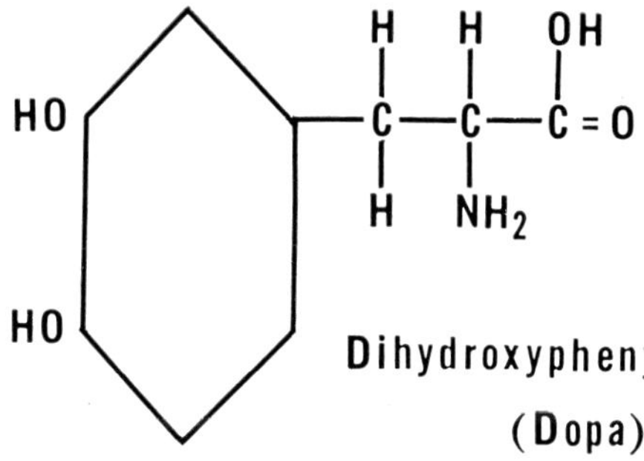

Dihydroxyphenylalanine
(Dopa)

L-DOPA and PARKINSONISM

Edited by

ANDRÉ BARBEAU, M.D.

Department de Neuro-biologie
Institut de Recherches Cliniques de Montreal
Montréal, Canada

FLETCHER H. McDOWELL, M.D.

Department of Neurology
Cornell University Medical College
New York, United States

F. A. DAVIS COMPANY
Philadelphia, Pa.

Manufactured in the United States of America.

Library of Congress Catalog Card Number 73-131190.

ISBN 0-8036-0595-1.

PREFACE

The discovery that a deficit in dopamine is present in Parkinson's disease has led to a completely new approach to the therapy of this incapacitating illness. L-3,4-Dihydroxyphenylalanine (L-Dopa, Levodopa) is now recognized as the drug of choice for alleviating most of the symptoms of parkinsonism. However, as with any new agent, side effects, difficulties and complications have been observed. The present book, and the research conference which gave it birth, came about in an effort to study and understand these side effects and to take a closer look at old physiological and biochemical concepts on which L-Dopa treatment is based.

The conference was the result of close cooperation between the American and Canadian Governments, their Food and Drug Administrations or Departments, and scientists from six countries. It was held from November 9th to 11th, 1969, at La Sapinère Hotel in Val David, Québec. The organizers and editors would like to acknowledge the all-important financial support of the National Institute of Neurological Diseases and Stroke (NS-09 170–01), Bethesda, Maryland, U.S.A., the Medical Research Council of Canada, and the W. Garfield Weston Charitable Foundation. Our thanks go also to our editorial secretary, Miss Nicole Guay, for her untiring efforts.

André Barbeau
Fletcher H. McDowell

PARTICIPANTS

N. E. Andén, M.D.
Associate Professor of Pharmacology
University of Goteborg
Goteborg, Sweden

R. D. Ansel, M.D.
Research Neurologist
University of California at Los Angeles
Los Angeles, California 90026

J. Arbit, M.D.
Assistant Professor of Neurology
Northwestern University
Chicago, Illinois 60611

André Barbeau, M.D.
Professor of Neurology
University of Montreal
110 ouest, avenue des Pins
Montreal (130), Québec, Canada

Aurèle Beaulnes, M.D.
Professor of Pharmacology and Therapeutics
McGill University
Montreal, Québec, Canada

Claude Bertrand, M.D.
Associate Professor of Surgery
University of Montreal
1560 est, rue Sherbrooke
Montreal (133), Québec, Canada

Gilles Bertrand, M.D.
Associate Professor of Neurosurgery
McGill University
3801 University Street
Montreal, Québec, Canada

Walter Birkmayer, M.D.
Professor of Neurology
University of Vienna
Blaasstrasse 28
Vienna XIX, Austria

Alan A. Boulton, Ph.D.
Department of Public Health
University of Saskatchewan
Saskatoon, Saskatchewan, Canada

Jacob A. Brody, M.D.
Division of Epidemiology
National Institute of Neurological Diseases and Stroke
Bethesda, Maryland 20014

D. B. Calne, M.D.
Division of Neurology
University College Hospital Medical School
University Street
London, W.C. 1, England

John Campbell, M.D.
Assistant Clinical Professor of Neurology
University of Cincinnati

125 William Howard Taft Road
Cincinnati, Ohio 45219

Arvid Carlsson, Ph.D.
Professor of Pharmacology
University of Goteborg
Goteborg, Sweden

William G. Clark, M.D.
Associate Professor of Biochemistry
University of California at Los Angeles
Los Angeles, California 90026

Jonathan Cole, M.D.
Department of Psychiatry
Harvard University
591 Morton Street
Boston, Massachusetts 02124

Mary P. Coleman, M.D.
Assistant Professor of Neurology
George Washington University
2125 13th Street, N.W.
Washington, D.C. 20009

I. S. Cooper, M.D.
Director, Department of Neurologic Surgery
St. Barnabas Hospital
4422 Third Avenue
Bronx, New York 10457

J. P. Cordeau, Ph.D., M.D.
Professor of Physiology
Universite de Montreal
C.P. 6128
Montreal, Québec, Canada

George C. Cotzias, M.D.
Senior Scientist
Brookhaven National Laboratory
Upton, L.I., New York 11973

Jacques de Champlain, M.D., Ph.D.
Assistant Professor of Physiology (Neuro-physiology)
Universite de Montreal
C.P. 6128
Montreal, Quebec, Canada

Henry B. Dinsdale, M.D.
Associate Professor of Medicine
Queens University
Kingston, Ontario, Canada

Arthur Dale Ericsson, M.D.
Assistant Professor of Neurology
Baylor University
Houston, Texas 77025

Judith Ericsson, M.D.
331 Duncaster
Houston, Texas 77024

Guy M. Everett, Ph.D.
Division of Neuro-pharmacology
Abbott Laboratories
Chicago, Illinois

Jacques Genest, M.D.
Professor of Medicine
University of Montreal
110 ouest, avenue des Pins
Montreal (130), Québec, Canada

F. Gerstenbrand, M.D.
Clinical Professor of Neurology
University of Vienna
Spitalgasse 23
1095 Vienna, Austria

G. L. Gessa, M.D.
Division of Chemical Pharmacology
National Heart and Lung Institute
Bethesda, Maryland 20014

L. I. Goldberg, M.D., Ph.D.
Professor of Medicine and Pharmacology
Emory University
69 Butler S.E.
Atlanta, Georgia 30303

Menek Goldstein, Ph.D.
Department of Biochemistry
New York University Medical Center
550 First Avenue
New York, New York 10016

Murray Goldstein, M.D.
Medical Director
National Institute of Neurological Diseases and Stroke
Bethesda, Maryland 20014

R. C. B. Graham, M.D.
Drug Advisory Bureau
Food and Drug Directorate
Department of National Health and Welfare
Tunney's Pasture
Ottawa (13), Ontario, Canada

Marian E. Hill, M.D.
Assistant Professor of Neurology
University of Toronto
101 College Street
Toronto (2), Ontario, Canada

Oleh Hornykiewicz, M.D.
Professor of Pharmacology
University of Toronto
Clarke Institute of Psychiatry
250 College Street
Toronto (2B) Ontario, Canada

Herbert H. Jasper, M.D., Ph.D.
Professor of Physiology (Neurophysiology)
University of Montreal
C.P. 6128
Montreal, Québec, Canada

Oscar Kofman, M.D.
Associate, Department of Medicine
University of Toronto
Suite 608
Ninety-nine Avenue Road
Toronto (5), Ontario, Canada

I. J. Kopin, M.D.
National Institute of Mental Health
5454 Wisconsin Avenue
Chevy Chase, Maryland 20015

K. Krnjevíc, Ph.D.
Professor of Physiology (Anesthesiology)
McGill University
McIntyre Medical Sciences Building
Montreal, Québec, Canada

Otto Kuchel, M.D.
Professor of Medicine
University of Montreal
Institut de Recherches Cliniques de Montreal
110 ouest, avenue des Pins
Montreal (130), Québec, Canada

Yves Lamarre, M.D.
Assistant Professor of Physiology (Neuro-physiology)
University of Montreal
C.P. 6128
Montreal, Québec, Canada

Charles H. Markham, M.D.
Associate Professor of Neurology
University of California at Los Angeles
Los Angeles, California 90024

Fletcher H. McDowell, M.D.
Professor of Neurology
Cornell University
1300 York Avenue
New York, New York 10021

Patrick L. McGeer, Ph.D.
Associate Professor of Medicine
University of British Columbia
Director, Kinsmen Laboratory of Neurological Research
Vancouver 8, B.C., Canada

Hugh McLennan, Ph.D.
Professor of Physiology (Neurophysiology)
University of British Columbia
Vancouver 8, B.C., Canada

Edward F. McNichol, M.D.
Director
National Institute of Neurological Diseases and Stroke
Bethesda, Maryland 20014

Alton Meister, M.D.
Professor of Biochemistry
Cornell University
1300 York Avenue
New York, New York 10021

Robert Y. Moore, M.D., Ph.D.
Associate Professor of Pediatrics, Medicine (Neurology) and Anatomy
University of Chicago
950 East 59th Street
Chicago, Illinois 60637

Sean O'Reilly, M.D.
Professor of Neurology
George Washington University
2150 Pennsylvania Avenue, N.W.
Washington, D.C. 20037

A. Pletscher, M.D.
Professor of Pathophysiology
University of Bern, Switzerland
F. Hoffmann-La Roche & Co. Ltd.
40002 Basel, Switzerland

Louis Poirier, M.D., Ph.D.
Professor of Anatomy
University Laval
Cite Universitaire
Québec, Québec, Canada

Merton Sandler, M.D.
Department of Chemical Pathology
Queen Charlotte's and Chelsea Hospital
London W. 6, England

Robert S. Schwab, M.D.
Clinical Associate Professor of Neurology
Harvard University
Massachusetts General Hospital
Boston, Massachusetts 02114

J. Siegfried, M.D.
Associate Professor of Neurosurgery
University of Zurich
Ramistrasse 100
8006 Zurich, Switzerland

T. L. Sourkes, Ph.D.
Professor of Biochemistry
McGill University
Allan Memorial Hospital
1033 Pine Avenue West
Montreal, Québec, Canada

A. M. Spiegel, M.D.
Department of Neurology (Emeritus)
Temple University
6807 Lawnton Avenue
Philadelphia, Pennsylvania 19126

E. A. Spiegel, M.D.
Professor Emeritus Experimental and Applied Neurology
Temple University
6807 Lawnton Avenue
Philadelphia, Pennsylvania 19126

R. R. Tasker, M.D.
Associate Professor of Surgery (Neurosurgery)
University of Toronto
Toronto General Hospital
Toronto, Ontario, Canada

R. Tissot, M.D.
Professor of Neurology
University of Geneva
1225 Chene-Bourg
Geneve (Bel-Air), Switzerland

Gertrude Tyce, M.D., Ph.D.
Department of Biochemistry
Mayo Foundation
Rochester, Minnesota

Arthur A. Ward, M.D.
Professor of Surgery, Head of Division of Neurosurgery
University of Washington
Seattle, Washington 98105

Melvin D. Yahr, M.D.
Professor of Neurology
Columbia College of Physicians
630 West 168th St.
New York, New York 10032

CONTENTS

PART 2

SIGNIFICANCE AND ETIOLOGY OF INDUCED ABNORMAL MOVEMENTS

PART 3

L-DOPA AND BLOOD PRESSURE REGULATION

PART 4

SIGNIFICANCE OF BEHAVIOR AND MENTATION CHANGES DURING THERAPY WITH L-DOPA

PART 5

HOW DOES L-DOPA WORK IN PARKINSONISM?

PART 6

PART 1

TREATMENT FAILURES WITH L-DOPA

CHAPTER 1

Clinical Implications of Treatment Failures

LIMITING FACTORS IN THE TREATMENT WITH DOPA *

George C. Cotzias

Under normal circumstances it would be a truism to state that lack of knowledge limits therapeutic success. Under the present crisis of priorities, however, it is necessary to stress that the successes which L-Dopa has brought us have also created new needs for basic research. Yet the support of basic research must now compete with other pressing needs in our society, and its priority is not always so high as we should like.

Perhaps the strongest voices in favor of basic neurological research are those of the afflicted, directly or indirectly, by various degenerations of the brain. These voices have been rendered effective by those among our colleagues who have placed behind L-Dopa their personal prestige as well as that of the great American universities. I am thinking of men like Fletcher McDowell and Melvin Yahr. Had they become frightened by the bad effects of Dopa, or had they honestly underrated its good effects, a study such as this would not have come about.

A detailed list of factors which may limit clinical success with Dopa would not be serviceable at this juncture. I would prefer to address myself instead to two phases of therapy, namely, the induction and the maintenance of therapeutic effects.

I can make one safe generalization with regard to the induction of improvement: my colleagues and I cannot predict with any degree of accuracy the

* Supported by the United States Atomic Energy Commission and in part by a grant (UI 00421–06) from the National Institutes of Health.

eventual response of a given case. We fail when predicting both the degree and the kind of improvement. We cannot even predict the months necessary before a given sign, such as tremor, will begin to respond. Furthermore, the toxicity which accompanies the induction of improvement has differed radically in our hands as opposed to the hands of others. For example, we do not have limiting problems with nausea, vomiting or anorexia. We wonder whether others do not condition their patients by increasing the dosage too fast. Among approximately 70 patients we have not had a single instance of cardiac arrhythmia attributable to Dopa. Only three individuals have developed hypotension, of whom only one required a dose reduction of the drug; a peripheral decarboxylase inhibitor was of no avail in correcting the hypotension in this case. We strongly suspect that differences in early toxicity reflect differences in the method of administration of Dopa. We have supportive evidence for this suspicion in two patients who did not tolerate Dopa elsewhere but tolerated large amounts with us.

Our inability to predict is even better illustrated by our efforts to maintain the therapeutic effect, once induced. All previous regimens for Parkinson's syndrome have been characterized by eventual loss of potency of the original agent, whether this was surgery or a drug. We were wrong when we expected the same to happen with L-Dopa. Worse, we had prepared our colleagues to face up to loss of potency of this drug. Surprisingly, the general trend has been toward lower therapeutic doses. Chief among the many reasons for diminishing the doses has been the late emergence of involuntary movements. We are apparently the first to have observed these movements, which we described in our initial publication.[1] We have stated that these dose-dependent, intermittent and reversible movements are encountered among patients whose parkinsonism has been remarkably improved. Since our first publication, these movements have been observed by others, and may now be considered part and parcel of the therapeutic success, even among small groups of parkinsonian patients treated with Dopa. This hypermobility can be regarded as the price of mobility itself.

Since we first observed these movements, we have been testing hypotheses as to their cause. We have not solved this problem but we have developed the conviction that dissociation of these movements from the therapeutic effects is going to prove possible. One of the many reasons for this conviction is the following: one of our patients exhibits involuntary movements while asleep, at a time when most of her parkinsonian symptoms have virtually disappeared and she is able to turn readily in bed.

The intermittency exemplified by these involuntary movements is not the only unexpected intermittency encountered during our chronic administration of L-Dopa. After three years of continuous administration we now see increasingly the sudden diurnal deviation from optimum performance which we described earlier.[2,3] This diurnal variation can still be corrected, at least in part, by rescheduling the drug. Further investigations of this progressive phenomenon are necessary, however. As we indicated before the recent World Congress of Neurological Sciences, injections of milligram quantities of apomorphine may show whether or not striatal neurons become refractory during these hypokinetic periods.

Some developments have actually refuted original expectations. Pyridoxine had been used earlier in the treatment of Parkinson's syndrome and it is known to activate the synthesis of dopamine from Dopa. Administration of pyridoxine has tended to cancel the effects of Dopa, however.

Another unexpected development is that something has happened to serotonin "on the way to the forum." Serotonin was the amine which opened this field of investigation. It was reported as low in reserpinized animals and low in the striatum of parkinsonian patients. Intravenous injections of its precursor, 5-OH tryptophan, were reported as beneficial to patients with parkinsonism before intravenous injections of Dopa had been abandoned. Some other neurological diseases are responding to 5-hydroxytryptophan these days, but one hears nothing about its role in Parkinson's syndrome. Perhaps we should reconsider this matter now that we are preoccupied with limiting factors in the treatment with L-Dopa.

From the foregoing, it seems to me that the basic findings available to us are not of predictive value. The success which the slow administration of Dopa to large final amounts has brought us has also created urgent needs for further physiological and biochemical investigations.

In summary, what holds true for any espousal may hold true for our wedding to Dopa: the reasons for the wedding are different from the reasons for remaining married. We stay married to Dopa because it has created an obvious, noncontrived need for new basic research. Basic research in this field is an urgent social need, and one which can be readily understood by all.

REFERENCES

1. Cotzias, G. C., Van Woert, M. H., and Schiffer, L.: Aromatic amino acids and modification of parkinsonism. New Eng. J. Med. *276:* 374, 1967.
2. (a) Cotzias, G. C., Papavasiliou, P. S., Gellene, R., and Aronson, R. B.: Parkinsonism and Dopa. Trans. Assoc. Amer. Physicians *81:* 171, 1968.
   (b) Papavasiliou, P. S., Gellene, R. and Cotzias, G. C.: Modification of Parkinsonism: Dyskinesias Accompanying Treatment with Dopa. Psychotropic Drugs and Dysfunctions of the Basal Ganglia; A Multidisciplinary Workshop. Workshop Series of Pharmacology Sect., N.I.M.H., No. 3, Proceedings of a Workshop, Bethesda, Nov. 1, 1968.
3. Cotzias, G. C., Papavasiliou, P. S. and Gellene, R.: Modification of parkinsonism: Chronic treatment with L-Dopa. New Eng. J. Med. *280:* 337, 1969.

TREATMENT FAILURES WITH L-DOPA

M. D. Yahr

Dr. Cotzias has characterized some of us as courageous in involving ourselves rapidly and extensively in this radically new approach to the treatment of parkinsonism with Dopa. However, as may often be the case, courage may in part be motivated by desperation. It has, after all, been a long time for those of us deeply involved in the care of patients with parkinsonism to have had any effective therapeutic approach to this devastating disorder.

One cannot escape the many problems inherent in clinical drug trials in general as well as those peculiar to parkinsonism. I need hardly point out the complexity, nor the fact that contradictory reports regarding the same therapeutic agent are more often the rule than the exception. We are sorely in need of agreement regarding terminology and methods of assessment in parkinsonism; when these are found some of the difficulties may be resolved. Leaving these problems aside, however, difficulties arise in choosing a representative patient population and a period of study of sufficient duration to balance the many variables in this disorder. I should like to point out a few of the problems which are intimately related to the characterization of treatment failures.

Table 1 summarizes the results of therapy in the same group of patients at two different time intervals—the first assessment carried out within the first two months, and the second some four to six months later. At the end of the initial phase of treatment with L-Dopa 21 patients showed improvement of 50 percent or more and 11 were classified as treatment failures. Our evaluation later (at 4 to 12 months) indicates that 34 patients ultimately enjoyed a degree of benefit above 50 percent and only four were classified as having an unsatisfactory result. It is quite apparent, therefore, that the time interval over which Dopa is administered is an important factor in assessing its full therapeutic efficacy. This incidentally is also the case in regard to its side effects, the nature and incidence of which vary with duration of treatment. This long latency, which is a unique therapeutic experience in treating parkinsonism and has been noted only with Dopa, raises a number of interesting questions regarding its mechanism of action. Are the beneficial and side effects a direct action of Dopa or does it have secondary effects in altering the effects of other amines? It is readily conceivable that the large Dopa load being administered for an extended period of time may well interfere with the transport, storage or appropriate utilization of other amines.

I should like to point out another aspect of treatment with Dopa which is

TABLE 1. Results of L-Dopa Therapy in Parkinsonism

Overall % Improvement	*Number of Patients* — *Initial Trial Period (4-8 wks)*	*Followup Trial Period (4-12 mos)*
80–100	1	9
50–79	20	25
20–49	28	16
0–19	11	4
Total number	60	54 *

* Six (6) patients were withdrawn from the study during initial trial period.

TABLE 2. Overall Percent Improvement in 116 Patients on Levodopa Treatment for 6 to 18 Months

% *Improvement*	*Number of Patients*
0–19	7
20–49	30
50–79	42
80–100	37

of some importance, namely, its sustained effects over an extended period of time. Table 2 summarizes 116 patients who were continuously treated for periods of six to eighteen months. Overall improvement of their parkinsonism of 50 percent or more occurs in almost 70 percent of this patient group. Further, almost half of the patients experience sufficient benefit to function at full capacity. These data are in marked contrast to those reported for other pharmacological agents used in parkinsonism. First, there is general agreement in the literature that no previous therapeutic agent has ever resulted in more than a 20 percent improvement and, by and large, this occurred within the first weeks or months of treatment. With continued usage the results usually became less impressive and in many instances almost imperceptible. Certainly the effects of Dopa are quite different.

One of the obvious determinants as to whether or not treatment might be successful is the degree of severity of the disease process. One might well assume from experimental data at hand that in instances of severe degeneration of the nigrostriatal system appropriate enzymes for metabolizing Dopa might well be lacking, negating any chance of raising brain dopamine. We, of course, have no way of clinically assessing this aspect of the disease except in setting up some guidelines regarding the degree of clinical involvement. We have previously reported a system of arbitrarily dividing the disease into five stages, Stage I indicating minimal involvement and Stage 5 involvement with complete incapacitation. In Table 3 a correlation of the results of treatment with the severity of the disorder is indicated. In general, patients with mild forms of the illness may benefit to a greater extent than those with severe involvement. However, an analysis of these data indicates that one is

TABLE 3. Distribution of 116 Patients on Levodopa Treatment Six Months or More by Percent Overall Improvement and Severity of Parkinsonism

Percent	*Severity of Parkinsonism*				
Improvement	*I*	*II*	*III*	*IV*	*V*
0–19	—	1	2	2	2
20–49	—	4	9	10	7
50–79	3	4	10	21	4
80–100	1	6	17	10	3
Totals	4	15	38	43	16

TABLE 4. Distribution of 116 Patients on Levodopa Treatment Six Months or More by Percent Overall Improvement and Dosage

Percent Improvement	*Daily Dose of Levodopa in Grams* 0–2.0	2.0–3.9	4.0–5.9	6.0–7.9	8.0–9.9
0–19	2	4	1	—	—
20–49	2	8	15	5	—
50–70	1	6	19	14	2
80–100	—	5	16	14	2
Totals	5	23	51	32	4

unable to predict from the degree of severity whether or not the patient will respond to the administration of Dopa. In the most severe form of this disorder—that classified as Stage 5, of which there were 16 patients—seven showed improvement of 50 percent or greater. Further, failures as well as successes occurred regardless of the degree of severity of the parkinsonism. After an experience of almost two years with more than 400 patients, I would agree with Dr. Cotzias that neither the type of symptomatology, the age of the patient nor any other factor that we have been able to assess bears a definite relationship to the degree of therapeutic response (Table 4).

I believe at this point that there are at least two positive factors that may explain nonresponsive cases. The first relates to drug tolerance. Nearly all of the patients in this series who have done poorly are those unable to tolerate an effective daily dose for an extended period of time. As indicated in Table 4, the seven patients that had less than 20 percent improvement were those in whom the dose had to be restricted because of side effects. Those tolerating 4 grams and above had a greater degree of improvement than the rest of the group. It would appear, therefore, that success of therapy is very intimately related to degree of tolerance for Dopa. The second factor that impresses me as an indicator of failure in Dopa treatment concerns atypical clinical manifestations. After all, parkinsonism must be looked at as a multi-etiological disorder in which there may be a host of underlying disease processes. It is my general impression at this time that atypical cases, which may represent pathological processes such as those described by Adams and Van Bogaert or by Steele-Richardson and Olszewski, are those that respond less well to treatment. This is not to say that some elements of the parkinsonism such as rigidity do not respond, for in fact they do, but rather that the total clinical picture is not reversed to a point that functional improvement results. I would say that the classical paralysis agitans and the postencephalitic variety are responsive to Dopa, but as yet one cannot identify specific features that are of predictive value in indicating responsivity.

TREATMENT FAILURES WITH L-DOPA

F. H. McDowell

I think it is clear that the definition of a treatment failure has to be preceded by an explanation of how a study is carried out. I would like to report our analysis of the treatment failures in the first 100 patients we treated, all of whom now have been under treatment for a period of at least six months and most for nine months or more.

When we originally began treating patients with L-Dopa we eliminated those who had mild Parkinson's disease. We did this mainly because we were not sure whether we could observe enough improvement in such patients to allow us to tell whether or not the drug was effective. We tended to load our series at the beginning with individuals who had severe parkinsonian involvement.

Our treatment program called for steadily increasing the dose of L-Dopa until the patient either responded or had side effects which caused us to stop the drug or to reduce the dose. Our method of analyzing improvement in parkinsonism consisted of two evaluation programs: (1) to examine the patient to determine his degree of tremor, rigidity and akinesia, and to score him on these findings; (2) to ask the patient about what he could do and score him on such activities as how well he could dress himself, get in and out of the bed, get in and out of a chair, bathe himself and use the toilet. We did not say that L-Dopa had improved a patient unless he showed a 20 percent improvement in his score on this evaluation program. The variations in the scoring system used were probably within that range. Those patients who did not show a 20 percent improvement were classed as treatment failures.

There were 13 patients who were considered treatment failures and who fall into four categories. First, there were those who by our standards did not show any evidence of improvement. Second, there were those who did show some response, evident especially to their family and friends, which was not demonstrable by our scoring system. Third, there were those who failed because of intolerable side effects from the drug, and fourth, there was one patient who was later found to have been misdiagnosed (Table 1).

TABLE 1. Treatment Failures

Response		*Patients*
No response to treatment		6
Dose range	4 to 8 gm	
Average dose	5.9 gm	
No scorable response		3
Dose range	5 to 7 gm	
Average dose	6 gm	
Side effects		3
Dose range	3 to 4.5 gm	
Wrong diagnosis		1

There were six patients who showed no response to treatment. Their L-Dopa dose range was 4 to 8 grams, with the average dose nearly 6 grams. These patients were severely involved with Parkinson's disease, with considerable disability; most of them were confined to bed or to a wheelchair. One patient in this group for reasons of her own refused to take enough medication. She once achieved a dose of 4 grams, but steadily, over a period of weeks before she ultimately died of pneumonia, took less and less L-Dopa and could not be persuaded to take more.

In the next category two patients with advanced parkinsonism come under the heading of no scorable response. The family in each instance thought that they could see some worthwhile improvement, but we could not. The dose range for these patients was 5 to 7 grams, with an average dose of 6 grams.

Three patients were considered treatment failures because of side effects: one because of hypotension, one because of severe nausea and vomiting, and one because of major mental disturbances which preceded L-Dopa treatment but were aggravated by it.

In the fourth category there was a patient who after several months of observation was found to have a disease other than parkinsonism.

In this study therefore, our rate of treatment failure was 13 percent.

MAJOR TREATMENT PROBLEMS IN L-DOPA THERAPY IN PARKINSON'S DISEASE

Charles H. Markham

Major treatment problems and treatment failures in L-Dopa therapy in Parkinson's disease are infrequent. In our series of 60 patients originally diagnosed as having Parkinson's disease (starting treatment from November 1, 1968 to July 1, 1969) eight fell into this category. They may be classified as: (1) one patient who could not tolerate the drug because of side effects; (2) two who lacked response for incapacitating parkinsonian symptoms; (3) two with lack of response probably due to misdiagnosis; and (4) three with unusual responses which limited the drug's usefulness. It should be noted that six of the eight patients are still taking L-Dopa because they consider that in spite of difficulties it is the best treatment available. Each of the six improved at least one stage (see Yahr et al. for stages of incapacity [1]).

The one patient with intolerable side effects was a woman of 45 years with postencephalitic akinesia, rigidity and severe swallowing difficulty. She had intractable nausea, anorexia and vomiting with even one gram per day of L-Dopa, and had to stop the drug.

The two patients with minimal therapeutic responses were males, 69 and 70 years of age, who exhibited severe akinesia with gait difficulty. They were in Stage 4 prior to therapy. One had had a unilateral thalamotomy with transient benefit. Both had great difficulty in initiating a single step. They would often be "frozen" in one position for a minute before they could move; after a few short steps, they would again become stuck. They would often lurch toward a chair or door only a few feet away. Both had severe postural instability and sometimes fell over backward. Hand and arm movements were less severely impaired. Both were intact intellectually; one was the head of his own engineering firm and the other was the owner and manager of a resort hotel. One tolerated 8 grams of L-Dopa a day without a trace of side effects and with modest improvement in walking and using his hands. The other became even more unsteady on his feet on 6 grams a day and was not helped at all by lesser doses.

Of the cases of probable misdiagnosis, one was a 70-year-old woman who had rapidly progressive parkinsonian akinesia and mild dementia over a year's time. L-Dopa helped her facial expression and use of her hands, but did little for her hesitancies in walking. She later developed increasing trouble in walking. In retrospect, we suspect she may have had an unusual form of parkinsonism, possibly with pallidal or striatal degeneration. She is still living.

The other patient in this group was a 56-year-old man with a four-year history of parkinsonian rigidity, tremor, moderate dementia and personality change. After continued progression of all symptoms over four months in spite of adequate L-Dopa dosage, he developed a fairly fixed dystonic posture of one arm. An electromyogram showed widespread denervation fibrillation. A clinical diagnosis of Jakob-Creutzfeld disease was made; we anticipate a cortical biopsy for slow virus studies.

Of the patients showing unusual responses, perhaps the most distressing was a 50-year-old rancher with severe truncal and limb akinesia secondary to postencephalitic Parkinson's disease. He was in Stage 4. He reached a maintenance dose of L-Dopa of 7 grams a day in the first two months of therapy and recovered from almost complete invalidism to the point that he could ski. After a few months he developed severe choreo-athetoid movements, and the therapeutic response to the L-Dopa became very erratic. After receiving L-Dopa for a year, he now takes 7.5 grams per day and fluctuates on an hour-to-hour basis between normality, parkinsonian akinesia and wild choreo-athetoid movements. Neither changes in dosage nor timing of administration has helped.

The two other individuals in this group are a 47-year-old woman and a 70-year-old man. Both responded well to L-Dopa, but after several months both had relapses over several weeks to almost their original state. Neither received vitamin B-6 or any dietary change. On the hypothesis that his intestinal flora might be utilizing the L-Dopa, the man received a week's course of neomycin; there was no change in his parkinsonism. After two more months the man has not improved on 5 to 8 grams of L-Dopa a day. In part because of her relapse, the woman became very depressed and tried to commit suicide. After several days in barbiturate coma, she recovered. While convalescing she showed striking improvement in her parkinsonian symptoms over several weeks; the improvement has persisted for six months. She has been taking 3 to 5 grams of L-Dopa constantly except for the time in coma.

These treatment failures should provoke our interest, for they may give us a better understanding both of diseases superficially mimicking parkinsonism and of variations in dopamine metabolism in cases with true Parkinson's disease.

REFERENCES

1. YAHR, M. D., DUVOISIN, R. C., HOEHN, M. M., SCHEAR, M. J., and BARRETT, R. E.: L- 23-4-dihydroxyphenylalanine (L-Dopa); its clinical effects in parkinsonism. Trans. Amer. Neurol. Assoc. *93:* 56, 1968.

FAILURES IN L-DOPA THERAPY

W. Birkmayer

We began using L-Dopa in the treatment of Parkinson's disease in the spring of 1961.[1] At the same time Barbeau, Murphy and Sourkes presented their results.[2] Our first administrations were both intravenous and oral. Over the last nine years we have treated 435 cases. Table 1 shows the age of patients; there was no correlation between age group and treatment failures. Table 2 gives the distribution of cases treated during the ensuing years.

TABLE 1. Age Distribution of Patients

Age	*Number of Cases*
40–50 years	21
50–60 years	69
60–70 years	189
70–80 years	142
80–90 years	14

TABLE 2. Patients Treated

Year	*Number of Patients Treated*
1961	42
1962	41
1963	43
1964	33
1965	40
1966	35
1967	43
1968	82
1969	76

Table 3 shows the scores of disability in patients under treatment with L-Dopa alone (235 cases), and the average improvement found; 53 cases (29 percent) were failures. Table 4 shows the scores of disability of patients treated with a combination of L-Dopa and Ro 4-4602 (200 cases); 22 cases (12 percent) in this group were failures. It can be seen that the amount of disability has no correlation with the kinetic effect and to treatment failure. There is no correlation between duration of treatment and failure rate (Table 5). The duration of the disease before L-Dopa treatment is 9.6 years and after Dopa treatment 18.3 years (Table 6). This difference is highly significant.

If a patient shows, after L-Dopa administration, a stimulation of the tendon reflexes, one can predict a positive kinetic reaction (Fig. 1). In a paper with Hornykiewicz [3] we demonstrated that patients with severe parkinsonism after reserpine or alpha-methyldopa administration show no decrease of akinesia. These patients fail to respond to L-Dopa administration. Patients with an increase of akinesia after treatment with reserpine show a positive kinetic effect after L-Dopa treatment.

Histological examination of the substantia nigra in patients with a positive kinetic effect after L-Dopa administration show some regions with melanin (Fig. 2). However, in patients counted as failures, the substantia nigra contains no melanin (Fig. 3). We assume that, for a positive kinetic effect of L-Dopa, traces of melanin in the substantia nigra or locus caeruleus have to be present.

TABLE 3. Disability * Before and After L-Dopa Treatment

*Average Degree of Disability * Before Treatment*	*Numbers in Each Group*	*Average Point Improvement in Disability **	*Number of Failures*
80	44	19.6	11
70	22	20.0	11
60	63	17.2	18
50	33	13.0	6
40	14	11.0	6
30	6	8.0	1

* *Disability Scores* (maximum disability per patient = 100 pts) are obtained by grading on a 1 to 10 scale (1 = normal) each of the following functions: speaking, walking, pushing, jumping, writing, dressing, posture, propulsion, miming and associated movements.

TABLE 4. Disability * Before and After Treatment with L-Dopa Combined With Ro 4-4602

*Average Degree of Disability * Before Treatment*	*Numbers in Each Group*	*Average Point Improvement in Disability **	*Number of Failures*
80	10	15.5	3
70	15	13.1	2
60	37	24.8	2
50	53	22.0	5
40	48	14.7	5
30	15	10.0	5

* For definition of *Disability Scores* see legend for Table 3.

TABLE 5. Duration of Treatment

Months	*With Dopa Alone Number of Patients*	*Dopa+Ro 4–4602 Number of Patients*
3	23	12
6	21	20
9	14	20
12	30	53
18	11	16
24	22	35
36	39	42
48	20	13
60	20	—
72	21	—
84	12	—
96	10	—
108	3	—

TABLE 6. Duration of Parkinson's Disease from Onset to Death

82 patients without L-Dopa treatment	mean 9.6 years
85 patients with L-Dopa treatment	mean 18.3 years

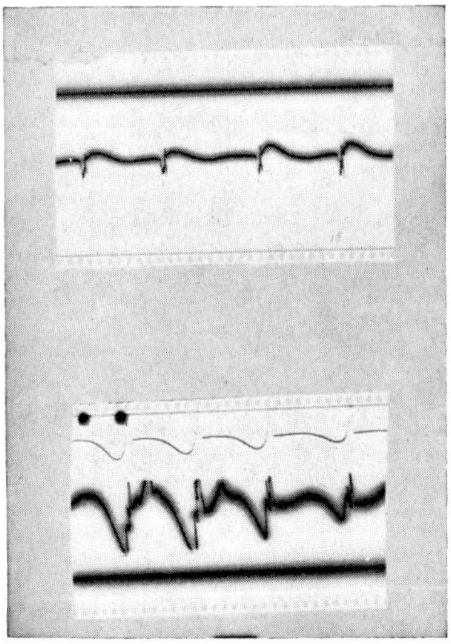

Fig. 1. Tendon reflex before and 30 minutes after an intravenous injection of 50 mg of L-Dopa. A stimulation of the gamma system demonstrates a positive kinetic effect.

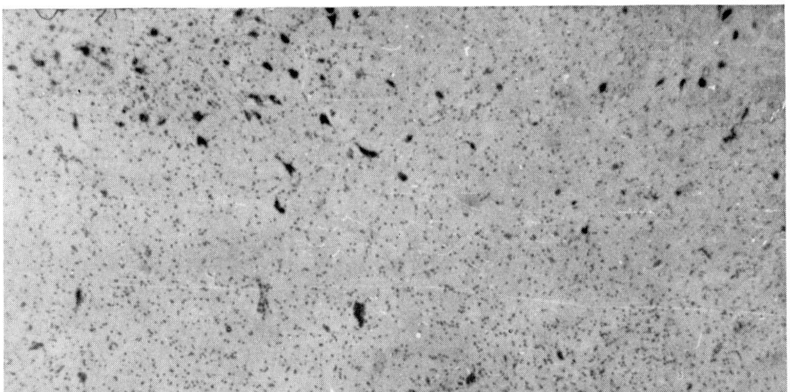

Fig. 2. Substantia nigra of a patient with idiopathic Parkinson's disease. KV×56. (Treatment a success.)

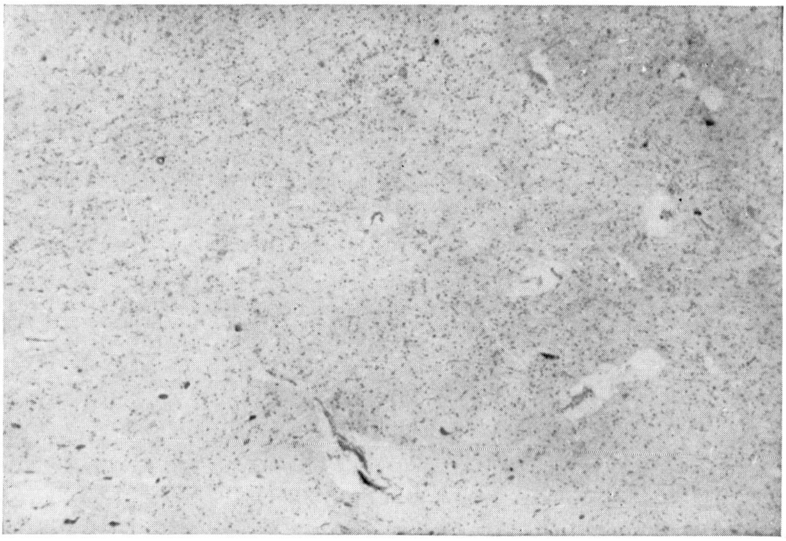

Fig. 3. Postencephalitic parkinsonism. Substantia nigra contains no melanin. KV×56. (Treatment a failure.)

REFERENCES

1. Birkmayer, W. and Hornykiewicz, O.: Der L-3,4-dioxyphenylalanin (Dopa)-effekt bei der Parkinson-akinese. Wien. Klin. Wschr. *73:* 787, 1961.
2. Barbeau, A., Sourkes, T. L., and Murphy, G. F.: Les catécholamines dans la maladie de Parkinson. *In:* Monoamines et système nerveux central, Symposium Bel-Air (J. de Ajuriaguerra, ed), Georg, Genève, 1961.
3. Birkmayer, W., and Hornykiewicz, O.: Der L-dioxyphenylalanin (= L-Dopa)-effekt beim Parkinson-syndrom des menschen: Zur pathogenese und behandlung der Parkinson-akinese. Arch. Psychiat. Nervenkr. *203:* 560, 1962.

FAILURES OF L-DOPA TREATMENT

F. Gerstenbrand

Failures with L-Dopa treatment can be divided into two groups. In the first group the treatment is interrupted because of insufficient effect. (Some patients decide to stop the medication by themselves.) Insufficient amelioration is observed in cases of severe tremor, especially postural tremor. Figures 1 and 2 show that L-Dopa influences not only rigidity but tremor as well, and especially passive tremor. Those cases with severe tremor are characterized by small differences from other parkinsonian patients; in the initial phase of Dopa therapy, postural tremor can increase.

If a stereotaxic operation is successful, L-Dopa therapy will also be successful postoperatively. In one of our patients Dopa was very effective after unilateral stereotaxic intervention greatly improved the tremor.

Another cause of lack of success with L-Dopa is advanced stage of disease. In patients with severe involvement it is thought that the extrapyramidal

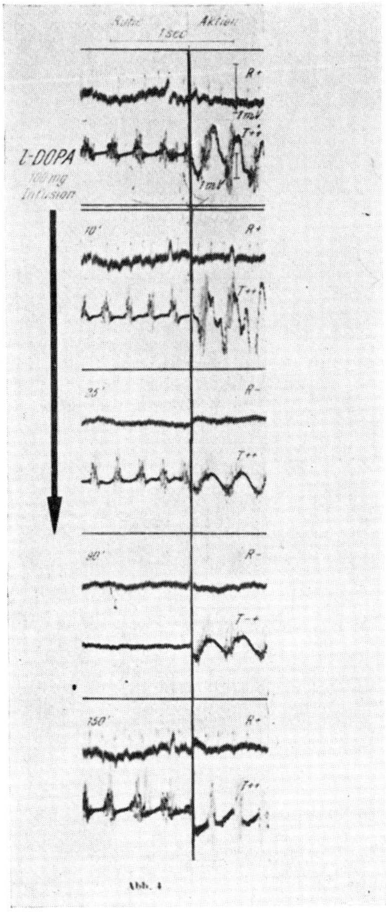

Fig. 1. Electromyogram of patient aged 53, postencephalitic parkinsonism, left-sided rigidity and right-sided tremor. EMG curve on the left during position of rest, on the right in action (light rise of the finger). *Upper curve* (m.bic.hum. coaxial needle electrode): rigidity unchanged in the resting position and in action. *Lower curve* (m.extens.indic. propr., coaxial needle electrode): resting tremor with a frequency of 5/s; on the right, postural tremor with a frequency of 6/s. The amplitude of the postural tremor is higher (6).

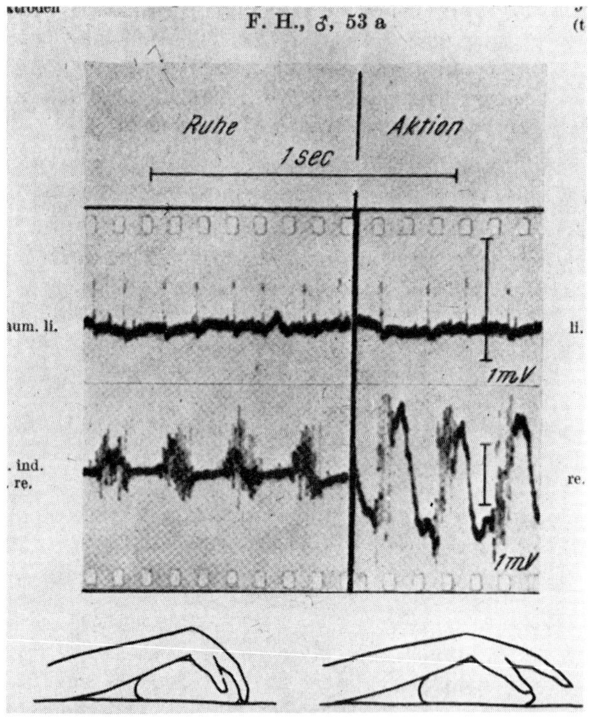

Fig. 2. An electromyogram of the patient before and for 150 minutes after 100 mg intravenous L-Dopa. Between 20 and 120 minutes there is interruption of the rigidity potential and between 40 and 100 minutes of the EMG-potentials of the resting tremor. There is no interruption of postural tremor (compare with description of Fig. 1).

structures, especially the substantia nigra, are damaged to a high degree and that the remaining nigra cells are not numerous enough to utilize the Dopa and to form dopamine from it. This supposition could be an explanation for the failure of L-Dopa in severe cases which show an initial improvement and then a brisk decompensation followed by coma.

Another subgroup of patients treated with L-Dopa shows other complications; this group comprises patients who after secondary traumatic midbrain lesions develop pseudo-parkinsonian symptoms (Fig. 3). In these patients (we call the complete clinical picture an "apallic syndrome") sometimes significant improvement is remarkable not only in the parkinsonian symptoms but in the "coma vigil" as well. The course of remission is accelerated. Our experience is confirmed by Dalle Ore and his collaborators (personal communication). For the full "apallic syndrome" we recommend first the intravenous administration of L-Dopa. The dose depends on the stage of the condition; we use doses of between 300 and 1000 mg per day. Failures are caused by the high irritability of the circulatory system and the exaggerated reaction of the autonomic system in "apallic" patients.

Failures can also be explained by the use of L-Dopa when indications do

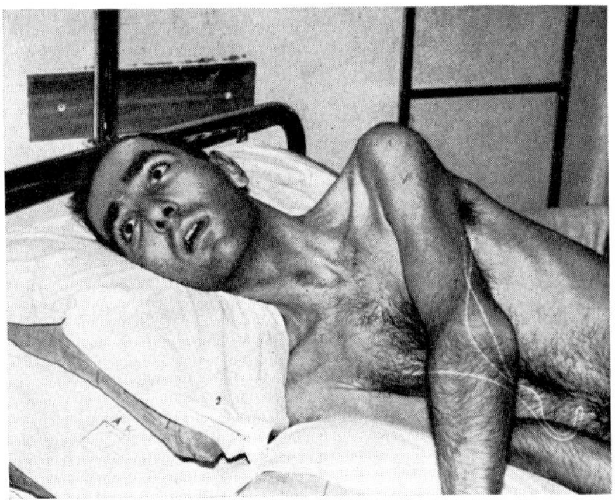

Fig. 3. A patient aged 27 with traumatic apallic syndrome with Parkinson symptoms.

not warrant it, such as with postencephalitic tic with slight parkinsonian symptoms, or the akinetic-rigid form of Huntington's chorea.

The second major group of failures is caused by side effects such as dyskinesia, hallucinations and a severe decrease of blood pressure.

The dyskinesias are very similar to the hyperkinesias which appear with treatment with high doses of Majeptil in psychotic or neurotic patients.[1] Calne and collaborators [2] confirm this opinion. From this analogy, the hyperkinetic movements might be explained as an overaction of extrapyramidal structures to a surplus of dopamine. A decrease in dosage stops the dyskinesias in nearly all cases. If the hyperkinetic movements persist with lower doses of L-Dopa, damage to the neostriate system is possible; the L-Dopa therapy should be cancelled.

Hallucinations are introduced by a period of restlessness and dysphoria. The hallucination state can change to a delirium with visual hallucinations. Hallucinatory patients often show signs of cerebral arteriosclerosis and abnormalities in the electroencephalogram; such abnormalities increase during treatment. Patients with distinct arteriosclerotic symptoms should be refused L-Dopa in order to avoid delirium and hallucinatory reaction. (Incidentally, hallucinations during L-Dopa treatment are very similar to hallucinations from Artane or belladonna.)

Occasionally L–Dopa therapy fails in patients with unstable hypertension because of persisting reactive hypotension. In these cases, L-Dopa cannot be administered in doses sufficient to be effective.

REFERENCES

1. Bruck, J., and Gerstenbrand, F.: Funktionelle Decerebration unter dem Bild eines apallischen Syndroms bei hochdosierter Majeptil-Behandlung. Nevenarzt *38:* 459, 1967.

2. Calne, D. B., Stern, G. M., and Laurence, D. R.: L-Dopa in postencephalitic parkinsonism. Lancet *1:* 744, 1969.
3. Gerstenbrand, F.: Das traumatische apallische Syndrom (Klinik, Morphologie, Pathophysiologie und Behandlung). Springer-Verlag, Wien, 1967.
4. Bruck, J., Gerstenbrand, F., Grundig, E., and Prosenz, F.: Stoffwechselveränderungen bei extrapyramidalen Syndromen und vorläufige therapeutische Konsequenzen. Fortschr. Neurol. Psychiat. *33:* 677, 1965.
5. Gerstenbrand, F., Pateisky, K., and Prosenz, P.: Erfahrungen mit L-Dopa in der Therapie des Parkinsonismus. Psychiat. Neurol. *146:* 246, 1963.
6. Gerstenbrand, F., and Pateisky, K.: Uber die Wirkung von L-Dopa auf die motorischen Störungen beim Parkinson-Syndrom. Wr. Ztschr. f. Nervenhk. *20:* 90, 1962.

PARKINSONISM AGGRAVATED OR UNIMPROVED BY L-DOPA. PARTICULAR CLINICAL SIGNS

R. Tissot

At examination we are unable to determine which patients will benefit from treatment with L-Dopa and which patients will not. At best we have perhaps obtained some indices which we would like to discuss here.

First, the etiological forms of parkinsonism, as far as one can identify them, do not seem important.

On the other hand, the three teams directed by Ajuriaguerra, Gauthier and myself, trying to evaluate the therapeutic effects of L-Dopa, came independently upon the same indices: patients aggravated or unimproved by L-Dopa had paradoxical kinesias. Akinesia (hypokinesia) was severe, but also peculiar. Difficulty in initiating movements was quite pronounced, lasting perhaps for several seconds. Once started, however, relatively rapid activity followed.

This latter manifestation was especially noted in walking: Blocked as if glued to the floor, patients start abruptly and quite rapidly, sometimes even running for several meters. Such was the case of a patient worsened by L-Dopa whose family was struck by this peculiarity. Stopped at the curbstone, he would take off into the street unable to stop and risking collision.

The "stick phenomenon," well known in the majority of patients with parkinsonism, is observed spontaneously in these patients. It is simply one index of an otherwise classic clinical picture.

These paradoxical kinesias must correspond to very particular modifications of muscle tone.

We hope to go beyond our strictly clinical observations by objectively noting changes in muscle tone with Paillard's method of comparative registration of H and T reflexes.

Finally, there are patients with classic parkinsonism who do not respond or respond badly to L-Dopa; certain dyskinesias other than parkinsonism are usually worsened by this treatment. This is well known in Huntington's chorea but perhaps less so in essential familial tremor where aggravation also occurs.

GENERAL DISCUSSION

Chairman Cotzias: Before we open the meeting for discussion I would like to recognize one man who is important to our conference and another who is important to me. It is an act of generosity on the part of Dr. Irving Cooper to have joined us today, and Dr. Melvin Van Woert has endured through our entire effort with DL-Dopa. You know his cardinal contributions to melanin biochemistry since he went to Yale.

The meeting is open for discussion, questions, and comments.

Dr. Cooper: I am very flattered to have been invited and I am delighted to be here. Although our surgical experience includes about 6,000 cases since 1952, we now believe L-Dopa to be the treatment of choice for most patients. I am extremely impressed with our own experience with L-Dopa as a therapeutic agent. We have treated almost 600 patients since George Cotzias presented his early results to us. Our statistics are about the same as those presented, and virtually every parkinsonian patient that comes to our service is put on the drug, barring contraindications. Very few patients with real parkinsonism come to surgery at the present time.

From a standpoint of interest in the problem rather than in the technique, I do not think there is any question of the immense value of the drug. Treatment failures are relatively insignificant if compared with value as a therapeutic agent.

It does fail totally in heredofamilial tremor, as mentioned by Professor Tissot. I think that the reason for this was demonstrated in the presentation of Professor Birkmayer by the fact that intravenous L-Dopa augmented deep tendon reflexes. Increase of deep tendon reflexes in cases of pure cerebellar tremor will usually exacerbate tremor.

If there is a cerebellofugal lesion, which there apparently is with familial tremor, we see a decrease of the deep tendon reflexes. If one performs a Jendrassik maneuver or augments the deep tendon reflexes in any way, one produces tremor. Since Dopa apparently does that, I think this is why it may not only fail in the treatment of heredofamilial tremor but actually make the condition worse.

Dr. Pletscher: I would like to ask the audience whether anyone has given 5-hydroxytryptophan (5-HTP) to patients who have been treated with Dopa after involuntary movements have appeared? We have noticed that in animals a combination of a peripheral decarboxylase inhibitor plus Dopa markedly diminishes the endogenous serotonin in the brain. It may well be that after repeated administration of Dopa alone a similar decrease of serotonin occurs. 5-Hydroxytryptophan or tryptophan, precursors of serotonin, may counteract the decrease of the endogenous serotonin and possibly also the abnormal movements.

Dr. Yahr: I believe Dr. Pletscher is anticipating a part of this program which will concern itself with induced involuntary movements. I can say at this time that we have administered 5-HTP in dosages up to a gram a day to patients receiving Dopa. First, it is tolerated well with L-Dopa; secondly, to date in a small series of cases it has not eliminated these movements.

My earlier presentation refers to Dr. Pletscher's work in showing a competitive mechanism for transport of amines. The concern we have had in parkinsonism is that there already exists a reduction in brain serotonin. The potential for further decrease of serotonin while administering a large Dopa load certainly exists.

Dr. Rajput: (1) Has anyone observed whether patients who have unilateral or bilateral thalamotomies require doses different from patients who have not had surgical treatment? (2) In long-term unilateral parkinsonism, what is the effect of treatment on the side involved and the side not involved, respectively? (3) Can anybody comment on what L-Dopa does to glaucoma? We know that we should not give anticholinergic drugs to patients with glaucoma.

Dr. McDowell: We have not been able to see any difference in the L-Dopa dose requirements of patients who had prior surgery of the basal ganglia or thalamus when compared to those not so treated. We have not had trouble with glaucoma. There are two patients in our series who developed glaucoma during the period they have been receiving L-Dopa, but we cannot say whether or not the glaucoma is related to the drug. One of the patients possibly had glaucoma before treatment with Dopa. Neither patient was receiving anticholinergic medication at the time glaucoma developed.

Dr. Schwab: My first remark concerns the glaucoma problem with anticholinergic drugs. We must distinguish between closed-angle and open-angle glaucoma. A closed-angle glaucoma requires surgery rapidly and is very liable to exacerbations with anticholinergic drugs. Luckily this involves only a small percentage of patients with glaucoma. Open-angle glaucoma victims can tolerate very high doses of anticholinergic drugs without any important increase in ocular pressure.

A very interesting patient I saw recently had unilateral rigidity and a resting tremor on his dominant side, while on the other side he had an action, or postural, familial tremor made worse over the years by a large intake of alcohol. Both the rigidity and tremor on his dominant side were vastly improved after treatment with L-Dopa, but the cerebellar tremor on the opposite side was made definitely worse. Will this asymmetry persist? Will the patient require stereotaxic surgery for his nondominant side? Time will tell.

Finally, I would like to emphasize that in the same parkinsonian patient you may well see both a resting and an intention tremor. It will be interesting to see upon which type of tremor L-Dopa proves to be effective in the long run.

Dr. Everett: Regarding the cases reported here, were they getting anti-parkinson agents in addition to L-Dopa?

Dr. Yahr: Our case material includes those to whom L-Dopa has been administered by itself as well as in combination with standard anti-parkinsonian agents. I would estimate that in about a third of the patients all drugs except L-Dopa were discontinued either prior to or during its administration. In some instances, we have continued anti-parkinson agents not necessarily because of their beneficial effects on parkinsonism but more for their control of side effects induced by Dopa. This particularly refers to the emetic action of L-Dopa.

Our initial study design enabled us to carry out comparative studies of standard anti-parkinson agents and L-Dopa. On a comparative basis, it is our opinion that L-Dopa is about three-and-one-half times more effective in controlling the symptoms of parkinsonism than any other pharmacological agent now available.

I should like to address a question to the group at large. In the two years that we have been administering L-Dopa, we have been trying to ascertain whether the natural history of parkinsonism is altered by its administration. Though the time interval is much too short to draw any definite conclusions, it is our impression that none of our patients has as yet developed new symptoms of this disorder. Since many participants in this discussion, especially Dr. Birkmayer, have had more extensive experience with its use, I wonder if he might wish to comment, specifically, in regard to whether the patients treated have developed any new symptoms of parkinsonism and whether the disease has progressed in its usual manner and at the expected rate.

Dr. Birkmayer: When we started treating our patients the mean duration of the illness was 9 years. Now, for the same patients receiving L-Dopa, it is 18.3 years. It would thus appear that in these patients there has been little progression of the illness.

Dr. Markham: We have one patient who had an excellent result, with relief of tremor, akinesia and rigidity over a year's time, but who in the last six months has developed a very soft resting parkinsonian tremor that gradually has been getting worse.

Dr. Stern: With respect to the factors limiting successful treatment, our experience [1,2] is generally in accord with the views already expressed, save on one issue. We found that patients with advanced postencephalitic parkinsonism may respond poorly because they can tolerate only very small amounts of L-Dopa, which contrasts of course with the staggeringly large amounts of anticholinergic drugs often consumed. We have not seen this therapeutic paradox in advanced, idiopathic cases and we wonder whether this is the experience of others?

REFERENCES

1. Calne, D. B., Stern, G. M., Lawrence, P. R., Sharkey, J., and Armitage, P.: L-Dopa in postencephalitic parkinsonism. Lancet *1:* 744, 1969.
2. Calne, D. B., Spiers, R. S. P., Stern, G. M., Lawrence, P. R., and Armitage, P.: L-Dopa in idiopathic parkinsonism. Lancet *2:* 973, 1969.

CHAPTER 2

Pharmacological and Physiological Implications of Treatment Failures

PHARMACOLOGICAL AND PHYSIOLOGICAL IMPLICATIONS OF TREATMENT FAILURES WITH L-DOPA *

André Barbeau

It is evident that L-Dopa does not give completely beneficial results to every patient under treatment. In our own series of 95 cases from three hospitals † there are 13 cases without satisfactory objective improvement, that is, improvement of more than 20 percent. In all of these cases one or more symptoms were modified, but never to a degree convincingly superior to a placebo effect. The reasons which justified discontinuation of the drug can be listed:

1. Psychotic reaction or aggravation of a preexisting senile dementia: 4 cases. It is noteworthy that in one case auditory and visual hallucinations persisted for nine weeks after cessation of L-Dopa, which had never exceeded 1.5 grams per day.
2. Atypical Parkinson's disease with large ventricles and pyramidal tract signs: 2 cases.
3. Complications of severe orthostatic hypotension: 2 cases. In both, localized cerebral infarcts (probably emboli) with focal minor seizures occurred.
4. In five other patients the only common factor was a severe underlying depression, manifested principally by a total lack of motivation or drive and made definitely worse by L-Dopa, with disproportionate anxiety.
5. No patient was removed from treatment because of nausea, vomiting or anorexia.

* Supported in part by Grant SP-1 from the National Research Council of Canada.

† Cases studied at Hôtel-Dieu, Maimonides or Jewish Convalescent Hospitals with the help of Drs. L. Gillo-Joffroy, H. Mars, A. Schwartz, and I. Libman.

These observations, with those of our colleagues in the literature, permit a few questions about underlying mechanisms.

The first question we should ask is: *Did we choose the proper patients?* It is obvious that the very design of our own study led us to accept the difficult problems (aged, invalid, long-term patients) and to exclude almost systematically the mild or even the moderately advanced cases. This being so, I think we can already place some limitations on the future selection of patients. I do not personally think that age, sex, or stage of the parkinsonian illness are major limiting factors, as long as we are dealing with a relatively pure extrapyramidal syndrome. The presence of severe, advanced, cerebral arteriosclerosis which would make the patient totally unresponsive to his environment, even if he were able to move with ease, should rule out prolonged Dopa therapy. However, it is not always easy to distinguish clearly between parkinsonian akinesia or apraxia-like confusion and arteriosclerosis.[1] Moreover, this distinction cannot be made until previous anticholinergic medication has been withdrawn for some time. I think it would be worthwhile to carry out studies which would permit the establishment of a battery of tests able to distinguish between these entities.

It does not appear that previous pallidal or thalamic surgery prevents good results unless there is consequent pyramidal damage. Moreover, speech modifications independent of the normal disease progression and subsequent to bilateral surgery are seldom helped by L-Dopa, whereas the monotony and low volume of parkinsonian speech are usually corrected. One further observation may be of interest. It is our experience that when thalamic surgery has resulted in the appearance of a permanent dyskinesia (athetosis, dystonia) this very symptom will often be made worse under L-Dopa therapy, even if total motility is considerably increased.

I think that parkinsonian patients who have a previous history of well-documented psychotic episodes should first be tried on a drug such as amantadine hydrochloride before being given L-Dopa. The same is true for severely depressed patients who appear to lack motivation for the treatment. A patient is not accepted on our program unless he himself asks for it. What the wife, the children or the rest of the family desire should be secondary in our decision.

Moving now from the implications of patient selection, I would like to ask: *May there not be a multitude of different diseases now classified under the term parkinsonism?* This concept, of course, is not new, and we can immediately think of parkinsonism secondary to encephalitis, to drugs, to carbon monoxide intoxication, and to manganese poisoning. We know that wherever we see akinesia (in Pick's presenile dementia, in juvenile Huntington's chorea, in manganese miners, in some case of Steele-Richardson-Olzewski syndrome) we can expect L-Dopa to produce symptomatic changes. The most difficult question is whether there exist within the so-called "idiopathic group" biochemical genetic differences which can help us in predicting the outcome of L-Dopa therapy. I can think of two specific ways to answer the question:

1. Is there a difference in Dopa metabolism between the reactors and nonreactors? This could be studied by examining the rate of increase of homovanil-

lic acid (HVA) in the cerebrospinal fluid of these patients after a Dopa load. It could also be inferred from studies with radioactive Dopa or dopamine. Such studies will be reported during this symposium, and perhaps the authors will take a second look at their data to see if there is a difference between reactors and nonreactors?

2. Another approach could make use of the methodology of pharmacogenetics. It seems to me that amantadine hydrochloride is a drug perfectly suited for such studies. In the human it is excreted totally unmetabolized, but nevertheless exerts a pharmacological action qualitatively similar to that of Dopa. It might be profitable to study the rate of excretion and the blood concentration of the drug in all patients to be later given L-Dopa and to correlate this with therapeutic results.

Finally there is one aspect of treatment failure which I think deserves a considerable amount of thought. That is the physiology and pharmacology of motivation. This word is difficult to define and at various times has implied the concepts of drive, of directed attention, of goal achievement, of aggression, and even has involved components of the wake-sleep system. I leave to others the detailed discussion of this factor, but I would like to stress the fact that it is a basic problem not only in the explanation of therapeutic failures but also in modern physiology. For many reasons having to do with the well-known studies of McLean regarding the limbic system, of Jouvet and others on the sleep cycle, of Brodie on depression, I would propose that we pay particular attention to serotonin and tryptamine metabolism. I think drugs such as the peripheral decarboxylase inhibitors Ro 4-4602 or MK-485 will be excellent tools for such studies. We already know from the experiments of Pletscher and his colleagues [2] that large doses of L-Dopa reaching the brain in this way may modify serotonin content. Such a mechanism acting upon a previously unbalanced system may aggravate a depressive state or even partially explain some of the hallucinations and paranoid manifestions. Studies along these lines should be encouraged.

REFERENCES

1. JOUBERT, M., and BARBEAU, A.: Akinesia in Parkinson's disease. An experimental study. *In:* Progress in Neuro-Genetics (A. Barbeau, and J. R. Brunette, eds.), Excerpta Medica Foundation, Amsterdam, 1969.
2. PLETSCHER, A., and GEY, K. F.: The effect of a new decarboxylase inhibitor on endogenous and exogenous monoamines. Biochem. Pharmacol. *12:* 223, 1963.

GENETIC CONSIDERATIONS IN PARKINSON'S DISEASE

Jacob A. Brody

A direct genetic role in Parkinson's disease has been postulated and several authors have suggested an autosomal dominant mode of inheritance.[1, 2] There are descriptions of familial aggregations of Parkinson's disease, and of the appearance of the disease in multiple generations. Although many physicians have the impression that there is a genetic component, the most recent studies by Duvoisin and Yahr failed to demonstrate familial patterns.[3] The reason for the inconclusiveness of the studies of genetic role in Parkinson's disease can be explained by at least three hypotheses: (1) There is no important genetic role in classical Parkinson's disease, and early reports dealt either with chance associations of a fairly common disease or, more probably, with the occurrence of hereditary diseases of the extrapyramidal system which were separate etiological entities. (2) There is a genetic role, but it is difficult or impossible to study clinically because Parkinson's disease does not represent a single etiological entity and because the disease occurs late in life and many potential patients die before they reach the age of susceptibility. (3) In light of the documentation that monoamine metabolism is abnormal in patients with Parkinson's disease, it is possible that there is a metabolic abnormality which is transmitted genetically, but which results in clinical illness only under special circumstances. Therefore, by using only clinical disease as the indicator, we see only weak and uninterpretable relationships within families.

If we accept the current limitation that the only readily available method for detecting Parkinson's disease is clinical, the most economical mechanism for determining the role of genetics is through the study of twins. Dr. Roswell Eldridge of my staff reviewed the Veterans Administration Twin Registry, which was last updated in 1962. Of the 16,000 registered twins, 7 twins (5 dizygotic and 2 monozygotic) had one member with diagnosed Parkinson's disease, but in no instance were both twins affected. We are also attempting a somewhat more costly approach, perhaps in collaboration with Dr. Barbeau. We are currently following the offspring of 70 Mormons in Utah who died with Parkinson's disease between 1930 and 1940; a similar approach utilizing a French Canadian population of Quebec would be possible. Although Mormons and French Canadians do not have much in common religiously, they do share certain propensities such as having large families which remain relatively intact. By selecting suitable control families, studies in Quebec and Utah should reveal if there is an increased rate of Parkinson's disease in the offspring of affected parents.

While clinical studies in carefully selected populations may provide information concerning a possible genetic role in Parkinson's disease, I am also intrigued by the possibility of conducting combined clinical and biochemical studies in defined populations. A fascinating report in the field of pharmacogenetics by Vesell[4] may provide a model. Vesell showed that the duration of phenylbutazone clearance from plasma varies from person to person. Among individuals with delayed clearance, the drug accumulates and quickly reaches toxic levels. Further, by studying twins it was shown that the inability to metabolize phenylbutazone was under genetic control. Clearly, no genetic pattern for drug toxicity had been suspected because no one had a series of sufficient size to make the clinical observation in sibships. Thus, a metabolic trait could be inherited but remain undetected in the absence of an exogenous factor.

Such a complex situation may exist in Parkinson's disease. In patients there is ample evidence of abnormal dopamine metabolism. Unfortunately, the most reliable mechanisms for demonstrating these abnormalities are studies of brain tissue or of cerebrospinal fluid. Both commodities are sufficiently unattainable as to make studies of dopamine metabolism in selected controls in the numbers necessary extremely difficult. However, it would be relatively easy to conduct a controlled study utilizing urinary excretion of monoamines and their catabolites; although the techniques are less sensitive and consistent, it is possible that meaningful genetic patterns would emerge through studies of urine.

We have not yet embarked on such investigations in idiopathic Parkinson's disease, but we have attempted to utilize the concept suggested by Vesell's work in the study of drug-induced extrapyramidal disease; we hope in the future to be able to apply this approach to populations receiving L-Dopa who do not respond or who develop specific secondary abnormal movements while under therapy. One of our studies (conducted by Drs. James A. Schnur and R. Michael Scott) was to compare the presence of parkinsonism-like movements among Negro and white schizophrenics receiving phenothiazines. Morbidity and mortality statistics indicate that Parkinson's disease occurs less frequently among Negroes than among whites. Undoubtedly, faulty diagnosis and reporting and shorter life expectancy among Negroes may partly explain these apparent racial differences. We have attempted to contact neurosurgeons in Africa and in medical centers in the United States where large numbers of elderly Negroes are seen, and it is our impression that Parkinson's disease indeed does not occur as frequently as would be expected among Negroes. Preliminary analysis of our data in phenothiazine-induced Parkinson's disease, however, reveals that the rate of disease is similar among whites and Negroes. These studies were conducted in mental institutions in Maryland and patients were matched to the best of our ability for age, sex, duration of mental illness, and drug histories.

In another study of drug-induced extrapyramidal disorders, we are examining identical twins with schizophrenia on high doses of phenothiazines to determine if the type of motion disorder induced is concordant. At present, our series is limited to four sets of identical twins. In one pair there were no extrapyramidal signs. In another pair one twin had dystonic movements, while the other had no motion disorder. In the remaining two sets dissimilar extrapyramidal signs were observed. Thus, although data are preliminary, it appears unlikely that genetics play an important role in the appearance or specific type of motion disorder in drug-induced parkinsonism.

Perhaps the most promising area to study a possible genetic mechanism in an extrapyramidal disorder is on Guam where the indigenous Chamorro population suffers from a disease we have called parkinsonism-dementia. At the onset, let me stress that the disease on Guam bears little resemblance to classical Parkinson's disease. The symptoms are those of severe, progressive, presenile dementia with massive bradykinesia and some rigidity. Fine tremor is not a prominent finding. Pathologically, there is marked cortical atrophy with neurofibrillary degeneration in the absence of senile plaques and marked cell loss, particularly in the basal ganglia and substantia nigra. On Guam, parkinsonism-dementia is the cause of death for approximately one in ten adults. Similarly, amyotrophic lateral sclerosis is the cause of death for approximately one in ten Guamanians. It is probable that these diseases are related in some way and that their high incidence in this population suggests that a genetic factor may be responsible.

Recently, in collaboration with Dr. Thomas N. Chase, Chief, Unit on Neurology, Laboratory of Clinical Sciences, National Institute of Mental Health, we studied the cerebrospinal fluid levels of homovanillic acid (HVA) and 5-hydroxyindoleacetic acid (5-HIAA) among Guamanian, parkinsonism-dementia patients, amyotrophic lateral sclerosis (ALS) patients, and controls.[5] The HVA levels and to a lesser extent the 5-HIAA levels were extremely low among patients with parkinsonism-dementia and comparable to levels encountered in Americans with classical paralysis agitans. The levels among ALS patients were higher than those in parkinsonism-dementia patients, but significantly lower than those in Guamanian controls. We also studied American ALS patients and found that levels were similar to those encountered among the ALS patients on Guam. The first three parkinsonism-dementia patients who were started on L-Dopa therapy have shown slight but definite improvement.

The observation which was perhaps most germane to the topic of the possible role of genetics in abnormal monoamine metabolism was the observation that in two of the six control patients HVA levels were as low as those encountered among the parkinsonism-dementia patients. We are aware, of course, of the limitations of the methods of analysis and the small numbers tested. It is possible, however, that these low values in controls indicate that an abnormality in monoamine metabolism occurs among apparently unaffected Guamanians. If this is the case, we must now determine how widespread this abnormality is and if it occurs in a definite genetic pattern. It is not possible to test this hypothesis on Guam using spinal fluid or brain tissue. Therefore, we are contemplating preliminary surveys of the urinary excretion of monoamines and their catabolites among parkinsonism-dementia patients and their families to determine the usefulness of this procedure.

REFERENCES

1. Mjones, H.: Hereditary extrapyramidal disease. Acta Neurol. Scand. *39:* Suppl. 14: 108, 1963.
2. Gudmundsson, K. R.: A clinical survey of parkinsonism in Iceland. Acta Neurol. Scand. *43:* Suppl. 33, 1967.
3. Duvoisin, R. C., Schweitzer, M. D., and Yahr, M. D.: Parkinsonism—the role of familial factors. *In* Progress in Neuro-Genetics (Barbeau, A. and Brunette, J. R., eds.). Excerpta Medica Foundation, Amsterdam, 1969.
4. Vesell, E. S., and Page, J. G.: Genetic control of drug levels in man: phenylbutazone. Science *159:* 1479, 1968.
5. Brody, J. A., Chase, T. N., and Gordon, E. K.: Depressed monoamine catabolite levels in cerebrospinal fluid of patients with parkinsonism-dementia of Guam. In press.

STRIATAL AND HYPOTHALAMIC EFFECTS OF TOPICALLY APPLIED L-DOPA AND VELVET BEAN EXTRACTS *

E. A. Spiegel, F. Escobedo, M. Spiegel-Adolf, D. Zivanovic, B. Rusy, and E. G. Szekely

The reason for administration of L-Dopa to patients with Parkinson's disease is chiefly based upon the deficiency of its derivative, dopamine, found in the striopallidum and substantia nigra of such patients.[1] The main symptom alleviated by L-Dopa medication is the brady- and akinesia of the Parkinson patient; it was attempted, therefore, to ascertain how the brady- and akinesia experimentally produced by chemical stimulation of the caudate nucleus [2] can be influenced by intrastriatal microinjection of L-Dopa. Since these experiments were previously reported,[3, 4] they are here summarized only briefly.

Slowing of locomotion and eventually more or less complete akinesia were produced in cats and rats by intrastriatal injection of alumina cream or tungstic acid gel. Injection of L-Dopa in the same caudate nucleus or of freeze-dried extracts of velvet beans containing about 13 percent L-Dopa in the opposite caudate nucleus counteracted the brady- or akinesia so that the original speed of locomotion reappeared (Fig. 1). These changes were statistically significant, while control experiments in which equal amounts of the solvent with the same pH as the Dopa or the velvet extract solution were injected failed to alter significantly the brady- or akinesia (Tables 1 and 2).

From these experiments it was concluded: (1) that chemical stimulation of the caudate nucleus may induce brady- or akinesia, so that an inhibitory influence of caudate impulses upon spontaneous movements, particularly locomotion, could be assumed; and (2) that dopamine, the derivative of L-Dopa, is able to reduce this inhibitory effect of striatal impulses upon spontaneous movements.

Electrographic studies seem to be in agreement with the second conclu-

* Aided by Grant NB 5316 from the National Institute of Neurological Diseases and Stroke, National Institutes of Health.

TABLE 1. Effect of Intracaudate Injections Upon Speed of Locomotion

	Tungstic Acid Effect (Speed Decrease)		*Dopa Effect (Speed Increase)*	
Series 1 N=13	Speed before minus speed after injection	Mean difference 2.56 ± 0.92 * t=10 $p<0.001$	Speed after minus speed before injection	Mean difference 2.35 ± 1.37 t=6.2 $p<0.001$
	Tungstic Acid Effect		*Effect of Solvent Without Dopa (Control)*	
Series 2 N=10	Speed before minus speed after injection	1.93 ± 0.95 t=6.4 $p<0.001$	Speed after minus speed before injection	0.3 ± 0.89 t=1.1 not significant

* Figures indicate number of rotations of pulley. One rotation corresponds to a locomotion of 8.5 cm on floor. t indicates mean difference divided by the standard error of the mean.

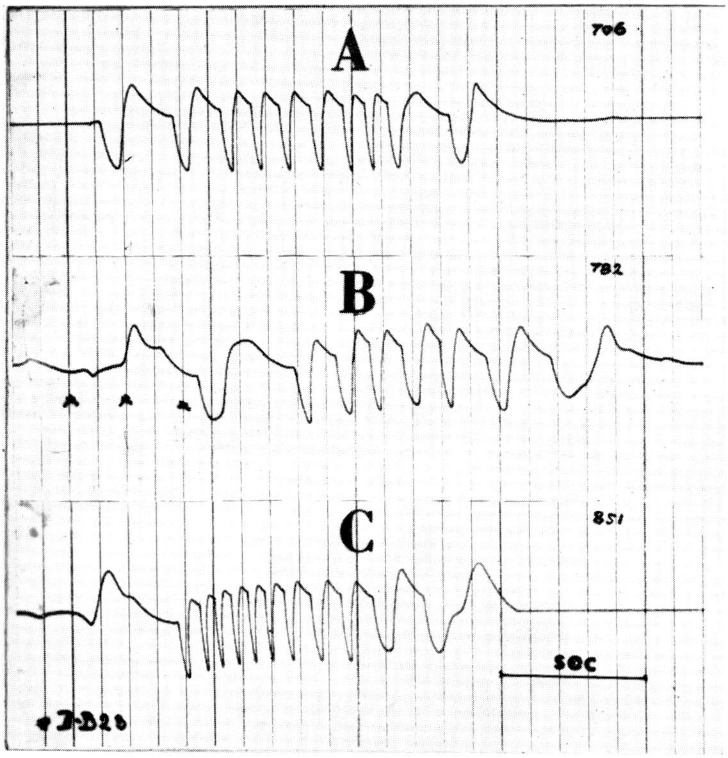

Fig. 1. Rat. *A:* Spontaneous running (1 deflection=8.5 cm on ground). *B:* 40 minutes after injection of 10 μl tungstic acid gel into left caudate nucleus. No spontaneous locomotion. After touching the body three times (signals), slow movements. *C:* 25 minutes after intracaudate injection of freeze-dried velvet bean extract containing 200 μg L-Dopa in 10 μl solution: spontaneous running.

TABLE 2. Effect of Intracaudate Injections Upon Speed of Locomotion

	Tungstic Acid or Amphogel Effect (Speed Decrease)		*Bean Extract (Speed Increase)*	
N=12	Speed before minus speed after inject.	Mean difference 4.39 ± 0.28 * t=15.44 p<0.001	Speed after bean inj. minus speed after amphogel inj.	5.63 ± 0.40 * t=14.04 p<0.001
	Tungstic Acid Effect		*Control (Effect of Solvent)*	
N=8	Speed before minus speed after inject.	4.32 ± 0.49 t=8.79 p<0.001	Speed after solvent — Speed before solvent (after tungstic acid)	0.00

* Number of rotations of pulley per sec induced by running.

sion. In studies of the cat's caudate nucleus with macroelectrodes (Spiegel in experiments with Szekely), the global effect of intravenous and intraperitoneal injection of large doses of L-Dopa resulted in long periods of electric silence in the caudate nucleus, interrupted by groups of slow waves.

Although microelectrode studies [5] revealed in some cells excitatory effects of dopamine, the majority of the units were inhibited.[6] On administration of large doses of L-Dopa, one may expect that its metabolites will affect not only the striatum but also other parts of the central nervous system.

Our attention was next directed to the hypothalamus, since bilateral hypothalamic lesions produced by Ranson [7] in monkeys induced a state of catalepsy, while stimulation of the hypothalamus or of the mesencephalic reticular formation restored the speed of locomotion of cats that were brady- or akinetic following chemical stimulation of the caudate nucleus.[8] We inferred that there are two antagonistic systems in the brain: a facilitating reticulohypothalamic and an inhibitory striatal system. In view of F. H. Lewy's findings of severe cell degeneration in the hypothalamus of patients with parkinsonism,[9] the assumption was made that the balance between these two antagonistic systems is shifted in favor of the striatal inhibition.

The involvement of the hypothalamus is also suggested by the numerous vegetative symptoms (profuse perspiration, seborrhea, drooling) found in parkinsonism and a possible action of L-Dopa on the hypothalamus by the beneficial effect of this drug upon these vegetative signs and symptoms. The suspicion that dopamine and/or its metabolite norepinephrine may affect hypothalamic areas, particularly on administration of large doses of L-Dopa, is further supported by the demonstration of monoaminergic neurons in the hypothalamus.[10, 11, 12] It may be recalled that, according to Fuxe, there is a high concentration of dopamine in the dorsomedial hypothalamic nucleus. Norepinephrine is abundant in the periventricular nuclei, in the external layer of the median eminence, in the vicinity of the ventral supraoptic decussation, and in the interstitial nucleus of the stria terminalis. Rinne and Sonninen [13] found norepinephrine neurons in the tuberohypophyseal system (dorsomedial, ventromedial, and arcuate nuclei). During various states of the reproductive cycle, Fuxe et al.[14] observed activity changes in the tuberoinfundibular dopamine neurons of the rat. Steroid-sensitive neurons of the basal hypothalamus were inhibited by dopamine in rats.[15] In the parkinsonian hypothalamus a reduction of the norepinephrine concentration was found by Hornykiewicz and his group.[16, 17]

In view of these considerations it seemed of interest to study the effects of L-Dopa and of velvet bean extract applied locally to the hypothalamus.

Methods and Materials

The experiments were performed in 22 cats. With the animal anesthetized with nembutal, a plastic screw, with two holes 5 to 6 mm distant from each other permitting insertion of 27-gauge needles, was fixed with its center in the midline of the skull. This fixation was stereotactically guided by a pin inserted at one end into the screw, at the other into the stereotactic apparatus.

Several days after this procedure, one needle was inserted into the left and later the other into the right hypothalamus of the conscious cat. In some of these cats spontaneous movements were reduced by intrastriatal injection of alumina cream or by subcutaneous injection of bulbocapnine hydrochloride (25 mg per kg). A solution (10 μl) containing 250 μg L-Dopa or 2.5 mg freeze-dried velvet bean extract with 325 μg L-Dopa was injected into the hypothalamus. The solution of L-Dopa or freeze-dried extract was stained by the addition of 1 drop of India ink to 0.5 cc solution.

In addition to observations of the general behavior, size of pupil, urination, and defecation, the cat's respiration was counted and locomotion was recorded photoelectrically as previously described. The electrocardiogram was recorded by means of clips fixed to the chest. The temperature was measured on the left forepaw by means of a Yellow Springs telethermometer. The blood pressure from the femoral artery was recorded. The site of the injection was checked macroscopically and in important cases microscopically in serial sections stained for fibers (Weil method) or for cells and fibers (Klüver method).

According to Marzcynski's review,[18] injection of solutions smaller than 20 μl yields reliable results that can be related to the site of the injection and are not vitiated by leakage of the fluid into the vicinity.

Results and Discussion

The location of the injections is shown in Figure 2 and the corresponding symptoms are summarized in Table 3.

It can hardly be expected that a compound such as L-Dopa or its derivatives will affect specifically only one of the many cell groups differentiated in the hypothalamus. The concept developed by Hess [19, 20] seems, therefore, useful; he distinguishes a dynamogenic, ergotropic region in the posterior hypothalamus extending into the lateral hypothalamus from a trophotropic-endophylactic region in the suprachiasmatic, preoptic and septal areas. The topical application of L-Dopa seems to influence both regions. The effects appear after a latent period of from one to several minutes. This seems to support the interpretation that the chemical effect of the compound slowly infiltrates the tissue while the mechanical effect of insertion of the needle, if any, appears within a few seconds and is of very short duration, as observed when recording the blood pressure.

The only syndrome elicited from the ergotropic region was evidenced by behavioral reactions suggestive of emotional excitement (rage, fear) and associated with running, sometimes circling, mostly to the opposite side with the head tilted to the same side, and climbing on the wall of the cage (pronounced after injections Nr. 6r, 91, 10m, 12r; see Fig. 3). Increased vocalization (crying) was observed after injection into either area. From the trophotropic zone increased locomotion, but without other sign of excitement, was induced.

It should be emphasized that injection into the ergotropic zone sometimes

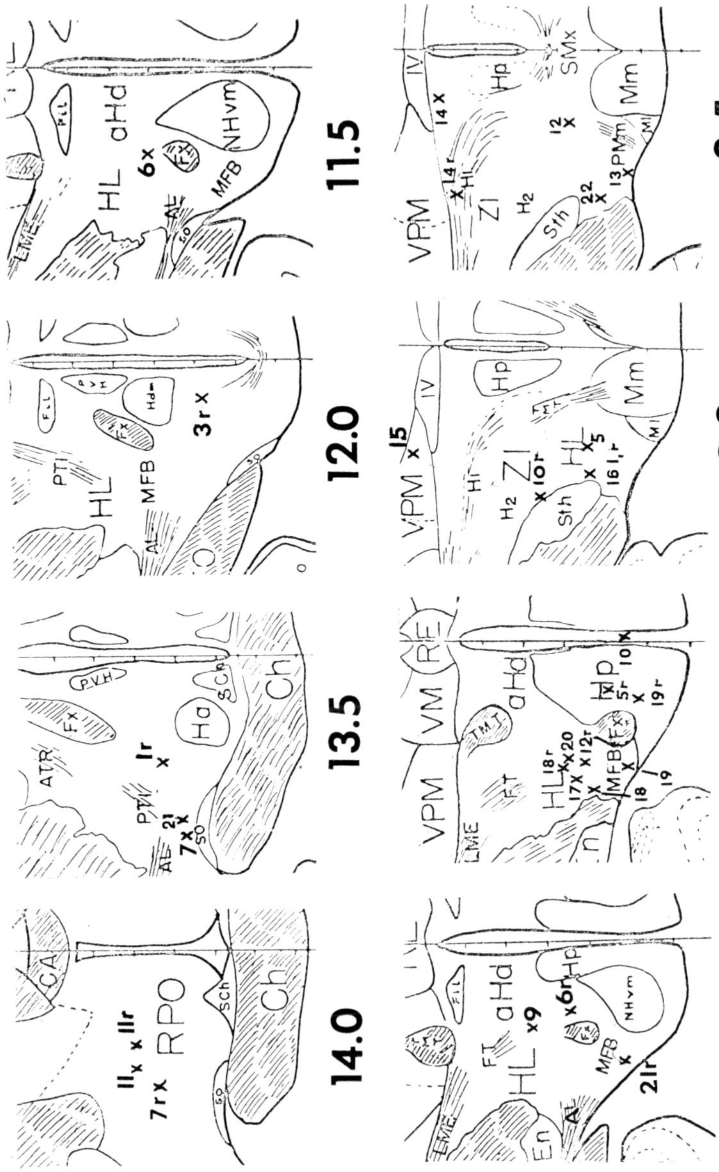

Fig. 2. Sites of injection shown in diagrams from Jasper and Ajmone-Marean's atlas. Right side indicated by r; the cat number alone indicates the left side.

TABLE 3. Effects of Hypothalamic Injections

Cat #	Inj.	*Respiration Av.* *Bef.*	*After*	*Heart Rate Av.* *Bef.*	*After*	*Temp. °C* *Bef.*	*After*	*Pupil.*	*Urinat.*	*Defec.*	*Loco-mot.*	*Behavior*
7 Bulbocap. akinetic	l.D.	25	41	174	147	30.5	28.5	sl. dil.	—	—	slow	constant vocaliz; quiet
7 Alumina bradykin.	r.D.	56	53	150	156	29.0	28.8	—	—	—	increas. circling both direct. later slow	constant vocaliz; later quiet
11 Alumina bradykin.	l.D.	36	34	207	228	28.8	28.0	sl. dil.	—	—	"frozen"	quiet sleepy
11 Alumina bradykin.	r.B.	34	104	228	253	28.0	32.5	dil. +++ aft. 15′ normal	+	+	increas. +++	crying
11 Alumina bradykin.	i.v.B. 300 μg	45	40	204	222	27.0	27.2	—	—	—	no spont.	quiet
1 Alumina in striat. akinetic	r.D.	24	56	127	192 (arrhyth)	28.7	32.0	aft. 10′ dil aft. 1½ hrs. mios.	+	—	increas. +++ after 2 hrs. no spont.	vocaliz. shivering
21 Bradykin. (untreated)	l.S.	27	26	150	156	32.0	32.0	—	—	—	no spont. locomot.	quiet sleepy
21	l.D.	26	33	164	177	31.0	32.5	slow light refl.	—	—	some spont. locomot.	quiet alert
6 Akinetic bulbocap.	l.D.	20	21	156	156	31.3	29.8	—	—	—	no spont.	sleepy
6 Alumina akinetic	r.D.	18	21	141	147	27.0	29.0	dil. +++	—	—	once spont. backward	rage on provocation. licking

6 Alumina bradykin.	i.p. D. 200 mg	17	40	129	152	29.0	30.0	sl. dil.	+	+	no locomot. even aft. stim.	vomit 2 x quiet
3 Bulbocap. akinetic	r.D.	42	40	156	171	27.0	26.5	dil. r>1	—	—	spont. walk for 20 min.	shivering later quiet
9	l.B.	28	64	153	243	29.2	36.6	after 2 hrs. miosis	+	+	very active +++ runs climbs on cage wall	excited much crying 20′ then quiet, sitting
10 Alumina in striat. slight bradykin.	m.B. part. in 3rd ventr.	25	73	175	234 (arrhyth)	28.5	35.0	dil. ad max.	—	+	runs, climbs on cage wall	continuous. vocalizat. excitement
10	r.D.	28	25	147	186	29.7	30.3	—	—	—	r. foreleg abnorm. posture; then spont. movements	much vocalizat. licking
12 Slow (untreat.)	l.D.	29	32	77	114	30.5	30.7	—	—	—	no spont.	sleepy quiet
12	r.B.	32	69	111	108	30.8	30.7	dil.+++	+	+	runs counter clockwise climbs cage wall	excited ++ rage like. vocalizat. ++
17	m.B.	Puncture produces short-lived rise in blood pressure, questionable effect of bean extract on pressure										
17	r.D.	200 μg no effect on blood pressure										
18	l.	Insertion of cannula. no effect on blood pressure, Dopa 250 μg: after 2 min. first rise then fall of blood pressure; 250 μg Dopa i.v. fall of blood pressure.										
18	r.	Insertion of cannula: no effect on blood pressure. Bean extr. no effect; i.v. slight depress.										

TABLE 3—Continued

Cat #	Inj.	*Respiration Av.* *Bef.*	*After*	*Heart Rate Av.* *Bef.*	*After*	*Temp. °C* *Bef.*	*After*	*Pupil.*	*Urinat.*	*Defec.*	*Loco-mot.*	*Behavior*
19 Alumina	l.S.	29	28	221	190	27.5	30.0	sl. dil.	—	—	no spont.	much vocalizat. sleepy quiet
19 Alumina	r.S.	32	24	166	162	29.5	32.0	sl. dil.	—	—	immediat. aft. inj. no spont. locomot., later spont.	some vocalizat. then sits quietly, sleepy
19	i.v. D. 250 μg	26	27	177	174	32.0	33.0	—	—	—	no locomot. akinetic	sleepy, sitting quietly eyes closed
20 Paralyz. (flaxedil)	l.D.	Insertion of needle: no effect on blood pressure 250 μg: after 3 min. definite rise of blood pressure, lasts 10 min.										
20	r.B.	Inj. 250 μg. Velvet bean extract, increase of blood pressure lasting 6 min.										
20	i.v. B.	Inj. 250 μg. mg. Velvet bean extract i.v. no effect on blood pressure										
21 Bradykin. (untreat.)	r.D.	32	37	134	126	30.5	31.5	sl. miosis lasts 30′	—	—	no locomot. even aft. body was hit	quiet, sleepy no attent. to sound stim.
22 Alumina bradykin.	l.D.	23	23	111	133	31.0	32.0	light refl. slower	—	—	no spont.	quiet, sleepy no reaction to sound
22	r.D. (above base behind mammil.)	23	22	133	150	31.0	31.0	slow light refl. 30′ lat. slight miosis	—	—	no locomot. even on painful stim.	sleepy, quiet

5 Spont. locomot. alumina part. in n. caud. part. in caps. int.	l.D.	47	60	126	144	28.5	25.5	slight dil. slow react.	—	—	decreas.	frozen
5	r.D.	53	29	126	159	25.5	25.7	slight dil.	—	—	decreas.	frozen
15 Paralyz. (flaxedil)	l.B.	Injection into ventral thalamus; slight increase in blood pressure										
	r.D.	Injection into ventral thalamus; slight increase in blood pressure										
16 Spont. locomot. (untreat.)	l.D.	27	21	207	190 (arrhyth)	28.0	30.0	refl. slower	—	—	no spont. locomot.	drinks milk-did not do before inj.; quiet
16	r.B.	21	25	190	255	29.0	30.5	slight dil.	—	—	some spont.	head tremor vocalizat. ++
16	r.S.	33	25	255	225	30.0	28.0	—	—	—	some spont.; then sits	vocalizat. ++ then sleepy
13 Spont. locomot. (untreat.)	r.D.	34	64	156	135	29.5	30.7	slight dil.	+	—	cataleptic	licking vocalizat. +++
14 Spont. locomot. (untreat.)	l.D.	46	37	141	150	27.0	29.0	slight dil.	—	—	slow; later slight cataton.	sleepy licking swallow
14	r.B.	40	32	135	130	26.5	29.2	slight dil.	—	—	slow; 20′ post inj. jumped into cage	quiet, frozen. sleepy

Explanation: The experiments are listed according to the frontal level of the injection site (oral levels first, caudal levels last). l = left, r = right, m = midline. D = injection of L-Dopa, B = injection of freeze-dried velvet bean extract, S = injection of solvent. i.p. = intraperitoneal (control), i.v. = intravenous (control). Injection in Cat 2 was outside diencephalon. Cat 4 died before hypothalamic injections.

A.

B.

C.

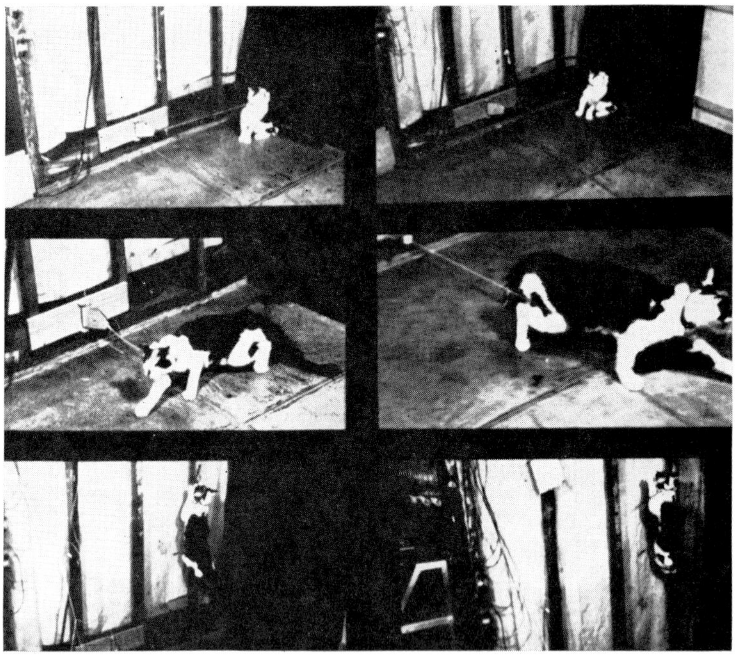

Fig. 3. Cat 12. *A:* before and *B, C* after injection of velvet bean extract in ergotropic zone. *B:* circling; *C:* climbing on wall of cage.

was followed by sleepiness or a cataleptic state (see No. 131). Hess observed, after 1 to 2 strong electric stimulations, a definite decrease of activity and a state of sleepiness.

Regarding respiratory and cardiac rates, changes in body temperature and pupil, definite differences between the responses to Dopa of the two regions could not be found. Six injections into the trophotropic zone elicited increased rate of respiration four times and increased heart rate four times; in one cat cardiac arrhythmia was noted. In one instance the heart rate dropped. Increase of body temperature was elicited in three cats and a decrease in one. The pupil reacted once with maximal, twice with slight dilation; in another instance the slight dilation was followed by miosis. Urination occurred twice and defecation once.

Injections into the ergotropic zone were performed thirteen times. They resulted seven times in an increase and once in a decrease of the respiratory rate, eight times in an increase and once in a decrease of the cardiac rate. The increased frequency of the latter was twice associated with arrhythmia (Fig. 4). The temperature rose seven times and dropped once. Dilatation of the pupil was on three occasions pronounced (Fig. 5), four times slight. Miosis appeared twice. Urination (seen twice) and defecation (observed three times) probably were the result of the general excitement of the animals.

If one compares the effect of the local application of L-Dopa with that of electrical stimulation of the hypothalamus, one finds only a partial parallelism,

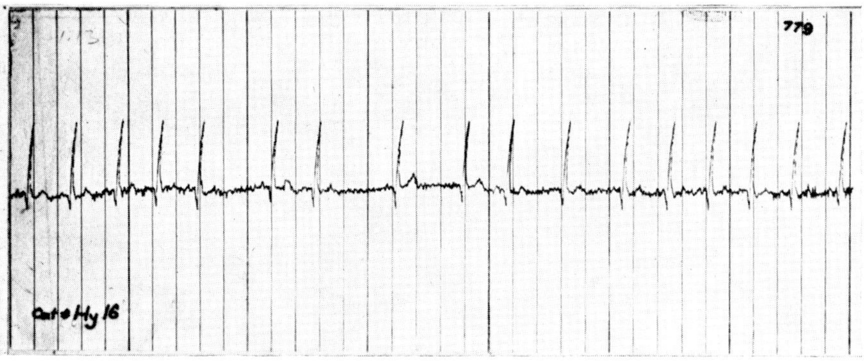

Fig. 4. Cat 16. Cardiac arrhythmia, 10 minutes after injection of 250 μg L-Dopa (10 μl solution) into left ergotropic zone.

e.g., the psychomotor excitement on injection into the ergotropic area corresponds to the effects of electrical stimulation. However, injection into the anterior hypothalamus and preoptic area failed to produce effects opposite to those seen after injection into the posterior hypothalamus, e.g., they induced only tachypnea, and the expected reduction of the respiration on injection into the trophotropic zone could not be observed.

It seems that, in the hypothalamus and septum, systems facilitating respiration and those inhibiting it are adjacent. For instance, Hess found a facilitatory

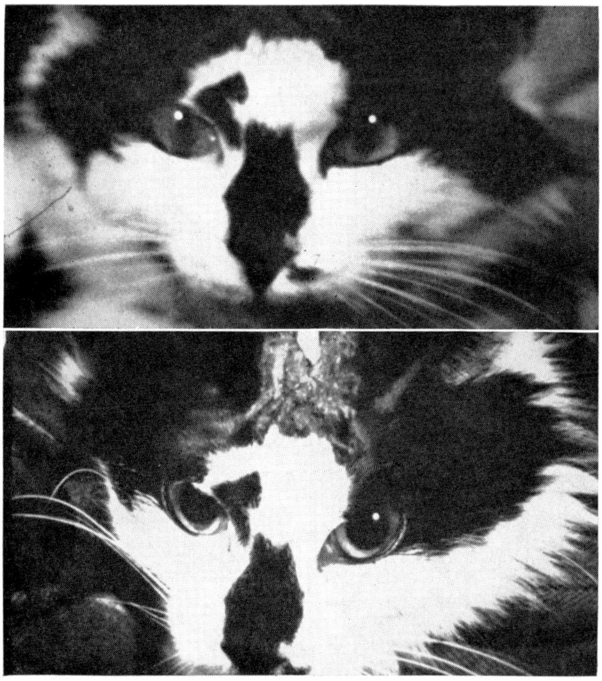

Fig. 5. Cat 12. Pupil before (*top*) and after (*bottom*) hypothalamic L-Dopa injection.

area between Meynert's tract and the posterior commissure. Tachypnea could also be induced by stimulation of the septum around the fornix, while inhibitory effects appeared on stimulation of the base of the columna fornicis descendens and from an area ventral, lateral and caudal to the perifornical activating center. Similarly Andersson [21, 22] observed panting on stimulation of an area ventral to the commissura anterior, but on stimulation of the septum, mediodorsal to the preoptic area, an inhibition of the polypnea.

In our experiments we demonstrated an inhibitory effect upon those areas that cause a reduction of the respiratory rate on electric stimulation. It should also be recalled that Yamaguchi, Marczynski and Ling [23] elicited an increase of the rate of respiration, tachycardia, and an electroencephalographic pattern of alertness on administration of the L-Dopa derivative norepinephrine to the preoptic region.

The effects of the bean extract injection were, as a rule, more pronounced and longer lasting than those of the injection of synthetic L-Dopa. This may partly be due to the fact that a somewhat higher amount of L-Dopa was injected with the bean extract; the possibility of other factors will have to be investigated.

It has already been pointed out that effects of the mechanical stimulus induced by the insertion of the needle, if they appeared at all, are noticed after a latent period of only a few seconds and last for seconds only, while the effects of the Dopa injections, e.g., the rise in blood pressure (Fig. 6) start after one to several minutes. The miosis (Nr. 21r) may last for half an hour and longer.

Another possible source of error is mechanical injury to the tissue by the injected fluid and the physicochemical effect of its acidity (pH 4.5 to 5.0). Therefore, in control experiments 10 μl of the solvent, i.e., of acidified water with a pH of 4.5 to 5.0, were injected. This caused, at least in some instances (No. 19l and 19r), a rise of body temperature and a slight dilatation of the pupil. Thus the increase of the body temperature observed after hypothalamic L-Dopa injection cannot be regarded with certainty as an effect of this com-

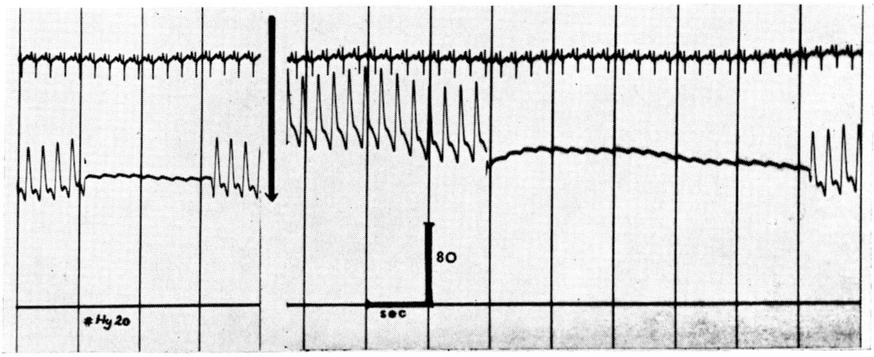

Fig. 6. Cat 20. Flaxedilized. Femoral blood pressure and EKG. *A:* After insertion of needle into left hypothalamus. *B:* Three minutes after injection of 10 μl solution containing 250 μg L-Dopa. Calibration 80 mm Hg and 1 sec.

pound, and the dilatation of the pupil should be considered as a Dopa effect only if very marked. The other changes induced by L-Dopa injection—excitation, increased locomotion, increased respiratory rate and other vegetative effects—should be attributed to this substance. Intravenous or intraperitoneal injection in amounts equal to those applied to the hypothalamus had no effect.

In view of the relatively small number of the present experiments the results are considered as preliminary only. Results on a larger series will be reported later.

It should be pointed out that the influence of topical application to the hypothalamus cannot be equated in all instances to the general effect of the substance. It is known that ingestion of large doses of L-Dopa induces orthostatic hypotensive episodes which are due, according to Barbeau et al.,[24] to a decrease of the plasma renin activity. In this instance, this antagonistic L-Dopa effect apparently outweighs its hypothalamic influence.

These preliminary observations indicate that the analysis of L-Dopa effects must take into consideration an action not only upon the striatum but also upon other cerebral areas such as the hypothalamus.

REFERENCES

1. Hornykiewicz, O.: Metabolism of brain dopamine in human parkinsonism: Neurochemical and clinical aspects. *In* Biochemistry and Pharmacology of the Basal Ganglia (Costa, E., L. J. Côté, and M. D. Yahr, eds.). Raven Press, Hewlett, New York, 1966.
2. Spiegel, E. A., and Szekely, E. G.: Prolonged stimulation of the head of the caudate nucleus. Arch. Neurol. *4:* 55, 1961.
3. Spiegel, E. A., Szekely, E. G., and Zivanovic, D.: The effect of intracaudate injection of L-Dopa upon the bradykinesia induced by caudate stimulation. 2nd Intl. Congr. Neurogenetics and Neuroophthalmology. Excerpta Med. Congr. Ser. *154:* 28. September 1967, Montreal.
4. Spiegel, E. A., Spiegel-Adolf, M., Wycis, H. T., Szekely, E. G., Zivanovic, D., and Schwartz, H. A.: Restoration of speed in parkinsonian and experimental bradykinesia and akinesia. Symposium on Parkinsonism and L-Dopa. 9th Intl. Congr. Neurology. September 23, 1969, New York City.
5. Bloom, F. E., Costa, E., and Salmoirhagi, G. C.: Anesthesia and the responsiveness of individual neurons of the caudate nucleus of the cat to acetylcholine, norepinephrine and dopamine administered by microelectrophoresis. J. Pharmacol. Exp. Ther. *150:* 244, 1965.
6. McLennan, H., and York, D. H.: The action of dopamine on neurons of the caudate nucleus. J. Physiol. *189:* 393, 1967.
7. Ranson, S. W.: Somnolence caused by hypothalamic lesions in the monkey. Arch. Neurol. Psychiat. *41:* 1, 1939.
8. Szekely, E. G., Spiegel, E. A., and Zivanovic, D.: Stimulation of hypothalamus, dorsomedial nucleus and reticular formation in experimental brady- and akinesia. Fed. Proc. *28:* 648, 1969.
9. Lewy, F. H.: Tonus und Bewegung. Springer, Berlin, 1923.
10. Dahlström, A., and Fuxe, K.: Evidence for the existence of monoamine containing neurons in the central nervous system. Acta Physiol. Scand. *62:* Suppl. 232, 1964.

11. Fuxe, K.: The distribution of monamine terminals in the central nervous system. Acta Physiol. Scand. *64:* Suppl. 247, 1965.
12. Shute, C. C. D., and Lewis, P. R.: Cholinergic and monoaminergic pathways in the hypothalamus. Brit. Med. Bull. *22:* 221, 1966.
13. Rinne, U. K., and Sonninen, V.: The occurrence of dopamine and NA in the tuberohypophyseal system. Experientia *24:* 177, 1968.
14. Fuxe, K., Hökfelt, T., and Nilsson, O.: Activity changes in the tubero-infundibular dopamine neurons of the rat during various states of the reproductive cycle. Life Sci. *6:* 2057, 1967.
15. Steiner, F. A., Pieri, L., and Kaufmann, L.: Effect of dopamine and ACTH on steroid sensitive single neurons in the basal hypothalamus. Experientia *24:* 1133, 1968.
16. Ehringer, H., and Hornykiewicz, O.: Verteilung von Noradrenalin u. Dopamin im Gehirn des Menschen u. ihr. Verhalten bei Erkrankungen des extrapyramidalen Systems. Klin. Wschr. *38:* 1236, 1960.
17. Bernheimer, H., Birkmayer, W., and Hornykiewicz, O.: Zur Biochemie des Parkinson syndroms des Menschen Einfluss der Monoaminoxydase Hemmer-Therapie auf die Konzentration des Dopamins, Noradrenalins und 5-Hydroxytryptamins im Gehirn. Klin. Wschr. *41:* 465, 1963.
18. Marczynski, T. J.: Topical application of drugs to subcortical brain structures and selected aspects of electrical stimulation. Ergebnisse der Physiologie *59:* 86, 1967.
19. Hess, W. R.: Vegetative Funktionen u. Zwischenhirn. B. Schwabe & Co., Basel, 1946.
20. Hess, W. R.: Beiträge zur Physiologie des Hirnstamms. II Das Zwischenhirn u. die Regulation von Kreislauf u. Atmung. Thieme, Leipzig, 1938.
21. Andersson, B., Granit, R., and Larsson, S.: Central control of heat loss mechanisms in the goat. Acta Physiol. Scand. *37:* 261, 1956.
22. Andersson, B.: Cold defense reactions elicited by electrical stimulation within the septal area of the brain in goats. Acta Physiol. Scand. *41:* 90, 1957.
23. Yamaguchi, N., Marczynski, T. J., and Ling, G. M.: The effects of electrical and chemical stimulation of the preoptic region and some nonspecific thalamic nuclei in unrestrained, waking animals. EEG Clin. Neurophysiol. *15:* 154, 1963.
24. Barbeau, A., Gillo-Joffroy, L., Boucher, R., Nowaczynski, W., and Genest, J.: Renin-aldosterone system in Parkinson's disease. Science *165:* 291, 1969.

SELECTION OF PATIENTS FOR L-DOPA OR SURGERY

C. Bertrand, S. N. Martinez, P. Molina-Negro, F. Velasco, and J. Hardy

We will discuss selection in the light of our experience in choosing more than 1000 cases for stereotaxic surgery during the past fifteen years.[1, 2] We wrote in 1963:[3] "Evidently surgery is not a cure for parkinsonism, but a means to abolish or at least to alleviate tremor, rigidity and to a much lesser degree bradykinesia." Although more data have been gathered, our selection has become more accurate, and the advent of L-Dopa has rendered us more selective, generally speaking this statement remains true.

The characteristics of the ideal candidate for surgery are:

1. predominance of tremor and/or rigidity.
2. unilateral motor signs involving mostly the upper limbs.
3. a satisfactory gait.
4. little or no ankylosis of the joints.
5. little or no alteration of speech.
6. a slowly progressive course.
7. age less than 65 years, or at least satisfactory cardiovascular and general condition.
8. mental alertness, little or no memory trouble, no confusion or hallucinations.
9. little or no alteration of the electroencephalogram.

The typical akinetic patient is definitely not a candidate for surgery. He can be described as follows:

1. a short history with rapid progression of symptoms.
2. very little tremor and little or no rigidity, but marked akinesia.
3. bilateral and almost symmetrical involvement of the limbs with tendency to propulsion and retropulsion.
4. lack of convergence of the eyes and even involvement of certain eye movements, possibly with a sign of Parinaud.
5. marked impairment of memory, occasional episodes of confusion or hallucinations especially under stress.
6. drooling, dysarthria or dysphagia, slow voluntary movements of tongue.
7. diffuse slow dysrhythmia on the electroencephalogram.

This malignant form of parkinsonism is relatively rare and the pure case of unilateral tremor or rigidity is not frequent.

Before discussing the factors involved in a decision for or against surgery, let us mention what surgery has to offer. It is true that results depend partly on the choice of patients and even more on the criteria used to assess improvement.[4, 5, 6, 7] However, it is well recognized that surgery can relieve or suppress tremor altogether in more than 85 percent of cases if precautions—such as accurate localization, stimulation, fine electrode recording which precisely determines the borders of the thalamus, and temporary suppression of functions (in our cases with a fine leukotome to allow radial exploration)—are used [3, 4, 5, 7, 8] (Figs. 1, 2).

We have found that minute lesions produce hypotonia, which can be increased almost at the target point (19 to 20 mm behind the anterior commissure; 13 mm lateral to the midline, and at or just below the intercommissural line in most subjects).[6] The hypotonia which occurs with relief of tremor in cases of heredofamilial tremor, where muscle tone is preoperatively normal, demonstrates that the areas mentioned have to do with the maintenance of muscle tone.

An important question concerns the usefulness of hypotonia. How much hypotonia is desirable? Diminution in muscle tone may render the patient slightly less aware of his limbs, at least at the beginning, and sometimes it may take him a few days to be sturdy on his feet. This change in tone is definitely beneficial for the upper limbs, since finger movements are speeded up and alternating movements useful for activities such as eating and writing are improved. In the lower limbs, fine movements are also enhanced but hypo-

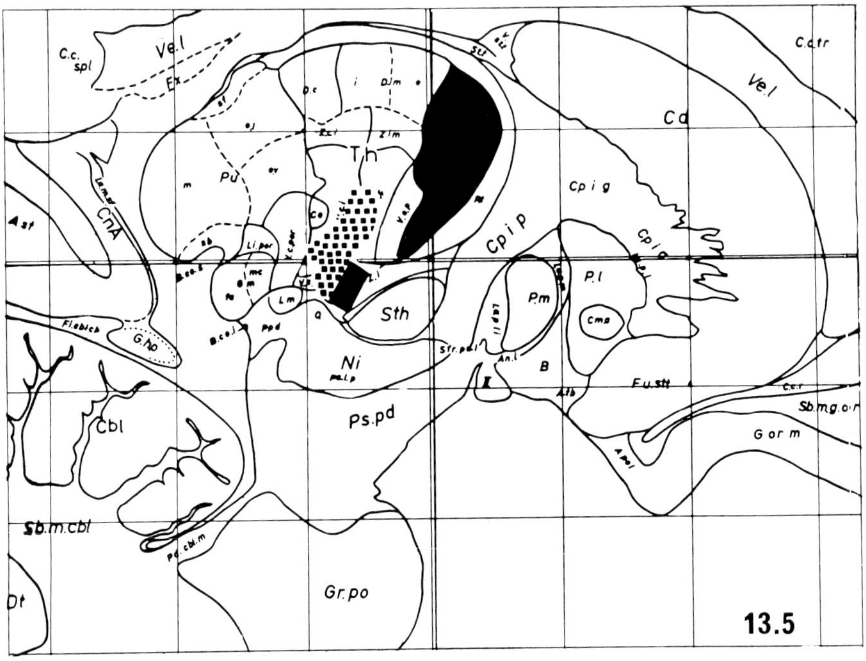

AUTONOMIC RHYTHMS ■ EVOKED RHYTHMS ⁞⁞⁞

Fig. 1. Sites where autonomous rhythms (black areas) and evoked rhythms (small black squares) were obtained.

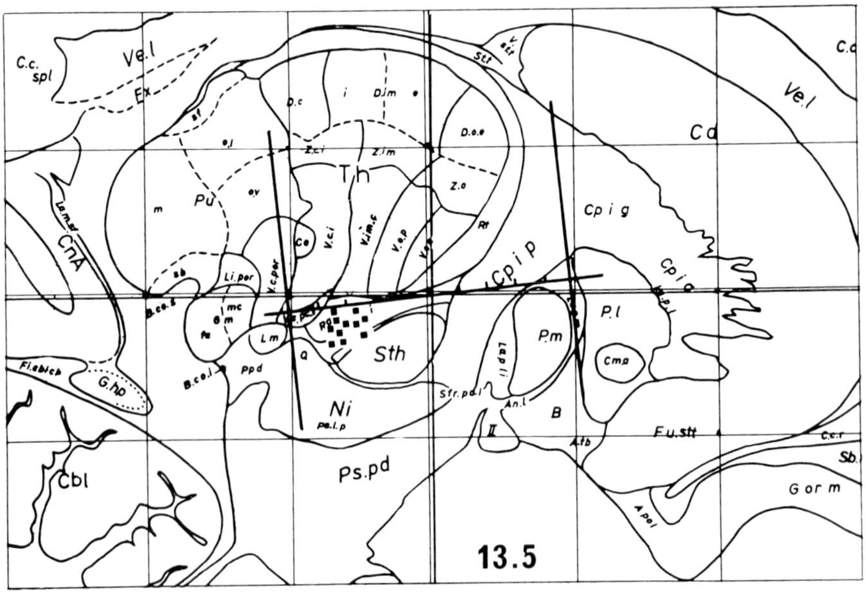

PARKINSON

Fig. 2. Points where tremor stopped on introduction of electrode. Note coincidence with lower sites of autonomous rhythms.

tonia does not help the gait unless the limbs are initially very rigid. In most cases, too much hypotonia in the lower limbs is undesirable. The presence of hypotonia has been a source of confusion in the interpretation of results and of indications for surgery. Surgery is directed at motor symptoms, and its indication depends on which symptoms are most troublesome to the patient. If his gait is the major problem, he should not be operated on unless careful studies show that his difficulty is due to rigidity of a lower limb. If the use of his hand is deficient, even if there is a slight gait problem, surgery is indicated.

The physician can form an impression as to the advisability for surgery by having the patient walk, perform rapidly alternating movements of the hand, and tap his fingers.

The volume of the voice is another point to be considered. A rigid vocal cord may be helped by surgery and speech may be improved in certain patients, but general volume is not improved and may occasionally diminish slightly with hypotonia, especially after bilateral surgical lesions. Only 20 percent of our cases have had bilateral surgery. The anterolateral quadrant of the thalamic nucleus on the second side is often spared, especially in the dominant hemisphere, to avoid the possibility of damaging the corticobulbar fiber which might exaggerate dysarthria. With stimulation and proper centering of the lesion this can be avoided. Using such safeguards, dysarthria or dysesthesia occur in less than 1 percent of patients. Hemiplegia or persistent choreic movements are not encountered, and mortality from hemorrhage is of the order of 0.5 to 1 percent.

Akinesia is a term which needs more precise definition. It is usually understood to describe difficulty in initiating movements apart from rigidity; however, one may encounter patients who can initiate movements quite rapidly, but are unable to do alternating movements in the absence of rigidity. They may be quite alert mentally, with no dysarthria and no difficulty with convergence of the eyes. In such patients surgery can produce very good results.

Surgery has established its value in treating tremor and rigidity. In following our patients over the years, it has become evident that, unless we are dealing with a rapidly progressive type of Parkinson's disease, most of the benefit remains once the first six months after surgery have passed and the results have stabilized. Even relief of tremor from the early pallidal lesions, thought to be effective primarily for hypotonia, persists in many instances.

REFERENCES

1. Bertrand, C.: Une nouvelle modification technique pour la chirurgie des mouvements involontaires. Un. Med. Canad. *84:* 150, 1955.
2. Bertrand, C.: A pneumotaxic technique for producing localized cerebral lesions. J. Neurosurg. *15:* 251, 1958.
3. Bertrand, C., Martinez, N., and Gauthier, C.: Surgical treatment of parkinsonism, the use of a pneumotaxic guide with recording and stimulation. Canad. Med. Assoc. J. *82:* 921, 1960.

4. BERTRAND, C., and MARTINEZ, N.: Localization of lesions, mostly with regard to tremor and rigidity. Confin. Neurol. *22:* 274, 1962.
5. BERTRAND, C., MARTINEZ, N., HARDY, J.: Localisation des fonctions lors de la chirurgie stéréotaxique par une voie d'abord frontale parasagittale. Neurochirurgie *10:* 389, 1964.
6. BERTRAND, C., HARDY, J., MOLINA-NEGRO, P. and MARTINEZ, S. N.: Optimum physiological target for the arrest of tremor. *In* Third Symposium on Parkinson's Disease (Gillingham, F. J. and Donaldson, I. M. L., eds.), E. & S. Livingstone, Edinburgh, 1969.
7. BERTRAND, C., MARTINEZ, N., POIRER, L., and GAUTHIER, C.: Une méthode de traitement chirurgical de la maladie de Parkinson. Un. Med. Canad. *89:* 297, 1960.
8. HARDY, J., BERTRAND, C., and MARTINEZ, N.: Activités cellulaires thalamiques liées au tremblement parkinsonien. Neurochirurgie, *10:* 449, 1964.

METABOLISM OF 3,4-DIHYDROXYPHENYLALANINE AFTER INHIBITION OF EXTRACEREBRAL DECARBOXYLASE

A. Pletscher and G. Bartholini

This paper will consider some changes in the metabolism of exogenous Dopa induced by inhibition of extracerebral decarboxylase (DC) of aromatic amino acids. The compound Ro 4–4602, which has been used to inhibit this enzyme, is shown in Figure 1.

Inhibition of DC

Ro 4–4602 added to tissue homogenates of heart and brain causes the same degree of DC inhibition in vitro (Fig. 2). Parenteral administration of Ro 4–4602 to the intact animal, however, inhibits DC more markedly in the heart than in the brain (Fig. 3). In rats, for example, 50 mg per kg of the drug cause an almost complete inhibition in the heart, but have no effect in the brain. Only doses of 300 mg per kg or larger of Ro 4–4602 inhibit the cerebral enzyme to a degree greater than 75 percent.

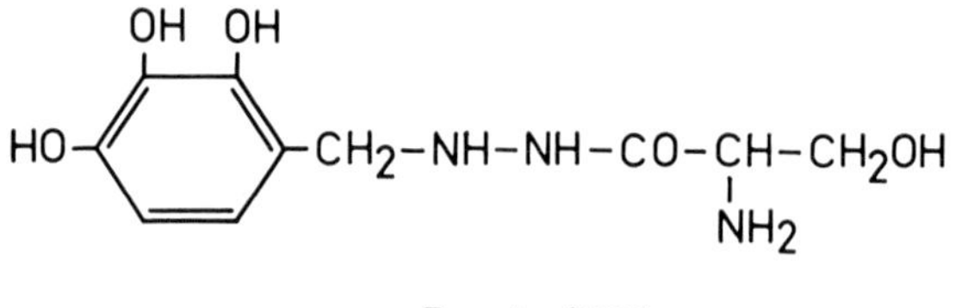

Ro 4-4602

Fig. 1. Formula of the decarboxylase inhibitor Ro 4-4602.

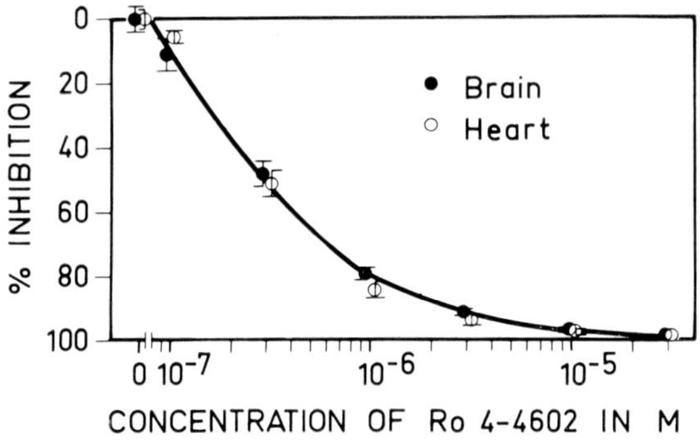

Fig. 2. Activity of decarboxylase of aromatic amino acids inhibited by Ro 4–4602 in vitro (rats). Total homogenates (2 ml) were incubated for 1 hour with 3 μg of ${}^{41}C$-Dopa. The points represent averages of 4 to 6 determinations ± S.E.. (G. Bartholini, W. P. Burkard, A. Pletscher, and H. M. Bates: Nature *215:* 852, 1967).

From these findings it can be concluded that Ro 4–4602 preferentially interferes with extracerebral DC. This preference does not seem to be due to differences in the enzyme of the extracerebral organs and the brain, but rather to a poor penetration of the inhibitor into the brain.

Dopa Metabolism in Animals

In the plasma and heart of rats, administration of relatively small doses of Ro 4–4602 (50 mg per kg) markedly enhances the Dopa-induced increase of the fraction of amino acids and reduces catecholamines and phenolcarboxylic acids, the main metabolites of the catecholamines (Table 1). This action of Ro 4–4602 is probably explained by inhibition of DC in extracerebral tissues, e.g., liver and kidney.

In the brain, 50 mg per kg Ro 4–4602 enhance the Dopa-induced rise of aromatic amino acids (mainly Dopa and 3–O-methyldopa), but in addition

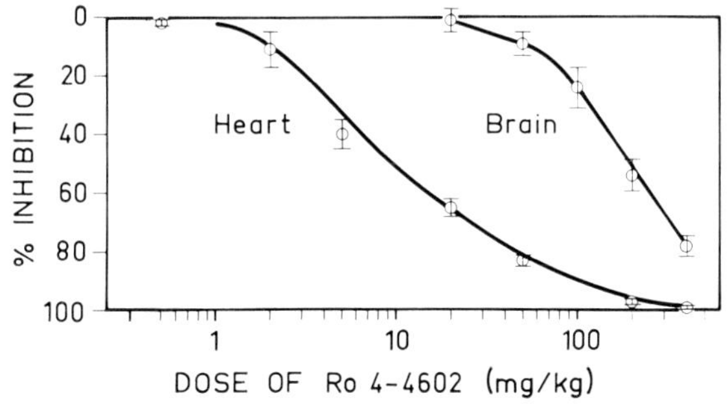

Fig. 3. Inhibition of decarboxylase of aromatic amino acids in brain and heart of rats 1 hour after i.p. injection of Ro 4–4602. Total homogenates (2 ml) were incubated 1 hour with 3 μg of ${}^{14}C$-Dopa. The points represent averages ± S.E. of 3 to 5 duplicate determinations of a pool derived from 2 to 5 animals (G. Bartholini, W. P. Burkard, A. Pletscher, and H. M. Bates: Nature *215:* 853, 1967).

TABLE 1. Distribution of the Radioactivity in the Different ^{14}C-Catechol Fractions in Blood Plasma and Heart 1 Hour after 16 mg/kg of ^{14}C-Dopa i.p. in Rats With or Without Pretreatment with 50 mg/kg of Ro 4–4602.*

^{14}C-Catechol Fraction	*Blood Plasma* ^{14}C-Dopa	*Blood Plasma* Ro 4-4602 + ^{14}C-Dopa	*Heart* ^{14}C-Dopa	*Heart* Ro 4-4602 + ^{14}C-Dopa
Amino acids	2.99±0.05	5.70 ±0.21	1.13±0.43	4.73±0.13
Catecholamines	0.03±0.005	0.005±0.00	0.34±0.03	0.06±0.01
Phenolcarboxylic acids	2.39±0.18	0.57 ±0.02	0.76±0.23	0.26±0.02

* Ro 4–4602 was administered i.p. 30 minutes before ^{14}C-dopa. The values are indicated in micromoles $\times 10^{-2}$ per ml of plasma or gram of heart and represent averages ± S.E. of two experiments. Each experiment was carried out with a pool of 4 to 6 animals (G. Bartholini and A. Pletscher, J. Pharmacol. Exp. Ther. *161,* 14, 1968).

the proportion of catecholamines (mainly dopamine and norepinephrine) is also increased. The quantity of phenolcarboxylic acids (mainly homovanillic acid and dihydroxyphenylacetic acid) also rises, but only after an initial decrease (Figs. 4 and 5).

In the catecholamine fraction, dopamine preferentially accumulates in extrapyramidal structures, e.g., the caudate nucleus (Table 2).

By histochemical investigations using the specific fluorescence method of Falck and Hillarp, it could be demonstrated that the accumulation of catechol-

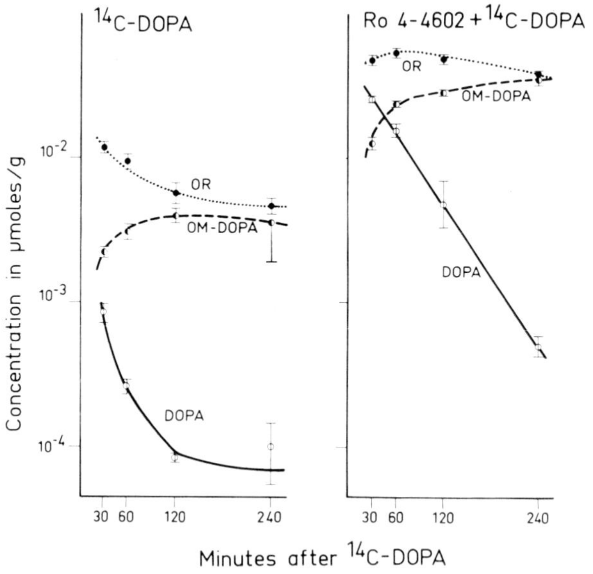

Fig. 4. Overall radioactivity (OR), concentration of ^{14}C-3-O-methyldopa (OM-Dopa) and of ^{14}C-3,4-dihydroxyphenylalanine (Dopa) at various time intervals after 16 mg per kg of ^{14}C-Dopa i.p. in brains of rats with and without pretreatment by 50 mg per kg of Ro 4–4602. Ro 4–4602 was administered i.p. 30 minutes before ^{14}C-Dopa. The values represent averages ± S.E. of 3 or 4 experiments. Each experiment was carried out with a pool of two brains (G. Bartholini and A. Pletscher: J. Pharmacol. Exp. Ther. *161:* 14, 1968).

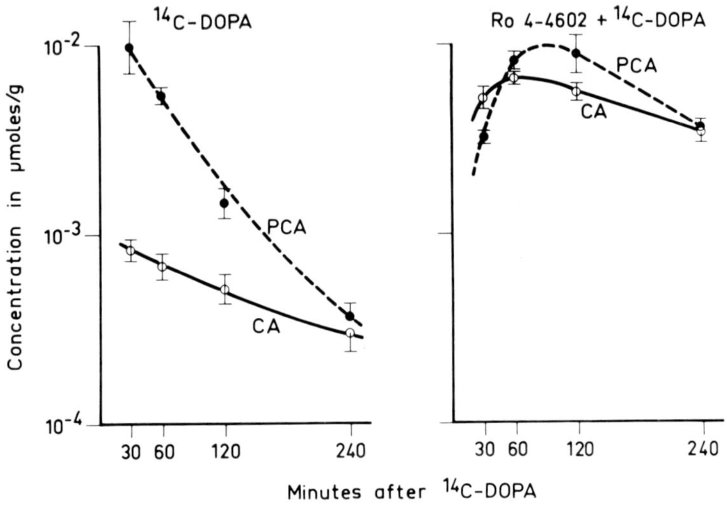

Fig. 5. Cerebral concentration of ¹⁴C-catecholamines (CA) and ¹⁴C-phenolcarboxylic acids (PCA) at various time intervals after ¹⁴C-Dopa i.p. in rats with and without pretreatment by 50 mg per kg of Ro 4–4602. Ro 4–4602 was administered i.p. 30 minutes before ¹⁴C-Dopa. The values indicate averages ± S.E. of 3 or 4 experiments. Each experiment was carried out with a pool of two brains (G. Bartholini and A. Pletscher: J. Pharmacol. Exp. Ther. *161:* 14, 1968).

amines after administration of Ro 4–4602 plus Dopa occurs in the parenchyma and not in the capillaries of the brain.

It may therefore be concluded that the Dopa-induced increase of catecholamines in the brain by small doses of Ro 4–4602 is due to inhibition of extracerebral DC. As a consequence, Dopa accumulates in the blood, and increased amounts of the amino acid penetrate into the brain parenchyma. This leads to an enhanced formation and accumulation of cerebral catecholamines, since the DC in the brain remains active.

With oral administration of Dopa alone, the various catechol fractions of the brain rise less markedly than after parenteral administration of the amino acid. After pretreatment with small doses of Ro 4–4602, however, oral and parenteral administrations of Dopa cause a similar increase of the cerebral catecholamines. This indicates that Ro 4–4602, probably an inhibition of the intestinal decarboxylation, increases the absorption of Dopa.

TABLE 2. Enhancement of the Dopa-Induced Increase of Catecholamines by Ro 4–4602 in the Caudate Nucleus of Rats.*

Treatment	*Dopamine*	*Norepinephrine*
Controls	6.36 ± 0.90	0.48 ± 0.03
Dopa	8.00 ± 0.30	0.51 ± 0.07
Ro 4-4602 + Dopa	18.56 ± 1.20	0.57 ± 0.04

* 75 mg per kg Dopa were administered i.p. 1 hour prior to sacrifice. Ro 4–4602 (50 mg per kg) was injected i.p. 30 minutes prior to L-Dopa. The results are expressed in µg per g and represent averages with S.E. of 3 to 9 experiments.

Dopa Metabolism in Humans

In the blood of humans, 16 to 24 mg per kg Ro 4–4602 or less also enhance the Dopa-induced rise of the aromatic amino acids and decrease that of catecholamines and phenolcarboxylic acids (Fig. 6). This indicates that small doses of the DC inhibitor interfere with the extracerebral decarboxylation of ^{14}C-Dopa. Whether or not the DC in the human brain remains unaltered after the above doses of Ro 4–4602 has still to be investigated. Based on the animal experiments with higher doses of Ro 4–4602 (50 mg per kg), however, it seems likely that the cerebral enzyme is not markedly inhibited in humans either.

Metabolism of Cerebral 5-HT

The combination of Ro 4–4602 plus Dopa also decreases endogenous 5-hydroxytryptamine (5-HT) of the rat brain. Doses as low as 25 mg per kg Ro 4–4602 plus 200 mg per kg Dopa have a significant effect, and with higher doses of Ro 4–4602 the 5-HT decreases by more than 60 percent. Alone, 200 mg per kg of Dopa do not significantly diminish the 5-HT content. The decrease of the cerebral 5-HT seems to be due to several mechanisms, e.g., displacement of 5-HT by dopamine, competition of Dopa with 5-HT precursors (e.g., 5-hydroxytryptophan) for penetration into the brain, competitive inhibition of the cerebral decarboxylation of 5-hydroxytryptophan by Dopa.

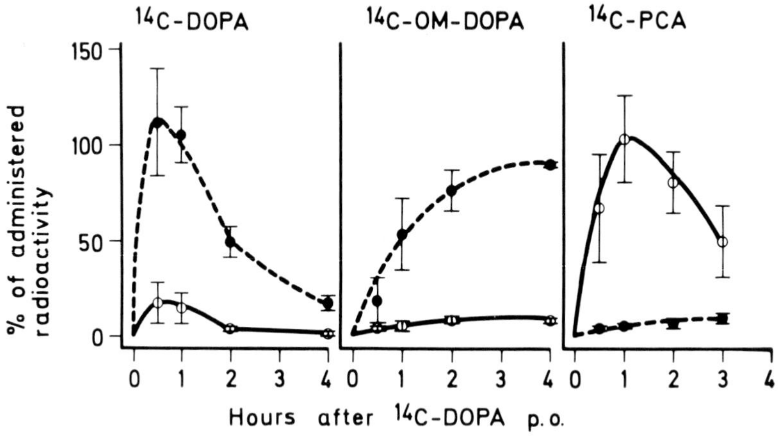

Fig. 6. Effect of oral Ro 4–4602 on the increase of ^{14}C-catechol metabolites in the blood plasma of humans after oral administration of 3 mg per kg ^{14}C-Dopa. 16 to 24 mg per kg Ro 4–4602 were administered in three divided doses within 15 hours prior to ^{14}C-Dopa. The ordinate indicates the radioactivity of the catechol fraction in percent of the total radioactivity administered per gram body weight. *Full line:* ^{14}C-Dopa alone; *dotted line:* Ro 4–4602 plus ^{14}C-Dopa. Each point represents an average with S.E. of 2 experiments. ^{14}C-OM-Dopa = ^{14}C-O-methyldopa; ^{14}C-PCA = ^{14}C-phenolcarboxylic acids (R. Tissot, G. Bartholini and A. Pletscher: Arch. Neurol. *20:* 187, 1969).

Conclusion

A combination of a DC inhibitor plus Dopa which selectively increases brain dopamine might have the following advantages over Dopa alone in the treatment of parkinsonism:

1. Greater specificity of action. Some side effects on Dopa (e.g., on the cardiovascular system), which are likely to be due to formation of catecholamines in the extracerebral tissues, may be reduced at least in part by inhibition of the extracerebral decarboxylation of Dopa.
2. Increased intestinal absorption of Dopa, since the degradation of the amino acid in the intestine is inhibited.
3. Reduction of the oral dose of Dopa as a consequence of a decreased degradation in the peripheral organs, e.g., intestine, liver and kidney.

With regard to the clinical action of Ro 4–4602 plus Dopa, it would seem that accumulation of cerebral dopamine is of importance. Whether a decrease in the endogenous brain serotonin may also be involved remains to be elucidated.

CLINICAL EFFECTS OF L-DOPA PLUS Ro 4–4602

W. Birkmayer

In 1964 we received from Dr. Pletscher an inhibitor of decarboxylase (Ro 4–4602) with the invitation to try this drug in cases of hypertension and affective disorders. We postulated that if Ro 4–4602 shows an effect ten times stronger than alpha-methyldopa it would be an ideal drug to quiet choreatic movements. However, a preliminary trial showed a paradoxical effect. After administration of 450 mg Ro 4–4602 for three days we saw an increase in choreatic movements. Because of the unexpected change in chorea, we tried Ro 4–4602 on patients with Parkinson's disease; we gave 450 mg of Ro 4–4602 and 50 mg L-Dopa intravenously daily, and later with 450 mg of L-Dopa orally. The kinetic effect was greater and longer lasting than after the administration of L-Dopa alone. No one, including Dr. Pletscher, could believe that an inhibitor of decarboxylase would produce an improvement in Parkinson akinesia. In certain cases Ro 4–4602 alone was effective. We also noted the side effects of L-Dopa treatment, such as confusions and hallucinations, after Ro 4–4602 administration. Table 1 summarizes the kinetic effects after L-Dopa alone and after the combination L-Dopa plus Ro 4–4602. The effect of the combined treatment is more pronounced and lasts longer than that of the Dopa alone. Three days after discontinuation of L-Dopa treatment the kinetic effect is lost, but three days after the combined treatment is stopped the kinetic effect is still present. The average decrease in disability after L-Dopa treatment is 11 points; the decrease in the group receiving combined treatment is 16 points. Our results[1,2] have been confirmed by J. Siegfried,[3]

TABLE 1. Kinetic Effects of L-Dopa Alone or With Ro 4-4602 (1 patient)

Treatment	*Kinetic Push* (*in Gs*—measured on Philips physiological acceleration transducer)
Before treatment	3.1
After 3 days of L-Dopa (450 mg/day)	3.6
After 3 days off Dopa	0.8
After 3 days of L-Dopa (450 mg/day) +Ro 4–4602 (450 mg/day)	4.8
3 days after discontinuation of treatment	7.0

R. Tissot [4] and A. Barbeau.[5] The studies of Pletscher [6] have also given support to these results.

REFERENCES

1. Birkmayer, W., and Mentasti, M.: Weitere experimentelle Untersuchungen über Catecholaminstoffwechsel bei extrapyramidalen Erkrankungen (Parkinson-und Chorea-Syndrom). Arch. Psych. u. Z. Neuro. *210:* 29, 1967.
2. Birkmayer, W.: Experimentelle Ergebnisse über die Kombinations-behandlung des Parkinson-Syndroms mit L-Dopa und einem Decarboxylasehemmer (Ro 4–4602). Wien. Klin. Wschr. *81:* 677, 1969.
3. Siegfried, J.: Traitement du parkinsonisme avec la L-Dopa associée à un inhibiteur de la décarboxylase. Médecine et Hygiène *27:* 543, 1969.
4. Tissot, R., Gaillard, J. M., Guggisberg, M., Gauthier, G., and de Ajuriaguerra, J.: Thérapeutique du syndrome de Parkinson par la L-Dopa "per os" associée à un inhibiteur de la décarboxylase (Ro 4–4602). Presse Méd. *77:* 619, 1969.
5. Barbeau, A., and Gillo-Joffroy, L.: Treatment of Parkinson's disease with L-Dopa and Ro 4–4602. *In* Excerpta Medica, Internat. Congr. Series 193, Amsterdam, 1969.
6. Bartholini, G., Burkard, W. P., and Pletscher, A.: Increase of cerebral catecholamines caused by 3,4-dihydroxyphenylalanine after inhibitor of peripheral decarboxylase. Nature *215:* 852, 1967.

ON THE NEUROPATHOLOGY OF PARKINSON'S DISEASE TREATED WITH L-DOPA

John B. S. Campbell, Charles W. True, and T. I. Mandybur

The effectiveness of L-Dopa in the treatment of a substantial percentage of patients with Parkinson's disease has turned attention to the basic sciences in an attempt to explain the mechanics of action of the drug. Considerable experimental information has been assembled but important questions remain, especially in relation to humans. In particular it is not known if histopathological changes occur in treated patients. Furthermore, one-third of the patients do not respond to treatment. It would be interesting to know if these latter patients have a histopathology different from those who improve.

These and other questions might be clarified by review of necropsy material from patients treated and untreated with L-Dopa. Considering the lack of pathological information, and in view of the increasing use of L-Dopa, it is important to report even a single necropsy in a treated case. The following account concerns a patient who died during L-Dopa treatment.

B.W., a 71-year-old man, developed idiopathic Parkinson's disease in 1963 with rapid progression in rigidity and bradykinesia. In 1968 he was unable to speak or swallow, marked tremor was present in the right hand, and rigidity was generalized. Shortly after admission in January 1969 to Good Samaritan Hospital for a trial with L-Dopa he developed pneumonia. Infection was treated successfully and treatment with L-Dopa was begun three weeks later. Because of difficulties in swallowing, the drug was administered by nasogastric tube. Slight improvement was noted after he received the first 100 grams of L-Dopa over a three-week period; the patient was able to talk a little and was more alert, but rigidity remained marked in all extremities. A clasp-knife rigidity noted in the upper extremities suggested a diffuse brain degeneration in addition to Parkinson's disease. A month later, after 180 grams, pneumonia again became a problem and L-Dopa was discontinued. It was restarted after one week and he received 120 grams in the ensuing three weeks. After this he was able to swallow fluids but further improvement did not occur. The treatment was terminated because of pneumonia five days prior to death.

The patient had received a total of 350 grams of L-Dopa in 80 days.

The necropsy revealed extensive bilateral bronchopneumonia with multiple lung abscess formation, chronic pyelonephritis and cystitis. The brain weighed 1230 grams. Slight cortical atrophy and moderate atherosclerosis of the circle of Willis were noted.

Microscopically the cerebral cortex showed diffuse senile plaques and neurofibrillary degeneration. There were no lesions in the hemispheric white matter except for slight generalized shrinkage. Several small cystic infarcts were found in the left putamen; there was also a moderate generalized spongy atrophy of the striatum, bilaterally. The substantia nigra showed severe degeneration with marked reduction of the number of the pigmented neurons and depigmentation of many of those remaining. Occasional Lewy's inclusion bodies were present in the neurons. The entire substantia nigra displayed severe spongy rarefaction associated with mild diffuse astrocytic gliosis. There were striking amounts of extraneuronal melanin granules scattered throughout the area, interstitially and in the glia. Small aggregations of perivascular melanin granules and occasional granules of hemosiderin were also found. The locus caeruleus showed similar changes but to a much lesser degree. Alzheimer's tangles were noted in an occasional neuron of the third cranial nerve, but not in the nigra or locus caeruleus. The dorsal nucleus of the tenth cranial nerve showed mild atrophy with a few neurons exhibiting Alzheimer's tangles. In addition, there was marked degeneration with gliosis in both inferior olives.

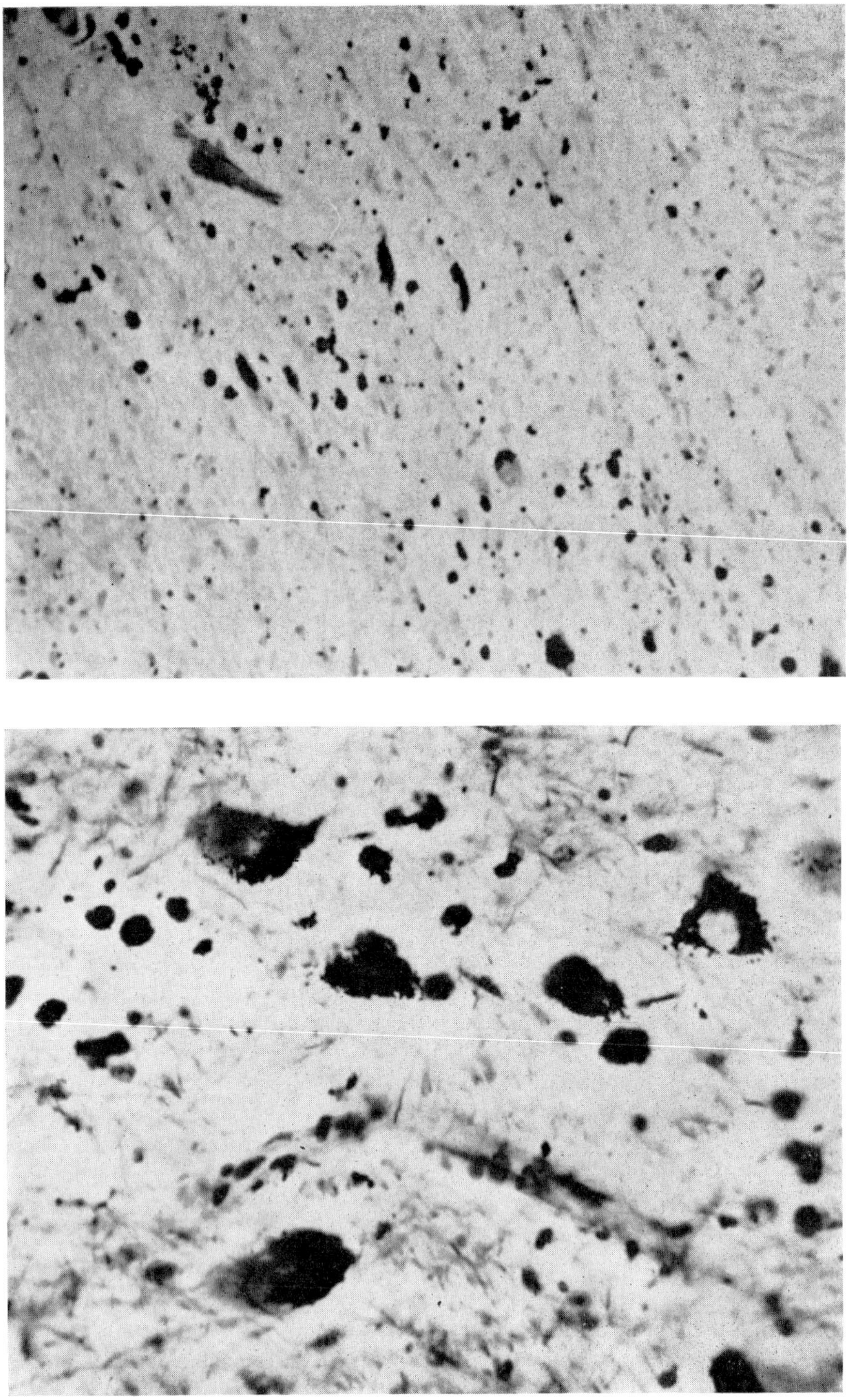

Figs. 1 and 2. Extraneuronal melanin in the substantia nigra. *Fig. 1,* low magnification; *Fig. 2,* high magnification.

Histopathological findings were compared with those of twenty cases of Parkinson's disease collected at Cincinnati General Hospital. The patient reported above differed from untreated instances only with respect to an *unusual amount of extraneuronal melanin* in the substantia nigra. The importance of this finding is not clear. (Figs. 1 and 2).

Extraneuronal-interstitial and perivascular melanin granules in Parkinson's disease are frequently found when the substantia nigra shows considerable degeneration;[1] however, this is usually not an important feature. Extraneuronal melanin is more abundantly found in brains with severe destruction of substantia nigra neurons, which is more common in postencephalitic forms of this disease. Therefore, it is our guarded opinion that the severe destruction of substantia nigra in the case reported, rather than the treatment, is responsible for the abundance of melanin seen interstitially.

In summary, there appeared to be no difference in these histopathological findings and those usually found in Parkinson's disease.[1, 2] In particular there was no appreciable repigmentation of neurons once depigmented. The moderate cerebral cortical degeneration of Alzheimer's type was considered to be coincidental.

It can be speculated, in regard to dopamine transfer in brain, that L-Dopa substitutes for a product of the nigra neurons that probably acts directly on striatum and pallidum. The fact that only slight clinical improvement was observed in this patient was perhaps explained by the severity of destruction of the nigra. However, since the restoration of functional integrity of the dopaminergic pathway by systemic administration of Dopa is not yet proven,[3] the advanced degeneration of striatum could be another reason for the lack of clinical response.

To explain the clinical improvement following L-Dopa treatment in patients with postencephalitic forms of Parkinson's disease, in which the destruction of substantia nigra usually is severe, other mechanisms of action of the drug should be considered than improvement of the function of the nigra neurons.

Summary

Histopathological examination of the brain of a patient with Parkinson's disease who died during therapy with L-Dopa revealed no differences when compared with untreated instances of this disease. The abundance of extraneuronal melanin in the substantia nigra which was found was thought to be related to the severe destruction of the pigmented neurons rather than to the treatment.

REFERENCES

1. NINDB Neurological Information Network. Parkinson's Disease. Present Status and Research Trends. 1968.
2. Hallervorden, J.: Paralysis Agitans. *In* Handbuch der speziellen pathologischer Anatomie und Histologie, Vol. 13. O. Lubarsch, F. Henke, and R. Rossle, eds. Springer, Berlin, 1957.

3. Yahr, M. D., Duvoisin, R. C., Schear, M. J., Barrett, R. E., and Hoehn, M. M. Treatment of parkinsonism with levodopa. Arch. Neurol. *21:* 343, 1969.

COMBINING L-DOPA AND AMANTADINE HYDROCHLORIDE (SYMMETREL)

Robert S. Schwab

We have used amantadine hydrochloride (Symmetrel) in Parkinson's disease for 18 months. This drug was designed to prevent influenza of the Asia Type II variety. Used for Parkinson's disease it becomes an experimental treatment and informed verbal consent is required from the patient. I have slowly added L-Dopa to the treatment of most of our patients on amantadine, some in the hospital, some in the office. The two drugs work together very well in 85 percent of the patients receiving the combination.

Patients exhibiting previous improvement with amantadine have a two-thirds chance of getting a good effect from L-Dopa. When amantadine fails, an approximately equal number do well on subsequent L-Dopa and the rest are not benefited. The combination of both drugs permits reduction of the L-Dopa dosage by 1 to 1½ grams daily.

If amantadine is withdrawn from patients receiving the combination, very promptly, because this drug is rapidly excreted, there is an exacerbation of symptoms (largely akinesia and rigidity). There has not been much reduction in tremor except in two patients. When amantadine treatment is resumed, the patients go back to their previous levels of improvement. We have done this experiment with a double-blind control, despite a published report to the contrary.

In conclusion, my feeling is that the combination of amantadine and L-Dopa is a very effective way to give L-Dopa. Using this combination in a group of 35 patients, we have had only one psychiatric complication and only one movement disorder. This may be due to the fact that we were using smaller doses of L-Dopa than we would have using L-Dopa alone.

GENERAL DISCUSSION

Chairman Cotzias: I would like to make two comments: (1) Under similar conditions to those faced by Dr. Campbell we have given L-Dopa by suppository. We found that the dosages given this way produced the same effects as oral ones. (2) I would like to mention a preliminary experiment opposite to that reported by Professor Pletscher. We have given parenteral dopamine to animals receiving a peripheral inhibitor prior to the administration of Dopa. This has protected the animals both from mortality induced by Dopa and from the stereotyped involuntary movements as well.

Dr. Poirier: In reference to Dr. Spiegel's paper which mentioned the Ransons' reporting of marked akinesia following bilateral hypothalamic lesions, I think it is interesting in the light of new data to attempt to get further insight into such results. As a matter of fact, we now know that lesions involving the dorsolateral area of the hypothalamus at the level of the mamillary bodies cause an interruption of numerous monoaminergic fibers. These lesions will lead to a complete disappearance of dopamine in the striata and to a decreased concentration of serotonin not only in the striatum but also in the thalamus and hypothalamus. On the other hand, they apparently do not interfere with the concentration of norepinephrine in the hypothalamus. Therefore we must keep in mind that some functions ascribed to the hypothalamus on the basis of lesions could, to a large extent, result from the loss of dopamine in the striatum and of serotonin in the striatum, thalamus, hypothalamus and other structures.

Dr. Spiegel: The lesions of Ranson affected, of course, not only ganglion cells but all fiber systems located in this region, chiefly the medial forebrain bundle that carries fibers in both ascending and descending directions.

Dr. Barbeau: I want to make clear my statement about the depression and motivation in patients treated with L-Dopa. I agree that Dopa is definitely an antidepressant in most of our patients, and there is an alerting effect which is quite evident.

I wanted to stress the fact that five patients we considered failures had severe depression, and did not get better during Dopa treatment. These patients were not treated simultaneously with antidepressant drugs, because this would have interfered with our metabolic studies.

Dr. Gerstenbrand: Two observations in Dr. Spiegel's report brought me to the following question: Have you seen turning movements which we call "Manège-movements"? Andén mentioned this pattern of turning movements in cats in his report.

In my opinion, such turning movements could be induced by an activation of the turning systems (Wendesubstrat) described by Hassler. In this case, we could suppose that these specific turning systems would use dopamine as a transmitter substance.

My second question is about the alerting reaction. In patients with apallic syndromes, Hassler performed stereotaxic stimulation in the ascending part of the reticular system. In these patients, who are in an akinetic-mutism state, it was observed that this procedure provoked an alerting reaction.

Dr. Spiegel: In answer to Dr. Gerstenbrand's question, we did not observe the circling movements in all animals. The reason for this remains to be studied. I am aware of the work of Hassler, of course. There apparently are ascending systems that go partly by way of the hypothalamus and the dorsomedial nucleus, partly by way of reticulothalamic fibers and the system of the nuclei of the lamina medullaris interna.

CHAPTER 3

Biochemical Implications of Treatment Failures

POSSIBLE BIOCHEMICAL REASONS FOR TREATMENT FAILURES WITH L-DOPA IN PARKINSON'S DISEASE

Patrick L. McGeer

The use of L-Dopa in Parkinson's disease is the end result of a series of research investigations spanning a number of fields. Among them are: the proposal of Carlsson [1] and others, based on the distribution of dopamine in brain, that this amine might be important to extrapyramidal function; the suggestion of Barbeau [2] that it might be deranged in Parkinson's disease; the finding of Hornykiewicz [3] of lowered dopamine in the basal ganglia of patients suffering from this condition; the production by Poirier and others of Parkinson-like symptoms in monkeys with ventromedial tegmental lesions; [4, 5, 6] the mapping out of nigrostriatal dopaminergic pathways by Anden, Dahlstrom, Fuxe [7, 8, 9] and their colleagues; and the partial reversal of reserpine-induced extrapyramidal reactions by Dopa by many investigators.

The fact that parkinsonian patients show a response to administration of L-Dopa is an exciting culmination of these studies, as well as a host of other ideas from independent pathological, biochemical, and neurophysiological investigations. For some years indifferent or even questionable responses were obtained with L-Dopa therapy.[10, 11, 12, 13] However, the more recent therapeutic regimens in which large doses of 3 to 8 grams of L-Dopa per day are administered [14, 15, 16, 17] seem to have removed any lingering doubts that L-Dopa will work.

It now seems probable that L-Dopa therapy will be widely tested in

Parkinson's disease. Emphasis is bound to shift from the excitement of those who obtain benefit to the disappointment of those whose conditions remain relatively refractory. As with all new treatments, the bright side of the coin is bound to be examined before the other side is seen. This may, therefore, be an appropriate time to attempt to set the therapy in perspective by considering possible reasons why response to this agent might be limited or absent. It may also be a time to consider the unsolved aspects of parkinsonism and to set new research objectives.

L-Dopa therapy focuses attention on the role of dopamine in the extrapyramidal system, but the extrapyramidal system appears to involve a complex interaction of a variety of neuronal pathways with a potential variety of neurotransmitter agents, receptor sites, and cell membrane characteristics. There is still no clue as to what causes Parkinson's disease. However, the same pathological studies and models of parkinsonism which suggested to many researchers that L-Dopa should be efficacious suggest as well, when viewed from other angles, that L-Dopa should have limitations and be less than a total answer.

Consider, for example, the reported pathology in the disease. Studies of postmortem materials from patients with idiopathic Parkinson's disease suggest that the pathology extends beyond the substantia nigra. Denny-Brown [18] considers loss of cells in the globus pallidus and loss or thinning of myelin sheaths in the globus pallidus and ansa lenticularis to be the most constant finding. In two of twenty-one postmortem cases, he failed to detect any abnormality in the substantia nigra. Other workers have drawn attention to cellular changes in the locus caeruleus, pontine tegmentum and corpus striatum in addition to the frequently noted pallor and cell loss in the substantia nigra.[19, 20, 21] There seems little doubt that an important part of the pathology of Parkinson's disease is either outright destruction or functional impairment of the dopaminergic neurons of the substantia nigra, but it also seems possible that changes beyond this structure are characteristic of the disease.

As to the biochemical pathology, Hornykiewicz and his colleagues have provided most of the limited information available. While a dramatically lowered dopamine content in the caudate, putamen, and globus pallidus remains the most prominent postmortem finding, both norepinephrine and serotonin values in the caudate, thalamus and hypothalamus were reported to be decreased to roughly half the normal values. A reduction in serotonin, but not in norepinephine, was reported for the globus pallidus.[22] Decreases in homovanillic acid and 5-hydroxyindoleacetic acid, the principle metabolites of dopamine, and of serotonin have been reported in the cerebrospinal fluid from living parkinsonian patients.[23] Bernheimer and Hornykiewicz report a lowered glutamic acid decarboxylase content in the caudate of parkinsonian patients,[24] but there is little further information regarding possible changes in GABA, acetylcholine and histamine because appropriate measurements do not appear to have been done. The action of drugs in Parkinson's disease, however, suggests the possible involvement of acetylcholine and histamine, and the distribution of both acetylcholine and GABA is suggestive of a relationship to the extrapyramidal system. Moreover, a miscellanea of reports such as that of abnormal glial ribonucleic acid in the globus pallidus of parkinsonian patients [25] suggests that other aspects of brain chemistry may also be involved.

The evidence suggests, therefore, that the pathology in Parkinson's disease goes beyond the nigrostriatal dopaminergic tracts. Serotonergic neurons seem particularly likely to be involved, but neurons whose biochemistry has not yet been determined may be involved as well. Since L-Dopa therapy is directed only at dopaminergic and noradrenergic paths, it cannot be expected to compensate for all of the suspected pathology.

A further reason for expecting that the scope of L-Dopa therapy may be limited comes from a consideration of the neurophysiological and neuropharmacological models of the disease.

There is no single lesion which perfectly duplicates the parkinsonian syndrome. For over a generation investigators have tried to duplicate the symptoms of parkinsonism with various midbrain lesions. The most productive area has always been in the midbrain region of the subthalamic nucleus, the red nucleus, the substantia nigra and the tracts in their vicinity. Lesions of the substantia nigra itself, or of that portion of the ventromedial tegmentum carrying the nigrostriatal tracts, are reported by Poirier and his colleagues to emphasize the hypokinetic aspects of parkinsonism. Lesions more medially placed, which are thought to involve serotonergic pathways arising from midline nuclei and the rubro-tegmento-spinal tract,[26] are more closely associated with tremor. Although this work produces evidence that both dopaminergic and serotonergic neurons are involved in duplication of the hypokinesis and tremor of Parkinson's disease, it must be remembered that it is difficult to delineate precisely the biochemical and anatomical pathways effected by even the smallest lesion. There is evidence, for example, suggesting that cholinergic pathways travel rostrally and caudally through the tegmental area.[27, 28] Some of these cholinergic pathways would be involved in midbrain lesions, and pathways whose biochemistry has not yet been determined are also bound to be involved.

Compensating lesions have also been investigated in an effort to correct the tremor and hypokinesia of the ventrotegmental lesions. Success has been obtained by making lesions in the globus pallidus and ventrolateral nucleus of the thalamus. These are the same regions selected for surgical correction in Parkinson's disease. They are also a main source of tracts to the gamma efferent system in the spinal cord. Little is known as yet about the biochemistry of these pallidal and thalamic neurons.

Within the ventrolateral nucleus of the thalamus, neurosurgeons define a ventralis oralis anterior and a ventralis oralis posterior nucleus.[29] A lesion of the former is thought to be effective in compensating for the rigidity of Parkinson's disease, while a lesion of the latter is regarded as effective in compensating for tremor. The neurophysiological models thus implicate more than dopaminergic and noradrenergic cells. Serotonergic and possibly other biochemical classes of neurons are involved both in the surgical production of parkinsonian symptoms and in their surgical alleviation.

Neuropharmacological models also suggest an involvement beyond dopaminergic neurons. Three different neuropharmacological models have been known for some years. Reserpine represents the amine-depletion model; certain phenothiazines are thought to represent an amine-blockade model; and tremorine represents a cholinergic overactivity model. Reserpine is regarded as the best of these biochemical models. It duplicates well the hypokinesia and rigidity of Parkinson's disease, but the tremor is apt to be of character somewhat different from the familiar pill-rolling of parkinsonism. In our experience with both humans and animals,[30, 31, 32, 33] Dopa does not completely reverse the reserpine syndrome. It is certainly more effective at this than is 5-hydroxytrytophan (5-HTP) but Dopa and 5-HTP together seem to work better than either of these amino acids alone. In animals, anticholinergic and antihistaminic drugs, used in the classical treatment of Parkinson's disease, bring some motor response but are relatively ineffective at reversing all of the reserpine-induced symptoms.[32] In humans,[32, 34, 35] the extrapyramidal reactions induced by reserpine are well controlled by the standard anti-parkinsonian drugs.

The phenothiazine model is more dramatic than the reserpine model. The symptoms can appear rapidly and duplicate in most respects idiopathic Parkinson's disease except again that the tremor tends to be different. Phenothiazines which are weakly antihistaminic and weakly sedative have a much greater tendency to produce parkinsonian reactions than do derivatives which are more strongly antihistaminic and more sedative in their action. Phenothiazines with a piperazine moiety in their side chains are particularly likely to produce extrapyramidal reactions. Dopa is only weakly active against this model, while anticholinergic and antihistaminic agents are dramatic in their effects.[32]

While the means by which tremorine produces its effects are not known, the most striking findings to date in tremorine- or oxotremorine-treated animals are increases in acetylcholine concentration,[36] decreases in norepinephrine concentration, and delayed increases in serotonin concentration.[37] Tremorine and its metabolite oxotremorine have been used for screening anti-parkinsonian drugs.[38, 39] Against tremorine itself, anticholinergic agents such as atropine have been disproportionately active, while against oxotremorine the balance of anticholinergic and antihistaminic action has been in closer keeping with clinical activity.[39]

Overdoses of anti-parkinsonian drugs produce a state of hypotonia, diminished tendon reflexes and choreiform movements. Choreiform movements, as well as the parkinsonian tremor, are interpreted by Denny-Brown as a failure in the normal rhythmic interplay between opposing postural reflexes.[18] A movement once initiated produces a lengthening in one muscle and a shortening in its antagonist. If the reflex lengthening is initially too great and is followed by an overcompensation, a rhythmic movement between two stereotyped postures takes place.

Although the drug models present a somewhat confusing picture, they do suggest involvement of more than just dopaminergic neurons in the production of Parkinson's disease and of more than just cholinergic neurons in the correction of its symptoms.

The pathological information, as well as the neurophysiological and neuropharmacological models of the parkinsonian syndrome, can perhaps be drawn together by assuming that the extrapyramidal system operates through a complex balancing mechanism. Dopaminergic and serotonergic neurons seem to be on one side of a teeter-totter, with cholinergic neurons on the other. Unidentified neuronal systems may be on either side, but the neuropharmacological models suggest that histamine may play a role on the cholinergic side of the teeter-totter.[32, 33] Evidence for histaminic pathways is extremely scant, but histamine may have some central or peripheral role in extrapyramidal function. Weighting of the teeter-totter against dopaminergic or serotonergic tone apparently produces the parkinsonian symptoms of akinesia, rigidity and tremor. This can result from destructive lesions of dopaminergic and serotonergic cell bodies in the brain stem, from depletion or blockade of the amines through drugs, or possibly from enhancement of the central cholinergic system through such agents as tremorine. Balancing of the teeter-totter can be brought about by amine precursors such as L-Dopa and 5-HTP, by amine potentiators such as imipramine or MAO inhibitors, by surgical lesions of the globus pallidus or ventrolateral thalamus, or by anticholinergic and antihistaminic drugs. Weighting of the teeter-totter against the cholinergic system can bring about a choreiform state characterized by hypotonia and chorea-like movements. The findings of Dalby[40] that tetrabenazine, a monoamine depletor, worsened parkinsonian tremors but improved choreiform and hemiballistic movements is in keeping with the hypothesis of balancing tone, although his results could also be interpreted as reflecting alterations only in dopaminergic pathways.

The hypothesis of a balancing tone between competing sets of pathways mediated by different amines suggests a number of therapeutic approaches. Since stimulation of dopaminergic tone and reduction of cholinergic tone are noncompetitive chemotherapeutic approaches, a combination of agents would seem to offer more promise than exclusive utilization of a single class. Dopaminergic tone can be increased by L-Dopa alone or in combination with an MAO inhibitor, a peripheral Dopa decarboxylase inhibitor or an amine uptake inhibitor. Cholinergic tone can be decreased by the standard anticholinergic agents.

The findings of lowered serotonin in the caudate and decreased 5-hydroxyindoleacetic acid in the cerebrospinal fluid in parkinsonian patients would suggest an impairment of serotonin pathways, and Poirier's results indicate that these pathways may be connected with the characteristic tremor of the disease. Further lowering of brain serotonin may be a consequence of L-Dopa therapy. Several years ago we found that large doses of Dopa given to cats or monkeys caused a decrease in brain serotonin[30, 31] and Bartholini et al. recently reported that combined administration of 50 mg per kg of Dopa with the peripheral decarboxylase inhibitor Ro 4–4602 caused a decrease in brain serotonin in rats.[41] Whether L-Dopa therapy, with or without a decarboxylase inhibitor, lowers serotonin concentrations in parkinsonian patients is not known. Bartholini et al.[41] take the view that such a lowering might be beneficial. They argue that since increasing brain serotonin by large doses of 5-HTP or intraventricular administration of serotonin causes tremors,[42, 43] a deficiency of serotonin might counteract tremors. On the other hand, Roos and Steg[44] found that both Dopa and 5-HTP reversed the rigidity and tremor induced by reserpine although 5-HTP, unlike Dopa, had no effect on the akinesia. Cotzias[15] and Yahr[14] both report that parkinsonian tremor responds to L-Dopa therapy, but even more beneficial results might be obtained if L-5-HTP or even large doses of L-tryptophan[46] were given with the L-Dopa. Barbeau[45] has already given a favorable preliminary report on this possibility.

The cholinergic pathways presumably involved in treatment of parkinsonism with anticholinergic drugs have not yet been identified. Definition of these might open up new avenues for surgical treatment of the disease. It could be that these pathways will correspond to the surgical targets in the thalamus and globus pallidus already defined. If that is the case, closure between neurosurgery and biochemistry will be brought about for this aspect of extrapyramidal function. If not, biochemical definition of the pathways in these nuclei might suggest new pharmacological approaches.

The most important aspect of parkinsonian research is, of course, elucidation of the causative mechanism. It could be due to a latent virus or the continued elaboration of a poison from some unidentified source in the body. It could simply be a consequence of the aging process with premature dropout of the particular cells affected in parkinsonism

Etiological models of parkinsonism do exist, but they are even less satisfactory than the neurosurgical or the pharmacological models. Carbon monoxide or manganese poisoning has been known for many years to produce parkinsonism.[19, 47] 6-Hydroxydopamine, which causes degradation of sympathetic nerve endings, has recently been reported to produce extrapyramidal reactions when locally applied.[48] A poison related to any of these is conceivable as a cause for parkinsonism but no specific suggestions have been made.

A viral origin is equally possible. The reason for the appearance of postencephalitic parkinsonism is uncertain, but the action of a latent virus seems conceivable in view of recent reports concerning kuru and other slowly developing viruses.[49] Little is known about the effect of various viruses on specific brain pathways, but a recent report on catecholamine and serotonin stimulation in brain by a herpes simplex encephalitis infection in mice[50] suggests that specific effects might be the consequence of specific virus infections.

In any event Parkinson's disease seems to be a relentlessly progressive degeneration of dopaminergic and other classes of neurons in the brain. The report of delayed responses to Dopa therapy and delayed regressions following Dopa withdrawal[14, 15] suggest that Dopa might actually retard the progression by altering the functional capability of neurons. A cautious attitude to this possibility should be kept until some definite proof can be obtained, either through model diseases or through metabolic studies on parkinsonian patients, that Dopa can influence the metabolism of cells in this favorable way.

For the moment the three most obvious reasons for lack of response to

Dopa therapy are: insufficient dosage, permanent destruction of dopaminergic cells, and destruction or impairment of nondopaminergic cells.

REFERENCES

1. Carlsson, A.: Occurrence, distribution and physiological rate of catecholamines in the nervous system. Pharmacol. Rev. *2:* 490, 1959.
2. Barbeau, A.: Dopamine and basal ganglia diseases. Arch. Neurol. *4:* 97, 1961.
3. Ehringer, H., and Hornykiewicz, O.: Verteilung von noradrenalin und dopamin (3-Hydroxytyramin) im gehirn des menschen und ihr verhalten bei erkrankungen des extrapyramidalen systems. Klin. Wschr. *38:* 1236, 1960.
4. Poirier, L. J., and Sourkes, T. L.: Influence of the substantia nigra on the catecholamine content of the striatum. Brain *88:* 181, 1965.
5. Ward, A. A., Magoun, H. W., and McCulloch, W. S.: Production of an alternating tremor at rest in monkeys. J. Neurophysiol. *11:* 317, 1948.
6. Vernier, V. G., and Unna, K. R.: The experimental evaluation of anti-Parkinson compounds. Ann. N.Y. Acad. Sci. *64:* 690, 1956.
7. Andén, N. E., Carlsson, A., Dahlstrom, A., Fuxe, K., Hillarp, N. A., and Larsson, K. Demonstration and mapping out of nigro-neostriatal dopamine neurons. Life Sci. *3:* 523, 1964.
8. Dahlstrom, A., and Fuxe, K.: Evidence for the existence of monoamine-containing neurons in the central nervous system. Acta Physiol. Scand. *62:* Suppl. 232, 1964.
9. Andén, N. E., Fuxe, K., Hamberger, B., and Hokfelt, T.: A quantitative study on the nigro-neostriatal dopamine neuron system in the rat. Acta. Physiol. Scand. *67:* 306, 1966.
10. Birkmayer, W., and Hornykiewicz, O.: Weitere experimentelle untersuchungen über L-dopa beim Parkinson-syndrom und reserpin-Parkinsonismus. Arch. Psychiat. Zeitsch. Neurol. *206:* 367, 1964.
11. Barbeau, A.: Les catécholamines dans la maladie de Parkinson. Proc. 7th Intern. Neurol. Congr. Roma. Societa Grafica Romana *2:* 925, 1961.
12. Fehling, C.: Treatment of Parkinson's disease with L-Dopa: double blind study. Acta Neurol. Scand. *42:* 367, 1966.
13. McGeer, P. L., and Zeldowicz, L. R.: Administration of dihydroxyphenylalanine to Parkinsonian patients. Canad. Med. Assoc. J. *90:* 463, 1964.
14. Yahr, M. D., Duvoisin, R. G., Schear, M. J., Barrett, R. E., and Hoehn, M. M.: Treatment of parkinsonism with levodopa. Arch. Neurol. *21:* 343, 1969.
15. Cotzias, G. C., Papavasiliou, P. S., and Gellene, R.: Modification of parkinsonism—chronic treatment with L-Dopa. New Eng. J. Med. *280:* 337, 1969.
16. Calne, D. B., Stern, G. M., and Laurence, D. R.: L-Dopa in postencephalitic parkinsonism. Lancet *1:* 744, 1969.
17. O'Reilly, S.: Dopa in Parkinson's disease. Lancet *1:* 984, 1969.
18. Denny-Brown, D.: Diseases of the basal ganglia: Their relation to disorders of movement. Lancet *2:* 1099, 1960.
19. Greenfield, J. G.: Neuropathology. Williams and Wilkins, London, 1958.
20. Lewy, F. H.: Zur pathologischen anatomie der paralysis agitans. Dtsch. Z. Nervenheilk. *50:* 50, 1913.
21. Greenfield, J. G., and Bosanquet, F. D.: Brain stem lesions in parkinsonism. J. Neurol. Psychiat. *16:* 213, 1953.
22. Bernheimer, H., Birkmayer, W., and Hornykiewicz, O.: Vorteilung des

5-hydroxytryptamins (Serotonin) im Gehirn des Menschen und sein Verhalten bei Patienten mit Parkinson-Syndrom. Klin. Wschr. *39:* 1056, 1961.

23. Gottfries, C. G., Gottfries, I., and Roos, B. E.: Homovanillic acid and 5-hydroxyindoleacetic acid in the CSF of patients with senile dementia, presenile dementia and parkinsonism. J. Neurochem. *16:* 1341, 1969.
24. Bernheimer, H., and Hornykiewicz, O.: Das verhalten einiger enzyme im gehirn normaler und Parkinson-kranker menschen. Arch. Exp. Path. Pharmak. *243:* 295, 1962.
25. Gomirato, G.: A biochemical glia error in the Parkinson disease. Brain *86:* 773, 1963.
26. Poirier, L. J., Sourkes, T. L., Bouvier, G., Boucher, R., and Carabin, S.: Striatal amines. Experimental tremor and the effect of harmaline in the monkey. Brain *89:* 37 1966.
27. Shute, C. C., and Lewis, P. R.: Cholinergic and monoaminergic systems of the brain. Nature *212:* 710, 1966.
28. Shute, C. C., and Lewis, P. R.: The ascending cholinergic reticular system: Neocortical olfactory and subcortical projections. Brain *90:* 497, 1967.
29. Cooper, I. S.: Involuntary Movement Disorders. Hoeber Med. Div., Harper and Row, New York, 1969.
30. McGeer, P. L., McGeer, E. G., and Wada, J. A.: Central aromatic amine levels and behavior. Arch. Neurol. *9:* 81, 1963.
31. Wada, J. A., Wrinch, J., Hill, D., McGeer, P. L., and McGeer, E. G.: Central aromatic amine levels and behavior. Arch. Neurol. *9:* 69, 1963.
32. McGeer, P. L., Boulding, J. E., Gibson, W. C., and Foulkes, R. G.: Drug-induced extrapyramidal reactions. Treatment with diphenhydramine hydrochloride and dihydroxyphenylalanine. J.A.M.A. *177:* 665, 1961.
33. McGeer, P. L.: Central amines and extrapyramidal functions. J. Neuropsych. *4:* 247, 1963.
34. Sarwer-Foner, G. J.: Recognition and management of drug-induced extrapyramidal reactions and "paradoxical" behavioural reactions in psychiatry. Canad. Med. Assoc. J. *83:* 312, 1960.
35. Kruse, W.: Treatment of drug-induced extrapyramidal symptoms. Dis. Nerv. Syst. *21:* 79, 1960.
36. Lundgren, G., and Malmberg, M.: Effects of oxotremorine on brain acetylcholine formation in vivo and in vitro. Biochem. Pharmacol. *17:* 2051, 1968.
37. Friedman, A. H., Aylesworth, R. J., and Friedman, G.: Tremorine: Its effect on amines of the central nervous system. Science *141:* 1188, 1963.
38. Ahmed, A. and Marshall P. B.: Relation between anti-acetylcholine and anti-tremorine activity in anti-parkinsonian and related drugs. Brit. J. Pharmacol. *18:* 247, 1962.
39. Leslie, G. B., and Maxwell, D. R.: Inhibition of tremorine by a group of non-anti-Parkinson phenothiazine derivatives. Nature *202:* 97, 1964.
40. Dalby, M. A.: Effects of tetrabenazine on extrapyramidal movement disorders. Brit. Med. J. *2:* 422, 1969.
41. Bartholini, G., Da Prada, M., and Pletscher, A.: Decrease of cerebral 5HT by 3,4-Dopa after inhibition of extracerebral decarboxylase. J. Pharm. Pharmacol. *20:* 228, 1968.
42. Udenfriend, S., Weissbach, H., and Bogdanski, D. F.: Biochemical findings relating to the action of serotonin. Ann. N.Y. Acad. Sci. *66:* 602, 1957.
43. Domer, F. R., and Feldberg, W.: Tremor in cats: The effect of administration of drugs into the cerebral ventricles. Brit. J. Pharmacol. *15:* 578, 1960.
44. Roos, B. E., and Steg, G.: The effect of L-3,4-dihydroxyphenylalanine and DL-5-hydroxytryptophan on rigidity and tremor induced by reserpine, chlorpromazine and phenoxybenzamine. Life Sci. *3:* 351, 1964.

45. Barbeau, A.: The pathogenesis of Parkinson's disease: A new hypothesis. Canad. Med. Assoc. J. *87:* 802, 1962.
46. Quay, W. P.: Effect of dietary phenylalanine and tryptophan on pineal and hypothalamic serotonin levels. Proc. Soc. Exp. Biol. Med. *114:* 718, 1963.
47. Lord Brain: Diseases of the Nervous System. 6th edition. Oxford Univ. Press, 1962.
48. Ungerstedt, U.: 6-Hydroxydopamine induced degeneration of central monamine neurons. Europ. J. Pharmacol. *5:* 107, 1968.
49. Gajdusek, D. C., Gibbs, C. J., and Asher, D. M.: Transmission of experimental kuru to the spider monkey (*Ateles geoffreyi*). Science *162:* 693, 1968.
50. Lycke, E., and Roos, B. E.: Effect on the monoamine metabolism of the mouse brain by experimental herpes simplex infection. Experientia *24:* 687, 1968.

EFFECTS OF L-DOPA IN NORMAL HUMANS

Robert D. Ansel and Charles H. Markham

The demonstration that the striatum of patients with Parkinson's disease has significantly decreased concentrations of dopamine (3, 4-dihydroxyphenylethylamine)[1] and that massive oral administration of its precurser, L-Dopa (3, 4-dihydroxyphenylalanine), dramatically improves the symptoms[2, 3] has led to widespread clinical trials of this agent.

Dosage levels and therapeutic effectiveness have been limited by side effects which include nausea and vomiting, postural hypotension, involuntary movements and mental disturbances.[2, 3]

In view of increasing numbers of patients with Parkinson's disease and other movement disorders taking L-Dopa, we would like to present our experience with four normal people who took the medication for one month.

Methods and Material

The subjects were three females and one male, ages 53 to 69, who were spouses of patients with Parkinson's disease. They took the medication as outpatients with full knowledge of and agreement with the nature and goals of the study. They were judged to be in good health by their private physicians and by the investigators. The subjects received general physical and neurological examinations prior to and at weekly intervals during the study. The following laboratory studies were obtained prior to and after completion of the study: blood, urine, cardiac, protein-bound-iodine, direct and indirect Coombs', blood-urea-nitrogen, alkaline phosphatase, fasting blood sugar, serum glutamic oxylate transaminase (SGOT), bilirubin, and total protein, calcium, and phosphorus.

The L-Dopa was given orally as 500-mg tablets, with an initial dosage of 1 gram per day. Every two days the daily dosage was increased by 0.5 gram to a maximum of 3 to 5 grams. Further adjustments were made according to the side effects (see Fig. 1). The medication was given with food. At the end of one month the medication was abruptly stopped.

Results

All four subjects had gastrointestinal symptoms (Fig. 1). In one patient (H.G.) there was mild nausea which started at a dosage of 4.0 grams and usually occurred within a half hour after taking the medication; there were no additional side effects. A second patient (R.T.) had mild nausea when the medication was raised to 3.0 and again to 3.5 grams. This was transient and did not recur as the dosage was further increased. R.B. developed a two-day episode of cramping abdominal pain and diarrhea while taking 4.0 grams. When examined, she did not have fever, chills, abdominal tenderness or other signs of systemic illness. Her spouse, also taking 4.0 grams, did not have similar symptoms at this time. When the L-Dopa was stopped for one day and started again at 3.0 grams per day, she had no further symptoms. The fourth patient (M.R.) developed persistent anorexia, eructations, and intermittent epigastric discomfort and vomiting, usually occurring or worsening after each dosage. When the medication was lowered from 3.5 to 3.0 grams, the symptoms became more tolerable.

Transient drowsiness was noted in two patients. They reported a desire to sleep much of the day and occasionally napped for several hours. They did not appear to be sedated. There was no reported change in their sleeping patterns and the other two patients denied any changes at all.

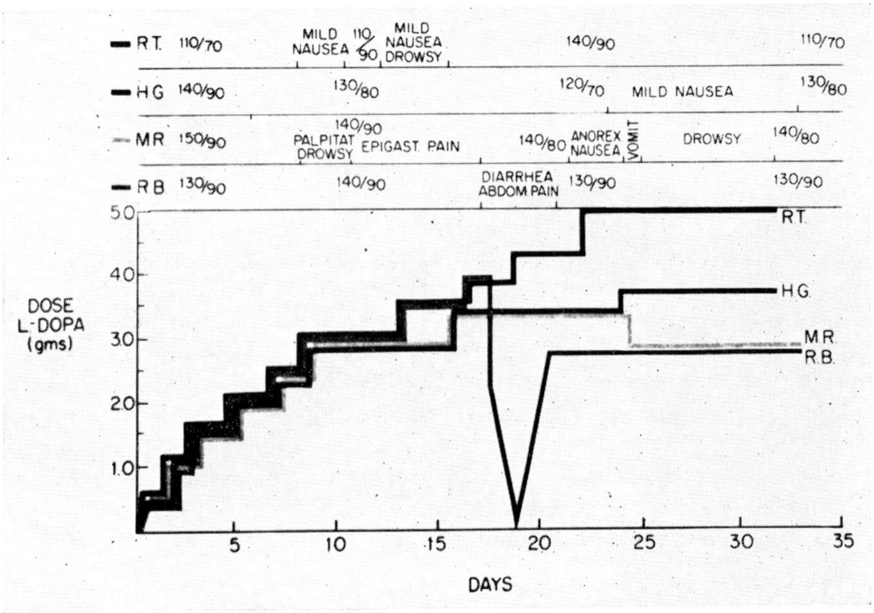

Fig. 1.

Blood pressure taken supine and standing at weekly intervals showed no significant changes. No patient complained of dizziness, nor were there any symptoms to suggest postural hypotension.

No subject reported any abnormal movements, nor were any observed. Fine motor control and coordination remained normal.

There were no changes in affect, behavior or intellectual function in any patient.

There was no significant change in any laboratory test.

Discussion

Factors limiting the dose and possibly the efficacy of L-Dopa are side effects of anorexia, nausea, vomiting and involuntary movements. In our series of 57 parkinsonian patients taking L-Dopa, 25 developed involuntary movements.[4] The schedule of dosage increase was approximately the same as that used in the present study. In nine patients involuntary movements became apparent in the first month of therapy as the dose was increased, and were found at dosages of 2.5 to 8.0 grams. It may be of some significance, therefore, that none of our normal controls developed involuntary movements at doses of L-Dopa capable of producing them in parkinsonian patients. Cotzias has found the most severe involuntary movements to be induced among patients whose disease had the longest duration.[2] This may suggest that some alteration of the striatal biochemical substrate is necessary to produce this phenomenon.

Gastrointestinal toxicity was commonly seen in all of the present subjects and to some degree in two-thirds of the 57 parkinsonian patients. In the latter group it was more frequent and more severe when the L-Dopa was increased rapidly in 0.5-gram increments every two to three days, as was done in the normal subjects. It is probable that gastrointestinal symptoms in the normal individuals could have been lessened if the L-Dopa had been administered more slowly. The observations that nausea and/or vomiting tended to occur, or worsen, a few minutes after taking an oral dose, depended on the amount taken, and could be inhibited by taking the medicine with food suggest a local gastrointestinal component to these symptoms. On the other hand, the persistent nausea has been noted in a few parkinsonian patients on high maintenance doses of L-Dopa (4 to 8 grams). This nausea may be reduced by reducing the L-Dopa dosage, or occasionally by use of phenothiazines. Such nausea may have a central nervous system component.

Summary

L-Dopa was administered to four normal adults in oral doses ranging from 3 to 5 grams per day for one month. No significant effects were noted in mentation, behavior or motor control. Gastrointestinal side effects developed in the four patients, but alterations in dosage permitted the medicine to be continued in all.

REFERENCES

1. Hornykiewicz, O.: Dopamine and brain function. Pharmacol. Rev. *18:* 925, 1966.
2. Cotzias, G. C., Papavasiliou, P. S., and Gellene, R.: Modification of parkinsonism —Chronic treatment with L-Dopa. New Eng. J. Med. *280:* 337, 1969.
3. Yahr, M. D., Duvoisin, R. C., Hoehn, M. M., Schear, M. J. and Barrett, R. E.: L-Dopa (L-3,4-dihydroxyphenylanine)—Its clinical effects in parkinsonism. Trans. Amer. Neurol. Assoc. *93:* 56, 1968.
4. Markham, C. H., and Ansel, R. D.: Unpublished data.

THE ROLE OF MINOR PATHWAYS OF DOPA METABOLISM

M. Sandler

Why some patients with parkinsonism fail to benefit from L-Dopa therapy cannot be understood without understanding the nature of the successful response. Throughout the extensive literature on this subject is the tacit assumption that clinical improvement derives from dopamine generation within the central nervous system. While it is probable that dopamine plays the leading part, it seems equally likely that there are other closely related biochemical factors of importance whose malfunction may provide the basis we seek for treatment failures. The probable supporting role of 5-hydroxytryptamine, for which evidence is accumulating,[1, 2] is germane to this thesis, but is dealt with in other papers in this volume.

It has recently been pointed out[3] that dopamine generation within the central nervous system is likely to be rapid, whereas the time course of the therapeutic response to orally administered Dopa is slow. One obvious explanation for this discrepancy lies in terms of drug absorption from the gastrointestinal tract—and, indeed, this aspect of the problem merits close attention. We need to know whether dietary interference from other amino acids modifies Dopa absorption. There is an alternative explanation, however, which deserves careful consideration: The clinical effect, and the delayed therapeutic response, may not (or not only) derive from dopamine replacement; it may result from a build-up in the central nervous sytem of some minor metabolite unconnected with the main route of Dopa degradation. A number of possible candidates for this role have been brought to light by the large doses of drug which are necessary to achieve therapeutic benefit.[3]

Versatile, quantitative gas chromatographic procedures[4, 5] show that urinary homovanillic acid (HVA) and dihydroxyphenylacetic acid (DOPAC) account for about 90 percent of the L-Dopa administered to parkinsonian patients, depending to some extent on the dose ingested. There is some evidence[3] to suggest that a small proportion of the HVA peak represents the positional isomer of HVA, 3-hydroxy-4-methoxyphenylacetic acid (iso-HVA). Ernst[6, 7] has provided some data in the experimental animal to suggest that *p*-methylation of dopamine may actually produce Parkinson-like phenomena. There are also indications in the literature that such *p*-methoxylated compounds tend to have an enhanced psychotomimetic activity.[8, 9] The relative production of iso-HVA in treatment successes compared with failures should obviously be investigated.

It seems worth mentioning here that despite earlier claims[10] homogentisic acid has been rigorously excluded as a Dopa metabolite.[11] After urinary HVA and DOPAC, quantitatively the most important acidic metabolite is 4-hydroxy-3-methoxyphenyllactic acid (vanillactic acid, VLA); excretion of this product of Dopa transamination[12] represents about 1 percent of the oral dose of the drug. The enzymes necessary for the generation of its chemically active precursor, 3,4-dihydroxyphenylpyruvic acid (DHPPA), are known to be present in brain.[13] It is possible that such an increased production of DHPPA may competitively inhibit *p*-hydroxyphenylpyruvate oxidase, an important enzyme of the main degradation route of tyrosine, for increased amounts of *p*-hydroxyphenylpyruvic acid were also excreted by patients under treatment.[3] If this interpretation were correct, DHPPA would itself be converted by an "NIH shift" mechanism to 2,4,5-trihydroxyphenylacetic acid. Shulgin et al.[8] have indicated that compounds of this type, with a 2,4,5-substitution pattern, appear to possess high psychotomimetic activity and that they may even be implicated in the pathogenesis of human psychoses.

A more direct route to the formation of such compounds, however, exists. It is well recognized that, under certain circumstances, dopamine itself can undergo 6-hydroxylation to form 6-hydroxydopamine (2,4,5-trihydroxyphenylehtylamine).[14] When this compound is administered to the experimental animal, depending on the route of administration, it can produce a "chemical sympathectomy"[15] or bring about degeneration of central neurons.[16, 17] Evidence of a sympathetic nervous system deficit should therefore be sought in patients on long-term Dopa therapy. (The occurrence of orthostatic hypotension might be relevant in this connection.) Aside from such treatment, it is even conceivable that 6-hydroxydopamine or a compound similar to it is generated endogenously in patients with idiopathic parkinsonism and causes neuronal degeneration in areas of high dopamine concentration such as the substantia nigra, thus providing a biochemical basis for the clinical features of the disease. Evidence of such a compound should be sought in the brains of parkinsonian patients who have not been treated with Dopa.

One other group of compounds on which we should keep a sharp eye is the *m*-hydroxylated amines. Calne et al.[3] noted a comparatively large increase in urinary *m*-hydroxyphenylacetic acid excretion after administration of oral L-Dopa. This compound almost certainly derives from *p*-dehydroxylation at some stage in the Dopa metabolic pathway, for small doses of neomycin brought about a quite significant decrease in output.[18] In view of the remarkable effects of metaraminol[19, 20] and the known pharmacological actions of *m*-tyrosine,[21, 22] one cannot rule out completely the possibility that this type of compound plays some part in the therapeutic action of Dopa or in the production of side effects. Perhaps treatment failures come from having the wrong kind of intestinal flora! To test this point, it will be necessary to treat drug responders with a larger dose of neomycin for a longer period of time to see if this gives rise to clinical deterioration.[18]

There is one other potential lead arising from the work of Calne et al.[3] which might merit closer attention: A shift to the oxidative pathway in the further metabolism of the intermediate aldehyde deriving from oxidative deamination of amine was noted; thus there was a proportionately greater increase in output of HVA at the expense of 4-hydroxy-3-methoxyphenylethanol (HMPE) but there was, if anything, an actual *decrease* in 4-hydroxy-3-methoxyphenylglycol (HMPG) output to match the increase in 4-hydroxy-3-methoxymandelic acid (VMA) excretion, marginal only even after massive doses of L-Dopa.

Another possibility deserves careful examination: Dopa decarboxylase, which is presumed to be responsible for the generation of dopamine from L-Dopa, may be quite atypical in the human. Very little or no activity could be detected in human brain by Robins et al.[23] who suggested that it might be the rate-limiting step in amine production. The finding of multiple forms of this enzyme,[24] which tends to have been neglected in recent years, should serve once more to focus attention on its probable key role in the successful treatment of parkinsonism with L-Dopa.

REFERENCES

1. SONNEVILLE, P. F.: Indirect action of dopamine. Acta Physiol. Pharmacol. Neerl. *15:* 419, 1969.
2. GOLDSTEIN, M., BATTISTA, A. F., NAKATANI, S., and ANAGNOSTE, B.: Drug-induced relief of the tremor in monkeys with mesencephalic lesions. Nature *224:* 382, 1969.
3. CALNE, D. B., KAROUM, F., RUTHVEN, C. R. J., and SANDLER, M.: The metabolism of orally administered L-Dopa in parkinsonism. Brit. J. Pharmacol. *37:* 57, 1969.
4. KAROUM, F., ANAH, C. O., RUTHVEN, C. R. J., and SANDLER, M.: Further observations on the gas chromatographic measurement of urinary phenolic and indolic metabolites. Clin. Chim. Acta *24:* 341, 1969.
5. KAROUM, F., RUTHVEN, C. R. J., and SANDLER, M.: Gas chromatographic measurement of phenolic acids and alcohols in human urine. Clin. Chim. Acta *20:* 427, 1968.
6. ERNST, A. M.: Phenomena of the hypokinetic rigid type caused by O-methylation of dopamine in the *para*-position. Nature *193:* 178, 1962.
7. ERNST, A. M.: Relation between the structure of certain methoxyphenylethylamine derivatives and the occurrence of a hypokinetic rigid syndrome. Psychopharmacologia *7:* 383, 1965.
8. SHULGIN, A. T., SARGENT, T., and NARANJO, C. Structure-activity relationships of one-ring psychotomimetics. Nature *221:* 537, 1969.
9. SMYTHIES, J. R., JOHNSON, V. S., BRADLEY, R. J., BENINGTON, F., MORIN, R. A., and CLARK, L. C., JR.: Some new behaviour-disrupting amphetamines and their significance. Nature *216:* 128, 1967.

10. Arras, M. J., and Bailey, G. W. H.: Metabolism of Dopa in parkinsonism. New Eng. J. Med. *278:* 280, 1968.
11. Sandler, M., Karoum, F., and Ruthven, C. R. J.: In preparation.
12. Cammarata, P. S., and Cohen, P. B.: Scope of transamination reaction in animal tissues. J. Biol. Chem. *187:* 439, 1950.
13. Fonnum, F., Haavaldsen, R., and Tangen, O.: Transamination of aromatic amino acids in rat brain. J. Neurochem. *11:* 109, 1964.
14. Senoh, S., Creveling, C. R., Udenfriend, S., and Witkop, B.: Chemical enzymatic and metabolic studies on the mechanism of oxidation of dopamine. J. Amer. Chem. Soc. *81:* 6236, 1959.
15. Thoenen, H., and Tranzer, J. P.: Chemical sympathectomy by selective destruction of adrenergic nerve endings with 6-hydroxydopamine. Arch. Pharmak. *261:* 271, 1968.
16. Ungerstedt, U.: 6-Hydroxydopamine-induced degeneration of central monoamine neurons. Europ. J. Pharmacol. *5:* 107, 1968.
17. Uretsky, N. J., and Iversen, L. L.: Effects of 6-hydroxydopamine on noradrenaline-containing neurons in the rat brain. Nature *221:* 557, 1969.
18. Sandler, M., Karoum, F., Ruthven, C. R. J., and Calne, D. B.: *m*-Hydroxyphenylacetic acid formation from L-Dopa in man: suppression by neomycin. Science, in press.
19. Shore, P. A.: The mechanism of norepinephrine depletion by reserpine, metaraminol and related agents. The role of monoamine oxidase. Pharmacol. Rev. *18:* 561, 1966.
20. Lundborg, P.: Studies on the uptake and subcellular distribution of catecholamines and their α-methylated analogues. Acta Physiol. Scand. Suppl. 302, 1967.
21. Mitoma, C., Posner, H. S., Bogdanski, D. F., and Udenfriend, S.: Biochemical and pharmacological studies on o-tyrosine and its meta and para analogues. J. Pharmacol. Exp. Ther. *120:* 188, 1957.
22. Blaschko, H., and Chrusciel, T. L.: The decarboxylation of amino acids related to tyrosine and their awakening action in reserpine-treated mice. J. Physiol. *151:* 272, 1960.
23. Robins, E., Robins, J. M., Croninger, A. B., Moses, S. G., Spencer, S. J. and Hudgens, R. W.: The low level of 5-hydroxytryptophan decarboxylase in human brain. Biochem. Med. *1:* 240, 1967.
24. Coulson, W. F., Bender, D. A., and Jepson, J. B.: Multiple electrophoresis peaks of rat liver decarboxylase for 3,4-dihydroxyphenylalanine and 5-hydroxytryptophan. Biochem. J., in press.

MONOAMINES IN HUNTINGTON'S CHOREA AND IN WILSON'S DISEASE *

T. L. Sourkes

The basal ganglia of the brain are characterized, among other things, by relatively high content of monoamines, in particular dopamine (3-hydroxytyramine) and serotonin (5-hydroxytryptamine). The relationships of these

* Supported by a grant from the Medical Research Council of Canada.

substances to functions of the basal ganglia have been dealt with by the author and others.[1, 2, 3, 4, 5] Knowledge of the metabolism of monoamines in Huntington's chorea and Wilson's disease (hepatolenticular degeneration) will be discussed in this paper.

Huntington's Chorea

Huntington's chorea is an inherited disease with pathology in dopamine-rich regions in the brain. If only on this account, it merits investigation in regard to the metabolism of Dopa, dopamine and related substances. However, despite the accumulation of much information about the monoamine content of particular regions of the brain in Parkinson's disease, as developed by Hornykiewicz and his colleagues, only three brains of patients dying with Huntington's chorea have been examined. These had normal catecholamine (dopamine and norepinephrine) content.[6] Serotonin was not estimated. We have found[4] subnormal concentrations of serotonin with normal levels of dopamine in two monkeys bearing brain lesions and exhibiting choreiform and ballistic movements of the limbs contralateral to the side of the lesions.

Because of the correlation between depleted dopamine stores of the brain and the subnormal excretion of this amine in the urine of patients with Parkinson's disease, it was of interest to determine whether dopamine excretion in Huntington's chorea would be normal, in agreement with the results of necropsy analyses. When we undertook such an investigation[7] we also attempted to determine if Huntington's chorea entails some defect in the metabolism of exogenous Dopa, the precursor of dopamine and the other catecholamines. It was thought that if there is a defect in the activity of an enzyme concerned with the metabolism of Dopa, then the administration of a large amount of this amino acid (100 mg) relative to the size of the body pool would result in an abnormal pattern of metabolites, perhaps even reflecting the particular point in intermediary metabolism that is defective. Our study included laboratory and other personnel at our Institute as the normal controls, four patients with Huntington's chorea, and four unaffected relatives of the patients as additional controls. Urines were collected on numerous occasions for a 24-hour period, and were analyzed for six metabolites.

Analysis of endogenous metabolites showed a few differences among the three groups. For example, the patients excreted more norepinephrine and less 3,4-dihydroxyphenylacetic acid (DOPAC) than the control group. Although these differences are statistically significant their pathophysiological meaning is not at all clear. In a previous study of five patients with Huntington's chorea from a different geographic population,[5] norepinephrine excretion was lower than in the controls; this may have stemmed from a continuing action of reserpine, which three of the five patients had been receiving chronically and from which they had been withdrawn only 24 hours before urine collection was begun.[8]

It was also found that the unaffected relatives of patients excreted significantly less dopamine and DOPAC than the control subjects, but all groups excreted approximately the same amounts of creatinine, epinephrine, and 5-hydroxyindoleacetic acid (5-HIAA). Others have shown that the output of homovanillic acid (HVA) is normal in Huntington's chorea.[9]

It is possible that some genetic difference between the patients and their relatives from a control population would account for the low amounts of DOPAC in the urine, but interpretation must await the accumulation of more data.

Studies of the metabolism of intravenously administered L-Dopa did not throw any light on the results with endogenous metabolites, for there were no differences between the four patients with Huntington's chorea and the normal subjects.

Perry and his colleagues [10] measured the concentration of many amino acids in the serum and cerebrospinal fluid of patients with Huntington's chorea. Among the few that were lower than normal in concentration was tyrosine, the precursor of L-Dopa. Oliphant, Evans and Forrest [11] found no abnormality in urinary indoles in three patients with Huntington's chorea as compared with two normals. Ruge and Oepen [12] examined urines of nine patients with Huntington's chorea and six controls for 26 phenolic acids and 17 indoles; again, there were no significant differences.

There do not seem to be any reports as yet on the content of HVA and 5-HIAA in the cerebrospinal fluid, although these compounds are currently being studied in relation to Parkinson's disease [13, 14, 15] and other diseases of the nervous system.

The administration of 5-hydroxytryptophan, the precursor of serotonin, to patients with Huntington's chorea has been tested as a potentially therapeutic measure. However, this amino acid causes an exacerbation of the motor symptoms of the disease.[16, 17]

Barbeau reported [18] that patients with Huntington's chorea excrete excessive amounts of an unknown urinary metabolite reacting with Lugol's solution. Curzon and Wald, who made a systematic study of this reaction (Simola test) and quantified it,[19] could not confirm that patients with Huntington's chorea have elevated concentrations of Simola-positive material in their urine. Their tests indicate that the reacting substance is not a catecholamine.

In seeking a defect in monoamine metabolism in this disorder, it is well to keep in mind that the striatal dopamine appears to be in the terminations of the nigrostriatal fibers, whereas the cell bodies observed in the striatum are probably serving cholinergic mechanisms. Hence, some other type of neurohumor, perhaps acetylcholine, may be in defective supply.

Wilson's Disease

Monoamine metabolism in hepatolenticular degeneration has received even less attention than in Huntington's chorea. In our study catecholamines, HVA, DOPAC and 5-HIAA were measured in the urines of five patients with hepatolenticular degeneration and in three persons with subclinical changes.[20] Some of the compounds were excreted in increased amounts, but this was not consistent for all subjects, nor even within subjects for whom there were more than a single 24-hour urine. No wholly consistent pattern of metabolite excretion applies to these cases, and a much larger series would be necessary to segregate the influence, if any, of factors such as age of patient, age at onset of illness, duration of illness at the time of the urine studies, severity of symptoms, and effect of drugs on the metabolism of monoamines. Those urinary abnormalities that appear to be present could well result from the hepatic cirrhosis in Wilson's disease.[21] The decreased amount of parenchymal tissue could lead to defective conjugation, so that relatively more of the free metabolites would be excreted.

Ceruloplasmin is able to oxidize serotonin in vitro,[22] a fact which suggests the possibility of a similar reaction in vivo. Warner and Sternlieb[23] have looked into this matter and have found no basis for the hypothesis of an interaction of these compounds under physiological conditions.

Again, dopamine interacts with ceruloplasmin and ^{II}Cu.[24] This has given rise to the view that such interactions may occur in vivo and play a role in the pathogenesis of Wilson's disease. Walaas and Walaas[25] postulate that "an effect of copper on structures containing aromatic amino acids and aromatic amines may also be of importance for the metabolism of nervous tissue." Under aerobic conditions the copper-amine complexes are "rather unstable and the ligands undergo oxidative transformation to indolic structures. The possibility of an altered metabolism of neurohormones due to accumulation of copper in the neural tissue therefore must be considered of importance in neural diseases." However, evidence for this view is lacking.

R. Papeschi in my laboratory has measured the concentration of HVA in lumbar fluid withdrawn from a patient with Wilson's disease: the value was 32 ng per ml. The normal range is variously given by other authors as 10 to 50 ng per ml,[26] 39 to 93 ng per ml,[13] and 57 ± 14 ng per ml.[14]

REFERENCES

1. Carlsson, A.: The occurrence, distribution and physiological role of catecholamines in the nervous system. Pharmacol. Rev. *11:* 490, 1959.
2. Sourkes, T. L.: Cerebral and other diseases with disturbances of amine metabolism. Prog. Brain Res. *8:* 186, 1964.
3. Sourkes, T. L.: Amines of the basal ganglia: biochemical, pharmacological and neurological studies. *In* Molecular Basis of Some Aspects of Mental Activity, Vol. 1 (O. Walaas, ed.). Academic Press, New York, 1967.
4. Sourkes, T. L., and Poirier, L. J.: Neurochemical bases of tremor and other disorders of movement. Canad. Med. Assoc. J. *94:* 53, 1966.
5. Hornykiewicz, O.: Dopamine (3-hydroxytyramine) and brain function. Pharmacol. Rev. *18:* 925, 1966.
6. Ehringer, H., and Hornykiewicz, O.: Verteilung von Noradrenalin und Dopamine (3-Hydroxytyramin) im Gehirn des Menschen und ihr Verhalten bei Erkrankungen des extrapyramidalen Systems. Klin. Wschr. *38:* 1236, 1960.
7. Sourkes, T. L., Pivnicki, D., Brown, W. T., Wiseman-Distler, M. H., Murphy, G. F., Sankoff I. and Saint-Cyr, S.: A clinical and metabolic

study of Dopa (3,4-dihydroxyphenylalanine) and methyldopa in Huntington's chorea. Psychiat. Neurol. *149:* 7, 1965.

8. Carlsson, A., Rasmussen, E. B., and Kristjansen, P.: The urinary excretion of adrenaline and noradrenaline by schizophrenic patients during reserpine treatment. J. Neurochem. *4:* 318, 1959.
9. Williams, C. M., Maury, S., and Kibler, R. F.: Normal excretion of homovanillic acid in the urine of patients with Huntington's chorea. J. Neurochem. *6:* 254, 1961.
10. Perry, T. L., Hansen, S., Diamond, S., and Stedman, D.: Plasma-amino acid levels in Huntington's chorea. Lancet *1:* 806, 1969.
11. Oliphant, J., Evans, J. L., and Forrest, A. D.: Huntington's chorea—some biochemical and therapeutic aspects. J. Mental Sci. *106:* 718, 1960.
12. Ruge, W., and Oepen, H.: Phenolsäuren und Indol im Urin bei Huntingtonscher Chorea. Humangenetik *1:* 296, 1965.
13. Bernheimer, H., Birkmayer, W., and Hornykiewicz, O.: Homovanillinsäure im Liquor cerebrospinalis: Untersuchungen beim Parkinson-Syndrom und anderen Erkrankungen des ZNS. Wien. Klin. Wschr. *78:* 417, 1966.
14. Guldberg, H. C., Turner, J., Hanieh, A., Ashcroft, G. W., Crawford, T. C., Perry, W. and Gillingham, F. J.: On the occurrence of HVA and 5-HIAA in the ventricular CSF of patients suffering from parkinsonism. Confin. Neurol. *29:* 73, 1967.
15. Papeschi, R., Molina-Negro, P., Sourkes, T. L., Hardy, J., and Bertrand, C.: Concentration of homovanillic acid in the ventricular fluid of patients with Parkinson's disease, including thalamotomized cases, attitudinal tremor and other diseases of the extrapyramidal system. In preparation.
16. Birkmayer, W., and Hornykiewicz, O.: Der L-Dioxyphenylalanin (=L-Dopa) —Effekt beim Parkinson-Syndrom des Menschen: zur Pathogenese und Behandlung der Parkinson-Akinese. Arch. f. Psychiat. *203:* 560, 1962.
17. Lee, D. K., Markham, C. H., and Clark, W. G.: Serotonin (5-HT) metabolism in Huntington's chorea. Life Sci. *7:* 707, 1968.
18. Barbeau, A.: Preliminary observations on abnormal catecholamine metabolism in basal ganglia diseases. Neurology *10:* 446, 1960.
19. Curzon, G., and Wald, I.: A quantitative Simola test and its application to the investigation of extrapyramidal syndromes. Clin. Chim. Acta *8:* 893, 1963.
20. Sourkes, T. L., Murphy, G. F., Sankoff, I., Distler, M. H. W., and Saint Cyr, S.: Excretion of dopamine, catecholamine metabolites and 5-hydroxyindoleacetic acid in hepatolenticular degeneration (Wilson's disease). J. Neurochem. *10:* 947, 1963.
21. Cooper, A. M., Eckhardt, R. D., Faloon, W. W., and Davidson, C. S.: Investigation of the aminoaciduria in Wilson's disease (hepatolenticular degeneration): demonstration of a defect in renal function. J. Clin. Invest. *29:* 265, 1950.
22. Eriksen, N., Martin, G. M., and Benditt, E. P.: Oxidation of the indole nucleus of 5-hydroxytryptamine and the formation of pigments: isolation and partial characterization of a dimer of 5-hydroxytryptamine. J. Biol. Chem. *235:* 1662, 1960.
23. Warner, R. R. P., and Sternlieb, I.: Study of a possible physiologic interaction between ceruloplasmin and serotonin. J. Lab. Clin. Med. *67:* 934, 1966.
24. Walaas, E., Walaas, O., Haavaldsen, S., and Pedersen, B.: Spectrophotometric and electron spin resonance studies of complexes of catecholamines with Cu(II) ions and the interaction of ceruloplasmin with catecholamines. Arch. Biochem. Biophys. *100:* 97, 1963.

25. WALAAS, E., and WALAAS, O.: Biochemical aspects of copper and aromatic amines in relation to the extrapyramidal system. Acta Neurol. Scand. Suppl. 1, *38:* 84, 1962.
26. DENCKER, S. J., MALM, U., ROOS, B. E., and WERDINIUS, B.: Acid monoamine metabolites of cerebrospinal fluid in mental depression. J. Neurochem. *13:* 1545, 1966.

L-DOPA AND DECARBOXYLASE INHIBITOR (IDC): BIOCHEMICAL AND CLINICAL IMPLICATIONS

R. Tissot

Study of the Enzymatic Barrier of Cerebral Capillaries

We will not review the general mechanisms of action of IDC Ro 4-4602 (Ro 4-46.02 N′(DL-seryl)-N^2-(2,3,4-trihydroxybenzyl)-hydrazine) and Ro 1756 (Ro 17.56 N′(glycyl)-N^2-(2,3,4-trihydroxybenzyl)-hydrazine). Let us remember simply that the phenomena which could be investigated in man gave results comparable to those obtained in the animal. After ingestion of Dopa ^{14}C alone, the rise in plasma levels of Dopa ^{14}C and ^{14}C methyldopa is modest, while that of the phenolcarboxylic derivatives is quite important. After ingestion of Dopa ^{14}C and Ro 4-4602, the inverse happens [1] (Fig. 1). This seems to indicate that the decarboxylase inhibitor facilitates the intestinal absorption of Dopa by hindering its degradation and lessening its extracerebral metabolism, as in the experimental animal. Evidently, these experiments do not prove that the third mechanism of action of IDC, namely, the decrease or elimination of the cerebrocapillary enzyme barrier, is actually functional in man nor that the cerebral catecholamine level rises. Perhaps one could be permitted to say so if therapeutic results can be credited. Lacking direct proof in man (which only cerebral biopsy can give) we approached the problem in the experimental animal and by indirect and conjectural methods in man. In the animal, as Bertler et al.[2, 3] have shown, Falck's [4] histofluorometric method allows visualization of the cerebral capillary barrier for precursors of monoamines the substrate of which is endothelial DC. In the rat, the administration of an IDC, Ro 4-4602 50 mg per kg, causes capillary fluorescence to disappear after Dopa 75 mg per kg and nialamid 250 mg per kg whereas the cerebral parenchyma becomes frankly fluorescent.[5]

Biochemical assays show that this parenchymal fluorescence, if partly related to Dopa, also denotes an increase of cerebral catecholamines.[6, 7, 8] This enzymatic barrier is not uniform for all the cerebral structures. Intense at the level of the cortex, where 10 mg per kg of Ro 4-4602 are required to make capillary fluorescence disappear, it is much less intense in the lower

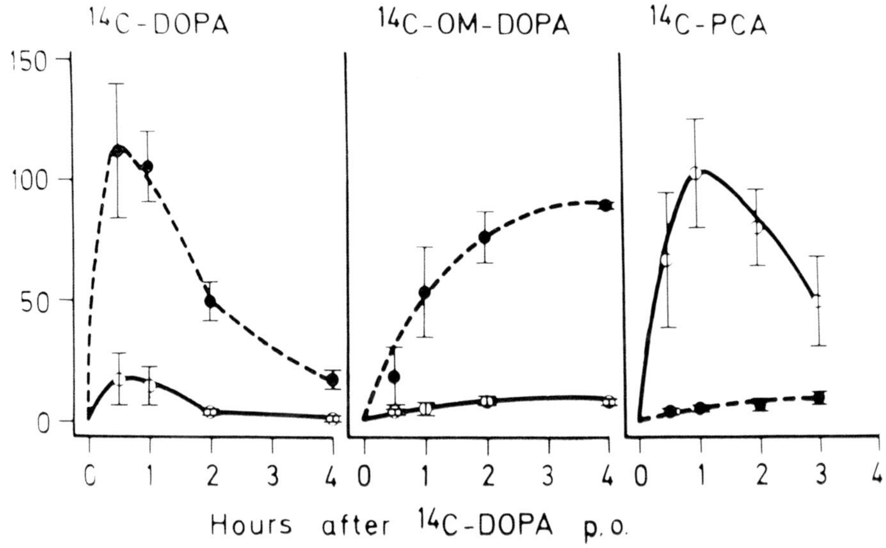

Fig. 1. Effect of Ro 4-4602 by mouth on the increase of ^{14}C.catechol metabolites in plasma of humans after administration of 3 mg/kg ^{14}C-Dopa by mouth. 16 to 24 mg Ro 4-4602 were administered by three divided doses within 14¾ hours prior to ^{14}C-Dopa. The ordinate indicates the radioactivity of the catechol fractions in plasma in percent of the total radioactivity administered per gram body weight. *Line:* ^{14}C-Dopa alone. *Dotted line:* Ro 4-46.02 + ^{14}C-Dopa. Each point indicates an average with standard error of three experiments. ^{14}C-OM-Dopa = ^{14}C-methyldopa; ^{14}C-PCA = ^{14}C-phenolcarboxylic acids.

hypothalamus and substantia nigra where capillary fluorescence disappears after 2 to 4 mg per kg of IDC.[9, 16] Nevertheless, this barrier cannot be nullified and its dissolution by IDC facilitates the penetration of Dopa in corresponding structures. This fact, tied to a better resorption of Dopa and to a smaller extracerebral metabolism, can in part explain the effectiveness of small doses of Dopa in the parkinsonian patient receiving IDC simultaneously. All of the catecholamine-rich structures did not have an enzymatic capillary barrier of low intensity. Indeed, the lateral ganglia of the pons and medulla oblongata require doses comparable to those necessary to eliminate capillary fluorescence after Dopa in the cortex, namely, 10 mg per kg.[10] Notwithstanding that, even without IDC, the injection of Dopa produces an increase of parenchymatous fluorescence at the level of these nuclei while no increase occurs at the cerebral cortex. Several hypotheses can be put forth to account for these observations: (1) Even a relatively elevated barrier is not sufficient to stop the passage of Dopa completely. In the brain-stem nuclei the small amount of Dopa passing the barrier is transformed into catecholamines which are fixed and stocked, their increase becoming visible. On the other hand, at the cerebral cortex, the Dopa that passed into the parenchyma is eliminated. (2) sites of predilection, "windows" in the enzymatic barrier, may exist for the passage of Dopa. This hypothesis is suggested by the existence of close contacts between capillaries and neurons in the substantia

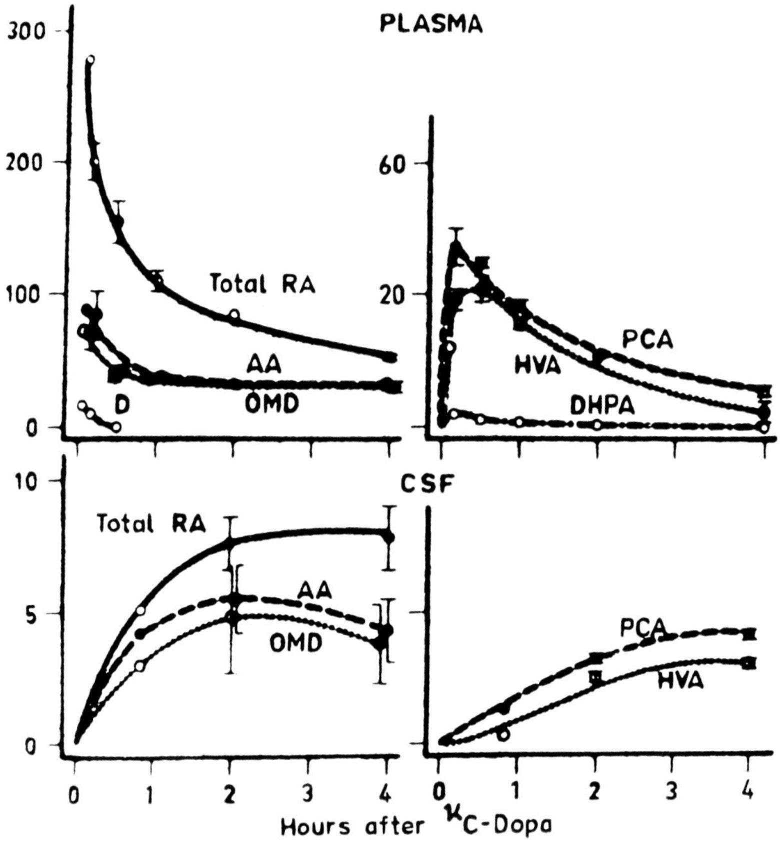

Fig. 2. Radioactivity of the various fractions of plasma and cerebrospinal fluid (CSF) of two patients injected with 48 g per kg ^{14}C-Dopa i.v. Each point represents an average ± S.E. of two corresponding values.

RA = radioactivity
AA = amino acids
OMD = 3-O-methyldopa
D = Dopa
PCA = phenylcarboxylic acids
HVA = homovanillic acid
DHPA = dihydroxyphenylacetic acid

nigra and above all by the activity of neurons of the median raphé. The capillary enzymatic barrier for Dopa is very intense at this level. Capillary fluorescence does not disappear completely after 50 mg per kg of IDC. Neurons normally loaded with serotonin accumulate catecholamines after the injection of Dopa, but they show odd prolongations which come in contact with the capillaries.[10] One can then ask if, along with the general regulation of the penetration of monoamine precursors into cerebral parenchyma, still other more subtle and specific mechanisms do not exist.

In man, previous experiments with Dopa ^{14}C have shown that the radioactivity of spinal fluid due basically to the phenolcarboxylic acids does not correspond at all to that in the blood and, in particular, that homovanillic acid (HVA) infused into the blood stream does not pass directly into the spinal fluid.[11, 12] Therefore it seems probable that the HVA of spinal fluid must come from the cerebral metabolism of catecholamines. After giving Dopa ^{14}C and an IDC (Ro 4-4602), we expected to see the level of HVA in the spinal fluid rise; on the contrary, the HVA level dropped.[13] These preliminary experiments must be verified in the experimental animal by the ventricular perfusion technique. If they are confirmed, they would lead us to think that the greatest part of spinal-fluid HVA comes directly from catecholamine metabolism at the level of cerebral capillaries and not from parenchymatous metabolism in the strict sense.

Without diminishing the value of studies showing that HVA levels in spinal fluid differ in the normal subject and in certain neurological and psychiatric diseases, particularly parkinsonism,[14, 15, 16, 17] these findings would open their significance to further question.

Preliminary Biological Studies on Treatment with L-Dopa 900 mg and Ro 4-4602 900 mg

Before treatment 24-hour urinary assays of monoamines and their metabolites gave us the results shown in Table 1. They seem to confirm certain earlier and controversial notions found in the literature.[18, 19, 20, 21, 22, 23, 24, 25] In parkinsonism dopamine metabolism seemed insufficient, while in depressive syndromes norepinephrine and epinephrine metabolism appeared abnormal, as did that of serotonin. After treatment with Dopa and Ro 4-4602, this contrast between parkinsonism and depressions is still seen but is less noticeably evident (Table 2).

Clinical Results

It is not the purpose of this symposium to discuss the effect of this therapy in depressive syndromes. In the space allotted us, we will consider what occurs in parkinsonism.

TABLE 1. 24-Hr Urinary Excretion of Monoamines and Their Metabolites Amines and Methylated Derivatives in μg, Acid Derivatives in mg

	Normal Subjects		*Parkinson's Disease*		*Depressions (Melancolia)*	
Dopamine	(17)	1122 ±1260	(5)	740 ±342	(9)	992 ±521
Methoxytyramine	(17)	477 ± 482	(5)	*176* ± 55	(9)	307 ±250
HVA	(20)	5.67± 2.6	(5)	*3.76*	(9)	6.04
VMA	(20)	3.56± 1.1	(5)	*6.02*± 4.03	(9)	*4.59*± 2.08
Adrenaline	(20)	20 ± 9	(5)	22 ± 18	(9)	46 ± 55
Metadrenaline	(10)	93 ± 70	(5)	106 ± 35	(9)	*223* ±192
Noradrenaline	(20)	131 ± 65	(5)	156 ± 93	(9)	*78* ± 50
Normetadrenaline	(10)	105 ± 74	(5)	187 ±117	(9)	*227* ± 98
Meta+normeta	(20)	198 ± 99	(5)	293 ±143	(9)	*440* ±248
5 HIAA	(20)	3.47± 1.1	(5)	3.66± 2.0	(9)	*4.84*± 1.6

TABLE 2. 24-Hr Urinary Execution of Monoamines and Metabolites After Treatment with Ro 4-4602 (900 mg) and L-Dopa (900 mg)

	Parkinson (4)	*Depression* (8)
Dopamine	12.000±10.161	20.937±14.180
Methoxytyramine	2.868± 2.178	4.205± 2.033
HVA	30.4 ±31.1	34.5 ±32.6
VMA	6.4 ± 1.0	7.6 ± 3.2
Adrenaline	34.5 ±24.8	18.9 ±21
Metadrenaline	97.3 ±57.9	136.6 ±95.9
Noradrenaline	120.0 ±42.5	170.8 ±85.2
Normetadrenaline	135.5 ±83.3	115.5 ±67.1
Meta+normeta	232.8	252.1
5 HIAA	*0.79* ± 0.6	*2.4* ± 1.5

After having treated 150 patients, our results seem to us equivalent to those obtained on this side of the Atlantic with Dopa alone.[26, 27, 28] With IDC added, the amount of Dopa required is at least ten times less, ranging from 450 to 1500 mg. As with Dopa alone, akinesia and hypertonicity are the symptoms which respond best and, above all, most rapidly. Resolution of tremors is very erratic, varying from no improvement to complete disappearance. We have the impression that the more regular the tremor the better it responds to treatment.

The advantages of the addition of IDC are not limited to a decrease in the dosage of Dopa. When Dopa and IDC are given in adequate doses, the therapeutic margin between amelioration of parkinsonism and the appearance of abnormal (involuntary) movements is wider. Other habitual side effects, such as nausea, vomiting and above all orthostatic hypotension, are more rare and less pronounced. In 150 patients we had one serious blood dyscrasia, reversed by cessation of treatment. Urinary retention at the start of treatment occurred in three elderly patients with prostatism. No sign of intolerance in large parenchymal organs was noted in our cases.

REFERENCES

1. Tissot, R., Bartholini, G., Pletscher, A.: Drug-induced changes of extracerebral Dopa metabolism in man. Arch. Neurol. *20:* 187, 1969.
2. Bertler, A., Falck, B., Owman, C., and Rosengren, C.: The localization of monoaminergic blood brain barrier mechanism. Pharmacol. Rev. *18:* 369, 1966.
3. Owman, C., and Rosengren, E.: Dopamine formation in brain capillaries, an enzyme blood-brain barrier mechanism. J. Neurochem. *14:* 547, 1967.
4. Falck, B., and Owman, C.: A detailed methological description of the fluorescence method for the cellular demonstration of biogenic monoamines. Acta Univ. Lund *2:* 7, 1965.
5. Constantinidis, J., Bartholini, G., Tissot, R., and Pletscher, A.: Accumulation of dopamine in the parenchyme after decarboxylase inhibition in the capillaries of brain. Experientia, *24:* 130, 1968.
6. Bartholini, G., Bates, H. M., Burkard, W. P., and Pletscher, A.: Increase of cerebral catecholamines caused by 3-4-dihydroxyphenylalanine after inhibition of peripheral decarboxylase. Nature *215:* 852, 1967.

7. Bartholini, G., and Pletscher, A.: Cerebral accumulation and metabolism of C14 Dopa after selective inhibition of peripheral decarboxylase. J. Pharmacol. Exp. Ther. *161:* 14, 1968.
8. Bartholini, G., Tissot, R., and Pletscher, A.: Selective increase of cerebral dopamine. A therapeutic possibility in parkinsonism. *In* Third Symp. on Parkinson Disease. Livingston, London, 1969.
9. Constantinidis, J., Torre, J. C. de la, Tissot, R., Geissbuhler, F.: La barrière capillaire pour la Dopa dans le cerveau et les différents organes. Psychopharmacologia *15:* 75, 1969.
10. Constantinidis, J., Bartholini, G., Geissbuhler, F., and Tissot, R.: La barrière capillaire enzymatique pour la Dopa au niveau de quelques noyaux du tronc cérébral du rat. In preparation.
11. Bartholini, G., Pletscher, A., and Tissot, R.: On the origin of HVA in the cerebrospinal fluid. Experientia *22:* 609, 1966.
12. Pletscher, A., Bartholini, G., and Tissot, R.: Metabolic fate of L-14 C-Dopa in cerebrospinal fluid and blood plasma of humans. Brain Res. *4:* 106, 1967.
13. Pletscher, A., Bartholini, G., and Tissot, R.: On the origin of homovanillic acid in the cerebrospinal fluid. In press.
14. Bernheimer, H., Birkmayer, W., and Hornykiewicz, O.: HVA in liquor cerebrospinalis. Untersuchungen beim Parkinson-Syndrom und anderen Erkrankungen des ZNS. Wien. Klin. Wsch. *78:* 417, 1966.
15. Bruck, H., Gerstenbrand, F., Grundig, E., and Teufelmayer, R.: Uber Ergebnisse von Liquoranalyse beim Parkinson-Syndrom. Acta Neuropath. *3:* 638, 1964.
16. Johanssen, B., and Roos, B. E.: 5-HIAA and HVA and levels in the cerebrospinal fluid of healthy volunteers and patients with Parkinson's syndrome. Life Sci. *6:* 1449, 1967.
17. Roos, B. E., and Sjöström, R.: 5-HIAA (and HVA) levels in the cerebrospinal fluid after Probenecid application in patients with manic-depressive psychosis. Pharmacol. Clinica *1:* 153, 1969.
18. Barbeau, A., Murphy, G. F., and Sourkes, T. L.: Excretion of dopamine in disease of basal ganglia. Science *133:* 1706, 1961.
19. Barbeau, A., and Jasmin, G.: Biochemistry of Parkinson's disease. Neurology *13:* 56, 1963.
20. Bergsman, A.: The urinary excretion of adrenaline and noradrenaline in some mental diseases. A clinical and experimental study. Acta Psychiat. Neur. Scand. *34:* Suppl. 133, 1959.
21. Burgermeister, J. J., Dick, P., Garrone, G., Guggisberg, M., and Tissot, R.: Excrétion urinaire de l'acide 5-hydroxyindolacétique (5-HIAA) chez 150 malades atteints de syndrome dépressif et d'agitation maniaque. Presse Médicale, *22:* 1116, 1963.
22. Greer, M., and Williams, C. M.: Dopamine metabolism in Parkinson's disease. Neurology *13:* 73, 1963.
23. Rinne, U. K., Laitinen, L. V., and Sonninen, V.: Excretion of homovanillic and vanilmandelic acids before and after thalamotomy for extrapyramidal disorders. Psychiat. Neurol. *153:* 345, 1967.
24. Strom-Olsen, R., and Weil-Malherbe, H.: Humoral changes in manic-depressive psychosis with particular reference to the excretion of catecholamine in urine. J. Ment. Sci. *104:* 696, 1958.
25. Tissot, R.: Perturbations du métabolisme des catécholamines dans les syndromes parkinsoniens. Presse Med. *77:* 617, 1969.
26. Cotzias, G. C.: L-Dopa for parkinsonism. New Eng. J. Med. *276:* 374, 1967.
27. Cotzias, G. C., Papavasiliou, P. S., Gellene, R., Aronson, R. B., and

MENA, I.: Long term effects of Dopa in parkinsonism. A proposal. *In* Third Symp. on Parkinson Disease. Livingston, London, 1969.
28. DUVOISIN, R., BARRETT, R., SCHEAR, M., HOHEN, M., and YAHR, M.: The use of L-Dopa in parkinsonism. *In* Third Symp. on Parkinson Disease. Livingston, London, 1969.

THE METABOLISM OF L-DIHYDROXYPHENYLALANINE-^{14}C BY ISOLATED PERFUSED RAT LIVER *

Gertrude M. Tyce

L-Dopa given orally is immediately available for uptake and metabolism by the liver. Since Dopa is given to patients with Parkinson's disease, we thought it important to study the hepatic phase of its metabolism. We used the isolated perfused rat liver to determine how quickly Dopa was cleared from the perfusate and also the nature and disposition of the metabolites.

Brauer's method [1] was used for the liver perfusion. In a temperature- and humidity-controlled cabinet, a liver was perfused with circulating oxygenated blood and the bile was collected.

L-Dopa-^{14}C (5 μc) labeled in the 3 position in the side chain was injected into the blood entering the portal vein; this dose was equivalent to 32 μg. Total ^{14}C was measured in aliquots of whole blood taken at intervals after Dopa injection and in aliquots of plasma and bile. After five hours of perfusion, the liver was removed and ^{14}C was measured in perchloric acid homogenates of the liver.

During the first hour of perfusion, ^{14}C in whole blood decreased to 50 percent of the dose; it then increased somewhat to a final value of 56 percent of the dose. In plasma, total ^{14}C decreased more rapidly than in whole blood; at 45 minutes 40 percent of the dose was in plasma, and then there was an increase to 49 percent after five hours. The difference between ^{14}C in whole blood and ^{14}C in plasma represents ^{14}C taken up by the cellular elements in the perfusate; this fraction represented 27 percent of the dose at five minutes after the Dopa injection, but decreased during the perfusion to 7 percent after five hours.

The ^{14}C in the liver at intermediate times was calculated from the difference between the total ^{14}C injected and the sum of total ^{14}C in whole blood and in bile. In the first 30 minutes, 36 percent of the injected ^{14}C was taken up by the liver; after one hour, ^{14}C in the liver steadily decreased to a final value of 6 percent of the dose.

Excretion of ^{14}C into the bile followed a sigmoid course. The cumulative excretion of ^{14}C in bile after five hours accounted for 32 percent of the dose.

* Supported in part by Grant NB-4004 from the National Institutes of Health, Public Health Service.

Dopa and metabolites in perchloric acid extracts of plasma, bile, and liver were separated by column chromatography using Dowex 50 and alumina by the methods of Bartholini and Pletscher.[2]

The disappearance of Dopa-^{14}C from plasma showed three components. The first phase was most rapid: after five minutes, only 31 percent of the dose remained in plasma as Dopa-^{14}C. From 30 minutes to two hours, the decrease in Dopa-^{14}C was linear with a half-life of about 27 minutes. At two hours, the Dopa-^{14}C remaining in plasma was only 1.7 percent of the dose. After this, the decrease was much slower; at five hours, less than 0.5 percent of the dose remained as unmetabolized Dopa.

The first phase in the disappearance of Dopa-^{14}C probably is largely accounted for by its uptake by the liver and by the cellular elements in blood, which together account for 50 percent of the dose. However, some metabolism of Dopa-^{14}C has already occurred in the first five minutes. During the second phase, Dopa is metabolized by the liver and the metabolites are released into bile and into plasma; more Dopa-^{14}C is then taken up from plasma by the active metabolic sites in the liver. During the final phase, either end-product inhibition was operative or the Dopa-^{14}C was so diluted in the 120 ml of perfusate that substrate concentration became limiting.

The most abundant fraction in plasma was the conjugate fraction, accounting for 36 percent of the dose after five hours.

The amount of ^{14}C-labeled O-methyldopa increased during the perfusion to about 4 percent of the dose.

Organic acids (homovanillic acid, dihydroxyphenylacetic acid) accounted for 4 percent of the dose early in the perfusion but only for 1.4 percent after five hours.

The amine fraction was a very minor one throughout the perfusion; this fraction accounted for 0.6 percent of the dose during the first 15 minutes and for less than 0.2 percent at the end of the perfusion. The shape of the amine fraction curve suggests that the liver very rapidly metabolized the amines produced.

Five compounds were found in bile. By elution and incubation with β-glucuronidase or with sulfatase, one of the major compounds was identified as the glucuronide of N-acetyl-3-methoxydopamine. Two other glucuronides were found in considerable amounts; one was the glucuronide of a compound which was volatile under acidic conditions and was not absorbed on alumina (this compound may result from cleavage of the catechol ring) and the other was the glucuronide of N-acetyldopamine.

After five hours of perfusion, the most abundant compounds in plasma were the sulfates of homovanillic acid (15.6 percent of dose) and dihydroxyphenylacetic acid (14.6 percent), and the glucuronide of N-acetyl-3-methoxydopamine (9.7 percent). Dopa accounted for 0.5 percent and methyldopa for 4.2 percent. Amines (about equal amounts of methoxy and dihydroxy amines) accounted for only 0.2 percent of the dose. Free homovanillic acid and free dihydroxyphenylacetic acid each accounted for less than 1 percent of the dose.

In liver, Dopa accounted for 0.1 percent, methyldopa for 1.3 percent, amines for 0.1 percent, and conjugated compounds for a total of 2.8 percent of the dose.

In bile there was the merest trace of Dopa and of amines. The glucuronides of methoxydopamine and homovanillic acid and the sulfates of homovanillic and dihydroxyphenylacetic acids each accounted for less than 1 percent of the dose. The glucuronide of N-acetyl-3-methoxydopamine accounted for 16.8 percent, the glucuronide of N-acetyldopamine for 5.4 percent, and the glucuronide of the unknown compound for 5.3 percent.

Thus, in the isolated perfused rat liver, L-Dopa-^{14}C is taken up rapidly by the liver and by the cellular elements in the perfusate. Metabolism by the liver is very rapid, and L-Dopa has a half-life of about 27 minutes in plasma. In bile, the most abundant compounds formed were the glucuronides of N-acetyl-3-methoxydopamine, N-acetyldopamine and an unknown compound. In plasma, conjugates of homovanillic acid, dihydroxyphenylacetic acid, and N-acetyl-3-methoxydopamine were the most abundant metabolites.

ACKNOWLEDGMENT

The author wishes to thank Dr. Eunice V. Flock and Dr. C. A. Owen, Jr., for advice and criticism during these investigations.

REFERENCES

1. Brauer, R. W., Pessotti, R. L., and Pizzoloto, P.: Isolated rat liver preparations: Bile production and other basic properties. Proc. Soc. Exp. Biol. Med. *78:* 174, 1951.
2. Bartholini, G., and Pletscher, A.: Cerebral accumulation and metabolism of C^{14}-Dopa after selective inhibition of peripheral decarboxylase. J. Pharmacol. Exp. Ther. *161:* 14, 1968.

INTERRELATIONSHIPS AMONG DOPAMINE, SEROTONIN AND ACETYLCHOLINE *

A. Orzeck and A. Barbeau

Reserpine has long been known to produce extrapyramidal symptoms in humans. Brodie and collaborators [1] first demonstrated that this drug depleted cerebral serotonin. Carlsson et al.[2, 19] added that it also depleted the brain, and particularly the extrapyramidal system of catecholamines. Both the clinical symptoms and the catecholamine depletion could be reversed by treatment with L-Dopa.[3] However, it is not yet certain that the symptomatol-

* Supported by Grants MA-3082 and SP-1 from the Medical Research Council of Canada.

ogy is directly related to a depletion in catecholamines or serotonin. Many other neurotransmitters may also be involved, and in 1962 Barbeau [4] suggested that the clinical syndrome resulted from an imbalance in a complex system involving dopamine, serotonin, acetylcholine and histamine. McGeer,[5] Everett [6] and Coleman [7] in this symposium have also proposed evidence for a balance between dopamine and serotonin. In the present paper we would like to summarize some evidence indicating that both dopamine and serotonin may act as modulators of acetylcholine (ACh) levels in the caudate nucleus.

The effect of reserpine on acetylcholine (ACh) levels is not clear. There is a divergence in results among several authors who found both an increase and a decrease in tissue levels of ACh in the brain of animals treated with reserpine.[8, 9, 10] This dichotomy can possibly be attributed to the selection of different brain areas as well as of different species. To avoid some of these difficulties, we selected the reserpinized rabbit as our experimental model. We also restricted ourselves to certain areas of the brain including the caudate nucleus, an extrapyramidal structure known to be rapidly depleted of catecholamines by reserpine and considered deficient in dopamine in the brain of parkinsonian patients.[11]

Material and Methods

Male rabbits weighing 2 kg received 2.5 mg per kg of reserpine through the marginal ear vein. They were observed in a quiet room at 21°C and were then killed by air embolism after four hours. Various drugs (listed in results and tables) were given intraperitoneally to the animals following different time schedules; control animals were injected with physiological saline. Control rabbits were killed concurrently with each set of experimental animals.

The caudate nuclei, the hypothalamus, the brain stem and remaining cerebrum were dissected out rapidly and extracted for total tissue ACh according to the method of Crossland.[12] ACh was estimated with the frog rectus-abdominus preparation adapted to a digital balance as described by Jasper and Koyama.[13] (Although this work is concerned with the estimation of ACh, it should be understood that it is ACh-like activity that we are measuring.)

Results

It was found that reserpine produced a selective and significant depletion of ACh levels as compared to controls in the caudate nucleus and not in the other structures studied (Table 1). Various drugs were administered intraperitoneally 30 minutes prior to sacrifice to try to correct this depletion. L-3,4-dihydroxyphenylalanine (L-Dopa 200 mg per kg) corrected the alteration in ACh stores with a significant elevation ($p<0.001$) (Table 2). A significant repletion in ACh levels ($p<0.001$) was also observed with γ-hydroxybutyrate (gamma-OH) (2000 mg per kg), a drug known to cause

TABLE 1. Acetylcholine Levels in Discrete Areas of Rabbit Brain 4 Hours After a Single Injection of Reserpine (2.5 mg/kg I.V.) Mean Acetylcholine Content (μg/g of wet tissue ± S.D.)

Brain Area	*Control*	*N*	*Reserpine*	*N*	*Excipient*	*N*
Caudate nuclei	3.86±0.09	(16)	1.94±0.60 *	(16)	4.88±0.37	(6)
Hypothalamus	3.46±0.85	(8)	3.13±0.68	(8)	3.29±0.57	(3)
Brain stem	2.23±0.34	(7)	2.10±0.25	(8)	2.60±0.18	(3)
Remaining cerebrum	1.16±0.31	(8)	0.09±0.08	(8)	1.30±0.25	(3)

an elevation in dopamine (DA) levels [14] (Table 3). Physostigmine (1.5 mg per kg), an anticholinesterase, was also examined and as expected was found significantly to increase the levels of ACh, particularly in the caudate (Fig. 1). However an aggravation rather than an amelioration of the behavioral state of the animal was observed in this case. This finding corresponds with clinical observations reported by Nashold [15] and Duvoisin [16] in parkinsonian patients having received physostigmine. This evidence provides additional proof, although indirect, of the role played by DA in reserpine sedation. Other authors have also discounted the participation of norepinephrine and serotonin as modulators of this action.

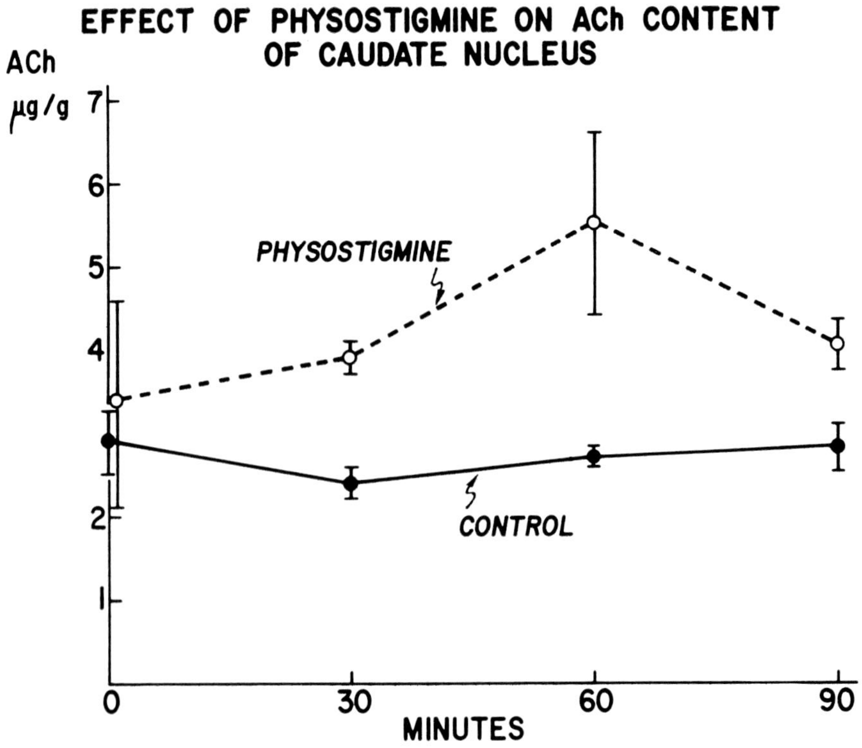

Fig. 1.

TABLE 2. Effect of Dopa on ACh Content of Caudate Nucleus in Saline Injected Rabbits and Rabbits Treated With Reserpine
Mean Acteylcholine Content (μg/g of wet tissue ± S.D.)

Brain Area	*Saline +Saline*	*N*	*Saline +Dopa*	*N*	*Reserpine +Saline*	*N*	*Reserpine +Dopa*	*N*
Caudate nuclei	4.30±1.02	(24)	4.91±1.38	(8)	1.76±0.64	(8)	5.32±2.22 *	(10)
Hypothalamus	3.76±0.62	(12)	4.38±0.82	(4)	4.34±0.73	(4)	3.88±0.47	(5)
Brain stem	2.77±0.49	(11)	2.58±0.47	(4)	2.92±0.46	(4)	2.62±0.19	(5)

TABLE 3. Effect of Gamma-OH on ACh Content in Saline Injected Rabbits and Rabbits Treated With Reserpine
Mean Acetylcholine Content (μg/g of wet tissue ± S.D.)

Brain Area	*Saline +Saline*	*N*	*Saline +Gamma-OH*	*N*	*Reserpine +Saline*	*N*	*Reserpine +Gamma-OH*	*N*
Caudate nuclei	4.30±1.02	(24)	4.25±0.53	(8)	1.76±0.64	(8)	3.28±0.60 *	(11)
Hypothalamus	3.76±0.62	(12)	3.61±0.21	(4)	4.34±0.73	(4)	4.46±0.38	(6)
Brain stem	2.77±0.49	(11)	2.67±0.41	(4)	2.92±0.46	(4)	2.82±0.57	(6)

TABLE 4. Effect of 5-HTP on ACh Content in Saline Injected Rabbits and Rabbits Treated With Reserpine
Mean Acetylcholine Content (μg/g of wet tissue ± S.D.)

Brain Area	*Saline +Saline*	*N*	*Saline +5-HTP*	*N*	*Reserpine +Saline*	*N*	*Reserpine +5-HTP*	*N*
Caudate nuclei	4.30±1.02	(24)	3.01±0.91 *	(8)	1.76±0.64 *	(8)	1.53±0.32 *	(10)
Hypothalamus	3.76±0.62	(12)	4.49±0.74	(4)	4.34±0.73	(4)	4.11±0.99	(5)
Brain stem	2.77±0.49	(11)	2.59±0.39	(4)	2.92±0.46	(4)	2.57±0.47	(5)

TABLE 5. Effect of 5-HTP on Elevated ACh Levels in Rabbits Pretreated With Physostigmine
Mean acetylcholine content (μg/g of wet tissue $\pm$ S.D.)

Brain Area	*Saline +Saline*	*N*	*Physostigmine +Saline*	*N*	*Physostigmine +5-HTP*	*N*
Caudate nuclei	2.75±0.37	(6)	4.63±1.00 *	(8)	3.11±0.94 *	(16)
Hypothalamus	3.01±0.91	(3)	5.24±1.00 *	(4)	4.11±0.63	(8)
Brain stem	1.93±0.15	(3)	2.88±0.58 *	(4)	2.67±0.75	(8)

5-hydroxytryptophan (5-HTP), 150 mg per kg, the direct precursor of serotonin, was found ineffective in repleting ACh stores (Table 4). However 5-HTP lowered significantly the ACh content of the caudate nucleus in animals with normal ACh levels, and prevented the increase in ACh levels produced by physostigmine (Table 5).

Discussion

Our results confirm the data of Beani and collaborators [10] of a reduction in ACh content under the influence of reserpine, an effect localized primarily to the caudate nucleus. This specificity in localization makes it likely that a direct effect of reserpine on ACh stores is not involved, since other areas with known ACh content are not affected. This assumption is corroborated by Beani and his group [10] who were unable to demonstrate any effect of reserpine upon the synthesis or metabolism of ACh.

From these studies it is apparent that both dopamine and serotonin can play a modulating role upon acetylcholine tissue levels in the caudate nuclei. Dopamine can correct a depletion in ACh levels, whereas serotonin will rectify abnormal increases. A model is proposed to exemplify this concept (Fig. 2). Antagonistic actions of DA and 5-HT have been reported by Viala and Buser [17] at the periphery, where Dopa injections increase extensor activity and 5-HTP increases flexor activity.

It thus becomes important to consider what happens to the level of other postulated neurotransmitters when L-Dopa is given. The present paper

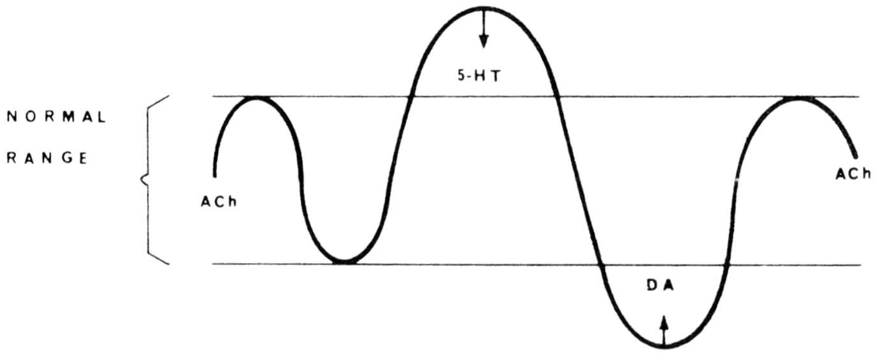

Fig. 2.

indicates that ACh levels can be influenced significantly. Experiments are in progress in our laboratory to look for a similar effect upon gamma-aminobutyric acid (GABA).

Summary

Injection of reserpine caused a specific decrease of acetylcholine levels in the caudate nucleus of the rabbit. This depletion was corrected by the administration of 3,4-dihydroxyphenylalanine, but not by 5-hydroxytryptophan. Excessive elevation of acetylcholine levels by physostigmine was prevented by administration of 5-hydroxytryptophan. Probable modulator functions of dopamine and serotonin in the regulation of normal levels of ACh in the caudate nucleus are discussed.

ACKNOWLEDGMENTS

The authors would like to thank Miss Lise Bergeron and Miss Francine Carrière for technical assistance. Preliminary results of this research were presented at the annual meeting of the "Club de Recherches Cliniques du Québec," September 27, 1969.[18]

REFERENCES

1. Shore, P. A., Silver, S. L., and Brodie, B. B.: Interaction of reserpine, serotonin and lysergic acid diethylamide in brain. Science *122:* 284, 1955.
2. Carlsson, A., and Hillarp, N. A.: Release of adrenaline from the adrenal medulla of rabbits produced by reserpine. Kgl. Fysiogr. Sällsk, Lund Förh *26:* 8, 1956.
3. Carlsson, A., Lindqvist, M., and Magnusson, T.: 3,4-dihydroxyphenylalanine and 5-hydroxytryptophan as reserpine antagonists. Nature *180:* 1200, 1957.
4. Barbeau, A.: The pathogenesis of Parkinson's disease: a new hypothesis. Canad. Med. Assoc. J. *87:* 802, 1962.
5. McGeer, P. L.: Central amines and extrapyramidal functions. J. Neuropsychiat. *4:* 247, 1963.
6. Everett, G. M., and Wiegand, R. G.: Central amines and behavioral states: a critique and new data. *In* Proceedings First Int. Pharmacol. Meeting, *8:* 85, 1962.
7. Coleman, M. P.: The L-DOPA effect in dystonia as a model of counterbalancing dopamine-serotonin systems in the CNS. *In* L-Dopa and Parkinsonism (Barbeau, A. and McDowell, F., eds.). F. A. Davis, Philadelphia, 1970.
8. Malhotra, C. L., and Pundlik, P. G.: The effect of reserpine on the acetylcholine content of different areas of the central nervous system of the dog. Brit. J. Pharmacol. *14:* 46, 1959.
9. Giarman, N. J., and Pepeu, G.: Drug-induced changes in brain acetylcholine. Brit. J. Pharmacol. *19:* 226, 1962.
10. Beani, L., Ledda, F., Bianchi, C., and Baldi, V.: Reversal by 3,4-dihydroxyphenylalanine of reserpine-induced regional changes in acetylcholine content of the guinea-pig brain. Biochem. Pharmacol. *15:* 779, 1966.
11. Ehringer, H., and Hornykiewicz, O.: Verteilung von noradrenalin und

dopamine (3-hydroxytyramine) im Gehirn des Menschen und ihr Verhalten bei Erkrankungen des extrapyramidalen Systems. Klin. Wschr. *38:* 1236, 1960.

12. Crossland, J.: Biologic estimation of acetylcholine. *In* Methods in Medical Research, Vol. 9 (J. H. Quastel, ed.), Year Book Medical Publishers, Inc., Chicago, 1961.
13. Jasper, H., and Koyama, I.: Use of a digital balance for bioassay of acetylcholine. Canad. J. Phys. Pharmacol. *46:* 692, 1968.
14. Gessa, G. L., Vargiu, L., Crabai, F., Boero, G. C., Caboni, F., and Camba, R.: Selective increase in brain dopamine induced by Gamma-hydroxybutyrate. Life Sci. *5:* 1921, 1966.
15. Nashold, B. S.: Cholinergic stimulation of globus pallidus in man. Proc. Soc. Exp. Biol. Med. *161:* 68, 1959.
16. Duvoisin, R. C.: The mutual antagonism of cholinergic and anticholinergic agents in parkinsonism. Trans. Amer. Neurol. Assoc. *91:* 73, 1966.
17. Viala, D., and Buser, P.: The effects of DOPA and 5-HTP on rhythmic efferent discharges in hind limb nerves in the rabbit. Brain Res. *12:* 437, 1969.
18. Orzeck, A., and Barbeau, A.: Effets de la sérotonine et de la dopamine sur la teneur en acetylcholine cérébrale chez le lapin. Un. Méd. Canad. *99:* 328, 1970.
19. Carlsson, A.: The occurrence, distribution and physiological role of catecholamines in the nervous system. Pharmacol. Rev. *11:* 490, 1959.

GENERAL DISCUSSION

Dr. Spiegel: The idea that the pallidum is involved in Parkinson's disease is over forty years old; it goes back to F. H. Lewy and his book *Tonus und Bewegung.* He studied about sixty cases of Parkinson's disease and found what he believed to be pathological changes in the pallidum. Because of these histological findings, Foerster described what he called the akinetic-rigid pallidum syndrome. It is now very difficult to accept these ideas, in view of the effects of stereotaxic operations on the pallidum. We have seen that when lesions were placed in the pallidum (before the operations on the ventrolateral thalamus were started) they reduced tremor and rigor. Thus the conclusions from the histopathological findings are contradicted by the results of stereotaxic operations.

I believe the trouble with these histopathological studies was that there were too few controls in the same age groups; if the authors had studied a sufficient number of senile brains, they would have found the same changes found in the parkinsonian pallida.

Dr. Coleman: Dr. McGeer speaks about the relationship of 5-HTP and its effect on dopamine levels; he comments that when he gave 5-HTP he lowered dopamine and norepinephrine, and when he gave Dopa he lowered serotonin.

This effect is probably at least partially due to the fact that these amino acids are all large neutral amino acids and share similar transport systems as they enter the brain, as well as some of the enzymes and coenzymes of their metabolic pathways. An experiment done with any member of this group of amino acids will indicate that when one amino acid is given the others are affected.

Dr. McGreer also points out that in patients with Parkinson's disease a large number of other biochemical modalities are affected. For example, serotonin was shown to be decreased by Dr. Hornykiewicz some years ago, and he now suggests that the GABA system may be affected.

If we study the brains of patients with the amino acidurias (the classical models we have for metabolic diseases), we would find that ten different amino acids are affected, and yet we know that in these disease entities there is only one enzyme defect.

Therefore, in treatment of Parkinson's disease, the fact that more than one biochemical modality is low does not necessarily indicate that all modalities need to be corrected for a good therapeutic result. Rather, it is best to try to concentrate on the one pathway, or the one enzyme defect.

Dr. Everett: Dr. McGeer's stimulating discussion in regard to the association of cholinergic and biogenic amine systems is an important concept for any of these discussions. He mentioned the parkinsonian model of tremorine and oxotremorine. These drugs increase brain acetylcholine by 30 percent. We have shown that decreases in norepinephrine occur after giving tremorine, and Friedman has shown decreases of dopamine in rats after tremorine.

We should also remember that this model is useful in reference to Dopa and dopamine because, if we give Dopa and raise the levels of brain dopamine, we can reverse a large number of the symptoms from oxotremorine. We believe this is due to a rebalancing of the systems by raising dopamine to match the increased levels of acetylcholine that have occurred after oxotremorine.

Dr. Hornykiewicz: Dr. McGreer and Dr. Barbeau were, as far as I know, the first to suggest that there was a kind of balance between dopaminergic and cholinergic brain mechanisms. Stimulated by this hypothesis, we have tried to look into this question in respect to parkinsonism. We estimated acetylcholine in the basal ganglia of normal and parkinsonian patients. We were completely unsuccessful; acetylcholine is so unstable that there was, upon postmortem, none remaining even in control brain material.

I would like to mention, however, two other findings which might be of some importance in this context. The first was published in 1952 by Weber in Switzerland, who found a lack of cholinesterase in the caudate nucleus of parkinsonian patients. Nobody has confirmed the finding—as a matter of fact, nobody has repeated the examinations—so nothing is known about its significance.

Another finding which might be relevant to the relationship between acetylcholine and parkinsonism is Duvoisin's report that there are no differences between the levels of acetylcholine in the cerebrospinal fluid of normal and parkinsonian patients. This observation might be significant.

Dr. Goldstein: Later in this symposium I will present some evidence which will support the view that a balance between the cholinergic and dopaminergic systems may play a role in the extrapyramidal disorders. With regard to Dr. Hornykiewicz's comments, I would like to point out that we analyzed cholinesterase and choline acetylase activities in the caudate nucleus of monkeys with unilateral nigral lesions and we could not find any significant differences between the two sides. However, in the globus pallidus the choline acetylase activity was slightly increased on the side of the lesion. Further studies are in progress to verify these results.

Dr. Poirier: We studied the distribution of cholinesterases following various lesions interrupting fibers from the substantia nigra, intralaminar thalamic nuclei, etc., ending in the striatum, and, up to now, we have not found any difference in the histochemical appearance of cholinesterases in the monkey and the cat. I would like also to raise another point which I will illustrate later. We must be concerned with the possibility that non-extrapyramidal structures may be involved in some of these symptoms, and thus one may have to take account of other factors underlying such symptoms.

Dr. Brody: I would like to comment on Dr. Ansel and Dr. Markham's work. As I understand the design of the study, the patients were given up to 3 to 5 grams per day for one month, and were of a selected age group. What is the predicted rate of extrapyramidal involvement after one month, in this age group, among parkinsonian patients?

Dr. Markham: Nine of our 57 parkinsonian patients exhibited involuntary movements during their first month of treatment on doses and dose schedules comparable to the controls, but these were not age related.

Dr. Moore: I would remark that it might be premature to neglect the globus pallidus entirely, since one need only go back to the classical paper of Richter on carbon-disulfide intoxication for the demonstration of a marked parkinsonian syndrome in monkeys in which the primary, if not only, abnormality was present in the globus pallidus.

Dr. McGeer: I would like to comment that I do not think we should write off any clues about Parkinson's disease, and of course carbon monoxide poisoning as well as carbon disulfide poisoning affect the globus pallidus and produce parkinsonism. To approach some of the biochemical correlations, glutamic acid decarboxylase was reported by Dr. Hornykiewicz to be decreased in Parkinson's disease. The highest levels of glutamic acid decarboxylase in the human brain are from the globus pallidus. We have made some measurements on normal human brain, and they confirm measurements that others have made previously. Therefore the possibility of disease in the globus pallidus is with us, and more measurements on glutamic acid decarboxylase in parkinsonian patients are needed.

One or two people have mentioned that as yet we cannot find anything wrong with acetylcholine. Although we have no information about acetylcholine, I can say one thing about the synthesizing enzyme choline acetylase. In our experience with other amines the synthesizing enzyme is very closely linked with the neurotransmitter amine itself.

In those ventrotegmental lesions which cause a dramatic decrease in tyrosine hydroxylase, the enzyme for synthesizing dopamine in the caudate, choline acetylase, remains normal. Clearly, identical pathways are not involved.

Dr. Sandler: Why does O-methyldopa stay around for such a long time?

Dr. Pletscher: The metabolite with greatest accumulation in the brain after repeated administration of Dopa is 3-O-methyldopa. In our experiments, 5 mg per kg of Dopa were administered intraperitoneally repeatedly to rats at 12-hour intervals, and there was a considerable build-up of O-methyldopa in the brain. The other metabolites, e.g., Dopa, catecholamines and phenolcarboxylic acids, accumulated to a much lesser extent than O-methyldopa. Whether these high levels of O-methyldopa interfere with the transport of amino acids into the brain, or whether they have still other effects, must be investigated. In the heart, O-methyldopa behaves similarly as in the brain, but catecholamines accumulate here also.

Experiments with ^{14}C-O-methyldopa carried out by Dr. Bartholini in our laboratories (intraperitoneal injections) indicate that the distribution of the compound is quite similar in blood, brain and heart. This indicates that O-methyldopa easily penetrates from the blood into the heart and the brain. The half-life of O-methyldopa in blood, brain and heart of rats is also quite similar, i.e., about 25 hours. This slow elimination probably explains the considerable accumulation of the compound after repeated administration.

Dr. Sandler: What about transaminase?

Dr. Pletscher: In preliminary experiments, Dr. Bartholini has found that O-methyldopa is slowly metabolized and that among various metabolites homovanillic acid is found. Whether this acid is formed by transamination of O-methyldopa remains to be investigated.

Dr. Hornykiewicz: I would like to show some additional results on brain dopamine and homovanillic acid in Huntington's chorea. These cases (we examined about ten brains) are still unpublished. There is, if any, very little change in the concentrations of striatal dopamine and homovanillic acid in Huntington's chorea as compared to control values.

I should also like to mention that we have published a short report on homovanillic acid concentration in the cerebrospinal fluid of patients with Huntington's chorea. We examined about four or five cases, as I remember, and the values for homovanillic acid were neither especially high nor especially low; it seems safe to say that they were somewhere within the normal range. The normal range for homovanillic acid in the cerebrospinal fluid is extremely wide, and the values for Huntington's chorea were near the lower range of normal.

Dr. Goldstein: I would like to add to Dr. Tyce's paper that we investigated the metabolism of Dopa several years ago and found that N-acetylation represents a metabolic pathway for inactivation of dopamine. However, we have not been able to demonstrate formation of N-acetyldopamine in the brain.

Dr. Hornykiewicz: Dr. Sandler mentioned that one of the smaller metabolites might be the phenylacetic acid metabolite of Dopa. I think I remember correctly that Gey injected dihydroxyphenylpyruvate or lactate in rats, and observed an increase of dopamine in the brain. Although the increase was small, Gey thought that there was the possibility of a conversion of the Dopa metabolites to dopamine.

Dr. Markham: I would like to comment on Dr. Sandler's suggestion about defective absorption of L-Dopa from the gut in resistant cases. In one of our patients who had a relapse after receiving L-Dopa for some months, we tried sterilizing the bowel with neomycin. There was no subsequent change in the state of relapse, but this approach was tried in one case only.

PART 2

SIGNIFICANCE AND ETIOLOGY OF INDUCED ABNORMAL MOVEMENTS

CHAPTER 4

Clinical Aspects of Abnormal Movements Induced by L-Dopa

ABNORMAL INVOLUNTARY MOVEMENTS INDUCED BY DOPA: CLINICAL ASPECTS *

Melvin D. Yahr

Among the several side effects occurring during the administration of levodopa to patients with parkinsonism, abnormal involuntary movements have emerged as the most limiting in achieving optimum therapeutic results. These movements, variable in type and intensity, may involve any or all segments of the body and occur regardless of the patient's age, the duration, severity, type of symptomatology or presumed etiology of the Parkinson process. They have, in addition, been induced in patients with dystonia musculorum deformans who have been treated with L-Dopa in a manner similar to the Parkinson patients. To date, we have not encountered these movements in patients receiving therapeutic trials with Dopa for other neurological disorders nor in those with presumably no structural disturbance of the nervous system. However, in these latter instances, though daily dosage of Dopa administered has been equivalent to that given Parkinson patients, the time intervals over which it has been administered have been relatively short. This latter factor is of considerable importance in regard to the appearance of involuntary movements.

The type of dyskinesia induced by Dopa has varied from generalized choreo-athetoid movements to localized dystonic movements and posture as well as myoclonic jerks limited to a single muscle group. They may occur

* Supported by Parkinson's Disease Foundation and the National Institutes of Health, NINDS, Grant NB 05184 and Contract 436454.

while the patient is at rest or be induced by volitional activity. The activities accentuating or inducing these movements need not be those involving the muscles which are a part of the dyskinesia. For example, it is not unusual to see bobbing movements of the head when the patient is performing an activity with one of his limbs.

The mode of onset has also been extremely variable. In some patients, the movements appear abruptly and with a degree of severity that markedly interferes with functional capacity. In others, their onset is more subtle and benign beginning with a general appearance of restlessness and fidgeting, stereotyped movement of a segment of the body, or merely overexaggeration of normal associated movements. In instances of this sort they are more apparent to an observer and rarely complained of by patients until brought to their attention. Not infrequently, symptoms such as jerking of a limb, curling of toes, or adventitious turning movements of the fingers have been present for weeks unbeknownst to the patient.

Of the many movements encountered, the most distressing are those involving the face, oral cavity, neck and respiratory apparatus. Not only are these disturbing to the patient and family from a cosmetic viewpoint but they seriously impair speech, feeding and breathing. Lesser difficulties have been encountered with those involving the limbs which rarely have become severe enough to interfere with function and have been remarkably well tolerated by patients. Paradoxically, the oral-facial movements resemble those seen with the use of neuroleptic agents. However, they differ in that they are intermittent rather than continuous and are usually activated by volitional motor activity.

In general, dyskinetic movements have occurred in patients who have experienced marked relief of their parkinsonian symptoms. The situation in these instances is a replacement of an akinetic-rigid state by a hyperkinetic-hypotonic one. In biochemical terms it might be considered as a deficiency of striatal dopamine being converted to an overabundance. However, I doubt if the situation is explainable on such a simplistic basis. Further, the clinical situation is hardly so clear-cut. A significant number of patients have had little or no resolution of their parkinsonism while acquiring abnormal movements. The coexistence of both is an unfortunate clinical reality for some patients.

The onset of these movements is not only dependent on the total daily dosage of L-Dopa but also on the duration of time over which it has been administered.

The following cases illustrate some of these points. In the patient whose record is shown in Figure 1, abnormal involuntary movements involving the pelvis made their appearance early, as soon as a daily dosage level of 4 grams of L-Dopa was reached. With reduction of the daily dosage to 3.5 grams a day, the movements subsided. In fact, at various times during his many months of treatment one could titrate the dose of Dopa against abnormal movement induction. The dose necessary to control his parkinsonism completely overlapped that which induced abnormal movements. Fortunately,

Case #8, H.B.

A. I. M.																
0	+	++	+	+	+	+++	++	++	+++	+−	+−	+−	+	++	++	−−
	0	++	0	+	+											
Park. Exam																
36	24	2	0	2	0	0	0	0	0	1	1	1	2	1	1	2
	21	2	0	0	0											
Dopa																
0	4.0	5.0	4.5	4.0	4.5	4.5	4.0	4.0	3.75	3.0	3.75	3.75	4.0	4.25	4.25	3.5
	3.5	4.5	3.75	4.5	4.5											
0	1	2	3	4	5	6	7	8	9	10	11	12	15	16	18	20

MONTHS

Fig. 1. Abnormal involuntary movements A.I.M. and parkinsonian symptoms with varying dosages of L-Dopa (A.I.M. scale 0 to +++, Park. Exam. scale 0 to 76, L-Dopa grams per day).

the movements were mild and tolerable to the patient and he preferred them to his Parkinson symptoms. With minor modification of dosage he has remained under treatment for some 20 months now, with good control of parkinsonism and mild movements.

In contrast, in the patient whose course is seen in Figure 2, abnormal movements made their appearance at the 4-gram level after a period of two months. They were severe, involving the face and neck, and did not disappear until the dose was reduced to 2.5 grams. Her movements were of such severity that they markedly interfered with her functional capacity. In this instance one had to settle for a less than optimum dosage of L-Dopa with resultant decreased benefit to her parkinsonism.

The marked latency with which these movements may make their appearance is indicated in Figure 3. For a period of seven months this patient received 6 grams of L-Dopa daily without evidence of abnormal movements. Their appearance was mild at onset but continued to increase in severity over a period of weeks. With reduction to a 5-gram level they abated and then, with an increase to 5.5 grams over the next few months, gave a satisfactory response without the recurrence of involuntary movements. It is, in fact, not infrequent for these movements to make their initial appearance only after many months on a fixed daily dosage. In addition, diurnal variations in the presence and severity of these movements have been in some, but not all, patients intimately correlated with the time of ingestion of a dose of Dopa. In such instances movements have been noted within one-half hour after a dose, persisting for an hour or so and then spontaneously subsiding. In others, their occurrence has been restricted either to the early part of the day following the first dose or to intervals during the evening after the major portion of the daily dose has been ingested. This capriciousness, with a random pattern of appearance, has made it difficult both to implicate and study causative factors and to evaluate therapeutic measures which have been attempted for their correction.

These many factors must be taken into consideration regarding any discussion of the incidence, causation and treatment of this side reaction. In regard to incidence, we have recently reviewed all of our patients who have been under continuous treatment for six months or more. The six-month interval was not chosen arbitrarily; it was based on our impression that it may require this length of time for the full therapeutic effect to develop and, more importantly, for abnormal movements to appear. (Actually, the six-month interval is not long enough to evaluate fully the prevalence of this side

Case #68, M.F.

A. I. M.	0	0	++	+++	++	0	+	+++	+
Park. Exam	34	12	5	3	9	22	18	4	12
Dopa	0	4.0	4.0	4.0	2.5	2.0	3.0	4 0	2.5
	0	1	2	3	4	5	6	7	13

MONTHS

Fig. 2. Abnormal involuntary movements and parkinsonian symptoms with varying dosages of L-Dopa (A.I.M. scale 0 to +++, Park. Exam. scale 0 to 76, L-Dopa grams per day).

Case #15, J.C.P.

A. I. M														
0	0	0	0	0	0	0	+	+++	++	0	0	0	0	0
Park. Exam														
54	34	27	26	20	18	19	22	17	17	16	19	19	14	12
Dopa														
0	5.0	6.0	6.0	6.0	6.0	6.0	6.0	6.0	6.0	6.0	5.0	5.5	5.5	5.5
0	1	2	3	4	5	6	7	8	9	10	11	12	13	20

MONTHS

Fig. 3. Abnormal involuntary movements and parkinsonian symptoms with varying dosages of L-Dopa (A.I.M. scale 0 to +++, Park. Exam. scale 0 to 76, L-Dopa grams per day).

reaction.) As a further prerequisite, all patients in this evaluation had an adequate trial on L-Dopa with the optimum dose determined by either a maximum therapeutic response, a limiting side effect or a daily dosage of 8 grams. In all, 116 patients fulfilled these criteria and the results are indicated in Table 1.

These data indicate the onset of involuntary movements at all time intervals up to a year. Most often they appear within the first six months. Since the data are additive for this period as they include all 116 cases, the overall incidence of these movements is 58 percent. However, one gets the impression of a rising incidence with prolonged Dopa administration. In those cases with one year follow-up these movements are encountered in almost three-quarters of the patients. However, only 71 cases were studied for a full year. Of interest is that the average daily dosage of Dopa is quite similar for each of the time intervals studied.

Regarding control, fortunately none of the movements has been permanent and all have subsided with reduction of dosage. However, dosage may have to be reduced to a level far below that previously tolerated. In some instances we have had to resort to less than one-half the daily dosage before the abnormal movements abated. With the reduced dosage, much of the beneficial therapeutic effect is sacrificed. In an effort to maintain full therapeutic doses various pharmacological agents (Table 2) have been tested, with a view to controlling the abnormal movements. To date, none has been fully satisfactory.

The anticholinergic agents used run the entire gamut of the so-called anti-parkinson drugs which are now available. We have noted no modification of abnormal movements when they have been administered with Dopa. However, in some instances the combined use has improved the Parkinson symptomatology, suggesting a synergistic action. Phenothiazine derivatives (particularly Stelazine) have also been administered and, though one can diminish or completely control the induced movements, an adverse effect on the Parkinson state concomitantly occurs. With Tofranil there has been no alteration in movements but some salutory effect on Parkinson symptoms. Because of the long-term usage of barbiturates and Dilantin in the control of choreo-athetoid movements, these were given a trial. No significant benefits were noted.

We have previously reported in detail about the effects of pyridoxine in

TABLE 1. Incidence of Abnormal Involuntary Movements during Dopa Administration

	Month of Dopa Therapy					
	1	*2*	*3*	*6*	*9*	*12*
No. patients	116	116	116	116	109	71
No. developing A. I. M.	20	20	13	9	9	6
Total no. with A. I. M.	20	40	53	62	71	52
	17%	34%	46%	53%	65%	73%
Dopa-avg. daily dose (gms/day)	5.0	4.4	4.3	4.7	4.8	4.8

TABLE 2. Drugs Administered for Control of Abnormal Involuntary Movements

	A. I. M.	*Parkinsonism*
Anti ACh	0	↓
Stelazine	↓	↑
Tofranil	0	↓
Barbiturates	0	0
Dilantin	0	0
Pyridoxine	↓	↑
L-Tryptophan	0	0
5-HTP	↓ ?	0
Glutamic acid	0	0

* 0 = no effect, ↑ = increase in symptoms, ↓ — decrease in symptoms.

Parkinson patients receiving Dopa. Its ability to wash out the effects of Dopa in parkinsonian patients is quite striking. It has been our impression that this results from catalyzing the conversion of Dopa to dopamine systemically and, hence, making less Dopa available to the brain. 5-HTP has been administered in dosage as high as 1½ grams per day. Our experience is limited to three or four patients, and we feel that we may have modified the movements to some extent. However, I am not impressed with this type of approach. L-Tryptophan has been used in dosage up to 15 grams per day with no effects on either abnormal movements or Parkinson symptomatology. Because of an ongoing interest in glutamic acid and the role its alteration in brain may play in parkinsonism, we have administered as much as 5 grams a day. No effect on any of the Parkinson symptoms or induced dyskinetic movements has been seen.

It would seem at the present time that the only effective way to control these movements is to reduce the total daily dose of Dopa. This requires careful titration, with the hope of reaching some balance in which Parkinson symptoms are effectively controlled and induced movements are at a minimum.

Our general impression is of a rising incidence with increased daily dosage. However, the range is wide, with some patients experiencing this side effect at the 2.5 gram level and others being able to tolerate 8 to 10 grams. In some individuals a critical dosage level has been encountered, with as small an addition as 250 mg of Dopa spelling the difference between the presence or absence of abnormal movements. In such individuals, minimal reduction in dosage results in alleviation of these movements. In others, once movements have been established, drastic reduction in dosage may be necessary before they subside and a period of a week or more may pass before they disappear. In fact, some patients are intolerant of doses previously taken without the occurrence of this side reaction.

These observations raise interesting and important questions regarding the role of Dopa and its metabolites as the direct and only causative factor in producing these movements. Certainly the alternate possibility—that the large Dopa load may interfere with the transport and storage of other amines essential for the maintenance of normal motor activity—must be considered. The time and dose relationships raise important pharmacodynamic considerations.

ACKNOWLEDGMENT

Clinical observations of all patients in this study were the joint responsibilities of Drs. Robert E. Barrett, Roger C. Duvoisin, Margaret M. Hoehn and Myrna J. Schear, to whom the author is extremely grateful.

DYSKINETIC MOVEMENTS

F. H. McDowell

Our experience has been virtually identical with that of Dr. Yahr. We have seen primarily oral, facial and cervical dyskinesias in our patients, and occasionally dyskinetic movements in the arms, legs or trunk.

In our 100 patients, 58 developed dyskinesias. We would probably be correct in assuming that, if every patient had a progressive increase in his dose of L-Dopa beyond the point of maximal improvement, we would be able to produce dyskinesias in 100 percent of the patients.

Several things stand out about the dyskinesia: In every instance when dyskinesia appeared the patient also showed improvement. In every instance in which dyskinetic movements appeared they were eliminated by a reduction in dose or cessation of the medication; there have been no permanent dyskinesias with L-Dopa.

In a few patients who developed dyskinetic movements, after the dose of L-Dopa was reduced and the dyskinetic movements were stopped, we have been able to increase the dose again without producing dyskinesias, and even to achieve higher doses than those given before dyskinetic movements appeared. Unfortunately, in other patients we have noted the occurrence of dyskinetic movements at lower and lower doses.

Table 1 shows the number of patients (58 out of 100) who had dyskinesia;

TABLE 1. Dyskinesia

Dyskinesia	*Patients*
Occurred in	58
Occurred again at lower doses in	12 (20%)

TABLE 2. Dyskinesias

Dose	*Patients At That Dose*	*Dyskinesias %*
0.5	100	0
1.0	100	1
1.5	100	4
2.0	100	9
2.5	99	12
3.0	98	12
3.5	89	13
4.0	86	17
4.5	74	27
5.0	64	28
5.5	52	31
6.0	43	33
6.5	26	23
7.0	20	25
7.5	9	44

in those 58, dyskinetic movements occurred at lower doses in 12 or 20 percent. We attempted to relate dyskinetic movements to the dose of L-Dopa. Table 2 indicates that the occurrence of dyskinesia is dose related; it shows the number of patients who achieved a particular dose and the percent of patients who exhibited dyskinetic movements at that dose. One patient showed dyskinetic movements on a 1-gram intake. There is a steady increase in the incidence of dyskinetic movements to the point where 44 percent of the patients who achieved a 7½ gram intake had abnormal movements.

As Dr. Yahr pointed out, many patients tolerate the abnormal movements because the reduction in Dopa which abolishes them frequently eliminates some improvement in their parkinsonism. However, other patients with abnormal movments have been so uncomfortable, or their families so distressed, that they have had to lose some of the improvement gained and take L-Dopa at a lower dose.

EXPERIENCE WITH DYSKINESIA IN TREATMENT OF 102 PATIENTS WITH L-DOPA OVER A PERIOD OF 2 TO 21 MONTHS

John Campbell

Of 102 patients taking L-Dopa in appropriate doses (2 to 8 grams, averaging 4.1 grams a day) over a 2 to 21 month period, 15 experienced episodes of mouthing and approximately the same number experienced dyskinesias of the extremities. The latter are commonly twisting and writhing of the foot or hand, and occasionally myoclonic or even choreiform movements.

The writhing movements may be rather noticeable to the observer and yet not noted by the patient, since his rigidity and bradykinesia are substantially improved. Akathisia (that is, a "need to move") has not been associated with the dyskinesias of the extremities or mouth.

Thalamotomies were performed on 20 percent of our patients before they received treatment with L-Dopa. Abnormal movements do not always occur on the side not operated upon, but in one patient at least they certainly did.

On several occasions, rapid respirations and, rarely, an apparent spasm of the diaphragm appeared with use of the drug. However, there was no evidence of hiccups or prolonged diaphragmatic irritation.

Most patients with dyskinesias experience them just prior to getting "better." There are short periods of dyskinesia and then a rather rapid release of rigidity and bradykinesia which may last one or two hours to six or eight hours, or until the next dose of medication.

All movements seem unique to L-Dopa and are unlike the face-mouth dyskinesia or athetotic dystonia induced by Thorazine or Serpasil. However, I have seen one patient with mouthing similar to that seen following L-Dopa administration which was associated with a small intake of Serpasil over a period of four years, and which persisted for over a year after the drug was withdrawn but has recently improved.

All patients had relief of the dyskinesias when the dosage of L-Dopa was reduced or discontinued. However, to date no other drug has been found to be especially beneficial in relieving the movements.

The following description is that of a physician-patient: "The dyskinesias are gradual in onset 3 to 3½ hours after the previous dose of Dopa, peak at 1 to 1½ hours, and gradually clear in another 1 to 1½ hours. The severity may vary; the timing of the dyskinesias remains the same. The sequence of the onset is as follows:

1. Internal rotation and inversion (occasionally with some clonus) of the left foot.
2. Hyperadduction and external rotation of the left humerus with 45° flexion at the elbow and with the hand supinated. Occasionally this occurs also on the right side, but never as markedly as on the left.
3. After the above, complete flexion of the ring and little fingers at the metacarpal and interphalangeal joints, with partial flexion and extension at the distal interphalangeal joints. The thumb becomes moderately flexed with distal phalanx resting at the distal end of the index finger.
4. Dysmetria and occasional overshooting of an object during this period.
5. Panting, if the above have also preceded.

Not in every episode does this entire sequence occur. The movements may abort any place along the sequence. The latter seems to be determined partially by the total dose or distribution of the total dose during the 24-hour period, as well as initiation of movement, loss of sleep, lack of rest, and the degree of pressure or tension at the time. Without any dyskinesias the maximum control of symptoms is not obtained. On most occasions 2 to 3 grams a day were necessary to produce the abnormal movements, but once they occurred with a smaller dose."

CLINICAL EXPERIENCE WITH INVOLUNTARY MOVEMENTS IN ASSOCIATION WITH THE ADMINISTRATION OF L-DOPA IN PATIENTS WITH PARKINSON'S DISEASE

Oscar Kofman

There is little need to reiterate what is already known about the incidence and character of the involuntary movement disorders observed in Parkinson patients receiving long-term intensive therapy with L-Dopa.

In our series of 46 patients, 33 have received L-Dopa for periods ranging from 2 to 11 months and 13 have been treated for shorter periods of time. Eleven have been observed to have clinically significant involuntary movements and, undoubtedly, others will develop such movements as therapy continues.

Generally these movements can be described as facial distortions, grimacing, chewing, masticatory movements, tongue dystonia, etc. In addition, there have been a variety of bizarre trunk and limb dyskinesias. One patient developed a severe form of hemiballismus which promptly cleared after L-Dopa therapy was discontinued. Other unusual patterns consisted of repetitive movements and spasms of both feet and toes, gross contractures of the muscular wall of the abdomen, and, in one patient, jerky respiratory movements. These irregularities appear to be dose dependent, as they generally occurred at levels over 3 grams daily and were reversible upon reduction of the dosage of L-Dopa—sometimes to levels considerably lower than the quantities taken at their onset.

Our particular interest has focused on a possible therapeutic means of controlling these distressing movements without compromising the dosage and effect of L-Dopa itself.

In view of the known effects of phenothiazine derivatives and Rauwolfia (reserpine) on producing extrapyramidal syndromes, we somewhat empirically considered the use of Stelazine (trifluoperazine), a potent phenothiazine. In theory we considered that this substance might influence the presumed excessive biochemical alteration which followed the administration of L-Dopa. Richter has suggested that active phenothiazines have a central effect that may be due to their interference with the transport of biogenic amines from the point of synthesis to the storage sites or to the receptor sites in the brain.

The dosage of Stelazine varied between 4 to 12 mg daily; generally 4 to 8 mg were used. Our observations are by no means consistent, but in three of the patients treated in this manner there was convincing evidence of reduction of involuntary movements without alteration in L-Dopa administration. In one of our patients (a physician), 6 to 8 mg of Stelazine daily stopped the facial grimacing and speech disturbance as well as the irregular jerking respiratory movements. While he continued with Stelazine we were able to raise the dosage of L-Dopa from 3 to 5 grams daily without recurrence of these movements. At no time did the administration of Stelazine appear to interfere with the observed beneficial effects of L-Dopa upon bradykinesia and rigidity.

In two other patients in whom Stelazine appeared to be of value, there was a significant reduction of the involuntary facial and tongue movements that had developed during L-Dopa therapy. In addition, it was noted that in both of these patients there appeared to be substantial improvement of the preexisting tremor that was a part of the parkinsonian disorder.

However, it must be noted that in three other patients in whom Stelazine was used there was no beneficial effect. In one of these patients the parkinsonian state became intensified, with a recurrence of bradykinesia and rigidity as well as tremor, all of which had been initially reduced with the administration of L-Dopa.

Several observers have noted the similarity of some of the induced involuntary movements that develop during L-Dopa administration to the extrapyramidal movements seen in Huntington's chorea.

Because of the known experience of others, together with our own favorable experience with the use of alpha-methyldopa (Aldomet) in the treatment of the choreiform movements of Huntington's chorea, we used this substance in conjunction with L-Dopa in patients with Parkinson's disease in an attempt to control the involuntary movements that had developed during therapy. We realize that the concept of alpha-methyldopa as developed implies a depletion of dopamine, as suggested by Sourkes and others. However, it should be noted that the clinical reports on the actions of alpha-methyldopa in Parkinson's disease in the "pre-Dopa" era are inconsistent and contradictory. We, therefore, cautiously administered alpha-methyldopa to four parkinsonian patients in whom serious involuntary movements had persisted.

In one patient, 250 mg of alpha-methyldopa four times daily in conjunction with the continued use of 6 grams of L-Dopa appeared to have a striking effect in reducing and eventually stopping the facial and tongue movements. In this particular patient, there was a pronounced beneficial effect upon the preexisting gross tremor of the upper limbs. This tremor in itself had been severe and incapacitating, with qualities sometimes associated with a so-called red nucleus tremor. Although the tremor had subsided slightly during administration of 6 to 7 grams of L-Dopa daily, a considerably more pronounced improvement occurred shortly after the addition of alpha-methyldopa.

In a second patient, the addition of 250 mg of alpha-methyldopa four times daily did not appear to be beneficial, and its use was therefore abandoned. This same patient had had an unfavorable response to the administration of Stelazine as well.

In a third patient the addition of alpha-methyldopa could not be properly evaluated, as the dosage of L-Dopa was simultaneously reduced because of the occurrence of a confusional state. In a fourth patient no change has been noted to date after a one-week trial.

There was no significant alteration of blood pressure during the administration of alpha-methyldopa to these four patients.

These limited and somewhat inconsistent observations of the potential effects of Stelazine and alpha-methyldopa upon the involuntary movements associated with the administration of L-Dopa therapy are intriguing, but they are obviously insufficient and inconclusive. With "dystonic tongue in cheek" we would suggest that the mechanism by means of which some of these responses may have occurred may be more sophisticated than Dopa decarboxylase inhibition, and thus a mere counterbalancing of the beneficial effects that were induced by dopamine.

Perhaps these preliminary observations may provoke further evaluation by others who are constantly challenged by this formidable problem.

DYSKINESIAS WITH L-DOPA

Sean O'Reilly

My experience has been a microcosm of the much larger series reported by others. We had 5 of 17 patients on long-term L-Dopa therapy who have shown involuntary movements, and they have fallen into the same three groups already described.

The first group show what might be best called multiple tics—repetitive little movements of the face, lips, and eyebrows. Unlike Dr. Yahr's patients, ours opened, rather than closed, their eyes.

The second type of involuntary movement is a frank choreo-athetosis quite similar in all respects to that of Huntington's chorea, with a stimulus-dependence of the athetotic posture as in athetosis from whatever cause.

The third type is exaggeration, or driving, of a previously existing dystonic posturing of a limb. The one patient who showed this had had a thalamotomy. and postoperatively the dystonic posturing was minimal. This was driven very markedly as soon as he attained a dosage of 2½ grams of L-Dopa per day; in fact, it became almost ballistic.

Pyridoxine in small doses (12½ mg a day) was sufficient to abolish the multiple tics in two of the patients so they were able to continue receiving increased doses of L-Dopa.

A second measure that was quite effective was either Stelazine alone, in quite small doses (1 mg), or Stelazine combined with Valium. Valium alone was better than Stelazine alone for multiple tics. However, Stelazine alone, in a dosage of 1 mg, was most effective in controlling the exaggerated dystonic posturing of one foot in the post-thalamotomy patient previously mentioned. In fact, this patient quickly learned to titrate his L-Dopa dosage against a small dose of Stelazine taken an hour or so later, and thus was able to control this undesirable side effect of the L-Dopa.

Fortunately, we did not have to resort to reduction of dosage in the treatment of any of these patients. It seems to me that a small dose of pyridoxine might be indicated while the Dopa dose is being increased; multiple tics did not appear in our patients who reached a maximum therapeutic dosage while taking pyridoxine. This of course may merely be attributable to our small number of patients.

One final remark concerning the appearance of multiple tics and the effect of pyridoxine on them: One patient had had quite severe parkinsonian tremor which was not affected by L-Dopa until he took 6 grams a day, at which dosage level multiple tics appeared. They didn't bother him but they bothered his wife, so we gave him some Stelazine which reduced them, added some Valium which reduced them a little more, but they were still present. We then stopped these drugs and gave him pyridoxine in a dosage of 12½ mg a day. The tics were eliminated, but his parkinsonian tremor reappeared and was worse than ever. I don't know the explanation of this, but it clearly separates parkinsonian tremor from other involuntary movements.

INDUCED ABNORMAL MOVEMENTS WITH L-DOPA AND Ro 4–4602 *

J. Siegfried, R. Klaiber, and W. H. Ziegler

The association of Ro 4–4602 to L-Dopa in treatment of Parkinson's disease decreases the amount of L-Dopa necessary to obtain a favorable clinical effect.[1, 2, 3, 4] Since Ro 4–4602 enhances the supply of L-Dopa to the brain by diminishing the formation of catecholamines in extracerebral tissues, due to inhibition of extracerebral decarboxylase [5] one can expect the same side effects of cerebral disturbance with Ro 4–4602 and L-Dopa as with L-Dopa alone. The percentage of induced abnormal movements with L-Dopa varies according to different authors, but can be as high as 75 percent with long-term treatment.[6]

A study of our material shows, surprisingly, a lower incidence of these complications.

In a series of more than 150 patients treated with L-Dopa and the inhibitor of decarboxylase Ro 4–4602, we had 100 patients treated for at least four months who were subjectively and objectively improved by the treatment. We described the involuntary movements in three different categories:

Grade I: Often not observed by the patient; transitory; in isolated muscle groups (mouth and/or proximal part of extremities); slight torsion of trunk. No indication for change in dose.

Grade II: Observed by patient; transitory; extensive, with amplitude of movements. Dosage must often be reduced.

Grade III: Often intolerable for patient; continuous; involving almost the entire body. The dose *has* to be reduced.

* Supported with the aid of grant No. 4801.3 of the Swiss National Foundation and also with funds of the Eric-Slack-Gyr Foundation, Zürich, and Hoffmann-La Roche, Basel.

The involuntary dystonic movements are often preceded by an unpleasant sensation in the muscles concerned. The side of the body which presents the most marked parkinsonian symptomatology often shows the strongest involuntary movements. Patients who underwent a prior stereotaxic thalamotomy do not have involuntary movements on the side operated upon.

We observed involuntary Grade I movements in 12 cases (12 percent), Grade 11 in 6 cases (6 percent) and Grade III in 5 cases (5 percent). The total percentage of complications is then 24 percent, but a reduction in dosage was made in 10 percent of the cases.

A relationship of complications with dose of the drug and/or length of treatment cannot always be found. Figure 1 shows the appearance of involuntary movements of slight grade with the increase of both drugs to 750 mg of each daily. The treatment was not interrupted because the troubles were discrete and the clinical effect excellent. Figure 2 shows the appearance of complications after two weeks of treatment with 900 mg of L-Dopa and 450 mg of Ro 4–4602. The involuntary movements appeared very suddenly and were very severe (Grade III). After reduction of L-Dopa to 450 mg daily, the troubles were still present. A reduction to 300 mg of each drug daily was not enough to cause a disappearance of the dystonic movements, and now the patient is receiving 150 mg Ro 4–4602 and 150 mg L-Dopa daily with a favorable, but not optimal, effect on the akinesia and cessation of involuntary movements. In Figure 3, we see a patient who needed more than 1 gram of L-Dopa with 450 mg of Ro 4–4602 to get a marked improvement of akinesia. We increased the L-Dopa to 1.5 gram a day and acute and strong dystonic movements appeared; these disappeared with reduction of the

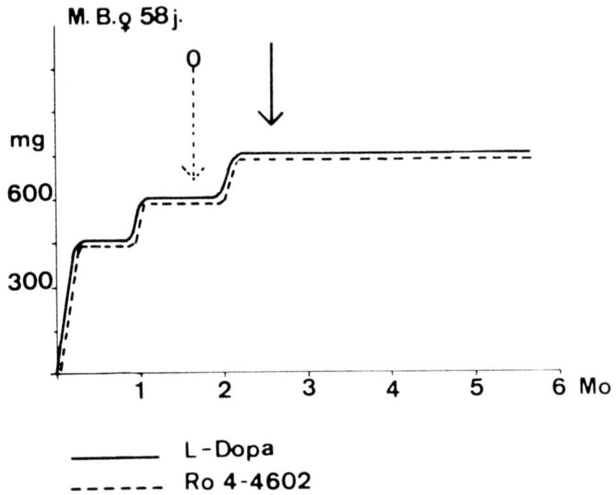

Fig. 1. Appearance of involuntary movement of slight grade after increasing Ro 4–4602 and L-Dopa to 750 mg each daily.

-------> No involuntary movements

———> Involuntary movements of slight grade (Grade I)

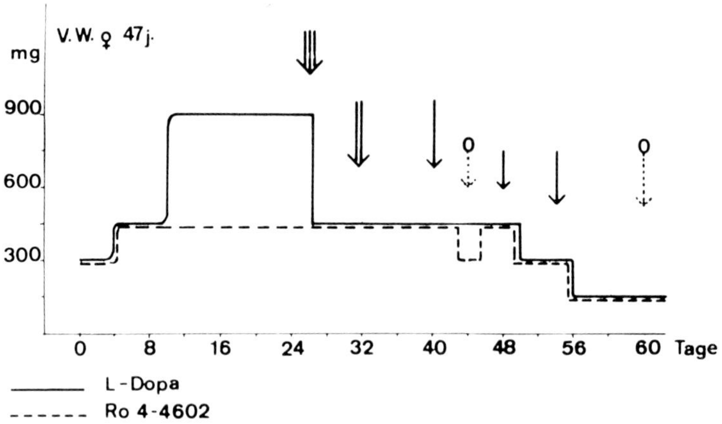

Fig. 2. Under a dose of 450 mg of Ro 4–4602 and 900 mg of L-Dopa for 2 weeks dystonic movements appeared in this case very suddenly. To obtain complete disappearance of these movements, Ro 4–4602 and L-Dopa were reduced to 150 mg each daily with a favorable, but not optimal, effect on the akinesia.

·········> No involuntary movements

———> Involuntary movements Grade I

═══> Involuntary movements Grade II

≡≡≡> Involuntary movements Grade III

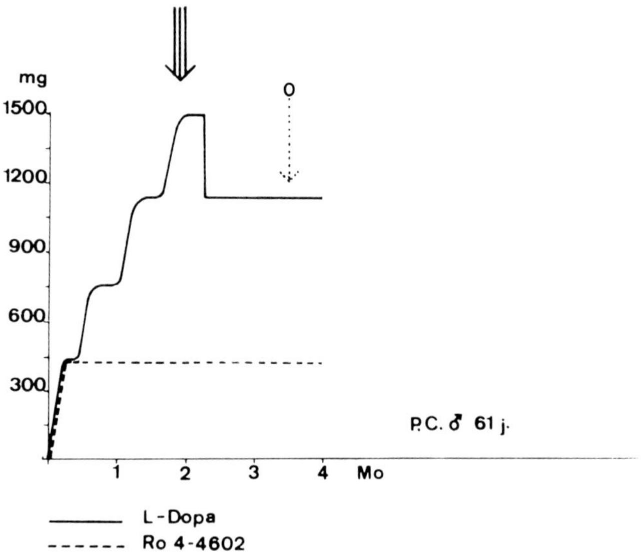

Fig. 3. By increasing progressively the dose of L-Dopa with 450 mg of Ro 4–4602, severe dystonic movements appeared with 1500 mg of L-Dopa.

·········> No involuntary movements

≡≡≡> Involuntary movements Grade III

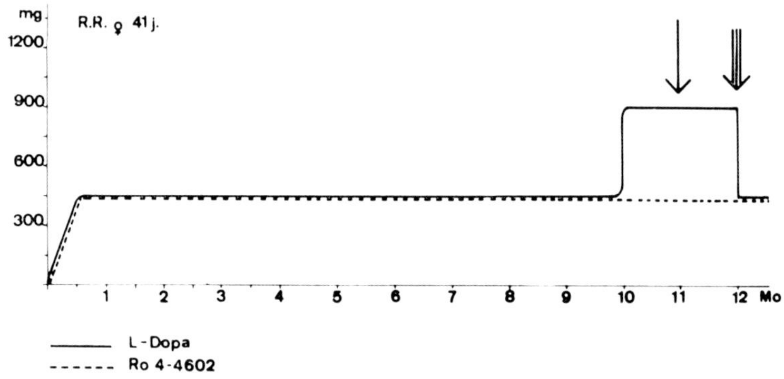

Fig. 4. By increasing L-Dopa to 900 mg a day after 10 months of treatment with 450 mg a day of Ro 4–4602 and L-Dopa, involuntary movements appeared.

→ Involuntary movements Grade I

⇛ Involuntary movements Grade III

dose to 1050 mg of L-Dopa daily in combination with 450 mg Ro 4–4602. Figure 4 shows the beginning of dystonic movements in one patient in whom we tried to increase the dose, aften ten months of treatment with 450 mg of L-Dopa and Ro 4–4602, to 900 mg of L-Dopa daily. After one month the dystonic disturbances began slowly and after two months it was necessary to reduce the dose to prior levels. Figure 5 shows an exceptional case of a woman with postencephalitic Parkinson's disease. As usual we started the treatment slowly with L-Dopa and Ro 4–4602, but after two days of treatment with 150 mg of each daily involuntary movements were observed. On increasing the dose to 300 mg of each, the troubles became very severe and the drugs

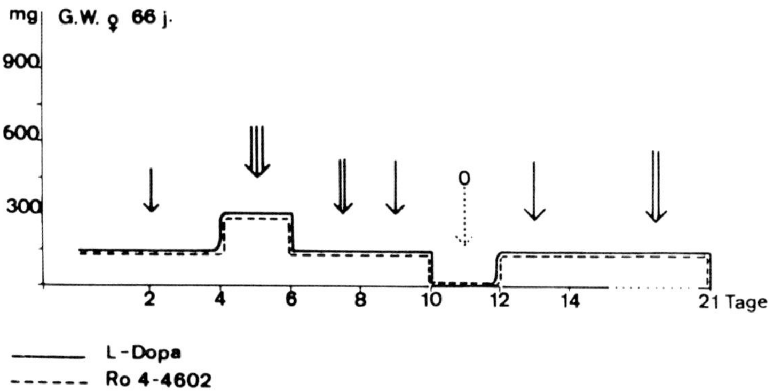

Fig. 5. Case of intolerance to Ro 4–4602 and L-Dopa with appearance of involuntary movements with very low dose of both drugs (150 mg of each daily).

⇢ No involuntary movements

→ Involuntary movements Grade I

⇉ Involuntary movements Grade II

⇛ Involuntary movements Grade III

were reduced; the disturbances were again marked after some days and we had to interrupt the medical treatment.

In conclusion, the treatment of Parkinson's disease with L-Dopa in combination with an inhibitor of decarboxylase (Ro 4–4602) produces in 25 percent of the patients involuntary movements which are classified in three grades. In 10 percent of our patients it was necessary to reduce the dose of drugs.

REFERENCES

1. Birkmayer, W., and Mentasti, M.: Weitere experimentelle Untersuchungen über den Catecholaminstoffwechsel bei extrapyramidalen Erkrankungen (Parkinson- und Chorea-Syndrom). Arch. Psychiat. Neurol. *210:* 29, 1967.
2. Siegfried, J.: Traitement du parkinsonisme avec la L-Dopa associée à un inhibiteur de la décarboxylase. Méd. Hyg. *27:* 543, 1969.
3. Siegfried, J., Kaiber, R., Perret, E., and Ziegler, W. H.: Behandlung des Morbus Parkinson mit L-Dopa in Kombination mit einem Decarboxylasehemmer. Dtsch. Med. Wschr. In press, 1969.
4. Tissot, R., Gaillard, J. M., Guggisberg, M., Gauthier, G., and Ajuriaguerra, J. de: Therapeutique du syndrome de Parkinson par la L-Dopa "per os" associée à un inhibiteur de la décarboxylase (Ro 4–4602). Presse Méd. *77:* 619, 1969.
5. Tissot, R., Bartholini, G., and Pletscher, A.: Drug induced changes of extracerebral Dopa metabolism in man. Arch. Neurol. *20:* 187, 1969.
6. Duvoisin, R., Yahr, M. D., Schaer, M., Hoehn, M. M., and Barrett, R. E.: The present state of L-Dopa in the treatment of parkinsonism. Excerpta Medica *193:* 170, 1969.

A PROPOSED CLASSIFICATION OF DOPA-INDUCED DYSKINESIAS *

A. Barbeau, H. Mars, L. Gillo-Joffroy, and A. Arsenault

Many sorts of induced abnormal movements have been observed during therapy with L-Dopa. The descriptions offered from many sources often lack precision in definition. Many terms which refer to symptoms observed in naturally occurring disorders—such as chorea, athetosis, dystonia—are used to describe the very different dyskinesias seen with L-Dopa. In our opinion these abnormal movements are rarely observed in spontaneous extrapyramidal disorders. Only the phenomena described during the acute phase of the epidemic of von Economo's encephalitis or the so-called "tardive dyskinesias" of long-term therapy with phenothiazine and butyrophenone can be compared to what we are now seeing. If a single component or move-

* Conducted with the help of grants MA-2530 and SP-1 of the Medical Research Council of Canada and a grant from L'Association Canadienne pour l'Avancement des Sciences Thérapeutiques.

ment is occasionally a classical choreic, myoclonic, athetotic or dystonic symptom, the total agglomeration of signs is something entirely new and different from what we are used to observe.

For this reason we thought it would be useful to present a descriptive classification based on the topography of the signs. This should be considered a working classification, but each one of the symptoms listed has been observed at least once in our series of 95 cases. Occasionally, as is frequent in basal ganglia disorders, the signs appear only under stress such as reading, intellectual effort (mental calculation), or movement. The movements can be classified under the following clinico-anatomical categories:

I. Cephalic Dyskinesias

A. *Ophthalmic*

1. Pseudo-exophthalmos: "wide-eyed" or astonished expression, with lifting of eyebrows.
2. Episodes of rapid blinking.
3. Sudden blepharospasms as if in intense concentration, often with hand gestures.
4. Sudden lateral deviations of eyes with or without rotation of head.
5. Internal strabismus of short duration with blurring of vision.

B. *Facial*

1. Unilateral rictus with or without short-duration hemifacial spasm.
2. Trismus.
3. Asymmetrical choreic or myoclonic movements of face (cheeks).

C. *Orobuccolingual*

1. Rapid, short-duration protrusion of tip of the tongue.
2. More prolonged protrusion with rhythmic and rolling movements of the tongue and licking motions.
3. Clicking and smacking of lips and tongue with or without gnawing.
4. Lateral or rotatory movements of lips and chin, sometimes like chewing, sometimes like rumination.
5. Rhythmic clicking of dentures (from tongue, cheek and jaw movements).
6. Rare palatal myoclonus.
7. Vocalization changes (nasal voice, slurring) with severe tongue movements.

D. *Cervical*

1. Latissimi contractions (with trismus).
2. Slow, wave-like, anteroposterior rocking or nodding of the head.
3. Much rarer lateral tremor of the head (similar to senile tremor).
4. Torsion dystonia (torticollis).
5. Short synchronous bilateral "shrugging" movements of shoulders.
6. Sudden massive unilateral contractions of shoulder muscles with bending of the head laterally toward the raised shoulder.

II. Truncal Dyskinesias

A. *Respiratory*

1. Hyperventilation, "air hunger" dyspnea.
2. Myoclonic jerks of diaphragm and intercostal muscles.

B. *Postural*

1. Whole-body rocking when sitting.
2. Oscillations while standing.
3. Oscillatory movements of the pelvis.
4. Compensatory lateral flexion (scoliosis), either occasional or constant and progressive.
5. Opisthotonic posture.
6. Acute akathisia anxiety reactions with inability to stay in one spot.

III. Upper-Extremity Dyskinesias

A. *Proximal*

1. Slow, wing-like flapping of entire arm, with elbow moving most.
2. Internal rotation with extension and retroversion of arm.
3. True ballistic movements ("arm shoots out").
4. Slow reptilian motion of arm with extension of hand and fingers alternating with flexion (athetosis).

B. *Distal*

1. Rapid jerking movements of fingers with or without extension and spreading.
2. Restlessness of hands and increased gesticulization.
3. Saccadic (3 per sec) lateral movement of whole hand from the wrist.
4. Tremor modification: increase in amplitude in initial phase of treatment with diminishing rigidity followed by decrease in rate (to 3 per sec) and amplitude.

IV. Lower-Extremity Dyskinesias

A. *When lying down*

1. Extension spasms of legs (kicking).
2. Restlessness (akathisia).

B. *When sitting*

1. Balancing motion of the leg from the knee: movement like organ player (lateral) or shuffling (back and forth) or mixture of both types.
2. Lateral oscillations of knee or ankle.
3. Rocking motion of foot resting on floor: alternate lifting of toes or heel.
4. Rhythmic spreading of toes.

C. *When standing*

1. Constant shifting from one foot to the other (akathisia).

D. *When walking*

1. Internal rotation of ankle progressing to whole limb internal rotation and circumvolution.
2. Slow wave-like movement of foot (athetosis).
3. More rapid, sudden propulsion of whole limb (ballismus).

GENERAL DISCUSSION

Dr. Cotzias: Dr. Yahr, do you have any personal impression as to whether involuntary movements are better or worse on the more afflicted side? or whether patients with previous thalamotomy show any differences in the lateralization of the involuntary movements?

Chairman Yahr: In our experience with the induction of these movements, it has been my impression that they occur in almost any muscle of the body, regardless of the previous degree of involvement of parkinsonism. We have analyzed 26 of our patients who developed abnormal movements and who previously had had thalamotomies.

In the first few cases we thought there was lateralization, but with time, as the number and intensity of the movements increased, we could find no further evidence for lateralization related to the previous surgery.

Dr. Barbeau: This is exactly our own experience, although the severity of the movements may be less on the side operated upon.

Dr. Coleman: Dr. Barbeau gave us a list of the various dyskinesias. I would like to add one. There is a buccolingual dyskinesia which occurs in retarded children and in particular in patients with Down's syndrome. This dyskinesia can be suppressed by elevating the low serotonin in these patients.

Originally we suppressed the dyskinesia by giving the patients 5-HTP, and recently we have had some success by giving pyridoxine. This is based on the work of Berman, who showed in 1965 that serotonin in the platelets of normal infants in the neonatal period can be doubled by giving pyridoxine.[1] We have duplicated that work in neonatal Down's syndrome, again showing that by giving pyridoxine serotonin levels can be raised. This may be one of the mechanisms of suppressing the buccolingual dyskinesia that is seen in Parkinson patients, since the two dyskinesias are clinically very similar and we know that by giving L-Dopa we are competing with the transport and enzyme systems of the tryptophan-serotonin pathways. B-6 helps to suppress the dyskinesia in both types of patients.

Dr. Everett: As a pharmacologist I would like to ask a question in regard to doses of 2, 3, or 4 grams per day. I am concerned about the unit of dosage and its distribution throughout the day. In the one case where the regime was mentioned, some of the side effects were apparently related to the time after the dose. Might it not be possible then to get rid of some side effects if smaller discrete doses were given throughout the day rather than in a few large doses?

Chairman Yahr: In regard to Dr. Everett's comment, these movements have been correlated with the total daily doses being administered regardless of the dose schedule. It is this type of variability in their appearance that makes evaluation of therapeutic measures for their control so difficult.

For example, particular patients have their movements sporadically throughout the day with varying degrees of intensity. In consequence, it is my feeling that any minor modification is probably of little consequence. The measure of effective therapy is really their total suppression over an extended period of time.

Dr. Calne: At various times it has been suggested that protein intake may modify both the therapeutic response to L-Dopa and the dyskinesias which this drug produces. Is this the current view? I would also like to ask whether apomorphine, with or without 5-hydroxytryptophan, induces involuntary movements?

Dr. Cotzias: Our work with apomorphine is still too preliminary for such interpretation. With regard to your other question, I subscribe to what Dr. Yahr says. We have quite often observed temporal dissociations between the therapeutic effects of L-Dopa and the involuntary movements. We have seen short-lived dissociation while manipulating the protein content of diets. We have had the same experience as Dr. Yahr with vitamin B-6. Following administration of some amino acids we have seen temporary reduction of involuntary movements.

Dr. Siegfried: I would like to review again the problem of involuntary movements due to overdoses of L-Dopa in stereotaxically operated patients. In our experience patients who underwent a thalamotomy with an excellent clinical result (disappearance of tremor and rigidity) did not show later the involuntary movements of L-Dopa overdose on the side operated upon. I wonder if Dr. Cooper observed the same effect?

Dr. Cooper: Yes, I would agree that some postoperative patients who are on Dopa and who develop involuntary movements do not have them on the operated side. I understood that Dr. Duvoisin and others have reported a similar finding in a patient after successful surgery, that is, when abnormal movements developed in the extremities they did not develop on the operated side. Moreover, it has been our experience that if a patient has postoperative dystonia of any kind it is invariably worsened by Dopa.

Chairman Yahr: I believe that Dr. Duvoisin and I are talking about the same case material. Initially, in the first cases we observed, it was our impression that there was a difference in movements between the operated and nonoperated sides. In the larger series of cases and with more opportunities to observe these movements, we find that they occur in operated as well as nonoperated patients, without regard to the side of operation. In fact, one of our patients with severe induced abnormal movements had a bilateral thalamotomy. Perhaps this relates to successful or unsuccessful surgery but we cannot make this distinction on clinical grounds alone.

REFERENCES

1. Berman, J. L., Justice, P., and Yi-Yung Hsia, D.: The metabolism of 5-hydroxytryptamine (serotonin) in the newborn. J. Pediat. *67:* 603, 1965.

CHAPTER 5

Pharmacological and Anatomical Aspects of Abnormal Movements Induced by L-Dopa

THE STRIOPALLIDAL SYSTEM AND ITS ACCESSORY MORPHOLOGICAL COMPONENTS: SOME FUNCTIONAL IMPLICATIONS

L. J. Poirier, A. Parent, C. Ohye, L. Larochelle, R. Boucher, and P. Bédard

Recent data from studies involving a number of approaches permit a reasonably accurate definition of the so-called extrapyramidal system and the role played by some of its components in the integration of motor and psychomotor activity.

The "extrapyramidal" or, more precisely, the striopallidal system may be regarded as being composed of a main center, represented by the neostriatum (caudate nucleus and putamen) and the paleostriatum (globus pallidus), and of accessory centers (substantia nigra, subthalamic nucleus and thalamic intralaminar nuclei) which are linked to the main gray masses by interconnecting pathways (Fig. 1).

The main efferent fibers of the neostriatum include striopallidal and strionigral projections.[1, 2, 3, 4] These two groups of fibers from the highly cholinergic caudate nucleus and putamen [5] are rich in acetylcholinesterase and, in this respect, represent the most characteristic intracerebrally coursing pathway.[6]

The acetylcholinesterasic striopallidal pathway contains important fascicles which constitute, with corresponding pallidal interconnections and pallidofugal fibers originating exclusively in the internal division of the pallidum,[1] the only well-established outflow by which the extrapyramidal system may influence

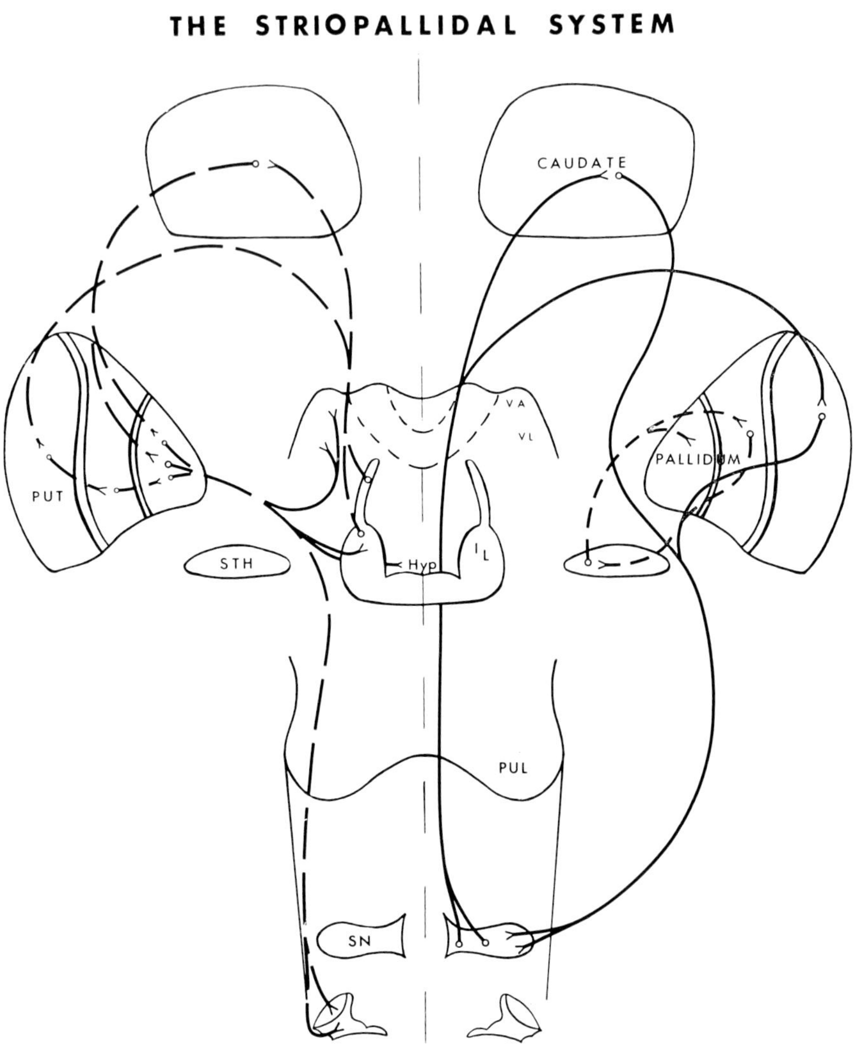

Fig. 1.

other nervous centers. The latter group of fibers (from the internal pallidum) project abundantly to areas of the ventral anterior and ventral lateral thalamic nuclei (pallidothalamic) and in smaller quantity to nucleus tegmentosus pedunculopontinus (pallidomesencephalic), on the ipsilateral side.[1] They probably include also pallidohypothalamic fibers (Fig. 1). Other fascicles from the internal pallidum ending in the thalamic intralaminar nuclei [1] constitute with the efferent fibers of the latter nuclei an intrinsic loop of the striopallidal system. As a matter of fact, the intralaminar nuclei project exclusively to the striatum.[7, 8] The acetylcholinesterasic strionigral fibers terminate in the substantia nigra,[1, 3, 4] the dopaminergic [9, 10] neurons of which project predominantly, if not exclusively, to the corresponding striatum.[11, 12] The latter two groups of fibers constitute a strionigrostriate loop by which the striatum apparently controls its own needs for dopamine.[13] The interruption of the nigrostriate pathway which travels within the medial forebrain bundle (MFB) [11, 14] causes a decrease of dopamine [10, 15, 16, 17] and its metabolite, homovanillic acid,[18] in the ipsilateral striatum, apparently as a consequence of the loss of the important enzymes tyrosine hydroxylase [19, 20] and Dopa decarboxylase [21, 22] involved in the synthesis of this amine.

Finally, a third intrinsic loop of the striopallidal system is represented by the pallidosubthalamic fibers from the external pallidum [1, 23] and the corresponding subthalamopallidal fibers ending predominantly in the internal segment [24] of the globus pallidus (Fig. 1).

Incoming fibers previously reported to end in the several structures of the striopallidal system (as defined above), in a general sense, are not firmly established. They include corticostriate, corticopallidal, corticosubthalamic, corticointralaminar and corticonigral fibers. Recent studies based on accurate silver staining methods, however, have cast some doubt on the existence of any important groups of corticopallidal,[25, 26] corticosubthalamic [25] and corticonigral fibers,[27] whereas the importance of the corticostriate fibers [26, 28] must be more precisely determined.[29] Important groups of fibers terminating in the relay nuclei of the striopallidal system are represented by the numerous fascicles which terminate in the intralaminar nuclei. They include corticointralaminar fibers from the motor [29] and premotor cortices [30] and spinothalamic [31] and reticulothalamic [32] fibers which reach the intralaminar nuclei by way of the ascending component of the central tegmental tract. Other ascending fibers ending in the same nuclei include cerebellothalamic fibers originating in the lateral cerebellar nuclei.[33] Finally, the origin and the course of the serotoninergic fibers ending in the striatum [9, 34] have not yet been precisely determined.[14]

As a corollary of the above data the striopallidal system is relatively autonomous except for important influences reaching it at the level of the intralaminar nuclei and, most probably, by way of the corticostriate fibers. On the other hand, this system exerts its influence predominantly on corticothalamocortical loops which involve the motor and premotor cortices and the ventral anterior and ventral lateral thalamic nuclei. Less important (in reference to the number of fibers) outflows of the striopallidal system include pallidohypothalamic and pallidomesencephalic fascicles.

The various motor and psychomotor disturbances encountered in the so-called "extrapyramidal" disorders may result not only from the impairment of the various components of the striopallidal system but also from the involvement of other nervous centers and pathways. Thus the impairment of the strionigrostriate loop seems to represent an important contributory factor in the appearance of various pathological, drug-induced and lesion-induced motor disturbances. The isolated interruption of this loop by unilateral sectioning of the nigrostriate fibers and the associated striatal decrease of dopamine, however, do not lead, per se, to any detectable disturbance in the monkey.[35] The production of a parkinsonian tremor in the monkey, for instance, involves, in addition to the striatal chemical imbalance, the concomitant impairment of the rubro-olivo-cerebello-rubral loop [36] which is not directly linked to the striopallidal system. On the other hand, harmaline, a monoamine oxidase inhibitor [37] which has a peculiar effect on brain dopamine and serotonin,[38, 39] exaggerates lesion-induced tremor [17, 35] and induces tremor in monkeys with lesions at one point or another of the rubro-olivo-cerebello-rubral loop.[36] This substance, however, does not bring about tremor after the isolated interruption of the striatal ascending monoaminergic pathways [35] but may exaggerate or cause tremor following lesions that simultaneously interrupt the latter mechanisms and other structures.[17, 35] These facts suggest that harmaline must exert some unknown effect such as activating the striatal cholinergic mechanisms.

In the light of the above, it was decided to investigate the effect of various catecholamine-blocking agents in monkeys with lesions of the rubro-olivo-cerebello-rubral loop which, however, spared the ascending monoaminergic pathways ending in the striatum. Such animals displayed unilateral tremor of the corresponding limbs within 10 to 15 minutes after the injection of harmaline (3 mg per kg), and again showed tremor following the injection of either α-methyldopa (100 mg per kg) or α-methyl-*p*-tyrosine (200 mg per kg) but after a delay of 4 to 5 hours or more.[40] In view of the fact that α-methyl-*p*-tyrosine, a known inhibitor of tyrosine hyroxylase,[41] prevents the formation of endogenous catecholamines but not the conversion of exogenous L-Dopa to dopamine in the presence of an intact nigrostriate pathway, the effect of L-Dopa (30 mg per kg) on tremor induced by α-methyl-tyrosine was tested in such monkeys. It was observed that under such conditions tremor was suppressed within 15 to 20 minutes and reappeared 45 to 60 minutes later. The same procedure repeated a few hours later gave the same results.[40] In view of the rapid effect of L-Dopa it is conceivable that under such circumstances the suppression of tremor is a consequence of the newly formed dopamine at the level of the striatum, where this amine is highly concentrated.

On the other hand, the administration of α-methyl-*p*-tyrosine in higher doses (1000 to 1500 mg intraperitoneally in two or more divided doses) to the same animals or to normal monkeys induced a catatonic state that developed after a delay of 18 to 24 hours and lasted 48 hours or more. The catatonic state, which resembles the behavior induced by reserpine, was associated with a disappearance of tremor in the lesioned monkey. Tremor, however, reappeared as the animal became more alert after 48 hours or more. The injection of L-Dopa (30 mg per kg) to the catatonic animal reversed the effect of α-methyl-*p*-tyrosine. The animal resumed normal behavior within 15 minutes after the injection of Dopa and returned to a catatonic state 45 to 60 minutes later. A reinjection of L-Dopa reproduced the same effect.[40] These facts suggest that the production of a catatonic state by α-methyl-*p*-tyrosine is dependent on a more profound disturbance of brain catecholamine metabolism than that necessary to induce tremor in lesioned monkeys. The rapid effect of L-Dopa to counteract the catatonic state induced by α-methyl-p-tyrosine, however, supports the idea that a repletion of brain dopamine represents the important factor in the transitory restoration to normal behavior, as in the case of tremor suppression.

The above data represent further evidence pointing out the importance of the strionigrostriate component of the striopallidal system in the control of motor and psychomotor activity. The rapid and transitory effect of L-Dopa on such abnormal disturbances, however, raises other questions. As a matter of fact these effects take place in the presence of intact nigrostriatal pathways in the monkey. This is to be opposed to the delayed beneficial effects obtained after long-term treatment with high doses of L-Dopa with or without the help of a peripheral decarboxylase inhibitor in parkinsonian patients [42, 43] whose striatal dopaminergic mechanisms are greatly impaired.[44]

Further evidence recently obtained in our laboratory might help to explain the above-mentioned apparent discrepancies. Since the interruption of the nigrostriate pathway greatly interferes with the conversion of Dopa to dopamine,[45] the study of the spontaneous activity of the striatal neurons deprived of their nigral afferents before and after the administration of L-Dopa was undertaken. Thus it was disclosed that the putaminal cells deprived of dopamine (as a consequence of the unilateral interruption of the corresponding nigrostriate pathway at different sites) fired at a higher frequency than the putaminal neurons of the intact side of the brain. Under such conditions L-Dopa (25 mg per kg intravenously) greatly enhanced the firing frequency in the putamen of the lesioned side but not in the putamen of the intact side.[46] It is obvious that under conditions where Dopa could not be converted to dopamine the partially denervated striatal neurons abnormally responded to Dopa itself, suggesting that this amine precursor acted as a false transmitter. These results, which should be completed by studies involving the chronic administration of L-Dopa to animals deprived of their nigrostriate mechanisms, open new avenues which may lead to a better understanding of the abnormal activities taking place in the striopallidal system and related nervous centers as a result of "extrapyramidal" disturbances.

Summary

The striopallidum (striatum and pallidum) is interconnected by three different loops to the corresponding subthalamic nucleus, substantia nigra and thalamic intralaminar nuclei, respectively. Therefore the latter three structures represent intrinsic components of the same system. In the light of such data it is proposed to substitute the term "striopallidal" for the misleading word extrapyramidal. The only well-established outflow of this relatively autonomous striopallidal system is represented by the pallidofugal fibers from the internal pallidum which terminate in the ventral anterior and ventral lateral thalamic nuclei, in the nucleus tegmentosus pedunculopontinus and, possibly, in the hypothalamus.

On the other hand the so-called extrapyramidal disorders involve not only the striopallidal system but also other extrinsic nervous mechanisms. The production of an experimental parkinsonian tremor in the monkey, for instance, involves the impairment of the rubro-olivo-cerebello-rubral loop in addition to disturbances (lesion- or drug-induced) of the monoaminergic mechanisms of the striopallidal system.

The importance of the catecholaminergic mechanisms in motor activity has been specifically demonstrated with the help of α-methyl-*p*-tyrosine, a tyrosine-hydroxylase inhibitor. The tremor and the catalepsy induced by this substance in lesioned and normal monkeys, respectively, are readily suppressed by L-Dopa, provided the nigrostriate pathways are intact.

Finally, L-Dopa apparently acts as a false transmitter at the level of the striatum in the absence of the nigrostriate pathway.

REFERENCES

1. Nauta, W. J. H., and Mehler, W. R.: Projections of the lentiform nucleus in the monkey. Brain Res. *1:* 3, 1966.
2. Rundles, R. W., and Papez, J. W.: Connections between the striatum and the substantia nigra in a human brain. Arch. Neurol. Psychiat. *38:* 550, 1937.
3. Szabo, J.: Topical distribution of the striatal efferents in the monkey. Exp. Neurol. *5:* 21, 1962.
4. Voneida, T. J.: An experimental study of the course and destination of fibers arising in the head of caudate nucleus in the cat and monkey. J. Comp. Neurol. *115:* 75, 1960.
5. MacIntosh, F. C.: The distribution of acetylcholine in the peripheral and the central nervous system. J. Physiol. *99:* 436, 1941.
6. Olivier, A., Parent, A., Simard, H., and Poirier, L. J.: Cholinesterasic striatopallidal and striatonigral efferents in the cat and the monkey. Brain Res. In press.
7. Mehler, W. R.: Further notes on the center median nucleus of Luys. *In* The Thalamus (Purpura, D., and Yahr, M.D., eds.). Columbia Univ. Press, New York, 1966.
8. Powell, T. P. S., and Cowan, W. M.: A study of thalamo-striate relations in the monkey. Brain *79:* 364, 1956.
9. Andén, N.-E., Carlsson, A., Dahlström, A., Fuxe, K., Hillarp, N.-A., and Larsson, K.: Demonstration and mapping out of nigro-neostriatal dopamine neurons. Life Sci. *3:* 523, 1964.
10. Poirier, L. J., and Sourkes, T. L.: Influence du locus niger sur la concentration des catecholamines du striatum. J. Physiol. *56:* 426, 1964.
11. Bédard, P., Larochelle, L., Parent, A., and Poirier, L. J.: The nigrostriatal pathway. A correlative study based on neuroanatomical and neurochemical criteria in the cat and the monkey. Exp. Neurol. In press.
12. Ferraro, A.: The connections of the pars suboculomotoria of the substantia nigra. Arch. Neurol. Psychiat. *19:* 177, 1928.
13. Poirier, L. J., Bédard, P., Boucher, R., Larochelle, L., Parent, A., and Singh, P:. The origin of different striato- and thalamo-petal neurochemical pathways and their relationship to motor activity. *In* Third Symp. on Parkinson's Disease. E. & S. Livingstone, London, 1969.
14. Parent, A., and Poirier, L. J.: The medial forebrain bundle (MFB) and ascending monoaminergic pathways in the cat. Canad. J. Physiol. Pharmacol. *47:* 781, 1969.
15. Goldstein, M., Anagnoste, B., Owen, W. S., and Battista, A. F.: The

effects of ventromedial tegmental lesions on the disposition of dopamine in the caudate nucleus of the monkey. Brain Res. *4:* 298, 1967.
16. Poirier, L. J., and Sourkes, T. L.: Influence of the substantia nigra on the catecholamine content of the striatum. Brain *88:* 181, 1965.
17. Poirier, L. J., Sourkes, T. L., Bouvier, G., Boucher, R., and Carabin, S.: Striatal amines, experimental tremor and the effect of harmaline in the monkey. Brain *89:* 37, 1966.
18. Sharman, D. F., Poirier, L. J., Murphy, G. F., and Sourkes, T. L.: Homovanillic acid and dihydroxyphenylacetic acid in the striatum of monkeys with brain lesions. Canad. J. Physiol. Pharmacol. *45:* 57, 1967.
19. Goldstein, M., Anagnoste, B., Owen, W. S., and Battista, A. F.: The effects of ventromedial tegmental lesions on the biosynthesis of catecholamines in the striatum. Life Sci. *5:* 2171, 1966.
20. Poirier, L. J., McGeer, E. G., Larochelle, L., McGeer, P. L., Bédard, P., and Boucher, R.: The effect of brain stem lesions on tyrosine and tryptophan hydroxylases in various structures of the telencephalon of the cat. Brain Res. *14:* 147, 1969.
21. Goldstein, M., Anagnoste, B., Battista, A. F., Owen, W. S., and Nakatani, S.: Studies of amines in the striatum in monkeys with nigral lesions. J. Neurochem. *16:* 645, 1969.
22. Lancaster, G., Larochelle, L., Bédard, P., Missala, K., Sourkes, T. L., and Poirier, L. J.: Effect of brain lesions and of harmaline on the dopa decarboxylase activity of striatum of the cat. J. Neurol. Sci. In press.
23. Ranson, S. W., and Ranson, S. W., Jr.: Efferent fibers of the corpus striatum. Res. Publ. Assoc. Res. Nerv. Ment. Dis. *21:* 69, 1942.
24. Carpenter, M. B., and Strominger, N. L.: Efferent fibers of the subthalamic nucleus in the monkey. A comparison of the efferent projections of the subthalamic nucleus, substantia nigra and globus pallidus. Amer. J. Anat. *121:* 41, 1967.
25. Petras, J. M.: Some fiber connections of the precentral and postcentral cortex with the basal ganglia, thalamus and subthalamus. Trans. Amer. Neurol. Assoc. *90:* 274, 1965.
26. Webster, K. E.: Cortico-striate interrelations in the albino rat. J. Anat. *95:* 532, 1961.
27. Rinvik, E.: The cortico-nigral projection in the cat. An experimental study with silver impregnation methods. J. Comp. Neurol. *126:* 241, 1966.
28. Carman, J. B., Cowan, W. M., Powell, T. P. S., and Webster, K. E.: A bilateral cortico-striate projection. J. Neurol. Neurosurg. Psychiat. *28:* 71, 1965.
29. Petras, J. M.: Some fiber connections of the precentral cortex (areas 4 and 6) with the diencephalon in the monkey (Macaca mulatta). Anat. Rec. *148:* 322, 1964.
30. Astruc, J.: Corticofugal fiber degeneration following lesions of area 8 (frontal eye field) in Macaca mulatta. Anat. Rec. *148:* 256, 1964.
31. Mehler, W. R., Feferman, M. E., and Nauta, W. J. H.: Ascending axon degeneration following anterolateral cordotomy. An experimental study in the monkey. Brain *83:* 718, 1960.
32. Nauta, W. J. H., and Kuypers, H. G. J. M.: Some ascending pathways in the brain stem reticular formation. *In* Reticular Formation of the Brain (Jasper, H., et al., eds.). Henry Ford Hospital Internat. Symposium. Little, Brown and Co., Boston, 1958.
33. Mehler, W. R., Vernier, V. G., and Nauta, W. J. H.: Efferent projections from the dentate and interpositus nuclei in primates. Anat. Rec. *130:* 430, 1958.

34. HELLER, A., HARVEY, J., and MOORE, R.: A demonstration of all fall in brain serotonin following central nervous system lesions in the rat. Biochem. Pharmacol. *11:* 859, 1962.
35. POIRIER, L. J., BOUVIER, G., BÉDARD, P., BOUCHER, R., LAROCHELLE, L., OLIVIER, A., and SINGH, P.: Essai sur les circuits neuronaux impliqués dans le tremblement postural et l'hypokinésie. Rev. Neurol. *120:* 15, 1969.
36. LAROCHELLE, L., BÉDARD, P., BOUCHER, R., and POIRIER, L. J.: The rubro-olivo-cerebello-rubral loop and postural tremor in the monkey. J. Neurol. Sci. In press.
37. UDENFRIEND, S., WITKOP, B., REDFIELD, B. G., and WEISSBACH, H.: Studies with reversible inhibitors of monoamine oxidase: harmaline and related compounds. Biochem. Pharmacol. *1:* 160, 1958.
38. POIRIER, L. J., SINGH, P., and BOUCHER, R.: Opposite effect of harmaline on serotonin and on dopamine and its metabolites, homovanillic acid and norepinephrine, in the brain of the cat. Canad. J. Physiol. Pharmacol. *46:* 585, 1968.
39. SINGH, P., POIRIER, L. J., and BOUCHER, R.: Effect of monoamine oxidase inhibitors on the concentration of dopamine and serotonin in the striatum of the cat with and without unilateral brain stem lesions. Canad. J. Physiol. Pharmacol. *45:* 897, 1967.
40. BÉDARD, P., LAROCHELLE, L., POIRIER, L. J., and SOURKES, L.: Reversible effect of L-Dopa on tremor and catatonia induced by α-methyl-*p*-tyrosine. Canad. J. Physiol. Pharmacol. In press.
41. NAGATSU, T., LEVITT, M., and UDENFRIEND, S.: The initial step in norepinephrine biosynthesis. J. Biol. Chem. *239:* 2910, 1964.
42. COTZIAS, G. C., VAN WOERT, M. H., and SCHIFFER, L. M.: Aromatic amino acids and modification of parkinsonism. New Eng. J. Med. *276:* 374, 1967.
43. DUVOISIN, R., BARRETT, R., SCHEAR, M., HOEHN, M., and YAHR, M.: The use of L-Dopa in parkinsonism. *In* Third Symp. on Parkinson's Disease. E. & S. Livingston, London, 1969.
44. HORNYKIEWICZ, O.: Dopamine (3-hydroxytyramine) and brain function. Pharmacol. Rev. *18:* 925, 1966.
45. POIRIER, L. J., SINGH, P., SOURKES, T. L., and BOUCHER, R.: Effect of amine precursors on the concentration of striatal dopamine and serotonin in cats with and without unilateral brain stem lesions. Brain Res. *6:* 654, 1967.
46. OHYE, C., BOUCHARD, R., LAROCHELLE, L., BÉDARD, P., BOUCHER, R., RAPHY, B., and POIRIER, L. J.: Effect of dorsal rhizotomy on postural tremor in the monkey. Exp. Brain Res. In press.

PHARMACOLOGICAL AND ANATOMICAL IMPLICATIONS OF INDUCED ABNORMAL MOVEMENTS WITH L-DOPA *

Nils-Erik Andén

When L-3, 4-dihydroxyphenylalanine (L-Dopa) is given, dopamine (DA) and norepinephrine (NE) are formed at many locations in the central nervous system. The amount of the two catecholamines synthesized from Dopa in

* Supported in part by the Swedish State Medical Research Council (B70-14X-502-06 B) and Svenska Livfürsäkringsbologs nämnd für Medicinsk Forskning.

the different regions is roughly proportional to L-Dopa decarboxylase activity and endogenous amine content.[1] Therefore the actions of Dopa may be evoked from many sites in the brain and the spinal cord. The demonstration and mapping out of central monoamine neuron systems [2] have been of importance in the analysis of the regional effects seen after treatment with Dopa. The function can most easily be studied in the places where the amines can be applied locally or where the afferent monoamine neurons can be severed selectively and completely. These conditions are found in the corpus striatum and in the spinal cord. Therefore, I will deal mainly with the L-Dopa actions in these two regions.

Corpus Striatum

About 80 percent of the rat brain DA occurs in nerve terminals of the caudate nucleus-putamen (neostriatum).[1, 3, 4] DA is also found in cell bodies of the substantia nigra, mainly in its pars compacta.[5, 6] The existence of a nigroneostriatal DA neuron system has been shown histochemically, i.e., in serial sections after an increase in the DA content in the nonterminal axons. Such an increase can be produced by removal of the terminals in the neostriatum.[7, 8] An even greater increase is observed after intrastriatal application of DA.[9] The DA axons leave the cell bodies in the substantia nigra and bend somewhat medially and dorsally toward the medial lemniscus and the red nucleus. Thereafter they aggregate and the fiber bundle ascends rostrally medial and dorsomedial to the ventral part of the crus cerebri. Some DA axons turn into the ventral part of the rostral crus cerebri. The others enter the retrolenticular part of the internal capsule from the H_2 field of Forel. Running rostrally and dorsally, the fibers diverge and enter the neostriatum via the fibrae capsulae internae. The DA terminals in the neostriatum are remarkably long and have a high content of amine in densely occurring enlargements (varicosities or boutons). Electronmicroscopically, Hökfelt [10] observed that the DA boutons make synaptic contact with cell bodies and dendrites in the neostriatum and that they constitute about 16 percent of all boutons in that nucleus.

The nigroneostriatal DA neurons are uncrossed. Therefore, the function of this neuron system and the influence of Dopa on it have been studied by observing asymmetries in rats after unilateral damage of the pathway. A clear-cut turning of the head and the tail to the side operated upon has been seen after selective damage of the nigroneostriatal DA axons on one side by 6-OH-DA [11] or by small electrolytic lesions. After large unilateral lesions of the ascending axons or of the terminals in the neostriatum, there is a tendency for the rats to turn to the operated side, but this finding is not constant (Fig. 1).[12] The absence of marked asymmetries after the large lesions may be due to destruction of efferent pathways from the corpus striatum or of a DA-antagonistic pathway to the corpus striatum. Obviously, rats with such a large unilateral lesion of the nigroneostriatal DA neuron system are useful for studies of the effects of drugs on the DA neurotransmission in the neo-

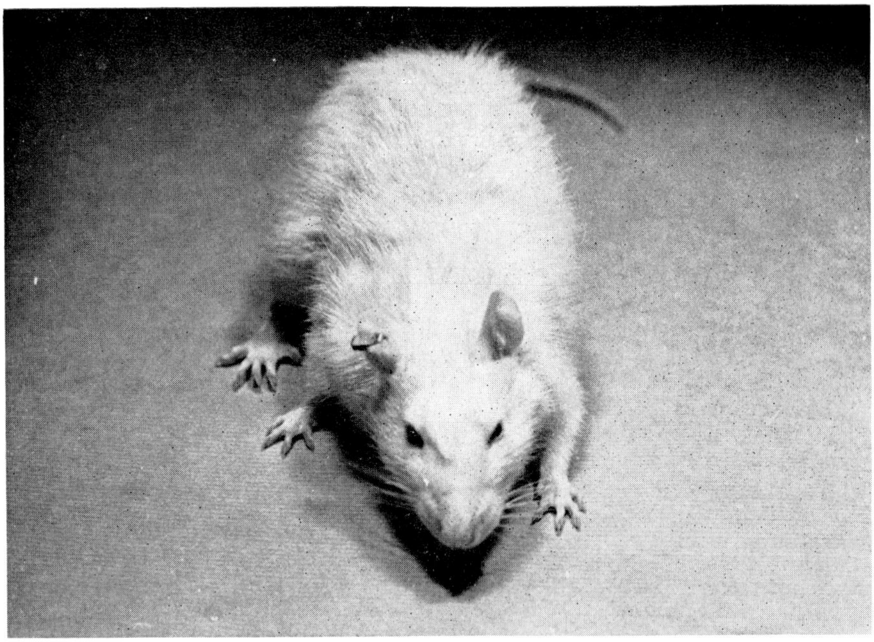

Fig. 1. Rat in which the left nigroneostriatal dopamine pathway was destroyed in the crus cerebri 30 days earlier (Andén, et. al.[12]).

striatum. Reserpine treatment of a surgically treated rat causes a DA loss on the unoperated side and a pronounced turning of the head and the tail to the unoperated side (Fig. 2).[12] Simultaneously, the extremities are adducted below the body on the unoperated side, possibly due to rigidity. The legs are abduced on the operated side. Similar asymmetries are observed after treatment with other neuroleptic drugs, such as the phenothiazine derivative chlorpromazine or the butyrophenone derivative haloperidol (Table 1).[12] Other butyrophenone compounds as spiroperidol and butyrophenone-like drugs as pimozide produce such an asymmetry in very small doses.[13] Blocking agents of adrenergic α-receptors (phenoxybenzamine) or β-receptors (propranolol) do not evoke any significant turning in rats with this type of lesion, nor is any asymmetry observed after treatment with sedatives lacking neuroleptic action as antihistamines (promethazine) or barbiturates (pentobarbital).

If the monoamine oxidase inhibitor nialamide is given before reserpine, the DA coming from the granules is neither degraded by means of the monoamine oxidase nor taken up again in the granules, but diffuses in large amounts to the receptors of the effector cells. In the case of a lesion of the nonterminal axons, it is necessary to wait for some days after the operation, until the DA terminals have degenerated, before giving drugs. Treatment with nialamide plus reserpine causes a unilaterally operated rat to turn to the side of the lesion (Fig. 3).[12] After such treatment, the animal is excited and often rotates vigorously. During the rotation, the legs on the unoperated side are more active. The same effect is seen in unilaterally striatotomized rats after treatment with

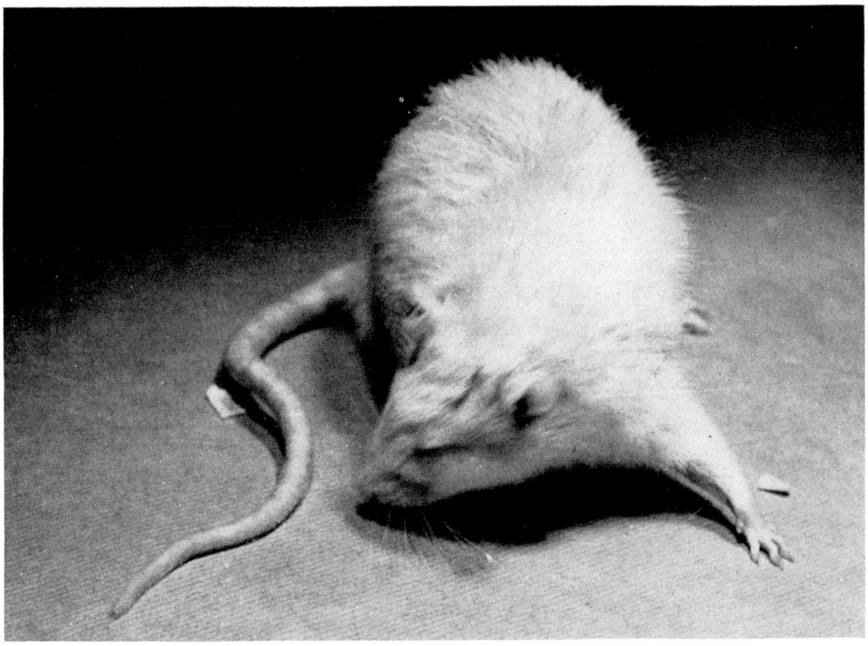

Fig. 2. Effect of reserpine (10 mg/kg i.p. 2¼ hrs before) on a rat in which the left nigroneostriatal dopamine pathway was destroyed in the crus cerebri 30 days earlier (Andén, et al.[12]).

TABLE 1. Drug-Induced Turnings of Heads and Tails in Rats with Large, Unilateral Lesions of the Nigroneostriatal Dopamine Pathway.

Treatment	*Effect*
No drug treatment	No turning
Drugs which may block monoamine transmission	
Reserpine (10 mg/kg i.p.)	Unoperated side
Chlorpromazine (5–10 mg/kg i.p.)	Unoperated side
Haloperidol (0.1–10 mg/kg i.p.)	Unoperated side
Spiroperidol (0.005–1 mg/kg i.p.)	Unoperated side
Pimozide (0.05–25 mg/kg i.p.)	Unoperated side
Phenoxybenzamine (20 mg/kg i.p.)	No turning
Propranolol (12 mg/kg i.p.)	No turning
Promethazine (20 mg/kg i.p.)	No turning
Pentobarbital (8–40 mg/kg i.p.)	No turning
Drugs which may cause stimulation of monoamine transmission	
Nialamide (100 mg/kg i.p.)+reserpine (10 mg/kg i.v.)	Operated side
Reserpine (10 mg/kg i.p.)+nialamide (100 mg/kg i.p.)+ L-Dopa (20–75 mg/kg i.v.)	Operated side
Reserpine (10 mg/kg i.p.)+nialamide (100 mg/kg i.p.)+ DL-5-HTP (25–50 mg/kg i.v.)	No turning
Reserpine (10 mg/kg i.p.)+α-methyltyrosine (500 mg/kg i.p.)+apomorphine (0.5–25 mg/kg s.c.)	Operated side
Reserpine (10 mg/kg i.p.)+α-methyltyrosine (500 mg/kg i.p.)+dexamphetamine (0.5–10 mg/kg i.p.)	No turning
Reserpine (10 mg/kg i.p.)+dexamphetamine (0.5–10 mg/kg i.p.)	Operated side

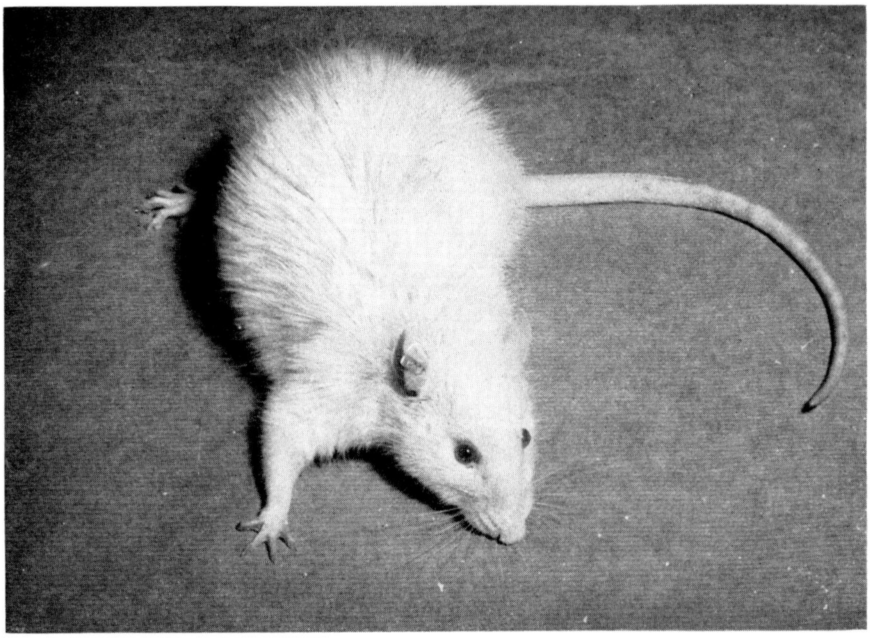

Fig. 3. Effect of nialamide (100 mg/kg i.p. 2½ hrs before) plus reserpine (10 mg/kg i.v. 30 min before) on a rat in which the left nigroneostriatal dopamine pathway was destroyed in the crus cerebri 30 days earlier (Andén, et al.[12]).

L-Dopa (Table 1).[12] Similar to other central effects of Dopa in the rat, this asymmetry will not appear if not preceded by monoamine oxidase inhibition or by blockade of the peripheral L-Dopa decarboxylase. If the surgically treated rat is pretreated with reserpine, treatment with Dopa changes the turning of the head and the tail from the unoperated to the operated side. The turning to the operated side is probably due to a facilitation of the DA transmission in the neostriatum on the intact side, since on the opposite side there are no neostriatal DA terminals from which DA can be released. This view has been strengthened by the observation that local application of DA after monoamine oxidase inhibition in the neostriatum on one side produces a turning of the rat to the opposite side.[9] The 5-hydroxytryptamine (5-HT) in the neostriatum probably acts differently from the DA since no asymmetry is observed in unilaterally operated rats after treatment with large doses of the precursor 5-hydroxytryptophan (5-HTP) (Table 1).[12]

The emetic agent apomorphine produces the same asymmetry as described above for Dopa after systemic treatment of unilaterally operated rats (Table 1)[14] or after unilateral application in the neostriatum.[9] This effect is seen also after pretreatment with reserpine plus α-methyltyrosine when all known stores of DA are depleted. Therefore it is likely that apomorphine directly stimulates the DA receptors in the neostriatum. Apomorphine does not cause emesis in rats but causes gnawing or other stereotyped movements. It has long been known that these actions are evoked from the corpus striatum.[15, 16] Ernst [17] was the first to propose a direct DA receptor stimulation by apomorphine from the finding that the stereotyped movements are not reduced by pretreatment with a synthesis inhibitor. Apomorphine causes a reduction of the DA turnover as seen from a retardation of the DA disappearance after synthesis inhibition,[14, 18] and from a decrease in the level of the main DA metabolite homovanillic acid.[19] This reduced DA turnover may be due to a negative feedback mechanism as a consequence of the DA receptor stimulation.

The same turning as produced by Dopa or apomorphine in reserpine-pretreated rats is seen also after injection of dexamphetamine. In contrast to apomorphine, dexamphetamine does not produce any asymmetry after pretreatment with both reserpine and α-methyltyrosine (Table 1).[14] Therefore it seems as if amphetamine acts in the neostriatum by release of DA from the nerve terminals. The same conclusion has previously been drawn from studies on the amphetamine-induced stereotypes.[17, 20]

The administration of Dopa, apomorphine or amphetamine to a reserpine-pretreated rat with a unilateral lesion of the nigroneostriatal DA pathway causes the rat to turn from the unoperated to the operated side. If chlorpromazine or haloperidol is given to such a rat, the animal again turns to the unoperated side.[12] Also the other neuroleptic drugs with a phenothiazine, butyrophenone or similar structure which, by themselves, cause a turning to the unoperated side can block the action of Dopa or apomorphine. On the other hand, phenoxybenzamine does not block this action, nor does it cause any clear-cut asymmetry in operated rats. The formation of DA from Dopa is not changed significantly by these drugs. Therefore it is most likely that these neuroleptics block the DA receptors of the effector cells in the neostriatum. Haloperidol and chlorpromazine also accelerate the turnover of brain DA as seen by an increase in the accumulation of methoxytyramine after monoamine oxidase inhibition,[21] by an elevation of the concentrations of the acid DA metabolites 3,4-dihydroxyphenylacetic acid and homovanillic acid [22] and by an acceleration of the DA disappearance induced by synthesis inhibition.[23] These biochemical effects can be explained by a positive feedback mechanism initiated by blockade of the DA receptors. In addition, haloperidol has been found to eliminate the effect of apomorphine on the DA turnover in the same way it blocks the effect of apomorphine on the turning.[14]

Treatment of rats with reserpine or chlorpromazine causes a rigidity which appears to be related to an increase in α-motoneuron excitability and a decrease in γ-motoneuron activity.[24, 25] These effects are probably evoked from the neostriatum since the changes are much more pronounced on the unoperated side of a reserpine-treated rat with a unilateral lesion of the nigroneostriatal DA pathway.[26] The turning may be due to a stronger reserpine-induced rigidity of the α-type on the unoperated side. The impulses from the neostriatum to the mononeurons are mediated via a pathway in the lateral funiculus of the spinal cord.[27] The reserpine-induced changes are eliminated by L-Dopa both in intact [25] and in surgically treated rats.[26]

Spinal Cord

The NE and 5-HT neurons to the spinal cord have their cell bodies in the lower brain stem and their terminals throughout the gray matter of the spinal cord.[28, 29, 30] The function of the monoamines in the spinal cord can conveniently be studied in a spinal animal. No nerve impulses reach the monoamine nerve terminals in the caudal part of the spinal cord; therefore, changes in the hindlimb reflexes can be studied without disturbing influences from the brain. Injection of L-Dopa causes formation of DA and NE also in the caudal part of the sectioned cord. Analogously, treatment with DL-5-HTP increases the concentration of 5-HT. The effects on the spinal reflexes after administration of Dopa or 5-HTP are central, since they cannot be reproduced by injection of NE or 5-HT. The amines formed are responsible for the effects observed, since decarboxylase inhibition blocks and monoamine oxidase inhibition potentiates them.[31] The effects of Dopa in the spinal cord can be considered to be due to stimulation of NE receptors since no DA neurons to this region have been detected.[28, 30]

Injection of Dopa induces in spinal animals a marked increase in the hindlimb flexor reflex evoked by pinching the legs [29] or by tetanic stimulation of the flexor reflex afferents (Fig. 4). This effect has been analyzed electrophysiologically and is probably due to inhibition of mechanisms which normally reduce the magnitude of the flexor reflex in the spinal animal.[32] Presumably,

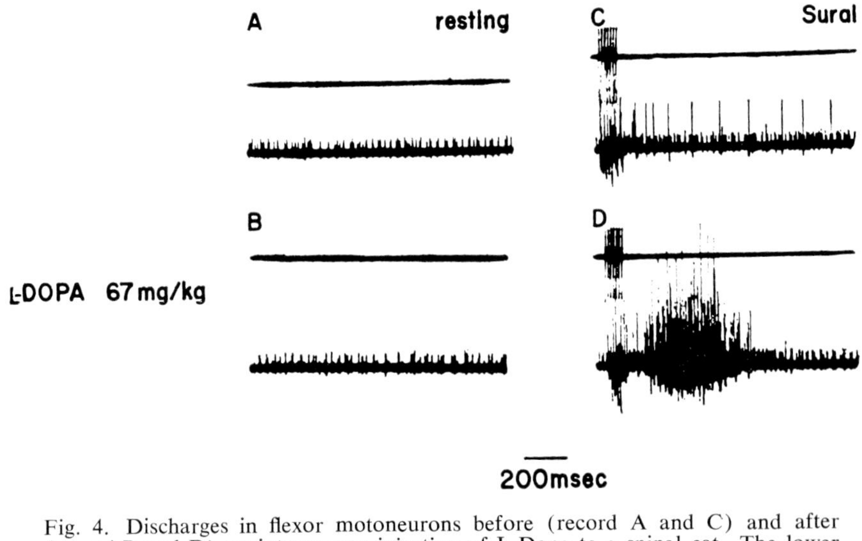

Fig. 4. Discharges in flexor motoneurons before (record A and C) and after (record B and D) an intravenous injection of L-Dopa to a spinal cat. The lower traces in each record are the efferent discharges in a nerve to a flexor (tenuissimus) and the upper traces are from the sciatic nerve (afferent). A and B are resting discharges; an ipsilateral, cutaneous nerve (sural) was stimulated by a train of volleys in C and D. Before treatment with Dopa (C) there was an early and short reflex discharge followed by an after-discharge in both α- and γ-efferents. After treatment with Dopa (D) there was a late, massive and long-lasting discharge in both α- and γ-efferents. Such a discharge is not seen in a nerve to an extensor (Andén, et al.[32]).

the Dopa treatment replaces the activity of bulbospinal NE neurons which in the intact animal facilitates the flexor reflex activity.

The Dopa-induced effects are blocked by the adrenergic α-receptor blocking agent phenoxybenzamine and the neuroleptic drugs haloperidol and chlorpromazine.[31, 33a] These drugs do not change the formation of DA and NE from Dopa. Therefore it is probable that phenoxybenzamine, haloperidol and chlorpromazine block the NE receptors mediating the increase in flexor reflex activity. On the other hand, pimozide does not block this Dopa effect and spiroperidol does it only in high doses.[13]

The same increase in flexor reflex activity as seen after treatment with Dopa is also seen with dexamphetamine. As in the neostriatum, the effect is blocked by pretreatment with reserpine and α-methyltyrosine indicating that the amphetamine effect is due to release of NE from the nerve terminals.[14] The DA receptor stimulating agent apomorphine has no effect on flexor reflex activity. The results with apomorphine, phenoxybenzamine and pimozide in the neostriatum and in the spinal cord show that the DA and NE receptors can be influenced specifically.

In the acute spinal rat, treatment with the 5-HT precursor DL-5-HTP does not evoke increased flexor activity like Dopa but produces hyperextension and athetoid movements of the hind legs.[33b] These effects are not significantly altered by haloperidol, chlorpromazine or phenoxybenzamine. The dissimilar actions after treatment with Dopa and 5-HTP and the unequal effect of receptor blocking agents on these actions indicate that 5-HT and the two catecholamines act on different receptors.

Effect of Dopa After Chronic Axotomy

After transection of the spinal cord, the L-Dopa decarboxylase activity is almost completely lost caudal to the transection (Fig. 5).[34] The time courses of the disappearance of NE (or 5-HT) and of the L-Dopa decarboxylase activity are about the same, indicating that the enzyme lost is localized in NE (or 5-HT) neurons. The percentage reduction is greater if the decarboxylase activity is determined in vitro than in vivo. In the former case, the levels of enzyme activity and of NA are reduced approximately to zero. The difference between the in vitro and the in vivo results may be explained by the small L-Dopa decarboxylase activity present in the pericytes and in the endothelial cells of the capillary walls. Decarboxylase activity in the capillary wall is small in comparison to the nerve terminals but may be of importance in vivo when the injected L-Dopa must pass through the capillary wall. In reflex studies, it has also been found that the effect of L-Dopa is not completely abolished even four weeks after transection of the spinal cord.[31] The effect is, however, markedly reduced, particularly the maximal effect. At the same time there is a slight formation of DA but not of NE. The DA may be formed in the capillary walls and diffuse to the receptors which are supersensitive to catecholamines due to the denervation. No NE formation is observed since the enzyme DA-β-hydroxylase is present only in NE neurons.[37]

Also in the neostriatum there is evidence for an action of Dopa after degen-

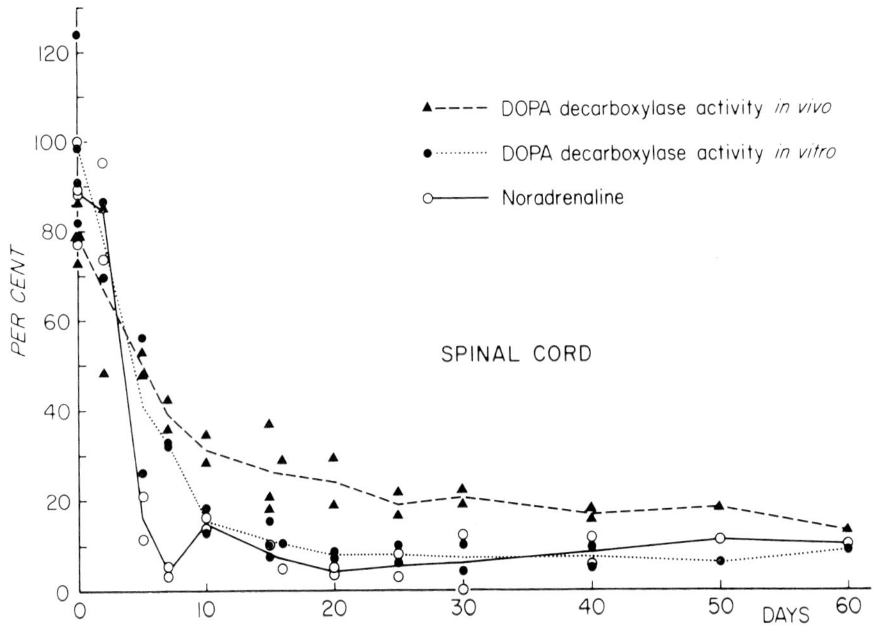

Fig. 5. The L-3,4-dihydroxyphenylalanine (Dopa) decarboxylase activity and the norepinephrine concentration in the caudal part of the rabbit spinal cord at various intervals after transection. The values are expressed as percent of those in the cranial part of the same animals (Andén, Magnusson and Rosengren [34]).

eration of the monoamine terminals. If L-Dopa is given to a rat more than one week after a unilateral section of the nonterminal axons of the nigro-neostriatal DA neurons, the animal rotates to the unoperated side.[38] After unilateral removal of the corpus striatum, the same treatment with L-Dopa causes a rotation to the operated instead of the unoperated side (see above). In chronically axotomized animals, there is a smaller accumulation of DA in the corpus striatum on the operated side, probably due to loss of neuronal L-Dopa decarboxylase activity. The small amount of DA formed in the corpus striatum on the operated side, probably in the remaining nerve terminals and in the capillaries, seems to have a stronger action than the greater amount of DA on the opposite side. As in the spinal cord, the phenomenon may be explained by a denervation supersensitivity of the amine receptors. Recently, Ungerstedt [39] has obtained further evidence for such a view. After almost complete loss of DA on one side after intracerebral injection of 6-OH-DA, apomorphine in small doses causes a pronounced turning or rotation of the rat to the unoperated side, i.e., to the opposite side of that seen after unilateral removal of the corpus striatum (see above).

Conclusions

Administration of Dopa causes changes in posture and movements which may be evoked from many regions in the central nervous system. The DA in the corpus striatum is of importance for the muscle tone whereas the NE in the spinal cord regulates the flexor reflex activity. By studies on these DA and NE mechanisms separately, it has been observed that DA and NE receptors can be selectively stimulated or blocked. A preferential effect on the neostriatal DA receptors is desirable in the Dopa treatment of parkinsonian patients.

Animal experiments with Dopa after chronic axotomy indicate that the small amount of DA formed may have a more pronounced effect than normal on the receptors due to denervation supersensitivity. A similar supersensitivity to DA is perhaps present in the neostriatum of parkinsonian patients in whom the DA formation from Dopa may be small, due to degeneration of the nigro-neostriatal DA neurons and accompanying reduction of L-Dopa decarboxylase activity.

ACKNOWLEDGMENTS

For generous gifts of drugs I thank the Swedish Ciba Ltd. (reserpine), the Swedish Pfizer Ltd. (nialamide), Leo Ltd. (chlorpromazine, phenoxybenzamine, haloperidol), and Janssen Pharmaceutica (pimozide, spiroperidol). For skillful technical assistance I am grateful to Miss Inger Karlsson and Miss Berit Nilsson.

REFERENCES

1. Bertler, Å., and Rosengren, E.: On the distribution in brain of monoamines and of enzymes responsible for their formation. Experientia *15:* 382, 1959.
2. Andén, N.-E., Dahlström, A., Fuxe, K., Larsson, K., Olson, L., and Ungerstedt, U.: Ascending monoamine neurons to the telencephalon and diencephalon. Acta Physiol. Scand. *67:* 313, 1966.
3. Fuxe, K.: Evidence for the existence of monoamine neurons in the central nervous system. IV. Distribution of monoamine nerve terminals in the central nervous system. Acta Physiol. Scand. 64, Suppl. 247: 37, 1965.
4. Andén, N.-E., Fuxe, K., Hamberger, B., and Hökfelt, T.: A quantitative study on the nigro-neostriatal dopamine neuron system in the rat. Acta Physiol. Scand. *67:* 306, 1966.
5. Bertler, Å.: Occurrence and localization of catecholamines in the human brain. Acta Physiol. Scand. *51:* 97, 1961.
6. Dahlström, A., and Fuxe, K.: Evidence for the existence of monoamine-containing neurons in the central nervous system. I. Demonstration of monoamines in the cell bodies of brain stem neurons. Acta Physiol. Scand. *62,* Suppl. 232: 1, 1964.
7. Andén, N.-E., Carlsson, A., Dahlström, A., Fuxe, K., Hillarp, N.-Å., and Larsson, K.: Demonstration and mapping out of nigro-neostriatal dopamine neurons. Life Sci. *3:* 523, 1964.
8. Andén, N.-E., Dahlström, A., Fuxe, K., and Larsson, K.: Further evidence for the presence of nigroneostriatal dopamine neurons in the rat. Amer. J. Anat. *116:* 329, 1965.
9. Ungerstedt, U., Butcher, L. L., Butcher, S. G., Andén, N.-E., and Fuxe, K.: Direct chemical stimulation of dopaminergic mechanisms in the neostriatum of the rat. Brain Res. *14:* 461, 1969.

10. HÖKFELT, T.: In vitro studies on central and peripheral monoamine neurons at the ultrastructural level. Z. Zellforsch. *91:* 1, 1968.
11. UNGERSTEDT, U.: 6-Hydroxy-dopamine induced degeneration of central monoamine neurons. Europ. J. Pharmacol. *5:* 107, 1968.
12. ANDÉN, N.-E., DAHLSTRÖM, A., FUXE, K., and LARSSON, K.: Functional role of the nigro-neostriatal dopamine neurons. Acta Pharmacol. *24:* 263, 1966.
13. ANDÉN, N.-E., BUTCHER, S. G., CORRODI, H., FUXE, K., and UNGERSTEDT, U.: Unpublished data, 1969.
14. ANDÉN, N.-E., RUBENSON, A., FUXE, K., and HÖKFELT, T.: Evidence for dopamine receptor stimulation by apomorphine. J. Pharm. Pharmacol. *19:* 627, 1967.
15. HARNACK, E.: Uber die Wirkungen des Apomorphines am Säugethier und am Frosch. Arch. Exp. Pathol. Pharmakol. *2:* 255, 1874.
16. AMSLER, C. Beiträge zur Pharmakologie des Gehirns. Arch. Exp. Pathol. Pharmakol. *97:* 1, 1923.
17. ERNST, A. M.: Mode of action of apomorphine and dexamphetamine on gnawing compulsion in rats. Psychopharmacologia *10:* 316, 1967.
18. BUTCHER, L. L., and ANDÉN, N.-E.: Effects of apomorphine and amphetamine on schedule-controlled behavior: reversal of tetrabenazine suppression and dopaminergic correlates. Europ. J. Pharmacol. *6:* 255, 1969.
19. ROOS, B.-E.: Decrease in homovanillic acid as evidence for dopamine receptor stimulation by apomorphine in the neostriatum of the rat. J. Pharm. Pharmaco., *21:* 263, 1969.
20. RANDRUP, A., and MUNKVAD, I.: Role of catecholamines in the amphetamine excitatory response. Nature *211:* 540, 1966.
21. CARLSSON, A., and LINDQVIST, M.: Effect of chlorpromazine or haloperidol on formation of 3-methoxytyramine and normetanephrine in mouse brain. Acta Pharmacol. *20:* 140, 1963.
22. ANDÉN, N.-E., ROOS, B.-E., and WERDINIUS, B.: Effects of chlorpromazine, haloperidol and reserpine on the levels of phenolic acids in rabbit corpus striatum. Life Sci. *3:* 149, 1969.
23. CORRODI, H., FUXE, K., and HÖKFELT, T.: The effect of neuroleptics on the activity of central catecholamine neurons. Life Sci. *6:* 767, 1967.
24. ROOS, B.-E., and STEG, G.: The effects of L-3,4-dihydroxyphenylalanine and DL-5-hydroxytryptophan on rigidity and tremor induced by reserpine, chlorpromazine and phenoxybenzamine. Life Sci. *3:* 351, 1964.
25. STEG, G.: Efferent muscle innervation and rigidity. Acta Physiol. Scand. *61:* Suppl. 225: 1, 1964.
26. ANDÉN, N.-E., LARSSON, K., and STEG, G.: Unpublished data, 1969.
27. ARVIDSSON, J., JURNA, I., and STEG, G.: Striatal and spinal lesions eliminating reserpine and physostigmine rigidity. Life Sci. *6:* 2017, 1967.
28. MAGNUSSON, T., and ROSENGREN, E.: Catecholamines of the spinal cord normally and after transection. Experientia *19:* 229, 1963.
29. CARLSSON, A., MAGNUSSON, T., and ROSENGREN, E.: 5-Hydroxytryptamine of the spinal cord normally and after transection. Experientia *19:* 359, 1963.
30. CARLSSON, A., FALCK, B., FUXE, K., and HILLARP, N.-Å.: Cellular localization of monoamines in the spinal cord. Acta Physiol. Scand. *60:* 112, 1964.
31. ANDÉN, N.-E., JUKES, M. G. M., and LUNDBERG, A.: The effect of Dopa on the spinal cord. 2. A pharmacological analysis. Acta Physiol. Scand. *67:* 387, 1966.
32. ANDÉN, N.-E., JUKES, M. G. M., LUNDBERG, A., and VYKLICKÝ, L.: The effect of Dopa on the spinal cord. 1. Influence on transmission from primary afferents. Acta Physiol. Scand. *67:* 373, 1966.
33a. ANDÉN, N.-E., CORRODI, H., FUXE, K., and HÖKFELT, T.: Increased impulse

flow in bulbospinal noradrenaline neurons produced by catecholamine receptor blocking agents. Europ. J. Pharmacol. *2:* 59, 1967.

33b. ANDÉN, N.-E., CORRODI, H., FUXE, K., and HÖKFELT, T.: Evidence for a central 5-hydroxytryptamine receptor stimulation by lysergic acid diethylamide. Brit. J. Pharmacol. *34:* 1, 1968.

34. ANDÉN, N.-E., MAGNUSSON, T., and ROSENGREN, E.: Occurrence of dihydroxyphenylalanine decarboxylase in nerves of the spinal cord and sympathetically innervated organs. Acta Physiol. Scand. *64:* 127, 1965.

35. BERTLER, Å., FALCK, B., and ROSENGREN, E.: The direct demonstration of a barrier mechanism in the brain capillaries. Acta Pharmacol. *20:* 317, 1964.

36. HAMBERGER, B.: Reserpine-resistant uptake of catecholamines in isolated tissues of the rat. Acta Physiol. Scand. Suppl. 295: 1, 1967.

37. CARLSSON, A., and WALDECK, B.: β-Hydroxylation of tyramine in vivo. Acta Pharmacol. *20:* 371, 1963.

38. ANDÉN, N.-E.: On the function of the nigroneostriatal dopamine pathway. *In* Mechanisms of Release of Biogenic Amines. Pergamon Press, Oxford, 1966.

39. UNGERSTEDT, U.: Behavioural registration of dopamine synaptic activity in the brain after 6-hydroxy-dopamine lesions. Acta Physiol. Scand. Suppl. 330: 117, 1969.

THE NIGROSTRIATAL PATHWAY: DEMONSTRATION BY ANTEROGRADE DEGENERATION

Robert Y. Moore

The analysis of the anatomical relations of the basal ganglia provides a continuing foundation for the understanding of extrapyramidal disease. A particular example of this is Parkinson's disease. The principal pathological changes in this disease affect the neurons of the substantia nigra. Associated with these there is a loss of the monoamine dopamine from both the substantia nigra and the neostriatum.[1, 2] From such observations it has been proposed that the pathophysiological process in Parkinson's disease affects dopamine-producing neurons with cell bodies in the substantia nigra and axon terminals in the neostriatal nuclei. The existence of a direct nigrostriatal projection was first suggested on the basis of retrograde degeneration studies. Ablation of the neostriatum results in degeneration of the cells of the substantia nigra, pars compacta, and the adjacent ventral tegmental area.[3] Attempts to demonstrate a nigrostriatal pathway by a variety of anterograde degeneration techniques, however, have thus far met with failure.[4, 5, 6, 7] Substantial experimental support for a nigrostriatal tract has come from a series of studies continuing the use of lesions with biochemical or histochemical analyses for dopamine in the neostriatum.[8, 9, 10, 11, 12, 13] The most definitive of these studies have been those using the histochemical method of Falck and Hillarp.[14, 15] A large number of dopamine-containing terminals are present in the rat neostriatum and these are lost following destruction of the dopamine-containing cells in the substantia nigra.[8] Following ablation of the neostriatum the course of the nigrostriatal tract can be followed by observing the accumulation of dopamine in the proximal part of the severed axons.[9, 10] Despite the elegance of these studies, some questions still remain concerning the origin, course and distribution of the nigrostriatal pathway which could be answered by a direct demonstration of the pathway using anterograde degeneration techniques. This would also be of interest in that it would establish that a monoaminergic pathway can be shown by conventional anatomical methods.

In this study lesions were placed in the brains of 13 adult cats under deep pentobarbital anesthesia (40 mg per kg) by stereotaxic placement of an insulated Nichrome wire electrode, bare at the tip, and passage of anodal direct current of 3 mA for 20 to 60 seconds. Stereotaxic coordinates were obtained from the atlases of Bleier [16] and Berman.[17] In 10 animals lesions were placed at the junction of the lateral hypothalamus and internal capsule at the level of the caudal border of the optic chiasm. This site was chosen because in previous experiments [11] we found similar lesions to cause marked decreases in caudate nucleus dopamine and degeneration of the cells of the substantia nigra, pars compacta and the adjacent ventral tegmental area (Table 1). One lesion was made by passing an electrode through the contralateral hemisphere at an angle to reach the area by an approach differing from the others. This provided a control for anatomical effects of passing the electrode through the same hemisphere in the other brains. In three animals, lesions were placed in the pars compacta of the substantia nigra. After a five-day survival period the animals were killed by transcardiac perfusion with 10 percent formaldehyde solution while deeply anesthetized with pentobarbital. Brains were removed and stored in 10 percent formaldehyde for two to four months. Frozen sections were cut at 30 μ in the frontal plane and prepared by one or both of the methods of Fink and Heimer [18] for the selective demonstration of degenerating axon terminals.

The lesions of the substantia nigra were centered in the pars compacta and partially destroyed this nucleus (Fig. 1, *F*). Each extended somewhat dorsally into the ventral tegmentum lying between the red nucleus and the substantia nigra, ventrally into the cerebral peduncle and medially into the ventral tegmental area of Tsai.[17] The electrode tract entered through the caudal cerebral

TABLE 1. Effect of Unilateral Hypothalamic Lesions on the Dopamine Content of the Caudate Nucleus in the Cat *

Lesion Location	*Dopamine Content*		*Percent Difference*
	Lesion Side	*Control Side*	
Medial hypothalamus	9.17±.44	9.72±.35	−5.1%
Medial forebrain bundle	8.31±2.33	10.97±1.24	−26.9%
Internal capsule	3.29±.46	8.72±.91	−62.4%
Medial forebrain bundle and internal capsule	0.78±.19	10.11±.83	−92.3%

* From Moore and Heller.[11]

cortex and white matter, the superior colliculus, mesencephalic reticular formation and red nucleus. Degenerating axons incidental to passage of the electrode and the dorsal tegmental extension of the lesion can be traced rostrally into the pretectal area, thalamus, mamillary bodies and the lateral hypothalamus. From the lesion, degenerating axons run forward in the ventral tegmentum below the red nucleus and enter the diencephalon in the ventral part of Forel's field H, dorsomedial to the subthalamic nucleus (Fig. 1, *E*). As the subthalamic nucleus disappears rostrally, the fibers lie medial and dorsal to the internal capsule (Fig. 1, *D*). From this position they pass ventrally and laterally until at rostral tuberal levels of the hypothalamus they run within the medial part of the internal capsule and the adjacent lateral hypothalamus (Fig. 1, *C*). Part of the fibers turn laterally at this level and pass through the entopeduncular nucleus and globus pallidus to enter the putamen. Few degenerating terminals are evident in either the entopeduncular nucleus or the globus pallidus. Similarly, very few degenerating terminals appear in the caudal putamen or the tail of the caudate nucleus; this is true regardless of the length of the postoperative survival period. As the caudate nucleus enlarges rostrally, degenerating fibers of passage spread out along the internal capsule (Fig. 1, *B*). In the rostral putamen and in the head of the caudate nucleus (approximately the rostral two-thirds of each nucleus), the terminal degeneration is dense (Fig. 1, *A* and *B;* Fig. 2). The distribution of degenerating terminals is homogeneous in the putamen (Fig. 1, *A* and *B*) whereas in the caudate nucleus more terminals are evident laterally than medially (Fig. 1, *A*). Neither this distribution nor the rostral-caudal distribution of terminals can be correlated with lesion size or placement. With the substantia nigra lesions and the internal capsule lesions (see below), the same pattern of degenerating terminals is evident. No degenerating axons or terminals are evident on the side contralateral to the lesion, indicating that the pathway does not cross the midline.

These observations from brains with substantia nigra lesions provide further direct evidence for a nigroneostriatal pathway. They are supported by the findings from internal capsule lesions. Following destruction of the medial border of the internal capsule at the level of the rostral tuberal hypothalamus, the distribution of degenerating terminals is essentially the same as that noted with a nigral lesion. The terminal degeneration appears more dense with the large internal capsule lesions (Fig. 2) but this is probably related to the fact that these completely transect the tract whereas the substantia nigra lesions each spare some of the pars compacta. If the lesion is placed medially, however, and is restricted to the lateral hypothalamus, sparing the medial internal capsule, there are many less degenerating terminals in the neostriatum. If the lesion is placed in the midportion of the internal capsule, lateral to the medial border, there are few, if any, degenerating terminals in the neostriatum.

Such cases support the view that the pathway is compact and largely located in the medial internal capsule at this level. They also indicate that passage of the electrode through the hemisphere will not, of itself, produce large numbers of degenerating terminals in the neostriatum. This is confirmed by a case in

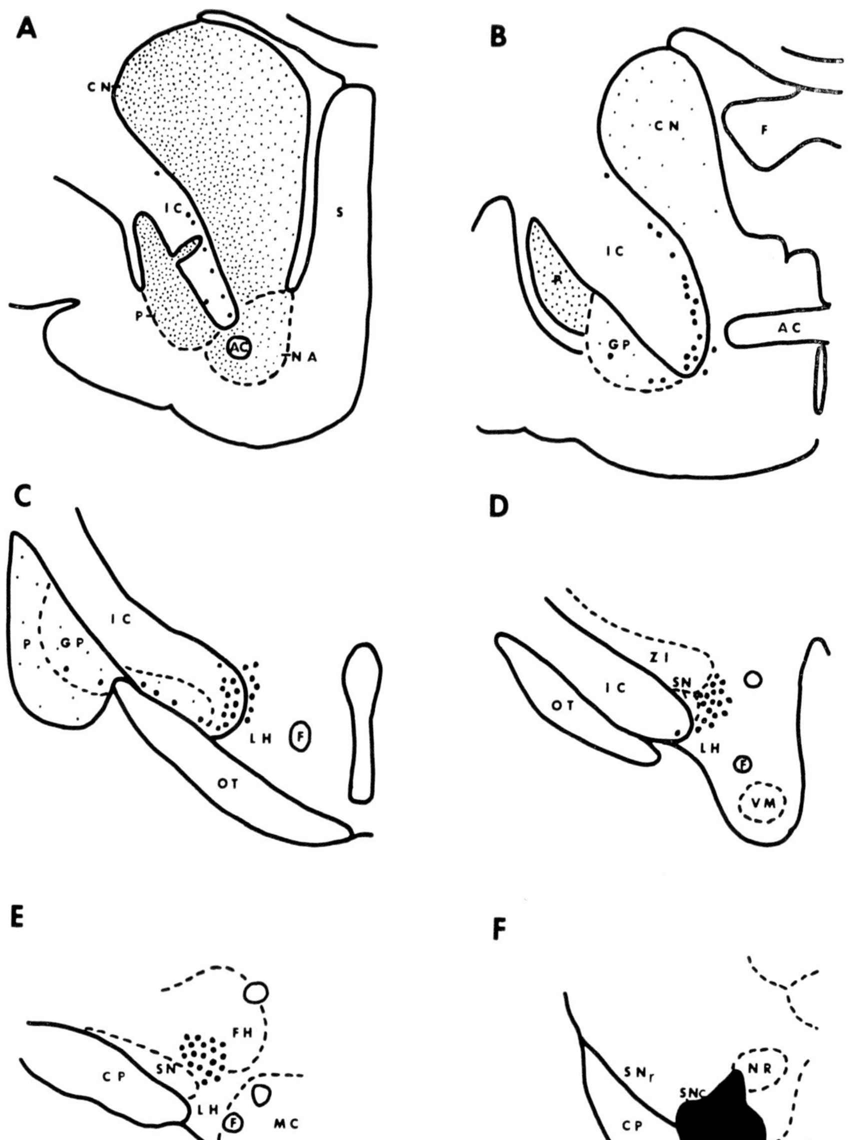

Fig. 1. Outline drawings of frontal sections from a brain with a substantia nigra lesion. The drawings are ordered from rostal to caudal (A through *F*). Large dots designate the location of degenerating fibers of passage. Small dots indicate the location of degenerating axon terminals. The area of the lesion in the substantia nigra and adjacent tegmentum is designated by the blacked-in region of drawing *F*. Abbreviations are: F, fornix; FH, field H of Forel; GP, globus pallidus; IC, internal capsule; LH, lateral hypothalamus; MC, mamillary complex; NA, nucleus accumbens; NR, red nucleus; OT, optic tract; P, putamen; SN, subthalamic nucleus; SNr, substantia nigra, pars reticulata; SNc, substantia nigra, pars compacta; ZI, zona incerta.

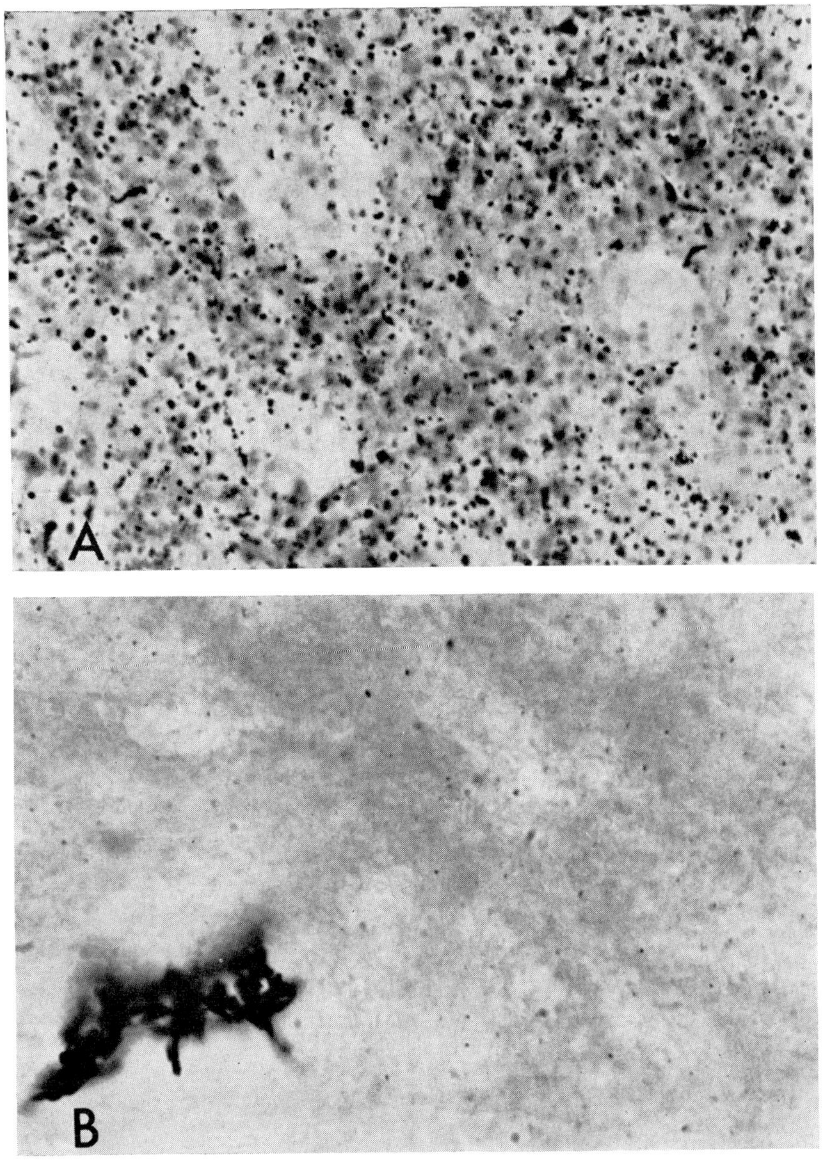

Fig. 2. Photomicrograph from an experimental brain. *A,* from the caudate nucleus ipsilateral to an internal capsule lesion showing numerous degenerating terminals. *B,* from the caudate nucleus contralateral to the lesion in *A* showing a few background silver grains and an impregnated blood vessel (arrow). Fink-Heimer, method 2. Magnified 960 times.

which the electrode was entered through the contralateral hemisphere to produce the lesion. As with the other medial internal capsule lesions, this resulted in degenerating terminals only in the neostriatum ipsilateral to the lesion.

The nigrostriatal projection, thus, appears to arise in the cells of the pars compacta of the substantia nigra and the adjacent ventral tegmentum and to traverse the ventral tegmental area, the subthalamus, the lateral hypothalamus and the internal capsule before distributing to the rostral putamen and the head of the caudate nucleus. With the data from the studies of Poirier and associates [12, 13] and those of Andén et al.,[8, 9, 10] the observations presented here form a firm morphological basis for current views on the pathophysiology of Parkinson's disease.

REFERENCES

1. Ehringer, H., and Hornykiewicz, O.: Verteilung von Noradrenalin und Dopamin (3-Hydroxytyramin) im Gehirn des Menschen und ihr Verhalten bei Erkrankungen des extrapyramidalen Systems. Klin. Wchsch. *38:* 1236, 1960.
2. Hornykiewicz, O.: Dopamin (3-Hydroxytyramin) im Zentralnervensystem und seine Bezeihung zum Parkinson-Syndrom des Menschen. Deusch. Med. Wchsch. *87:* 1807, 1962.
3. Ferraro, A.: The connections of the pars suboculomotoria of the substantia nigra. Arch. Neurol. Psychiat. *19:* 177, 1928.
4. Ranson, S. W., and Ranson, S. W., Jr.: Efferent fibers of the corpus striatum. Res. Publ. Assoc. Res. Nerv. Ment. Dis. *21:* 69, 1942.
5. Carpenter, M., and McMaster, R. E.: Lesions of the substantia nigra in the rhesus monkey. Efferent fiber degeneration and behavioral observations. Amer. J. Anat. *114:* 293, 1964.
6. Cole, M., Nauta, W. J. H., and Mehler, W. R.: The ascending efferent projections of the substantia nigra. Trans. Amer. Neur. Assoc. *89:* 74, 1964.
7. Adinolfi, A.: Observations on the fine structure of the feline caudate nucleus. Anat. Rec. *157:* 203, 1967.
8. Andén, N. E., Carlsson, A., Dahlstrom, A., Fuxe, K., Hillarp, N. A., and Larrson, K.: Demonstration and mapping out of nigrostriatal dopamine neurons. Life Sci. *3:* 523, 1964.
9. Andén, N. E., Dahlstrom, A., Fuxe, K., and Larrson, K.: Further evidence for the presence of nigroneostriatal dopamine neurons in the rat. Amer. J. Anat. *116:* 329, 1965.
10. Andén, N. E., Dahlstrom, A., Fuxe, K., Larsson, K., Olson, L., and Ungerstedt, U.: Ascending monoamine neurons to the telencephalon and diencephalon. Acta Physiol. Scand. *67:* 313, 1966.
11. Moore, R. Y., and Heller, A.: Subcortical systems participating in the control of caudate nucleus dopamine content. *In* Progress in Neurogenetics (A. Barbeau and J. R. Brunette, eds.). Excerpta Medica Foundation, Amsterdam, 1969.
12. Poirier, L. J., and Sourkes, T. L.: Influence of the substantia nigra on the catecholamine content of the striatum. Brain *88:* 181, 1965.
13. Poirier, L. J., Singh, P., Boucher, R., Bouvier, G., Olivier, A., and Larochelle, P.: Effect of brain lesions on striatal monoamines in the cat. Arch. Neurol. *17:* 601, 1967.

14. FALCK, B., HILLARP, N. A., THIEME, G., and TORP, A.: Fluorescence of catecholamines and related compounds condensed with formaldehyde. J. Histochem. Cytochem. *10:* 348, 1962.
15. FALCK, B.: Observations on the possibilities of the cellular localization of monoamines by a fluorescence method. Acta Physiol. Scand. *56*, Suppl. 197: 1962.
16. BLEIER, R.: The Hypothalamus of the Cat. Johns Hopkins Univ. Press, Baltimore, 1961.
17. BERMAN, A. L.: The Brain Stem of the Cat: A Cytoarchitectonic Atlas with Stereotaxic Coordinates. Univ. of Wisconsin Press, Madison, 1968.
18. FINK, R. P., and HEIMER, L.: Two methods for the selective silver impregnation of degenerating axons and their synaptic endings in the central nervous system. Brain Res. *4:* 369, 1967.

SOME OBSERVATIONS ON THE EFFECTS OF SHORT-CHAIN POLYPEPTIDES ON PERIPHERAL AND CENTRAL DOPAMINE

Aurèle Beaulnes

The elegant work of Poirier and his co-workers on dopaminergic pathways in the brain and, in fact, this symposium on the central effects of L-Dopa and the role of dopamine in extrapyramidal diseases serve to stress one point in my mind: our appalling lack of knowledge of the role and actions of peripheral dopamine. This will be discussed later in the symposium, but a few short comments may be in order here. One of the important side effects of L-Dopa may well be the approach it has provided toward understanding the role of dopamine in brain function. It is hoped that this will also create incentive for the study of the peripheral effects of dopamine.

During the last few decades, there has been perhaps too much emphasis on norepinephrine and epinephrine. In adrenergic nerves, a small percentage of the amine content consists of dopamine, and its role at this level is probably that of a precursor. However, the lungs, liver and gastrointestinal tract contain important stores of dopamine localized in non-nervous cells. In some species, the lungs contain 100 times more dopamine than norepinephrine. What is the role of dopamine in such organs? Is it involved in the local regulation of vascular tone? Can it act systemically? What are the mechanisms of release of dopamine from such stores? Are these stores depleted in Parkinson's disease?

Some of the vascular effects of exogenous dopamine can be blocked by adrenergic blocking agents, but its important vasodepressor component cannot be antagonized by β-blockers. In many respects, dopamine shows interesting similarities with the prostaglandins. We are presently investigating the role of these two vasoactive agents in the pressor and depressor effects of angiotensin and bradykinin. We have recently shown that angiotensin, administered in doses many million times smaller than the lowest doses administered up to now, can induce clear-cut increases or decreases in blood pressure in the rat, the cat, the rabbit and, to a smaller extent the dog. Bradykinin, the potent vasodilator polypeptide, administered in very minute doses, has also been shown to produce hypertension in the same species. These effects are apparently not mediated through adrenergic or noradrenergic mechanisms, since they are not prevented by adrenergic blockade. Ganglionic blockade potentiates such responses. We believe that these effects might be brought about by release of prostaglandin or dopamine.

My second point deals with the central effects of short-chain polypeptides. We have observed recently that, while angiotensin does not penetrate inside cells, it has a surprising affinity for some cell surfaces; a survey of the amounts of radioactive-labeled angiotensin taken up by the major organs and tissues shows a preferential uptake by the brain and the liver.

More relevant to the topic of this symposium is work initiated in our laboratories on the effects of angiotensin and other short-chain polypeptides on brain acetylcholine. We have already shown that a species-specific increase in total brain acetylcholine can be produced by intravenous injections of angiotensin and bradykinin. More recently, we have become interested in the effects of these polypeptides on brain amines, and we have shown that angiotensin can alter the concentration of amines in various parts of the cat brain. In the caudate nucleus, however, no clear-cut change in dopamine content occurred following the administration of large (10 μg per kg) doses of angiotensin and bradykinin. Our control values of 12.5 μg per g of wet tissue, measured with Sourkes and Murphy's method, are reduced to 11.6 with angiotensin and 10.5 with bradykinin, but the differences are not statistically significant.

I will end by asking whether dopamine-β-hydroxylase inhibitors have been considered as a means to increase brain dopamine and enhance tremors. I am thinking about disulfiram, which is Antabuse, the drug used in alcoholism. Both disulfiram and its active catabolite diethyldithiocarbamate have been shown markedly to increase the dopamine content of various organs, including the brain, as demonstrated by Goldstein and Nakajima. I am well aware of course that the major pathway for the disposal of dopamine in brain is not through hydroxylation but through oxidative deamination.

CHAPTER 6

Physiological Aspects of Abnormal Movements Induced by L-Dopa

PHYSIOLOGICAL IMPLICATIONS IN THE DYSKINESIAS

A. A. Ward

The Motor System

Before it is possible meaningfully to discuss drug actions on abnormalities of movement, it is useful to have a general model of how the central nervous system controls motor activity. It is increasingly apparent that motor control is dependent on central processing of sensory information, i.e., on feedback loops. In addition to simple loops at the segmental levels in the spinal cord, it is clear that motor control involves complex hierarchies of feedback loops superimposed one on the other which become increasingly complex at the levels of the central nervous system ascending to the brain stem, thalamus, basal ganglia and cortex.

Since we know most about the detailed synaptic function at the segmental level in the spinal cord, it is logical to discuss these feedback loops before considering the suprasegmental influences on such loops or considering what little is known regarding some of the interacting feedback loops located at higher levels of the central nervous system.

In the ventral horn of the cord are large neurons known as *alpha* motoneurons which innervate the peripheral striated muscles. These cells are the final common path for driving the power plant. Although higher mammals appear to be able to activate this final common path directly in certain instances, direct activation of the alpha system is used only for producing gross and often powerful motor activity which usually involves proximal muscle groups. This system cannot generate the fine control required in dextrous motor performance. As is true in engineering, fine control of power plants is best achieved by servomechanism; this is accomplished in mammals by the *gamma* system. Thus in the ventral horn are small motor neurons whose gamma efferent fibers innervate the polar contractile elements of the muscle spindles. Activation of these fibers causes contraction of the polar contractile elements of the muscle spindle, stretching the annulospiral ending between them. This sensory transducer responds to stretch but cannot distinguish whether it is passively stretched by pull of the entire muscle mass or whether it is stretched by contraction of its polar contractile elements in the absence of gross muscle contraction. When the spindle is stretched, the annulospiral ending generates a train of discharges which are conducted back to the cord over large, rapidly conducting fibers which project monosynaptically on the alpha motoneuron. Thus stretch of the spindle generates signals which excite the anterior horn cells of that particular muscle causing that muscle to contract. This is the myotatic reflex, of which the knee jerk is the classical example.

That the gamma system plays an important role in muscle control is suggested by the fact that one-third of the efferent ventral root fibers are in the gamma category. The role of the muscle spindle in motor activity represents a fairly recent addition to neurophysiology [1] and even this relatively simple feedback loop is still only partially understood.[2, 3] The muscle spindle may be considered to be an exquisite sensing element of a proprioceptive servosystem. It registers the difference in length between the main muscle mass and itself, whereupon the servo loop acts to reduce that difference.[4] Hunt [5] and Eldred et al.[6] showed that spinal and supraspinal reflexes are heralded by an initial heightened gamma efferent discharge which in turn generates, via the loop, the alpha motoneuron discharge and motor contraction. Thus it may be that, when fine motor control is required, we give our orders for "voluntary" motor activity to the gamma neurons and they in turn control the power plant. For that reason, those muscles most involved in dextrous motor activity (such as the intrinsic muscles of the hand) are also those most richly endowed with muscle spindles. Conversely, those muscles over which we have little voluntary control (such as the latissimus dorsi) are sparsely supplied with muscle spindles. This would be consistent with the observation of Fidone and Preston [7] indicating that corticomotor coupling from cortex to fusimotor neurons is more effective than to *alpha* motoneurons.

As a generalization, it appears that the suprasegmental control is characterized by a linked coordinated activity between alpha and gamma motor systems. Motor regulation is achieved by mechanisms of excitation and inhibition for each of the two systems. These functions of brain stem, cerebellum, basal ganglia and cortex are projected onto segmental mechanisms in the cord not only over the pyramidal tract but also over nonpyramidal, multisynaptic pathways.

Parkinsonism

As a preliminary model, we proposed in 1963 [8] that the pathological alteration of motor control in parkinsonism is the result of an altered alpha-gamma balance characterized by depressed gamma activity and augmented alpha activity.

It is proposed that the lesion in the ventromedial tegmentum that induces a parkinsonian syndrome in monkeys [9, 10] partially interrupts the excitatory circuit leaving the inhibitory input to the *gamma* neuron intact. Thus, the gamma neuron is effectively inhibited. The relevant lesion also is postulated to "release" the suprasegmental control of the *alpha* system with resultant augmented alpha activity.[8] In addition to production of a Parkinson-like state in monkeys by tegmental lesions, a clinical syndrome can be induced in experimental animals by the intravenous injection of reserpine [11] which is perhaps the best drug-induced model of the disease. As Steg [11] has shown, this drug-induced animal model is also characterized by inactivity of the gamma efferents while the slow motor units are tonically hyperactive. Both of these models not only simulate the clinical signs of parkinsonism but also exhibit the biochemical correlates of the human disease as expressed by reduced dopamine in the striatum.[12, 13] Finally, chlorpromazine, in appropriate doses, inhibits the gamma system [14] and induces augmented alpha drive.[15] That extrapyramidal syndromes resembling parkinsonism can be produced in some patients receiving large doses of the phenothiazine drugs is well documented.

If the physiological substrate of the dyskinesia in the human consists of an altered alpha-gamma balance characterized by depressed gamma activity and augmented alpha activity, it should be possible to relate the symptoms of bradykinesia, tremor and rigidity to this substrate.[16]

There are pathways from the brain which excite the gamma efferent neuron and others which inhibit it. The model proposes that the relevant lesion partially interrupts the excitatory circuit leaving the inhibitory input to the gamma neuron intact. Thus, the gamma neuron in the cord has an inhibitory bias. Some confirmation for this proposal is provided by the observation of Jung and Hassler [17] that the Jendrassik reinforcement of deep tendon reflexes is absent in patients with parkinsonism, and that in cases of hemiparkinsonism it is absent only on the affected side. If the Jendrassik reinforcement maneuver is due to excitation of the gamma system [18] this would indicate that the ability to excite the gamma neurons is compromised in patients with parkinsonism. This inhibitory bias of the gamma system might account for the bradykinesia exhibited by parkinsonian patients. If the patient has lost a major part of the circuits which he normally utilizes in fine, dextrous movement via activation of his gamma neurons, he now has to drive the remaining circuits unusually forcefully to achieve movement at all. Patients with this dyskinesia state that enormous effort is often required to induce fine-finger movement repetitively and that the circuits tire rapidly. Since the gamma loop is maximally involved in fine movements and to a minimal degree in gross movements involving the proximal musculature, one would predict that fine movements would be compromised to a much greater degree. Such is the case in the dyskinesias.

In addition to depression of the gamma system, the model also proposes that suprasegmental excitation of the alpha motoneuron is augmented. Rhythmic discharges of single neurons in the thalamus have been recorded in the monkey model of parkinsonism [19] as well as in the human with this disease.[20] In both, there is evidence that such rhythmic discharges in some thalamic neurons are related to the generation of the tremor. It has been proposed that the rate of the tremor is determined by the properties of an oscillating feedback loop between the nucleus ventralis lateralis and the motor cortex.[21] Since dorsal rhizotomy does not block the tremor in human patients [22] or in the monkey model,[23] the data would appear to indicate that the tremor in parkinsonism is a consequence of direct pulsed activation of the alpha motoneuron.

Finally, the model proposes that the rigidity is also a consequence of increased suprasegmental facilitation of motoneurons. Since all muscle spindle activity is not abolished but only depressed, it would be expected that the residual spindle input, playing on alpha motoneurons whose activity is increased, should be a factor in the rigidity which is observed. Thus, while the alpha rigidity of the ischemic decerebrate preparation is not modified by dorsal rhizotomy, Pollock and Davis [22] have shown that dorsal rhizotomy in a patient with parkinsonism resulted in modifications of muscle tone in which the initial decrease in tone after operation was replaced by a "contracture" in flexor muscles at elbow and wrist which, interestingly, disappeared during sleep. Although desirable documentation is not available on patients, the data presented by Steg [11] clearly showed that the rigidity in his experimental model was related to tonic hyperactivity of slow motor units.

Thus there is modest support for the hypothesis that the pathological alteration of motor control in parkinsonism is a result of an altered alpha-gamma balance characterized by depressed gamma activity (accounting for the bradykinesia) and augmented alpha activity (tonic hyperactivity of motoneurons accounting for the rigidity, phasic alpha drive accounting for the tremor).

Physiological Basis of Therapy

On the basis of this model, one would predict that a part of the imbalance in feedback control would be rectified by restoring the circuits which activate the gamma neurons. Obviously, pathways which have been destroyed by disease cannot be replaced. Although it is not possible to increase the activation of the gamma system, it is possible to reduce the inhibitory input on these cells, thereby permitting the remaining facilitatory signals to activate these cells more effectively. One of the major sources of inhibitory input to the gamma system appears to arise in the VL of the thalamus.

The investigation of the supraspinal control of the gamma system was initiated by Granit and Kaada.[24] They obtained gamma facilitation from stimulation of pontile and mesencephalic tegmentum, cerebellar cortex, dorsal hypothalamus, midline nuclei of thalamus, motor cortex and pyramidal tract. Gamma inhibition was observed after stimulation of culmen, certain sites in brain stem reticular formation, orbital and motor cortex. Some years ago we investigated the influence of the nucleus ventralis lateralis on this system.[25]

In the cat, one can monitor the activity of a single muscle spindle in an extensor muscle by electrically recording its discharge from a single dorsal root fiber. Under light static stretch, a stable rate of firing of a single muscle spindle can be recorded. The characteristic effect of stimulation of VL upon the contralateral muscle spindle discharge was profound inhibition.[25] A five second, 60 cps stimulation of VL could abolish the spindle discharge for as long as 50 seconds. The integrity of the sensorimotor cortex was necessary for this inhibitory effect which would be consistent with the known powerful projections from VL to cortex. These observations on the suprasegmental control of the gamma system have been carried out in animals under barbiturate anesthesia. It is clear that the levels of anesthesia can profoundly alter many supraspinal mechanisms of motor control.[15, 26]

To obviate this variable, studies have been undertaken in unanesthetized man where the myotatic reflex has been used as an index of gamma activity. This assumption is valid providing there is no alteration of excitability of the alpha motoneuron. Electrical stimulation of VL in unanesthetized patients with a variety of dyskinesias evokes inhibition or complete suppression of the contralateral knee jerk.[8] Thus it would appear that the more direct data reported in the cat are not artifacts of anesthesia. Since one of the major sources of inhibitory input to the gamma system thus appears to arise in VL, this may account for the improvement of bradykinesia that follows surgical lesions directed at the anterosuperior part of this thalamic target.

It is presumed that therapeutically effective drugs act in the same fashion. As Arvidsson, Roos and Steg [27] have pointed out, administration of drugs inhibiting cholinergic transmission or drugs facilitating monoamine transmission such as L-Dopa will increase gamma neuron activation evoked by dorsal root stimulation. They proposed that this augmentation of gamma activity is a consequence of drug action at the level of the neostriatum. Unfortunately, there is no definitive data regarding the site of drug action. A recent study does indicate that biogenic amines can also act segmentally at the spinal level. Wooley [28] noted that L-Dopa increased the monosynaptic reflex and decreased presynaptic inhibition, and he postulated that dopamine may be the presynaptic inhibitory transmitter in the cord. It is also known that these compounds can act on neurons in the neostriatum, since direct chemical stimulation of dopaminergic circuits in the neostriatum of the rat can induce postural responses.[29] However direct evidence of the effects of activation of striatal neurons upon the gamma system are currently lacking. Thus, although the site of action is not yet clear, the fact that L-Dopa increases gamma neuron activation would be consistent with the model that this is accomplished by a decrease of inhibitory input.

Granit, Holmgren and Merton [30] proposed the general concept that there are two routes of muscle innervation from motor cortex which they designate the alpha route and the gamma route. The classical suprasegmental pathway of the alpha route is the pyramidal tract projecting monosynaptically or, more commonly, through an interneuron on anterior horn cells. Although an anatomically discrete pathway, it is now apparent that the pyramidal tract, strictly defined, represents a relatively small part of the total suprasegmental control of alpha motoneuron activity. Stern and Ward [15] have shown that stimulation of the contralateral or homolateral globus pallidus produces facilitation of the alpha motoneuron as monitored by the monosynaptic reflex evoked by dorsal root stimulation in the non-narcotized, curarized cat. This facilitation, or alpha drive, is not blocked by ablations of the sensorimotor cortex and may be mediated by direct projections from pallidum to the mesencephalic tegmentum.[31, 32] Furthermore, stimulation of the contralateral VL also facilitates the alpha system, but this is dependent upon the integrity of the sensorimotor cortex. The alpha drive evoked by VL or pallidal stimulation is abolished or reversed by barbiturates.[15] Utilizing intracellular recording in alpha motoneurons, Sasaki and Tanaka [33] have confirmed that activation of VL evokes excitatory postsynaptic potentials in motoneurons as does excitation of the pyramidal decussation, the reticular formation of pons and medulla, and the dentate nucleus. Some of these types of alpha drive are rapidly conducted to motoneurons while others are conducted more slowly.

If a part of the disorder of parkinsonism is a consequence of augmented alpha activity, it is not surprising that surgical lesions should be beneficial if directed at a target such as VL, which is a principal relay station in the generation of alpha drive. The clinical improvement which may follow the administration of L-Dopa would also be consistent with this model since Arvidsson, Roos and Steg [27] have shown that this monoaminergic drug decreases alpha motoneuron activation.

It is not unreasonable that circuits should be concentrated in the lateral thalamus which excite the alpha system and inhibit the gamma system, or that a drug such as L-Dopa should have similar reciprocal effects. Calma and Kidd [34] evolved the concept of linked coordinated activity between alpha and gamma motor systems. They proposed that motor regulation is achieved by a mechanism of excitation and inhibition for each of the two systems. As in all servosystems, the requirements for accurate control are best met by servo loops containing negative feedback, in which the signals inducing muscular movement (alpha system) also retard the gamma system which is a positive feedback loop at segmental levels of the cord. Thus, at one extreme, negative feedback is provided by recurrent collateral inhibition involving Renshaw cells in the local circuit at segmental cord levels, while similar negative feedback is also a property of the suprasegmental control circuits within the brain. Thus, in the central nervous system as in the steam engine, activation of the mechanism tends to shut off the power which drives it. Alterations of the balance between positive and negative feedback can lead to pathological activity in physical servosystems, and it is not unreasonable that similar events should occur in biological control systems. In both man and machine, therapy is thus directed at the restoration of stability in such servo loops.

It is more difficult to form hypotheses to explain why the administration of L-Dopa can induce involuntary motor activity in patients. Such choreiform movements seem to be induced by the higher doses of L-Dopa. In our very limited experience, we infrequently see this complication of L-Dopa therapy in patients who have undergone bilateral thalamotomies. In patients who have had a unilateral thalamotomy, the drug-induced dyskinesia has always been most prominent on the side of the body homolateral to the lesion. I would propose the speculative hypothesis that the mechanisms operating in the improvement which follows the administration of L-Dopa involve a site of action other than the striatum, as is the case with surgical lesions of the lateral thalamus.

Ungerstedt et al.[29] have shown that localized, unilateral intrastriatal injection of dopamine in rats induces an asymmetric posture with turning to the opposite side, and that this response can be blocked by chlorpromazine (which is assumed to block dopaminergic receptor sites). If one makes the assumption that such movements are mediated by direct alpha drive and that these movements are related to those occasionally seen in parkinsonian patients receiving large doses of L-Dopa, one could perhaps ascribe the Dopa-induced dyskinesia to activation of these dopaminergic circuits in the striatum. On the other hand, systemically administered L-Dopa is known to decrease alpha motoneuron activation [27] and this would satisfactorily account for the improvement of function in parkinsonian patients based on the alterations of alpha-gamma balance which have been proposed as the physiological substrate for this disease. One can only reconcile these conflicting drug actions by assuming different loci of action with differing thresholds. It may be that there are dopaminergic circuits at lower levels in the central nervous system which have a lower threshold to dopamine than those in the striatum. Activation of these unspecified circuits would then account for the improvement in function noted following L-Dopa therapy, rather than activation of striatal circuits. This might also explain, to a degree, the somewhat anomalous finding that the dopamine deficiency in the striatum of monkeys with tegmental lesions is not influenced by lesions of lateral thalamus which result in abolition of tremor.[35]

Thus it may be that the effectiveness of L-Dopa therapy in parkinsonism is not related to the dopamine deficiency in the striatum in these patients. If that is the case (which remains to be proven), this would be another example to add to the history of medicine where the right therapy is initially undertaken for the wrong reasons!

REFERENCES

1. LEKSELL, L.: The action potential and excitatory effects of the small ventral root fibers to skeletal muscle. Acta Physiol. Scand. Suppl. 31, *10:* 1, 1945.
2. BARKER, D.: Symposium on Muscle Receptors. Hong Kong Univ. Press, Hong Kong, 1962.
3. MATTHEWS, P. B. C.: Muscle spindles and their motor control. Physiol. Rev. *44:* 219, 1964.
4. MERTON, P. A:. Speculations on the servo control of movement. *In* The Spinal Cord, a Ciba Foundation Symposium (J. L. Malcolm et al., eds.). Little, Brown and Co., Boston, 1953.
5. HUNT, C. C.: Reflex activity of mammalian small-nerve fibers. J. Physiol. *115:* 456, 1951.
6. ELDRED, E., GRANIT, R., and MERTON, P. A.: Supraspinal control of the muscle spindles and its significance. J. Physiol. *122:* 498, 1953.
7. FIDONE, S. J., and PRESTON, J. B.: Patterns of motor cortex control of flexor and extensor cat fusimotor neurons. J. Neurophysiol. *32:* 103, 1969.
8. STERN, J., and WARD, A. A., JR.: The relationship of the alpha and gamma motor systems to the efficacy of the surgical therapy of parkinsonism. J. Neurosurg. *20:* 185, 1963.
9. WARD, A. A., JR., McCULLOCH, W. S., and MAGOUN, H. W.: Production of an alternating tremor at rest in monkeys. J. Neurophysiol. *11:* 317, 1948.
10. POIRIER, L. J.: Neuroanatomical study of an experimental postural tremor in monkeys. J. Neurosurg. *24:* 191, 1966.
11. STEG, G.: Efferent muscle innervation and rigidity. Acta Physiol. Scand. Suppl. 225, *61:* 1, 1964.
12. SOURKES, T. L.: Effect of brain stem lesions on the concentration of catecholamines in basal ganglia of the monkey. J. Neurosurg. *24:* 194, 1966.
13. BARBEAU, A.: Some biochemical disorders in Parkinson's disease—a review. J. Neurosurg. *24:* 162, 1966.
14. HENATSCH, H. D., and INGVAR, D.: Chlorpromazin und Spastizität. Eine experimentelle elektrophysiologische Untersuchung. Arch. Psychit. Nervenkr. *195:* 77, 1956.
15. STERN, J., and WARD, A. A., JR.: Supraspinal and drug modulation of the alpha motor system. Arch. Neurol. *6:* 404, 1962.
16. WARD, A. A., JR.: The function of the basal ganglia. *In* Handbook of Clinical Neurology (Vinken, P. J. and Bruyn, G. W., eds.). North-Holland Publ. Co., Amsterdam, 1968.
17. JUNG, R., and HASSLER, R.: The extrapyramidal motor system. Handbook Physiol. *2:* 863, 1960.
18. PAILLARD, J.: Réflexes et régulations d'origine proprioceptive chez l'homme. Etude neurophysiologique et psychophysiologique. Librairie Arnette, Paris, 1955.
19. CORDEAU, J. P.: Microelectrode studies in monkeys with postural tremor. Rev. Canad. Biol. *20:* 147, 1961.
20. GUIOT, G., HARDY, J., and ALBE-FESSARD, D.: Délimitation précise des structures sous-corticales et identification de noyaux thalamiques chez l'homme par l'électrophysiologie stéréotaxique. Neurochirurgia *5:* 1, 1962.
21. LEBLANC, F. E., and CORDEAU, J. P.: Modulation of pyramidal tract cell activity by ventrolateral thalamic regions. Its possible role in tremorogenic mechanisms. Brain Res. *14:* 255, 1969.
22. POLLOCK, L. J., and DAVIS, L.: Muscle tone in parkinsonian states. Arch. Neurol. Psychiat. *23:* 303, 1930.

23. POIRIER, L. J., OHYE, C., and BOUCHARD, R.: Effect of dorsal rhizotomy on postural tremor in the monkey. Physiologist, *12:* 329, 1969.
24. GRANIT, R., and KAADA, B.: Influence of stimulation of central nervous structures on muscle spindles in cat. Acta Physiol. Scand. *27:* 130, 1952.
25. STERN, J., and WARD, A. A., JR.: Inhibition of the muscle spindle discharge by ventrolateral thalamic stimulation. Arch. Neurol. *3:* 193, 1960.
26. SEGUNDO, J. P., MIGLIARD, E. F., and ROIG, J. A.: Effect of striatal and claustral stimulation upon spinal reflex and strychnine activity. J. Neurophysiol. *21:* 391, 1958.
27. ARVIDSSON, J., ROOS, B. E., and STEG, G.: Reciprocal effects on α- and γ-motoneurons of drugs influencing monoaminergic and cholinergic transmission. Acta Physiol. Scand. *67:* 398, 1966.
28. WOOLEY, D. F.: A possible role for biogenic amines in presynaptic inhibition. Physiologist *12:* 399, 1969.
29. UNGERSTEDT, U., BUTCHER, L. L., BUTCHER, S. G., ANDÉN, N., and FUXE, K.: Direct chemical stimulation of dopaminergic mechanisms in the neostriatum of the rat. Brain Res. *14:* 461, 1969.
30. GRANIT, R., HOLMGREN, B., and MERTON, P. A.: Two routes for excitation of muscle and their subservience to the cerebellum. J. Physiol. *130:* 213, 1955.
31. JOHNSON, T. N., and CLEMENTE, C. D.: An experimental study of the fiber connections between the putamen, globus pallidus, ventral thalamus, and midbrain tegmentum in cat. J. Comp. Neurol. *112:* 83, 1959.
32. NAUTA, W. J. H., and MEHLER, W. R.: Projections of the lentiform nucleus in the monkey. Brain Res. *1:* 1, 1966.
33. SASAKI, K., and TANAKA, T.: Phasic and tonic innervation of spinal alpha motoneurones from upper brain centers. Jap. J. Physiol. *14:* 56, 1964.
34. CALMA, I., and KIDD, G. L.: The action of the anterior lobe of the cerebellum on alpha motoneurons. J. Physiol. *149:* 626, 1959.
35. BATTISTA, A. F., GOLDSTEIN, M., NAKATANI, S., and ANAGNOSTE, B.: The effects of ventrolateral thalamic lesions on tremor and the biosynthesis of dopamine in monkeys with lesions in the ventromedial tegmentum. J. Neurosurg. *31:* 164, 1969.

SIGNIFICANCE AND ETIOLOGY OF INDUCED ABNORMAL MOVEMENTS: PHYSIOLOGICAL IMPLICATIONS

Ronald R. Tasker

This paper is divided into three parts. First, an attempt is made to relate the experience in our laboratory with drug-induced tremor to the mechanism of the 4 to 6 cps rest tremor of parkinsonism. Second, the action of L-Dopa is considered in relation to tremorogenesis. Finally, comments are made pertinent to the problem of L-Dopa dyskinesias.

Our interests in the mechanisms of drug-induced tremor began with the study of tremorine in the rat. The fine 10 to 20 cps tremor induced by this

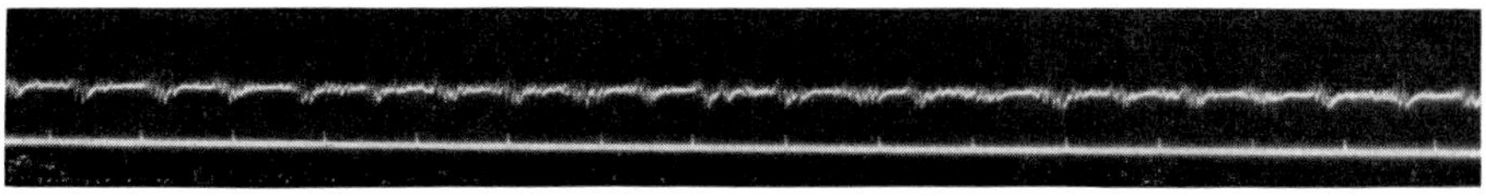

Fig. 1. Recording of action potentials in a portion of rat ventral root 15 minutes after 25 mg/kg harmaline i.p. Time scale division 100 msec.

drug occurs both at rest and during activity and involves the whole body. We showed that this tremor was prevented neither by deafferentation nor by supracollicular neuraxis transection, but that it was completely prevented by more caudal neuraxis transection, by cord transection below the level, and by bilateral midbrain tegmental lesions. Neuraxis transection was ineffective.[1]

We then went on to show that the quality, timing, and physiological characteristics of the tremor were identical to those of the synchronous bursts of neuronal activity induced by tremorine in the rat's ventral root fibers,[2] and in the midbrain reticular formation as well.[3]

Our attention then turned to harmaline, a drug which produced in animals, including monkeys, a type of tremor almost identical to that of tremorine. To our surprise, the findings described above for tremorine also held for this substance (Figs. 1 and 2).

This suggested that the site of tremorogenic action for both harmaline and tremorine was the midbrain, where synchronous activity was induced in the reticular neurons and transmitted bilaterally to motoneurons presumably via the reticulospinal system.

At this stage we were struck by an observation of Poirier's group [4] confirmed by others.[5] These workers attempted to produce a 4 to 6 cps rest tremor on the side of a monkey's body contralateral to certain tegmental lesions. In some animals in which the expected tremor was not seen injection of harmaline caused the 4 to 6 cps tremor to appear. On the side ipsilateral to the lesion only the usual rapid harmaline tremor was seen.

This synergistic action of drug plus lesion suggests at least a dual mechanism for the tremor, which closely resembles that of parkinsonian patients. That is, the lesion appears to modify the harmaline effect, converting it from a fine rapid tremor to a 4 to 6 cps parkinsonian-like activity.

It is tempting to consider that the tegmental lesion damages a diencephalic circuit, which possibly includes the ventral lateral nucleus of thalamus, which in turn alters the harmaline's reticulospinal activation. Further speculation

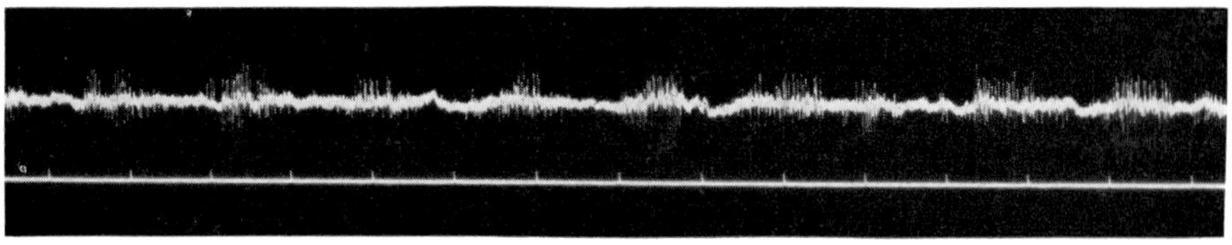

Fig. 2. Unit discharges in rat mesencephalic tegmental reticular formation 15 minutes after 25 mg/kg harmaline i.p. Time scale as above.

awaits experimentation, since neither the mechanism of action of harmaline nor the physiological basis of parkinsonian tremor is yet known.

This brings us to the attempt to relate the action of L-Dopa to involuntary movement. In the Parkinson's Disease Clinic at Toronto General Hospital we have been following several hundred patients since the 1950s in an attempt to assess the long-range effects of surgery and now of L-Dopa on the natural history of the disease.

Figure 3 documents our records on a typical patient. This 66-year-old man first noted right-hand tremor in 1961. Under our observations he maintained a fairly constant degree of tremor and rigidity over a period of seven years despite conventional therapy. During this time, however, he deteriorated slowly until he was forced to retire from business and required round-the-clock attendance. L-Dopa therapy was instituted in April of 1969. There was dramatic improvement in locomotion and the patient again became quite independent. Tremor and rigidity, however, showed only modest improvement. Figure 4 is the record of a second patient showing similar features. The striking well-maintained reduction in tremor and rigidity following thalamotomy is clearly shown. The point is that, despite the postoperative improvement in tremor and rigidity, the patient deteriorated steadily while after L-Dopa therapy he improved dramatically without significant change in tremor or rigidity.

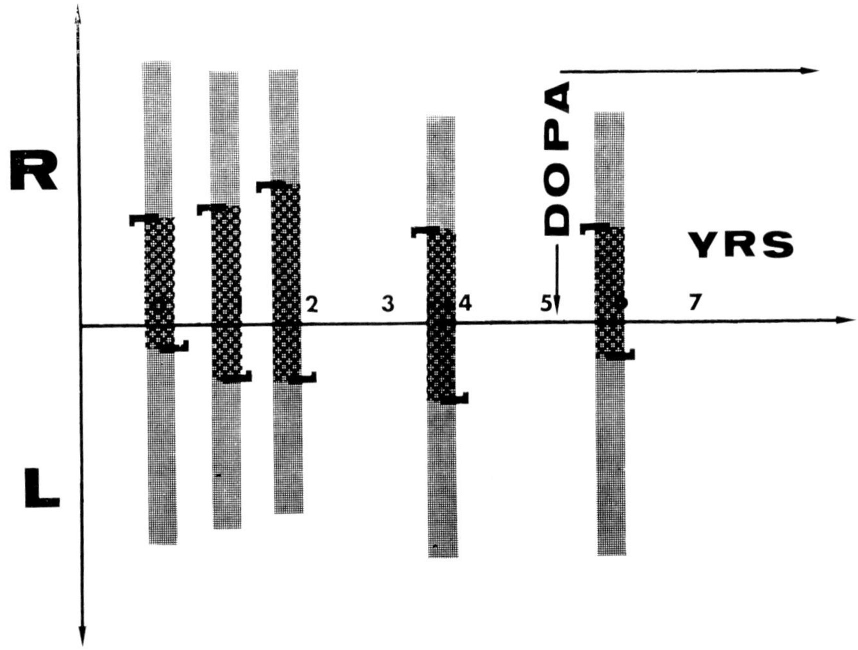

Fig. 3. 66-year-old man followed for seven years. Light hatched bars indicate semiquantitative degree of rigidity, heavy hatched bars with hooks, that of tremor. Right side of body indicated above, left side below the horizontal line. Institution of Dopa therapy gradually building up to the maximum tolerable dose of 5.0 gm, later 3.5 gm, daily indicated by arrow.

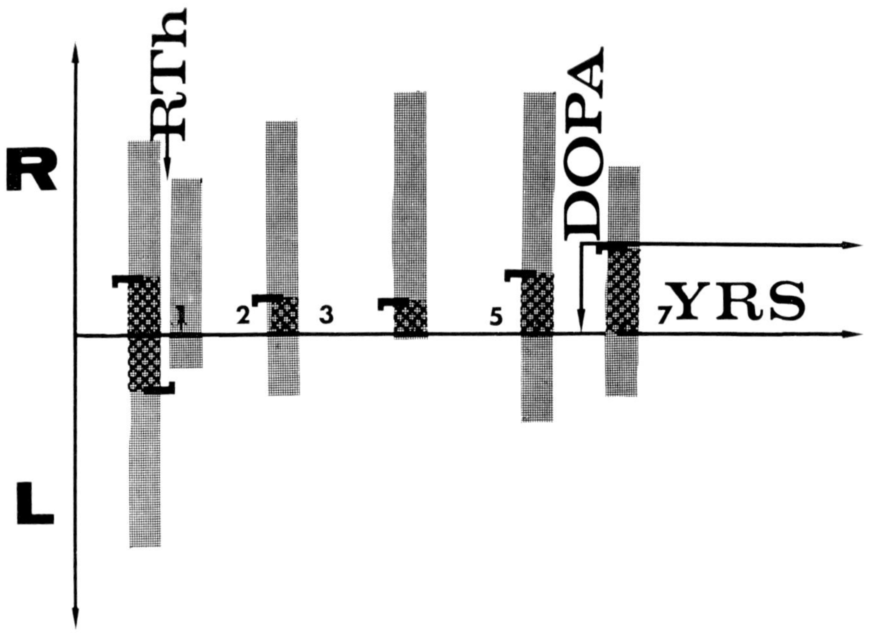

Fig. 4. 54-year-old man. Details as in Fig. 3. Stereotactic lesion placed in right VL thalamus as indicated by arrow. Dopa therapy institution marked by arrow building up to 6.5 gm daily.

Our observations are quite clear-cut. L-Dopa dramatically improves the general capabilities and locomotion of the patient by presumably reversing akinesia, while exerting at best a modest improvement in tremor or rigidity. The reverse is true for surgery. It would appear then that L-Dopa and dopamine are not directly involved in the mechanism of tremor.

Finally, a few comments on Dopa-induced involuntary movement. Table 1 lists some of our patients treated to date with L-Dopa. They are divided into three groups: those with unilateral, those with bilateral, and those with no previous surgery. The data suggest that a previous operation for Parkinson's disease protects the patient from Dopa hyperkinesia. Thus, since this is the

TABLE I.

Previous Surgery	*Incidence of Dopa Hyperkinesis*
Unilateral 12 patients	83% All 12 patients * 50% Ipsilateral to previous surgical lesion * 8% Contralateral to previous surgical lesion
Bilateral 3 patients	0%
None 12 patients	92%

* Only hyperkinesis in limbs considered.

same lesion which counteracts tremor and rigidity, it suggests that part of the tremorogenic mechanism may be common to that of Dopa hyperkinesis.

In summary, then, we have presented evidence for at least a dual tremorogenic mechanism, one part of which may involve the midbrain reticular formation. It would appear that Dopa is not directly involved in this mechanism of tremorogenesis. It is suggested that Dopa dyskinesia is dependent on the integrity of the same thalamic circuit as is tremor and rigidity.

REFERENCES

1. TASKER, R. R., and KERTESZ, A.: The physiology of tremorine-induced tremor. J. Neurosurg. *22:* 449, 1965.
2. FORCIER, P. and TASKER, R. R.: Participation of the ventral root in drug-induced tremor (tremorine). J. Neurosurg. *30:* 30, 1969.
3. PETER, J., and TASKER, R. R.: In preparation.
4. POIRIER, L. J., SOURKES, T. L., BOUVIER, G., BOUCHER, R., and CARABIN, S.: Striatal amines, experimental tremor and the effect of harmaline in the monkey. Brain *89:* 37, 1966.
5. BATTISTA, A. F., GOLDSTEIN, M., NAKATANI, S., ANAGNOSTE, B.: Drug induced changes of abnormal movements in monkeys with central nervous system lesions. Confin. Neurol. *31:* 135, 1969.

THALAMIC UNIT ACTIVITY IN MONKEY WITH EXPERIMENTAL TREMOR *

Yves Lamarre and Alfred J. Joffroy

A postural tremor can be produced in macaque monkeys by placing lesions in the ventromedial portion of the midbrain.[1] The mean frequency of this tremor is 4 to 6 cps and its clinical and electromyographic (EMG) characteristics are very similar to those of parkinsonian tremor. Previous investigations carried out in our laboratory have produced arguments in favor of a supraspinal pacemaker responsible for this type of tremor.[2] One of these arguments was the recording, in the curarized monkey, of rhythmic unit activity at the tremor frequency in the sensorimotor cortex and in the corticospinal tract. The animals being paralyzed, the possible involvement of a rhythmic afferent input of peripheral origin could thus be eliminated. The purpose of our present work is further to investigate the involvement, in this experimental dyskinesia, of the various cortical and subcortical structures concerned with motor function. This report presents some results on microelectrode recordings in the lateral ventral portion of the thalamus in the unanesthetized monkey, whether curarized or not.

* Supported by the Medical Research Council of Canada Grant MA-2863.

Some data have been obtained after intravenous administration of harmaline. Poirier and his collaborators [3] have shown that this substance is able to enhance or provoke the tremor in the monkey with midbrain lesions, and has no such effect in the normal monkey.

Methods

The recording technique used is modified from one described by Mountcastle and collaborators.[4] Briefly, the animals were prepared in the following way: At least two days before the first recording session, and under general anesthesia, a plastic chamber is stereotaxically attached to the skull, at the margins of a trephine hole above the subcortical structure to be explored. On the day of recording, the animal is placed in a special restraining chair. The chamber is attached to a metal bridge supporting the microdrive. The animal is thus held exclusively by the plastic chamber attached to the skull with the head oriented in Horsley-Clarke stereotaxic planes. A steel microelectrode is vertically inserted into the brain.

Recordings were also carried out in paralyzed unanesthetized monkeys. In this case, on the day of recording the animal is briefly anesthetized with Fluothane, intubated, and artificially ventilated after being paralyzed with Gallamine. The chamber is then inserted in the metal bridge, with the animal lying on a table. At the end of the session, the Gallamine infusion is stopped and after complete motor recovery the animal is placed back in its restraining chair.

Three to five recording sessions are usually carried out on each animal, during a period of two to three weeks.

Results

Effect of harmaline on tremor. Stereotaxic operations were performed on 17 monkeys for unilateral placement of lesions in the medioventral portion of the midbrain tegmentum. Two of these monkeys showed a sustained spontaneous contralateral tremor. In five other animals it was possible to induce a tremor of the same type by intravenous or intramuscular injection of harmaline. Harmaline is a potent, short-acting, reversible monoamine oxidase inhibitor. It exerts a direct action on the central nervous system, reaching its maximal concentration in the brain 30 minutes after parenteral administration.[5]

Figure 1 (upper curve) shows the time-course of the harmaline-induced tremor. The following procedure was employed in measuring the amount of tremor. Two electronic clocks were operated simultaneously, one (0.1 cps) to measure exactly successive 10-second timing periods, the other (10 cps) activated only when definite tremor was observed in the extremities, head, or trunk. The system allowed reading afterward of the number of pulses delivered by the second clock during each successive 10-second period. If, for instance, the animal is trembling continuously, the corresponding 10-second periods con-

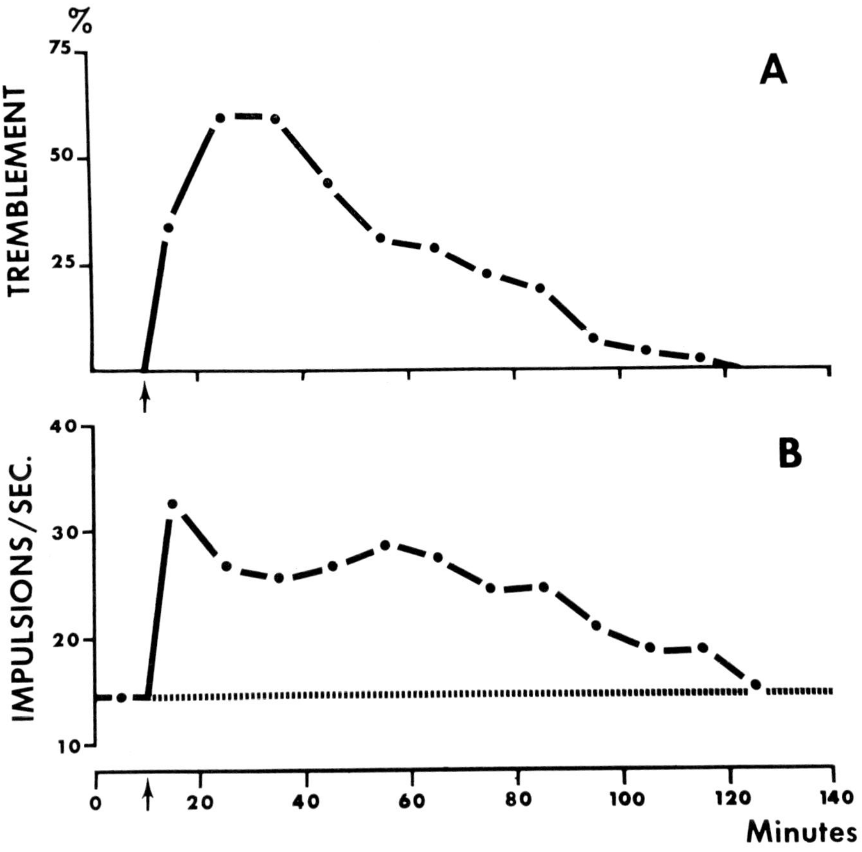

Fig. 1.

tain 100 counts, that is 100 percent. In this way, we obtain direct reading of the percent of the 10-second timing period during which tremor is noted. In Figure 1, the arrow indicates the time of harmaline injection (5 mg per kg, intravenously). The tremorogenic effect is maximal around 30 minutes after injection and lasts for about two hours.

Thalamic unit recordings. In both normal monkey and monkey with tremor the occurrence of two firing patterns is characteristic of many units in the lateral ventral portion of the thalamus. Periods of sustained activity (15 to 30 per second) alternate with periods of relative silence, the activity being then always in the form of short-duration, high-frequency bursts. These bursts occur in rhythmic sequences at, or very close to, 5 per second. This rhythmic activity, which at times can be quite sustained, is always contemporary to a state of drowsiness and to the appearance of slow waves on the electrocorticogram. Under harmaline, these rhythmic bursts disappear and the mean frequency of discharge is increased. This effect on thalamic units has a time course similar to the induced peripheral tremor, as shown in Figure 1. The upper curve depicts the time course of harmaline-induced tremor. The lower curve shows the effects of harmaline on the mean discharge frequency of a thalamic unit of the same animal recorded while curarized with Gallamine. This central effect of harmaline was also observed in the normal monkey and parallels clearly, in the unparalyzed animal, the arousal provoked by the drug.

In the monkey with tremor, in a more posterior region of the lateral ventral portion of the thalamus, we found other units which display a sustained rhythmic bursting not related to the level of arousal. Furthermore, this rhythmic activity does not disappear under harmaline but, on the contrary, is often enhanced. Figure 2 illustrates such units. The recordings in *A* and *B* were obtained from a monkey having a spontaneous tremor. They were recorded with the animal completely paralyzed and without harmaline.

The trace in record *A* illustrates the recruitment and then the waning of the activity often displayed by these rhythmic units. Record *B* shows two units. Both discharge rhythmically at the same frequency as the contralateral tremor which had been previously recorded by electromyography in the same non-curarized animal. The recordings in *C* and *D* were obtained from another monkey having a spontaneous tremor, but in this case the animal was not paralyzed. In each record, the upper trace is a monopolar recording on the precentral cortex; the middle trace, the electromyogram from the left biceps, and the lower trace show unit activity from the right thalamus. The animal was completely awake. The rhythmic bursting is quite sustained, either with (*D*) or without (*C*) contralateral tremor. In this case, the frequency of the bursting (3.5 cps) is quite different from the frequency of tremor (5 cps).

With autocorrelation analysis we can measure precisely the period of a rhythmic activity for several seconds of recording. In Figure 3, the correlogram of a thalamic unit recorded in the curarized animal (upper curve) is here

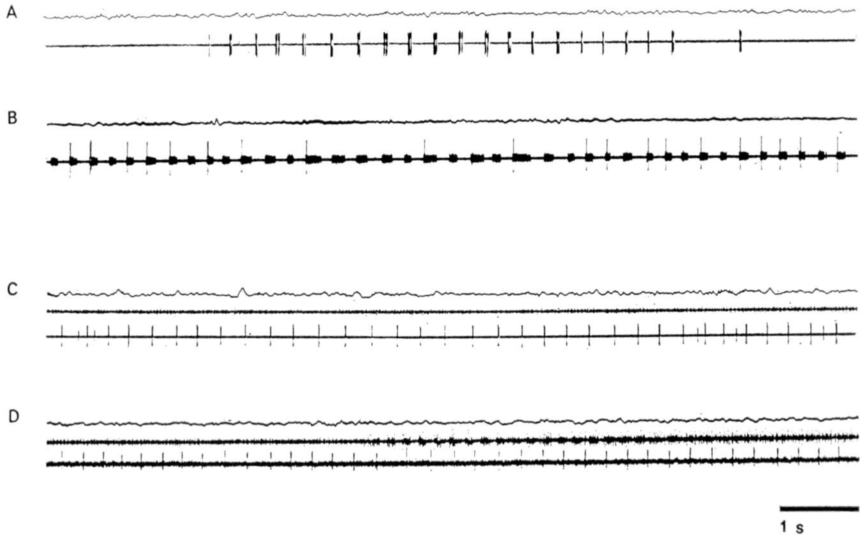

Fig. 2.

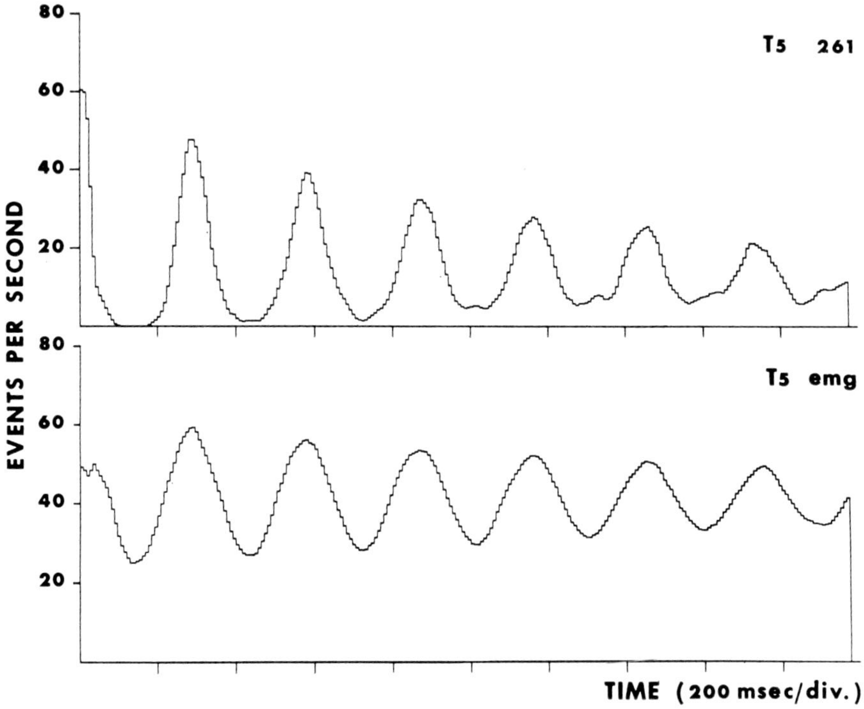

Fig. 3

compared with the correlogram of a sequence of tremor previously recorded by electromyography in the same animal (lower curve). The period is 270 msec in both cases, corresponding to a frequency of 3.7 per second. However, we would like to point out that a number of other rhythmic units in the same thalamic region do not beat at the tremor frequency, but at a somewhat lower frequency (as shown in Figures 2, *D* and 5, *C*).

In some cases this sustained rhythmic bursting stops during a voluntary movement as shown in Figure 4, *A* and *B*. The experimental conditions are

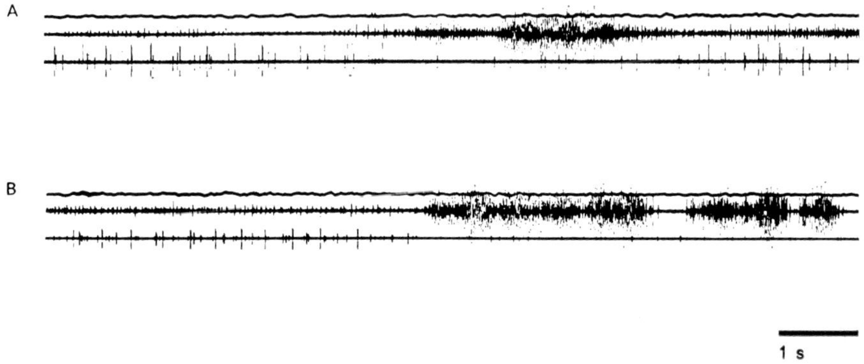

Fig. 4

the same as in Figure 2, *C, D*. In record *A*, the animal tries to reach the edge of its chair with the left hand. In *B*, the animal scratches its right shoulder with the left hand. In both cases, the unit bursting stops completely, or nearly completely, during the movement of the contralateral arm. This type of rhythmic firing unit could never be activated by passive movements or by any other type of peripheral sensory stimuli.

Intermingled with these units showing sustained rhythmic activity, we found other units which fired phasically in relation with tremor. However, their rhythmic activity occurs only during tremor and usually only for short periods of time. These units were activated by voluntary movements but not by passive movements. Such a unit is illustrated in Figure 5, *B*. The unit (lower trace) is first silent during tremor (EMG, middle trace), then fires phasically with tremor in a rather irregular way. Later the cell shows a long burst of activity during a voluntary movement, and then stops firing while the animal maintains a fixed posture (end of tracing).

More posteriorly in the thalamus, i.e., in the specific joint and touch receptor zones, we found tremor-synchronous cells which could always be activated by gentle passive movements. This is illustrated in the two records of Figure 5, *A*. In each record, the upper trace is the unit activity and the lower trace the EMG of the contralateral biceps. The upper record shows rhythmic activity during a few episodes of tremor and also rhythmic activity in the absence of EMG tremor (the animal was nevertheless trembling in

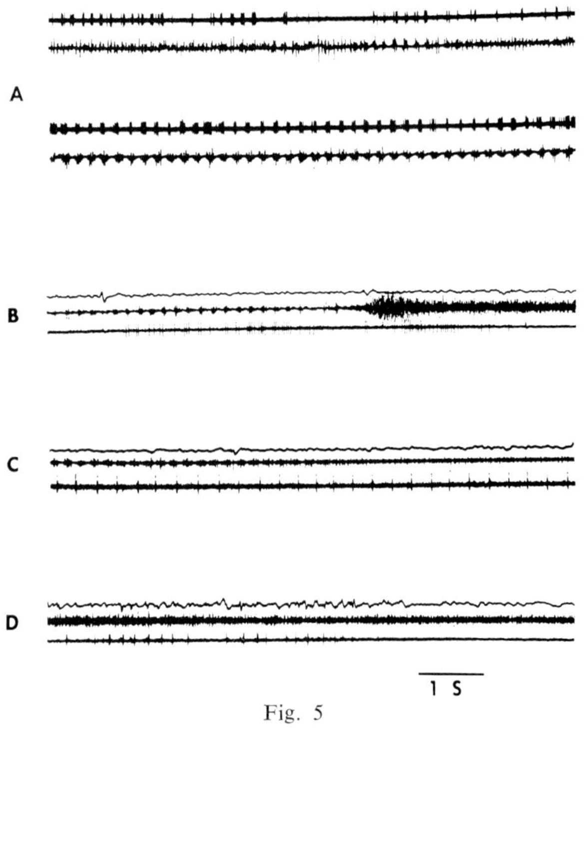

Fig. 5

other muscles of the same limb). The record underneath shows the unit responses to rhythmic passive movements of the contralateral arm.

Discussion

To summarize the results, four main types of rhythmic firing at or near the tremor frequency can be recognized in the lateral ventral portion of the thalamus. They are illustrated in Figure 5:

1. Tremor-synchronous activity recorded in the more caudal parts of the lateral ventral portion of the thalamus, i.e., in the specific joint and touch receptor zones (Fig. 5, *A*).
2. Rhythmic activity at 5 cps occurring in relation to a decreased level of arousal and to appearance of slow waves in the electrocorticogram (Fig. 5, *D*). This type of activity could be recorded in all thalamic regions investigated, in the lesioned as well as in the intact monkey.
3. Episodes of rhythmic activity contemporary to the presence of tremor but not locked precisely with each period of tremor (Fig. 5, *B*). The units showing this type of activity are activated during voluntary movements of the contralateral limbs but not by passive movements.
4. Very sustained rhythmic activity not related to the actual occurrence of tremor (Fig. 2, *C* and *D,* and Fig. 5, *C*). In general, those units fire at a somewhat lower frequency (3 to 5 cps) than the peripheral tremor (5 to 6 cps).

The clearly evoked rhythmic activity of type 1 and the rhythmic bursting of type 2 can hardly be thought of as part of a central mechanism responsible for the rhythmic activation of segmental neurons. However, harmaline provokes a considerable increase of the mean discharge frequency of these cells, together with a strong arousal effect. This kind of effect could well have an "energizing" influence responsible in part for the induction of tremor by harmaline.

Rhythmic activities of types 3 and 4 occurring only in the monkey with tremor and recorded in a restricted part of the thalamus, just in front and above the sensory relay nuclei, are more likely candidates for an abnormal central activity involved in the genesis of tremor. Additional arguments support this hypothesis, such as the influence of voluntary movements on the firing of these cells, the enhancement of the rhythmic activity in some of these cells under harmaline, and the decrease or disappearance of tremor which can be observed after several microelectrode trajectories in the thalamic region where these cells are recorded.

There are, as we have shown, major differences in the behavior of type 3 and type 4 cells. Although only type 3 cells show definite temporal relations with the tremor, we cannot exclude a possible tremorogenic action from the type 4 cells displaying sustained, clock-like, rhythmic activity.

Assuming that the experimental tremor that we study in monkey is valid as a parkinsonian "model",[6, 7] the abnormal rhythmic activity found in this animal should be examined more intensively, since this neuronal activity appears to be located in a region homologous to the thalamic area where in parkinsonian patients some neuron discharges are closely associated with tremor.[8, 9, 10]

REFERENCES

1. POIRIER, L. J.: Experimental and morphological study of midbrain dyskinesias. J. Neurophysiol. *23:* 534, 1960.
2. LAMARRE, Y., and CORDEAU, J. P.: Etude du mécanisme physiopathologique responsable, chez le singe, d'un tremblement expérimental de type parkinsonien. Actualités Neurophysiol. 7, 1967.
3. POIRIER, L. J., SOURKES, T. L., BOUVIER, G., BOUCHER, R., and CARABIN, S.: Striatal amines, experimental tremor and the effect of harmaline in the monkey. Brain *89:* 37, 1966.
4. MOUNTCASTLE, V. B., TALBOT, N. H., SAKATA, H., and HYVÄRINEN, J.: Cortical neuronal mechanisms in flutter-vibration studied in unanesthetized monkeys. Neuronal periodicity and frequency discrimination. J. Neurophysiol. *32:* 452, 1969.
5. VILLENEUVE, A., and SOURKES, T. L.: Metabolism of harmaline and harmine in the rat. Rev. Canad. Biol. *25:* 231, 1966.
6. CORDEAU, J. P., and LAMARRE, Y.: Further studies on patterns of central unit activity in relation with tremor. J. Neurosurg. *24:* 213, 1966.
7. CORDEAU, J. P., GYBELS, J., JASPER, H. H., and POIRIER, L. J.: Microelectrode studies of unit discharges in the sensorimotor cortex: investigations in monkeys with experimental tremor. Neurology *10:* 591, 1960.
8. ALBE-FESSARD, D., GUIOT, G., LAMARRE, Y., and ARFEL, G.: Activation of thalamocortical projections related to tremorogenic processes. *In* The Thalamus (D. Purpura and M. D. Yahr, eds.). Columbia Univ. Press, New York, 1966.
9. BATES, J. A. V.: The significance of tremor phasic units in the human thalamus. *In* Third Symposium on Parkinson's Disease (F. J. Gillingham and I. M. L. Donaldson, eds.). E. & S. Livingston, London, 1969.
10. JASPER, H. H., and BERTRAND, G.: Thalamic units involved in somatic sensation and voluntary and involuntary movements in man. *In* The Thalamus (D. Purpura and M. D. Yahr, eds). Columbia Univ. Press, New York, 1966.

CLINICAL PHYSIOLOGY OF ABNORMAL MOVEMENTS

I. S. Cooper

It seems pertinent in considering the clinical physiology of induced involuntary movement disorders in man to compare those induced by infliction of surgical lesions within the brain to those precipitated iatrogenically by drugs. In our personal experience in 500 cases of chemopallidectomy there was only one case of transient contralateral choreiform movement; in 2,300 cases of chemothalamectomy 4 percent demonstrated some type of induced involuntary movement, while in almost 4,000 thalamic lesions produced by freezing the prevalence was 1.4 percent.

Fig. 1. Postsurgical dystonic posture of arm.

Some examples of various types of involuntary movement produced by surgical lesions are shown in Figures 1 to 3. These movements were produced almost exclusively in male patients, median age 55, with clinical evidence of severe generalized parkinsonism believed to be postencephalitic.

In two cases the site of the lesion was verified anatomically. The lesion shown in Figure 4 in the inferior medial portion of the ventrolateral nucleus (VL) of the thalamus induced hemiballismus. The subthalamic nucleus was unaffected. The incoming rubrothalamic tract was not interrupted. The thalamic lesion shown in Figure 5 produced severe hemiballismus, starting one hour after infliction, in a 38-year-old man with severe postencephalitic parkinsonism. Again there is lack of involvement of subthalamic nucleus and failure to interrupt the rubrothalamic tract. The induced movement disorder was abolished by the pallidal lesion demonstrated in Figure 6 approximately six months later.

The L-Dopa induced movements are similar to those induced by surgical lesions. These have occurred in 35 percent of our series of 600 patients receiving therapy of dosages as small as 500 mg t.i.d. It was noted that such movements were invariably accompanied by restoration of muscle tone to

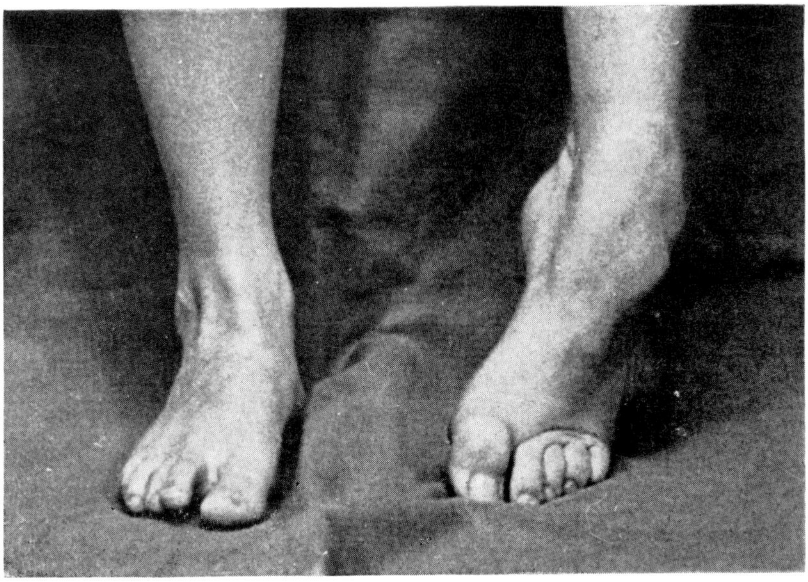

Fig. 2. Postsurgical dystonic posture of foot.

normal or reduced tonality, a fact also noted in the group whose lesions were surgically induced. Movements were invariably reversible when the drug was stopped or dosage decreased or could be abolished by Haldol while L-Dopa therapy was continued. However, Haldol invariably produced akinetic parkinsonian symptoms even with L-Dopa. Elavil, when given as an antidepressant to patients receiving L-Dopa, increased the incidence of involuntary movements.

Fig. 3. Postsurgical dystonic posture of hand.

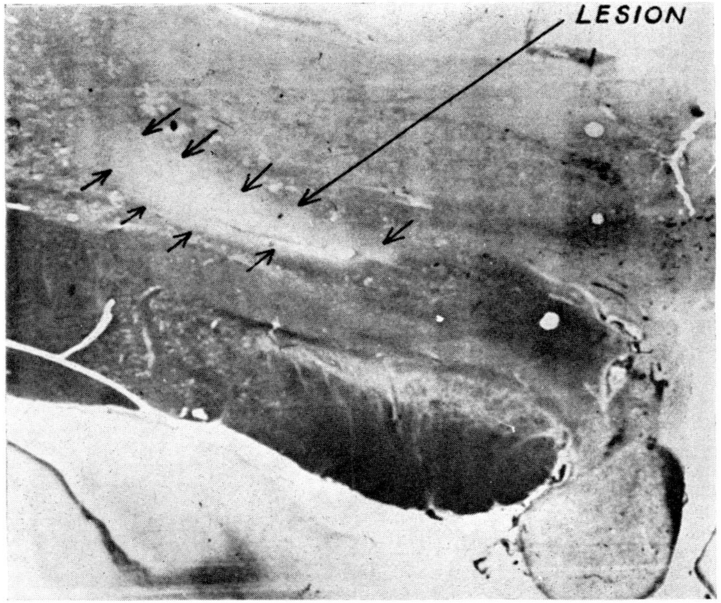

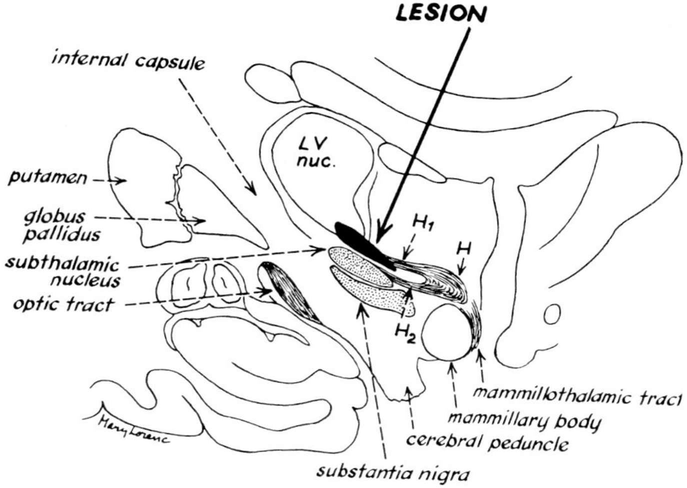

Fig. 4. Surgical lesion confined to thalamus and zona incerta which resulted in hemiballismus.

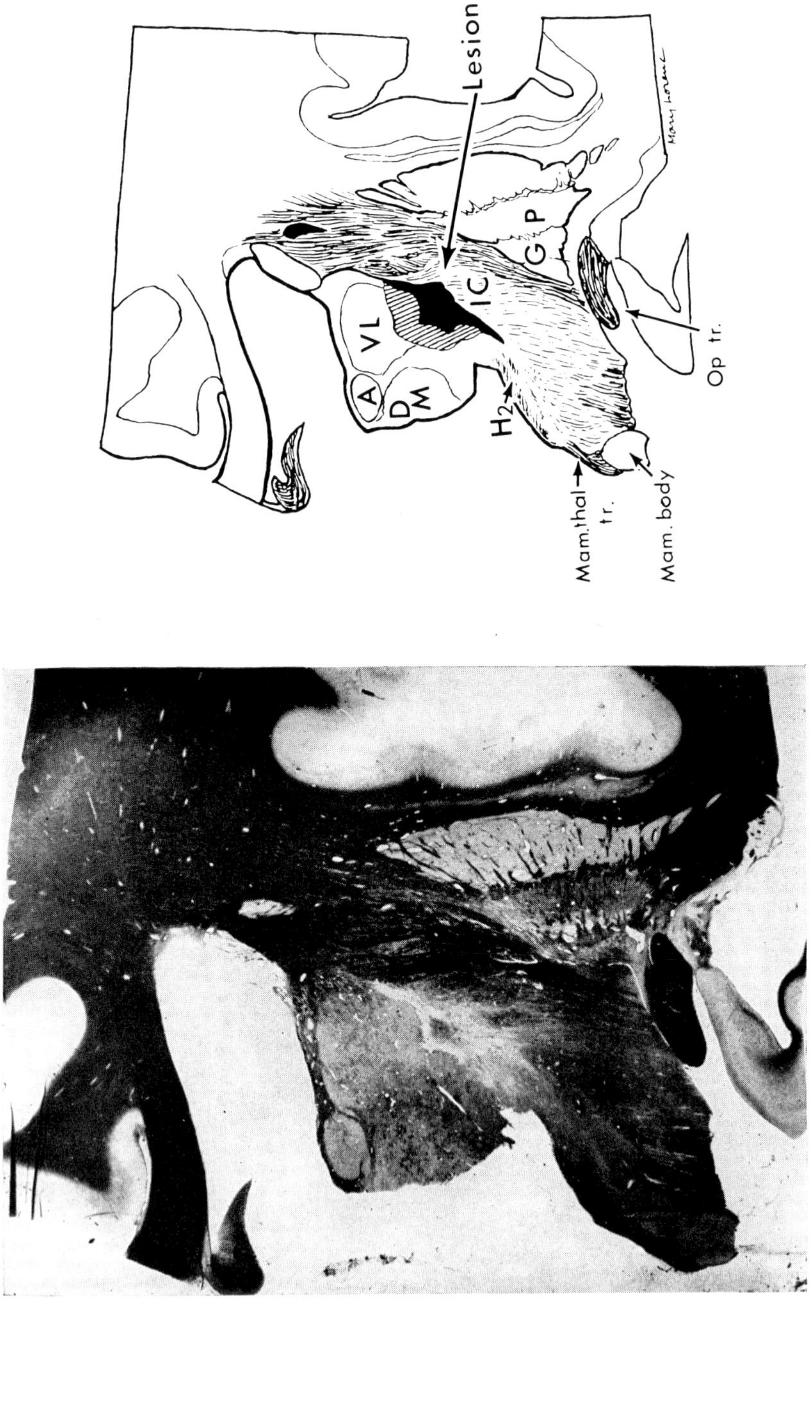

Fig. 5. Thalamic lesion which resulted in hemiballismus.

On the basis of laboratory, clinical and surgical evidence summarized elsewhere, I have postulated that the classic basal ganglia disease symptoms result from pathological conflict between pallidofugal and cerebellofugal pathways (in the modulation of VL and centrum median activity). Rigidity results principally from striatal (principally putamen) deficiency, classic postural or intention tremor from involvement of cerebellofugal pathways between the dentate nucleus and nucleus ruber, while parkinsonian tremor is produced by stimultaneous involvement of the dopaminergic nigrostriatal pathway and the ventral portion of the adjacent cerebellofugal fibers in the mesencephalon. This hypothesis is supported by the recent work of Poirier and associates.

Dystonia also appears to result from an imbalance of modulation of VL and centrum median by these two anatomical systems, lesions having been described in both striatum and cerebellum as well as centrum medianum.

Hemiballism and hemichorea cannot be ascribed to a specific neuropathological lesion. Rather, such movements seem to result from a release of complex motor mechanisms caused by variable lesions in subcortical structures which process sensory information.

Dystonia, hemichorea and hemiballism can be relieved by a lesion in the posterior one-third of the ventrolateral nucleus overlapping slightly into VPL. Moreover, in patients operated upon unilaterally and subsequently treated with L-Dopa, this lesion usually prevents appearance of Dopa-induced movements even though they may appear on the opposite, nonoperated side.

The seeming paradox that this same lesion can, when inflicted on some parkinsonian brains, produce movements identical to those produced by L-Dopa therapy may be explained by the fact that the thalamic lesions shown earlier, which induced involuntary movements, were laterally placed and did not interrupt the cerebellofugal and pallidofugal pathway. They simply produced a different abnormal balance between these two systems at the thalamic level, rather than virtually abolishing this interaction as the more complete lesion demonstrated here apparently does. Other possible explanations include interruption of thalamopallidal or thalamosubthalamic circuits.

On the basis of present knowledge it appears that normal motor activity demands an interaction of dopaminergic and cholinergic function in striatum. Excessive dopaminergic modulation, superimposed on some brains already pathologically involved, results in abnormally lessened muscle tone, involuntary movements and dystonic posturing. The fact that thalamic and zona incerta lesions may also produce identical symptoms indicates the possibility that Dopa-induced movements are produced at least in part by alteration of dopaminergic-cholinergic control of the striopallidofugal pathway to VL, affecting the pallidofugal-cerebellofugal interplay at this level. In some instances striatal control of the cortex is undoubtedly affected as well. The thalamic lesions which have resulted in similar abnormalities could readily have produced a similar imbalance by producing either cerebellofugal or pallidofugal dominance, depending on the location of the lesion. Of paramount

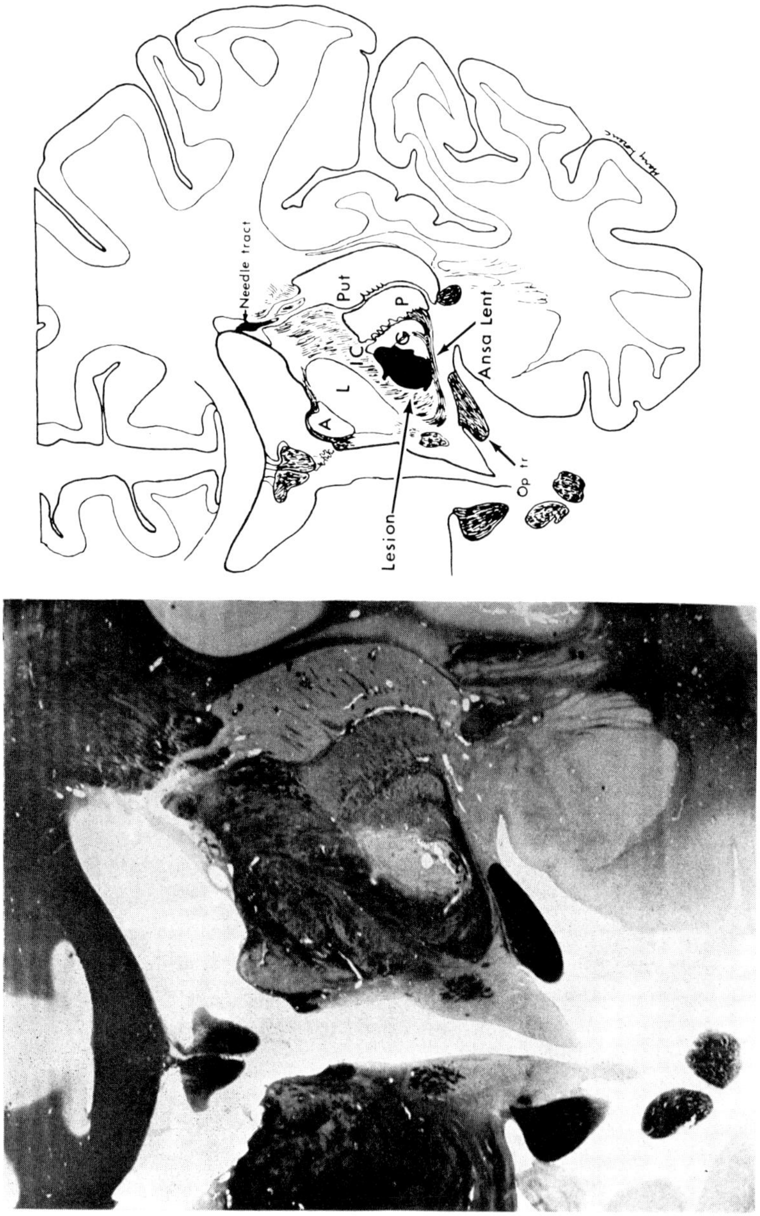

Fig. 6. Pallidal lesion which alleviated postsurgical hemiballismus.

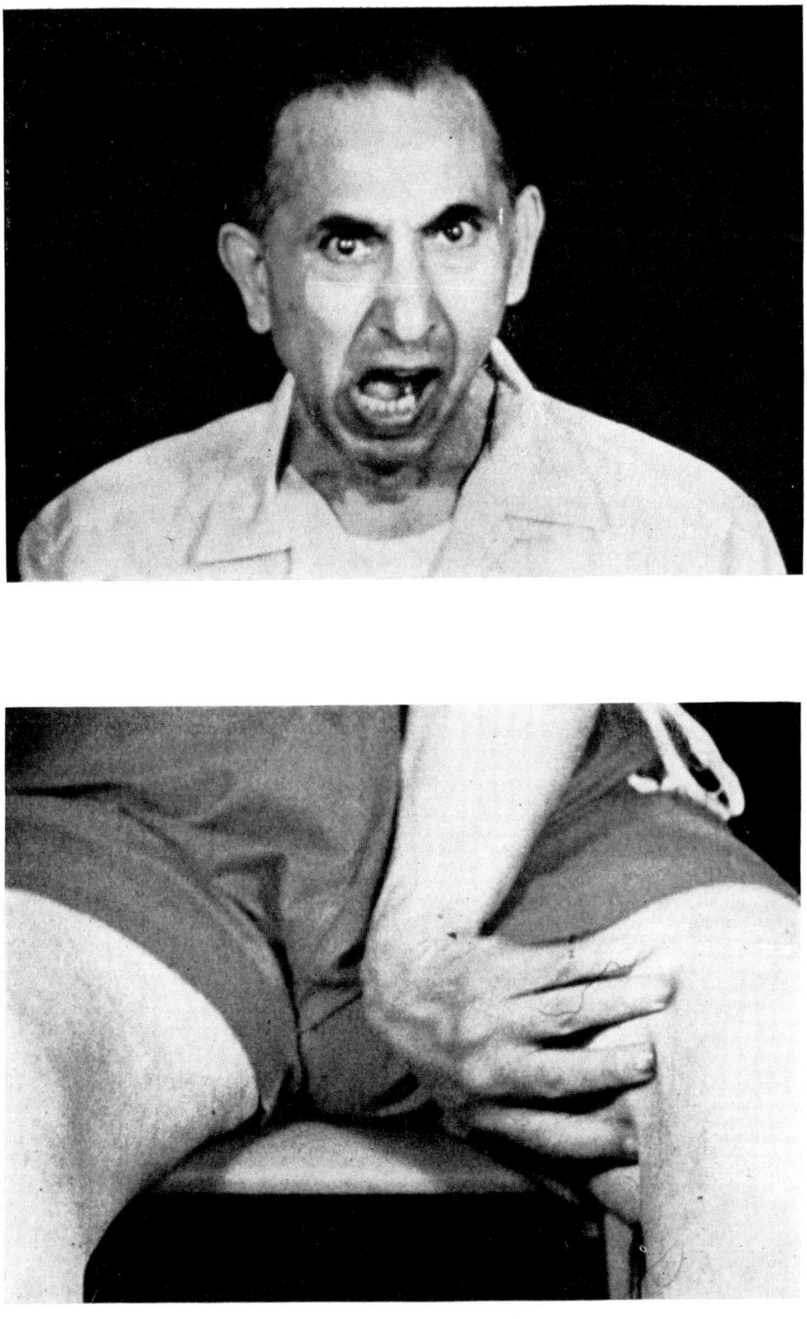

Figs. 7 and 8. Involuntary movements produced by L-Dopa.

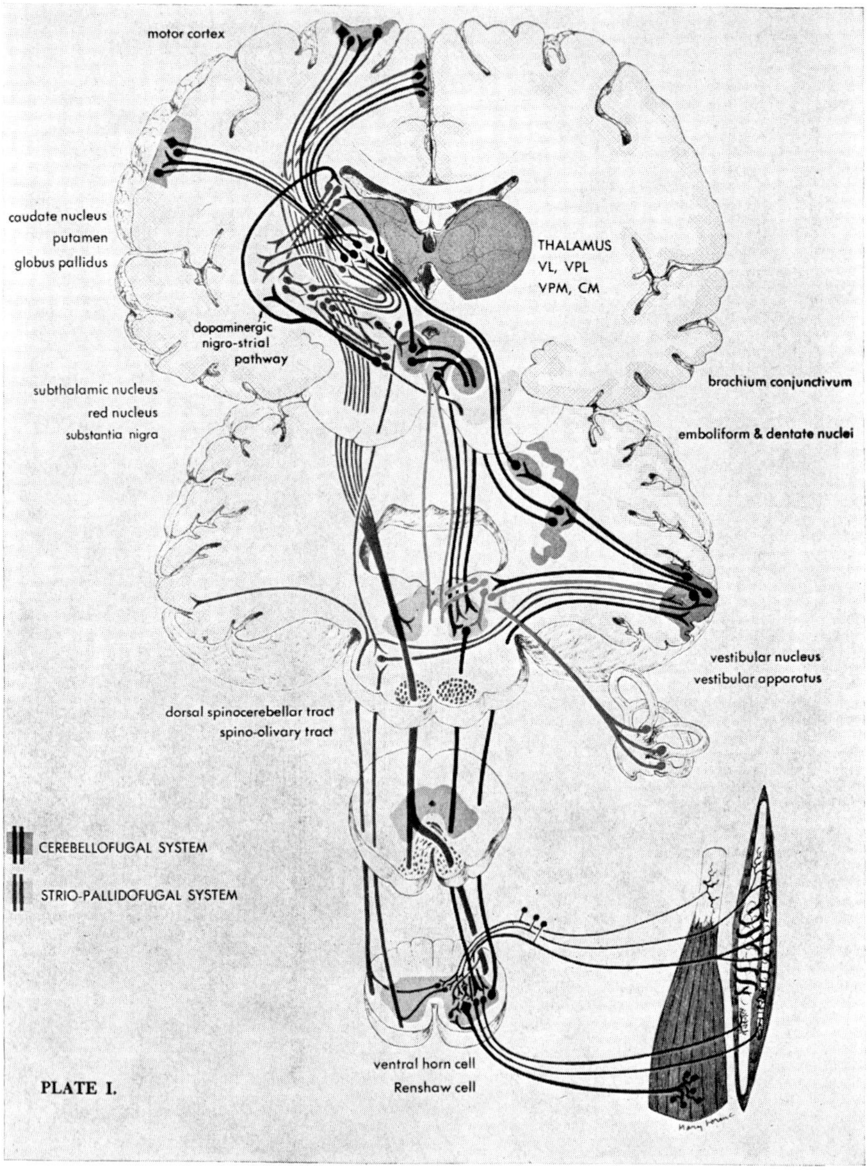

Fig. 9. Diagramatic representation of cerebellofugal and striopallidofugal systems. Pathological conflict between these systems results in involuntary movement disorders.

importance is the fact that L-Dopa or the surgical subcortical lesion inducing involuntary movements appears to require a brain in which *abnormality* of dopamine-cholinergic balance or pallidofugal-cerebellofugal interaction already exists.

THALAMIC RHYTHMIC UNITS AND PARKINSONIAN TREMOR *

Gilles Bertrand

Microelectrode recordings carried out to determine some of the physiological landmarks necessary for accurate placement of therapeutic thalamic lesions have permitted the identification of many types of neurons in the ventralis lateralis and the ventralis posterior nuclei. Three main cell types have been recognized easily by all who have used variants of this method.[1, 2, 3, 4, 5, 6, 7, 8, 9, 10, 11, 12]

Cells responding to light tactile stimuli in small circumscribed cutaneous or mucosal areas (Fig. 1) are grouped somatotopically in the 5 or 6 mm just above the intercommissural plane and slightly below. Their anterior limit, however, varies considerably [13, 14] in relation to the posterior commissure,

* Work carried out with the collaboration of Dr. Herbert Jasper, Dr. Andrew Wong, and Dr. George Mathews.

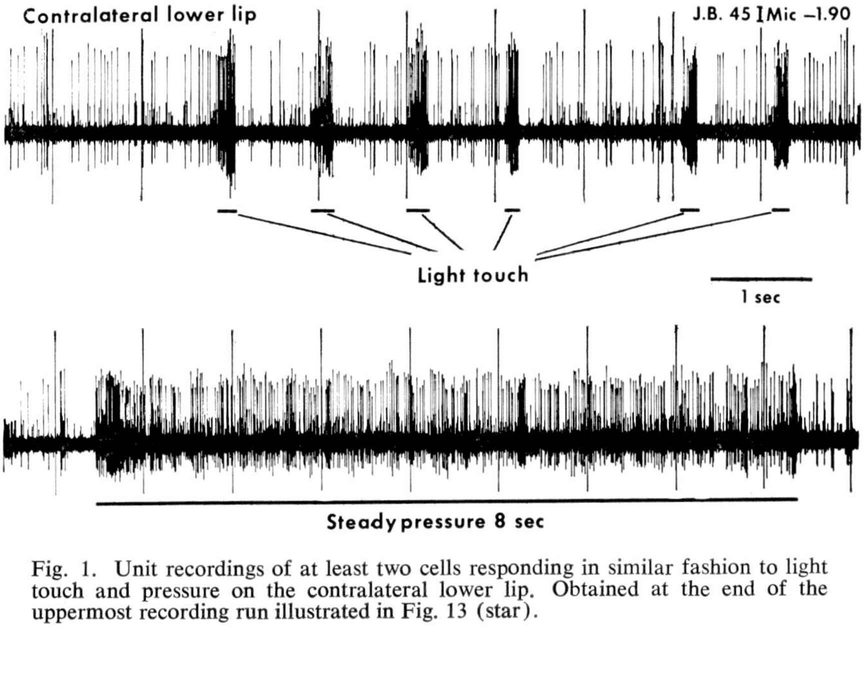

Fig. 1. Unit recordings of at least two cells responding in similar fashion to light touch and pressure on the contralateral lower lip. Obtained at the end of the uppermost recording run illustrated in Fig. 13 (star).

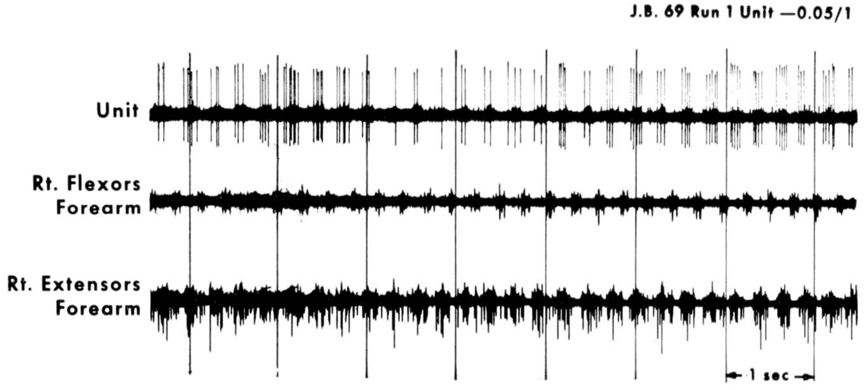

Fig. 2. Rhythmic firing of a neuron related to movements of the contralateral elbow joint: note constant phase relationship with tremor recorded in forearm muscle EMG. Obtained at end of recording run illustrated in Fig. 14.

and knowledge of this limit is of practical importance since encroachment on this zone by the therapeutic lesion can produce permanent sensory deficit and paresthesia. They constitute part of the nuclei ventralis posterolateralis and posteromedialis.

Immediately in front of this tactile group, and also to some extent above it, one finds a fairly broad zone in which a certain proportion of large cells exhibit a striking rhythmic synchronous activity when tremor is present. Analysis quickly indicates that in these cells the rhythmic firing is evoked by peripheral tremor. A more detailed study permits differentiation between cells responding to joint movements (Figs. 2, 3, 4) and others responding to muscle contraction, stretch and palpation (Figs. 5, 6). These are most certainly part of the

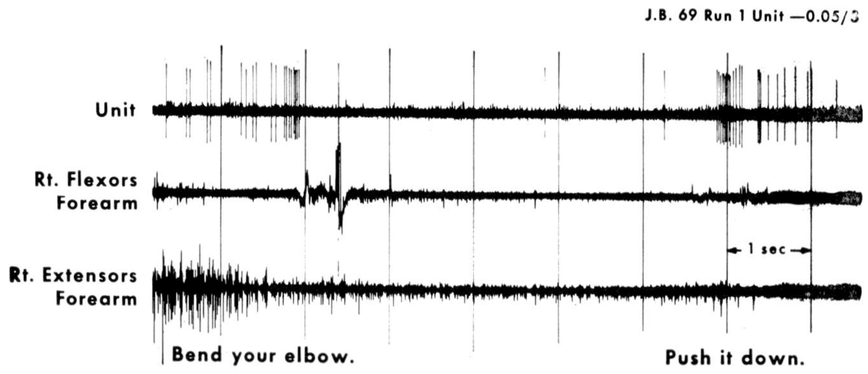

Fig. 3. Same unit showing disappearance of rhythmic activity and activation in a burst of spikes at the time of active extension of the elbow, when the order is given to the patient: "push it down" (arm). The brief burst which comes with "bend your elbow" was not constant in other tests and was due to a small extension jerk the patient gave just before he flexed the elbow (2nd trace EMG artefact).

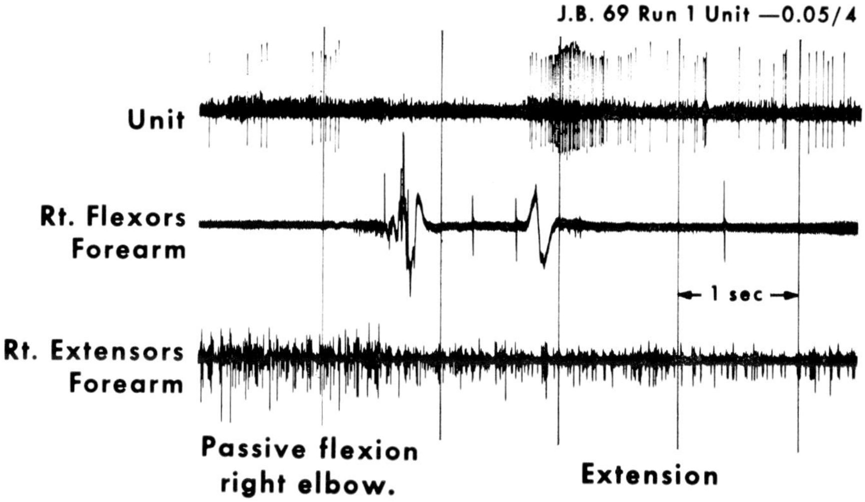

Fig. 4. Same unit responds also to passive extension of the elbow but not to passive flexion.

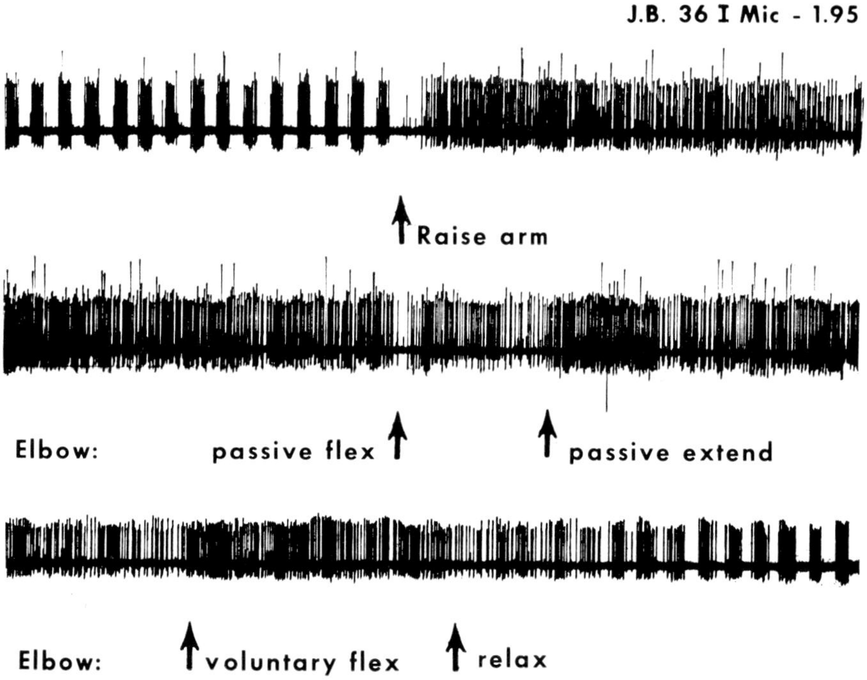

Fig. 5. Rhythmic tremor synchronous unit responding to biceps muscle contraction (voluntary flexion of the elbow) and stretch (passive extension). Relative inhibition occurs during voluntary extension and passive flexion, both of which relax the biceps.

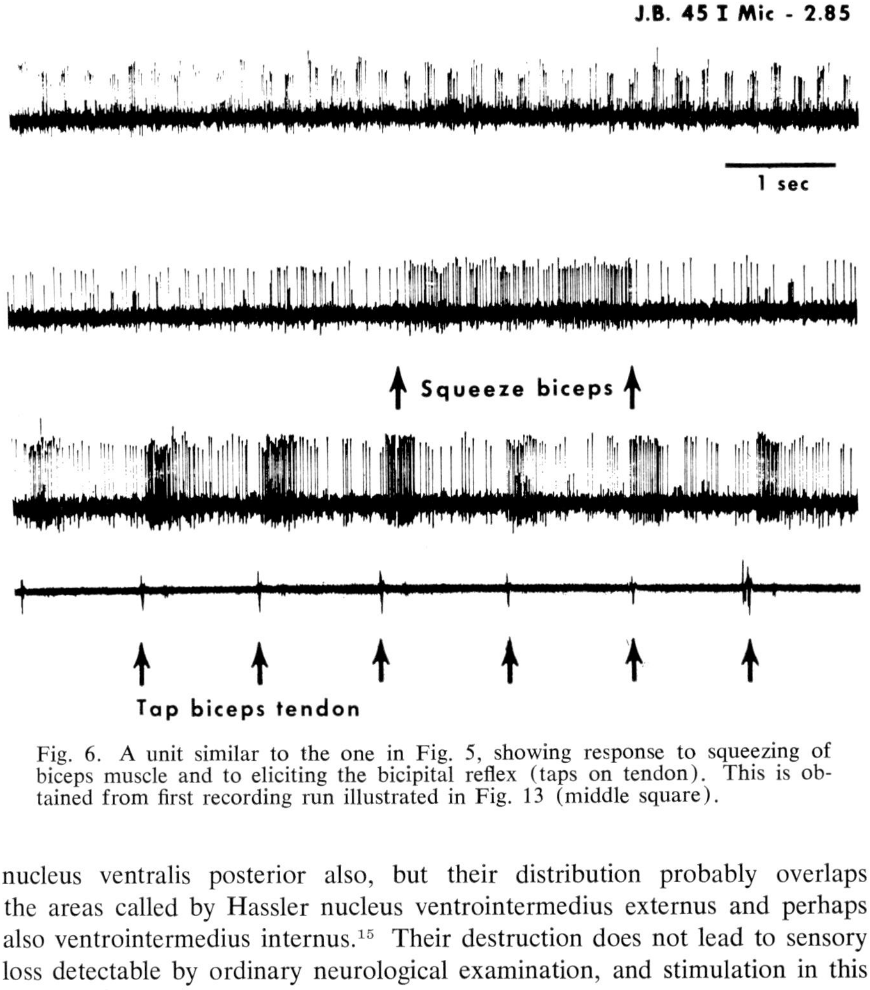

Fig. 6. A unit similar to the one in Fig. 5, showing response to squeezing of biceps muscle and to eliciting the bicipital reflex (taps on tendon). This is obtained from first recording run illustrated in Fig. 13 (middle square).

nucleus ventralis posterior also, but their distribution probably overlaps the areas called by Hassler nucleus ventrointermedius externus and perhaps also ventrointermedius internus.[15] Their destruction does not lead to sensory loss detectable by ordinary neurological examination, and stimulation in this zone with currents adequate to produce paresthesia when applied to the more posterior tactile area produces nothing noticeable to the patient. There is good evidence that many of these cells project to the motor cortex.[16, 17, 18]

A third group of cells is found more anteriorly still in the thalamus, somewhat higher up, as high as the nucleus dorsalis externus of Hassler. These cells also exhibit striking rhythmic discharge but they are not evoked by peripheral movements.[19] The bursts of spikes are less regularly spaced than those of the preceding group and the spikes within a burst are less numerous. In some of these cells the rhythmic activity is completely independent of the presence of tremor, in other it is inhibited by any maneuver which stops peripheral tremor (Figs. 7, 8).

During the course of such recordings, before the introduction of the leukotome used to make the lesion, we have often been struck by the sudden decrease in the amplitude of tremor as the fine tungsten microelectrode traversed zones where rhythmic cells and particularly joint and muscle

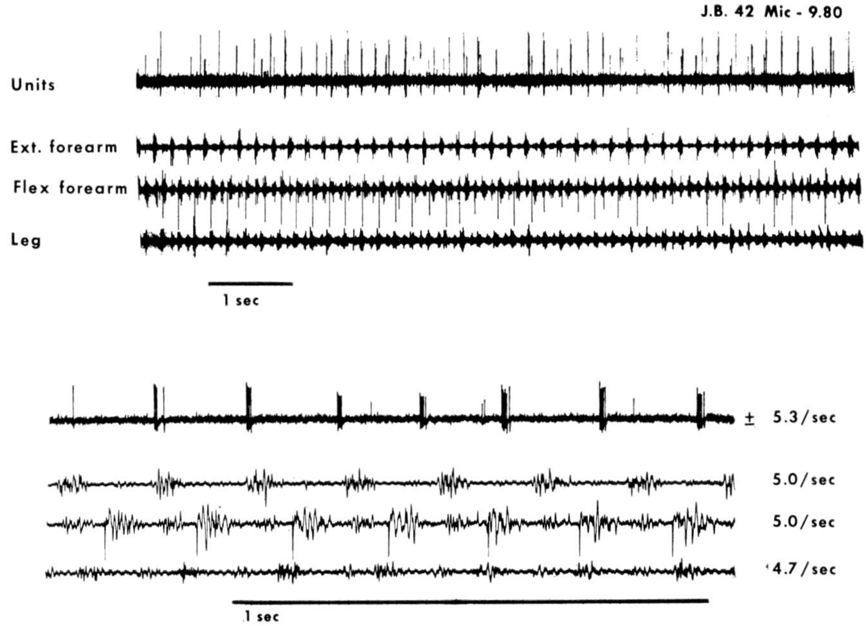

Fig. 7. Rhythmic firing of a cell from nucleus ventralis lateralis, anterior and above area where joint and muscle cells are encountered. Note slight irregularity of intervals between bursts and lack of synchrony with tremor recorded in arm or leg. This unit was not affected by movements or by presence or absence of tremor.

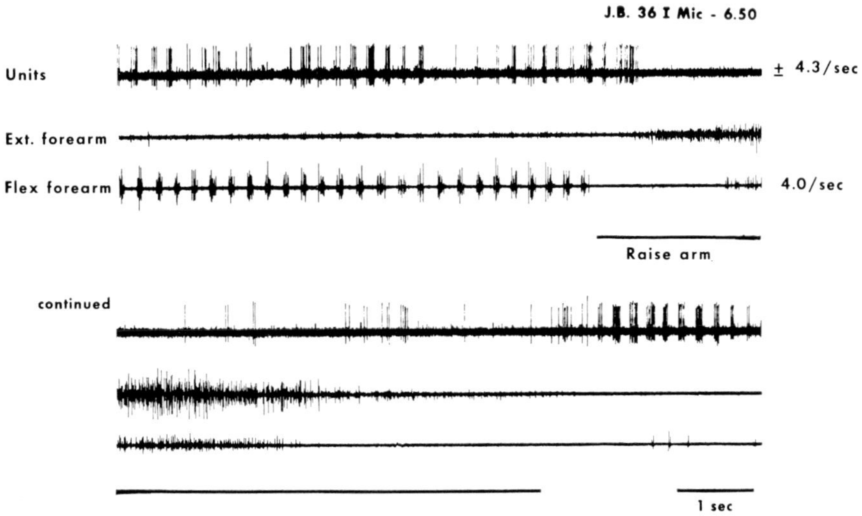

Fig. 8. Rhythmic unit, slightly irregular in frequency and not synchronous with the forearm tremor yet dramatically inhibited by movements which also stop tremor. The rhythmic activity starts again when the arm is relaxed. This unit could not be driven by passive movements. It was obtained 5 mm above the intercommissural plane, in front of others which were clearly proprioceptive.

stretch afferents had been localized (Fig. 9). This phenomenon was confined to the extremity from which afferents had been recorded; elsewhere, tremor continued as before.

In twelve cases in which a recurrence of the parkinsonian tremor necessitated a reoperation a few weeks or months after an initial good result, abolition of the tremor was always accomplished by an extension of the lesion posteriorly, close to the tactile zone of the nucleus ventralis posterior and destroying more completely the area where muscle and joint afferents had been found in front of this tactile zone. These lesions were often enlarged downward as well, below the thalamus, into the zona incerta (Figs. 9, 10, 11, 12) as described by Claude Bertrand.[19] In other cases where the initial lesion had included this low "subthalamic" area and tremor had recurred, an additional effective lesion was placed higher, well within the thalamus but again more posterior than the first (Figs. 13, 14). These posteriorly placed lesions obviously destroy many cells and fibers which are not muscle and joint afferents and which we

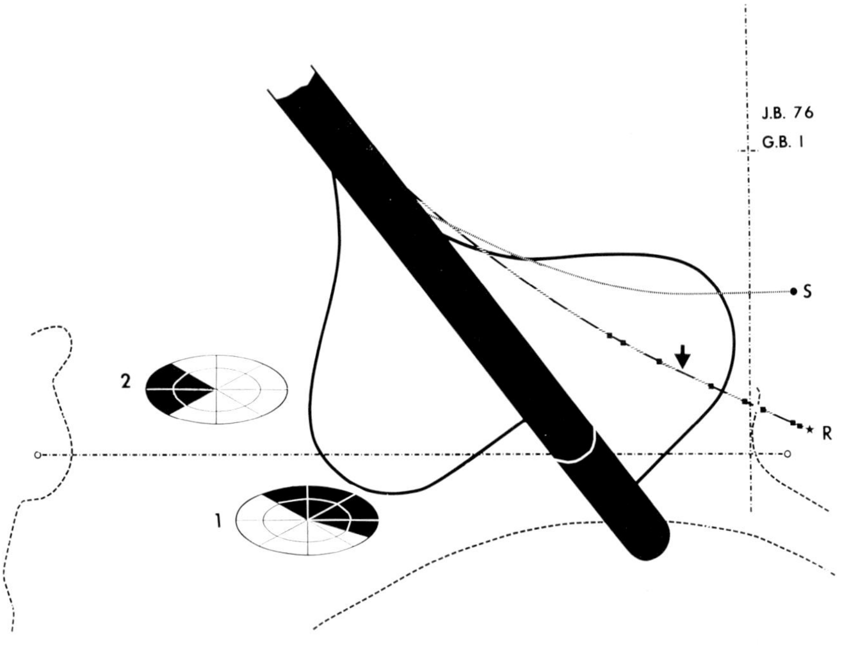

Fig. 9. Diagram of the leukotome "lesion" (solid black lines) and of the recording (R) and stimulating (S) electrode tracts as they appeared on lateral radiographs and superimposed on an outline of the third ventricle. Leukotome reached intercommissural plane at 15 mm from midline of third ventricle. The squares along the recording tract indicate points where tremor-synchronous and proprioceptive cells were encountered. The star represents the position of a light touch sensitive cell. The circle on the stimulation run indicates the point where 2 msec square waves, 60 per sec, 2 volts, 0.5 mA produced tingling in the contralateral hand. The arrow indicates position of the recording electrode tip when considerable decrease in arm tremor occurred. The blackened sectors in the ellipses correspond to those in which the leukotome loop was opened. No lesion was made in white sectors. The numbers indicate the chronological order of the various parts of the lesion.

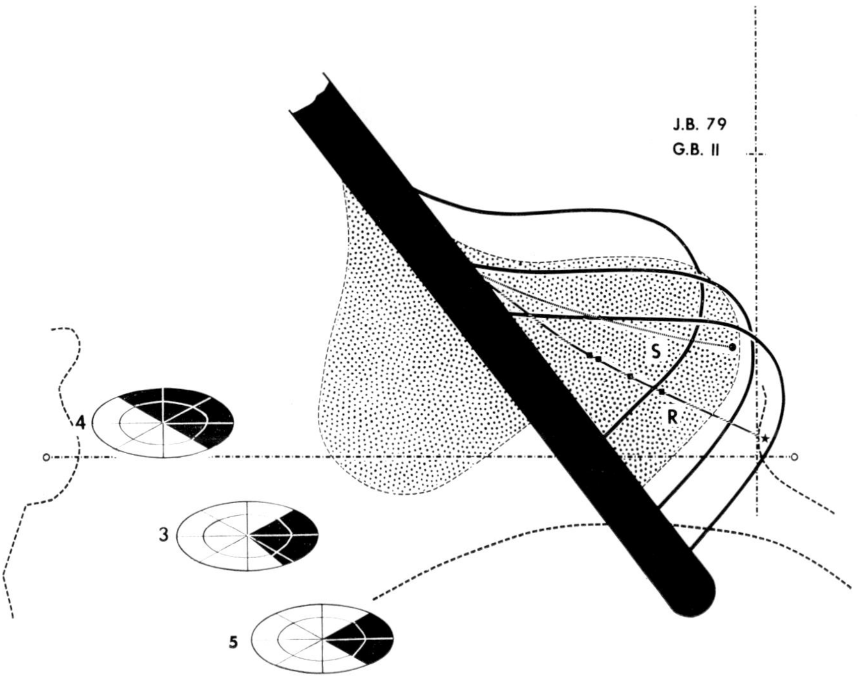

Fig. 10. Diagram of leukotome and electrode positions at reoperation two weeks later. Stippling indicates area involved in first procedure. This time the leukotome was introduced 2 mm more medially at 13 mm from midline. The new lesion is therefore more medial as well as lower and more posterior but the tremor ceased only when sectors 5 were cut, posteriorly and down.

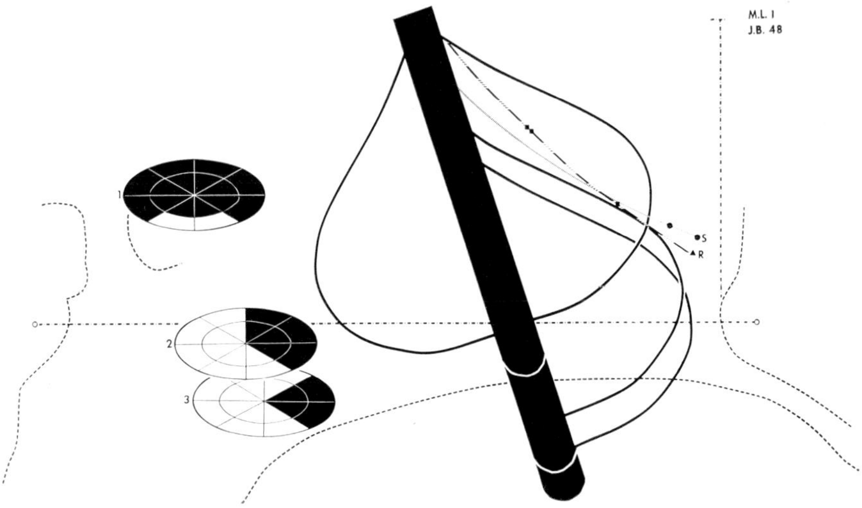

Fig. 11. The leukotome was inserted at 15 mm from the midline in this first operation and a rather large lesion was created (leukotome loop extends 6.5 mm from shaft). Light touch cells were not reached but a cell responding to deep pressure over the contralateral cheek was found at the end of the recording run (black triangle).

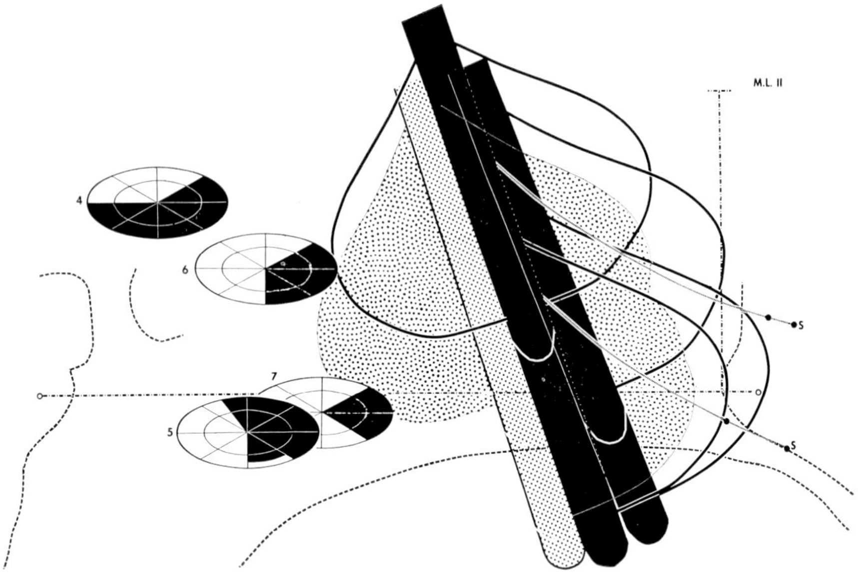

Fig. 12. Same patient, reoperated 18 days later. Leukotome 1 mm more medial, at 14 mm from midline and further back. Sectors 4 and 5 brought no improvement in tremor, sectors 6 and 7 arrested it permanently. Stimulation points marked by circles produced tingling in the hand at 1 volt or less (0.25 mA or less).

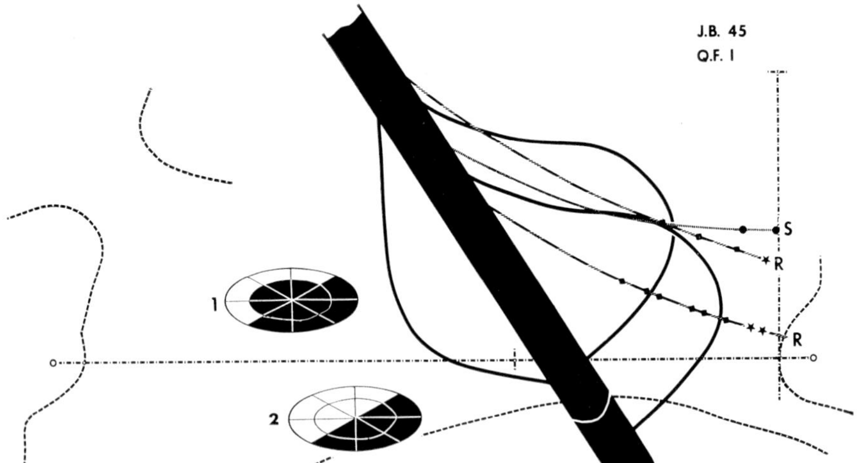

Fig. 13. Leukotome crosses the intercommissural plane 14 mm from midline. Tremor was abolished after sectors 2 were cut.

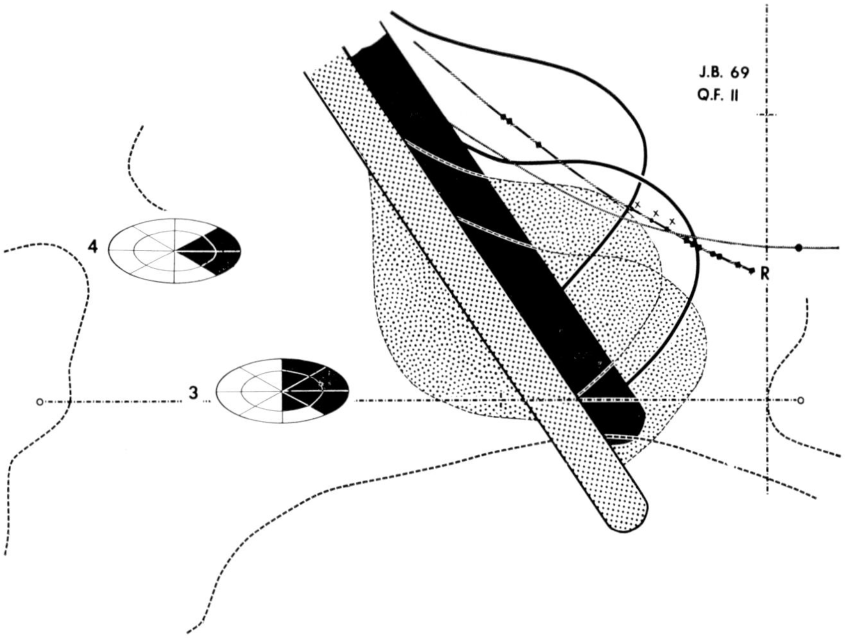

Fig. 14. Same patient as in Fig. 13, reoperation one year later, tremor and rigidity diminished after cutting sectors 3, tremor abolished after section of sectors 4 (also 14 mm from midline). In this case, the extension of the lesion is posterior, up and slightly more medial than the original one.

have not learned to recognize in our recordings; for instance, fibers of cerebellar origin must terminate there. The lesions certainly involve these "movement" cells in large numbers, thus depriving the motor cortex, where they undoubtedly project, of a powerful rhythmic input.

Since one knows that peripheral rhythmic stimuli such as rapidly alternated passive movements of a joint or elicitation of a clonus at a frequency close to that of the spontaneous tremor can induce such a tremor to start again at a time when a partially made therapeutic lesion had apparently abolished it, one cannot help but think that these rhythmic cells have some role to play in the mechanism of tremor.

The classical report of Pollock and Davis [20] and the more recent experiments of Poirier, Ohye and Bouchard,[21] in which extensive dorsal rhizotomies failed to abolish parkinsonian and experimental tremors, clearly indicate that peripheral afferents are not necessary to the production of tremor although they play a more important role in rigidity.

Conversely, Cordeau and Lamarre [22] have shown that tremor-like rhythms persist in neurons of the precentral cortex and nucleus ventralis lateralis of monkeys even after the experimentally induced tremor is completely abolished by a curarizing agent.

Thus, the true pacemaker of parkinsonian tremor is probably a cortico-thalamic loop. The nonsensory rhythmic neurons found in the nucleus ventralis lateralis of parkinsonian patients probably represent a link in such an oscillating system, the abnormal firing of which is the result of disease-induced loss of inhibition.

The facts that tremor can be markedly reduced by a small lesion in the proprioceptive portion of the nucleus ventralis posterior, and that this tremor can be made to reappear by rhythmic peripheral stimuli until a more complete destruction of this same portion of the thalamus has been achieved, lead us to believe that the rhythmic feedback from the trembling extremity, although it is not essential to the genesis of tremor, is nevertheless a powerful reinforcing and synchronizing agent on the abnormal activity of cortical and thalamic neurons, much like the tramp of their own boots on a hard pavement keeps marching soldiers in step.

REFERENCES

1. Albe-Fessard, D., Arfel, G., Derome, P., and Guilbaud, G.: Thalamic unit activity in man. Electroenceph. Clin. Neurophysiol. *25:* 132, 1967.
2. Albe-Fessard, D., Arfel, G., Guiot, G., Hardy, J., Vourc'h, G. Hertzog, E., Aléonard, P., and Derome, P.: Dérivations d'activités spontanées et évoquées dans les structures cérébrales profondes de l'homme. Rev. Neurol. *106:* 89, 1962.
3. Bates, J. A. V.: The significance of tremor phasic units in the human thalamus. *In* Third Symposium on Parkinson's Disease (F. J. Gillingham and I. M. L. Donaldson, eds.). E. & S. Livingstone, London, 1969.
4. Bates, J. A. V.: The electrophysiology of the human thalamus. J. Roy. Coll. Phys. *1:* 118, 1967.
5. Bertrand, G., and Jasper, H.: Microelectrode recording of unit activity in the human thalamus. Confin. Neurol. *26:* 205, 1965.
6. Bertrand, G., Jasper, H., Wong, A., and Mathews, G.: Microelectrode recording during stereotactic surgery. Clin. Neurosurg. *16:* 328, 1969.
7. Gaze, R. M., Gillingham, F. J., Kalyanaraman, S., Porter, R. W., Donaldson, A. A., and Donaldson, E. M. L.: Microelectrode recordings from the human thalamus. Brain *7:* 691, 1964.
8. Guiot, G., Albe-Fessard, D., Arfel, G., and Derome, P.: Représentation et organisation de la somesthésie dans le noyau ventral postérieur de l'homme. Rev. Neurol. *109:* 465, 1963.
9. Guiot, G., Albe-Fessard, D., Arfel, G., and Derome, P.: Dérivations d'activités unitaires en cours d'interventions stéréotaxiques. Neurochirurgie *10:* 427, 1964.
10. Hardy, J., Bertrand, C., and Martinez, N.: Activités cellulaires thalamiques liées au tremblement parkinsonien. Neurochirurgie *10:* 449, 1964.
11. Jasper, H., and Bertrand, G.: Stereotaxic microelectrode studies of single thalamic cells and fibers in patients with dyskinesia. Trans. Amer. Neurol. Assoc. *89:* 79, 1964.
12. Jasper, H., and Bertrand, G.: Thalamic units involved in somatic sensation and voluntary and involuntary movements in man. *In* The Thalamus (D. Purpura and M. Yahr, eds.). Columbia Univ. Press, New York, 1966.
13. Bertrand, G., Jasper, H., and Wong, A: Microelectrode study of the human thalamus: functional organization of the ventrobasal complex. Confin. Neurol., *29:* 81, 1967.
14. Taren, J., Guiot, G., Derome, P., and Trigo, J. C.: Hazards of stereotaxic

thalamectomy. Added safety factor in corroborating x-ray target localization with neurophysiological methods. J. Neurosurg. *29:* 173, 1968.
15. SCHALTENBRAND, G., and BAILEY, P. (eds.): Introduction to Stereotaxis with an Atlas of the Human Brain. G. Thieme, Stuttgart, 1959.
16. ALBE-FESSARD, D., GUIOT, G., LAMARRE, Y., and ARFEL, G.: Activation of thalamo-cortical projections related to tremorogenic processes. *In* The Thalamus (D. Purpura and M. Yahr, eds.). Columbia Univ. Press, New York, 1966.
17. MALIS, L., PRIBRAM, K., and KRUGER, L.: Action potentials in "motor" cortex evoked by peripheral nerve stimulation. J. Neurophysiol. *16:* 161, 1953.
18. WOOLSEY, C. N., CHANG, A. T., and BARD, P.: Distribution of cortical potentials evoked by electrical stimulation of dorsal roots in Macaca mulatta. Fed. Proc. *6:* 230, 1947.
19. BERTRAND, C., HARDY, J., MOLINA-NEGRO, P., and MARTINEZ, S. N.: Optimum physiological target for the arrest of tremor. *In* Third Symposium on Parkinson's Disease (F. J. Gillingham and I. M. L. Donaldson, eds.). E. & S. Livingstone, London, 1969.
20. POLLOCK, L. S., and DAVIS, L.: Muscle tone in parkinsonian states. Arch. Neurol. *23:* 303, 1930.
21. POIRIER, L. J., OHYE, C., and BOUCHARD, R.: Effect of dorsal rhizotomy on postural tremor in the monkey. Physiologist *12:* 329, 1969.
22. CORDEAU, J. P., and LAMARRE, Y. Further studies on patterns of central unit activity in relation with tremor. J. Neurosurg. *24:* 213, 1966.

DOPAMINE, ACETYLCHOLINE, AND EXCITATORY AMINO ACIDS IN NIGROSTRIATAL TRANSMISSION *

K. Krnjević

The abundance and rapid metabolic turnover of dopamine in the nigrostriatal pathway [1, 2, 3] have led to the general acceptance of the idea that dopamine is an important neurotransmitter in this system. An ascending cholinergic pathway is suggested by the presence of large numbers of acetylcholinesterase-containing fibers linking the substantia nigra and the striatum.[4] However, it is perhaps not sufficiently realized that there is still only very inadequate evidence of a significant action of dopamine or acetylcholine (ACh) on the *relevant* striatal neurons.

Unlike L-Dopa, which is mainly excitatory,[5, 6] dopamine when applied directly by microiontophoresis tends to depress neuronal firing in various parts of the central nervous system.[5, 7, 8, 9, 10] The first studies on striatal neurons—by Bloom, Costa and Salmoiraghi,[11] McLennan and York,[12] Connor,[13] and Herz and Zieglgänsberger [14]—all agreed in finding primarily depressant effects of dopamine. ACh, on the other hand, has an excitatory effect on many striatal neurons.[11, 12b] This evidence appeared to support the hypothesis of a double nigrostriatal projection, with an excitatory cholinergic

* Supported in part by the Canadian Medical Research Council.

component and an inhibitory dopaminergic one. The functional state of the striatum would thus depend on the balance between these two ascending systems. However, the tests mentioned above were not made on cells shown to be directly innervated by a nigral projection.

The results of more recent experiments throw some doubts on the above hypothesis. According to York [15] some cells in the putamen receive a slow-conducting, but apparently direct, excitatory input from the substantia nigra; these cells are excited by dopamine. A careful analysis of direct connections between the substantia nigra and the caudate nucleus by Feltz and MacKenzie [16] has revealed a probably monosynaptic excitatory input on certain caudate neurons. Subsequent tests,[17] however, have shown these cells to be quite insensitive to dopamine. As they are also not affected by acetylcholine, but depolarized by glutamate, the nigrocaudate excitatory pathway is probably neither dopaminergic nor cholinergic but could be glutaminergic. Certain other caudate neurons which are much more active spontaneously are clearly inhibited by nigral stimulation, though after a somewhat longer delay. They are depressed by dopamine. In some cases, cells inhibited by nigral stimulation were recorded very close to cells excited by the same stimulation. This suggests the possibility that the directly excited cells may be intracaudate inhibitory interneurons, and that they inhibit other caudate cells by releasing dopamine. In keeping with this idea is the fact that some other caudate neurons, which are not clearly responsive to nigral stimulation, are also depressed by dopamine.

In conclusion, the evidence from the most recent electrophysiological studies does not favor the hypothesis of direct cholinergic excitatory or dopaminergic inhibitory projections from the substantia nigra to the striatum. The role of ACh in nigrostriatal connections is not clear. However, a cholinergic projection from the striatum to the pyramidal tract cells of the cerebral cortex [6, 18, 19] may be of importance in determining motor tone and activity.

REFERENCES

1. Poirier, L. J., and Sourkes, T. L.: Influence of the substantia nigra on the catecholamine content of the striatum. Brain *88:* 181, 1965.
2. Fuxe, K.: The distribution of monoamine terminals in the central nervous system. Acta Physiol. Scand. *64:* Suppl. 247: 37, 1965.
3. Cheramy, A., Besson, M. J., and Glowinski, J.: Effect of neuroleptics on synthesis and release of dopamine from striatal dopaminergic terminals in the rat. Europ. J. Pharmacol., in press.
4. Shute, C. C. D., and Lewis, P. R.: The ascending cholinergic reticular system: neocrortical, olfactory and subcortical projections. Brain *90:* 497, 1967.
5. Krnjević, K., and Phillis, J. W.: Acetylcholine-sensitive cells in the cerebral cortex. J. Physiol. *166:* 296, 1963.
6. Krnjević, K., and Phillis, J. W.: Actions of certain amines on cerebral cortical neurones. Brit. J. Pharmacol. *20:* 471, 1963.
7. Curtis, D. R., and Davis, R.: Pharmacological studies upon neurones of the lateral geniculate nucleus of the cat. Brit. J. Pharmacol. *18:* 217, 1962.

8. PHILLIS, J. W., TEBECIS, A. K., and YORK, D. H.: The inhibitory action of monoamines on lateral geniculate neurons. J. Physiol. *190:* 563, 1967.
9. GALINDO, A., KRNJEVIĆ, K., and SCHWARTZ, S.: Microiontophoretic studies on neurones in the cuneate nucleus. J. Physiol. *192:* 359, 1967.
10. BRADLEY, P. B.: Synaptic transmission in the central nervous system and its relevance for drug action. Int. Rev. Neurobiol. *11:* 1, 1968.
11. BLOOM, F. E., COSTA, E., and SALMOIRAGHI, G. C.: Anesthesia and the responsiveness of individual neurones of the caudate nucleus of the cat to acetylcholine, norepinephrine and dopamine administered by microelectrophoresis. J. Pharmacol. Exp. Ther. *150:* 244, 1965.
12a. MCLENNAN, H., and YORK, D. H.: The action of dopamine on neurons of the caudate nucleus. J. Physiol. *189:* 393, 1967.
12b. MCLENNAN, H., and YORK, D. H.: Cholinergic mechanisms in the caudate nucleus. J. Physiol. *187:* 163, 1966.
13. CONNOR, J. D.: Caudate unit responses to nigral stimuli: evidence for a possible nigro-neostriatal pathway. Science *160:* 899, 1968.
14. HERZ, A., and ZIEGLGÄNSBERGER, W.: The influence of microelectrophoretically applied biogenic amines, cholinomimetics and procaine on synaptic excitation in the corpus striatum. Int. J. Neuropharmacol. *7:* 221, 1968.
15. YORK, D. H.: Dopaminergic transmission in mammalian central nervous system. Brain Res., in press.
16. FELTZ, P., and MACKENZIE, J. S.: Properties of caudate unitary responses to repetitive nigral stimulation. Brain Res. *13:* 612, 1969.
17. FELTZ, P.: Dopamine, amino-acids and caudate unitary responses to nigral stimulation. J. Physiol., in press.
18. KRNJEVIĆ, K., and SILVER, A.: A histochemical study of cholinergic fibers in the cerebral cortex. J. Anat. *99:* 711, 1965.
19. KRNJEVIĆ, K., and SILVER, A.: Acetylcholinesterase in the developing forebrain. J. Anat. *100:* 63, 1966.

THE EFFECTS OF DRUGS ON MOTOR DISTURBANCES IN MONKEYS WITH TEGMENTAL LESIONS *

M. Goldstein, A. F. Battista, S. Nakatani, and B. Anagnoste

During the past decades, considerable progress has been achieved in the understanding of the physiopathology and biochemistry of Parkinson's disease. Since it was reported that the basal ganglia contain high concentrations of dopamine and serotonin,[1] attention has been focused on the possible role of monoamines in extrapyramidal dysfunction. The observation by Ehringer and Hornykiewicz [2] that the content of monoamines in the striatum in patients dying with parkinsonism is low provided the first experimental evidence for the association of extrapyramidal disorders with abnormal monoamine levels in the striatum.

Monkeys with unilateral lesions in the ventromedial areas of the brain stem

* Supported by USPHS Grants MH-02717 and NB-06801 and by NSF Grant GB-8465.

exhibit neurological and neurochemical deficits which are similar to those observed in Parkinson's syndrome.[3]

A close relationship between motor disturbances and a cell loss in the substantia nigra, with a concomitant decrease in striatal monoamines, was found in patients with parkinsonism as well as in the experimental monkeys. The data in Table 1 summarizes the neuropathological and neurochemical deficits in Parkinson's syndrome and in monkeys with mesencephalic lesions. Studies in the experimental monkeys have shown that the reduction of monoamines in the striatum is due to an impairment in storage and biosynthesis of the corresponding monoamines.[4, 5] Thus monkeys with mesencephalic lesions were found to be useful models in studies of the role of monoamines in extrapyramidal dysfunction. This paper describes the effects of centrally acting drugs on motor disturbances in monkeys with mesencephalic lesions and correlates these effects with monoamine changes.

Materials and Methods

Green monkeys (Cercopithecus sabaeus) were used and unilateral radiofrequency lesions were induced in the ventromedial tegmental region of the brain stem as previously described.[3, 5] The lesions were induced at least one month prior to the beginning of the experiments. Hypokinesia of the contralateral extremities appeared immediately afterward. In some monkeys a resting tremor (4 to 6 cycles per sec) developed five to seven days later. Tremograms were obtained by means of a transducer attached to the extremities and recorded on an electroencephalograph. The effects of centrally acting drugs were investigated in monkeys with ventromedial tegmental lesions with a slight spontaneous resting tremor of the contralateral extremities and in monkeys exhibiting hypokinesia of the contralateral extremities but no resting tremor.

Localization of the lesion in each brain was verified by gross examination and by histological study. The locus and the extent of the lesions were similar to those reported previously.[3] Gross examination revealed that the lesions in the monkeys which developed tremor and hypokinesia were larger, involved more medial structures and extended further into the medial substantia nigra

TABLE 1. The Neuropathological, Neurological and Neurochemical Deficits in Human Parkinsonism and in Experimental Monkeys with Mesencephalic Lesions

Human Parkinsonism	*Monkeys with Mesencephalic Lesions*
Degeneration of the melanin-containing nerve cells of the pars compacta of the substantia nigra	Degeneration of the melanin-containing nerve cells of the pars compacta of the substantia nigra
Rigidity, akinesia and tremor	Hypokinesia and tremor
Striatal amine deficiency: depletion of dopamine and serotonin	Striatal amine deficiency: depletion of dopamine and serotonin

and the dorsomedial peduncle areas than the lesions in monkeys which developed hypokinesia but no tremor.

Results

Monkeys with ventromedial tegmental lesions exhibiting hypokinesia and postural tremor of the contralateral extremities. The data in Table 2 show the effects on tremor of Dopa and 5-hydroxytryptophan (5-HTP) administered alone and in combination with atropine. The intravenous administration of DL-Dopa at a dose of 30 mg per kg or of DL-5-HTP at a dose of 25 mg per kg resulted in disappearance of the tremor. The administration of smaller doses of DL-Dopa or DL-5-HTP did not produce any marked effect on the intensity of the tremor. The combined administration of DL-Dopa and DL-5-HTP had an additive but not a synergistic effect. However, the effective dose of DL-Dopa or DL-5-HTP varies in different monkeys and seems to be dependent on the duration and size of the lesion. A typical tremogram obtained from a monkey with a tegmental lesion prior to and following the administration of DL-5-HTP is shown in Figure 1.

Atropine at a dose of 0.18 mg per kg intravenously slightly relieved the tremor. Following the combined administration of DL-Dopa or DL-5-HTP with atropine the tremor disappeared for 30 minutes to two hours even when the concentration of each of these amino acids was lowered to 15 mg per kg. Thus, the combined administration of these drugs has a much more pronounced effect on the relief of the tremor than the administration of each of these drugs alone.

The relief of the tremor by the administration of Dopa decarboxylase inhibitors in combination with either Dopa or 5-HTP was investigated. The data presented in Table 3 show that the extracerebral acting inhibitor DL-α-methyldopa-hydrazine (MK-485) reduces the effective dose of DL-Dopa or of DL-5-HTP. The peripherally and centrally acting decarboxylase inhibi-

TABLE 2. The Effect of Separate and Combined Treatment with DL-Dopa, DL-5-HTP and Atropine on Postural Tremor in Monkeys with Mesencephalic Lesions

Drug *	*Pharmacological Responses on the Activities of the Contralateral Limbs* **
None	Sustained hypokinesia and postural tremor
DL-Dopa (30) or DL-5-HTP (25)	Diminished tremor or completely absent for 30 min. to 2 hrs.
DL-Dopa (15) or DL-5-HTP (15)	No effect
Atropine (0.18)	Slightly diminished tremor
DL-Dopa (15) + Atropine (0.18)	Tremor absent for 30 min. to 2 hrs.
DL-5-HTP (15) + Atropine (0.18)	Tremor absent for 30 min. to 2 hrs.

* Each drug was administered (i.v.) to two monkeys. The drugs were tested in each monkey 3 times.

** The tremor is described as diminished when the amplitude of the tremor was decreased and the tremor was absent for short periods of time (5 to 10 min.).

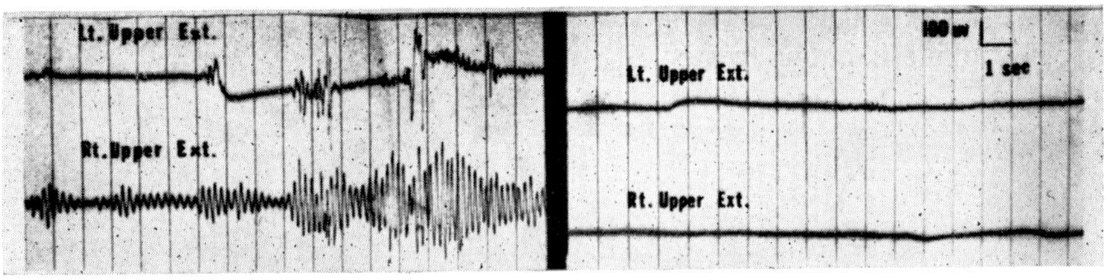

Fig. 1. Tremogram recorded from the upper limbs of a monkey with a mesencephalic lesion. Record on the left taken prior to the administration of DL-5-HTP. Record on the right taken following administration of DL-5-HTP (25 mg/kg, i.v.).

tor DL-α-methyldopa diminishes the relief of the tremor produced by the administration of DL-Dopa.

Monkeys with ventromedial tegmental lesions exhibiting hypokinesia but no postural tremor of the contralateral extremities. In monkeys exhibiting hypokinesia but no resting tremor, the administration of harmaline evoked a marked resting tremor of the extremities contralateral to the tegmental lesion (Table 4). The frequency was 4 to 6 cycles per sec and the tremor resembled that which spontaneously develops after a ventromedial tegmental lesion.

We have also investigated the effects of specific inhibitors of amine synthesis on harmaline-induced tremor. As shown in Table 4, administration of α-methyl-*p*-tyrosine, a known inhibitor of tyrosine hydroxylase, had no tremogenic effect. When *p*-chlorophenylalanine, an inhibitor of serotonin synthesis, was administered for eleven days, a very fine resting tremor of low amplitude and 4 to 5 cycles per sec appeared on the eighth day in the extremities contralateral to the tegmental lesion. It seems unlikely that the resting tremor developed spontaneously since the administration of *p*-chlorophenylalanine was started six months after the lesion was placed.

To determine whether the action of harmaline is mediated by striatal amines, harmaline was given to a monkey pretreated with both inhibitors of amine

TABLE 3. The Effects of Dopa Decarboxylase Inhibitors on Postural Tremor in Monkeys with Mesencephalic Lesions

*Drug ** *(Dose in mg/kg)*	*Pharmacological Responses on the Activities of the Contralateral Limbs*
None	Sustained hypokinesia and postural tremor
DL-Dopa (30) or DL-5-HTP (25)	Diminished tremor or completely absent for 30 min. to 2 hrs.
DL-Dopa (15) or DL-5-HTP (15) or MK-485 (10)	No effect
DL-Dopa (10) or DL-5-HTP (10) + MK-485 (10)	Diminished tremor or completely absent for 30 min. to 2 hrs.
DL-α-methyl Dopa (15)	No effect
DL-α-methyl Dopa (15) + Dopa (30)	Postural tremor

* MK-485 was given i.p. 1 hr. prior to the administration of Dopa or 5-HTP and DL-α-methyldopa was given i.v. 30 min. prior to the administration of Dopa.

TABLE 4. The Effects of Centrally Acting Drugs on Harmaline Induced Tremor in Monkeys with Mesencephalic Lesions

Drug *	*Pharmacological Responses on the Activities of the Contralateral Limbs*
None	Sustained hypokinesia
Harmaline	Coarse tremor
MT	No effect
PCPA	Slight tremor (low amplitude)
Harmaline + MT + PCPA	Coarse tremor
5-HTP + harmaline	No tremor or markedly diminished tremor (tremor of the ipsilateral extremities occasionally observed)

* MT, α-methyl-*p*-tyrosine; PCPA, *p*-chlorophenylalanine.
Harmaline (5 mg/kg), DL-5-HTP (25 mg/kg), MT (80 mg/kg) were injected intravenously.
PCPA (80 mg/kg) was administered orally every day for a period of eleven days.
DL-5-HTP was given 20 min prior to the administration of harmaline.

synthesis. Production of tremor by harmaline was not abolished by lowering the amine levels with *p*-chlorophenylalanine and α-methyl-*p*-tyrosine (Table 4). Thus, the action of harmaline does not appear to be mediated by striatal amines, but could be due to the direct action of harmaline on the receptors.

In other experiments we have investigated whether the increase in brain serotonin and dopamine levels affects tremor production by harmaline. The experimental monkeys were treated with 5-HTP or Dopa prior to the administration of harmaline. As shown in Table 4, administration of 5-HTP protects the monkeys from tremors induced by harmaline. The administration of Dopa has no effect on the harmaline-induced tremor. Since harmaline and serotonin are both indoles, it is conceivable that harmaline affects the functions of the central nervous system by interaction with serotonin receptors.

The effects of various centrally acting drugs on motor disturbances in monkeys with tegmental lesions and in rats with lesions in the nigroneostriatal dopamine pathway. The data in Table 5 show that in rats with lesions in the nigroneostriatal dopamine pathway the administration of amantadine (25 mg per kg) causes a potentiation of the Dopa-induced turning behavior.[6] It is also evident from Table 5 that in monkeys with tegmental lesions the administration of amantadine at a dose of 10 mg per kg intravenously causes a potentiation of the lesion-induced tremor.

TABLE 5. The Effects of Centrally Acting Drugs on Tremor in Monkeys and on the Turning Behavior in Rats

Drugs	*Pharmacological Responses*
Dopa + amantadine	Potentiates the Dopa-induced turning behavior in rats.*
Amantadine	Potentiates the tremor in monkeys.
Apomorphine	Produces a transient relief of the tremor in monkeys.

* Fuxe and Goldstein.[6]

The administration of apomorphine (1 mg per kg subcutaneously) to monkeys with tegmental lesions produces a transient relief of the tremor (Table 5). Apomorphine also potentiates the Dopa-induced relief of the tremor.

Various cholinergic blocking agents and antihistamines used for treatment of parkinsonism were found to inhibit the uptake of dopamine in the neostriatum.[6] The administration of benztropine (Cogentin), procyclidine (Kemadrin), atropine and benzhydryloxy (Rigidyl) in a dose of 50 mg per kg intraperitoneally produces a 50 percent inhibition of dopamine uptake in the neostriatum of rats. These drugs also potentiate the Dopa-induced turning behavior in rats with striatal lesions.

Discussion

In many aspects the extrapyramidal defects in monkeys with mesencephalic lesions resemble the human syndrome first described by James Parkinson in 1817 and generally known as Parkinson's disease. The present findings illustrate, at least in one primate species, an apparent role of a deficit of serotonin and dopamine in experimental parkinsonism. The therapeutic effect of L-Dopa in parkinsonian patients has found confirmation in the experimental monkeys. The reduction of the effective dose of Dopa or of 5-HTP when given in combination with the extracerebral decarboxylase inhibitor MK-485 shows that the relief of the tremor is associated with the increase of the corresponding amine concentrations in the brain. The conclusion that the amines are the active agents responsible for the relief of tremor and other symptoms is also supported by the finding that the cerebral acting decarboxylase inhibitor methyldopa diminishes the beneficiary action of Dopa on tremor. In support of this conclusion it could also be cited that the effective dose of Dopa or 5-HTP seems to depend on the decarboxylase activity in the ipsilateral striatum.

Although Dopa was effective in the relief of the tremor in the experimental monkeys, it is noteworthy that 5-HTP was also effective. The response to Dopa or to 5-HTP differs in various monkeys: in some monkeys with the tegmental lesion placed for over one year, only 5-HTP but not Dopa was effective in relief of tremor. The response to Dopa or 5-HTP seems to be dependent on striatal decarboxylase activity, and in monkeys with very low activity of this enzyme in the ipsilateral striatum the response is minimal or completely abolished.

It is conceivable that different biochemical deficits are associated with various neurological manifestations. While Dopa appears to be most effective in the relief of parkinsonian akinesia and rigidity, the relief of tremor appears only after prolonged treatment with the drug. The relief of tremor might not necessarily be due to a direct action of dopamine. It is conceivable that prolonged treatment with Dopa causes an activation of serotoninergic and other monoaminergic neurons which contribute to relief. It is noteworthy that the urinary dopamine excretion is reduced in patients with akinesia and rigidity but not in patients with tremor.[7] Of considerable interest also are the findings by Cotzias (personal communication) that in patients with manganese poisoning 5-HTP was found to be effective in the relief of the symptomatology of the syndrome.

Preliminary studies in our laboratory have shown that intracerebral injection into the substantia nigra of 6-hydroxy-dopamine (6-OH-DA), a compound which causes selective degeneration of sympathetic nerve terminals, produces posture changes and hypokinesia but no tremor in monkeys. Further studies with 6-OH-DA are now in progress to obtain additional information on the association of dopamine and serotonin with motor disturbances.

Although considerable attention has been directed toward the role of dopamine and serotonin in extrapyramidal functions, cholinergic mechanisms should also be considered. Anticholinergic drugs are widely use in the treatment of the symptomatology of Parkinson's disease. The present findings that atropine and Dopa or atropine and 5-HTP have a synergistic effect on tremor relief in experimental monkeys suggests interaction of both sympathomimetic and cholinergic systems in extrapyramidal motor functions. Thus the idea that an imbalance between the sympathomimetic and cholinergic systems results in motor disturbances has found some further support. A hypothetical scheme in Figure 2 outlines the imbalance in the experimental

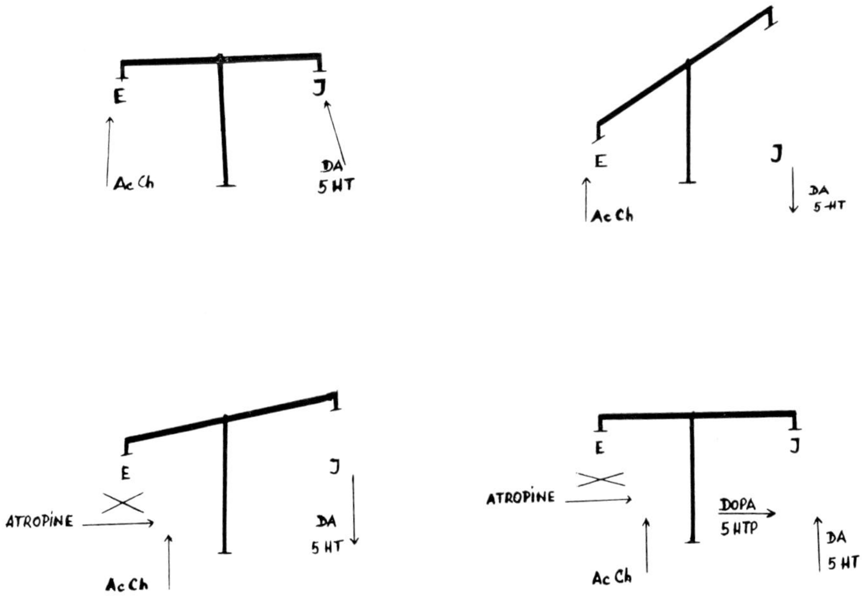

Fig. 2. The imbalance between the sympathomimetic and cholinergic systems in experimental parkinsonism.
Upper left: Equilibrium between the excitatory (E) and inhibitory (I) systems maintained by the ratio of the acetylcholine (AcCh) to dopamine (DA) and serotonin (5-HT) levels.
Upper right: The shift in equilibrium due to the reduction of DA and 5-HT levels.
Lower left: Partial restoration of the equilibrium by administration of atropine.
Lower right: Restoration of the equilibrium by the administration of atropine plus Dopa or 5-HTP.

parkinsonism and shows the effects of Dopa or 5-HTP and atropine on the restoration of balance.

Another mechanism for the synergistic effect of anticholinergic drugs with Dopa or 5-HTP has also to be considered. The recent findings that various cholinergic blocking agents inhibit the uptake of dopamine in the neostriatum [6] suggest that these drugs might potentiate the Dopa effect by blocking the dopamine uptake which consequently will increase the amine concentration at the receptor site.

Apomorphine, which is known to stimulate the dopamine receptors, produces a transient relief of the tremor and potentiates the Dopa-induced relief of the tremor. The effects of amantadine are still somewhat obscure. The potentiation of the Dopa-induced turning behavior in rats points to a possible membrane pump blockade of the dopamine neurons by amantadine. On the other hand, amantadine potentiates the lesion-induced tremor, and it is conceivable that amantadine might produce both relief and potentiation of the tremor and that this effect might be dose dependent.

Summary

The effects of centrally acting drugs on motor disturbances were investigated in monkeys with ventromedial tegmental lesions exhibiting hypokinesia or hypokinesia and tremor. The administration of DL-Dopa or of DL-5-HTP causes a transient relief of the tremor. The relief of the tremor is effective at a lower dose of Dopa or 5-HTP when given in combination with anticholinergic drugs or with extracerebral decarboxylase inhibitors. The potentiation by atropine of the Dopa or 5-HTP induced relief of the tremor points to an imbalance between the sympathomimetic and cholinergic systems in parkinsonism. The synergistic effect of atropine and Dopa could also be due to the inhibition by the anticholinergic drugs of dopamine uptake in the striatum.

The tremor in monkeys is relieved by administration of apomorphine. Amantadine potentiates the Dopa-induced turning behavior in rats and potentiates the lesion-induced tremor in monkeys. 5-HTP protects the monkeys from harmaline-induced tremor. These results suggest that the tremor produced by harmaline might be related to its interaction with receptors for serotonin.

REFERENCES

1. Carlsson, A.: The occurrence, distribution and physiological role of catecholamines in the nervous system. Pharmacol. Rev. *11:* 490, 1959.
2. Ehringer, H., and Hornykiewicz, O.: Verteilung von noradrenalin und dopamin (3-hydroxytyramin) im gehirn des menschen und ihr verhalten bei erkrankungen des extrapyramidalen systems. Klin. Wschr. *38:* 1236, 1960.
3. Poirier, L. J., and Sourkes, T. L.: Influences of the substantia nigra on the catecholamine content of the striatum. Brain *88:* 181, 1965.
4. Goldstein, M., Anagnoste, B., Battista, A. F., Owen, W. S., and Nakatani, S.: Studies of amines in the striatum in monkeys with nigral lesions. J. Neurochem. *16:* 645, 1969.

5. GOLDSTEIN, M., ANAGNOSTE, B., OWEN, W. S., and BATTISTA, A. F.: The effects of ventromedial tegmental lesions on the disposition of dopamine in the caudate nucleus of the monkey. Brain Res. *4:* 298, 1967.
6. FUXE, K., and GOLDSTEIN, M.: Unpublished data.
7. BARBEAU, A.: Parkinson's disease as a systemic disorder. *In* Third Symposium on Parkinson's Disease (F. J. Gillingham and I. M. L. Donaldson, eds.) E. & S. Livingston, London, 1969.

PHYSIOLOGICAL CHANGES INDUCED IN TREMOR BY L-DOPA

J. Arbit

Let me begin by speaking of the concept of "motivation" which has been previously introduced. By this term we may be able to bridge the gap between what were referred to as therapeutic failures and the material I will present.

We are not only interested in terms of a patient's specific motivation for therapy but also in other aspects of his needs and their relation to treatment with L-Dopa. For example, we find that in a series of purposive acts we ask people to perform, such as throwing a large ball or walking a certain course, the ability shown is very much a function of the social interactions occurring at the time. The patient being treated with L-Dopa who is asked to carry out these acts with a group of other patients similarly treated will do significantly better than when these tasks are performed alone. There is thus not a basic mechanism or basic behavioral measure that one can call the measure of the L-Dopa effect, because it is related not only to the patient's individual motivations and needs but also to the conditions of the measurement situation itself.

Dr. Tasker noted that in his group of patients tremor apparently was not affected by L-Dopa. I should like to report a series of patients in which the results seem quite the contrary. These data were obtained from 36 patients undergoing treatment at Northwestern University Medical School and the Chicago Wesley Memorial Hospital. The mean age was 65; there were 23 men and 13 women in the study.

Figure 1 presents the results of L-Dopa effect upon tremor frequency in the right and left hand during hand extension and at rest. In each instance the increase is significant; in each instance the findings after treatment with L-Dopa are close to what we consider the normal range.

Figure 2 represents the effect of L-Dopa upon tremor amplitude, again right and left hand during extension and at rest. In each case the comparisons before and after drug treatment are significant.

Figure 3 presents our findings using as a measure the percent of time during a fixed time interval in which the tremor frequency is in the "organic" range (less than 8 cycles per second). Once again, the measure was obtained under

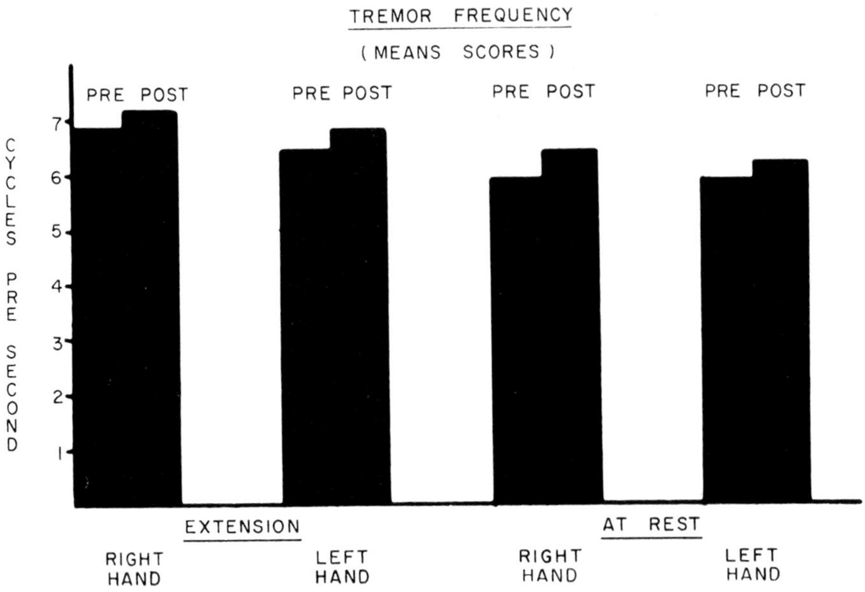
TREMOR FREQUENCY
(MEANS SCORES)
PRE POST
PRE POST
PRE POST
PRE POST
CYCLES PRE SECOND
7
6
5
4
3
2
1
EXTENSION
AT REST
RIGHT HAND
LEFT HAND
RIGHT HAND
LEFT HAND

Fig. 1

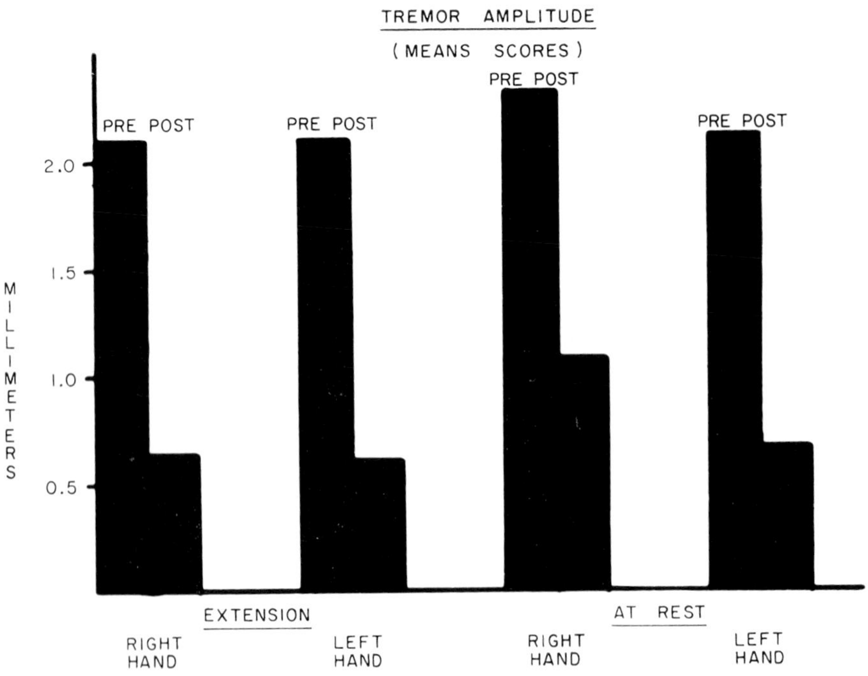
TREMOR AMPLITUDE
(MEANS SCORES)
PRE POST
PRE POST
PRE POST
PRE POST
MILLIMETERS
2.0
1.5
1.0
0.5
EXTENSION
AT REST
RIGHT HAND
LEFT HAND
RIGHT HAND
LEFT HAND

Fig. 2

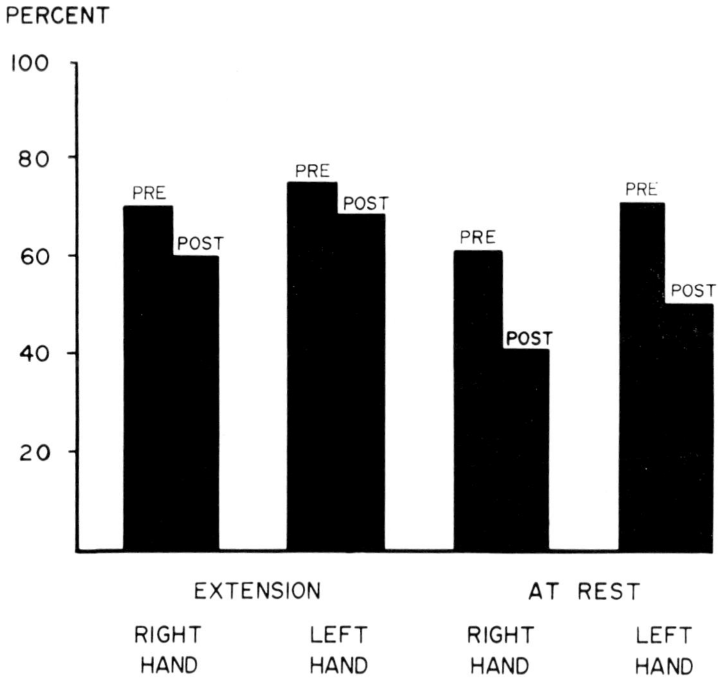

Fig. 3

extension and rest, right and left hand. As with the results already noted, these too show significant differences.

Figure 4 shows results obtained with a 70-year-old woman with an eight-year history of Parkinson's disease. The initial measure of tremor frequency was obtained in September, 1964. In January, 1969, a measure was made immediately preceding treatment with L-Dopa, and in February, 1969, tremor

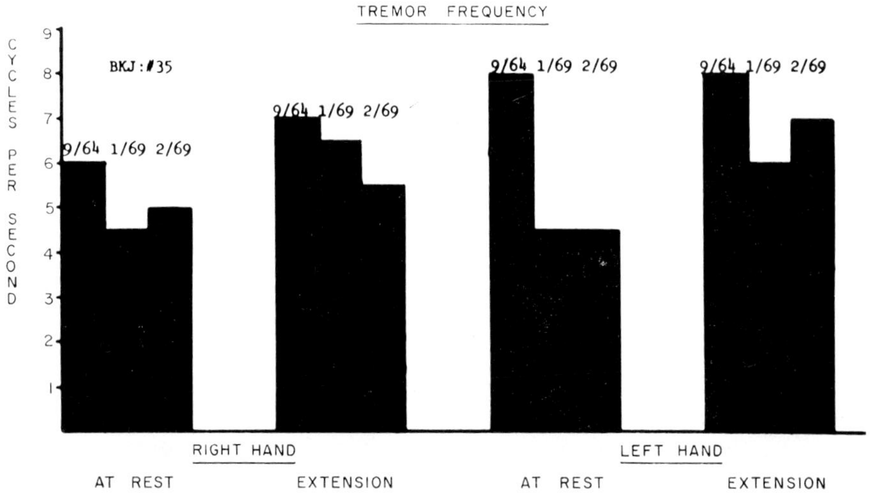

Fig. 4

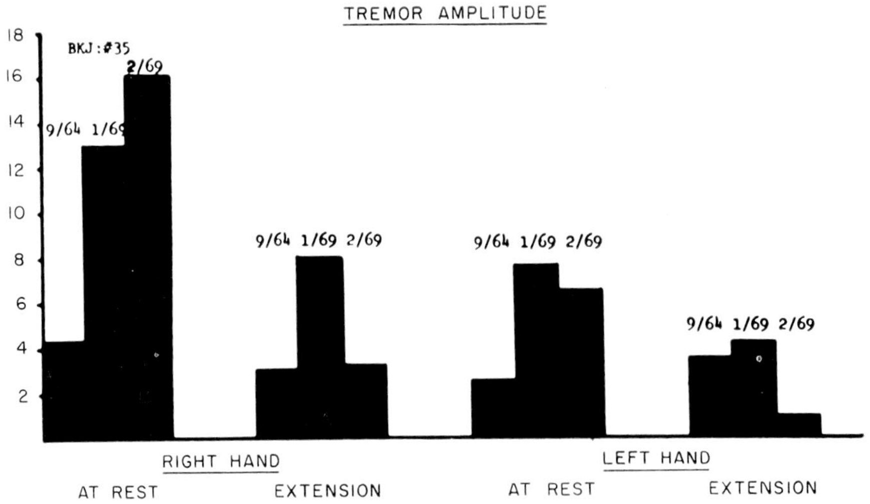

Fig. 5

frequency was recorded approximately one month after beginning treatment. The changes are representative of those we have observed in patients undergoing L-Dopa treatment.

Figure 5 presents tremor amplitude measures with this same woman; these findings again appear representative of the changes observed as a result of L-Dopa treatment.

GENERAL DISCUSSION

Dr. Sourkes: I think there is something more to be said about alpha-methyldopa. The program that led to its introduction into therapeutics was the search for decarboxylase inhibitors.[1, 2] After many years of testing such inhibitors, methyldopa proved to be the only one that acted as an antihypertensive agent. The others were discarded except for some limited experiments in biochemical pharmacology. Now they have been brought out again so that the original search for decarboxylase inhibitors, while not working out generally in hypertension, may yet provide useful drugs potentiating Dopa in Parkinson's disease.

There are occasional occurrences of Parkinson's disease-like symptoms in patients treated with methyldopa. If one accepts Dr. Poirier's concept of the rubro-olivo-cerebello-rubral loop, interruption of any part of which might lead to tremor after harmaline or methyldopa, for example, one could expect that in a very small percentage of the large number of patients who are hypertensive there might be some minimal brain damage in parts of this loop. When insult is added to this cerebral pathology by methyldopa, which interferes with monoamine synthesis and storage, the signs of parkinsonism may then occur.

Dr. Poirier: Some of Dr. Goldstein's findings, namely, those related to the beneficial effect of Dopa and of 5-HTP in monkeys with nigral lesions, are difficult to correlate with our own results. As a matter of fact, a few years ago Dr. Sourkes and I tested the effect of relatively high intravenous dosages of Dopa in monkeys with lesion-induced (not harmaline-induced) spontaneous tremor. We found very little action except that the weak effect corresponded to the relaxing state of the animal with spontaneous tremor.

If we consider my results concerning the effect of this tyrosine hydroxylase inhibitor (α-methyl-*p*-tyrosine), I may mention that several investigators, and especially Corrodi and Hanson, have shown that this substance does not interfere with the concentration of serotonin. As I pointed out, this substance interferes with the concentration of dopamine and norepinephrine, but not with the concentration of serotonin in the brain, and causes tremor in the lesioned monkey. This is, I think, quite clear-cut. If the biochemical evidence is correct, this means that a typical tremor may be induced without a change in the concentration of 5-HT.

Dr. Everett: I would like to comment concerning 5-HT in monkeys showing a Parkinson-like state produced by giving reserpine. If one gives 5-HTP, the precursor of serotonin, abduction of the limbs and tremor develop. Thus serotonin can produce and enhance some Parkinson-like signs and symptoms. These effects are also observed in mice. In other words, increasing the brain levels of 5-HT can produce a Parkinson-like picture in animals.

Dr. O'Reilly: Dr. Sandler asked if L-Dopa had any effect on menstruation. I have a young woman of 23 with parkinsonism who has shown no change whatsoever in her menstrual rhythm and is symptom-free on 1.6 grams of L-Dopa per day. On that dosage, L-Dopa has not affected her menstruation.

REFERENCES

1. SOURKES, T. L.: Inhibition of dihydroxyphenylalanine decarboxylase by derivatives of phenylalanine. Arch. Biochem. Biophys. *51:* 444, 1954.
2. SOURKES, T. L.: The action of alpha-methyldopa in the brain. Brit. Med. Bull. *21:* 66, 1965.

CHAPTER 7

Biochemical Aspects of Abnormal Movements Induced by L-Dopa

BIOCHEMICAL IMPLICATIONS OF DOPA-INDUCED ACTIONS ON THE CENTRAL NERVOUS SYSTEM, WITH PARTICULAR REFERENCE TO ABNORMAL MOVEMENTS *

Arvid Carlsson

In experimental animals L-Dopa causes a syndrome characterized mainly by certain actions on peripheral autonomically innervated organs and by various signs of central stimulation. Particularly striking is the ability of Dopa to reverse the reserpine syndrome.[1] It was, incidentally, during our attempts to analyze this action that we discovered that dopamine (DA) is a normal brain constituent occurring in even greater amounts than norepinephrine (NE) and with a distribution entirely different from that of NE, the main part occurring in the neostriatum. As reserpine, a drug capable of inducing the parkinsonian syndrome, caused DA to disappear, and since L-Dopa was able again to raise the DA level in the brain, we concluded that DA may be involved in the control of extrapyramidal motor functions and in the production of parkinsonian and choreic syndromes.

Since then, numerous studies have been carried out in which the central actions of L-Dopa have been further characterized in various species. There is little doubt that the actions are essentially caused by the DA, and partly also the NE, formed from Dopa, and that they involve the activation of specific DA and NE receptors in the central nervous system.[2, 3, 4]

All drugs capable of activating central DA receptors directly (e.g., topically

* Supported by grants from the Swedish Medical Research Council (B70-14X-155-06A, B70-14X-2157-04B).

applied DA, topically or systemically administered apomorphine) or indirectly (L-Dopa, which acts via DA and NE, or amphetamine, which causes release of catecholamines into receptors) have certain features in common. They cause central excitation and induce stereotype hyperkinesia of various types—gnawing, licking, sniffing, etc.[5]

There are thus strong reasons for implicating central monoamines, notably DA, in stereotype hyperkinesia.

However, the biochemical pharmacology of L-Dopa is complex. Theoretically, the central actions induced by this agent might be attributed to one or more of the following alternatives (Fig. 1).

A. L-Dopa itself; it is pharmacologically inert, but may possibly interfere with other amino acids in the body, e.g., with their intestinal absorption. Mutual interaction is possible.

B. Dopamine (DA) formed from L-Dopa
  1. in DA neurons; released onto DA receptors.
  2. in NE neurons; here DA may
     a. be converted to NE, which may be released onto NE receptors.
     b. displace endogenous NE onto NE receptors.
     c. be released as DA and activate NE receptors.
  3. in 5-HT neurons (5-hydroxytryptophan decarboxylase and Dopa decarboxylase are closely related or identical)
     a. DA released onto 5-HT receptors would probably have no effect.
     b. DA released onto DA receptors, e.g., of neighbor DA synapse, may possibly be of some importance in parkinsonism.
     c. DA displacing 5-HT, which in turn is released onto 5-HT receptors.
  4. in capillaries/pericytes.[6, 7, 8] Release may occur onto adjacent DA receptors: Dopa decarboxylase in these cells is low compared to intraneuronal decarboxylase, but may possibly be important in parkinsonism, where many monoamine neurons may have degenerated.
  5. in unknown structures such as hypothetical epinephrine neurons and other monoamine neurons.

C. Other metabolites.
  1. O-methylated amines; they are relatively inactive; COMT inhibition potentiates the actions of L-Dopa (see below).

Fig. 1. Fate of L-dopa in the central nervous system.

The amino acid, unlike its decarboxylation product dopamine (DA), is capable of passing through the capillary endothelium, probably by means of a specific carrier mechanism. The capillary/pericyte system appears to contain Dopa decarboxylase and monoamine oxidase (MAO); thus part of the L-Dopa is decarboxylated and the DA formed is at least partly deaminated within the capillary/pericyte system. The penetration of DA from this system into the brain tissue is counteracted by a desipramine-sensitive but reserpine-resistant pump mechanism, indicated; when the pump is blocked this penetration appears to be facilitated. By means of the decarboxylase inhibitor Ro 4–4602 it is possible to block the capillary decarboxylase, while only partially inhibiting the neuronal decarboxylase in the brain. Thus the enzymatic barrier is put out of function, allowing the L-Dopa to penetrate without loss (Constantinidis, de la Torre, Tissot and Geissbuhler[23]).

The Dopa penetrating into the parenchyma may reach decarboxylase-containing neurons (DA, NA or 5-HT neurons), or other structures as indicated. For further explanation, see text.

Fig. 1A serves to illustrate the general case, Fig. 1B the events taking place in 5-HT (left) and norepinephrine (right) neurons.

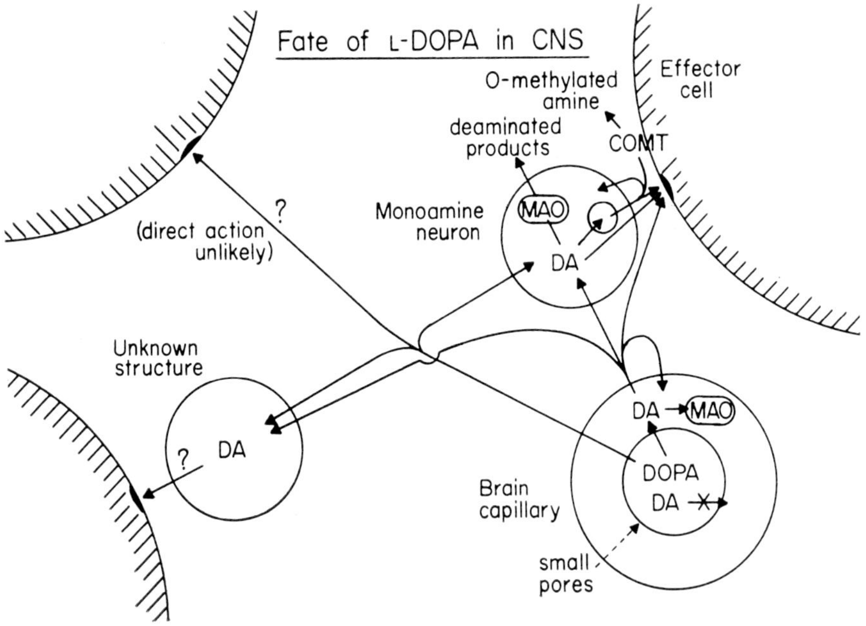
Fate of L-DOPA in CNS
Effector cell
O-methylated amine
deaminated products
COMT
?
(direct action unlikely)
Monoamine neuron
MAO
DA
Unknown structure
?
DA
DA
MAO
DOPA
DA
Brain capillary
small pores

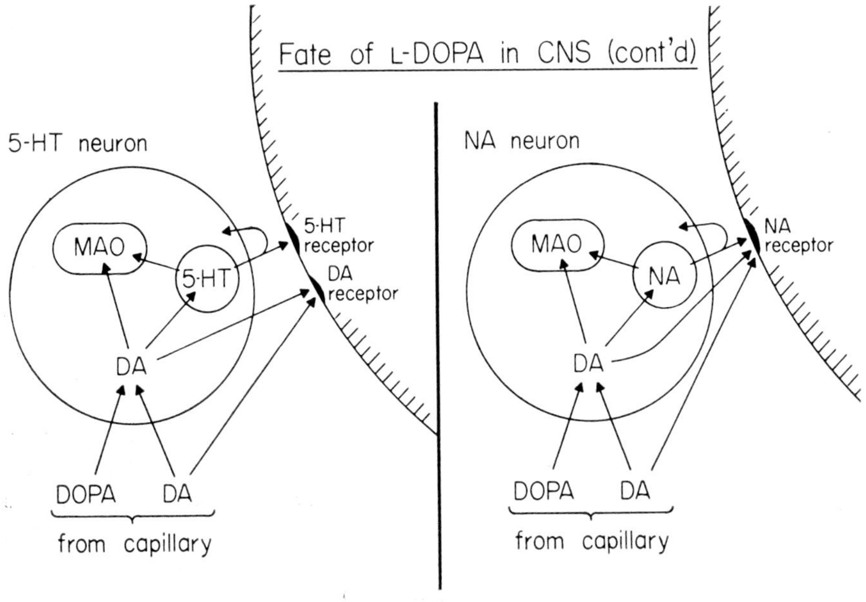
Fate of L-DOPA in CNS (cont'd)
5-HT neuron
MAO
5-HT
5-HT receptor
DA receptor
DA
DOPA
DA
from capillary
NA neuron
MAO
NA
NA receptor
DA
DOPA
DA
from capillary

2. deaminated metabolites; they are generally inactive; MAO inhibition potentiates the action of L-Dopa (see below).

When these various alternatives are considered, it may be concluded that mechanisms leading to activation of DA and NE receptors in the brain are most likely to be involved in the central actions of L-Dopa. An interaction with 5-HT neurons, e.g., might possibly contribute to some extent to the overall picture.

Parkinsonian Lesions and Overall Sensitivity to L-Dopa

We have to consider at least two factors: (1) loss of Dopa decarboxylase as a consequence of neuronal degeneration and (2) denervation supersensitivity. These two factors obviously work in opposite directions. There is an experimental basis for the existence of both phenomena in central dopaminergic synapses. We do not know their relative importance in parkinsonism. We have to consider the possibility that both factors vary considerably in different patients and also in different parts of the brain in the same patient. In a given patient a high degree of supersensitivity with relatively well-preserved Dopa decarboxylase would of course provide optimum conditions for a good therapeutic response with few side effects. However, suppose that in a given patient the balance between the two factors is different in different parts of, for example, the nigroneostriatal DA pathway or in different monoaminergic pathways. We then have conditions for the simultaneous occurrence of signs of overdosage and insufficient dosage with respect to different functions, as often seen in parkinsonian patients undergoing Dopa treatment. With respect to abnormal movements, which probably should be looked upon as signs of overdosage at certain monoaminergic synapses, it is interesting to note that the pattern observed generally appears to be characteristic for the individual patient (head movements in one patient, leg movements in another, buccolingual movements in a third patient, etc.). This may reflect differences in overall sensitivity to L-Dopa in different parts of a given monoaminergic system, most probably an uneven loss of nigroneostriatal DA neurons and thus of Dopa decarboxylase.

Apart from Dopa decarboxylase activity and supersensitivity, several other factors may influence sensitivity to Dopa, e.g., the activities of the various catecholamine-inactivating mechanisms—the so-called membrane pump, MAO and COMT. Needless to say, extracerebral factors such as absorption and metabolism have to be considered.

It appears that hyperkinesia may sometimes be induced by L-Dopa at a higher integrative level than usual. We observed a parkinsonian patient who responded well to L-Dopa treatment apart from persistent amimia and aphonia; therefore the dose was elevated somewhat. Shortly afterward the patient developed a strange kind of hyperkinesia, with apparently compulsive hammering, day and night putting up nails on doors and walls. When his wife tried to stop him he threatened to throw her out. After the dose had been lowered to the earlier level the compulsive hyperactivity and aggressiveness disappeared.[9]

It is interesting to note that amphetamine, an agent causing central stimulation via release of catecholamines, may also cause hyperkinesias at different levels of integration. Best known is the so-called punding, where an addict can go on for hours with some stereotype activity such as taking apart and putting together a watch, sorting out a handbag or tidying up a flat. This appears to approach the hammering just mentioned after Dopa treatment. However, stereotype movements of a more elementary type have also been described in amphetamine addicts. For example, the mouth may be drawn sideways or rounded, or the tongue may move in and out. Grinning and chewing may also occur, or there may be jerking movements of the arms.[10]

There is thus strong evidence that activation of catecholamine receptors may result in stereotype hyperkinesia. If we assume that the elementary stereotype movements which may occur without mental involvement are due to synaptic activation of the nigroneostriatal DA pathway, it is tempting to suggest that some other pathway(s) is (are) responsible for the more complex patterns, e.g., the DA and/or NE pathways belonging to the limbic system. Analogous considerations may be applied to the sexual hyperactivity and the various mental changes observed after both amphetamine and (occasionally) L-Dopa administration.

The Hyperkinetic and Choreic Syndromes Induced by Long-Term Treatment with Neuroleptic Drugs; the Possible Mechanism Involved

The neuroleptic drugs—chlorpromazine, haloperidol and reserpine—are generally known to cause actions which in many respects are opposite to those of Dopa, e.g., parkinsonism and other syndromes with hypokinesia and increased muscle tone. The mechanism of these effects is probably a deficiency in catecholamines (reserpine) or blockade of catecholamine receptors (chlorpromazine and haloperidol). However, these drugs sometimes produce hyperkinetic and choreic syndromes which are remarkably similar to L-Dopa-induced abnormal movements. These syndromes generally set in insidiously and may persist for a long time after cessation of medication.

How can we explain the paradox that more or less the same syndrome may be induced by excess and deficit of catecholamines at receptor sites? A tentative explanation may be as follows. Deficiency of catecholamines at receptor sites appears to activate a trans-synaptic feedback mechanism by means of which the release and turnover of the DA and NE in the brain are accelerated.[4] If the condition persists changes can be seen histochemically in the cell bodies of the catecholamine neurons, indicating an increased formation of storage granules, etc. Biochemically, an increased synthesis of tyrosine hydroxylase has been detected.[11] As these various neuronal components are transported down the axons to the terminal system, the capacity for synthesis, storage and release of transmitter will further increase. At the same time the deficiency of catecholamines at receptor sites will induce denervation supersensitivity. During long-term therapy these changes may ultimately reach such a degree that whenever the concentration of the neuroleptic agent in the brain tends to go down these compensatory changes are unmasked, resulting in a syndrome that is essentially the opposite of the initial action of the drug.

Admittedly this explanation is hypothetical. It seems, however, to be substantiated by the available experimental and clinical data. It is thus possible to explain both the Dopa- and the neuroleptic-induced hyperkinetic and choreic syndromes as being due to excess activation of DA receptors.

The existence of feedback mechanisms would imply that during L-Dopa medication the activity and transmitter turnover of catecholamine neurons would decrease. It is difficult to predict the long-term consequences of this change. Enzyme and granule production might be reduced. On the other hand the possibility exists that the unloading of damaged neurons may stimulate processes of recovery. If this is true L-Dopa medication should begin at an early stage of the disease in order to retard the progress of neuronal degeneration.

L-Dopa and Parkinsonian Tremor

It is generally recognized that tremor responds less promptly than akinesia and rigidity to L-Dopa. In certain cases it is hard to detect a clear-cut effect on the tremor in spite of a good therapeutic response otherwise. How can we explain this? I should like to discuss three possibilities.

1. The tremor may primarily be due to loss of DA neurons and is resistant to Dopa because the dopaminergic synapses involved in tremor may have a low sensitivity to L-Dopa; the Dopa decarboxylase activity remaining may be low. There are two strong arguments in favor of this interpretation: (1) in a certain proportion of cases L-Dopa has a clear-cut action on the tremor, and (2) characteristic parkinsonian tremor can be induced by phenothiazines and butyrophenones, which selectively block DA (and NE) receptors but do *not* block 5-HT receptors in the brain.[4]
2. The dopaminergic synapses involved in tremor may respond to L-Dopa, but the effect is counteracted by an epinephrine-like effect of the circulating catecholamines; the ability of epinephrine in low dosage to enhance parkinsonian tremors already is well known. The weakness of this interpretation is that adrenalinemia is hardly produced by L-Dopa and it is doubtful if circulating DA can aggravate parkinsonian tremor. The tremor-aggravating effect of epinephrine is β-adrenergic; thus β-adrenergic blocking agents might be tried in cases where tremor persists in spite of L-Dopa treatment.
3. The tremor is induced by some mechanism other than DA deficiency. For example, 5-HT deficiency due to loss of 5-HT neurons has been implicated. Administration of 5-hydroxytryptophan would then be useful, but there are some difficulties involved in the simultaneous use of L-Dopa and 5-hydroxytryptophan owing to competition between the two.

Some attention should be drawn to the fact that 5-HT formation is low in parkinsonism [12] and may possibly be further inhibited by L-Dopa. L-Dopa may possibly interfere with 5-HT formation by different mechanisms: (1) competition with tryptophan for carrier mechanisms in the intestinal mucosa and blood-brain barrier, (2) inhibition of the 5-HT synthesizing enzymes tryptophan hydroxylase and 5-HTP decarboxylase, which is identical with—or closely related to—Dopa decarboxylase. Examination of 5-hydroxyindole acetic acid in the cerebrospinal fluid will show whether this interference with 5-HT formation does in fact exist.

In conclusion, the weight of evidence in my opinion suggests that deficiency of DA at receptor sites in the brain is an important factor in parkinsonian tremor. Other factors, e.g., deficiency of 5-HT at central receptor sites, may possibly contribute, however.

Possible Ways to Modify the Therapeutic Response to L-Dopa

I should like to discuss three main alternatives: (1) Dopa-potentiating agents, (2) Dopa analogues, and (3) direct DA receptor-stimulating agents.

Potentiation of the central action of Dopa can be obtained by several types of agent:

1. Peripheral decarboxylase inhibitors [13, 14, 15]
2. MAO inhibitors [1]
3. COMT inhibitors [16]
4. Amphetamines [17]
5. Membrane-pump blocking agents (e.g., desipramine)[18]
6. Anticholinergic agents.[19]

Peripheral decarboxylase inhibitors will reduce Dopa dosage and side effects due to circulating catecholamines. The will hardly be expected to solve the problem of stereotype hyperkinesia.[14]

It is impossible to predict whether groups 2–5 in addition to potentiation will modify the response, and, if so, if this will be favorable or unfavorable; they may be worth trying, even though several of these agents have serious side effects. The advantages of combination with anticholinergic therapy are of course already utilized routinely.

Certain Dopa analogues are readily decarboxylated by Dopa decarboxylase, and their decarboxylation products possess physiological activity.[20] It is impossible to predict whether they would induce a more or less favorable therapeutic response than L-Dopa. Meta-tyrosine (40 mg intravenously) did not counteract akinesia in parkinsonian patients.[21]

The only compound penetrating into the brain and stimulating DA receptors directly is apomorphine; structurally it can be looked upon as a DA analogue.[4, 22] It seems to have certain unfavorable properties: low oral relative to parenteral activity, a strong emetic action, and central depressant properties.

As mentioned, the sometimes unfavorable pattern of the response induced by L-Dopa, e.g., resistance of certain symptoms such as aphonia and tremor or appearance of side effects such as choreiform movements at a low dosage, may be due to an unfavorable distribution of the Dopa decarboxylase remaining in the brain. Insofar as this hypothesis is correct, direct DA-receptor stimulating agents would have something to offer. Here the medicinal chemists may help us to develop orally active drugs with high selectivity.

Conclusions

Both the antiparkinsonian and the hyperkinetic actions of L-Dopa are probably mainly due to the activation of neostriatal dopamine receptors.

The release of endogenous dopamine onto receptors is normally under rigorous feedback control. In contrast, the parkinsonian patient undergoing L-Dopa treatment is at the mercy of the largely uncontrolled concentration of dopamine at receptor sites.

The chief determinants of the neostriatal response to L-Dopa may be the concentration of Dopa in the neostriatum, the activity of the remaining neuronal Dopa decarboxylase in the neostriatum, and finally the degree of supersensitivity developed.

The simultaneous occurrence of parkinsonian symptoms and hyperkinesias during Dopa therapy may be due to an uneven pattern of Dopa-receptor activation in different parts of the neostriatum. The individual response pattern may depend on the topography of the neostriatal lesions.

REFERENCES

1. Carlsson, A., Lindqvist, M., and Magnusson, T.: 3,4-Dihydroxyphenylalanine and 5-hydroxytryptophan as reserpine antagonists. Nature *180:* 1200, 1957.
2. Carlsson, A.: Evidence for a role of dopamine in extrapyramidal functions. Acta Neuroveg. *26:* 484, 1964.
3. Carlsson, A.: Morphologic and dynamic aspects of dopamine in the central nervous system. *In* Biochemistry and Pharmacology of the Basal Ganglia (E. Costa, L. J. Coté and M. D. Yahr, eds.). New York, 1966.
4. Andén, N.-E., Carlsson, A., and Häggendal, J.: Adrenergic mechanisms. Ann. Rev. Pharmacol. *9:* 119, 1969.
5. Randrup, A., and Munkvad, I.: Behavioural stereotypes induced by pharmacological agents. Pharmakopsychiat. Neuro-Psychopharmakol. *1:* 18, 1968.
6. Bertler, A., Falck, B., Owman, Ch., and Rosengren, E.: The localization of monoaminergic, blood-brain barrier mechanisms. Pharm. Rev. *18:* 369, 1966.
7. Hamberger, B.: Reserpine-resistant uptake of catecholamines in isolated tissues of the rat. A histochemical study. Acta Physiol. Scand. Suppl. *295:* 1, 1967.
8. Fuxe, K., and Ungerstedt, U.: Histochemical studies on the effect of (+)-amphetamine, drugs of the imipramine group and tryptamine on central catecholamine and 5-hydroxytryptamine neurons after intraventricular injection of catecholamines and 5-hydroxytryptamine. Europ. J. Pharmacol. *4:* 135, 1968.
9. Andén, N.-E., Carlsson, A., Kerstell, J., Magnusson, T., Olsson, R., Roos, B.-E., Sten, B., Steg, G., Svanborg, A., Thieme, G., and Werdinius, B.: Oral L-Dopa treatment of parkinsonism. Acta Med. Scand., in press.
10. Rylander G.: Clinical and medico-criminological aspects of addiction to central stimulating drugs. *In* Abuse of Central Stimulants (F. Sjöqvist and M. Tottie, eds.). Stockholm, 1969.
11. Mueller, R. A., Thoenen, H., and Axelrod, J.: Increase in tyrosine hydroxylase activity after reserpine administration. J. Pharmacol. *169:* 74, 1969.
12. Olsson, R., and Roos, B.-E.: Concentrations of 5-hydroxyindole-acetic acid and homovanillic acid in the cerebrospinal fluid after treatment with probenecid in patients with Parkinson's disease. Nature *219:* 502, 1968.
13. Bartholini, G., and Pletscher, A.: Cerebral accumulation and metabolism of C^{14}-dopa after selective inhibition of peripheral decarboxylase. J. Pharmacol. *161:* 14, 1968.
14. Cotzias, G. C., Papavasiliou, P. S., and Gellene, R.: Modification of parkinsonism: Chronic treatment with L-Dopa. New Eng. J. Med. *280:* 337, 1969.

15. BUTCHER, L. L., and ENGEL, J.: Peripheral factors in the mediation of the effects of L-Dopa on locomotor activity. J. Pharm. Pharmacol. *21:* 614, 1969.
16. CARLSSON, A., LINDQVIST, M., FILA-HROMADKO, S., and CORRODI, H.: Synthese von Catechol-O-methyl-transferase-hemmende Verbindungen. In den Catecholaminmetabolismus eingreifende Substanzen 1. Helv. Chim. Acta *45:* 270, 1962.
17. HANSON, L. C. F.: Evidence that the central action of (+)-amphetamine is mediated via catecholamines. Psychopharmacologia *10:* 289, 1967.
18. EVERETT, G. M., WIEGAND, R. G., and RINALDI, F. V.: Pharmacologic studies of some nonhydrazine MAO inhibitors. Ann. N.Y. Acad. Sci. *107:* 1068, 1963.
19. SCHEEL-KRÜGER, J.: Pharmacological studies on a counterbalancing adrenergic-cholinergic system in the brain. Acta Physiol. Scand. Suppl. *330:* 66, 1969.
20. BLASCHKO, H., and CHRUŚIEL, T. L.: The decarboxylation of amino acids related to tyrosine and their awakening action in mice. J. Physiol. *151:* 272, 1960.
21. BIRKMAYER, W.: Experimentelle Befunde und neue Aspekte bei extrapyramidalen Erkrankungen. Wien. Z. Nervenheilk. *23:* 128, 1966.
22. ERNST, A. M.: Mode of action of apomorphine and dexamphetamine on gnawing compulsion in rats. Psychopharmacologia *10:* 316, 1967.
23. CONSTANTINIDIS, J., DE LA TORRE, J. C., TISSOT, R., and GEISSBUHLER, F.: La barrière capillaire pour la dopa dans le cerveau et les différents organes. Psychopharmacologia *15:* 75, 1969.

TYROSINE METABOLISM IN PARKINSON'S DISEASE *

André Barbeau, André Arsenault, and John Donaldson

The metabolism of tyrosine has been implicated in the pathogenesis of Parkinson's disease for many reasons. It is well known, of course, that dopamine is found in reduced concentration in the brain of parkinsonian patients.[1] A similar decrease in striatal dopamine has been demonstrated in monkeys with specific brain stem lesions.[2, 3] From such studies possible explanations for the low dopamine concentrations were immediately considered: the first proposed were a decrease in the total number of nigrostriatal dopaminergic fibers and a defect in the actual synthesis of dopamine. There is evidence for and against both concepts.

From human Dopa-loading experiments, Barbeau and collaborators [4] postulated in 1961 a defect in Dopa-decarboxylase. Soon afterwards, Hornykiewicz and his collaborators [5] indicated that the actual measurement of this enzyme in autopsied material, although very low, was not significantly different from control brains. These autopsies were carried out more than six hours after death. Recently, however, Metzel and collaborators,[6] using biopsy material obtained during stereotaxic surgery, clearly demonstrated a lower level of this enzyme in parkinsonian thalamus than in other types of dyskinesias.

* Supported by Grants SP-1 and MA-2530 from the Medical Research Council of Canada and a grant from the Huntington's Chorea Foundation.

Other evidence, reviewed elsewhere,[7,8] implicated in a more direct fashion the next step above, i.e., the transformation of tyrosine to Dopa. Tyrosine hydroxylase, as is well known, is the rate-limiting step in the synthesis of catecholamines and therefore could easily be the most vulnerable enzyme.

Studies in experimental animals (monkeys and cats) by Moore and Heller,[9] Poirier,[2] McGeer et al.,[10] and Goldstein and collaborators[11] have confirmed the fact that when a lesion is strategically placed along the nigrostriatal dopaminergic pathway there is a decrease in the concentration of tyrosine hydroxylase in the striatum. Other enzymes such as Dopa decarboxylase can also be involved, as demonstrated by Goldstein.[11] It seems evident from these studies that both synthesis *and* storage of DA are involved in the experimental animals.

The concentration of blood tyrosine is usually considered normal in Parkinson's disease[12,13,14] except in advanced akinetic patients where it is low or at the lower limits of normal.[13] Tyrosine load tests, taking into account the known diurnal variation in blood concentrations of this substance,[15] can be considered normal.[13] These studies have recently been confirmed.[14]

Despite these normal results, some further evidence exists for a possible abnormal metabolism of tyrosine in Parkinson's disease. In 1965, Barbeau and collaborators,[16] using the Friedhoff technique of extraction on paper chromatography, had demonstrated an increase in 85 percent of parkinsonians of a substance with a Rf similar to that of 3,4-dimethoxyphenylethylamine. This was confirmed by others but the identity of the substance was challenged and the whole problem became lost in the confusion generated around the so-called "pink spot."[17] We now know, with the aid of more refined methods, that at least 17 substances, including some drug metabolites, can migrate in the initial system at a similar Rf.[18,19] However Boulton[20] and Kuehl[21] have produced strong evidence, backed by mass spectrography, that some of these substances can be definitely identified as paratyramine and/or tryptamine and that both are found in increased quantity in parkinsonian urines. Smith and Kellow have recently confirmed the increase in excretion of both paratyramine and tryptamine in parkinsonian urines.[22] Again the conclusion to be drawn from such studies would seem to be that a block may exist at the tyrosine hydroxylase level in Parkinson's disease.

The metabolism of tyrosine is extremely complex and few studies have considered the total picture. O'Reilly studied the metabolism of ^{14}C tyrosine in a few parkinsonians, but could not conclusively demonstrate a clear-cut anomaly.[23] Other authors have reinvestigated this aspect before and after L-Dopa therapy in parkinsonian patients. Cotzias first indicated an occasional increase in PBI but carefully stated that no conclusions could yet be drawn from this observation.[24] It now appears that this may have been due to contamination from a certain amount of iodine contained in the capsules used for L-Dopa. Our group could not demonstrate changes in T_3 and T_4 before or after L-Dopa, and this is the general finding of most authors. Cotzias demonstrated a clinical worsening of the patient's condition when given phenylalanine.[7] This we also confirm.

TABLE 1. Tyrosine and Phenylalanine Before and After L-Dopa Treatment

	Phenylalanine (mg/100 ml)	*Tyrosine (μM/100 ml)*
Number of cases	12	12
Before Dopa	14.2±0.9	3.7±0.2
After Dopa	14.7±0.7	5.1±0.4 *

* $p < 0.05$.

In a series of parkinsonian patients we could not demonstrate a change in blood phenylalanine concentration after treatment with L-Dopa. On the other hand, tyrosine levels as measured increased slightly but significantly over the base-line period (Table 1). It must be remembered, however, that the method used for tyrosine determination in the blood (Udenfriend and Cooper: α-nitroso-naphthol fluorescence in the presence of nitric acid) also detects p-tyramine, and we therefore cannot be certain of the absolute identification of tyrosine in this case. Nevertheless, whether it is tyrosine or p-tyramine, these findings would indicate that, in the presence of large amounts of L-Dopa, the tyrosine normally utilized for catecholamine synthesis could overflow toward other pathways.

Such a possibility would have theoretical interest, if confirmed, because some of the metabolites of tyrosine could be implicated in the production of abnormal movements. Parahydroxyphenylpyruvic acid (pHPPA) and homogentisic acid (HGA) are the first two that we considered.

As seen in Table 2, pHPPA (determined through the method of La Du which measures the reaction of phenolic groups with mercuric ions to give a colored chelate) appears elevated in Parkinson's disease and in Huntington's chorea. The specificity of this method, however, leaves much to be desired and further conclusions at this stage are not warranted, especially since homogentisic acid levels in the same urines were normal.

These discrepancies in our preliminary results, which have not yet been resolved, lead us to be extremely cautious. We should not repeat here the errors of the "pink spot" controversy. Therefore all we can conclude is that there is some evidence to indicate that an abnormal tyrosine metabolism may be involved in Parkinson's disease. Further studies are in progress to delineate this possible defect and to clarify its relationship, if any, to symptoms before and after L-Dopa therapy.

TABLE 2. Parahydroxyphenylpyruvic Acid (pHPPA) in Parkinson's Disease and Huntington's Chorea

Number of Subjects		*pHPPA (mg/24 hrs±S.D.)*
Control subjects	39	5.9±2.8
Parkinson		
• before Dopa	25	7.9±0.7
• after Dopa	27	15.5±1.8
Huntington's chorea	3	10.9±0.6

ACKNOWLEDGMENTS

The authors would like to thank Miss C. Simard, Miss N. Poirier, and Miss M. Giroux for technical assistance.

REFERENCES

1. EHRINGER, H., and HORNYKIEWICZ, O.: Verteilung von noradrenalin und dopamine (3-hydroxytyramin) im Gehirn des Menschen und ihr Verhalten bei Erkrankungen des Extrapyramidalen Systems. Klin. Wschr. *38:* 1236, 1969.
2. POIRIER, L. J., and SOURKES, T. L.: Influence of the substantia nigra on the catecholamine content of the striatum. Brain *88:* 181, 1965.
3. ANDÉN, N. E., CARLSSON, A., DAHLSTRÖM, A., FUXE, K., HILLARP, N. A., and LARSSON, K.: Demonstration and mapping out of nigroneostriatal dopamine neurons. Life Sci. *3:* 523, 1964.
4. BARBEAU, A., SOURKES, T. L., and MURPHY, G. F.: Les catécholamines dans la maladie de Parkinson. *In* Monoamines et Système nerveux central (J. de Ajuriaguerra, ed.). Masson & Cie, Paris, 1962.
5. BERNHEIMER, H., and HORNYKIEWICZ, O.: Das Verhalten eineger Enzyme im Gehirn normales und Parkinson Kranker Menschen. Arch. Exp. Path. Pharmak. *243:* 295, 1962.
6. METZEL, E., WEINMANN, D., and RIECHERT, T.: A study of the enzymes of Dopa metabolism in parkinsonism from biopsies of the basal ganglia. *In* Third Symposium on Parkinson's Disease (F. J. Gillingham and I. M. L. Donaldson, eds.). E. & S. Livingston, London, 1969.
7. COTZIAS, G. C., PAPAVASILIOU, P. S., and GELLENE, R.: Modification of parkinsonism: chronic treatment with L-Dopa. New Eng. J. Med. *280:* 337, 1969.
8. BARBEAU, A.: Dopamine and dopamine metabolites in Parkinson's disease—A review. Proc. Aust. Assoc. Neurol. *5:* 95, 1968.
9. MOORE, R. Y., and HELLER, A.: Subcortical systems participating in the control of caudate nucleus dopamine content. *In* Progress in Neuro-Genetics (A. Barbeau and J. R. Brunette, eds.). Excerpta Medica Foundation, Amsterdam, 1969.
10. MCGEER, P. L., WADA, J. A., GIBSON, S., LOESER, C., and MCGEER, E. G.: Tyrosine hydroxylase levels in the cat caudate and basal forebrain following selective mid-brain lesions. *In* Progress in Neuro-Genetics (A. Barbeau and J. R. Brunette, eds.). Excerpta Medica Foundation, Amsterdam, 1969.
11. BATTISTA, A. F., GOLDSTEIN, M., NAKATANI, S., and ANAGNOSTE, B.: The effects of ventrolateral thalamic lesions on tremor and the biosynthesis of dopamine in monkeys with lesions in the ventromedial tegmentum. J. Neurosurg. *31:* 164, 1969.
12. BRUCK, J., GERSTENBRAND, F., GRUNDIG, E., and PROSENZ, F.: Stoffwechselveränderungen bei extrapyramidalen Syndromen und vorläufige Therapeutische Konsequenzen. Fortschr. Neurol. Psychiat. *33:* 677, 1965.
13. MONTPLAISIR, J., and BARBEAU, A.: Serum tyrosine in Parkinson's disease. *In* Progress in Neuro-Genetics (A. Barbeau and J. R. Brunette, eds.). Excerpta Medica Foundation, Amsterdam, 1969.
14. BRAHAM, J., SAROVA-PINHAS, I., CRISPIN, M., GOLAN, R., LEVIN, N., and SZEINBERG, A.: Oral phenylalanine and tyrosine tolerance tests in parkinsonian patients. Brit. Med. J. *2:* 552, 1969.
15. WURTMAN, R. J., CHOU, C., and ROSE, C. M.: Daily rhythm in tyrosine concentration in human plasma: persistence on low-protein diets. Science *158:* 660, 1967.

16. Barbeau, A., DeGroot, J. A., Joly, J. G., Raymond-Tremblay, D., and Donaldson, J.: Urinary excretion of a 3,4-dimethoxyphenylethylamine-like substance in Parkinson's disease. Rev. Canad. Biol. *22:* 469, 1963.
17. Barbeau, A.: The "pink spot," 3,4-dimethoxyphenylethylamine and dopamine. Relationship to Parkinson's disease and to schizophrenia. Rev. Canad. Biol. *26:* 55, 1967.
18. Bartlett, M. F., Ulshaffer, P. R., Korzum, B. P., Ox, H., Hunt, M., Hackett, V., and Barbeau, A.: A chemical investigation of the urinary bases in Parkinson's disease. *In* Progress in Neuro-Genetics (A. Barbeau and J. R. Brunette, eds.). Excerpta Media Foundation, Amsterdam, 1969.
19. Pushpathadam, J. J., and Barbeau, A.: The "pink spot" and Parkinson's disease. *In* Progress in Neuro-Genetics (A. Barbeau and J. R. Brunette, eds.). Excerpta Medica Foundation, Amsterdam, 1969.
20. Boulton, A. A.: The pink spot in parkinsonism. *In* Progress in Neuro-Genetics (A. Barbeau and J. R. Brunette, eds.). Excerpta Medica Foundation, Amsterdam, 1969.
21. Kuehl, F. A., Jr., Van den Heuvel, W. J. A., and Ormond, R. E.: The pink spot and Parkinson's disease. *In* Progress in Neuro-Genetics (A. Barbeau and J. R. Brunette, eds.). Excerpta Medica Foundation, Amsterdam, 1969.
22. Smith, I., and Kellow, A. H.: Aromatic amines and Parkinson's disease. Nature *221:* 1261, 1969.
23. O'Reilly, S., Urban, M., and Smekal, D.: The intermediary metabolism of tyrosine in basal ganglia. *In* Progress in Neuro-Genetics (A. Barbeau and J. R. Brunette, eds.). Excerpta Medica Foundation, Amsterdam, 1969.
24. Cotzias, G. C., VanWoert, M. H., and Schiffer, L. M.: Aromatic amino acids and modification of parkinsonism. New Eng. J. Med. *276:* 374, 1967.

TYROSINE METABOLISM IN MOVEMENT DISORDERS

Arthur Dale Ericsson, Nansie S. Sharpless, Edward Honos, Walter E. Szpunar, and D. S. McCann

Tyrosine metabolism is intrinsically interwoven into the neurochemistry of movement disorders. Although it is the only precursor of dopamine (DA), there are other vital metabolic pathways in which tyrosine is preeminent. Nevertheless, only two of the major metabolic routes of tyrosine metabolism will be considered here: (1) Transamination, from which p-hydroxyphenylpyruvic acid (pHPPA) is formed; this route continues via homogentisic acid to the Krebs cycle. (2) Ring hydroxylation, by which 3,4-dihydroxyphenylalanine (Dopa) is formed; this route leads to the biologically active catecholamines.

These two pathways are interrelated. Normally, the rate of tyrosine transamination in the brain is 100 times faster than that of hydroxylation. There is evidence that depletion of brain catecholamine levels leads to a decrease in the activity of brain tyrosine transaminase, which may serve as a mechanism by which tyrosine is shunted into the hydroxylation pathway.[1] Repletion of

Dopa and DA by the in vivo administration of L-Dopa returns brain tyrosine transaminase activity to normal.

The 24-hour urinary excretion of pHPPA in "normal" control subjects is of the order of 9.2 ± 3.2 * mg per 24 hrs.[2] Essentially the same value was obtained for a series of hospitalized subjects with a variety of diseases. Significant increases in urinary pHPPA levels were found in liver disease, connective tissue diseases,[2] and in parkinsonism.[3] Interestingly, it was the patients belonging to the latter group who showed the highest average pHPPA values. Initially, our finding of elevated pHPPA levels in 20 patients with parkinsonism with a mean of 16.6 mg per 24 hrs and a standard deviation of 4.5, and in a larger, more controlled series of hospitalized patients with a mean of 22.9 mg per 24 hrs, seemed to be somewhat of a paradox to the concept that tyrosine transaminase might regulate the synthesis of DA. Abundant evidence is available confirming that DA is localized in the globus pallidus, substantia nigra, and caudate nucleus of the brain.[4, 5] Ehringer and Hornykiewicz [6] demonstrated that DA was virtually absent from the caudate nucleus and putamen of patients who had died with parkinsonism. The concentration of homovanillic acid (HVA) in these regions was also found to be subnormal [7] in patients with parkinsonism. Furthermore, the magnitude of the elevation of pHPPA in the urine of parkinsonism patients appeared inexplicable on the basis of alterations in brain tyrosine transaminase activity alone.

A search of the literature revealed that in 1963 Kusch and Heinrich [8] found elevation of liver tyrosine transaminase activity in animals treated with chlorpromazine and reserpine. Adrenalectomy, per se, made no difference; furthermore, the same two drugs still elevated tyrosine transaminase activity in adrenalectomized animals. The authors were interested in ACTH and adrenocortical steroids; consequently no attempt was made to interpret these findings in terms of concentration of catecholamines in the central nervous system. Black and Axelrod in 1968 [9] did a very similar, though rather more elegant, series of experiments in search of evidence that hepatic tyrosine transaminase activity might be controlled by humoral or neural mechanisms. Their studies demonstrated that pharmacological depletion of central nervous system catecholamines increased hepatic tyrosine transaminase and that the phenomenon was reversible with L-Dopa. It appears, then, that the elevation of pHPPA in the urine of the parkinsonian subject is due to abnormally high hepatic tyrosine transaminase activity secondary to the well-established pathological depletion of catecholamines in the basal ganglia of these patients.

Tyrosine metabolism via the Dopa pathway to the catecholamines has been studied intensively by Goodall and Alton [10] in normal subjects. Utilizing radioactive DA (which does not cross the blood-brain barrier), they demonstrated that only 25 percent of the tracer dose was converted to norepinephrine (NE) and excreted as such or as its metabolic products including epinephrine (E), dihydroxymandelic acid, vanillylmandelic acid (VMA), normetanephrine, or metanephrine and their conjugates. Seventy-five percent of the administered DA was catabolized directly, the principal excretion products

* Standard deviation.

being 3-methoxy-dopamine, 3,4-dihydroxyphenylacetic acid (DOPAC), 3-methoxy-4-hydroxyphenylethanol, 3,4-dihydroxyphenylethanol, HVA, and their conjugates. The predominant single product (19 percent of the dose) resulting from exogenous administration of DA was HVA.

The urinary concentration of DA itself in parkinsonism has been reported variously as depressed,[11] as elevated,[12] and as normal.[13] HVA has been reported as depressed [14, 15] although the statistical significance of the depression has been questioned.[13] In our hands, urinary HVA levels of 4.0 ± 1.7 * mg per 24 hrs appear to be on the low side of the normal range. The statistical significance of this value is questionable, however, until our series of normal patients is larger than at the time of this writing. VMA values of 6.8 ± 2.4 * mg per 24 hrs (normal range 1.8 to 7.6 mg per 24 hrs with a mean of 5.2 mg per 24 hrs) appears to be slightly elevated. Most of these determinations were made on parkinsonism patients just hospitalized and removed for two to three days from all medication. This may have resulted in sufficient stress to have increased epinephrine synthesis and, consequently, VMA excretion. Under the influence of orally administered L-Dopa, both HVA and VMA rose in direct proportion to the administered dose (Figs. 1 and 2). The fact that HVA rises much more precipitously than VMA indicates that relatively little of the administered L-Dopa is converted to E and NE.

HVA in these studies was determined by the method of Sato.[16] VMA was

* Standard deviation.

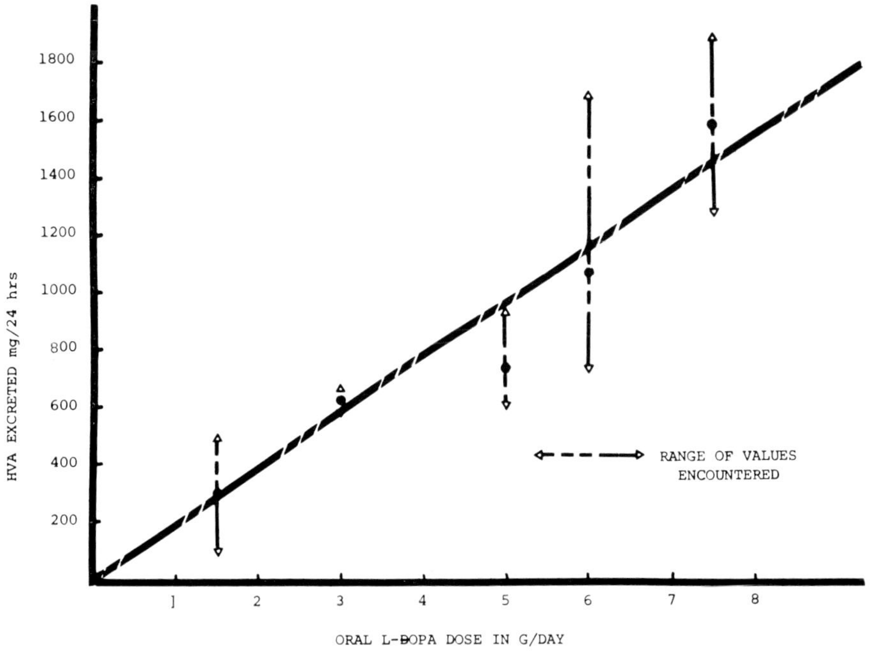

Fig. 1. Urinary HVA excretion versus oral L-Dopa dose. Arrows indicate the range.

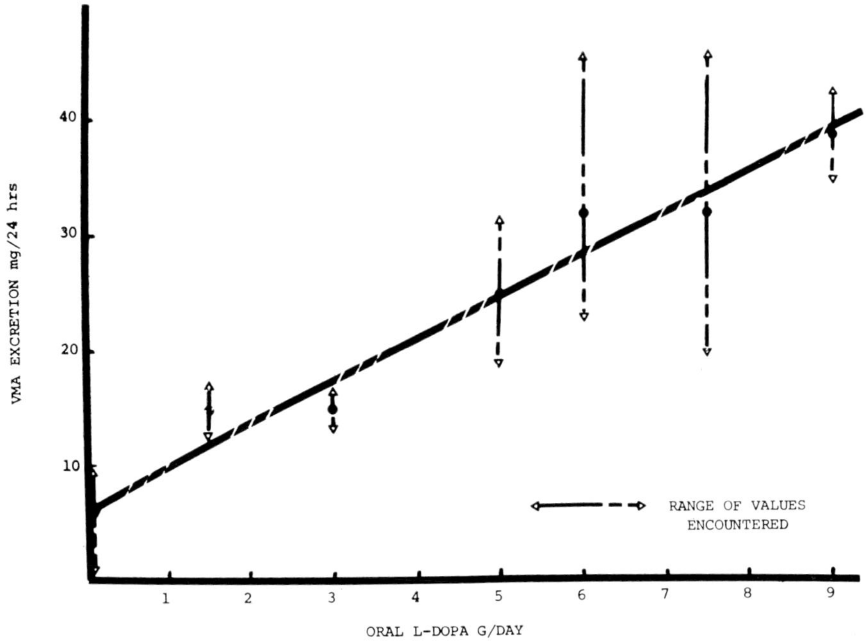

Fig. 2. Urinary vanilmandelic acid (VMA) excretion versus oral L-Dopa dose. Arrows indicate the range.

quantitated by the method of Wybenga and Pileggi,[17] which depends on preliminary treatment of the urine with ion exchange chromatography, oxidation of the VMA to vanillin which is condensed with indole to form a chromophore. In our hands, the method was found to be satisfactory provided certain precautions were observed. It proved necessary to run a standard curve with each set of determinations. Solutions of VMA in a concentration of 6 mg per 100 ml, which were used as stock standards, were renewed every two weeks as they proved unstable if stored for a longer period, even at 4 °C. The color reaction was found to follow Beer's law only over the range 10 μg to 90 μg per ml, and therefore analysis volumes had to be adjusted so that determinations fell within this range. A urine blank, which had been omitted by the authors of the method, was found to be essential, and therefore was included for every specimen. Finally, timing and temperature during color development had to be monitored closely.

A study of the concentrations of Dopa, HVA, DOPAC, NE, and DA in cerebrospinal fluid (CSF) was undertaken in 31 parkinsonism patients (Table 1) to learn more about the central nervous system metabolism of these compounds, both in the disease state per se and also during oral therapy with L-Dopa.[18] This project offered an opportunity to evaluate the possibility of a correlation between clinical and biochemical findings. Preliminary work had confirmed reports of others [19, 20, 21] that the low basal ganglia DA levels seen in parkinsonism are reflected by low CSF levels of HVA. In addition, a pronounced HVA concentration gradient in CSF taken from different regions

TABLE 1. Distribution of 31 Patients in Whom Simultaneous CSF Chemical Analysis and Clinical Assessment Were Performed

	Female	*Male*
Number of patients	11	20
Age range	49–76	44–77
Weight range	95–185	119–225
Mean weight	134.4	168.6
L-Dopa dosage/day		
Range	2.5–6.0 gm/day	4.0–12.0 gm/day
Mean	4.86 gm/day	5.84 gm/day

of the central nervous system was observed. Since the principal dopaminergic neurons are located adjacent to the ventricles, the findings of much higher HVA concentrations in CSF taken from this region as opposed to the lumbar region added to the evidence that the HVA in the CSF may normally be viewed as an indicator of DA metabolism in the brain. The introduction of chronic massive oral doses of L-Dopa into the therapeutic regimen may alter this assumption. In the studies with patients receiving oral L-Dopa, CSF Dopa and HVA levels both correlated positively with the L-Dopa dose in a manner essentially identical (Fig. 3). This was interpreted as indicating that under these conditions (i.e., during administration of L-Dopa) much of the CSF HVA arises from the systemic circulation. Prior to administration of L-Dopa, CSF levels of Dopa averaged 12 ± 8 * ng per ml, of HVA 35 ± 27 * ng per ml, and of DOPAC 21 ± 13* ng per ml. As was true of urine, HVA levels increased most precipitously, often to more than 20 times their original levels, when patients were receiving an average L-Dopa dose of 5.23 gm per day, while CSF Dopa levels rose from 5 to 6 times their pretreatment values on the same dose. DOPAC levels remained essentially unchanged, while NE and DA concentrations, even on a dose of 12 gm of L-Dopa per day, remained too low to permit measurement.

In our method, DA could have been measured accurately, if present, in CSF in a concentration of 0.02 μg per ml and its presence detected at a concentration lower than this. NE could have been measured at a CSF concentration of 0.01 μg per ml and detected at a much lower concentration.

Sympathomimetic amines were found to affect CSF levels of Dopa and HVA in patients receiving L-Dopa therapy, but did not affect clinical improvement as measured by an objective neurological point score system. Concurrent administration of these amines with L-Dopa doubled the expected concentration of this amino acid in CSF, while HVA levels were somewhat lower than in patients receiving other or no supportive therapy.

It was possible to establish correlations between (1) treatment CSF Dopa levels and the clinical response of the patients, as measured by a neurological score change, and (2) treatment CSF HVA levels and the pretreatment incidence of dementia, defined as a loss of mental capabilities, measured in an independent neuropsychological study.[22]

* Standard deviation.

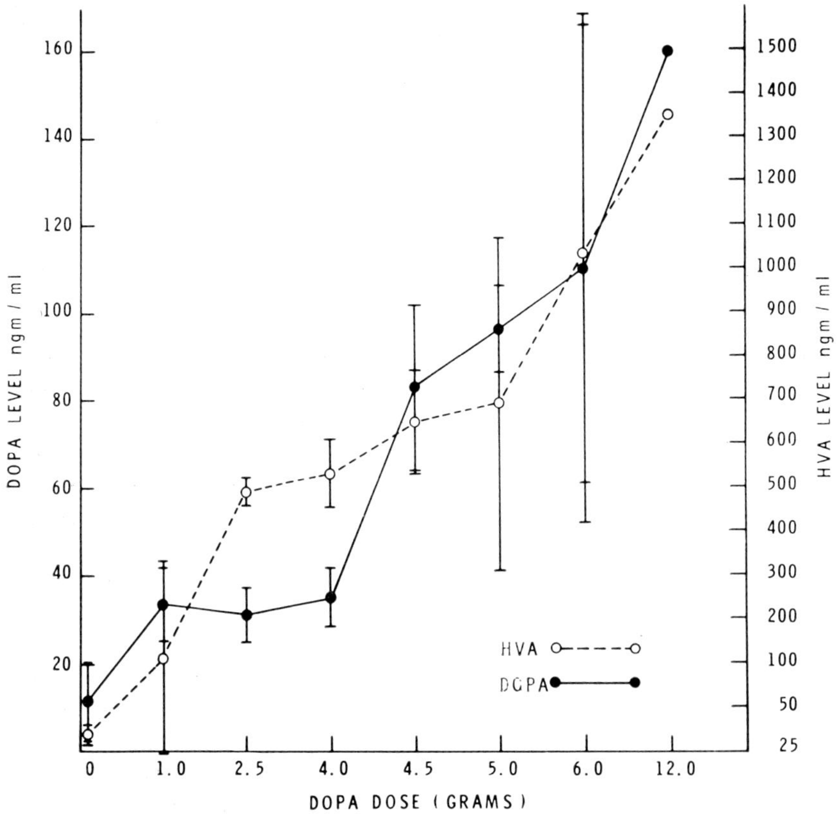

Fig. 3. Relation of Dopa and HVA levels in cerebrospinal fluid with oral dose of L-Dopa in gm per day.

It was possible on the basis of CSF Dopa levels to divide the patients into three categories:

Group A: Those whose CSF Dopa levels during treatment with L-Dopa did not exceed 50 ng per ml. The mean L-Dopa dosage here was 4.1 gm per day.

Group B: Those whose CSF Dopa levels during treatment with L-Dopa exceeded 120 ng per ml. The mean dose in this group was 6.25 gm per day.

Group C: Those whose CSF Dopa levels on L-Dopa therapy fell between these two extremes. The mean dose in this group was 5.5 gm per day.

Clinical evaluation of these patients before and during L-Dopa therapy was based on a numerical neurological scoring system which took into account the disability imposed by rigidity, tremor, and dyskinesia as well as impairment of mentation and perception. Weighted scores were assigned to each functional measurement such that a total of 100 points would be obtained in the absence of any disability. The patients in group A showed the best clinical improvement with an average score change of +22.3 points, while group

C patients showed the poorest response with an average score change of +15.7 points.

The incidence of dementia appeared to be related to CSF HVA levels but not to CSF Dopa levels. Thus, patients with dementia tended to have higher CSF HVA concentrations during treatment with L-Dopa than similarly treated patients in whom dementia was absent. It was also noted that the patients with dementia had, in general, lower neurological scores than patients without dementia both before and during treatment with L-Dopa, although the response to L-Dopa, as measured by a change in score, was the same for the two groups. If one plots the before- and during-treatment CSF HVA concentrations against the corresponding scores, this difference can be seen clearly (Fig. 4). The lines connecting the before and during points of dementia patients are seen to lie predominately in the left portion of the figure, while those connecting points of patients without dementia fall in the right portion of the figure.

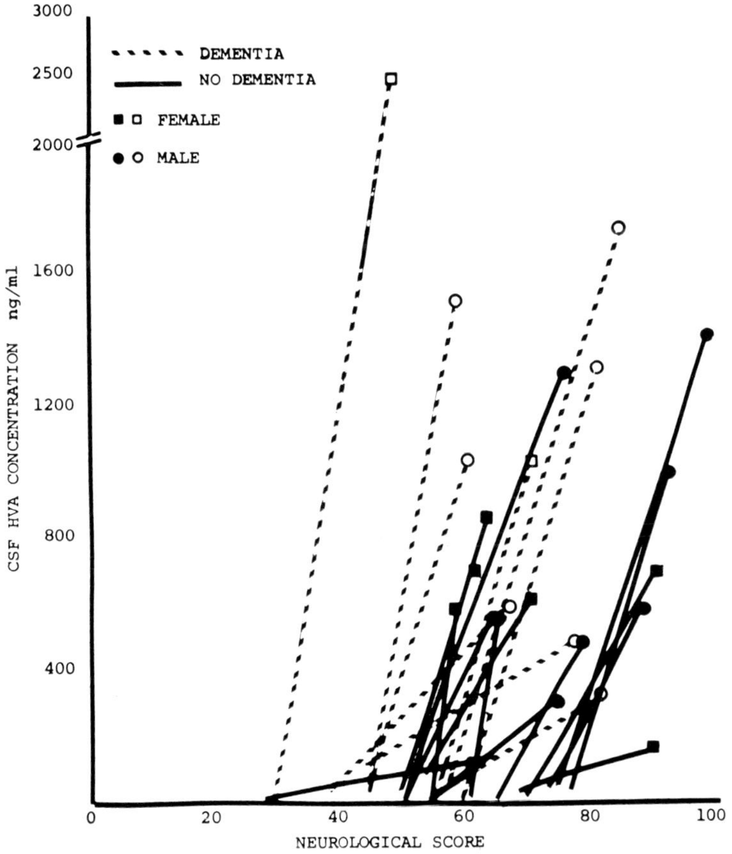

Fig. 4. Correlation of initial neurological score and CSF HVA levels with final neurological score and CSF HVA levels. Final values are indicated by the large dot (males) or square (females).

TABLE 2. Cerebral Blood Flow and Metabolism in Parkinsonism Patient Before and After L-Dopa Administration (Preliminary Data).

	HBF (ml/100g brain/min)	*pO_2 (mmHg)*	*SO_2*	*pCO_2 (mmHg)*	*pH*	*O_2 Content*	*Glucose (mg/dl)*	*Pyruvate (mg/dl)*	*Lactate (mg/dl)*
Steady State									
Right	36.7	27.5	41.3	45.5	7.248	7.48	127.0	0.355	6.05
Left	31.5	35.7	64.0	43.8	7.340	11.56	126.0	0.450	7.30
Artery	—	69.3	93.7	32.3	7.389	17.04	132.5	0.310	6.00
Hexobendine									
Left	40.0	35.2	64.2	36.8	7.357	11.61	121.7	0.480	6.50
Artery	—	69.8	94.2	35.0	7.410	17.16	129.0	—	5.40
L-Dopa									
Left	42.4	31.9	64.0	39.0	7.382	11.60	141.0	0.500	7.40
Artery	—	68.8	93.5	37.3	—	17.08	147.0	0.380	6.80

It is evident, then, that there are gross differences in tyrosine metabolism such as increased pHPPA excretion in specific pathological conditions. In addition, there are more subtle differences among individuals apparently suffering from the same movement disorder. At least some of the differences in catecholamine metabolism of the parkinsonism patient appear to relate to clinical groupings—for example, the presence or absence of dementia.

The interrelationships between Dopa metabolism and general metabolism have hardly been touched as yet. A relationship between a disturbance in catecholamines of the extrapyramidal system and general brain metabolism has been reported. Cerebral blood flow and oxidative metabolism are reduced in patients with "arteriosclerotic" parkinsonism but not in patients with paralysis agitans.[23, 24] Preliminary investigations into (1) the nature of overall metabolic disturbance in parkinsonism and (2) the value of L-Dopa and/or its metabolites in altering this disturbance have been undertaken in our laboratory by the methods developed by Meyer, et al.[25, 26, 27] Initial data support the hypothesis that in bilateral extrapyramidal disease there is a difference in hemispheric reduction of blood flow and metabolism. Replacement therapy (200 mg L-Dopa intravenously) over a short interval of time reversed the clinical symptoms and significantly altered hemispheric blood flow. Initial arterial and jugular venous elevations of blood sugar levels (12 to 14 mg per 100 ml) suggest a direct effect on hepatic glucose stores. Preliminary hemispheric blood-flow data on a parkinsonism patient before and after administration of L-Dopa are shown in Table 2. It is obvious that interdisciplinary utilization of in vivo biochemical and physiological techniques will offer a fecund approach for further investigation.

REFERENCES

1. Gibb, J. W., and Webb, J. G. : The effects of reserpine, alphamethyltyrosine, and L,-3,4-dihydroxyphenylalanine on brain tyrosine transaminase. Proc. Nat. Acad. Sci. *63:* 364, 1969.
2. McCann, D. S., Mitchell, J. G., Keech, M. K., and Boyle, A. J.: Tyrosine metabolism in connective tissue disease. J. Chronic Dis. *20:* 781, 1967.
3. Honos, E., Ericsson, A. D., and McCann, D .S.: Parahydroxyphenylpyruvic acid excretion in parkinsonism. Life Sci., in press.
4. Bertler, A.: Occurrence and localization of catecholamines in the human brain. Acta Physiol. Scand. *51:* 97, 1961.
5. Sano, I., Gamo, T., Kakimoto, Y., Taniguchi, K., Takesada, M., and Nishinuma, K.: Distribution of catechol compounds in human brain. Biochim. Biophys. Acta *52:* 586, 1959.
6. Ehringer, H., and Hornykiewicz, O.: Verteilung von Noradrenalin und Dopamin (3-hydroxytyramin) im Gehirn des Menschen und ihr Verhalten bei Erkrankungen des Extrapyramidalen Systems. Klin. Wschr. *38:* 1236, 1960.
7. Bernheimer, H., and Hornykiewicz, O.: Herabgesetzte Konzentration der Homovanillinsäure im Gehirn von parkinsonkranken Menschen als Ausdruck der Störung des zentralen Dopaminstoffwechsels. Klin. Wschr. *43:* 711, 1965.
8. Kusch, T., and Heinrich, I.: Die Aktivitäts veränderungen der Leber Tryptophan-Pyrrolase und Tyrosina α-Keto glutarat-Transaminase durch Chlor-

promazin und Reserpin unter Berücksichtigung der Funktionen der Nebennieren. Acta Biol. Med. Germanica *10:* 554, 1963.
9. Black, I., and Axelrod, J.: Elevation and depression of hepatic tyrosine transaminase activity by depletion and repletion of norepinephrine. Proc. Nat. Acad. Sci. *59:* 1231, 1968.
10. Goodall, M., and Alton, H.: Metabolism of 3-hydroxytyramine (dopamine) in human subjects. Biochem. Pharmacol. *17:* 905, 1968.
11. Barbeau, A., Murphy, G. F., and Sourkes, T. L.: Excretion of dopamine in diseases of basal ganglia. Science *133:* 1706, 1961.
12. Smith, I., and Kellow, A. H.: Aromatic amines and Parkinson's disease. Nature *221:* 1261, 1969.
13 Rinne, Urpo, K., and Sonninen, V.: Dopamine and Parkinson's disease. CA *71:* 1599e, 1969.
14. Barbeau, A.: Effect of phenathiazines on dopamine metabolism and biochemistry of Parkinson's disease. Agressologie *9:* 195, 1968.
15. Simek, J.: Excretion of 3-methoxy-4-hydroxyphenylacetic (homovanillic) acid in patients with Parkinson's disease. CA *70:* 104555w, 1969.
16. Sato, T. L.: The quantitative determination of 3-methoxy-4-hydroxyphenylacetic acid (HVA) in urine. J. Lab. Clin. Med. *66:* 517, 1965.
17. Wybenga, D., and Pileggi, V. J.: Quantitative determination of 3-methoxy-4-hydroxymandelic acid (VMA) in urine. Clin. Chim. Acta *16:* 147, 1967.
18. Sharpless, N. S., Ericsson, A. D., and McCann, D. S.: Clinical and cerebrospinal fluid changes in parkinsonism patients treated with L-3,4-dihydroxyphenylalanine (L-Dopa). In press.
19. Johansson, B., and Roos, B. E.: Acid monoamine metabolites in the cerebrospinal fluid of patients with Parkinson's syndrome. *In* Proceedings of the 8th International Neurological Congress, Wien, 1965. Compt. Rend. *4:* 141, 1965.
20. Bernheimer, H., Birkmayer, W., and Hornykiewicz, O.: Homovanillinsäure im liquor cerebrospinalis: Untersuchungen beim Parkinson-syndrom und anderen Erkrankungen des ZNS. Wien. klin. Wschr. *78:* 417, 1966.
21. Guldberg, H. C., Turner, J. W., Hanieh, A., Ashcroft, G. W., Crawford, T. B. B., Perry, W. L. M. and Gillingham F. J.: On the occurrence of homovanillic acid and 5-hydroxyindolyl-3-acetic acid in the ventricular C.S.F. of patients suffering from parkinsonism. Confin. Neurol. *29:* 73, 1967.
22. Ericsson, A. D., and Schwartz, M.: In preparation.
23. Gotoh, F., Meyer, J. S., and Ebihara, S.: Continuous recording of cerebral blood flow and metabolism. Med. Res. Eng. *5:* 13, 1966.
24. Gotoh, F.: Cerebral blood flow and metabolism in parkinsonism. Personal communication, 1968.
25. Meyer, J. S., Gotoh, F., Akiyama, M., and Yoshitake, S.: Monitoring cerebral blood flow, oxygen, and glucose metabolism. Circulation *36:* 197, 1967.
26. Meyer, J. S., Toyoda, M., Shinohara, Y., Kitamura, A., Ryu, T., Wiederholt, I., and Guiraud, B.: Regional cerebral blood flow (carotoid perfusion) measured by clearance of hydrogen from cerebral venous blood. Trans. Amer. Neurol. Assoc. *93:* 246, 1968.
27. Meyer, J. S., Wiederholt, I., Toyoda, M., Ryu, T., Shinohara, Y., and Guiraud B.: A new method for continuous sampling of cerebral venous blood without extracranial contamination in man. Bilateral transbrachial catheterization of the cerebral venous sinuses. Neurology *19:* 353, 1969.

URINARY EXCRETION AND CEREBRAL DISTRIBUTION OF P-TYRAMINE *

Alan A. Boulton, Lillian Quan, and J. R. Majer

The spontaneous urinary excretion of free *p*-tyramine has been shown to be above normal in about 50 percent of patients suffering with Parkinson's disease.[1, 2] The excretion levels of conjugated *p*-tyramine and free and conjugated *p*-hydroxyphenylacetic acid (the principal metabolite of *p*-tyramine) do not seem markedly different from normal values, although there are pronounced variations from individual to individual; this is probably only a reflection of the dietary intake of *p*-tyramine. Smith and Kellow [3] have claimed that a relationship exists between the urinary concentration of free *p*-tyramine and the severity (severe indicating both tremor and rigidity, mild meaning tremor or rigidity but not both) and extent of symptoms. From our own results,[4] however, it was not possible to observe any particular relationship beyond the general one that high excretion levels may be related to actively progressing disease.

Feeding cheese rich in *p*-tyramine did not increase the urinary excretion of free *p*-tyramine [5] either in normal people or in patients who on previous occasions had been heavy excreters. The sum of the conjugated *p*-tyramine and the free and conjugated *p*-hydroxyphenylacetic acid, however, increased so as to account almost stoichiometrically for the ingested amine.

Eradication of most of the intestinal flora with neomycin or succinylsulphathiazole considerably increased both free and conjugated amines and acid.[5]

Increased spontaneous urinary excretion of free *p*-tyramine has been reported [1] in about 10 percent of a schizophrenic population. Decreased urinary excretion of both free and conjugated *p*-tyramine has been described in migraine.[6] It has also been shown that orally ingested *p*-tyramine can precipitate the onset of migraine attack.[7]

Because *p*-tyramine is an interesting amine possessing complex physiological properties and because it seems to be implicated in some neurological diseases, we have investigated its presence and regional distributions in normal rat brain. Spector et al.,[8] using a relatively nonspecific technique based upon the formation of the 1-nitroso-2-naphthol fluorophore, claimed that *p*-tyramine was present in rat brain at about 2 ng per gram wet tissue. Gunne and Jonsson,[9] using an ion exchange procedure, claimed that in rabbit brain if *p*-tyramine was present at all it would have to be at a concentration below 10 ng per gram wet tissue. Anagnoste and Goldstein,[10] using intraventricular injections of radioactive *p*-tyramine during monoamine oxidase blockade, found in rat brain that a labeled substance isographic with synthetic acetylated *p*-tyramine was present in the amine fractions following acetylation and paper chromatographic separation. They also showed that the *p*-tyramine

* Supported in part by the Psychiatric Services Branch of the Province of Saskatchewan and the Medical Research Council of Canada.

was principally located in the striatum. In nonenzyme-inhibited brain tissue no *p*-tyramine could be detected by this technique.

In normal rat brains we have unambiguously identified *p*-tyramine and investigated its regional distribution.

Methods and Results

Full details of the techniques used have already been published[11, 12,13] these publications should be consulted for the precise methodology. Briefly, rats (Wistat, approximately 200 grams) were stunned and decapitated, their brains were removed, and various regions were dissected out. After washing in ice-cold saline solution, whole brains or the various regions were homogenized at 2°C in 0.4 N perchloric acid. After centrifugation the supernatant was neutralized with 1N sodium hydroxide and the phenolic amine fraction isolated on a Bio-Rad AG50W-X2 H^+ form column as described by Kakimoto and Armstrong.[14] The *p*-tyramine content of aliquots (5 μl) of the dried amine extract dissolved in 50 μl of ethanol:ammonia 70:30 (v/v) was obtained by placing the aliquot on a probe and inserting the probe directly into a mass spectrometer near to the ionizing beam. The integrated ion current at a selected m/e value (in the case of *p*-tyramine this is 108 or 137) was recorded on a chart as the amine (in the solid state since the solvents quickly evaporate) evaporated from the probe. Measurement of the area (planimetry or tracing and weighing) enclosed under the ion current envelope gave the amount of the selected ion in the aliquot by reference to a calibration curve obtained by similarly evaporating samples of the synthetic amine.

The calibration curve is linear at least in the range 10^{-12} to 10^{-5} grams. Other amines, contaminants and metabolites do not interfere with the analysis or recovery of the selected ion and in the case of *p*-tyramine the recovery is better than 90 percent. Although it would be rare for two or more ions to be produced at any selected m/e value they can be distinguished by precise mass analysis using the peak matching facilities of the double focusing mass spectrometer. It has been shown that at m/e values of 137 and 108 a urine extract contains only one ion with precise molecular weights of 137.0841 (mol. wt. of *p*-tyramine) and 108.0545 (major fragment of *p*-tyramine) respectively. In brain extracts there are two ions at both the m/e values selected; 108 and 137. At 108.0575 we found $C_7H_8O^+$ (from *p*-tyramine), at 108.0939 $C_8H_{12}^+$ (presumably a lipid fragment passing through the extraction procedure), at 137.0841 $C_8H_{11}ON^+$ (*p*-tyramine), and at 137.1330 $C_{10}H_{17}^+$ (presumably the lipid fragment). The integrated ion current in both these latter cases exhibits a double envelope so that analysis of either the tyramine or the hydrocarbon moiety was possible.

Using this technique we have shown conclusively and unambiguously that *p*-tyramine is a normal constituent of rat brain. The concentration in whole brain is 188 $\pm$ 35 ng per gram (n = 8) wet tissue. This concentration is below the level of sensitivity of most chromogenic and fluorogenic chromatographic procedures. The so-called DANSYL (1-dimethylaminonaphthalene-5-sulphonyl chloride) derivatives, however, as a consequence of their intense fluorescence, may be used in certain susceptible cases [15] for qualitative and quantitative analysis. *p*-Tyramine may be detected down to a level of 5 ng. For comparative purposes we used this technique coupled with extensive prefractionation and thin-layer chromatographic separations to indicate that *p*-tyramine in whole rat brain is present and at a concentration of 192±24 ng per gram (n=7) wet tissue.

The regional distribution of *p*-tyramine as obtained using the integrated ion circuit technique is shown in Table 1.

Discussion

It is apparent from Table 1 that the distribution of *p*-tyramine in rat brain resembles that of norepinephrine. Neither the significance of this distribution nor the cerebral function of *p*-tyramine is presently understood.

The principal metabolites of *p*-tyramine seem to be *p*-hydroxyphenylacetic acid [10, 16, 17] and octopamine.[10, 16, 18, 19] Both of these metabolites have been identified in urine.[5, 17, 20] *p*-Tyramine can be converted to catecholamines and catecholamine metabolites both in vivo [19, 21, 22] and in vitro.[23, 24] Although the amount of catecholamines arising from the nonclassical route is small as a proportion of total synthesis, this does not preclude the possibility of an alternative or abnormal route in some specific location and/or in some disease states. Further studies on the identification and quantitative distribution of some tyramine metabolites and on other amines and their metabolites are under way.

Summary

A technique—the so-called integrated ion current technique—is described for the precise, unambiguous and quantitative analysis of metabolites in neurochemical specimens. Using this technique *p*-tyramine has been shown to be a normal constituent of rat brain. *p*-Tyramine concentration (ng per gram wet tissue) in whole brain and in various regions is: whole brain 188±35, hypothalamus 1712±43, caudate nucleus 636±107, cerebellum 280±51, stem 192±9, and other parts 55±1.

TABLE 1. Regional Distribution of *p*-Tyramine (ng per gram) in Rat Brain

	Experiment		
	1	2	*3*
Caudate nucleus	735	650	523
Hypothalamus	1670	1711	1756
Cerebellum	229	280	330
Stem	200	183	192
Other	55	54	55
Whole brain	188±35 (n=8)		

REFERENCES

1. BOULTON, A. A., POLLITT, R. J., and MAJER, J. R.: Identity of a urinary 'pink spot' in schizophrenia and Parkinson's disease. Nature *215:* 132, 1967.
2. BOULTON, A. A., and FELTON, C. A.: The pink spot: A red herring? Lancet *2:* 964, 1966.
3. SMITH, I., and KELLOW, A. H.: Aromatic amines and Parkinson's disease. Nature *221:* 1261, 1967.
4. BRODIE-HUGHES, E.: Personal communication.
5. BOULTON, A. A.: The pink spot in parkinsonism. Progress in Neurogenetics, Excerpta Medica *175:* 437, 1969.
6. HANINGTON, E., SMITH, I., and KELLOW, A. H.: Tyramine metabolism in dietary migraine. Migraine Symposium, London, 1969.
7. HANINGTON, E.: The role of tyramine in the aetiology of migraine, and related studies on the cerebral and extracerebral circulations. Migraine Symposium, London, 1969.
8. SPECTOR, S., MELMON, K., LOVENBERG, W., and SJOERDSMA, A.: The presence and distribution of tyramine in mammalian tissues. J. Pharmacol. Exp. Ther. *140:* 229, 1963.
9. GUNNE, L. M., and JONSSON, J.: On the occurrence of tyramine in the rabbit brain. Acta Physiol. Scand. *64:* 434, 1965.
10. ANAGNOSTE, B., and GOLDSTEIN, M.: The metabolism of tyramine-H^3 in different regions of the CNS. Life Sci. *6:* 1535, 1967.
11. MAJER, J. R., and BOULTON, A. A.: Absolute, unambiguous, ultramicro analysis of metabolites present in complex biological extracts. Nature *225:* 658, 1970.
12. BOULTON, A. A., and QUAN, L.: The presence and distribution of *p*-tyramine in rat brain. Neurochemical Meeting, Milan, 1969.
13. BOULTON, A. A., and MAJER, J. R.: Mass spectrometry of crude biological extracts: Absolute quantitative detection of metabolites at the submicrogram level. J. Chromatogr., in press.
14. KAKIMOTO, Y., and ARMSTRONG, M. D.: The phenolic amines of human urine. J. Biol. Chem. *237:* 208, 1962.
15. BOULTON, A. A.: The automated analysis of absorbent and fluorescent substances separated on paper strips. *In* Methods of Biochemical Analysis (E. Glick, ed.). Interscience, New York, 1968.
16. BREESE, G. R., CHASE, T. N., and KOPIN, I. J.: Metabolism of tyramine-^3H and octopamine-^3H by rat brain. Biochem. Pharmacol. *18:* 863, 1969.
17. BOULTON, A. A.: Unpublished observations.
18. CHIDSEY, C. A., KAISER, G. A., and LEHR, B.: The hydroxylation of tyramine in the isolated canine heart. J. Pharmacol. Exp. Ther. *144:* 393, 1964.
19. FISHER, J. E., MUSACCHIO, J., KOPIN, I. J., and AXELROD, J.: Effects of denervation on the uptake and β-hydroxylation of tyramine in the rat salivary gland. Life Sci. *3:* 413, 1964.
20. KAKIMOTO, Y., and ARMSTRONG, M. D.: On the identification of octopamine in mammals. J. Biol. Chem. *237:* 422, 1962.
21. CREVELING, C. R., LEVITT, M., and UNDENFRIEND, S.: An alternative route for biosynthesis of norepinephrine. Life Sci. *1:* 523, 1962.
22. LEMBERGER, L., KLUTCH, A., and KUNTZMAN, R.: The metabolism of tyramine in rabbits. J. Pharmacol. Exp. Ther. *153:* 183, 1966.
23. AXELROD, J.: Enzymatic formation of adrenaline and other catechols and monophenols. Science *140:* 499, 1963.

24. LEMBERGER, L., KUNTZMAN, R., CONNEY, A. H., and BURNS, T. J.: Metabolism of tyramine to dopamine by liver microsomes. J. Pharmacol. Exp. Ther. *150:* 292, 1965.

AROMATIC AMINO ACID METABOLISM DURING L-DOPA THERAPY OF PARKINSON'S DISEASE *

Melvin H. Van Woert and Malcolm B. Bowers, Jr.

Subnormal levels of dopamine and serotonin have been found at autopsy in the brains of parkinsonian patients.[1] The acid metabolites of these amines, homovanillic acid (HVA) and 5-hydroxyindoleacetic acid (5-HIAA) are present in measurable amounts in the cerebrospinal fluid (CSF).[2] In Parkinson's disease the CSF HVA and 5-HIAA are decreased,[3] thus suggesting that the CSF levels may reflect the metabolism of brain catechol and indole amines.

The demonstration of decreased dopamine synthesis in the parkinsonian brain has led to the successful use of its precursor dihydroxyphenylalanine (Dopa) in the treatment of this disease.[4, 5, 6] In the brain, Dopa is decarboxylated to dopamine by the enzyme aromatic L-amino acid decarboxylase which also acts on several other aromatic amino acids, including 5-hydroxytryptophan and tyrosine. L-Dopa might be expected to compete with other aromatic amino acids (e.g., 5-hydroxytryptophan and tyrosine) for membrane transport, other enzymes and coenzymes.

This paper reports the results of studies of the effect of L-Dopa on the concentrations of HVA, 5-HIAA and tyrosine in the cerebrospinal fluid and plasma of patients with Parkinson's disease. The effect of L-Dopa on the conversion of phenylalanine to tyrosine following an oral load of tyrosine was also examined.

Methods

HVA was measured according to Andén et al.[7] by an extraction with ethyl acetate instead of ether and a more dilute solution of $K_3Fe(CN)_6$ as described previously.[2] Values were calculated from internal standards prepared by adding known amounts of HVA to CSF and plasma. CSF 5-HIAA was assayed by the method of Ashcroft and Sharman [8] and tyrosine according to Waalkes and Udenfriend.[9] The addition of Dopa and dopamine to CSF and plasma were found not to affect HVA, 5-HIAA or tyrosine assays.

Plasma and CSF were obtained from patients with Parkinson's disease before and one to two months after starting L-Dopa treatment. The blood and CSF samples were deproteinized and stored at $-20°C$ until analyzed (analysis

* Work supported by USPHS Grants NB-07542-02 and CA-08341.

was done within a week). L-Dopa was administered orally in four equal doses (after meals and at 9:00 p.m.). The total daily dose of L-Dopa for each patient was determined individually according to therapeutic response and side effects, and ranged from 4 to 7.5 gm. All CSF and blood samples were obtained between 9:00 and 10:00 a.m. after a twelve-hour fast (except during the phenylalanine load tests) and prior to the morning dose of L-Dopa. Control CSF for HVA and 5-HIAA determinations was obtained from psychiatric inpatients who were not taking medication. Control samples of CSF and blood for tyrosine assay were obtained from hospitalized patients without hepatic, renal or endocrine disease, and between the ages of 56 and 78 years.

A phenylalanine conversion test was determined on eight of the parkinsonian patients before and during L-Dopa therapy and ten of the control patients. The phenylalanine conversion test consisted of feeding L-phenylalanine (100 mg per kg body weight) mixed in applesauce. Blood samples for tyrosine determinations were obtained before and ½, 1, 2, 3, and 4 hours after phenylalanine feeding.

Results

The concentrations of HVA and 5-HIAA in the lumbar CSF of 14 patients with Parkinson's disease were significantly lower than those of the control group (Table 1). These results are similar to those of Johansson and Roos [3] and are consistent with the findings that dopamine, HVA and serotonin are decreased in the parkinsonian brain.[1] The plasma HVA was undetectable in all but one of the parkinsonian patients before Dopa administration. Similarly, only an occasional control patient had measurable levels of plasma HVA.

During L-Dopa administration (4 to 7.5 gm per day, orally), the CSF HVA showed a mean 16-fold increase when measured 12 hours after the last dose. The mean concentration of plasma HVA was increased to 7 times the level in the CSF (obtained at the same time as the plasma).

The CSF 5-HIAA decreased in 11 of 13 parkinsonian patients during L-Dopa therapy. This mean decrease is highly significant ($p<0.01$). CSF tyrosine was normal in parkinsonian patients and did not change during L-Dopa therapy. There was no correlation with the levels of HVA, 5-HIAA and tyrosine before or during L-Dopa administration and neurological findings, age of patient, duration of illness, dose or response to therapy.

Fasting plasma tyrosine concentrations were measured in 16 parkinsonian patients. The mean plasma tyrosine was normal in the parkinsonian patients before L-Dopa but there was a slight decrease during therapy (Table 2, zero time) ($p<0.001$). The reduction in plasma tyrosine suggested the possibility that L-Dopa was inhibiting phenylalanine hydroxylase in these patients, as has been described in animals.[10, 11] Therefore, the conversion of phenylalanine to tyrosine was determined in 10 control subjects and in 8 parkinsonian patients before and during L-Dopa therapy (Table 2). There was no differ-

TABLE 1. Plasma HVA, CSF HVA, 5-HIAA and Tyrosine Before (B) and During (D) L-Dopa Therapy. Patients Grouped According to Predominant Symptom: Tremor, Akinesia and Rigidity or a Combination of These (Mixed).

Patient	*Age and Sex*	*Stage and Duration of Illness (Yrs)*		*Improvement*	*Plasma HVA ng/ml*		*CSF HVA ng/ml*		*CSF 5-HIAA ng/ml*		*CSF Tyrosine µg/ml*	
					B	*D*	*B*	*D*	*B*	*D*	*B*	*D*
Tremor												
1	63F	4	3	2	—	—	<10	152	22	6	1.51	1.45
2	86F	4	18	1	—	1125	<10	392	19	3	1.79	1.23
3	55M	2	9	2	<20	370	<10	170	9	20	1.32	1.25
4	55M**	2	6	4	—	300	22	78	25	16	1.21	1.24
Akineto rigid												
5	67M*	4	14	4	—	1300	—	350	—	10	2.33	2.51
6	53M**	4	17	2	160	770	<10	105	24	—	1.34	—
7	57M***	3	25	3	<20	3130	<10	155	27	14	1.96	1.56
8	44M**	3	8	2	<20	820	<10	185	24	14	—	—
9	59F	3	2	1	—	1464	11	156	14	7	1.77	1.57
10	46M	3	6	4	—	410	—	136	—	25	1.18	1.18
Mixed												
11	65M	2	11	3	—	1730	<10	126	25	13	2.30	1.99
12	73F	2	12	2	<20	1000	<10	145	11	17	1.54	1.41
13	74F	5	20	3	—	—	11	146	26	9	1.30	1.10
14	58M	4	10	4	<20	1080	<10	70	34	20	1.32	1.48
15	59M*	2	10	1	—	—	<10	128	9	6	1.69	1.87
16	53M	4	21	1	<20	1500	<10	75	37	28	0.97	1.19
Mean±SEM of parkinsonian patients					<20	1154±199	<10	161±21	22±2	14±2	1.57±0.10	1.50±0.10
Mean±SEM of controls					<20 (7)		23±2 (31)		30±2 (27)		1.63±0.20 (18)	

Clinical stages: 1—unilateral involvement, 2—bilateral involvement, 3—impaired postural and righting reflexes, 4—fully developed severe disease, 5—confinement to bed or wheel chair (Hoehn and Yahr [16]). Improvement during L-Dopa therapy recorded as: 1 = 0–24%, 2 = 25–49%, 3 = 50–74%, 4 = 75–100% overall improvement. * = right thalamotomy, ** = left thalamotomy, *** = bilateral thalamotomy. Number in parenthesis indicates the number of controls.

TABLE 2. Phenylalanine Conversion Tests *

	Plasma tyrosine concentrations (μg/ml)					
Time: hours	0	½	1	2	3	4
Control subjects	10.2±0.7 (16)	17.8±1.1 (10)	24.0±1.2 (10)	29.4±1.3 (10)	29.8±1.4 (10)	28.1±1.4 (10)
Parkinsonian patients						
before Dopa therapy	10.6±0.3 (16)	18.4±1.0 (8)	22.5±1.3 (8)	26.0±1.5 (8)	25.3±1.7 (8)	21.4±1.4 (8)
during Dopa therapy	8.7±0.4 (16)	14.9±1.4 (8)	22.6±2.3 (8)	25.4±1.8 (8)	26.1±2.3 (8)	23.1±1.5 (8)

* Values are mean ± S.E.M. Numbers in parentheses are numbers of patients.

ence in the changes in plasma tyrosine concentration with time in the control and parkinsonian patients. L-Dopa administration had no effect on the phenylalanine conversion tests.

Discussion

Bernheimer et al.[12] and Bartholini et al.[13] demonstrated that only minor amounts of HVA are found in the CSF after a continuous intravenous infusion of HVA. They concluded that the principle source of CSF HVA is from the metabolism of brain dopamine. The absence of correlation of the levels of plasma HVA with the CSF HVA in our patients supports this conclusion.

The increase in CSF HVA during therapy is compatible with the hypothesis that L-Dopa replenishes the dopamine concentration in the parkinsonian brain. However, the quantitative increase in CSF HVA does not reflect the therapeutic response in the individual patient. Perhaps this is because the clinically important conversion of L-Dopa to dopamine in the basal ganglia is only a small fraction of the total increase in dopamine synthesis. In our patients, L-Dopa therapy increases the mean CSF HVA 7-fold greater than the normal non-parkinsonian concentration. It would be expected that the decarboxylation of L-Dopa to dopamine and subsequently HVA occurs throughout the brain as well as in the wall of the cerebral blood vessels. An increased conversion of dopamine to HVA in the parkinsonian brain, as suggested by the work of Bernheimer and Hornykiewicz,[14] might also contribute to the supernormal levels of CSF HVA during L-Dopa treatment.

The decrease in CSF 5-HIAA induced by L-Dopa is probably due to a decrease in synthesis of brain serotonin. Dopa and dopamine are potent inhibitors of brain tryptophan hydroxylase,[15] the rate-limiting enzymic step in serotonin biosynthesis. This may be the mechanism producing the decrease in CSF 5-HIAA induced by L-Dopa therapy. However, L-Dopa could also competitively inhibit cerebral decarboxylation of 5-hydroxytryptophan or interfere with the transport of serotonin precursors into the brain.

It is of interest that clinical improvement with L-Dopa therapy is associated with a further reduction in an already subnormal 5-HIAA in the CSF. Clinical improvement in the 75 to 100 percent range was seen in all three categories (tremor, akinesia and rigidity, and mixed) in spite of this further reduction in 5-HIAA. Our results suggest that the reduction in serotonin levels in the brain may have little or no relationship to the clinical symptomatology. However, the reduction in serotonin synthesis produced by L-Dopa may be pertinent to some of the side effects of this drug.

Summary

The concentrations of HVA and 5-HIAA were found to be significantly lower in the spinal fluid of patients with Parkinson's disease than in an adult control group. The administration of L-Dopa to patients with Parkinson's disease produced an increase in CSF HVA and a decrease in CSF 5-HIAA and plasma tyrosine. CSF tyrosine was normal in parkinsonian patients and did not change during L-Dopa therapy. The conversion of phenylalanine to tyrosine following an oral load of phenylalanine was normal in parkinsonian patients and unaffected by L-Dopa administration.

ACKNOWLEDGMENTS

The authors are indebted to Miss Amanda Smith, Miss Angelica Rozitis and Mrs. Audrey Lindsey for technical assistance.

REFERENCES

1. EHRINGER, H., and HORNYKIEWICZ, O.: Verteilung von noradrenalin und dopamin (3-hydroxytyramin) im gehrin des menschen und ihr verhalten bei erkrankungen des extrapyramidal systems. Klin. Wschr. *38:* 1236, 1960.
2. GERBODE, F. A., and BOWERS, M. B.: Measurement of acid monoamine metabolites in human and animal cerebrospinal fluid. J. Neurochem. *15:* 1053, 1968.
3. JOHANSSON, B., and ROOS, B.-E.: 5-hydroxyindoleacetic and homovanillic acid levels in the cerebrospinal fluid of healthy volunteers and patients with Parkinson's syndrome. Life Sci. *6:* 1449, 1967.
4. BARBEAU, A.: Biochemistry of Parkinson's disease. *In* Proceedings of the Seventh International Congress, Societa Grafica Romana, Vol. 2. Rome, 1961.
5. BIRKMAYER, W., and HORNYKIEWICZ, O.: Der L-3,4-dioxyphenylalanin (=DOPA)-effekt bei der Parkinson-akinese. Wien. Klin. Wschr. *73:* 787, 1961.
6. COTZIAS, G. C., VAN WOERT, M. H., and SCHIFFER, L. M.: Aromatic amino acids and modification of Parkinsonism. New Eng. J. Med. *276:* 374, 1967.
7. ANDÉN, N.-E., ROOS, B.-E., and WERDINIUS, B.: On the occurrence of homovanillic acid in brain and cerebrospinal fluid and its determination by a fluorometric method. Life Sci. *7:* 448, 1963.
8. ASHCROFT, G. W., and SHARMAN, D. F.: Drug-induced changes in the concentration of 5-OR indolyl compounds in the cerebrospinal fluid and caudate nucleus. Brit. J. Pharmacol. *19:* 153, 1962.
9. WAALKES, T. P., and UDENFRIEND, S.: A fluorometric method for the estimation of tyrosine in plasma and tissues. J. Lab. Clin. Med. *50:* 733, 1957.
10. DEGRAW, J. I., CORY, M., SKINNER, W. A., THEISEN, M. C., and MITOMA,

C.: Experimentally induced phenylketonuria. I. Inhibitors of phenylalanine hydroxylase. J. Med. Chem. *10:* 64, 1967.

11. Ross, S. B., and Haljasmaa, O.: Inhibitors of the enzymatic hydroxylation of phenylalanine and tryptophan by rat liver extract *in vitro*. Life Sci. *3:* 579, 1964.
12. Bernheimer, H., Birkmayer, W., and Hornykiewicz, O.: Homovanillic acid in the cerebrospinal fluid in Parkinson's syndrome and other disorders of the CNS. Wein. Klin. Wschr. *23:* 417, 1966.
13. Bartholini, G., Pletscher, A., and Tissot, R.: On the origin of homovanillic acid in the cerebrospinal fluid. Experientia *22:* 609, 1966.
14. Bernheimer, H., and Hornykiewicz, O.: Herabgesetzte konzentration der homovanillinsaure im gehirn von parkinsonkranken menschen als ausdruck der storung des zentralen dopaminstoffwechsel. Klin. Wschr. *43:* 711, 1965.
15. Jequier, E., Robinson, D. S., Lovenberg, W., and Sjoerdsma, A.: Further studies on tryptophan hydroxylase in rat brain stem and beef pineal. Biochem. Pharmacol. *18:* 1071, 1969.
16. Hoehn, M., and Yahr, M.: Parkinsonism: onset, progression and mortality. Neurology *17:* 427, 1967.

HOMOVANILLIC ACID IN THE CEREBROSPINAL FLUID OF PARKINSON'S AND OTHER EXTRAPYRAMIDAL DISEASES *

R. Papeschi, P. Molina-Negro, T. L. Sourkes, C. Bertrand, and J. Hardy

Several investigators have postulated an abnormality of dopamine (DA) metabolism or storage, or both, in Parkinson's disease.[1, 2, 3, 4, 5] As the most consistent lesion of Parkinson's disease is pathological damage of the substantia nigra,[6] some of the efferent connections of this structure may play a physiopathological role in determining some of the symptoms of Parkinson's disease.

The axons and nerve endings of the nigrostriatal pathway are rich in DA.[7, 8, 9, 10] Recently Llamas and Reinoso-Suarez [11] have demonstrated submicroscopic unmyelinated nigrostriatal fibers by anterograde degeneration methods, among other more numerous efferents of the substantia nigra to the thalamus, the mesencephalic tegmentum, the pallidum and the cortex.

DA is a likely candidate as an inhibitory neurotransmitter in the nigrostriatal pathway.[12, 13, 14, 15] No data are available on the nature of the other efferents of the substantia nigra to the thalamus and the cortex, but in view of the low concentration of DA in these structures it seems unlikely that these fibers are dopaminergic. Thus we should expect that damage to cells of the substantia nigra would produce symptoms derived from or related to a decrease of DA. Similarly, the terminal metabolite of DA, homovanillic acid (HVA), might be involved in this relation.

* Supported by research grants from the Medical Research Council of Canada.

Investigators have reported a decrease of HVA in the lumbar cerebrospinal fluid (CSF) of parkinsonian patients in comparison with neurologically normal people or neurological patients not affected by extrapyramidal diseases.[16, 17, 18] This finding was confirmed by Guldberg et al.[4] for the ventricular CSF in a large series of parkinsonian patients compared with a group of neurological patients with no signs of extrapyramidal disorder.

In view of the unsolved questions about the pathogenesis of Parkinson's disease, we considered it of interest to compare the concentration of HVA in the ventricular CSF of such patients with that in the CSF of patients affected by striatal syndromes such as essential (attitudinal) tremor,[19] dystonia, and choreo-athetosis. In the latter conditions the nigrostriatal pathway is presumably not involved. Some of our parkinsonian patients presented an additional therapeutic lesion of the thalamus or subthalamus.

Material and Methods

Sixty patients were investigated on the occasion of neurosurgery at Notre Dame Hospital, Montreal. All patients discontinued medication for at least two to three days prior to surgical intervention.

The diagnoses were assessed by three of us (P.M.-N., C.B., J.H.). The degrees of tremor and rigidity of parkinsonian patients were evaluated on a conventional 0 to 3 scale for each separate limb or head muscle, and then summed up for each patient. Akinesia was considered severe when, in absence of pyramidal, cerebellar, or vestibular signs, there was: slowing of movements, especially of associated ones; difficulty in walking, balance and speech not accounted for by the gravity of the tremor or the rigidity; slow electroencephalographic activity; disturbances of memory and reduced alertness and interest in the surroundings. Akinesia was considered mild when there was only some slowing of the associated movements.

Specimens of ventricular CSF were collected at the time of the operation and immediately frozen. After two to four weeks the determinations of HVA were carried out following the extraction procedure of Sharman.[20] The final step consisted of a fluorescent reaction [21] and measurement of the intensity of the product in an Aminco-Bowman spectrophotofluorometer, at 400 nm (activation at 310 nm). Internal standards of HVA were carried through the whole procedure for each sample, to assess losses during the method. Recovery varied between 40 and 70 percent.

The exchange of information regarding clinical and biochemical data was performed only after the respective investigations were completed. The clinical data on age, sex and duration of illness in the various diagnostic groups are presented in Table 1. Of 39 parkinsonian patients, only two were classified as postencephalitic. Seven had undergone previous thalamotomy. There was a difference in the mean age of the various groups, the parkinsonian patients being older than all the other groups. The patients with attitudinal tremor showed the longest duration of illness.

TABLE 1. Age, Sex and Duration of Illness of Patients Investigated, Divided in Diagnostic Groups

Condition	*Sex*	*No. of Pts.*	*Age (years)**	*Duration of Illness (years) **
Parkinson's disease	M	28		
" "	F	11		
" "	total	39	57.5±1.5	5.8±0.8
Attitudinal tremor	M	7		
" "	F	5		
" "	total	12	51.1±5.1	20.5±4.9
Attitudinal dystonia	M	1		
" "	F	3		
" "	total	4	30.7±10.9	2.4±1.0
Dyst. musc. deformans	F	2	10;22	5;2
Double athetosis	M	1	19	19
" "	F	1	21	21
Senile chorea	M	1	69	2.5

* Mean ± S.E.

Results

Table 2 presents the mean concentration of HVA in the ventricular CSF of the various groups of patients studied. The parkinsonian patients had a mean concentration of HVA lower than the patients affected by other extrapyramidal diseases. The two parkinsonian patients whose disease was of postencephalitic nature (not represented separately in the table) had a very low concentration of HVA (16 and 31 ng per ml, respectively).

An analysis of variance was conducted to assess the statistical significance of the difference between the means for the groups of patients with Parkinson's disease, dystonia, attitudinal tremor and double athetosis (Table 3). These means were significantly different from one another ($P<0.01$). The comparison among individual groups showed that the parkinsonians had a significantly lower concentration of HVA than the patients with attitudinal tremor or double

TABLE 2. Concentration of HVA in Ventricular CSF of Patients Affected by Various Extrapyramidal Diseases

Condition	*No.*	*No. of Determinations*	*HVA * (ng/ml)*
Parkinson's disease total	39	40 **	154±19
Parkinson's disease operated	7	7	192±52
Parkinson's disease unoperated	32	33	146±22
Attitudinal tremor	12	13 **	248±33
Dystonia	6	6	231±48
Double athetosis	2	2	445±91
Senile chorea	1	1	127
Total non-parkinsonian patients	21	22 **	264±24

* Mean ± S.E.
** Two samples of CSF were obtained on different occasions from one patient.

TABLE 3. Analysis of Variance of HVA Concentration in the CSF of Patients Affected by Parkinson's Disease, Attitudinal Tremor, Dystonia or Double Athetosis

Sources of Variation	*Degrees of Freedom*	*Mean Square*	*F Ratio*
Total	60		
Between groups	3	78,952	5,658 *
Within groups	57	13,954	

* $P < 0.01$.

athetosis (Table 4). The patients with Parkinson's disease who had been operated upon presented a slightly higher concentration of HVA than the unoperated patients, but the difference was not statistically significant.

Because the parkinsonian patients were older than the other groups, we calculated a regression equation for each of the main groups studied to see whether age had any effect on HVA concentration (Table 5). The slopes of the various diagnostic groups were not significantly different from zero. The same held true for the duration of the disease process in our parkinsonian patients.

We looked for a possible correlation of the HVA level with the symptoms of Parkinson's disease. Because thalamotomy, as a rule, improves postural tremor and rigidity but often worsens akinesia, we did not include the previously thalamotomized patients in the correlational study. The mean concentration of HVA in the patients showing a degree of tremor or rigidity ranked as low (arbitrary units 0 to 5, as defined in Methods) was not significantly different from that of patients ranked as high (6 to 12). The parkinsonian patients presenting a severe akinesia had a mean value of HVA lower than the nonakinetic or mildly akinetic patients, but the difference was not convincingly significant (Table 6).

Discussion

The finding that parkinsonian patients have a significantly lower concentration of HVA in the ventricular CSF than patients affected by other extrapyramidal diseases is a further extension of the report by Guldberg et al.,[4] in which parkinsonian patients were compared with neurological patients not affected by extrapyramidal disorders. In our study the non-parkinsonian pa-

TABLE 4. Comparison of the Mean Concentration of HVA Between Groups of Patients with Various Extrapyramidal Diseases

Comparison	*t Value*	*Degrees of Freedom*	*P*
Parkinson/attitudinal tremor	2.49	51	<0.02
Parkinson/dystonia	1.48	44	N.S.
Parkinson/double athetosis	3.39	40	<0.01
Parkinson operated/unoperated	0.93	38	N.S.

TABLE 5. Regression Equations of HVA Concentration on Age or Duration of Illness in Various Extrapyramidal Diseases

Independent Variable = X	*Condition*	*Y * = a + b X*	*Significance of Difference of b from Zero*
Age	Parkinson's disease	Y = 156.3 − 0.1 X	P = NS
	Attitudinal tremor	Y = 263.2 − 0.04 X	P = NS
	All non-parkinsonian patients	Y = 314.3 − 1.2 X	P = NS
Duration of illness	Parkinson's disease	Y = 153 − 0.37 X	P = NS

* Y = HVA, ng/ml.

tients were affected by striatal syndromes, and, presumably, in no instance was the nigrostriatal pathway involved. The mean value of HVA in this group was markedly lower than that reported by the Edinburgh authors in their control group, whereas the two values for the parkinsonian patients are remarkably close. Thus the possibility exists that our extrapyramidal patients exhibited a subnormal value, although significantly higher than the parkinsonian patients.

However, it appears that pathological lesions of the striatum or of other extrapyramidal centers, not preferentially involving the substantia nigra, are not sufficient to produce a decrease of HVA in the ventricular CSF comparable to that induced by a lesion of the substantia nigra.

The validity of the assumption that the level of HVA in the CSF closely represents the metabolic situation of DA in the CNS is supported by the observations of Eccleston and others.[22, 23, 24] They observed parallel changes of the concentrations of the acid metabolites of brain amines in the CSF and in discrete brain areas after injections of amine precursors.

The parkinsonian patients previously operated upon by thalamotomy presented a slightly higher concentration of HVA in the CSF than the unoperated patients, but the difference was not statistically significant. An increased urinary excretion of DA but not of HVA was reported by Westlake and Tew [25] after thalamotomy; this finding seemed to the authors themselves unlikely to reflect alterations in the cerebral metabolism of DA.

TABLE 6. Concentration of HVA in the Ventricular CSF of Parkinsonian Patients, According to Severity of Symptoms

Severity of Symptoms	*No. of Pts.*	*No. of Dtms.*	*HVA * ng/ml*	*t Test*
Tremor 0–5	21	21	131±22	P=NS (Tremor 0–5 vs. 6–12)
Tremor 6–12	11	12	165±51	
Rigidity 0–5	20	21	150±24	P=NS (Rigidity 0–5 vs. 6–12)
Rigidity 6–12	12	12	130±45	
1—Severe akinesia	5	5	56±29	
2—Mild akinesia	6	6	227±79	$t_{1-2}=2.39$; $P<0.05$
3—No akinesia	22	23	144±22	$t_{1-3}=1.61$; P=NS

* Mean ± S.E.

We examined the different variables present in our group of patients to see whether the level of HVA in the CSF was influenced by any particular factor. In agreement with most authors, we could not detect any correlation between age and HVA in the different diagnostic categories investigated. Only Bowers and Gerbode [26] have described such a correlation in the lumbar CSF of psychiatric patients; however, their patients were not divided in diagnostic groups and some of them were taking phenothiazines, and it is thus possible that the correlation ascribed to age was in fact due to the last two factors.

Guldberg et al.[4] have reported that the parkinsonian patients with a longer duration of illness tended to present a lower concentration of HVA in the CSF; we could not confirm this observation in our correlational study.

Among the factors that could influence the concentration of HVA, we took into consideration the presence and the degree of the symptoms of Parkinson's disease. The rationale underlying this correlational study will be discussed in the next few paragraphs.

Tremor of the parkinsonian type may be produced in monkeys by lesions of the tegmentum of the mesencephalon involving the rubro-tegmento-spinal tract and serotonin-containing fibers, besides interrupting the nigrostriatal pathway.[27] Moreover, tremor is beneficially influenced by thalamotomy in humans, a fact that indicates a dependence of parkinsonian tremor upon the release of mechanisms distinct from those operating through the nigrostriatal connections, where DA is contained. The mechanism operating in the pathogenesis of postural tremor could involve a "tremorogenic circuit" released by the interruption of the connections of the substantia nigra with the thalamus, the pallidum and the cortex,[28] as well as by the interruption of other descending mesencephalic outputs to the spinal cord. These pathways do not seem to contain DA, so that there is no reason to expect a correlation of the HVA level with postural tremor. Our parkinsonian patients did not show any such correlation.

Could rigidity be correlated with DA metabolism in the brain? The favorable effect of Dopa treatment on this symptom would appear to support this theory, but several other arguments can be brought against it. First, lesions of DA-containing fibers in the monkey have never induced disturbance of muscular tone; second, rigidity benefits from thalamotomy as well as from Dopa treatment; third, there are some lines of evidence in favor of a role of hyperactivity of the static gamma neurons in the pathogenesis of rigidity,[29] although not all authors agree with this opinion. The relative excitability of dynamic and static gamma neurons would be regulated at a spinal level by a norepinephrine-containing bulbospinal pathway.[30, 31] It is thus possible that the effect of Dopa on rigidity is mediated by its conversion to norepinephrine rather than through DA; in fact, the reports of the presence of DA in the spinal cord have not been confirmed.[32] The expectation would therefore again be negative for a correlation between HVA concentration in the CSF of parkinsonian patients and degree of rigidity. Our statistical analysis is compatible with this.

The question arises as to whether akinesia can be related to a deficit of DA metabolism in the brain. Although this point is debatable, the following facts support the hypothesis: the improvement of akinesia in parkinsonian patients treated with L-Dopa; the lack of response or even the negative effect of thalamotomy on akinesia; the function of the caudate nucleus, main terminus of the DA-containing fibers, as an inhibitory structure over the motility initiated in the cerebral cortex; [33] the inhibition in turn of caudate nucleus cells by iontophoretic application of DA,[13, 15] or by stimulation of the substantia nigra.[14] It would thus be conceivable that akinesia can result from release of the neostriatum following the interruption of the nigrostriatal fibers. Against such an hypothesis is the possibility of observing akinesia in man without damage of the pars compacta of the substantia nigra, as in progressive supranuclear palsy or in some cases of Parkinson's disease.[34]

In our study the patients who presented a clear-cut akinesia had a mean concentration of HVA in the CSF lower than patients with mild or no akinesia, but the difference was not quite statistically significant. Given the small number of truly akinetic patients studied thus far, we propose to extend our observations before trying to answer the question of a correlation of the CFS HVA level with akinesia.

Summary

Sixty patients with Parkinson's disease, attitudinal tremor, dystonia, double athetosis or senile chorea have been investigated for concentration of homovanillic acid (HVA) in the ventricular cerebrospinal fluid (CSF). The parkinsonian patients had a significantly lower mean concentration of HVA than the non-parkinsonian group as a whole, and as compared to the patients with attitudinal tremor and double athetosis.

No influence on the concentration of HVA in the ventricular CSF was found to be exerted by previous thalamotomy, age or duration of illness. Among the parkinsonian symptoms, the severity of tremor, rigidity or akinesia was not correlated with HVA concentration. However, the patients with severe akinesia showed a lower concentration of HVA than the patients with mild or no akinesia.

REFERENCES

1. Ehringer, H., and Hornykiewicz, O.: Verteilung von Noradrenalin und Dopamin (3-Hydroxytyramin) im Gehirn des Menschen und ihr Verhalten bei Erkrankungen des extrapyramidalen Systems. Klin. Wschr. *38:* 1236, 1960.
2. Barbeau, A., and Sourkes, T. L.: Some biochemical aspects of extrapyramidal diseases. Rev. Canad. Biol. *20:* 197, 1961.
3. Barbeau, A., Sourkes, T. L., and Murphy, G. F.: Les catecholamines dans la maladie de Parkinson. *In* Monoamines et Système Nerveux Central (J. de Ajuriaguerra, ed.). Georg, Geneva; Masson, Paris, 1962.
4. Guldberg, H. C., Turner, J. W., Hanieh, A., Ashcroft, G. W., Crawford, T. B. B., Perry W. L. M., and Gillingham, F. J.: On the occurrence of homovanillic acid and 5-hydroxyindol-3-ylacetic acid in the ventricular CSF

of patients suffering from parkinsonism. Third International Symposium Stereoencephalography, Madrid. Confin. Neurol. *29:* 73, 1967.
5. Cotzias, G. C., Van Woert, M. H., and Schiffer, L. M.: Aromatic amino acids and modification of parkinsonism. New Eng. J. Med. *276:* 374, 1967.
6. Hassler, R.: The extrapyramidal motor system. *In* Handbook of Physiology (J. Field, H. W. Magoun and V. E. Hall, eds.). Amer. Physiol. Soc., Washington, 1960.
7. Sourkes, T. L., and Poirier, L. J.: Effect of brain stem lesions on the concentration of catecholamines in basal ganglia of the monkey. *In* Second Symposium on Parkinson's Disease, Washington, 1963. J. Neurosurg. Suppl. II: 194, 1966.
8. Andén, N.-E., Dahlström, A., Fuxe, K., and Larsson, K.: Further evidence for the presence of nigroneostriatal dopamine neurons in the rat. Amer. J. Anat. *116:* 329, 1965.
9. Sourkes, T. L., and Poirier, L. J.: Influence of the substantia nigra on the concentration of 5-hydroxytryptamine and dopamine of the striatum. Nature *207:* 4993, 202, 1965.
10. Dahlström, A., and Fuxe, K.: Evidence for the existence of monoamine-containing neurons in the central nervous system. I. Demonstration of monoamines in the cell bodies of brain stem neurons. Acta Physiol. Scand. *62:* Suppl. 232, 1964.
11. Llamas, A., and Reinoso-Suarez, F.: Projections of the substantia nigra and ventral tegmental mesencephalic area. Third Symposium Parkinson's Disease (F. J. Gillingham and I. M. L. Donaldson, eds.). E & S Livingston, London, 1968.
12. McLennan, H. The release of dopamine from the putamen. Experientia *21:* 725, 1965.
13. McLennan, H., and York, D. H.: The action of dopamine on neurons of the caudate nucleus. J. Physiol. *189:* 393, 1967.
14. Connor, J. D.: Caudate unit responses to nigral stimuli: Evidence for a possible nigroneostriatal pathway. Science *160:* 899, 1968.
15. Herz, A., and Zieglgänsberger, W.: The influence of microelectrophoretically applied biogenic amines, cholinomimetics and procaine on synaptic excitation in the corpus striatum. Int. J. Neuropharmacol. *7:* 221, 1968.
16. Johansson, B., and Roos, B.-E.: Acid monoamine metabolites in the cerebrospinal fluid of patients with Parkinson's syndrome. Eighth International Congress Neurology, Vienna, 1965.
17. Bernheimer, H., Birkmayer, W., and Hornykiewicz, O.: Homovanillinsäure im Liquor cerebrospinalis: Untersuchungen beim Parkinson-Syndrom und anderen Erkrankungen des ZNS. Wien. Klin. Wschr. *23:* 417, 1966.
18. Gottfries, C. G., Gottfries, I., and Roos, B.-E.: Homovanillic acid and 5-hydroxyindoleacetic acid in the cerebrospinal fluid of patients with senile dementia, presenile dementia and Parkinsonism. J. Neurochem. *16:* 1341, 1969.
19. Molina-Negro, P., Robert-Robillard, F., and Hardy, J.: The pathology of attitudinal tremor. In preparation.
20. Sharman, D. F.: A fluorimetric method for the estimation of 4-hydroxy-3-methoxyphenylacetic acid (homovanillic acid) and its identification in brain tissue. Brit. J. Pharmacol. *20:* 204, 1963.
21. Andén, N.-E., Roos, B.-E., and Werdinius, B.: On the occurrence of homovanillic acid in brain and cerebrospinal fluid and its determination by a fluorometric method. Life Sci. *7:* 448, 1963.
22. Eccleston, D., Ashcroft, G. W., Moir, A. T. B., Parker-Rhodes, A., Lutz, W., and O'Mahoney, D. P.: A comparison of 5-hydroxyindoles in

various regions of dog brain and cerebrospinal fluid. J. Neurochem. *15:* 947, 1968.
23. Andersson, H., and Roos, B.-E.: 5-Hydroxyindoleacetic acid in cerebrospinal fluid after administration of 5-hydroxytryptophan. Acta Pharmacol. Toxicol. *26:* 293, 1968.
24. Guldberg, H. C., and Yates, C. M.: Some studies of the effects of chlorpromazine, reserpine, and dihydroxyphenylalanine on the concentrations of homovanillic acid, 3,4-dihydroxyphenylacetic acid and 5-hydroxyindol-3-ylacetic acid in ventricular cerebrospinal fluid of the dog using the technique of serial sampling of the cerebrospinal fluid. Brit. J. Pharmacol. Chemother. *33:* 457, 1968.
25. Westlake, R. J., and Tew, J. M.: Urinary amines in patients undergoing thalamotomy for Parkinson's disease. Neurology *16:* 619, 1966.
26. Bowers, M. B., and Gerbode, F. A.: Relationship of monoamine metabolites in human cerebrospinal fluid to age. Nature *219:* 1256, 1968.
27. Poirier, L. J., Sourkes, T. L., Bouvier, G., Boucher, R., and Carabin, S.: Striatal amines, experimental tremor and the effect of harmaline in the monkey. Brain *89:* 37, 1966.
28. Hardy, J., Molina-Negro, P., Bertrand, C., and Martinez, N.: Le tremblement essentiel d'attitude. Etude physiopathologique. Un. Med. Canad. *99:* 499, 1970.
29. Rushworth, G., Some aspects of the pathophysiology of spasticity and rigidity. Clin. Pharmacol. Ther. *5:* 828, 1964.
30. Andén, N.-E., Jukes, M. G. M., Lundberg, A., and Vyklicky, L.: The effect of Dopa on the spinal cord. I. Influence on transmission from primary afferents. Acta Physiol. Scand. *67:* 373, 1966.
31. Bergmans, J., and Grillner, S.: Reflex activation and regulation of spontaneous activity in static and dynamic γ-motoneurons. Brain Res. *5:* 114, 1967.
32. Laverty, R., and Sharman, D. F.: The estimation of small quantities of 3,4-dihydroxyphenylethylamine in tissues. Brit. J. Pharmacol. *24:* 538, 1965.
33. Mettler, F. A., Ades, H. W., Lipman, E., and Culler, E. A.: The extrapyramidal system. An experimental demonstration of function. Arch. Neurol. Psychiat. *41:* 984, 1939.
34. Denny-Brown, D.: The Basal Ganglia. Oxford University Press, London, 1962.

BLOOD AND URINE FINDINGS IN PATIENTS TREATED WITH L-DOPA *

Gertrude M. Tyce and Manfred D. Muenter

In 17 patients with Parkinson's disease treated with L-Dopa (L-dihydroxyphenylalanine), measurements were made of the urinary concentrations of 3-methoxy-4-hydroxyphenylacetic acid (homovanillic acid, HVA) by the method of Mellinger.[1] As an index of the amount of L-Dopa metabolized

* Supported in part by Grant NB-4004 from the National Institutes of Health, Public Health Service.

to norepinephrine and epinephrine, we also measured urinary concentrations of 3-methoxy-4-hydroxymandelic acid (vanillylmandelic acid, VMA), the most abundant metabolite of these amines. VMA concentrations were measured by the paper chromatographic method of Armstrong and co-workers.[2] Preliminary experiments had indicated that, in patients ingesting Dopa, determination of VMA by methods involving oxidation to vanillin [3] yielded values much higher than those obtained by paper chromatography.

Urinary concentrations of HVA in each of the 17 patients were expressed as a function of the dose of L-Dopa, and lines of best fit were drawn by the method of least squares. In 13 of the patients, HVA accounted for a fairly constant 20 to 25 percent of the dose as the dose of L-Dopa was increased. In 4 of the patients there was a significant decrease in the percentage of Dopa excreted as HVA as the dose of L-Dopa increased.

In one patient the dose of L-Dopa was gradually increased to 8 gm per day and held at that level for eight days; a placebo period of seven days followed. During this entire period, daily urine collections were made. On the first day of the placebo period the amount of HVA in urine decreased precipitously, from more than 2 gm per day to 23 mg per day; for the rest of the placebo period, urinary HVA excretion was between 7 and 9 mg per day. This very rapid decrease in HVA excretion suggests that the body does not build up stores of L-Dopa. It was noteworthy that the beneficial effects of the L-Dopa continued for some days after administration of the drug was discontinued and while the HVA excretion was in the normal range (up to 9 mg per day).

Urinary concentrations of VMA in each of the 17 patients were also expressed as a function of L-Dopa dose, and lines of best fit were drawn. The percentage of the dose of L-Dopa converted to VMA varied considerably among patients, from 0 in one patient to the highest value of 0.13 percent.

In an attempt to determine if peripheral decarboxylase was saturated, urinary concentrations of 5-hydroxyindoleacetic acid (5-HIAA) were measured by the method of Macfarlane and co-workers.[4] Aromatic L-amino acid decarboxylase catalyses the decarboxylation of both 5-hydroxytryptophan (5-HTP) and Dopa. Large doses of Dopa could competitively inhibit 5-HTP decarboxylation and therefore inhibit 5-hydroxytryptamine (5-HT) and 5-HIAA formation. In most of the patients receiving L-Dopa, amounts of 5-HIAA in urine were markedly decreased. 5-HIAA in urine is largely formed from 5-HT in the gastrointestinal mucosa. During treatment with L-Dopa, this gastrointestinal 5-HT is apparently depleted; perhaps this may contribute to the undesirable side effects of the drug.

We also measured concentrations of Dopa in plasma of patients receiving L-Dopa. Dopa was isolated from perchloric acid extracts of plasma by adsorption and elution from alumina. Dopa in the eluate was oxidized by iodine to a fluorophore with activating and fluorescing wavelengths distinct from those of the catecholamines.

The maximal concentration of Dopa in plasma of patients occurred between two and three hours after the capsule of drug was taken. The amount of Dopa in plasma was low; the maximal concentrations reached were between 1 and 4 μg per ml. At this concentration, the Dopa in the total plasma volume accounted for only 0.5 percent of the dose.

In summary, as the dose of L-Dopa was increased HVA in urine accounted for a fairly constant 25 percent of the dose in 13 of 17 patients with Parkinson's disease so treated. In 4 of the 17 patients the fraction of the dose excreted as HVA decreased as the dose of Dopa increased.

There was considerable variation among patients in the amount of the dose of L-Dopa metabolized to VMA but in none did this exceed 0.14 percent of the dose.

5-HIAA concentrations in urine were markedly decreased in most patients receiving L-Dopa.

The concentrations of Dopa in plasma of patients receiving this drug were low. The maximal concentrations, between 1 and 4 μg per ml, were reached two to three hours after the drug was taken.

ACKNOWLEDGMENT

The authors thank Dr. Eunice V. Flock and Dr. C. A. Owen, Jr., for their advice and criticism during these investigations.

REFERENCES

1. Mellinger, T. J.: Spectrofluorometric determination of homovanillic acid in urine. Amer. J. Clin. Path. *49:* 200, 1968.
2. Armstrong, M. D., McMillan, A., and Shaw, K. N. F.: 3-Methoxy-4-hydroxy-D-mandelic acid, a urinary metabolite of norepinephrine. Biochim. Biophys. Acta *25:* 422, 1957.
3. Wybenga, D., and Pileggi, V. J.: Quantitative determination of 3-methoxy-4-hydroxy-mandelic acid (VMA) in urine. Clin. Chim. Acta *16:* 147, 1967.
4. Macfarlane, P. S., Dagliesh, C. E., Dutton, R. W., Lennox, B., Nyhus, L. M., and Smith A. M.: Endocrine aspects of argentaffinoma with special reference to the use of urinary 5-hydroxyindoleacetic acid estimations in diagnosis. Scot. Med. J. *1:* 148, 1956.

AMINO ACIDS AFTER L-DOPA

F. Gerstenbrand and E. Gründig

This paper will present the results of our recent research on the amino acid pattern in cerebrospinal fluid (CSF) and serum in Parkinson's disease before and after L-Dopa administration. There is no brain-CSF barrier for amino acids; therefore, CSF values can be taken to reflect changes in brain metabolism.

Table 1 shows a statistically significant increase in glycine, serine, threonine, cysteine + cystine and methionine in parkinsonism patients as compared to control subjects. Glutamic acid is significantly lower in the parkinsonians than in the controls. However, in serum the amino acids are all within the normal range.

Ten schizophrenic patients with phenothiazine-induced extrapyramidal symptoms were similarly studied and compared to the parkinsonians. In both groups the same pattern was found: glycine, serine, threonine, cysteine + cystine and methionine were increased, and glutamic acid decreased or not detected at all. This alteration is still more striking in phenothiazine-induced parkinsonism.

To investigate a possible metabolic disorder within the brain induced by these phenothiazines, we determined the amino acid level of some brain regions of phenothiazine-treated rats. In these animals, with clinical akinesia, rigidity, and tremor, we found the same change in amino acid content in cortex, striatum, thalamus, and cerebellum. Glycine, serine and threonine content was higher, but glutamic acid content lower than in the control animals; in addition, γ-amino butyric acid was increased two- or threefold.

The CSF was also investigated in brain-injured patients with a midbrain lesion and with distinct parkinsonian symptoms. We know that, in these cases, the histology usually shows a lesion in the substantia nigra on one or both sides. Again in this post-traumatic parkinsonian syndrome, the total amount of amino acids was in the normal range. Glycine was increased (as compared to controls) by about 50 percent, serine 160 percent, threonine 330 percent, cysteine + cystine 80 percent, and methionine 150 percent; glutamic acid was decreased.

From these results, we believe that in the parkinsonism syndrome a disturbance in the metabolism of glycine is highly important, because a change in the turnover of glycine will definitely provoke a change in the turnover of serine, threonine, cysteine and methionine. Thus a disturbance in glycine metabolism may be present with the well-known deficiency in dopamine content of some brain structures, and these disorders may be related to each other.

This hypothesis is supported by the following results: when 100 mg of L-Dopa are injected intravenously to patients with Parkinson's disease, the glycine content of the CSF returns to a normal range while the tyrosine level rises. In normal controls, glycine content increases for 12 hours while tyro-

TABLE 1. CSF Amino Acids in Parkinsonism and Control Subjects (average values in mg percent)

	Control Subjects	*Parkinsonism*
Glycine	0.32	0.62
Serine	0.36	0.76
Threonine	0.21	0.56
Cyteine + cystine	0.19	0.47
Methionine	0.07	0.16
Glutamic acid	0.90	0.10

sine levels remain unchanged. Other amino acids, except lysine, do not show great differences between parkinsonian patients and controls.

In cases of Huntington's chorea, where we found a similar alteration in the amino acid content of CSF, the glycine level is increased fourfold after L-Dopa administration.[1]

We may thus conclude that

1. Every administration of L-Dopa provokes in parkinsonian patients and in normal persons an initial, short-lasting increase in the total amount of amino acids in the CSF and in the serum. This change is accompanied by increases in the blood sugar level.
2. Catecholamine synthesis starts from tyrosine to L-Dopa. After administration of L-Dopa in parkinsonian patients a feedback inhibition of tyrosine hydroxylation can be observed. In our investigation, we could find that a single L-Dopa administration produced a distinct increase in CSF tyrosine as well as in the serum, an effect which we could not observe in normal persons. Thus in Parkinson's disease there exists a disturbance in tyrosine hydroxylation.[2]
3. In parkinsonism the glycine level in CSF is increased. A mechanism probably exists which partly compensates the dopamine deficiency by an increased production of "inhibitor" amino acids such as glycine and gamma-amino butyric acid. After L-Dopa administration, the production of dopamine increases and an enhanced glycine synthesis is no longer necessary. The rise in tyrosine and decrease in glycine after L-Dopa administration confirms the close relationship between these amino acids.

In the CSF of parkinsonian patients, the glutamic acid content is very low. In our experience, this is not characteristic for the Parkinson syndrome. The same deficiency can be found in other neurological diseases, and seems to be a consequence of an insufficient oxygen consumption by brain tissues.[3]

We finally would like to report on our study of the amino acid level in the CSF of parkinsonians with long-term (high-dose) L-Dopa therapy. During this study, six patients were given 1.5 to 4 grams of L-Dopa for several weeks. The high values of glycine and serine were reduced after eight to ten days. In preliminary experiments it would seem that the glycine and serine levels, as well as the content in methionine, threonine and cysteine, remain normal after long-term L-Dopa therapy. The low glutamic acid content was unchanged. The other amino acids have been unchanged as well. An interruption of high-dose L-Dopa caused an increase in glycine and serine.

We think that the finding of typical changes in amino acid content in organic Parkinson syndromes, as well as a similar reversible disturbance of the amino acid pattern in cases of phenothiazine-induced parkinsonism, could improve our knowledge about the biochemistry of Parkinson's disease.

The influence of L-Dopa upon tyrosine and glycine seems to be especially important. It indicates a disturbance in specific enzymatic systems. We actually cannot give an exact explanation. Success of a treatment with L-Dopa could possibly be evaluated through changes in CSF metabolism.

In these investigations amino acids were measured by column chromatography, paper chromatography and paper electrophoresis.

REFERENCES

1. Bruck, H. Gerstenbrand, F., Gnad, H., Grundig, E., and Prosenz, P.: Uber Veränderungen der Zusammensetzung des Liquor cerebrospinalis beim choreatischen Syndrom. J. Neurol. Sci. *5:* 257, 1967.
2. Bruck, H., Gerstenbrand, F., Grundig, E., and Teuflmayr, R.: Uber Ergebnisse von Liquoranalysen beim Parkinson-Syndrom. Acta Neuropath. *3:* 638, 1964.
3. Grundig, E., Gerstenbrand, F., Bruck, H., Gnad, H., Prosenz, P., and Teuflmayr, R.: Der Einfluβ der Verabreichung von Aminosäuren, speziell von L-Dopa und α-Methyldopa, auf die Zusammensetzung des Liquor cerebrospinalis bei extrapyramidalen Syndrome. I. Veränderungen der Liquorzusammensetzung nach L-DOPA-Gaben bein Parkinson-Patienten und Gesunden. Dtsch. Ztschr. f. Nervenhk. *196:* 236, 1969.

EFFECT OF DOPA AND JB516 ON GLUCOSE METABOLISM IN BRAIN *

Gertrude M. Tyce

We became interested in the effect of large doses of dihydroxyphenylalanine (Dopa) on glucose metabolism. Since glucose is the chief source of energy in the central nervous system, we were particularly interested in effects of Dopa on glucose metabolism in the brain.

Glucose-U-^{14}C (20 μc, 0.24 mg) was injected intravenously in control rats and in rats which 2¼ hours earlier had been injected with the monoamine oxidase inhibitor β-phenylisopropylhydrazine (JB516) (20 mg per kg) and 15 minutes earlier with Dopa (100 mg per kg). Fifteen minutes after glucose-^{14}C administration the rats were anesthetized with Nembutal (20 mg) and the brains were frozen in situ by immersing the rat in liquid nitrogen. The frozen brains were chiseled out. Concentrations of glucose and lactate were measured in protein-free extracts of plasma and of brain by enzymatic procedures.[1, 2] Concentrations of amino acids in picric acid extracts of brain were measured on ion exchange columns in an amino acid analyzer. Radioactivity in glucose and metabolites in the effluent from the column was measured in a liquid scintillation counter.

At 15 minutes after the administration of glucose-U-^{14}C, the total ^{14}C in brain as unmetabolized glucose was 30 percent in the treated (Dopa and JB516) rats and less than 10 percent in the control rats. There was no difference in the percentages of total ^{14}C in lactate and alanine in the two groups of animals. The proportions of ^{14}C in aspartate, glutamine, glutamate,

* Supported in part by Grant NB-4004 from the National Institutes of Health, Public Health Service.

and GABA were significantly less in the brains of the treated rats than in the brains of the controls.

The concentration of glucose in plasma was 17.90 μmoles per ml in treated rats and 9.28 μmoles per ml in controls. In brain, the glucose concentrations were 2.89 μmoles per gm in the treated animals and 0.87 μmole per gm in the controls. Specific activity of plasma glucose, at 14 minutes (23.8×10^3 dpm per μmole) and as an average from 0 to 14 minutes (42.7×10^3 dpm per μmole), was less in treated rats than in control rats (45.2×10^3 and 64.9×10^3 dpm per μmole, respectively). Total ^{14}C in brain was 230×10^3 dpm in treated rats and 369×10^3 dpm in control rats.

Expressing total ^{14}C in brain as a function of the average specific activity of plasma glucose from 0 to 14 minutes gives the plasma glucose equivalent in brain. The plasma glucose equivalent in brain is an index of glucose uptake by brain; this was the same in treated rats (230×10^3 dpm per gm per 42.7×10^3 dpm per μmole $= 5.39$ μmoles per gm) and in control rats (369×10^3 dpm per gm per 64.9×10^3 dpm per μmole $= 5.69$ μmoles per gm).

The percentage of the total ^{14}C in brain in the combined fraction, glutamic acid + aspartic acid + glutamine + γ-aminobutyric acid, was 38.5 percent in treated rats and 61.2 percent in control rats. The plasma glucose equivalents in these amino acids was derived by multiplying the percentage in this fraction by the plasma glucose equivalent in brain. The plasma glucose equivalent in this combined fraction was less in the treated rats (38.5×5.39 μmoles per gm $= 2.08$ μmoles per gm) than in controls (61.2×5.69 μmoles per gm $= 3.48$ μmoles per gm). Apparently, metabolism of glucose to the amino acids derived from the tricarboxylic acid cycle intermediates was decreased after Dopa and JB516 treatment.

Similar decreased conversion of glucose to amino acids in brain has previously been noted in our laboratory in rats subjected to total hepatectomy [3] or ethanol intoxication,[4] by Gaitonde and Richter [5] in immature animals, and by Cremer[6] in rats during triethyltin intoxication.

ACKNOWLEDGMENT

The author thanks Dr. Eunice V. Flock and Dr. C. A. Owen, Jr., for their advice and criticism during these investigations.

REFERENCES

1. Slein, M. W.: D-Glucose: Determination with hexokinase and glucose-6-phosphate dehydrogenase. *In* Bergmeyer, H.-U.: Methods of Enzymatic Analysis. New York, Academic Press, Inc., 1963.
2. Scholz, R., Schmitz, H., Bücher, Th., and Lampen, J. O.: Über die Wirkung von Nystatin auf Bäckerhefe. Biochem Z. *331:* 71, 1959.
3. Flock, E. V., Tyce, G. M., and Owen, C. A., Jr.: Regulatory effects of insulin and liver on brain glucose metabolism. Endocrinology *84:* 392, 1969.
4. Flock, E. V., Tyce, G. M., and Owen, C. A., Jr.: Effects of ethanol on glucose utilization in rat brain. Submitted for publication.

5. GAITONDE, M. K., and RICHTER, D.: Changes with age in the utilization of glucose carbon in liver and brain. J. Neurochem. *13:* 1309, 1966.
6. CREMER, J. E.: Amino acid metabolism in rat brain studied with ^{14}C-labelled glucose. J. Neurochem. *11:* 165, 1964.

GENERAL DISCUSSION

Dr. Tyce: In our studies on the metabolism of L-Dopa-^{14}C in the isolated perfused rat liver we found that the amount of O-methyldopa-^{14}C in the perfusate increased with time after Dopa injection; after five hours, O-methyldopa-^{14}C accounted for about 6 percent of the injected dose. The shape of the curve of O-methyldopa-^{14}C in the perfusate suggested that this compound was not itself being metabolized by the liver. In the perfusate of the isolated rat liver, total ^{14}C in platelet-rich plasma and in platelet-poor plasma were found to be essentially similar, indicating that platelets did not take up Dopa-^{14}C or its metabolites. Dr. W. M. Manger (personal communication) has found, by histofluorometric methods, that Dopa is taken up by leukocytes in blood.

Dr. Sandler: I would like to say a word about *p*-tyramine excretion in parkinsonism. In the late 1950s, Sjoerdsma and his colleagues [1] proposed that urinary excretion of tryptamine, or sometimes of tyramine, would be useful as an in vivo index of monoamine oxidase activity. And indeed, if inhibitors of this enzyme are given to human subjects, increased outputs of tryptamine and tyramine are observed. As this finding provides a possible basis for the increased urinary output of tyramine in parkinsonian patients,[2] we measured platelet activity in a small series of affected subjects [3] using a microfluorescence assay with kynurenine as substrate.[4] There was a significant decrease in activity in these patients when compared with control values. We are not ready to draw far-reaching conclusions about these results, however, because our patients were severely affected postencephalitic parkinsonians and it did not appear justifiable to withdraw their anticholinergic drugs. Therefore these observations must be repeated on idiopathic parkinsonians not on drug treatment. If there were a defect of amine inactivation, this might well be of importance in the pathogenesis of the disease; my colleagues and I have recently shown [5] that intravenous tyramine administration in man brings about dopamine release in a similar manner to the way it brings about norepinephrine release.[6]

Dr. Kopin: Tyramine can competitively interfere with the β-hydroxylation of dopamine. Might that explain the homovanillic acid increase?

Dr. Sandler: I doubt it, but do not really know. Another possible explanation is that tyramine itself is β-hydroxylated to dopamine, but I do not think so because this is only a very minor pathway, at least in the rat.[7]

Dr. Kopin: What dose of tyramine did you use?

Dr. Sandler: This was relatively high: 5 mg intravenously. We also got a similar effect with 2 mg.

Dr. Krnjevic: It seems to me that Dr. Gerstenbrand's demonstration of a large drop in CSF glutamate could be interpreted as substantial confirmation of my suggestion that the nigrostriatal pathway is in fact glutaminergic.

REFERENCES

1. Sjoerdsma, A.: Techniques for measuring monoamine oxidase inhibiting activity in man. J. Neuropsychiat. *2:* 159, 1961.
2. Boulton, A. A., Pollitt, R. J., and Majer, J. R.: Identity of a urinary "pink spot" in schizophrenia and Parkinson's disease. Nature *215:* 132, 1967.
3. Sandler, M., Karoum, F., Ruthven, C. R. J., Southgate, J., and Calne, D. B.: Metabolism of L-Dopa in Parkinsonism. Abstr. Fourth International Congress Pharmacology, Basel, 1969.
4. Kraml, M.: A rapid microfluorimetric determination of monoamine oxidase. Biochem. Pharmacol. *14:* 1684, 1965.
5. Bonham Carter, S. B., Karoum, F., Sandler, M., and Youdim, M. B. H.: In preparation.
6. Sandler, M., and Youdim, M. B. H.: The effect of tyramine, reserpine and other drugs on catecholamine metabolism in man. Brit. J. Pharmacol. *34:* 224, 1968.
7. Creveling, C. R., Levitt, M., and Udenfriend, S.: An alternative route for biosynthesis of norepinephrine. Life Sci. *1:* 523, 1962.

PART 3

L-DOPA AND BLOOD PRESSURE REGULATION

CHAPTER 8

Clinical Effects of L-Dopa Upon Blood Pressure

BLOOD PRESSURE IN PARKINSONIAN PATIENTS RECEIVING L-DOPA

Charles H. Markham, Leo Treciokas, and Robert D. Ansel

Treatment of Parkinson's disease with L-Dopa has brought, in addition to beneficial effects, certain ancillary problems. From the patient's standpoint, these range from curious to serious. Alterations in sense of taste might fall into the first category, while fracture of an osteoporotic femur in a patient who could walk for the first time in years would be in the second. Alterations in blood pressure could belong in either group. Blood pressure changes are worth noting, from the standpoints both of understanding the physiology and of dealing with a possible serious side effect.

Our observations were made on 58 patients with Parkinson's disease who were treated from 4 to 12 months with L-Dopa. Thirty-seven were male and 21 female. They ranged in age from 41 to 79 years (mean age 62.4). The Parkinson's disease was idiopathic in 51 and postencephalitic in 7. Using Yahr et al.'s system of degrees of incapacity,[1] none were in Stage 1 (mild unilateral symptoms), 16 were in Stage 2 (bilateral symptoms), 19 in Stage 3 (moderate incapacity), 18 in Stage 4 (significant akinesia and needing major assistance), and 5 were in Stage 5 (complete invalidism). Thirty-one of the 58 had had stereotaxic thalamotomies.

The 58 patients were in reasonably good health aside from the parkinsonism. Two had old myocardial infarctions. Three had had congestive heart failure but had compensated well and were taking no cardiac supportive drugs except digitalis. Four had had mild hypertension, but at the start of the study had diastolic blood pressures no higher than 95 mm Hg without hypertensive medication.

Adjunctive medications such as Artane (trihexyphenidyl hydrochloride), Cogentin (benztropine mesylate) and Kemadrin (procyclidine hydrochloride) were being administered to 46 of the 58 patients (80 percent) before and in the early stages of the L-Dopa trial. After three months, 33 (57 percent) were continuing to take anticholinergic medications. A trial of withdrawal had been carried out on most of the others; this led to a worsening of their parkinsonian symptoms. Another withdrawal trial will be attempted later.

L-Dopa was started at a level of 1.0 grams a day and was increased 0.5 gram every two to three days. At a level of 3 or 4 grams, or earlier if there were significant side effects such as nausea or vomiting, increases of 0.5 grams were added every five to ten days up to 4 or 5 grams. Further increases, if any, were made over longer intervals depending on the response of the parkinsonian symptoms and the presence of side effects such as nausea, anorexia, choreo-athetoid movements, or, less frequently, insomnia, nervousness, mental confusion or low blood pressure. The average optimal therapeutic dose was 4.6 grams per day.

Blood pressure and heart rate were taken in each patient supine (or occasionally sitting) and then after standing for about a minute. The same arm was used on each occasion. Blood pressures were taken at least two to three times over a two-week period before the L-Dopa trial was begun. Many of the patients had been followed for one or more years, and in these individuals numerous blood pressure readings were taken while sitting. These observations were the base line. After L-Dopa was started and increased as noted above, blood pressures and heart rates were observed at weekly intervals for a month and thereafter at one-month intervals. Our first eight patients were hospitalized for the initial two to three weeks of the study, and in these individuals blood pressures and heart rates were taken four times a day, supine and standing.

Considering the group as a whole, alterations in blood pressure did not constitute much of a problem. We had no patients who developed significant blood pressure elevations on L-Dopa, as previously reported.[2] We did observe modest falls in blood pressure in the first one to two months in many of the patients. Many also showed a further small reduction in the blood pressures taken while standing. Heart rates remained remarkably stable, both in the supine and in the erect positions. These blood pressure changes are reflected in the slight falls in the mean systolic and diastolic pressures at one month and at three months, as may be seen in Figure 1. These data were obtained from our first 45 patients. Yet it is noteworthy that, comparing blood pressure changes during the course of therapy as well as changes from supine to standing by means of the Student method, all p values were in the range of, or greater than, 0.4, indicating that the changes were not statistically significant.

Six of the 58 patients did have marked falls in blood pressure, as may be seen in Figures 2 and 3. Typically, the supine pressures in this group fell

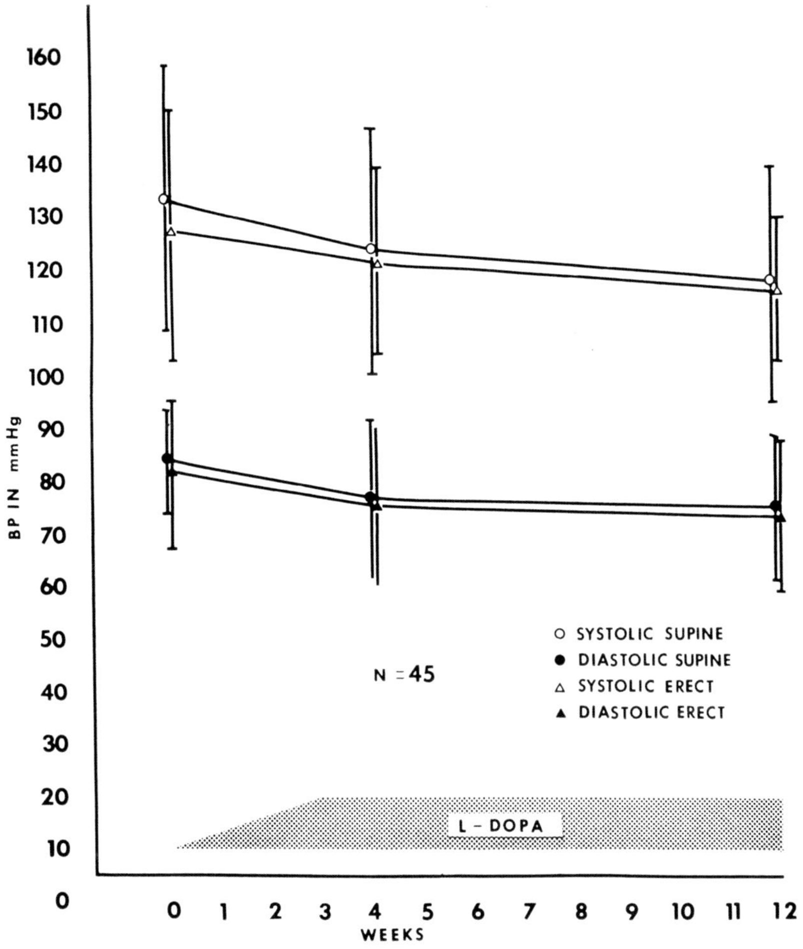

Fig. 1. Means of blood pressure, supine and standing, taken before, one month after and three months after beginning of L-Dopa therapy in 45 patients with Parkinson's disease. The vertical bars through the means represent one standard deviation on each side of the mean.

from about 130/90 to about 100/70 mm Hg. On standing, the pressure tended to fall to the 90/60 range. Heart rates during the course of the L-Dopa treatment and in the supine and standing positions were not significantly changed; certainly there were no marked increases, even when the lowest pressures were noted. The falls in both the supine and standing pressures seemed to parallel the period of maximal improvement in akinesia and rigidity. At this time there was some increased mobility, but, because of these patients' passivity, dementia or depression (see below), the actual increase in activity was modest. In two patients (J.M. and M.D.) the physical improvement took place in the first month and the low blood pressures coincided with this period. Two patients (B.B. and H.J.) had significant improvement in rigidity and akinesia only after four months; their low pressures were found at this

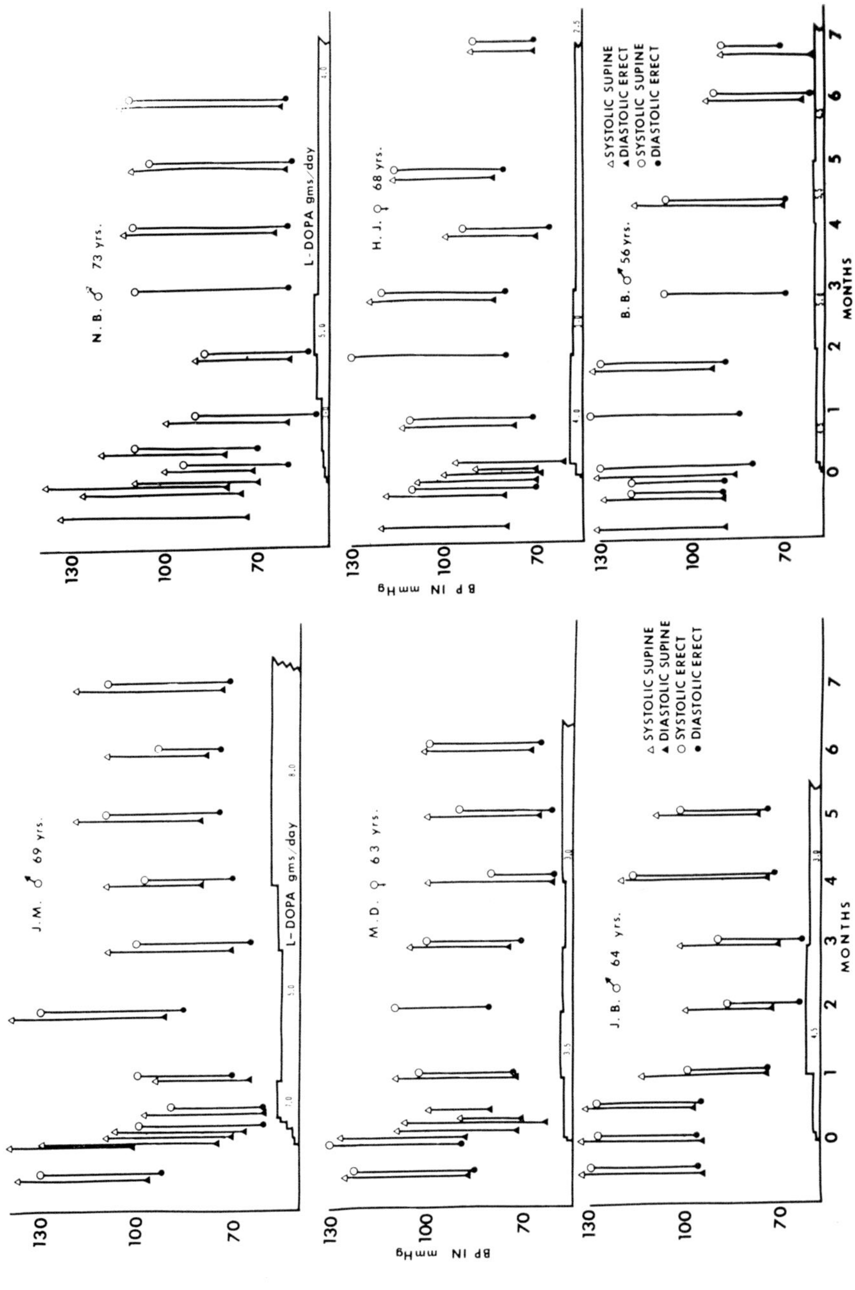

Figs. 2 and 3. Blood pressure observations, supine and standing, on six patients with Parkinson's disease during L-Dopa therapy. These patients showed marked asymptomatic fall in blood pressure at times when they were showing the most rapid clinical improvement.

time. Similarly, the remaining two patients (J.B. and N.B.) had major clinical improvement and blood pressure fall in the second month. As may be seen from Figures 2 and 3, several patients showed rise in blood pressure after several months in spite of increases in L-Dopa dosage.

It is surprising that few symptoms resulted from standing systolic pressures of 90 or even 80. The patients would stand easily and without complaint when such pressures were being recorded. Each of the six patients or their families reported instances at home of being "dizzy" or lightheaded on standing or getting up from a chair quickly. Yet in most instances such reports were only gained on direct questioning by the doctor. Two patients (J.B. and B.B.) had periods of increased confusion while standing. Yet they were both mildly demented and were also intermittently confused while lying in bed. No patient had overt syncope, or even pallor, sweating or other signs of an impending faint. Thus, while more severe hypotensive symptoms have been reported with parkinsonian patients on L-Dopa therapy,[1, 3] we did not observe them.

All six patients reported they sweated in hot weather. The four men have been impotent for years. This is probably due to their moderately severe akinesia, plus mild depression and/or dementia, and not a primary defect of the autonomic nervous system.

We might ask ourselves how these six patients compared to the entire group of 58. Their general health was on a par with the rest. Their duration of illness averaged 9.0 years as compared to 11.2 for the group. Their average age was 65.1 as compared to 62.4 for the entire group. These six patients had degrees of incapacity which put them in Stages 3, 4, and 5, where the majority of our patients fell. L-Dopa therapy resulted in significant reduction in rigidity and akinesia and all improved at least one stage. While in these respects and in others it is difficult to separate this hypotensive group from the rest, there are three clinical findings which, at least taken together, tend to set them apart. First, all had moderate to severe akinesia. Secondly, each had either mild dementia or was depressed, or both. Lastly, they were even less active, both mentally and physically, than seemed accounted for by the above symptoms. Even when the akinesia had been moderately improved by the L-Dopa, they tended to be quiet, passive and inactive.

It is commonly recognized but poorly documented that patients who are confined to bed or chair, who are elderly, and who are not active even in these confines often have low blood pressure. It may fall even lower on standing. When healthy young subjects are placed in the upright posture following prolonged bed rest, they have a marked fall in blood pressure[4, 5] accompanied by a brisk increase in heart rate.[5] When patients in their 50s and 60s with recent myocardial infarction were tilted to 70° after strict bed rest for 9 to 24 days, they also had impaired orthostatic tolerance.[6] In this last study, most had a heart rate increase of less magnitude than seen in healthy subjects and two patients had reduced heart rate and vasovagal symptoms.[6] Relative immobility may contribute to hypotension, especially the orthostatic element, by several different means. Among these are: (1) The aortic and carotid sinus baroreceptors respond not only to mean fluid pressure but also to changes in pressure. It has also been shown that perfusion of baroreceptors with a pulsatile flow causes greater reflex effects on systemic blood pressure than does perfusion with a steady flow at the same mean pressure.[7] It seems possible that the slow postural adjustments of the akinetic parkinsonian patient may cause only sluggish changes in intra-aortic and intracarotic pressures and thus provide a less than adequate stimulus to the baroreceptors. (2) There may be a defective reflex cardiac response to orthostatic venous pooling in central and peripheral veins and decreased right heart filling. It is noteworthy that in our akinetic patients there was no significant change in heart rate even when the systolic pressure fell below 90 on standing. (3) It is known that there is a rapid increase in plasma renin when the subject is tilted or put in the upright posture.[8] Could infrequent and slow assumptions of the upright posture lead to a diminished renin release in the normal individual? Barbeau et al.[3] have shown that renin levels are low in patients with akinetic parkinsonism; it is not known whether the relative postural immobility is the cause of the low renin, or whether there are other causes (see below). Whatever the underlying mechanism, it seems likely that postural immobility and infrequent assumption of the upright posture accounts for some of the hypotension in certain parkinsonian patients.

Considering the parkinsonian patients in a different light, we might ask whether there is anything to Parkinson's disease itself which makes for low blood pressure. Since the bulk of our parkinsonian patients do not have hypotension either lying down or standing, it is likely that the primary pathophysiological substrate of Parkinson's disease, degeneration of the nigrostriatal dopamine pathway and low striatal dopamine, is not a factor in the disturbed blood pressure regulation which occurs in a minority of the patients. However, a few cases with parkinsonian syndromes have other (or additional) pathological changes in the globus pallidus,[9] the striatum,[10] the centrum medianum of the thalamus,[11] the hypothalamus, pretectal region, midbrain tegmentum,[12] and elsewhere. Cortical atrophy and neuronal loss may also be found, particularly in those individuals with dementia. Parkinsonian cases with ancillary lesions should be compared to the following problems. In a certain number of cases of idiopathic orthostatic hypotension there are "signs of basal ganglia

TABLE 1. Pertinent Information on Six Patients Who Showed Significant Hypotension During L-Dopa Therapy

	J.M.	*M.D.*	*J.B.*	*N.B.*	*H.J.*	*B.B.*
Age	69	63	64	72	69	56
Sex	M	F	M	M	F	M
Stage before & after Rx.	3→2	3→2	4→3	5→3	5→4	3→2
Main symptom	Akinesia	Akinesia	Akinesia	Akinesia	Akinesia	Akinesia
Amt. of activity on feet	Little	Little	Little	None	None	Little
Mentation	Normal	Mild dementia	Dementia	Mild dementia	Normal	Mild dementia
Mood	Depressed	Normal	Normal	Normal	Depressed	Depressed

disease"[13] and reduced aldosterone secretion. Some patients with mass lesions in the posterior commissural and posterior hypothalamic regions have diminished aldosterone excretion on sodium restriction.[14] Further, some patients with akinetic Parkinson's disease have reduced aldosterone secretion.[3] Considering all this, it may be that a minority of parkinsonian patients also have atypical lesions, possibly in the pretectal region, which are responsible for low aldosterone secretion from the adrenals and thus result in low arterial blood pressure. Needless to say, facts are fewer than hypotheses in this chain of reasoning.

The final consideration we would like to discuss is the effect of L-Dopa on blood pressure. It is surprising that L-Dopa rarely produces pressor or sympathomimetic effects, since it seems that about 25 percent of radioactivity following a single ^{14}C-labeled dose of dopamine is found in the urine as norepinephrine or norepinephrine metabolites.[15] Yet the only such effect we have seen is that three of our present 58 patients developed widely dilated pupils, nearly fixed to light. All developed this pupillary abnormality in the initial stages of L-Dopa administration. It lasted only about an hour and occurred four or five times in each of two patients. The third patient has had varying degrees of pupillary dilation for three months while on L-Dopa and uses dark glasses as a protection against photophobia. None of these patients nor any of our other patients has had any blood pressure or heart rate increases which we could attribute to L-Dopa. On the other hand, it is interesting that each of our six patients with significant hypotension developed it at a time when the L-Dopa was working its maximal improvement on the parkinsonian rigidity and akinesia.

Summary

Measurements of blood pressure and heart rate were obtained in 58 parkinsonian patients before and during L-Dopa therapy. Six had asymptomatic hypotension in supine and upright postures. There were no significant changes in heart rate. These six patients seemed to differ from the rest in that they had a combination of akinesia, depression and/or dementia, and little activity on their feet, even when the akinesia was much improved. The paucity of physical activity, especially in the upright posture, is postulated as one of the main causes of the hypotension.

REFERENCES

1. Yahr, M. D., Duvoisin, R. C., Hoehn, M. M., Schear, M. J., Barrett, R. E.: L- -3-4-dihydroxyphenylalanine (L-Dopa); its clinical effects in parkinsonism. Trans. Amer. Neurol. Assoc. *93:* 56, 1968.
2. Cotzias, G. C., Papavasiliou, P. S., and Gellene, R.: Modification of parkinsonism—chronic treatment with L-Dopa. New Eng. J. Med. *280:* 337, 1969.
3. Barbeau, A., Gillo-Joffroy, L., Boucher, R., Nowaczynski, W., and Genest, J.: Renin-aldosterone system in Parkinson's disease. Science *165:* 291, 1969.

4. Deitrick, J. E., Whedon, G. D., and Shorr, E.: Effects of immobilization upon various metabolic and physiologic functions of normal men. Amer. J. Med. *4:* 3, 1948.
5. Miller, P. B., Johnson, R. L., and Lamb, L. E.: Effects of four weeks of absolute bed rest on circulatory functions in man. Aerospace Med. *35:* 1194, 1964.
6. Fareeduddin, K., and Abelmann, W. H.: Impaired orthostatic tolerance after bed rest in patients with myocardial infarction. New Eng. J. Med. *280:* 345, 1969.
7. Bard, P., ed.: Medical Physiology, 11th ed. C. V. Mosby Co., St. Louis, 1961.
8. Vander, A. J.: Control of renin release. Physiol. Rev. *47:* 359, 1967.
9. Denny-Brown, D. : The Basal Ganglia and Their Relation to Disorders of Movement. London, Oxford Univ. Press, 1962.
10. Adams, R. D., Van Bogaert, L., and Eecken, H. V.: Striato-nigral degeneration. J. Neuropath. Exp. Neurol. *23:* 584, 1964.
11. Markham, C. H., Brown, W. J., and Rand, R. W.: Stereotaxic lesions in Parkinson's disease; Clinico-pathological correlations. Arch. Neurol. *15:* 480, 1966.
12. Greenfield, J. G., Blackwood, W., McMenemey, W. H., Meyer, A., and Norman, R. M.: Neuropathology. Edward Arnold, Ltd., London, 1958.
13. Slaton, P. E., and Biglieri, E. G.: Reduced aldosterone excretion in patients with autonomic insufficiency. J. Clin. Endocr. *27:* 37, 1967.
14. Krieger, D. T., and Krieger, H. P.: Aldosterone excretion in pretectal disease. J. Clin. Endocr. *24:* 1055, 1964.
15. Goodall, McC., and Alton, H.: Metabolism of 3-hydroxytyramine (dopamine) in human subjects. Biochem. Pharmacol. *17:* 905, 1968.

BLOOD PRESSURE AND L-DOPA TREATMENT

G. C. Cotzias

In our experience, we have not found a significant problem with blood pressures. As I reported earlier, we have had only three cases of hypotension out of 70 parkinsonian patients treated with L-Dopa.

In one patient we had to diminish the daily dose of L-Dopa. When we gave this patient a peripheral Dopa-decarboxylase inhibitor we were able to control her parkinsonism with 20 percent of the original dose of Dopa, but her orthostatic hypotension was not improved. Only our cutting down the L-Dopa from 6 grams to 4.5 grams per day (without the administration of a decarboxylase inhibitor) improved the hypotension.

I wonder whether the differences in numbers of hypotensive patients are not due to differences in dose rate. Eventually we give very high doses, but we do so by slowly increasing dose rates. Our individual doses are very much smaller than those used by others. They are given frequently throughout the day, and a full therapeutic dose is reached very slowly.

A second point I would like to make is that, in our preliminary experiences with apomorphine, the patients with Parkinson's syndrome are the ones who have developed most episodes of hypotension and other peripheral vascular difficulties. The effects on the vascular system which we have found are identical to those described by Dr. Schwab many years ago.

The final point I would make is that we have had one case of carotid sinus syndrome from L-Dopa. This seems to make it pertinent to postulate a role for serotonin in the complications of some individuals receiving the drug.

ORTHOSTATIC HYPOTENSION

F. H. McDowell

One should begin by defining what one means by orthostatic hypotension. In our analysis we have used as a criterion a drop in systolic blood pressure of 30 mm Hg or more on going from the lying to the standing position, or a drop of systolic pressure to the level of 80 mm Hg. When we analyzed our patients on this basis, we had 25 who demonstrated orthostatic hypotension. Many of these patients had low blood pressures and some of them demonstrated an orthostatic drop in blood pressure prior to receiving L-Dopa.

Table 1 shows the 25 patients who had orthostatic hypotension. We were surprised to note that low blood pressures were tolerated well by patients in the age group we were dealing with. Only six had symptoms of a change in cerebral blood flow. Among those, two had syncope and four complained of light-headedness.

In our group of patients, 20 of the 25 no longer have orthostatic hypotension, and indeed, surprisingly enough, some of these 20 patients would be considered hypertensive. The diastolic pressures in some range from 90 to 110 mm Hg. In three patients the orthostatic hypotension remains intermittent, with an unexplained appearance of hypotension sometimes with symptoms. Orthostatic hypotension continues in two patients. Both of these patients have mild Parkinson's disease, have responded fairly well to Dopa, are quite physically active, are employed, and use public transportation daily. One of them is now being studied in more detail because she has demonstrated evidence of denervation hypersensitivity, intermittent Horner's syndrome, and some difficulty with urination. We are investigating the possibility

TABLE 1. Orthostatic Hypotension in 25 of 100 Patients

Symptomatic	6
Syncope	2
Lightheadedness	4
No longer present	20
Intermittent	3
Still present	2

TABLE 2. Postural Hypotension

Dose	*Patients at That Dose*	*Postural Hypotension*	
		Asymptomatic %	*Symptomatic* %
0.5	100	2	1
1.0	100	3	2
1.5	100	5	2
2.0	100	6	2
2.5	99	6	1
3.0	98	7	0
3.5	89	6	0
4.0	86	8	1
4.5	74	4	0
5.0	64	5	0
5.5	52	6	0
6.0	43	7	0
6.5	26	0	0
7.0	20	0	0
7.5	9	0	0
8.0	5	0	0

that there may be some patients who have the syndrome of autonomic insufficiency described by Shy and Drager [1] and are restudying our group to see whether there are others with more generalized evidence of autonomic insufficiency. I am not aware of any study of patients with Parkinson's disease which has included a careful survey of the pathology of the autonomic nervous system, but this is something that should be done.

We investigated whether orthostatic hypotension was dose related. In Table 2 it can be seen that orthostatic hypotension tends to occur rather early in treatment, and that it does not rise in prevalence with increasing doses of L-Dopa.

REFERENCE

1. SHY, G. M., and DRAGER, C. A.: A neurological syndrome associated with orthostatic hypotension; a clinical-pathological study. Arch. Neurol. *2:* 511, 1960.

PERSONAL EXPERIENCE WITH BLOOD PRESSURE CHANGES

M. Hill

Much has been written about the many side effects of L-Dopa, including those relating to blood pressure change. This change is important in both those with normal blood pressure and those who are hypertensive.

Of 29 parkinsonian patients treated with L-Dopa at the Toronto General Hospital for more than eight weeks, only two (6.8 percent) developed significant alterations in their blood pressures. These two men experienced marked postural hypotension; however, tolerance developed with time. With an adjustment in the dose of L-Dopa and the temporary use of tensor bandages, they were able to continue L-Dopa therapy in smaller doses. It is of interest that other troublesome side effects, namely, facial grimacing, tongue and limb dystonia, and retrocollis, were most distressing at the time the hypotension was marked.

In older persons a blood pressure fall of the magnitude seen in these two men, in the order of 50 mm Hg systolic and 30 diastolic, could be potentially hazardous.

Several other patients receiving L-Dopa therapy have complained of faintness; however, no blood pressure changes could be objectively documented.

One young man with Parkinson's disease and severe hypertension (diastolic pressure over 110 mm Hg) readily attained normal blood pressure control with methyldopa (Aldomet). The blood pressure remained within normal limits when Aldomet was stopped and replaced by L-Dopa. This man's blood pressure does fall—but only slightly—when he assumes the erect position, but he does not experience symptoms. It is of passing interest that he is certain that his parkinsonian symptoms, which had been confined to one side, became generalized shortly after methyldopa therapy was instituted in the treatment of his hypertension. It is probable that the development of bilateral disease was coincidental to the initiation of Aldomet therapy, and simply represented the natural progression of the disease.

A second man with Parkinson's disease and hypertension saw his parkinsonian symptoms improve with L-Dopa while there was no significant change in his blood pressure.

Thus our experience has been comparable to other reported series in that troublesome hypotension is an uncommon side effect of L-Dopa therapy. That it can be troublesome, however, is indicated in two cases of a group of 29.

GENERAL DISCUSSION

Dr. Barbeau: (Parenthetically, I would like to make two comments on Dr. Markham's paper. In our paper [1] we did mention the postural changes, and these will be elaborated on later in this symposium. Secondly, we never claimed that the excretion rate of aldosterone was low; it is the *secretion* rate. Excretion is normal, as I will show later. This makes a big difference in theoretical thinking.)

I would like to say that, although hypotension is never responsible for cessation of treatment, it can be potentially dangerous. I will give two examples: Both cases were patients who had been inactive for a number of years because of Parkinson's disease, and who under Dopa were doing well, except that they remained at the lower limits of normal blood pressure (between 100 and 110 mm Hg systolic, and about 70 diastolic). Both these men decided that they felt well enough to carry out some major outside activity.

One of them, 61 years old, played tennis for the first time in fifteen years and managed beautifully. The second time he tried it, he played three sets and then had a heart attack.

The other patient, 65 years old, had been unable to do anything at home, was walking with much difficulty with a cane, and most of the time was in a wheelchair; after Dopa treatment he was doing well enough to tend his garden and his lawn. Then he went out to cut a tree in his back yard. This happened to be a fairly big tree which he managed to cut completely. After he finished that, he had a stroke and has remained hemiplegic. In both these patients I had the distinct impression that there was a certain degree of frontal-lobe "insouciance."

We have to remember that these patients are often elderly and that, even though they feel, and are, much better with L-Dopa, they are now at the lower limits of normal blood pressure or hypotensive. Added activity, not expected normally from a 60- or 65-year-old person, can then become a dangerous occurrence. This point must be emphasized, for we may see more of these episodes as we go on.

Dr. Cotzias: We also have gained the impression that parkinsonian patients under treatment with L-Dopa may have less of a sense of physical danger than one would expect. Several cases of accidents were mentioned earlier. One of our patients suddenly walked right into the traffic when coming out of a movie. We have the impression (which we cannot substantiate) that lessening of the normal sense of physical danger may be induced in some patients by L-Dopa. I would like to suggest to our colleagues that some of these accidents might indicate that these people are deprived of some alerting mechanism. The only way I can describe it is deprivation of a sense of impending danger, namely, the opposite of "impending doom." We ask our patients not to drive.

Dr. Yahr: I should like to make two comments. First, parkinsonian patients as a group run lower blood pressures than a group of normal individuals matched for age and sex. Further, in the parkinsonian population, the incidence of hypertension is far below the expected rate. Second, in regard to Dr. McDowell's comment, we would confirm the fact that initially the administration of Dopa tends to lower blood pressure. However, with continued administration there is a tendency for blood pressures to regain pretreatment levels. We have proposed the assumption that the mechanism by which Dopa reduces blood pressure is replacement of the normal stores of norepinephrine in vasomotor nerve fibers, and the result is a loss of vasomotor tone due to "a false transmitter effect" with resultant vasodilation. In this regard, it is of interest that patients with varicosities in their lower limbs are more subject to blood pressure lowering during Dopa administration, and more susceptible to postural change. It has been possible in many patients to obtain a tolerable blood pressure level by using tight elastic stockings or bandages on the lower limbs, as well as abdominal binders. In some instances, expanding blood volume by administering 1 to 2 grams of sodium chloride a day has also been helpful. Certainly hypotension is a problem with the use of Dopa, and restrictive in those instances where one suspects occlusive arterial disease involving heart or brain.

Dr. Schwab: I would like to make two points, the first one about apomorphine. When we tried to use it as a therapeutic tool, we produced low blood pressures only when we gave it intravenously. There was no change in blood pressure when it was taken by mouth.

The second point concerns blood pressure. We have just completed a study of 200 patients with Parkinson's disease. We followed blood pressure in some of these patients for as long as fifteen years. We have analyzed the whole group and compared them with controls of the same age. There is a highly significant difference, both in the diastolic and systolic readings and in the response to assuming upright posture. This difference is of the order of 20 to 25 mm Hg.

I have not been able significantly to change blood pressures with such things as ephedrine sulphate three times a day, and I would like to ask what other drugs with stronger effects we could use when we have persistent hypotension with L-Dopa?

Dr. Yahr: We really found that the most effective approach is 2 grams of sodium chloride a day.

Dr. Cotzias: Salt did not work in our study of one case.

Dr. Calne: We have observed orthostatic hypotension in parkinsonian patients on conventional therapy. In addition, we have found that L-Dopa lowered the systolic blood pressure when erect by 19.3 mm Hg (mean) in a group of 20 patients taking part in a therapeutic trial.

Dr. Coleman: For the record, we have treated eight patients with dystonia musculorum deformans with L-Dopa, and in two of these patients, one aged 10 years and one aged 8 years, the early stages of treatment have been limited by the development of hypertension with diastolic pressures around 100.

Dr. Kopin: I wonder if Dr. Cotzias would clarify his comments on the euphoriant effect of L-Dopa. As you know, we have been using Dopa for the treatment of depression.

Dr. Cotzias: I am ignorant of the textbook definition of euphoria. The way I understand it, euphoria means an uncalled-for optimism, such as that seen sometimes in patients with multiple sclerosis.

Dr. Kopin: Do you see anything that would associate L-Dopa with a patient's being happier?

Dr. Cotzias: Happy, yes; unduly happy, no.

Dr. Cole: One further question on euphoria: Do these patients have bad judgment in areas other than walking in traffic? Do they show bad judgment in personal relations?

Dr. Cotzias: Let me give you an example: One lawyer in our group is constantly winning difficult criminal cases, which takes a lot of judgment. I do not believe he can be fully trusted to get his car into the garage without hitting the sides, because he goes in too fast. I do not know how to define this in psychiatric terms.

Dr. Yahr: In summary: Parkinson patients tend to run blood pressures below normal. The underlying mechanism is at present not known. Whether it is due to some generalized disturbance in catecholamines, or one localized to the central nervous system, or in any way related to disturbed metabolism, are questions not yet determined. Studies such as Dr. Markham's in correlating clinical aspects of the Parkinson syndrome are an important beginning in an understanding of the interrelationship of these two phenomena. If hypotension is more profound in the akinetic as well as in patients in the late phases of parkinsonism, this might be a clue to its causation. At any rate, it makes one cautious about treating patients of this type with Dopa.

There appears to be general agreement that the oral administration of Dopa at the higher dosage levels results in further lowering of blood pressure. Fortunately, only a small percentage become symptomatic and some of these can be controlled by the usual means of treating orthostatic hypotension. The mechanism whereby Dopa lowers blood pressure has not been defined. Of some interest in this regard is the fact that in earlier studies, where small amounts of Dopa were given intravenously, precipitious elevations in blood pressure occurred—in contrast to the large oral doses which lower blood pressure levels. In regard to the latter there apparently is a mechanism of readaptation, since in some cases the initial lowering of blood pressure seems to disappear with continued administration of the drug.

REFERENCE

1. Barbeau, A., Gillo-Joffroy, L., Boucher, R., Nowaczynski, W., and Genest, J.: Renin-aldosterone system in Parkinson's disease. Science *165:* 291, 1969.

CHAPTER 9

Pharmacology and Biochemistry of Blood Pressure Regulation

THE ROLE OF THE SYMPATHETIC NERVOUS SYSTEM IN EXPERIMENTAL HYPERTENSION *

Jacques de Champlain

Despite numerous studies, the role of the sympathetic nervous system in the pathogenesis and maintenance of human and experimental hypertension has not been clearly defined.[1] Nevertheless, the increased sensitivity to infused catecholamines and the hypotensive effect of numerous drugs altering the functions of the sympathetic nervous system in several forms of human and experimental hypertension constitute strong indirect evidence that this system may play a role in the pathogenesis and maintenance of hypertension.

Due to the numerous mechanisms involved in the proper functioning of the sympathetic nervous system, more direct means of study are required to investigate the various functions of the sympathetic nerve fibers during the development and evolution of hypertension. Normally, the norepinephrine synthesized in the nerve cell is stored in an inactive form within storage vesicles or granules localized at the nerve endings. Circulating catecholamines, or norepinephrine released by nerve impulses, are mainly inactivated by uptake or reuptake across the neuronal membrane and subsequently by uptake and binding in the storage vesicles. Moreover, part of the free norepinephrine is metabolized extraneuronally by the action of catechol-O-methyl transferase and intraneuronally by the action of monoamine oxidase. An impairment of

* Supported in part by Grant MA 3131 of the Medical Research Council of Canada and Grant ME 3138 of le Conseil de la Recherche Médicale du Québec.

any of these mechanisms or functions could result in a change in the concentration of physiologically active norepinephrine at the receptor site.

These various mechanisms and functions were studied in one form of experimental hypertension in the rat. The animals were made hypertensive by the administration of desoxycorticosterone (DOCA) (10 mg per week) and by drinking saline (NaCl 1 percent). With such treatment, systolic blood pressure started to increase by the tenth day of treatment and continued to rise progressively in the following weeks to reach levels of 180 to 220 mm Hg after six weeks. The systolic blood pressure was recorded in unanesthetized animals by means of a pulse transducer applied to the tail.[2, 3]

In hypertensive animals, but not in animals receiving either saline or DOCA alone, the endogenous norepinephrine levels were decreased significantly in the heart, spleen, intestine, kidney and skeletal muscle, while found to be normal in organs containing mainly parenchymatous innervation, such as the salivary glands and the vas deferens (Fig. 1). The neuronal uptake of tritiated norepinephrine, in vivo and in isolated perfused hearts, was found

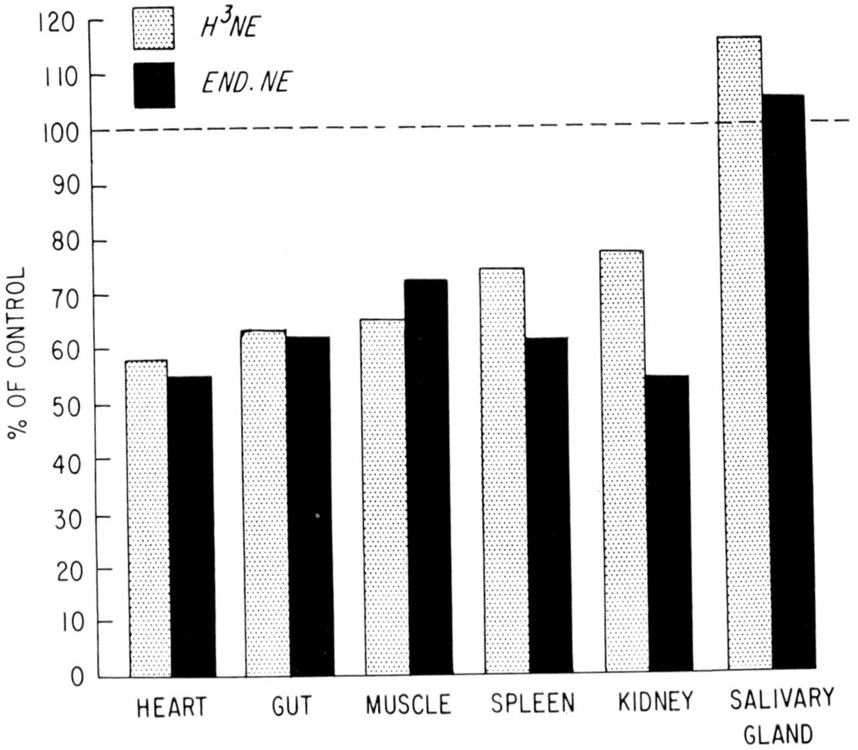

Fig. 1. Endogenous norepinephrine levels (END.NE) in various tissues of hypertensive animals and the retention of ^{3}H-norepinephrine (H^3NE) 1 hour after intravenous injection of 25 μC of the tritiated amine in the same organs. Results expressed as the percentage of the values found in organs of normotensive control rats. Each bar represents the mean of 10 to 16 values. (From de Champlain et al.[2])

normal or slightly increased in the heart of hypertensive animals.[2] However, one hour after intravenous injection of tritiated norepinephrine, the retention of the amine was markedly decreased in the heart and other organs (Fig. 1). A highly significant inverse relationship was also observed between systolic blood pressure and the endogenous norepinephrine levels or the capacity to retain tritiated norepinephrine.[2] The decrease in capacity to retain tritiated norepinephrine could be explained by changes in the turnover of norepinephrine in endogenous pools of nerve fibers.

An increase in the turnover of norepinephrine was observed in the heart and other organs of the hypertensive animals when studied after intravenous injection of tritiated norepinephrine,[4] after intravenous injection of tritiated dopamine,[5] or after blockade of tyrosine hydroxylase by α-methyl-*p*-tyrosine (Fig. 2). Moreover, it was found that the changes in turnover preceded the appearance of hypertension, thus indicating that they were primary rather than secondary to the elevation of blood pressure.[6] In association with the increased turnover, an abnormal subcellular distribution of norepinephrine was found in the heart.[4] In that organ, the amount of "free" or soluble norepinephrine was increased in relation to the amount bound to the storage

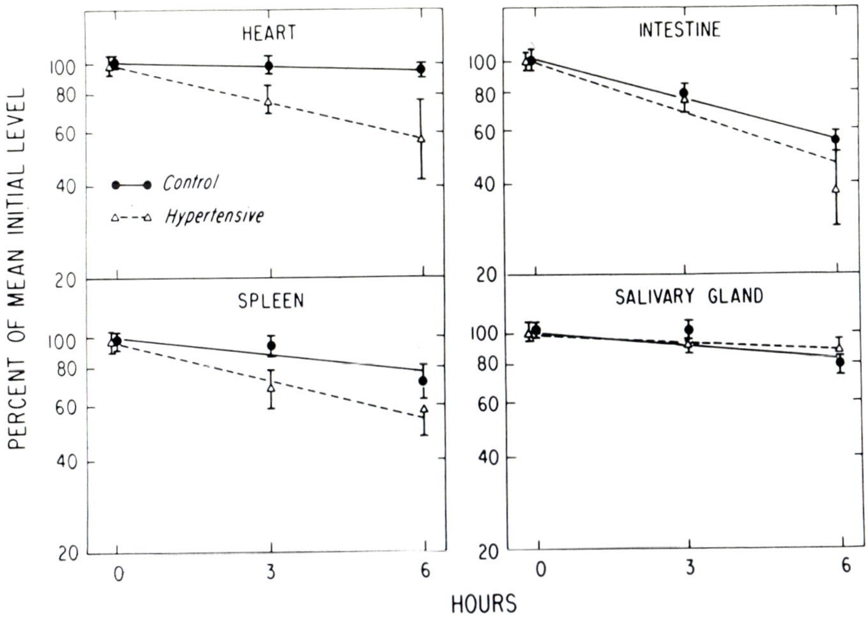

Fig. 2. Decline in endogenous norepinephrine levels in tissues of normotensive and hypertensive animals following inhibition of tyrosine hydroxylase. Each point represents the mean of 6 to 13 individual values+SEM, expressed as the percentage of the initial value. The animals were injected intraperitoneally with α-methyl-*p*-tyrosine (250 mg/kg every 3 hours) and were killed 0, 3, or 6 hours after the first injection. The rats killed 3 or 6 hours after administration of α-methyl-*p*-tyrosine had body weights and systolic blood pressure comparable to those of the initial group of rats. As illustrated, the rate of decline of endogenous norepinephrine was increased in the heart and spleen of hypertensive rats, slightly increased in the intestine but normal in the salivary glands. (From de Champlain et al.[5])

vesicles. Moreover, it appeared that more physiologically active norepinephrine might be liberated by the sympathetic nerves of hypertensive animals, as was suggested by the finding of a marked increase in the urinary excretion of tritiated norepinephrine and O-methylated metabolites 4 to 24 hours following injection of tritiated norepinephrine.[1] These findings are indicative of abnormal activity of the sympathetic nervous system in this form of experimental hypertension. Therefore, these studies suggest that in this condition more physiologically active norepinephrine is liberated by the nerve endings and is available to react with the receptors.

Various factors could contribute to the development of the abnormal sympathetic function in this form of hypertension. Since numerous forms of experimental and clinical hypertension are often accompanied by a change in the ionic content of the tissues,[7, 8] it is possible that in DOCA and sodium hypertension, which is accompanied by a marked sodium retention, ionic disturbance at the site of the nerve ending might influence the storage of norepinephrine. The withdrawal of sodium from the diet of hypertensive rats, for two weeks, resulted in lowering the blood pressure to normotensive levels and in restoring the norepinephrine turnover and the storage capacity of the nerve fibers to normal (Fig. 3). Moreover, sodium depletion and restriction in normotensive animals caused a slight decrease in blood pressure while the norepinephrine storage capacity increased above normal levels in the nerve endings.[6] Also, the subcellular distribution of norepinephrine was modified by decreasing the amount of "free" norepinephrine and by increasing the amount of bound norepinephrine, thus probably decreasing the amount of norepinephrine available to react with the receptors.[6] These findings indicate that the capacity of the sympathetic nerve granules to bind and store norepinephrine may be influenced by the state of sodium balance and can be closely correlated with the levels of systolic blood pressure.

Ionic disturbance could either affect the binding of norepinephrine directly in the storage granules or modify binding capacity indirectly by changing the sympathetic tone. The effect of ganglion blockade by chlorisondamine indicated that sympathetic nervous tone was indeed increased in the hypertensive animals. Within 24 hours after the beginning of treatment with that drug, the blood pressure of the hypertensive rats was reduced to normotensive levels and the norepinephrine storage capacity as well as the norepinephrine turnover were restored to normal levels (Fig. 4).

To determine whether blood pressure was maintained elevated by the effect of sympathetic activity at the level of the adrenal glands or at the level of the cardiovascular system, the effect of chemical sympathectomy and adrenalectomy was studied on the blood pressure of hypertensive animals. Treatment with 6-OH-dopamine, which was described to induce specific degenerescence of the adrenergic nerves,[9, 10] decreased the systolic blood pressure of the hypertensive rats by about 40 mm Hg while it decreased the blood pressure of normotensive animals by about 30 mm Hg (Fig. 5). Adrenalectomy alone induced a similar decrease in blood pressure of the hypertensive animals. This indicated that neither the sympathetic innervation of the vas-

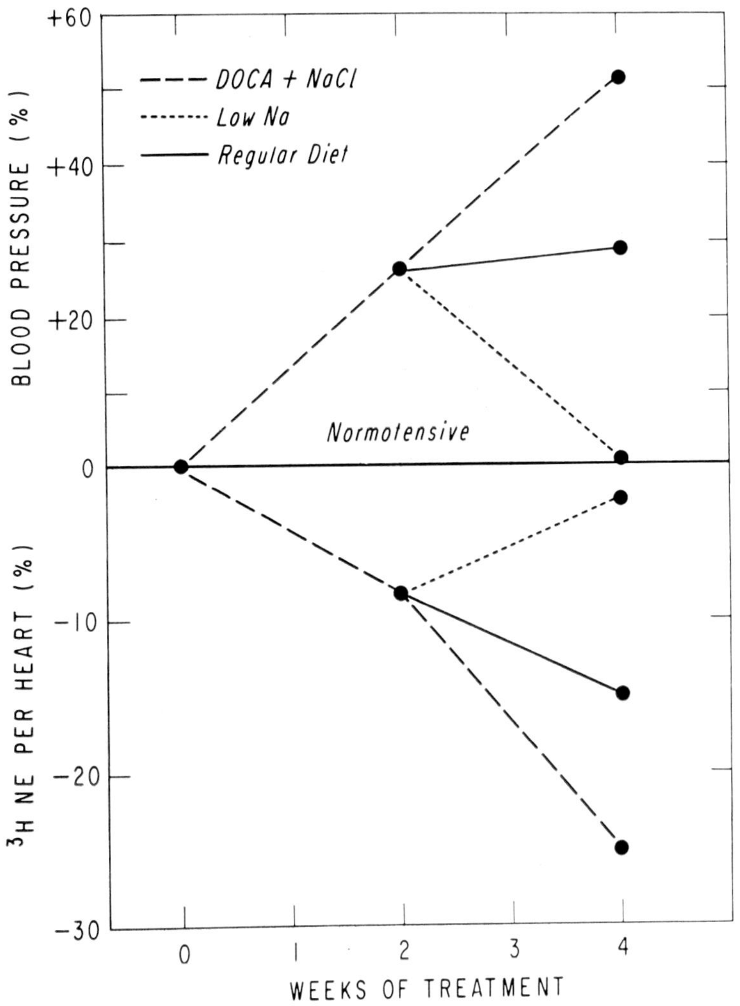

Fig. 3. Comparative effects of sodium intake on blood pressure and on the capacity of the heart to retain norepinephrine in various groups of animals. Each point is the mean of 6 to 8 values expressed as the percentage of increase or decrease compared to values in normotensive untreated animals. Upper part of graph illustrates change in systolic blood pressure. After 2 weeks of treatment with DOCA and saline (long dash lines), the blood pressure was significantly increased. At this point the rats were divided in 3 groups. One group continued to receive DOCA and saline for the following 2 weeks and the blood pressure continued to rise progressively. In the 2 other groups, DOCA and saline treatment was stopped. One of these was given a normal laboratory diet (full line) and the blood pressure was still elevated 2 weeks later. Finally, the other group of hypertensive rats was given a sodium-free diet (short dash line) and the blood pressure returned to normotensive levels within 2 weeks after the start of that diet. Lower part of graph illustrates the retention of tritiated norepinephrine in the hearts of the same groups of animals 4 hours after intravenous injection of 25 μC of the tritiated amine. The changes in the capacity to retain norepinephrine vary in the opposite direction to the changes in blood pressure. The retention capacity is the lowest when the blood pressure is the highest and this abnormality can be restored to normal simultaneously with the blood pressure in hypertensive rats treated with a sodium deficient diet for two weeks.

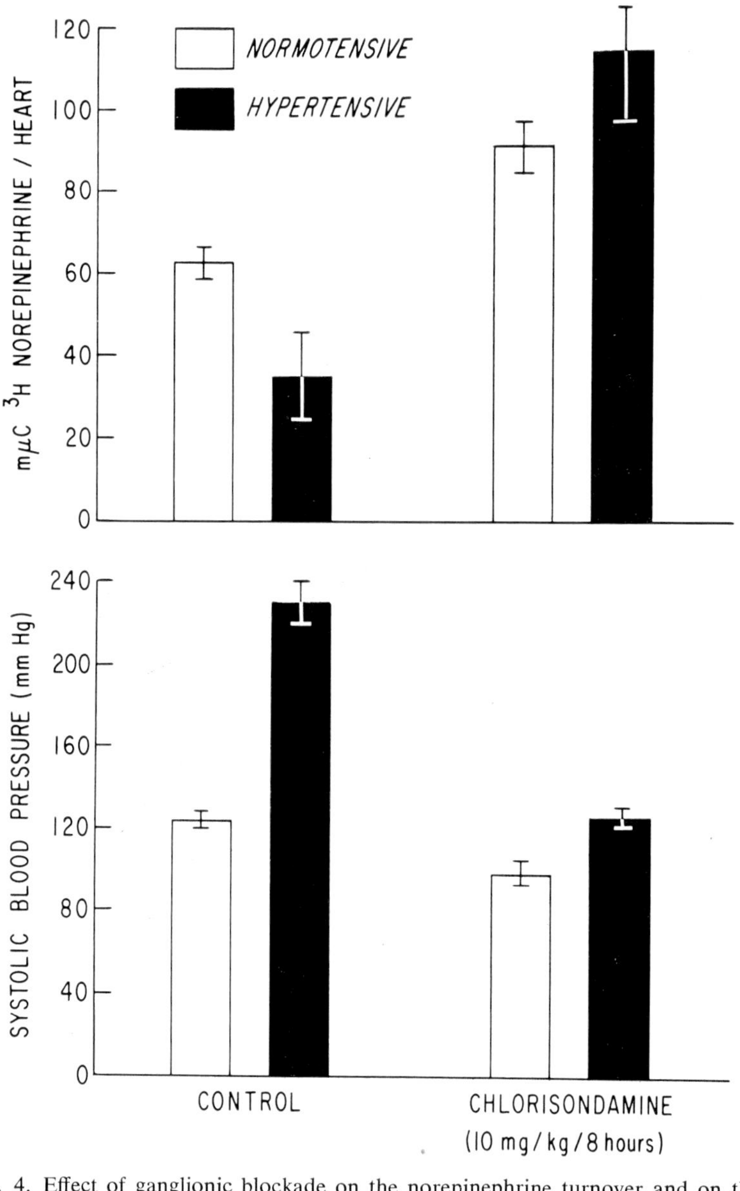

Fig. 4. Effect of ganglionic blockade on the norepinephrine turnover and on the blood pressure of normotensive and hypertensive rats. The animals were first injected intravenously with 25 μC of tritiated norepinephrine; 15 minutes after this injection, treatment with chlorisondamine (10 mg/kg every 8 hours) was started. The rats were killed 24 hours after the beginning of treatment. During treatment, the blood pressure of the hypertensive rats fell rapidly within the normotensive range while the normotensive rats became slightly hypotensive. Simultaneously, the norepinephrine turnover decreased markedly as shown by the greater retention of tritiated norepinephrine in the hearts of the treated animals. Whereas the retention of norepinephrine in hearts of hypertensive animals was half that of normotensive rats before treatment, the norepinephrine retention became even greater than that of normotensive rats in hypertensive animals treated with the ganglion blocker for 24 hours. Each group contains 7 to 8 rats and the results are expressed as the mean ± SEM. (From de Champlain et al.[6])

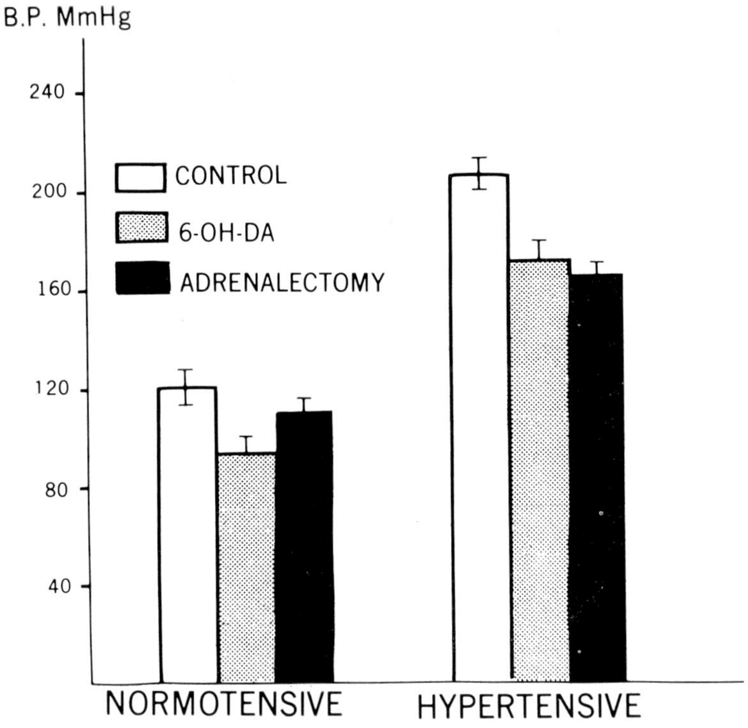

Fig. 5. Effect of chemical sympathectomy and adrenalectomy on the systolic blood pressure of normotensive and hypertensive animals. One week after intravenous injection of 6-OH-dopamine (100 mg/kg) or one week after adrenalectomy, the blood pressure of the hypertensive animals was significantly decreased to about the same level. In normotensive animals, the hypotensive effect of the treatment with 6-OH-dopamine was greater than the effect of adrenalectomy alone.

cular system nor the adrenal medulla alone could account entirely for the elevated blood pressure in the hypertensive rats. However, when both procedures were used in the same hypertensive animals, a marked fall in blood pressure was observed. Animals previously treated with 6-OH-dopamine showed a sharp and profound drop in blood pressure minutes after removal of the adrenal glands (de Champlain, J. and Lacroix, A.: unpublished data). It therefore appears that, in this form of hypertension, there is an increased central sympathetic activity, which through the synergetic action of the vascular sympathetic nerves and the adrenal medulla is responsible for the elevated blood pressure.

Impairment in the function of the sympathetic nervous system is not exclusive to DOCA and sodium hypertension, since similar abnormalities have been recently observed in other forms of experimental hypertensions. The increased norepinephrine turnover in DOCA and sodium hypertension was confirmed recently.[11] An increased turnover of norepinephrine was also observed by others in the hearts of animals made hypertensive by denervation of the carotid sinus,[12] by experimental perinephritis,[13] and by kidney infarction.[14] Similar findings were observed in animals made hypertensive by renal artery stenosis and by treatment with cadmium (de Champlain, J.: unpublished data). In contrast to other forms of hypertension, the turnover of norepinephrine was decreased in the heart of one strain of genetic hypertensive rats,[15] indicating that in this latter form the changes in norepinephrine turnover were secondary to the increase in blood pressure.

Although the findings made in experimental hypertension are difficult to extrapolate to human hypertensive diseases, it is of interest to note that abnormalities of similar nature have been observed in human hypertensive patients. Recent studies have shown that more norepinephrine and epinephrine [16] as well as more normetanephrine [17] are excreted in urine of patients with normal renal function. Using tritiated norepinephrine, Gitlow et al.[18] were able to obtain indirect evidence that the storage of norepinephrine is decreased in essential hypertension. Finally, Portnoy et al.[19] have reported recently that circulating plasma catecholamine levels were doubled in patients with essential hypertension.

Our studies and those of others in recent years have brought support for a role of the sympathetic nervous system in the pathogenesis of various hypertensive diseases. However, many more studies are required before we obtain a comprehensive evaluation of the participation of the peripheral and central nervous system in mechanisms of hypertension and before we understand the interrelationships of neurogenic tone, sodium metabolism, adrenal function, receptor sensitivity and blood pressure.

REFERENCES

1. DE CHAMPLAIN, J., KRAKOFF, L. R., and AXELROD, J.: Interrelationship of sodium intake, hypertension and norepinephrine storage in the rat. Circ. Res. *24:* 1, 1969.
2. DE CHAMPLAIN, J., KRAKOFF, L. R., and AXELROD, J.: Catecholamine metabolism in experimental hypertension in the rat. Circ. Res. *20:* 136, 1967.
3. DE CHAMPLAIN, J., KRAKOFF, L. R., and AXELROD, J.: Reduction in the accumulation of H^3-norepinephrine in experimental hypertension. Life Sci. *5:* 2283, 1966.
4. KRAKOFF, L. R., DE CHAMPLAIN, J., and AXELROD, J.: Abnormal storage of norepinephrine in experimental hypertension in the rat. Circ. Res. *21:* 583, 1967.
5. DE CHAMPLAIN, J., MUELLER, R. A., and AXELROD, J.: Turnover and synthesis of norepinephrine in experimental hypertension in rats. Circ. Res. *25:* 285, 1969.
6. DE CHAMPLAIN, J., KRAKOFF, L. R., and AXELROD, J.: Relationship between sodium intake and norepinephrine storage during the development of experimental hypertension. Circ. Res. *23:* 479, 1968.
7. TOBIAN, L.: Interrelationship of electrolytes, juxtaglomerular cells and hypertension. Physiol. Rev. *40:* 280, 1960.
8. CIER, J. F., and FROMENT, A.: Le sodium dans les hypertensions expérimentales. Path. Biol. *13:* 1052, 1965.
9. TRANZER, VAN, J. P., and THOENEN, H.: Ultramorphologische Veronderungen der sympathis chen Nervendigungen der Kotze nach Vorbehandlung mit 5-und 6-Hydroxy-Dopamin. Arch. Pharmacol. Exp. Path. *257:* 342, 1967.

10. MALMFORS, T., and SACHS, C.: Degeneration of adrenergic nerve produced by 6-hydroxy-dopamine. Europ. J. Pharmacol. *3:* 89, 1968.
11. KAZDA, ET AL.: Norepinephrine content of tissues in DOCA-hypertensive rats. Amer. J. Physiol. *216:* 1472, 1969.
12. DEQUATTRO, V., NAGATSU, T., MARONDE, R., and ALEXANDER, N.: Catecholamine synthesis in rabbits with neurogenic hypertension. Circ. Res. *24:* 545, 1969.
13. VOLICER, L., SCHEEN, E., HISLE, H., and WISWESWARAM, D.: The turnover of norepinephrine in the heart during experimental hypertension in rats. Life Sci. *7:* 525, 1968.
14. HENNING, M.: Noradrenaline turnover in renal hypertensive rats. J. Pharm. Pharmacol. *21:* 61, 1969.
15. LOUIS, W. J., SPECTOR, S., TABEI, R., and SJOERDSMA, A.: Noradrenaline in the heart of the spontaneously hypertensive rats. Lancet 1013, 1968.
16. IKOMA, T.: Studies on catechols with reference to hypertension. Jap. Circ. J. *29:* 1269, 1965.
17. STOTT, A. W., and ROBINSON, R.: Urinary normetadrenaline excretion in essential hypertension. Clin. Chim. Acta *16:* 249, 1967.
18. GITLOW, S. E., MENDLOWITZ, M., KRUK-WILK, E., WILK, S., WOLF, R. L., and NAFTCHI, N. E.: Plasma clearance of dl-H^3-norepinephrine in normal human subjects and patients with essential hypertension. J. Clin. Invest. *43:* 2009, 1964.
19. PORTNOY, B., ENGELMAN, K., and WYATT, R.: Plasma catecholamines in hypertensive and psychiatric disorders. Clin. Res. *17:* 258, 1969.

CATECHOLAMINES IN BLOOD PRESSURE REGULATION

I. J. Kopin

The mutifactorial control of blood pressure in the servomechanism for maintaining a steady level of arterial resistance and cardiac output is probably simpler and more fully understood than those involved in muscle tone, movement and coordination. There are, perhaps, some useful analogies which might help to clarify our thinking in regard to neuronal control of smooth and skeletal muscles. It might helpful to mention a few of the possibilities.

The biochemical basis for the disturbance of control of blood pressure is, of course, unknown but, regardless of the etiology of blood pressure elevation, interference with adrenergic function appears to be an effective means of lowering abnormally elevated blood pressure. The drugs which alter adrenergic function may provide insight into the means by which any neurochemical process may be altered. The processes of synthesis, storage, release, receptor interaction and termination of activity of the neurochemical transmitter should be considered.

There is a transport system which brings the amino acid precursor tyrosine into the cell. Then, by the action of tyrosine hydroxylase, tyrosine is converted to Dopa. There seems to be feedback inhibition of norepinephrine synthesis at the tyrosine hydroxylase step, presumably by the amount of cytoplasmal free norepinephrine or dopamine present. When norepinephrine is released by nerve impulses from vesicles, the empty binding sites take up some of the soluble amines in the cytoplasm; the free catecholamine levels in the cell decrease and tyrosine hydroxylase activity increases. Since this is the rate-limiting step, there is an increase in the rate of norepinephrine synthesis. This more rapid rate of synthesis results in a replenishment of the site from which the catecholamine was released to a greater extent than other less accessible stores. When the next impulse arrives, there is a selective release of the newly synthesized norepinephrine. Once released, the transmitter interacts with the receptor to produce a response, which may be only an alteration in the state of excitability. The major means of inactivation appears to be reuptake into the neuron from which the amine was released. COMT (catechol-O-methyl transferase) appears to play a role in the termination of the action of norepinephrine at some sites, however.

In addition to these local factors, which exist at any nerve terminal, there is also a system of neuronal feedback. Adrenergic function may be decreased by altering any of these processes in a manner that might be expected to diminish the amount of receptor activation. For example, synthesis of the transmitter might be decreased. Alpha-methyltyrosine, of course, is an excellent inhibitor of tyrosine hydroxylase. Although this compound has not been particularly effective as a treatment for hypertension, it is useful in the surgical preparation of patients who have pheochromocytoma: by inhibiting catecholamine synthesis in the tumors it decreases the amount of epinephrine and norepinephrine available, and diminishes the problem of excessive elevation of blood pressure. Evidently, at the doses used, inhibition of this enzyme in normal individuals is not sufficiently complete to result in depletion of the store of norepinephrine available for release from the nerve endings, either in the central nervous system or in the peripheral nervous system.

Another means of decreasing synthesis would be to inhibit Dopa decarboxylase, if one could get a good Dopa-decarboxylase inhibitor. There doesn't seem to be one available. Alpha-methyldopa was first introduced because it was thought it might be effective. Carlsson and Lindqvist [1] showed that α-methylnorepinephrine and α-methyldopamine could be formed from this compound and replace norepinephrine and dopamine. Dr. Carlsson first introduced the term "false transmitter" in the adrenergic neuron to explain the actions of α-methyldopa.

Alpha-methyldopa also has other effects. It forms α-methyldopamine which is not destroyed by monoamine oxidase and can remain in the soluble fraction. This compound is a catecholamine and can inhibit tyrosine hydroxylase. Alpha-methyldopa may indirectly inhibit synthesis of catecholamines by that means.

In relation to the action of Dopa, it is very strange that one can restore the actions of a cell that is not there. We have seen that in Parkinson's disease there is destruction of neurons, but Dopa apparently reverses the effects of these lesions. I would like to think of alternative explanations than the fact that the 5 percent of the neurons that remain take over the function of the 95 percent that are missing by supplying a little bit more transmitter. One of these alternatives, of course, is that there is the possibility of amine substitution. Thoa, Eccleston, and Axelrod [2] have shown that, in the guinea pig vas deferens, serotonin can be taken up into adrenergic nerves and released by nerve stimulation. This acts like a "false transmitter." In the pineal gland, Axelrod and others have provided evidence that the pinealocytes make serotonin, and that the adrenergic nerve may acquire serotonin from the pinealocytes and release it along with norepinephrine. If two aminergic cells are relatively close to each other, the amine from one may reach the other. I wonder whether this possibility doesn't exist in the central nervous system as well. Administered Dopa could form dopamine in serotonergic or other decarboxylase-containing cells. If serotonergic cells are complementary and opposite in function to dopaminergic cells, perhaps destruction of dopaminergic cells leads to an imbalance; too many serotonergic cells function.

The possibility of amine substitution, rather than amine overflow, as a mechanism of Dopa activity should be considered. If this mechanism operates, administration of 5-hydroxytryptophan might be expected to reverse the effects of Dopa. I do not know of any such trials.

False transmitters might explain how monoamine-oxidase inhibitors diminish elevated blood pressure. Octopamine is normally present in neurons; it normally can be released during nerve stimulation. It is derived from the beta-hydroxylation of tyramine. When monoamine oxidase is inhibited, there is a two-fold increase in the norepinephrine, but there is a 50-fold increase in the octopamine content of the peripheral sympathetic nerves. Replacement of norepinephrine by octopamine might explain some of the actions of monoamine oxidase inhibitors.

Another way, of course, of altering the synaptic efficacy is to alter the levels of enzymes concerned with transmitter synthesis. Chemical sympathectomy, with 6-hydroxydopamine or reserpine, produces an elevation of the tyrosine hydroxylase in the adrenal medulla by a neuronally mediated mechanism.[3] In our laboratory it was shown that repeated immobilization of rats for about 2½ hours a day results in elevation of adrenal catecholamine synthesizing enzymes. Initially, immobilization produces a depletion of adrenal epinephrine with increased excretion in the urine. With repeated immobilization tyrosine hydroxylase and dopamine beta-hydroxylase are elevated. There is a slight elevation of tyrosine hydroxylase six hours after immobilization for 2½ hours; after one week there is a three-fold elevation of tyrosine hydroxylase. There is also a doubling of the dopamine beta-hydroxylase. Hypophysectomy markedly diminishes the levels of these enzymes in unstressed animals, but the nerve impulses still produce an increase in tyrosine hydroxylase. Repeated utilization of a transmitter may result in the production of more of the transmitter synthesizing enzymes.

Another point which should be mentioned is the differences among patients. One area of treatment failure investigated in our laboratory has been the drug-induced movement disorders. Drs. Chase, Scherr and Gordon studied levels of homovanillic acid (HVA) and 5-hydroxyindoleacetic acid (5-HIAA) in spinal fluid in three groups of patients. Patients in Group I had not been treated with phenothiazines; those in Group II were treated with phenothiazines but did not develop movement disorders; those in Group III were treated with phenothiazines and developed movement disorders.

The authors found an elevation of HVA in response to treatment with phenothiazine derivatives, presumably, as Carlsson has explained, because of blockage of the receptor with a neuronal feedback so that there is an increase in the formation of dopamine. However, certain patients (those in Group III) evidently could not respond as well and did not have elevated HVA levels; these patients developed movement disorders. We do not know why one patient differs from another in this respect.

There is one other point I would like to raise. Elevated circulating levels of epinephrine, referred to by de Champlain, have certainly been confirmed in hypertensive patients. However, depressed patients have higher levels of circulating norepinephrine, about six times normal, and do not have hypertension.[4] It would therefore appear that mechanisms other than catecholamines are involved in hypertension.

REFERENCES

1. Carlsson, A., and Lindqvist, M.: In vivo decarboxylation of o-methyl-m-tyrosine. Acta Physiol. Scand. *54:* 87, 1962.
2. Thoa, N. B., Eccleston, D., and Axelrod, J.: The accumulation of C^{14}-serotonin in the sympathetic nerves of the guinea pig vas deferens. J. Pharmacol. Exp. Ther. *169:* 68, 1969.
3. Thoenen, H., Mueller, R. A., and Axelrod, J.: Trans-synaptic induction of adrenal tyrosine hydroxylase. J. Pharmacol. Exp. Ther. *169:* 249, 1969.
4. Engelman, K.: Personal communication.

THE POSSIBLE EFFECTS OF DOPA ON CATECHOLAMINE BIOSYNTHESIS *

M. Goldstein, B. Anagnoste, A. F. Battista, and S. Nakatani

The most common side effects during treatment with L-Dopa are nausea, vomiting and anorexia, postural hypotension, cardiac arrythmias, and pronounced psychic manifestations. While the gastrointestinal side effects have proved to be less significant with further experience in the treatment, hypotension, involuntary movements and psychic side effects have remained difficult problems in treatment of some Parkinson's patients.[1]

* Supported by USPHS Grant NB-06801 and by NSF Grant GB-8465.

The purpose of this paper is to describe some possible alterations in the catecholamine biosynthesis produced by administration of Dopa and to relate these changes to some side effects observed during treatment with L-Dopa.

Material and Methods

Unilateral radio frequency lesions were placed in the ventromedial tegmental region of the rostral pons in monkeys by previously described procedures.[2, 3] Tyrosine hydroxylase activity was determined by the isotopic procedure of Nagatsu et al.,[4] and Dopa decarboxylase activity by a modification of the procedures described by Bertler et al.[5] Labeled catecholamines were isolated as acetylated derivatives and separated by paper chromatography.[6]

Results

The biosynthetic and catabolic pathways of dopamine. The main pathway for the biosynthesis and catabolism of dopamine is presented in Figure 1. The formation of dopamine from L-tyrosine requires two enzymatic steps, namely, the conversion of tyrosine to Dopa which is catalyzed by tyrosine hydroxylase and the conversion of Dopa to dopamine which is catalyzed by Dopa decarboxylase. In sympathetic innervated tissues and in the norepinephrine-containing regions of the brain, dopamine is further converted to

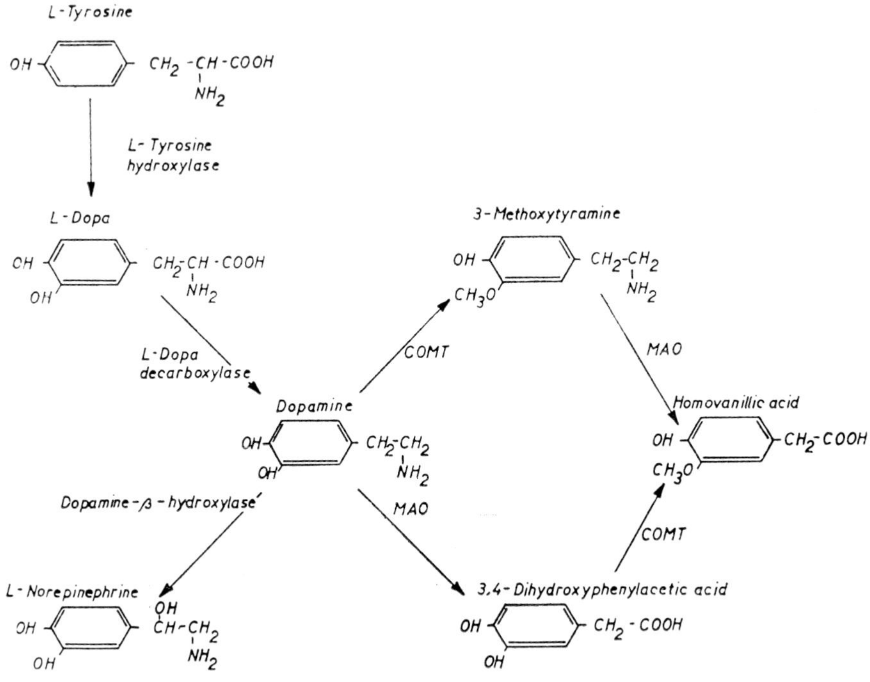

Fig. 1. The biosynthetic and metabolic pathway of dopamine.

norepinephrine and this enzymatic reaction is catalyzed by dopamine-β-hydroxylase. However, in the striatum the activity of dopamine-β-hydroxylase is very low and the conversion of dopamine to norepinephrine occurs only to a limited extent.[3]

The major pathways in the metabolism of dopamine are deamination catalyzed by monoamine oxidase (MAO) and O-methylation catalyzed by catechol-O-methyl transferase (COMT). The aldehyde formed from dopamine by the action of MAO is further metabolized by the action of aldehyde dehydrogenase to the corresponding acids and alcohols. The metabolic breakdown products of dopamine are homovanillic acid (HVA) and 3,4-dihydroxyphenylacetic acid (HPA) as well as 3,4-dihydroxyphenylethanol (DHPE) and 3-methoxy-4-hydroxyphenylethanol (MHPE).[7] The formation and accumulation in the brain of the dopamine aldehyde generated by MAO cannot be excluded, especially in patients treated for a long period with Dopa. Some evidence has been obtained that the oxidative pathway which leads to the formation of acids occurs primarily extraneuronally while the reductive pathway which leads to the formation of alcohols occurs intraneuronally.[8] It is therefore conceivable that the ratio of excreted acids to alcohols will be higher in subjects treated with Dopa.

Dopamine can also undergo N-acetylation and this metabolic pathway is enhanced following inhibition of monoamine oxidase.[9] Following administration of Dopa, N-acetylation might represent an important route for the inactivation of dopamine in the liver and other peripheral organs.

The effects of ventromedial tegmental lesions on tyrosine hydroxylase and Dopa decarboxylase activities in the striatum of monkeys. The effects of tegmental lesions on tyrosine hydroxylase and Dopa decarboxylase activities in the striatum are summarized in Table 1. Following tegmental lesions the tyrosine hydroxylase activity was reduced 40 to 75 percent on the lesion side below that of the corresponding intact side. The Dopa decarboxylase activity was also significantly lower on the lesion side than on the intact side. The extent of reduction of these two enzyme activities varied in different monkeys and seemed to be dependent upon the size of the lesion and on the time which elapsed after the lesion was placed. It is noteworthy that the effective dose of Dopa to relieve the tremor varies in monkeys with tegmental lesions and seems to be dependent on Dopa decarboxylase activity in the ipsilateral striatum. The percentage loss of enzyme activities could be correlated with

TABLE 1. The Effects of Tegmental Lesions on Tyrosine Hydroxylase and Dopa Decarboxylase Activities in the Striatum

	*Activity **	
Lesion Duration	*Tyrosine Hydroxylase*	*Dopa Decarboxylase*
2–8 days	40–60	50–70
2–5 months	25–50	30–50

* The values are expressed as percentage of enzyme activity in the lesion side as compared to the intact side. The figures represent the range of values obtained from three animals in each group.

the loss of the cells in the pars compacta of the substantia nigra. No significant changes in MAO activity were observed on the lesion side of the striatum. Also the activities of cholinesterase and of choline acetylase were not significantly different on the lesion side of the striatum than on the intact side.

The rate of catecholamine biosynthesis from ^{14}C-tyrosine in brain cortex slices. The time course of ^{14}C-dopamine and ^{14}C-norepinephrine formation from ^{14}C-tyrosine was investigated in brain cortex slices of rats. The data presented in Figure 2 show that after 40 minutes of incubation time the rate of ^{14}C-dopamine synthesis decreases. The rate of ^{14}C-norepinephrine synthesis is initially slow and increases after 20 minutes of incubation time. The rate of ^{14}C-norepinephrine synthesis increases with increased concentrations of ^{14}C-dopamine in the slices.

Discussion

Evidence has been presented that in monkeys with centromedial tegmental lesions the storage and synthesis of dopamine and serotonin is impaired in the ipsilateral striatum.[3] The finding that Dopa decarboxylase activity is decreased in the striatum ipsilateral to the lesion is of importance in view of the therapeutic use of L-Dopa in treatment of Parkinson's disease. In patients treated with L-Dopa the formation of dopamine depends on Dopa decarboxylase activity. It is therefore possible that in some Parkinson's patients

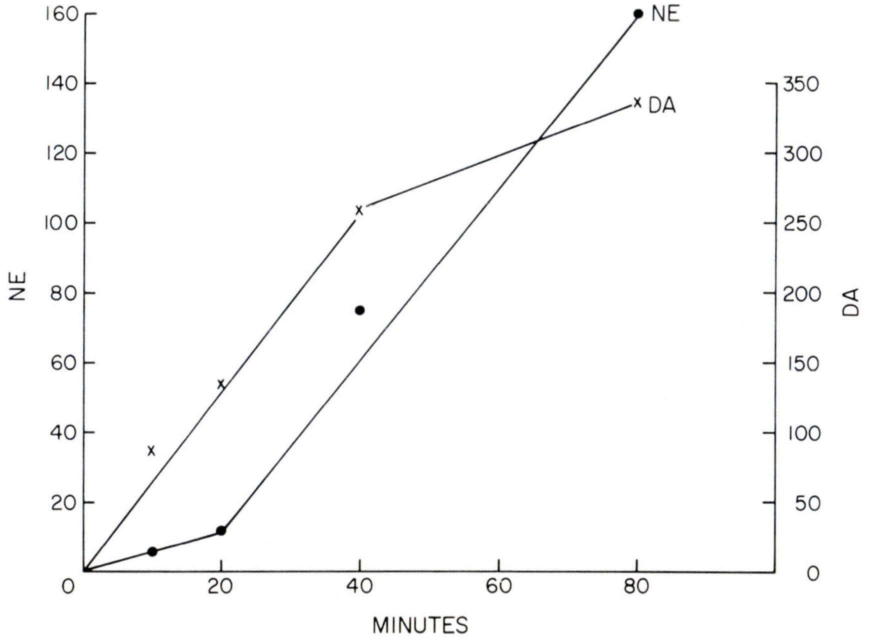

Fig. 2. The time course of ^{14}C-dopamine (DA) and of ^{14}C-norepinephrine (NE) formation in brain cortex slices of rats. The values are expressed as percentage of ^{14}C-tyrosine ($\times 10^{-3}$) per 340 mg of slices.

the formation of dopamine from Dopa is limited; this might explain the lack of response to Dopa therapy in some individuals.

A diagram showing the enzymatic formation of catecholamines from tyrosine and from Dopa is shown in Figure 3. The formation of catecholamines from tyrosine is regulated by the feedback inhibition at the tyrosine hydroxylase stage.[4, 10] The formation of catecholamines from Dopa is not controlled by the feedback inhibition. However, the dopamine formed from Dopa will potentiate the feedback inhibition at the tyrosine hydroxylase stage. This will result in a decrease of catecholamine synthesis from tyrosine and the newly formed dopamine will originate mainly from Dopa. As a result of such an alteration, tyrosine which was not utilized for the catecholamine formation might be further metabolized by other pathways. Consequently the formation of tyramine and octopamine from tyrosine might be enhanced. It is conceivable that the decrease in tyrosine hydroxylase activity enhances tryptophan hydroxylase activity, since both enzymes require the same tetrahydropteridine as a cofactor.

The absence of an increase in norepinephrine levels after administration of Dopa indicates that the norepinephrine synthesis is not only regulated at the tyrosine hydroxylase stage. The penetration of dopamine into the vesicles containing dopamine-β-hydroxylase might be responsible for the regulation of norepinephrine synthesis from dopamine. The apparent K_M value of dopamine-β-hydroxylase for dopamine is in the order of 5×10^{-3} M, which suggests that a rather high concentration of dopamine is required at the enzyme site for the maximal conversion to norepinephrine.

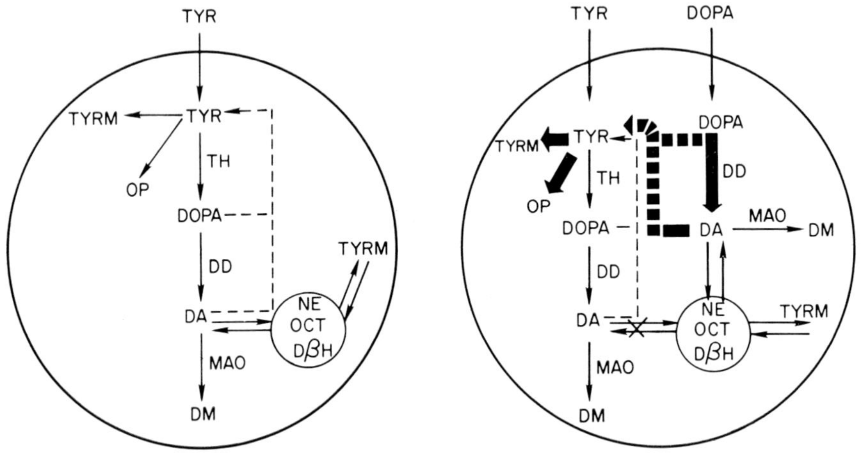

Fig. 3. Diagram showing the enzymatic formation of catecholamines from tyrosine and Dopa, and the possible effect of Dopa on the regulation of synthesis. (Left-synthesis from tyrosine; right-synthesis from tyrosine and from Dopa).

Abbreviations used: TYR, tyrosine; TYRM, tyramine; OCT, octopamine; OP, other products; DA, dopamine; NE, norepinephrine; DM, deaminated metabolites; TH, tyrosine hydroxylase; DD, Dopa decarboxylase; DβH, dopamine-β-hydroxylase; MAO, monoamine oxidase.

The enhanced formation of tyramine from tyrosine following Dopa administration might inhibit the conversion of dopamine to norepinephrine. Tyramine has a direct effect on dopamine-β-hydroxylase, competing with the β-hydroxylation of dopamine,[9] and has an indirect effect competing with dopamine for the accumulation in the vesicles containing dopamine-β-hydroxylase.

It is evident from the present data that ^{14}C-dopamine accumulates to a great extent in brain tissue cortex slices and is not further converted to norepinephrine. In the brain tissue slices the biosynthesis of catecholamines is not governed by functional neuronal activity and an analogous situation might exist after prolonged administration of Dopa. The activity of the catecholamine neurons might be diminished since dopamine formed from Dopa might accumulate at the receptors and also replace norepinephrine at the receptor sites. It is possible that the Dopa-induced hypotension and other observed side effects are caused by accumulation of dopamine and other sympathomimetic amines which act as false transmitters. Disulfiram or other dopamine-β-hydroxylase inhibitors [11] could be used as tools for pharmacological and clinical studies to investigate the effects of dopamine accumulation on the production of side effects following Dopa administration.

Summary

Some side effects observed during treatment with L-Dopa were related to possible alterations in the catecholamine biosynthesis. The accumulation of dopamine and other sympathomimetic amines which might function as false transmitters might account for some observed side effects during prolonged treatment with L-Dopa.

REFERENCES

1. Duvoisin, R., Yahr, M. D., Schear, M., Hoehn, M. M., and Barrett, R. E.: The present status of L-Dopa in the treatment of parkinsonism. Excerpta Medica, International Congress Series *193:* 170, 1969.
2. Poirier, L. J., and Sourkes, T. L.: Influences of the substantia nigra on the catecholamine content of the striatum. Brain *88:* 181, 1965.
3. Goldstein, M., Anagnoste, B., Battista, A. F., Owen, W. S., and Nakatani, S.: Studies of amines in the striatum in monkeys with nigral lesions. J. Neurochem. *16:* 645, 1969.
4. Nagatsu, T., Levitt, M., and Udenfriend, S.: Tyrosine hydroxylase: the initial step in norepinephrine biosynthesis. J. Biol. Chem. *239:* 2910, 1964.
5. Bertler, A., Carlsson, A., and Rosengren, E. : A method for the fluorimetric determination of adrenaline and noradrenaline in tissues. Acta Physiol. Scand. *44:* 273, 1958.
6. Goldstein, M.: Cerebral metabolism of Dopa-C^{14} in rats. Int. J. Neuropharmacol. *3:* 37, 1964.
7. Goldstein, M., and Gerber, H.: Phenolic alcohols in the brain after administration of Dopa-C^{14} or dopamine-C^{14}. Life Sci. *2:* 97, 1963.
8. Goldstein, M., Anagnoste, B., Yamamato, A., and Felch, W. C., Jr.: Re-

gional distribution and metabolism of H^3-tyramine in the rat brain. J. Pharmacol. Exp. Ther., in press.

9. GOLDSTEIN, M., and CONTRERA, J. F.: The substrate specificity of phenylamine β-hydroxylase. J. Biol. Chem. *237:* 1898, 1962.
10. COSTA, E., and NEFF, N. H.: Isotopic and non-isotopic measurements of the rate of catecholamine biosynthesis. *In* Biochemistry and Pharmacology of the Basal Ganglia. Proceedings of the Second Symposium of the Parkinson's Disease Information and Research Center, College of Physicians and Surgeons of Columbia University (Costa, E., Cote, L. J., and Yahr, M.D., eds.). Raven Press, New York, 1965.
11. GOLDSTEIN, M.: Inhibition of norepinephrine biosynthesis at the dopamine-β-hydroxylation stage. Pharmacol. Rev. *18:* 77, 1966.

RENIN, DOPAMINE AND PARKINSON'S DISEASE *

André Barbeau, Lise Gillo-Joffroy, and Yves Brossard

Previous papers have reviewed some of the evidence leading to the conviction that there is a relationship between the sympathetic nervous system and the pathogenesis of essential hypertension. In summary, the main arguments are, as partially outlined by De Quatro and Sjoerdsma: [1]

1. Some hypertensive patients exhibit sympathetic "hyperactivity" measured by an increase in urinary excretion of norepinephrine. However, these studies are still very controversial.
2. Hypertensive subjects have an enhanced vascular responsiveness to injections of norepinephrine.
3. Treatment with antiadrenergic drugs is effective, as Dr. Kopin has earlier pointed out.
4. Vascular wall metabolism of norepinephrine in animals is made hypertensive with DOCA plus sodium, as shown by de Champlain and others.[2]
5. Robertson and others have recently demonstrated an increased concentration of dopamine in the caudate and of norepinephrine in the forebrain of genetically hypertensive rats.[3] This form of hypertension would not seem to be of renal origin.

Our own approach to this problem was motivated by the following observations:

1. Akinetic parkinsonian patients have a lower blood pressure (mean of 200 observed cases: 126/83 mm Hg) than control subjects of similar age (140/88 mm Hg) and are subject to occasional episodes of postural hypotension. In fact, derangements in postural regulation are frequent symptoms.
2. Treatment of these patients with L-Dopa occasionally produced a marked drop in blood pressure, not always symptomatic. In some patients this postural hypotension can be accompanied by complications: myocardial infarcts or cerebrovascular accidents.[4]
3. A defect in the sympathetic nervous system has been postulated in essential postural hypotension [5] and in some patients with this disease extrapyramidal

* Supported by Grants MA-2530 and SP-1 of the Medical Research Council of Canada.

signs have been observed. In one case a depigmentation of the substantia nigra was reported at autopsy.

4. There is a growing amount of evidence indicating that the sympathetic nervous system is involved in some part of the mechanism of renin release. The studies of Vander and collaborators [6] are reinforced by the observation, with fluorescence methods and electron microscopy, of catecholamine terminals near the juxta glomerular apparatus.[7]
5. There is now some direct evidence obtained by Winer and his colleagues [8] that many forms of stimulation of renin secretion are mediated through adrenergic receptors in the kidney. Blockade of these receptors by either phentolamine, an alpha-adrenergic blocking agent, or by propanolol, a beta-adrenergic agent, markedly suppressed the increase in plasma renin activity (PRA) produced by upright posture, diazoxide, ethacrynic acid or theophylline. It is interesting to note that this effect could be separated from any action upon sodium metabolism.

We therefore studied renin activity in the plasma by the method of Boucher in 35 normotensive control subjects and 31 parkinsonian patients.[9] Under an unrestricted salt intake, plasma renin activity was significantly lower in patients that in control subjects (Table 1). This difference persisted when these patients were given a diet of 135 mEq of sodium per day (potassium being limited at 90 mEq per day) for three days. Of the 48 measurements made, only 9 were above 5 ng per liter per minute and 28 had undetectable amounts.

This initial study has now been confirmed in over 60 patients with Parkinson's disease, 26 of whom received a complete metabolic investigation (Table 2). It can be seen that these patients have low urinary excretion of dopamine, with borderline low excretion of norepinephrine. The blood pressure of the group as a whole is below normal. Blood cortisol, urinary 17-hydroxy- and 17-keto-steroids are within normal limits, as are thyroid function tests. Kidney function is not impaired. Despite this, PRA is considerably diminished in the group as a whole.

Treatment with a progressively higher dose of L-Dopa uniformly decreased plasma renin activity to undetectable levels.[9] Recently Mohammed and collaborators [10] demonstrated a similar effect of α-methyldopa upon PRA.

Parkinsonian patients react to stimuli that normally enhance renin secretion, although possibly in a more sluggish way than normal controls. In six patients thus studied, there was a clear-cut elevation from 2.5 to 13.1 ng per

TABLE 1. Plasma Renin Activity in Parkinson's Disease and in Normotensive Healthy Control Subjects Expressed in Nanograms of Angiotensin Liberated per Liter per Minute (Mean ± S.D.) (From Barbeau, A. et al.[9])

Na Diet	*Number Patients*	*Angiotensin* (ng/1/min)
	Control	
Unrestricted	27	9.5± 6.7
135 mEq/day	8	9.8±12.7
	Parkinson	
Unrestricted	31	2.4± 4.1
135 mEq/day	17	2.7± 2.7

TABLE 2. Metabolic Studies of the Sympatho-reno-adrenal System in Parkinson's Disease (26 patients)

P.R.A.	
· Total group	2.6± 2.8 ng/1/min
· Na diet 135 mEq/li (N=7)	2.1± 0.8 ng/1/min
· Na diet 10 mEq/li (N=7)	11.2± 4.4 ng/1/min
Cortisol	16.8± 6.1 mcg %
17 OH (THE)	4.1± 1.6 mg/24 hrs
17 KETO	9.4± 3.1 mg/24 hrs
P.B.I.	6.3± 2.5 μg %
T_3	31.4± 6.8 μg/100 cc
T_4	7.0± 2.5 μg/100 cc
Plasma aldosterone	14.8±10.0 mμg %
Aldosterone secretion rate	69.9±40.2 μg/24 hrs
Aldosterone excretion rate	8.4± 3.5 μg/24 hrs
Dopamine (urine)	256.8±26.0 μg/gm of creat.
Noradrenaline (urine)	23.4± 3.3 μg/gm of creat.
Adrenaline (urine)	17.3± 1.8 μg/gm of creat.
H.V.A. (urine)	3.7± 0.3 mg/24 hrs
Blood pressure (mean)	125.7± 3.8/86.6±2.1
Creatinine clearance	83 ±12.8 ml/min

liter per minute after three days of sodium diet at 10 mEq per day.[9] This new level is still lower than the one observed in age-matched normal controls.

Upright posture also produces an increase in renin secretion, but to levels still subnormal.[9] Thus we can say that in Parkinson's disease the renin-angiotensin system is responsive, but very slowly, to stimuli mediated through the sympathetic nervous system, while it is almost normally responsive to sodium-mediated and other stimuli.

This can be better illustrated by the aldosterone system. Plasma aldosterone is usually normal, as is urinary excretion of aldosterone. However, the rate of secretion of the hormone is at the lower limits of normal (Table 2). It thus appears that, at least in Parkinson's disease, there are many systems controlling aldosterone in addition to renin and angiotensin and that renin secretion and aldosterone do not always vary in parallel. Moreover it is evident that in Parkinson's disease the suppression of renin activity is not due to a state of hyperaldosteronism nor to increased plasma volume (in fact, in many cases this volume is decreased). Thus, by exclusion, it appears that the low PRA in Parkinson's disease is a consequence of hypoactivity in the sympathetic nervous system.

The evidence thus obtained led us to ask a number of questions relating to the central nervous system control of blood pressure regulation and the role specifically played by dopamine in this mechanism, both peripherally and centrally.

Peripheral Role of Dopamine

As will be seen from Dr. Kuchel's paper in this volume, dopamine urinary excretion decreases significantly upon assuming upright posture, thus lowering the DA/NE ratio. The relative action of these two amines upon the adrenergic receptors in the kidney will be discussed.[11]

Central Role of Dopamine

Our evidence is thus that dopamine plays an important function in this postural system at it does in other postural neural mechanisms in the brain. However, since dopamine is not the main catecholamine in the brain stem or hypothalamus, but rather in the striatum, we suspected that there may be a more rostral center for blood pressure regulation than those usually postulated. We approached this hypothesis through a series of experiments with a mild hypotensive drug called Catapres,* alone or in combination with a potent natriuretic, furosemide (Lasix).

Recent studies from many authors [12, 13] had indicated that the site of action of Catapres is in the brain stem. Indeed, most authors agreed that it produced a decrease in central sympathetic outflow coupled to an increase in central parasympathetic outflow.[13] Clinically furosemide potentiates these effects.

In rabbits, Catapres, and more so Catapres plus furosemide, caused a marked decrease in the concentration of norepinephrine in the brain stem and also in the caudate nucleus (Figs. 1 and 3). In this latter area there was no change in total dopamine concentration during the period of observation, but a marked increase in dopamine turnover was observed through a significant increase in HVA (Fig. 2). Thus the caudate, by increasing the turnover of dopamine, seems to favor a decreased sympathetic outflow as evidenced by low norepinephrine contents in the usual vasomotor centers of the brain stem. This would appear to be a situation opposite to that de Champlain now reports in hypertensive animals.

Parallel with this control of norepinephrine synthesis, Catapres or Catapres plus Lasix caused a marked elevation in serotonin content of the caudate and brain stem (Figs. 1 and 3). Because of the postulated association between serotonin and acetylcholine, this mechanism could account for the increase in parasympathetic outflow.[14]

Conclusion

New evidence is presented to support an important role for dopamine in the postural regulation of blood pressure, not only at the periphery (particularly at the level of the adrenergic renal receptors important in renin secretion) but also in the central nervous system (mainly in the caudate nuclei). These nuclei could play a role as monitors of changes in composition of the internal milieu (ionic or hormonal) and as modulators of the blood pressure centers in the brain stem and hypothalamus. A feedback mechanism is probably active. The baroreceptors in the carotid sinus, known to have a high content of dopamine, may be an important part of this feedback loop. Here again dopamine may be playing a role in determining the "set" of sensitivity.

---

* 2-(2,6-Dichlorophenylamino)-2-imidazoline hydrochloride (ST-155).

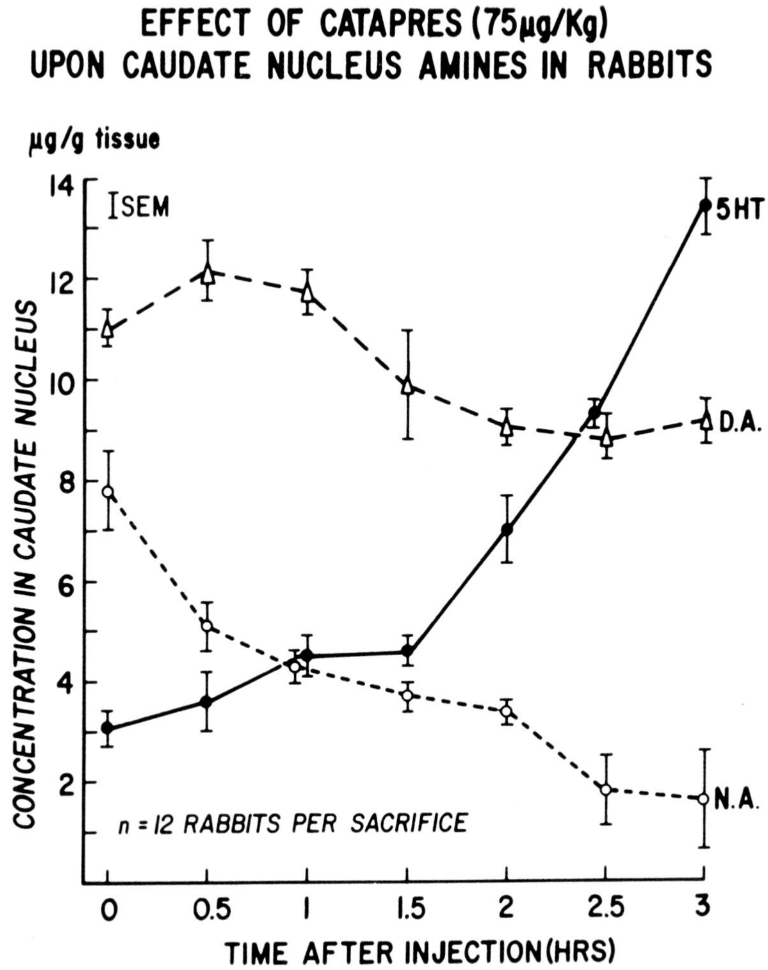

Fig. 1. Concentration of various amines in the caudate nuclei of rabbit following Catapres. For norepinephrine scale $\times 10$.

At the periphery, PRA responses during postural modifications are under the influence of the sympathetic nervous system, while, centrally, ionic or hormonal changes will influence the content and relative output of the caudate upon lower brain stem centers. In all of these areas, dopamine is a key component. Parkinson's disease, which is characterized by a deficiency in dopamine, therefore becomes an ideal human model for the study of such mechanisms.

ACKNOWLEDGMENTS

The authors would like to thank Drs. R. Boucher, O. Kuchel, W. Nowaczynski, and J. Genest, and Miss G. Fullum, R.N., for their collaboration during these investigations.

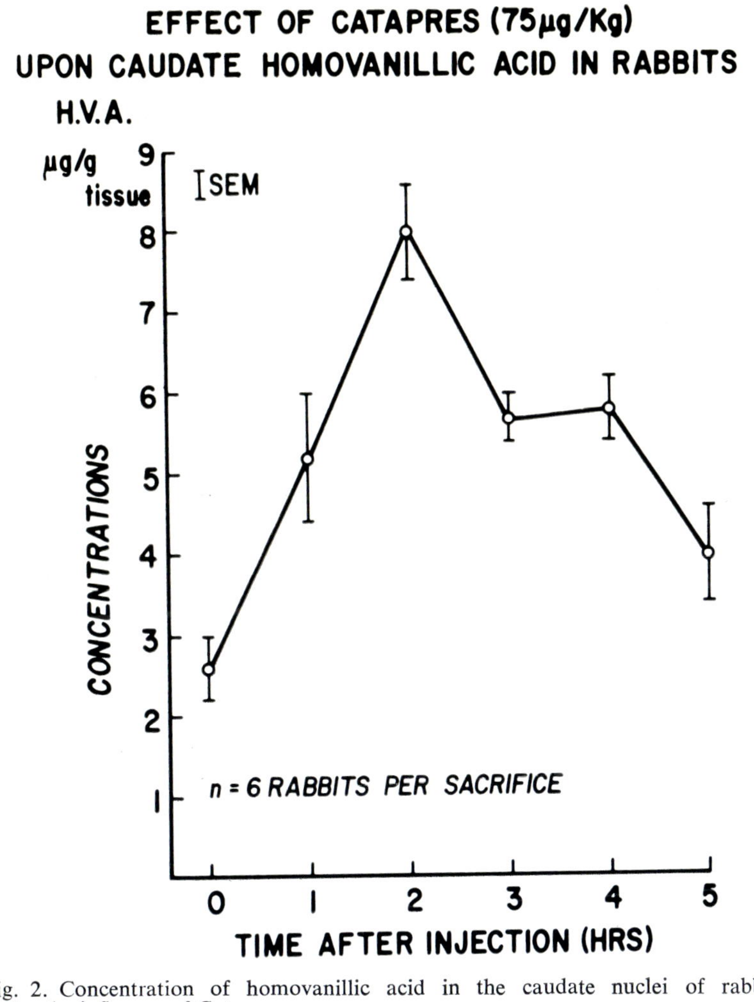

Fig. 2. Concentration of homovanillic acid in the caudate nuclei of rabbit under the influence of Catapres.

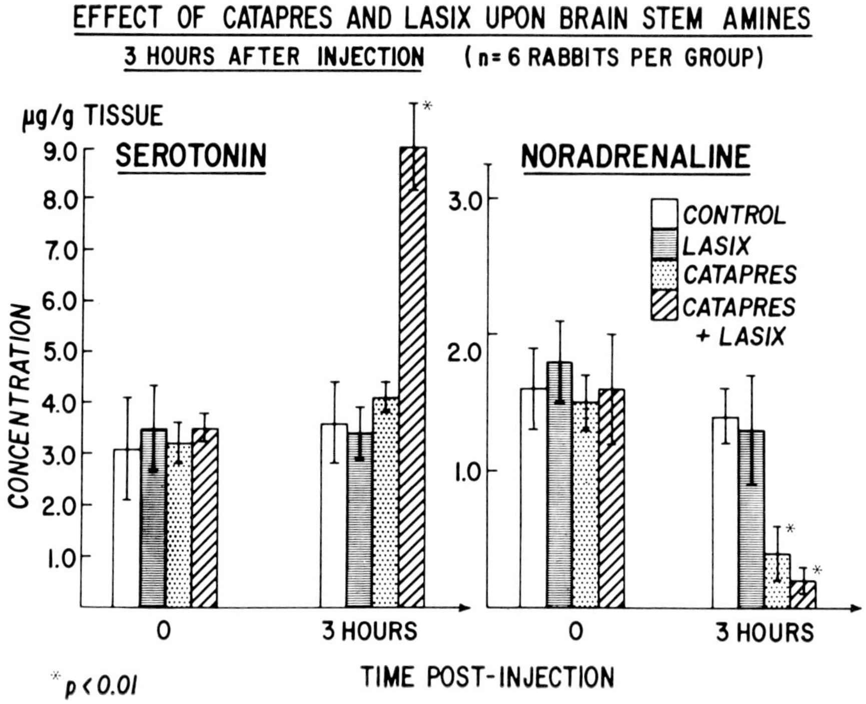

Fig. 3. Concentration of various amines in the brain stem of rabbit under the influence of Catapres alone or of Catapres plus Lasix.

REFERENCES

1. DeQuatro, V., and Sjoerdsma, A.: Catecholamine turnover in normotensive and hypertensive man: effects of antiadrenergic drugs. J. Clin. Invest. *97:* 2359, 1968.
2. DeChamplain, J., Krakoff, L. R., and Axelrod, J.: Catecholamine metabolism in experimental hypertension in the rat. Circ. Res. *20:* 136, 1967.
3. Robertson, A. A., Hodge, J. V., Laverty, R., and Smirk, F. H.: Tissue catecholamine levels in normotensive, genetically hypertensive and renal hypertensive rats. Aust. J. Exp. Biol. Med. Sci. *46:* 689, 1968.
4. Barbeau, A.: L-Dopa therapy in Parkinson's disease: a critical review of nine years' experience. Canad. Med. Assoc. J. *101:* 791, 1969.
5. Slaton P. E. Jr., and Biglieri, E. G.: Reduced aldosterone excretion in patients with autonomic insufficiency. J. Clin. Endocr. *27:* 37, 1967.
6. Vander, A. J.: Control of renin release. Physiol. Rev. *47:* 359, 1967.
7. Rojo-Ortega, J. M., Hatt, P. Y., and Genest, J.: A propos de l'innervation des cellules juxtaglomérulaires. Etude au microscope électronique dans diverses conditions expérimentales chez le rat. Path. Biol. *16:* 497, 1968.
8. Winer, N., Chokshi, D. S., Yoon, M. S., and Freedman, A. D.: Adrenergic receptor mediation of renin secretion. J. Clin. Endocr. *29:* 9, 1969.
9. Barbeau, A., Gillo-Joffroy, L., Boucher, R., Nowaczynski, W., and Genest, J.: Renin-aldosterone system in Parkinson's disease. Science *165:* 291, 1969.
10. Mohammed, S., Fasola, A. F., Privitera, P. J., Lipisky, R. J., Martz, B. L.,

and GAFFNEY T. E.: Effect of methyldopa on plasma renin activity in man. Circ. Res. *25:* 543 1969.
11. KUCHEL, O., CUCHE, J. L., BARBEAU, A., BOUCHER, R., and GENEST, J.: The urinary ratio of dopamine/norepinephrine (DA:NE) and of norepinephrine/epinephrine (NE:E) upon assuming upright posture in normo and hypertensive subjects. Clin. Res. *17:* 645, 1969.
12. SHERMAN, G. P., GREGA, G. J., WOODS, R. J., and BUCKLEY, J. P.: Evidence for a central hypotensive mechanism of 2-(2,6-dichlrophenylamino)-2-imidazoline (Catapressan, St-155). Europ. J. Pharmacol. *2:* 326, 1968.
13. SCHMITT, H., SCHMITT, H. MRS., BOISSIER, J. R., GIUDICELLI, J. F., and FICHELLE, J.: Cardiovascular effects of 2-(2,6-dichlorophenylamino)-2-imidazoline hydrochloride (St-155). II—Central sympathetic structures. Europ. J. Pharmacol. *2:* 340, 1968.
14. BARBEAU, A.: Dopamine and blood pressure regulation. Clin. Res. *17:* 634, 1969.

THE ROLE OF DOPAMINE, NOREPINEPHRINE AND EPINEPHRINE IN THE ORTHOSTATIC REGULATION OF RENIN IN NORMOTENSIVE AND HYPERTENSIVE SUBJECTS

O. Kuchel, J. L. Cuche, A. Barbeau, M. Brecht, R. Boucher, and J. Genest

A change from recumbency to upright posture results in pooling of the blood below the heart, which leads to a decrease in venous return and "effective" blood volume and a decrease in cardiac output.[1] The hemodynamic changes are compensated for by two types of responses: (1) a rapid one consisting of an increase in sympathetic activity, leading to an immediate homeostatic response (cardiac acceleration, venoconstriction, increased myocardial contractility and peripheral resistance, all combining to maintain arterial blood pressure);[2] and (2) the more sluggish mechanisms of the renin-angiotension-aldosterone system, which expand extracellular fluid volume by retention of sodium and thus support the blood pressure in another way.[3]

Previous studies from our group and others presented evidence for adrenergic innervation of the juxtaglomerular apparatus,[4] for the stimulating effect of norepinephrine upon renin secretion,[5] and for the fact that the sympathetic nervous system, in addition to controlling the almost instantaneous adaptation of the cardiovascular system to upright posture, also plays an important role in mediating the more sluggish renin-aldosterone response to this change in posture.[6] Many mechanisms thus operate in a coordinated fashion and the sympathetic nervous system plays a key integrative role.

Normal subjects respond to the assumption of upright posture by parallel increases in the urinary excretion of catecholamines (epinephrine plus norepinephrine), in plasma renin activity and in the urinary excretion of aldosterone. A decrease in sympathetic discharge (e.g., in autonomic insufficiency[7] or Parkinson's disease[8]) is followed by a deficient rise in renin and aldosterone, and orthostatic hypotension. If the deficient sympathetic discharge is corrected in both the above entities by an infusion of norepinephrine and epinephrine, one observes a normal increase in renin and aldosterone.[7, 8]

However, there is still some controversy about the role of catecholamines in stimulating renin production during postural changes. For example, adequate renin production can occur in the upright posture without a concomitant increase in catecholamine excretion.[9] Moreover, the renin-stimulating effect of norepinephrine may not be wholly dependent on renal nerves, since human transplanted kidneys show a normal response to stimuli which provoke renin release.[10]

On the other hand, the recent experiments of Winer[11] presented evidence that many, if not all, stimuli which provoke renin release act either through adrenergic receptor sites or at a step distal to the receptor site. Such receptors are probably renal in origin and not dependent on renal nerves. There is also growing evidence from experiments in vitro[12] and in vivo[13] that cyclic adenosine monophosphate (AMP), a typical "second messenger" of norepinephrine action in target tissues, mediates renin secretion in the kidney.

Since the final effect of catecholamines is believed to be dependent on the interaction of these various substances upon many specific receptors, or upon competition of the amines for the same receptor site,[14] we wondered if the discrepancies mentioned above could not be caused by the fact that in most studies only norepinephrine plus epinephrine were measured. Experiments in Parkinson's disease indicated for the first time that some alterations in renin secretion may also be related to a defect in dopamine metabolism.[8]

The present study represents our attempt to go a step further by measuring not only total catecholamines but also qualitative changes in urinary catecholamines and plasma catecholamines and the correlation of both with plasma renin activity during postural adaptation. Since some evidence suggests differences in postural-induced catecholamine excretion between normotensive and hypertensive patients,[15] we have compared the two groups and further divided essential hypertensive patients into those with a normal postural blood pressure response and those with orthostatic hypotension or orthostatic hypertension. We were especially interested in finding qualitative catecholamine changes within that subpopulation of essential hypertensive patients (approximately 25 percent of all essential hypertensives) who show suppressed renin secretion.[16]

Material and Methods

Six control subjects and nine essential hypertensive patients were put on a diet containing 135 mEq of sodium and 90 mEq of potassium per day at least three days before the experiment in an attempt to produce a stable electrolyte balance. While they were on the diet, we studied the effect of posture on the fourth and fifth days in the following way: For the first morning the subjects remained strictly recumbent from 8:00 a.m. to noon. At 8:00 a.m.

a blood sample was drawn for the determination of plasma renin activity (PRA) and plasma catecholamines, and urine was collected for the 4-hour period (8:00 a.m. to 12:00) for the measurement of urinary catecholamines, sodium, potassium, and creatinine. On the second morning the subjects remained recumbent to 8:00 a.m. and thereafter were up and about, standing or sitting, until noon. Urine was again collected for the 4-hour period. At noon, a blood sample was drawn for determination of plasma renin and catecholamines. Plasma renin activity (PRA) was assayed by the method of Boucher et al.[17]; dopamine, norepinephrine and epinephrine were determined in urine by the fluorometric method of Sourkes and Murphy,[18] and the results were expressed in μg per gram of creatinine. Plasma catecholamines were assayed by our own modification of the method of Anton and Sayre.[19] Blood pressure and pulse rate in the recumbent and upright positions were measured every hour during the studies.

The hypertensive patients were selected for this study after a careful clinical examination to exclude all those with known causes of hypertension, and all patients had renal arteriograms to exclude the possibility that renovascular hypertension was present. The age of the hypertensive patients was 26 to 66 years; six were women, three men. Control subjects were healthy normotensive volunteers between 21 and 30 years of age.

Results

Table 1 summarizes the results and Figure 1 shows graphically the statistically significant differences between a group of seven hypertensive patients with normal postural response in blood pressure (while two others with abnormal postural response are described later) and six control subjects. Urinary dopamine excretion (expressed in μg per gram of creatinine) fell consistently in every control and hypertensive subject when the upright posture was assumed. Urinary norepinephrine excretion (per gram of creatinine) rose only slightly in the upright position, but the base-line value in the recumbent position was higher in essential hypertension ($p<0.01$). The urinary dopamine/norepinephrine ratio decreased in both groups, mainly because of the large decrease in dopamine excretion. This decrease was significant in the control subjects ($p<0.05$). Plasma norepinephrine did not change when control subjects assumed the upright posture but increased significantly when hypertensives did ($p<0.05$). PRA increased significantly in both groups upon assuming upright posture.

In Figure 2, we compare the blood pressure response of control subjects and hypertensive patients to the upright posture. Hypertensive patients, who have a "normal" postural blood pressure response, show slightly greater fall in systolic blood pressure than normotensive controls. However, they show no increase in diastolic blood pressure. These differences from normals are well recognized and are probably due to a greater than normal fall in plasma volume when hypertensives assume the upright posture.[20]

Not all essential hypertensive patients behave in this way.[21] In postural

TABLE 1. Blood Pressure, Urinary Catecholamine Excretion, Plasma Catecholamines in Normal and Essential Hypertensive Subjects in Recumbent and Upright Position *

	Normal Subjects		*Essential Hypertension*					
			Postural Hypotension		*Normal Postural Response*		*Postural Hypertension*	
	Recumbent ↓	*Upright* ↓	↓ *Recumbent*	↓ *Upright*	*Rec.* ↓	*Up.* ↓	*Rec.* ↓	*Up.* ↓
Blood pressure								
Systolic mmHg	100±8	99.5±8	167	121	144±21	136±24	130	230
	n=4				*n*=7			
Diastolic mmHg	65±5	73±5	88	115	94±10	94±11	80	160
	n=4							
Urine								
Dopamine γ/gr creat	290±113	187±70	408	275	491±331	232±90	168	60
	n=6							
	↓ p<0.001 (Normal Recumbent vs. Normal Postural Response Rec.)				↓			
Norepinephrine γ/gr creat	25.1±9.7	31.3±14.9	8	40	39±19	37±15	29	45
	n=6							
Epinephrine γ/gr creat	4.9±1.1	6.4±2.5	33	32	42±12	41±27	4	6
	n=6							
	↓ p<0.05 ↓							
DA/NE ratio	13.6±7.3	7.6±4.5	49.1	6.8	12.6±6.8	6.2±1.5	5.8	1.3
	n=6							
NE/E ratio	4.5±3.7	4.6±3.5	0.25	1.20	0.94±0.44	0.98±0.17	7.6	6.8
	n=6							
Blood								
PRA ng/l/hr.	2.9±4.8	11.4±1.8	0	0	1.7±2.3	6.7±4.6	5.5	21
	↑ p<0.05 ↑				↑ p<0.05 ↑			
	n=5							
Plasma Eγ/1	0	0	0.05	0.15	0.02	0.04	0	0
	n=4							
Plasma NEγ/1	0.21±0.1	0.23±0.04	0.24	0.51	0.21±0.09	0.35±0.15	0.51	1.17
					↑ 0.02<p<0.05 ↑			

* Values represent means ± standard error of mean (SEM), n = number of subjects, statistically different changes are marked by p.

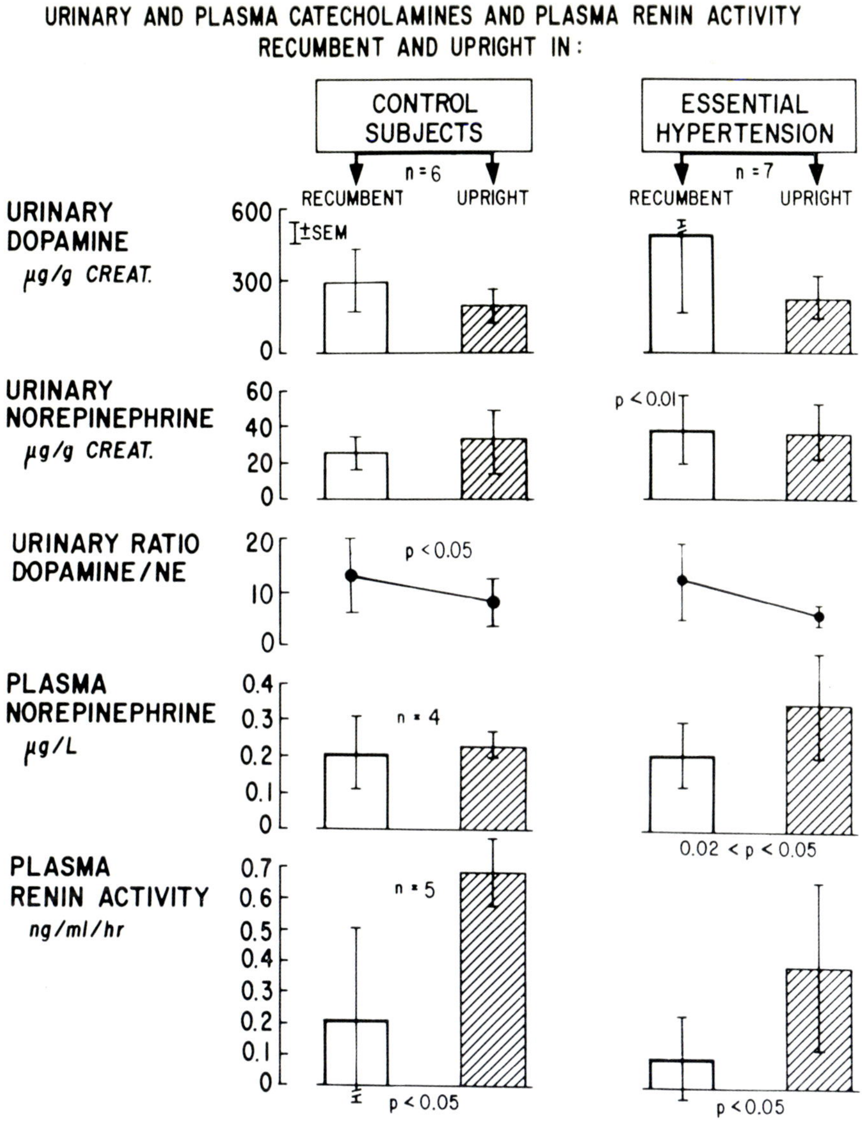

Fig. 1. Urinary and plasma catecholamines and PRA recumbent and upright in normal subjects and in essential hypertension. Symbols as on Table 1.

hypotension there is an exaggerated decrease in the systolic blood pressure and excessive increases in diastolic blood pressure and pulse rate (Fig. 2; this was observed in one patient in whom a drug-induced orthostatic hypotension could be positively excluded). Orthostatic hypertension, with an excessive increase in both systolic and diastolic pressure in response to upright posture, was observed in another patient in our series. This patient presents a special problem since postural hypertension is associated in his case with some manifestations of beta-adrenergic hyperreactivity.[22] Four

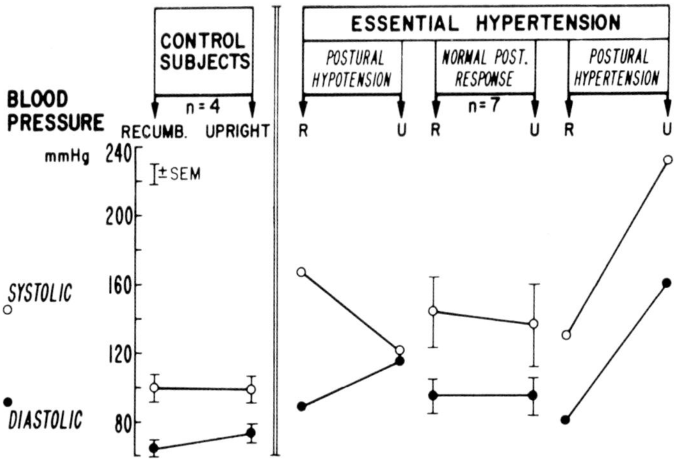

Fig. 2. Blood pressure recumbent and immediately after the subjects assumed upright position in normal subjects and essential hypertension.

members of the same family, three sisters and the mother, present similar symptoms. In three of the four members we have observed an intolerance to salt depletion, such depletion leading rapidly to a change from orthostatic hypertension to orthostatic hypotension. The syndrome is well controlled by beta-adrenergic inhibitors. This syndrome, apparently genetically determined, is described in detail elsewhere.[23]

In Figure 3 we compare urinary and plasma components (recumbent and upright) in the three types of postural responders in essential hypertension.

In the patient with orthostatic *hypotension* there was a very low recumbent excretion of norepinephrine in the presence of normal dopamine and higher epinephrine excretion. This leads to an excess of dopamine and epinephrine over norepinephrine. Plasma norepinephrine is in the normal range, plasma epinephrine is higher than normal, and PRA undetectable. In the upright posture, urinary norepinephrine excretion rises to high normal values, while dopamine falls and epinephrine is unchanged. This is associated with a marked increase in plasma norepinephrine and epinephrine and a concomitant rise in the diastolic blood pressure and pulse rate, but PRA is still undetectable. It should be stressed that this patient maintains a suppressed renin even after more potent stimulation by low sodium diet and diazoxide.[16]

In orthostatic *hypertension,* however, the recumbent values of urinary dopamine are less than half normal. Those of urinary epinephrine and norepinephrine are in the normal range. Plasma norepinephrine values are also high and PRA is at the upper limit of the normal range. On assuming upright posture, the disproportion between dopamine and norepinephrine is further accentuated. Urinary dopamine falls to about a third of the already low base-line level, epinephrine does not change, and norepinephrine rises to almost twice its normal value. Plasma norepinephrine also rises very substantially and PRA increases more than in other patients.

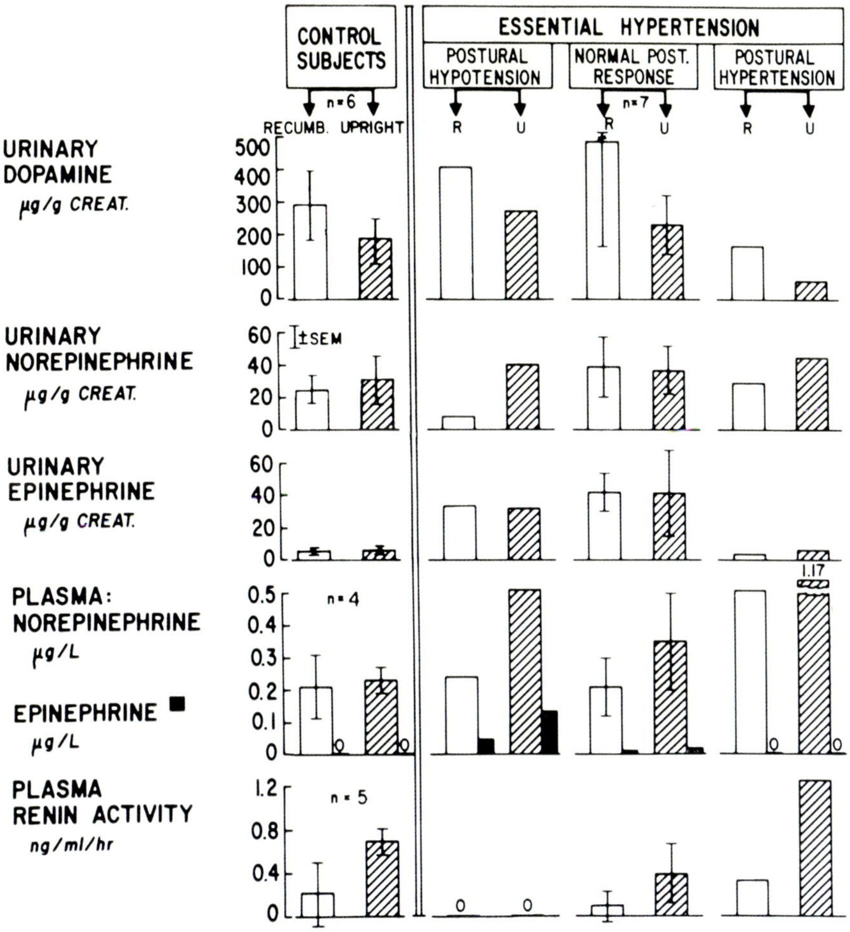

Fig. 3. Urinary and plasma catecholamines and PRA recumbent and upright in normal subjects and essential hypertension with normal postural response, postural hypotension and postural hypertension. Symbols as in Table 1.

Discussion

Our data indicate that a significant qualitative change in the urinary excretion of the three main catecholamines reflects the sympathetic discharge associated with adaptation to upright posture and that the ratio of those three catecholamines is correlated with the posturally-induced increase in plasma renin activity.

In control subjects and in most hypertensive patients, upright posture is associated with a decrease in urinary dopamine excretion, while norepinephrine and epinephrine excretion change only very slightly. The increase in plasma norepinephrine is not significant in control subjects, but plasma renin rises significantly.

If the only noteworthy change associated with an increase in renin during upright posture is a decrease in dopamine excretion, the norepinephrine-like sympathetic effect (increase in pulse rate and diastolic blood pressure) may be a reflection of a decrease in dopamine production. This amine is not only a precursor to norepinephrine but also a mediator of the adrenergic nervous system.[24] The biological effects of dopamine are completely different from those of norepinephrine. Dopamine increases renal and splanchnic blood flow and urinary sodium excretion; it has no effect on peripheral resistance and exerts only a slight pressor effect. It may even have a depressor effect. Since renal adrenergic receptors seem to be critically involved in regulation of renin,[11] the differences between the effects of dopamine and those of norepinephrine upon renal plasma flow and urinary sodium excretion are of particular interest. In addition, Goldberg recently described specific dopaminergic receptors in the kidney.[25] Norepinephrine seems to have a direct stimulating effect on the renin-producing cells in the kidneys since this effect is reproducible in vitro.[12] Upon the mixed renal receptors, a competitive inhibition between norepinephrine and dopamine may exist. A decrease only in urinary dopamine with unchanged urinary norepinephrine could be translated into a relative hyperactivity of norepinephrine, even without changes in the concentration of this substance. Since an effect on renal blood flow is the main difference between norepinephrine and dopamine, it may be that the sympathetic activation of renin secretion is mediated through renal arteriolar constriction from unopposed norepinephrine action, leading to a decrease in renal blood flow and thus to stimulation of renin.[26] Indirect evidence for this link between dopamine and renin is given by recent studies from our laboratories on the diurnal rhythm in dopamine, norepinephrine and epinephrine excretion in recumbent subjects receiving a controlled diet.[27] Although there was no evident oscillation in norepinephrine and epinephrine excretion under these conditions, there was an evident peak in dopamine excretion in the afternoon between 4:00 and 8:00 p.m. This peak coincides with the peak in renal plasma flow and glomerular filtration rate examined under similar conditions by Wesson[28] and with the lowest point in the diurnal variation of plasma renin activity studied independently of postural factors by Gordon et al.[29] The basic physiological diurnal rhythm of dopamine thus may be causally linked to the diurnal rhythm in renal plasma flow and renin, independently of postural variations. Upright posture then becomes a superimposed stimulus. The changes we have described are even more significant since they occur in the opposite direction to those resulting from the diurnal rhythms. In the period between 8:00 a.m. and noon we have observed a decrease in dopamine excretion, while in the diurnal rhythm dopamine excretion tends to increase at that time. Renin, on the other hand, increases whereas normally it decreases during this part of the day.[29]

These results indicate that the measurement of catecholamine excretion as usually carried out (i.e., norepinephrine plus epinephrine) cannot give meaningful information on sympathetic activity. It appears that not only dopamine but also epinephrine play the role of endogenous antagonists to norepinephrine. This is evident at the renal level where epinephrine and dopamine exert a natriuretic effect in opposition to the anti-natriuretic effect of norepinephrine.[30] The ratios of dopamine to norepinephrine and norepinephrine to epinephrine thus reflect not only the conversion rates of one catecholamine to another but also the final biological effect of a spectrum of catecholamines upon receptor sites.

Our data in hypertensive patients add further indirect evidence to the hypothesis that the ratio of dopamine and/or epinephrine to norepinephrine is crucial for the adrenergic effectiveness of norepinephrine and its renin-stimulating property. We were unable to confirm Serrano's observation that hypertensive patients have a lower excretion of dopamine than normotensive subjects.[31] This is probably because Serrano did not express his values per gram of creatinine and thus had no correction for the lower glomerular filtration rate in hypertensive patients as a group. We have found a clearly higher norepinephrine excretion per gram of creatinine in hypertensive patients, in agreement with earlier reports.[32]

In the same line of thought is the significantly higher level of plasma norepinephrine in essential hypertension, a finding recently confirmed by Engelman and Sjoerdsma [33] with double isotope techniques. It may be inferred that this more evident sympathetic discharge in response to upright posture in essential hypertensive patients compensates for the more significant decrease in plasma volume.[20] This is not reflected, however, in postural increase in diastolic blood pressure. Probably an already high peripheral resistance is not further increased by a sympathetic discharge which is only sufficient to modify normal levels of resistance.

In spite of higher norepinephrine values, PRA in our hypertensive patients was generally lower than in normotensive controls, a finding in accordance with our previous results.[34, 35] Our preliminary data indicate that there is a negative correlation between urinary DA/NE ratio and PRA in normotensive subjects. No such correlations can be found in essential hypertensive subjects. It appears that in these patients there is a basic disturbance in the functional relationship between catecholamines and renin. This disturbance can play an important role in the development of the final hypertensive process, especially in the initial stages.

The extreme cases of postural hypo- and hypertension seem to be important for the understanding of those alterations.

In our patient with orthostatic hypotension there is, during recumbency, an extreme excess of dopamine and epinephrine over norepinephrine, associated with suppressed renin activity. This constellation was not observed in other cases. The chronic suppression of renin is associated in this patient with low plasma aldosterone (6, 7 ng percent) and hypovolemia (approximately 15 percent decrease). The base-line rate of conversion of dopamine to norepinephrine seems to be low, assuming upright posture leads to an additional fall in plasma volume and to a marked drop in blood pressure, two potent stimuli for an increase of sympathetic activity. A rapid homeostatic response, the increased conversion of dopamine to norepinephrine, takes place, and plasma norepinephrine and epinephrine rise even more than in control subjects. This is reflected by an unusual increase in diastolic blood pressure and by tachycardia. This reaction, however, is of short duration and ineffective in the absence of further, more sluggish, adaptive reactions: increase in renin, aldosterone, sodium retention and volume expansion.

In our patient as well as in other cases of essential orthostatic hypotension it is not clear whether suppression of renin is primarily located in the juxtaglomerular cells or whether it is the consequence of a previous long-standing adrenergic inhibition. Upright posture is probably not a stimulus effective enough to overcome this inhibition. Since patients with suppressed renin tend to have a higher total exchangeable sodium,[36] and since suppressed renin may be corrected in hypertensives by long-term treatment with spironolactones,[37] it seems that prolonged saluretic stimulation is necessary to correct a suppressed PRA. Only further investigation will show whether this type of patient may be a model for more general patterns in essential hypertensive patients with suppressed renin in the absence of elevated aldosterone. Such a possibility would be supported by the fact that an overwhelming majority of patients with suppressed renin have orthostatic hypotension.[38]

In orthostatic hypertension a low dopamine excretion, with norepinephrine at the upper limits of normal, presents just the opposite situation. This disequilibrium in favor of norepinephrine is further accentuated when the upright position is assumed: There is an additional fall in dopamine to the lowest observed value; norepinephrine rises, especially in the plasma where it reaches concentrations at least 5 to 6 times as high as in control subjects. In our patient renin rises more than in any other control or hypertensive subject studied.

In this patient, probably an example of a new familial syndrome with hyperkinetic circulation [23] (three further members of the family similarly affected), there is an evident tendency to higher than normal conversion rates of dopamine to norepinephrine upon postural stimulation. We may be dealing with a new inborn error of metabolism. This change in conversion rate may explain the excessive responsiveness of the adrenergic mechanisms to all kinds of stimulation (upright posture, cold pressure test, low sodium diet, etc.), but at the same time it may cause rapid exhaustion of the system if another stimulus is added. In our patient fainting and dizziness occured after only two days on a low sodium diet and 25 mg of Aldactazide, and orthostatic hypertension was soon reversed to hypotension. In this syndrome renin rises normally in response to a low sodium diet in the recumbent position, but if the stimulus of upright posture is added a paradoxical decrease in PRA occurs. Three of four patients studied from this family presented this intolerance to sodium depletion.

These findings show that qualitative analysis of all urinary catecholamines under a standard postural challenge may provide a dynamic approach to functional relationships that we would be unable to prove under static conditions. The adaptive process to upright posture discloses a link between hemodynamics and the endocrine system. This is a link between a rapid-acting homeostatic system (nervous system and catecholamines) important for acute situations and a longer-acting, sluggish adaptive system (renin-angiotensin-aldosterone) for more chronic adaptive reactions. The relationship between the sympathetic nervous system and renin secretion seems to be bidirectional. Angiotensin may play the role of a feedback link in the system by accelerating the biosynthesis of catecholamines [39] and thus changing the ratio of the catecholamines competing for the same receptor site. The time may come when we will be thinking in terms of a unified sympatho-reno-adrenal system.

REFERENCES

1. WATERFIELD, R. L.: The effects of posture on the circulating blood volume. J. Physiol. *72:* 110, 1931.
2. WRIGHT, S.: Applied Physiology. London, Oxford University Press, 1961.
3. COHEN, E. L., CONN, J. W., and ROVNER, D. R.: Postural augmentation of plasma renin activity and aldosterone excretion in normal people. J. Clin. Invest. *46:* 418, 1967.
4. ROJO-ORTEGA, J. M., HATT, P. Y., and GENEST, J.: A propos de l'innervation des cellules juxtaglomérulaires. Etude au microscope électronique dans diverses conditions expérimentales chez le rat. Path. Biol. *16:* 497, 1968.
5. VANDER, A. J.: Effect of catecholamines and the renal nerves on renin secretion in anesthetized dogs. Amer. J. Physiol. *209:* 659, 1965.
6. GORDON, R. D., KUCHEL, O., ISLAND, D. P., LIDDLE, G. W.: Role of the sympathetic nervous system mediating renal and adrenocortical secretory responses to upright posture. J. Clin. Invest. *45:* 44, 1966.
7. GORDON, R. D., KUCHEL, O., LIDDLE, G. W., and ISLAND, D. P.: Role of the sympathetic nervous system in regulating renin and aldosterone production in man. J. Clin. Invest. *46:* 599, 1967.
8. BARBEAU, A., GILLO-JOFFROY, L., BOUCHER, R., NOWACZYNSKI, W., GENEST, J.: Renin-aldosterone system in Parkinson's disease. Science *165:* 291, 1969.
9. HEDELAND, H., DYMLING, J. F., HOKFELT, B.: Catecholamines, renin and aldosterone in postural hypotension. Acta Endocr. *62:* 399, 1969.
10. BLAUFOX, M. D., LEWIS, E. J., JAGGER, P., LAULER, D., HICKLER, R., and MERRILL, J. P.: Physiologic responses of the transplanted human kidney. Sodium regulation and renin secretion. New Eng. J. Med. *280:* 62, 1969.
11. WINER, N., CHOKSHI, S., YOON, M. D., and FREEDMAN, A. D.: Adrenergic receptor mediation of renin secretion. J. Clin. Endocr. *29:* 1168, 1969.
12. MICHELAKIS, A. M., CAUDLE, J., and LIDDLE, G. W.: In vitro stimulation of renin production by epinephrine, norepinephrine and cyclic AMP. Proc. Soc. Exp. Biol. Med. *150:* 748, 1969.
13. WINER, N., BURCH, B. H., CHOKSHI, D. S., and FREEDMAN, A. D.: Cyclic AMP stimulation of renin secretion. Endocrine Society, 51st Meeting, June 1969.
14. FURCHGOTT, R. F.: The receptors for epinephrine and norepinephrine (adrenergic receptors). Pharmacol. Rev. *11:* 429, 1959.
15. SUNDIN, T.: The effect of body posture on the urinary excretion of adrenaline. Acta Med. Scand. *161:* Suppl. 336, 1958.
16. KUCHEL, O., FISHMAN, L. M., LIDDLE, G. W., and MICHELAKIS, A.: Effect of diazoxide on plasma renin activity in hypertensive patients. Ann. Intern. Med. *67:* 791, 1967.
17. BOUCHER, R., VEYRAT, R., DECHAMPLAIN, J., and GENEST, J.: New procedure for measurement of human plasma angiotensin and renin activity levels. Canad. Med. Assoc. J. *90:* 194, 1964.
18. SOURKES, T. L., and MURPHY, G. F.: Determination of catecholamines and

catecholamino-acids by differential spectrophotofluorometry. Meth. Med. Res. *9:* 147, 1961.

19. GENEST, J., SIMARD, S., ROSENTHAL, J., and BOUCHER, R.: Norepinephrine and renin content in arterial tissue from different vascular beds. Canad. J. Physiol. Pharmacol. *47:* 87, 1969.
20. EISENBERG, S., and WOLF, P. C.: Plasma volume after posture change in hypertensive subjects. Arch. Intern. Med. *115:* 17, 1965.
21. FROHLICH, E. D., TARAZI, R. C., ULRYCH, M., DUSTAN, H. P., and PAGE, I. H.: Tilt test for investigating a neural component in hypertensives: Its correlation with clinical characteristics. Circulation *36:* 387, 1967.
22. FROHLICH, E. D., TARAZI, R. C., and DUSTAN, H. P.: Hyperdynamic β-adrenergic circulatory state. Arch. Intern. Med. *123:* 1, 1969.
23. KUCHEL, O., CUCHE, J. L., BARBEAU, A., BRECHT, M., BOUCHER, R., and GENEST J.: In preparation.
24. MCDONALD, R. H., JR., GOLDBERG, L. I., MCNAY, J. L., and TUTTLE, E. P., JR.: Effects of dopamine in man: augmentation of sodium excretion, glomerular filtration rate, and renal plasma flow. J. Clin. Invest. *45:* 1116, 1964.
25. GOLDBERG, L. J.: Dopaminergic Receptors in the Kidney. Val David, November 1969.
26. SKINNER, S. L., MCCUBBIN, J. W., and PAGE, I. H.: Control of renin secretion. Circ. Res. *15:* 64, 1964.
27. CUCHE, J. L., KUCHEL, O., BARBEAU, A., BRECHT, M., BOUCHER, R., and GENEST, J.: In preparation.
28. WESSON, L. G., JR.: Electrolyte excretion in relation to diurnal cycles of renal function. Plasma electrolyte concentrations and aldosterone secretion before and during salt and water balance changes in normotensive subjects. Medicine *45:* 547, 1964.
29. GORDON, R. D., WOLFE, L. K., ISLAND, D. P., and LIDDLE, G. W.: A diurnal rhythm in plasma renin activity in man. J. Clin. Invest. *45:* 1587, 1966.
30. RUBENSTEIN, H. A., MORRIS, R. P., CLIFT, G. V., SCHLETTER, F. E., SPELLER, P. J., and STREETEN, D. H. P.: Analysis of renal action of epinephrine (EPI). Clin. Res. *15:* 369, 1969 (abstract).
31. SERRANO, P. A., FIGUEROS, G., TORRES, M. G., and DELANGEL, A. R.: Adrenaline, noradrenaline and dopamine excretion in patients with essential hypertension. Amer. J. Cardiol. *13:* 484, 1964.
32. NESTEL, P. J., and DOYLE, A. E.: The excretion of free noradrenaline and adrenaline by healthy young subjects and by patients with essential hypertension. Aust. Ann. Med. *17:* 295, 1968.
33. ENGELMAN, K., and SJOERDSMA, A.: Double isotope measurements of plasma catecholamines concentrations in patients with hypertension. The Council for High Blood Pressure Research Meeting, Cleveland, October 1969.
34. FISHMAN, L. M., KUCHEL, O., LIDDLE, G. W., MICHELAKIS, A. M., GORDON, R. D., and CHICK W. T.: Incidence of primary aldosteronism uncomplicated "essential" hypertension. JAMA *205:* 497, 1968.
35. GRANGER, P., BOUCHER, R., and GENEST, J.: L'aldostéronisme primaire. Un. Med. Canada *97:* 1226, 1968.
36. WOODS, J. W., HILL, C., LIDDLE, G. W., STANT, E. G., MICHELAKIS, A. M., and BRILL A. B.: Effect of an adrenal inhibitor in hypertensive patients with suppressed renin. Arch. Intern. Med. *123:* 366, 1969.
37. CRANE, M. G.: International Workshop on Adrenal Cortex, Aldosterone and Hypertension. San Diego, February 1969.
38. WEINBERGER, M. H., DOWDY, A. J., NOKES, G. W., and LUETSCHER, J. A.: Plasma renin activity and aldosterone secretion in hypertensive patients dur-

ing high and low sodium intake and administration of diuretic. J. Clin. Endocr. *28:* 359, 1968.
39. Boadle, M. C., Hughes, J., and Roth, R. H.: Angiotensin accelerates catecholamine biosynthesis in sympathetically innervated tissues. Nature *222:* 987, 1969.

RENAL VASODILATING ACTIONS OF DOPAMINE

Leon I. Goldberg

Dopamine exerts several interesting cardiovascular and renal actions which may be of importance to patients receiving L-Dopa. We first became interested in dopamine because the amine displayed an unusual constellation of actions in the open-chest, vagotomized, anesthetized dog.[1, 2] When a dose response curve was constructed it was observed that in the dose range between 2 to 15 μg per kg dopamine increased cardiac contractility and lowered diastolic blood pressure without affecting heart rate. When larger doses of dopamine were administered diastolic pressure and heart rate increased (Fig. 1). The cardiac stimulation produced by dopamine was completely blocked by dichloroisoproterenol, but the reduction in blood pressure was not affected by the beta-adrenergic blocking agent. Studies in the perfused hind limb indicated that intravenous injections of dopamine caused vasodilation in the limb but intra-arterial injections caused vasoconstriction. The vasodilation was prevented by transection of the nerves to the limb suggesting that it was the result of a neurogenic mechanism.[2]

Hemodynamic studies in normal subjects indicated that intravenous infusions of dopamine at rates ranging from 2 to 11 μg per kg per minute increased cardiac output and decreased total peripheral resistance without significantly changing heart rate.[3] On the basis of these studies dopamine was administered to patients with congestive heart failure and edema in hopes of improving renal function by increasing cardiac output. When dopamine was administered to 11 patients an increase in sodium excretion occurred in each of them.[4, 5] In order to explore the mechanism for this natriuresis, renal hemodynamic studies were carried out in normal subjects and in patients with congestive heart failure. Renal blood flow in both the normal subjects and patients increased without significant change in arterial pressure suggesting that dopamine was active as a renal vasodilator.[6] Changes in glomerular filtration rate, estimated renal plasma flow and sodium and potassium excretion produced by dopamine in normal subjects are shown in Figure 2.

Subsequent studies in the dog demonstrated that injections of dopamine directly into the renal artery caused vasodilation in the injected kidney.[7–10]

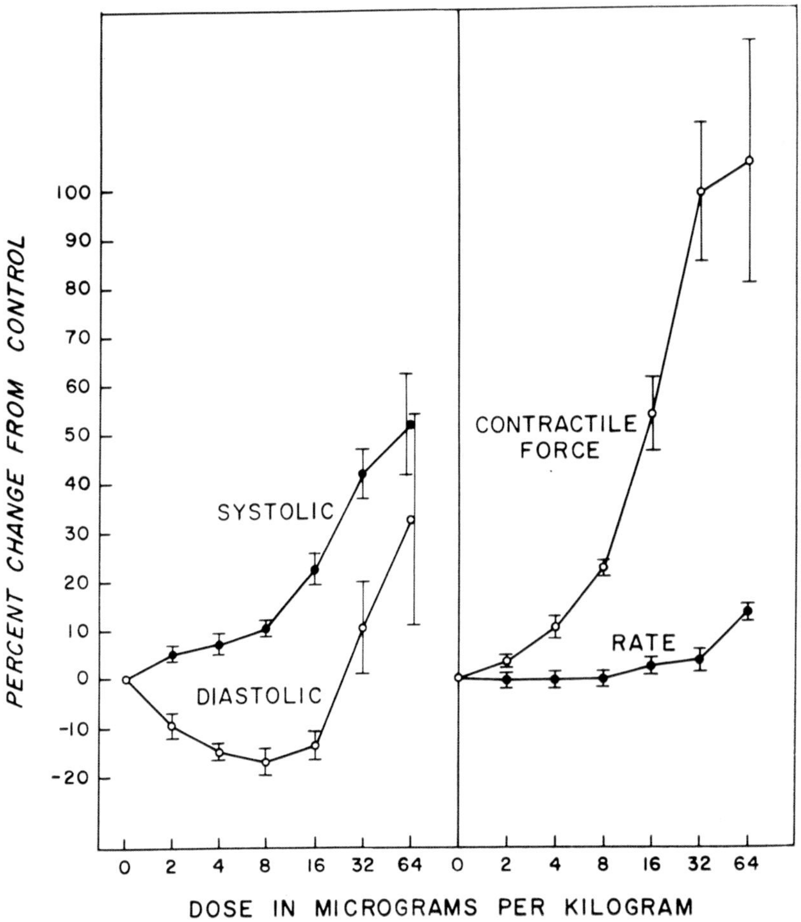

Fig. 1. Average effects (±S.E.) of increasing doses of dopamine on systolic and diastolic blood pressure, and heart rate and contractile force in 16 anesthetized dogs (McDonald and Goldberg [2]).

The vasodilation was not blocked by atropine, beta-adrenergic blocking agents, or by pretreatment of the animal with reserpine, monoamine oxidase inhibitors or the histamine depleting agent 48–80.[9, 10]

Injection of dopamine into the femoral artery caused only vasoconstriction of that vascular bed. After administration of phenoxybenzamine, the vasoconstriction to the limb was blocked indicating that this effect was due to an action on alpha-adrenergic receptors. Vasodilation such as was observed in the kidney was never found in the limb. Much larger doses of dopamine injected in the femoral artery, after phenoxybenzamine, produced vasodilation which differed from the renal vasodilation since it was blocked by beta-adrenergic blocking agents.[9, 11] Eble found that dopamine produced similar vasodilation in the mesenteric vascular bed.[12]

Investigations of 44 analogues of dopamine indicated that dopamine-like

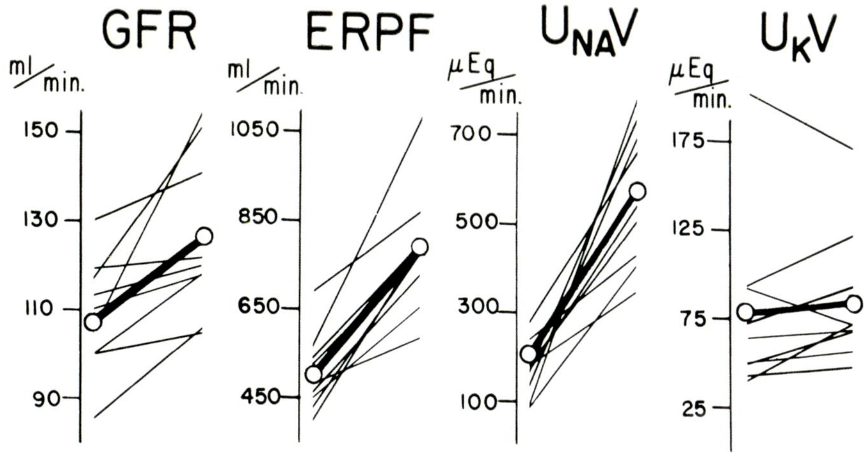

Fig. 2. Effects of intravenous infusions of dopamine on glomerular filtration rate (GFR), estimated renal plasma flow (ERPF), sodium excretion ($U_{Na}V$) in nine normal subjects. The starting point of each line designates the average of three control determinations during dopamine infusion. The heavy line is the average for the nine subjects. All values are corrected to 1.73 m^2 BSA (McDonald et al [6]).

renal vasodilation is extremely specific.[13] Only the N-methyl analogue, epinine, exerted similar renal vasodilation. Interestingly, apomorphine, which is structurally related to dopamine, exerted a similar vasodilation. The renal vasodilation produced by dopamine and epinine in the dog are shown in Figure 3. As can be seen, the initial vasoconstriction produced by epinine was blocked by phenoxybenzamine. However, the renal vasodilation produced by both dopamine and epinine was not affected by propranolol in doses which were sufficient markedly to attenuate the vasodilating actions of isoproterenol.

In 1966, Van Rossum suggested that haloperidol and spiramide might be dopamine blocking agents and demonstrated that these neuroleptic drugs attenuated the reduction in blood pressure produced by dopamine in the yohimbine-treated cat.[14] We investigated the actions of haloperidol on dopamine-induced renal and mesenteric vasodilation in the dog.[15] After considerable investigation we found that within a relatively narrow dose range haloperidol was capable of selectively attenuating the renal and mesenteric vasodilation produced by dopamine. Figure 4 illustrates dose response curves of the renal vasodilation produced by bradykinin, isoproterenol and dopamine with and without addition of haloperidol in a concentration of 1.4×10^{7} moles added to the solution of the vasodilating agents. As noted the dose response curve of dopamine was significantly shifted to the right, whereas the dose response curves of isoproterenol and bradykinin were not affected. We have recently investigated the effects of chlorpromazine and several of its analogues and found that these drugs also can attenuate selectively the renal vasodilation produced by dopamine when used in concentrations approximately twice that of haloperidol.[16] We feel that these investigations support the concept that dopamine is acting on a specific receptor in the

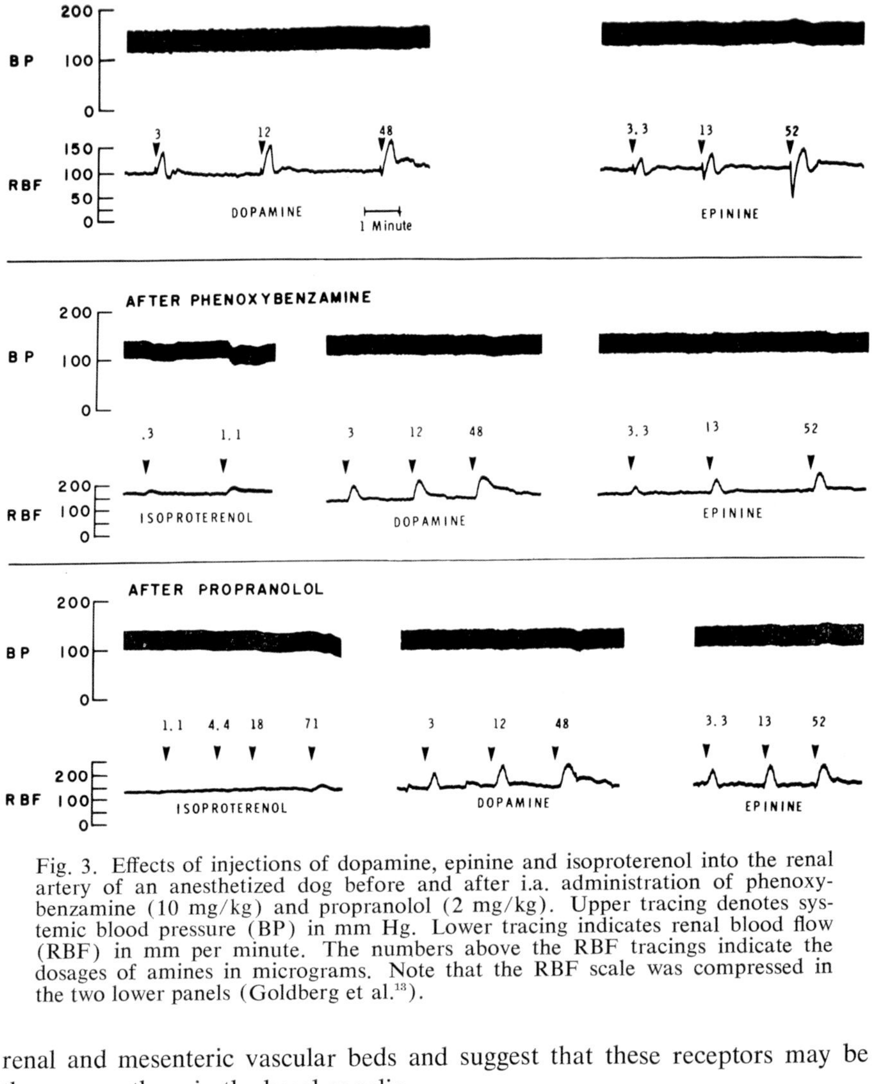

Fig. 3. Effects of injections of dopamine, epinine and isoproterenol into the renal artery of an anesthetized dog before and after i.a. administration of phenoxybenzamine (10 mg/kg) and propranolol (2 mg/kg). Upper tracing denotes systemic blood pressure (BP) in mm Hg. Lower tracing indicates renal blood flow (RBF) in mm per minute. The numbers above the RBF tracings indicate the dosages of amines in micrograms. Note that the RBF scale was compressed in the two lower panels (Goldberg et al.[13]).

renal and mesenteric vascular beds and suggest that these receptors may be the same as those in the basal ganglia.

The question can be asked whether dopamine-induced renal and mesenteric vasodilation is the cause of the hypotension occurring with administration of L-Dopa in patients with Parkinson's disease. We have not observed hypotension in normal subjects receiving infusions of dopamine, but these subjects were in the supine position and we did not determine whether orthostasis occurred. On the other hand, we have observed that a reduction in supine blood pressure occurs both in dogs [17] and in hypertensive patients [18] when dopamine is administered after phenoxybenzamine, suggesting that the mesenteric and renal vasodilation produced by dopamine is masked by alpha-adrenergic vasoconstriction in other vascular beds.

We are currently investigating the cardiovascular and renal effects of

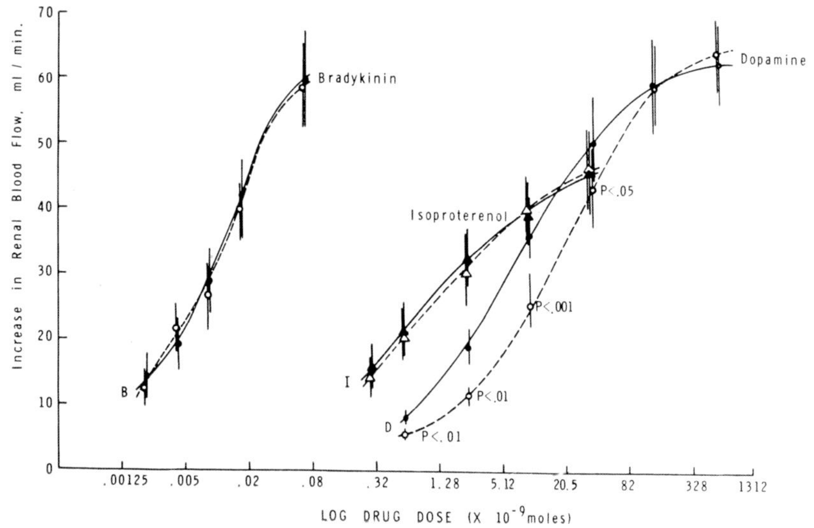

Fig. 4. Mean increments in renal blood flow produced by i.a. injections of several doses of bradykinin (B), isoproterenol (I) and dopamine (D) alone and in combination with haloperidol (1.4×10^{-7} mol). Results were obtained from eight experiments in anesthetized dogs. Solid lines and closed symbols represent data obtained with the vasodilators alone; broken lines and open symbols represent data obtained when the vasodilators were injected at the same time as haloperidol. Values for P were obtained by the paired t test. Vertical lines indicate the standard error (Yeh et al.[15]).

L-Dopa to determine whether renal and mesenteric vasodilation produced by dopamine is the cause of the hypotension. During the course of these studies we found that there appears to be a tolerance to the cardiac stimulating actions of L-Dopa after administration of the drug for more than seven weeks.[20] We have also recently observed that intravenous injections of 29 mg per kg of L-Dopa in the anesthetized dog and 40 mg per kg in the anesthetized cat attenuate the increase in blood pressure resulting from bilateral carotid artery occlusion.[21] Thus, more than one factor may be responsible for the orthostatic hypotension.

Conclusion

Dopamine causes vasodilation in the renal and mesenteric vascular beds by an unusual mechanism. This vasodilation is selectively attenuated by administration of haloperidol and chlorpromazine. L-Dopa produces similar vasodilation. Additional studies are necessary to determine whether this phenomenon is responsible for the hypotension observed in patients with Parkinson's disease.

When L-Dopa is injected intravenously in the anesthetized dog, cardiac contractile force and renal blood flow increase in a manner similar to that which occurs with intravenous infusions of dopamine.[19]

REFERENCES

1. GOLDBERG, L. I., and SJOERDSMA, A.: Effects of several monoamine oxidase inhibitors on the cardiovascular actions of naturally occurring amines in the dog. J. Pharmacol. Exp. Ther. *127:* 212, 1959.
2. MCDONALD, R. H., JR., and GOLDBERG, L. I.: Analysis of the cardiovascular effects of dopamine in the dog. J. Pharmacol. Exp. Ther. *140:* 60, 1963.
3. HORWITZ, D., FOX, S. M., and GOLDBERG, L. I.: Effects of dopamine in man. Circ. Res. *10:* 237, 1962.
4. GOLDBERG, L. I., MCDONALD, R. H., JR., and ZIMMERMAN, A. M.: Sodium diuresis produced by dopamine in patients with congestive heart failure. New Eng. J. Med. *269:* 1060, 1963.
5. ZIMMERMAN, A. M., GOLDBERG, L. I., MCDONALD, R. H., JR., and MCNAY, J. L.: Increased sodium excretion produced by dopamine in patients with congestive heart failure. Circulation *28:* 828, 1963 (abstract).
6. MCDONALD, R. H., JR., GOLDBERG, L. I., MCNAY, J. L., and TUTTLE, E. P., JR.: Effects of dopamine in man: Augmentation of sodium excretion, glomerular filtration rate, and renal plasma flow. J. Clin. Invest. *43:* 1116, 1964.
7. MCNAY, J. L., MCDONALD, R. H., JR., and GOLDBERG, L. I.: Comparative effects of dopamine on renal and femoral blood flows. Pharmacologist *5:* 269, 1963 (abstract).
8. MCNAY, J. L., MCDONALD, R. H., JR., and GOLDBERG, L. I.: Direct renal vasodilation produced by dopamine in the dog. Circ. Res. *16:* 510, 1965.
9. MCNAY, J. L., and GOLDBERG, L. I.: Comparison of the effects of dopamine, isoproterenol, norepinephrine and bradykinin on canine renal and femoral blood flow after POB. J. Pharmacol. Exp. Ther. *151:* 23, 1966.
10. MEYER, M. B., MCNAY, J. L., and GOLDBERG, L. I. : Effects of dopamine on renal function and hemodynamics in the dog. J. Pharmacol. Exp. Ther. *156:* 186, 1967.
11. GOLDBERG, L. I.: Discussion. *In* Second Symposium on Catecholamines. Pharmacol. Rev. *18:* 539, 1966.
12. EBLE, J. N.: A proposed mechanism for the depressor effect of dopamine in the anesthetized dog. J. Pharmacol. Exp. Ther. *145:* 64, 1964.
13. GOLDBERG, L. I., SONNEVILLE, P. F., and MCNAY, J. L.: An investigation of the structural requirements for dopamine-like renal vasodilation: phenylethylamines and apomorphine. J. Pharmacol. Exp. Ther. *163:* 88, 1968.
14. ROSSUM, J. M. VAN: The significance of dopamine-receptor blockade for the mechanisms of action of neuroleptic drugs. Arch. Int. Pharmacodyn. Ther. *160:* 492, 1966.
15. YEH, B. K., MCNAY, J. L., and GOLDBERG, L. I.: Attenuation of dopamine renal and mesenteric vasodilation of haloperidol: Evidence for a specific dopamine receptor. J. Pharmacol. Exp. Ther. *168:* 303, 1969.
16. GOLDBERG, L. I., and YEH, B. K.: Specific blockade of dopamine receptors in the renal vascular bed by chlorpromazine. Proceedings Fourth International Congress of Pharmacology, Basel, 1969.
17. MCNAY, J. L., and GOLDBERG, L. I.: Hemodynamic effects of dopamine in the dog, before and after alpha-adrenergic blockade. Circ. Res. *18:* i-110, 1966.
18. MCNAY, J. L., BOGART, M., and GOLDBERG, L. I.: Hypotensive effects of dopamine after phenoxybenzamine in experimental animals and man. Proceedings Third International Pharmacological Congress, 1966 (abstract).
19. WHITSETT, T. L., and GOLDBERG, L. I. : Unpublished data.
20. WHITSETT, T. L., MCKINNEY, A. S., and GOLDBERG, L. I.: Tolerance to the

cardiac effects of L-Dopa in patients with Parkinson's disease. Clin. Res. *18:* 28, 1970 (abstract).
21. WHITSETT, T. L., HALUSHKA, P. V., RYAN, H. and GOLDBERG, L. I.: Reduced carotid sinus reflex after L-Dopa in cats and dogs. Fed. Proc., in press (abstract).

THE ACTIONS OF DOPAMINE, L-DOPA AND L-TYROSINE ON HUMAN ADRENERGIC NERVE ENDINGS

D. B. Calne and A. S. D. Spiers

Application of drugs to the human eye affords useful information on the pharmacology of peripheral adrenergic nerve endings and receptors. Simple methods may be employed to record the responses of the dilator pupillae muscle and high local concentrations of catecholamines can be achieved without the adverse cardiovascular reactions which result from systemic administration.

Sneddon and Turner [1] have carried out extensive studies of human pupillary responses to sympathomimetic drugs and also investigated their interaction with agents which block adrenergic function. They conclude that the pharmacological responses of the dilator pupillae muscle indicate the presence of alpha-adrenergic receptors homologous to those of the blood vessels.[2] We have undertaken an investigation of human pupillary responses to ocular instillation of L-Dopa in an attempt to elucidate the mechanisms of cardiovascular disturbance which occurs in parkinsonian patients treated with high oral doses of this drug. Patients receiving L-Dopa excrete large quantities of homovanillic acid,[3] the major metabolite of dopamine; we therefore administered dopamine eyedrops in our initial studies, some of which have been reported elsewhere.[4] We have also investigated the effect of L-tyrosine, the immediate precursor of L-Dopa in the synthesis of norepinephrine.

Methods and Materials

Observations have been made on 41 adult volunteers. Twenty-nine experiments with dopamine were performed on 27 subjects, 14 experiments with L-Dopa on 14 subjects, and 14 experiments with L-tyrosine on 14 subjects. In all cases 2 to 3 drops were instilled hourly in one eye, and a 'blind' observer recorded the size of both pupils approximately every hour, using the pupillometer of a standard ophthalmoscope. Results have been expressed as the difference obtained by subtracting the pupillary diameter of the untreated eye (in mm) from that of the treated eye. Guanethidine eyedrops were administered at 12-hour intervals for 1 to 3 days before L-Dopa or dopamine was given.

Dopamine was given in 10 percent aqueous solution buffered to pH 6.5. L-Dopa was made up in 1 percent solution buffered to pH 7.0. Guanethidine was administered in a 5 percent commercial preparation (Ismelin eyedrops, Ciba). L-tyrosine was given in 0.5 percent aqueous solution buffered to pH 7.5.

Results

Ocular administration of dopamine consistently led to pupillary dilatation. It usually took 4½ hours for maximal response to be achieved. Observations on 10 subjects receiving dopamine are shown on Table 1. In the early stages the pupil frequently became eccentric or elliptical. Ultimately these irregularities disappeared to result in a regular, central mydriasis.[4] Later mydriasis sometimes declined despite continuing instillation of dopamine. Diminishing mydriasis was not due to fatigue of the dilator pupillae muscle since the administration of 10 percent phenylephrine resulted in a brisk increase in pupillary diameter.

Pretreatment with guanethidine eyedrops reduced or completely blocked the response to dopamine. Subsequent application of phenylephrine led to marked enlargement of the pupil.

Administration of L-Dopa eyedrops resulted in pupillary dilatation which was very similar to that with dopamine. Mydriasis induced by L-Dopa may be extreme (Fig. 1). The time course of a typical series of observations is recorded in Figure 2. It can be seen that in several cases the mydriasis waned despite continued administration of L-Dopa. The responses to L-Dopa in 14 subjects are summarized in Table 2. The similarity between the actions of L-Dopa and dopamine is evident. The response to L-Dopa also was subject to inhibition by pretreatment with guanethidine. As with dopamine, this blockade did not always occur. Reasons for this inconsistent action of guanethidine have been discussed elsewhere.[4] Neither L-Dopa nor dopamine had any significant effect on accommodation. L-tyrosine eyedrops had no action on the dilator pupillae muscle.

TABLE 1. Dilatation with Ocular Administration of Dopamine *

	Blue Eyes (6)	*Brown Eyes* (4)	*All Subjects* (10) SD
Mean time to onset of dilatation (hr)	2.37	1.56	2.05 (1.71)
Mean time to maximum dilatation (hr)	4.21	5.19	4.60 (2.57)
Mean maximum dilatation observed (mm)	2.9	3.5	3.15 (2.05)

* From Spiers and Calne.[4]

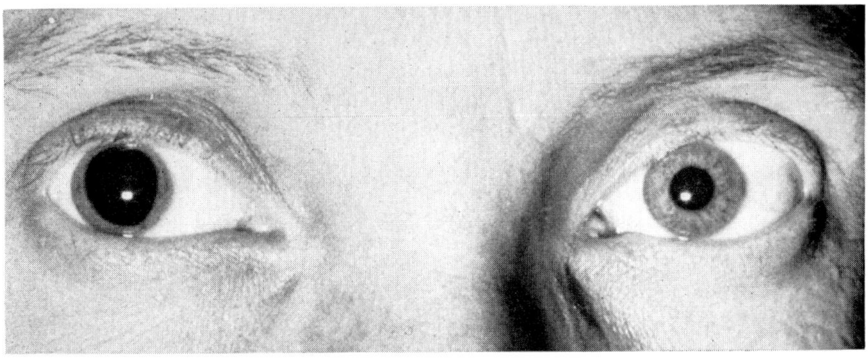

Fig. 1. An example of the response to L-Dopa eyedrops (1 percent, 2 drops per hour instilled into the right eye).

Discussion

These observations indicate that high local concentrations of L-Dopa and dopamine lead to activation of the dilator pupillae muscle. Since the effect is suppressed by guanethidine which does not block adrenergic receptors, neither L-Dopa nor dopamine can be acting directly at the latter site. The mydriatic action of these agents must therefore be due to an effect upon the

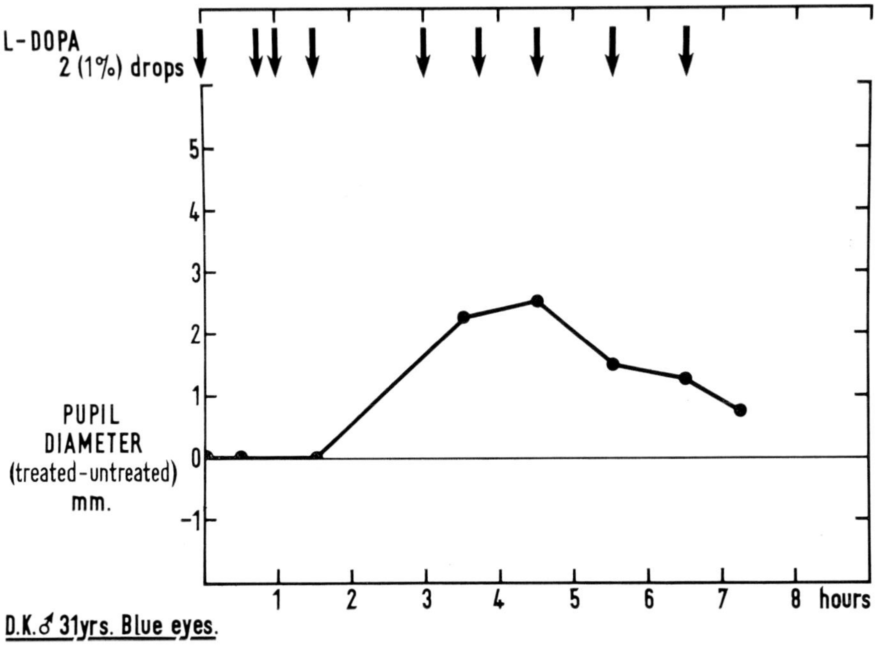

Fig. 2. The time course of a typical response to L-Dopa eyedrops, dilatation being expressed as the difference in pupillary diameter between the treated and the untreated eye.

TABLE 2. Response of the Iris to Instillation of L-Dopa (1 percent) into the Conjunctival Sac

	Blue Eyes (*11*)	*Brown Eyes* (*3*)	*All Subjects* (*14*) SD
Mean time to onset of dilatation (hr)	3.63	3.66	3.20 (1.76)
Mean time to maximum dilatation (hr)	4.72	4.66	4.80 (1.38)
Mean maximum dilatation observed (mm)	2.4	2.0	2.25 (1.82)

adrenergic nerve endings. It has been shown that both L-Dopa and dopamine may release norepinephrine from adrenergic nerve endings in the iris of the rat.[5] A similar mechanism may be responsible for the action of these agents on the human eye.

It has been suggested that sympathomimetic drugs fall into two categories: those which elicit their response by exciting adrenergic receptors (direct acting: e.g., phenylephrine) and those which exert their effect by releasing norepinephrine from adrenergic nerve endings (indirect acting: e.g., tyramine). This is probably an oversimplification[6] but it does provide the basis for a useful classification, in which both L-Dopa and dopamine would clearly appear as indirect acting drugs.

It has not been established that L-Dopa itself is exerting the pharmacological reactions which have been reported. It is quite possible that decarboxylation of L-Dopa to dopamine is so rapid that all of the effects described are in fact induced by dopamine.

Hydroxylation of L-tyrosine to L-Dopa is the rate-limiting step in the synthesis of norepinephrine. It is evident that this transformation—the addition of a *meta*-hydroxyl group—is necessary to elicit the mydriatic response under investigation.

Parkinsonian patients receiving L-Dopa may develop hypertension[7] or hypotension,[8] the latter being more common. Calne et al.[9] found a mean fall in the erect systolic blood pressure of 19.3 mm Hg in 20 patients taking L-Dopa. It is possible that both the hypertension and the hypotension induced by L-Dopa result from an action upon vascular adrenergic nerve endings similar to that seen following local application of L-Dopa and dopamine to the eye. Thus the hypertension might be consequent upon the release of norepinephrine from nerve endings, while the hypotension might result from subsequent depletion of norepinephrine stores. These changes could correspond to the initial marked pupillary dilation induced by L-Dopa and dopamine, sometimes followed by waning mydriasis in spite of continued drug administration (Fig. 2).

It has been demonstrated that dopamine may reduce levels of norepinephrine in the guinea-pig heart.[10] This indicates that it is possible for the rate of release of norepinephrine by dopamine to exceed the rate of synthesis of norepinephrine from dopamine. The roles of dopamine as both a releaser and a precursor of norepinephrine may have important therapeutic implications. This aspect of the pharmacology of dopamine could have a bearing on the mechanism of production of several of the adverse effects of L-Dopa. It is possible that the contraindications to use of L-Dopa may be the same as those of other indirect-acting sympathomimetic drugs such as ephedrine and tyramine.

Summary

Installation of L-Dopa and dopamine eyedrops leads to pupillary dilatation which is inhibited by pretreatment with guanethidine. L-tyrosine eyedrops are inactive.

It is concluded that an indirect action of L-Dopa and dopamine, involving the release of norepinephrine from adrenergic nerve endings, may play an important part in the pharmacogenesis of some clinical effects of L-Dopa therapy, such as alterations in blood pressure.

ACKNOWLEDGMENTS

We wish to thank Professor D. R. Laurence and Dr. R. A. Webster for helpful discussion. Mr. T. B. French kindly formulated the eyedrops. The figures were prepared by Mr. V. K. Asta and the Photographic Department of University College Hospital Medical School.

REFERENCES

1. Sneddon, J. M., Turner, P.: Structure-activity relationship of some sympathomimetic amines in the guanethidine-treated human eye. J. Physiol. *192:* 23P, 1967.
2. Turner, P., and Sneddon, J. M.: Alpha receptor blockade by thymoxamine in the human eye. Clin. Pharmacol. Ther. *9:* 45, 1968.
3. Calne, D. B., Spiers, A. S. D., Stern, G. M., Laurence, D. R., Armitage, P.: Lancet, in press.
4. Spiers, A. S. D., Calne, D. B.: Brit. Med. J., in press.
5. Eränkö, O., Räisänen, L.: Mechanisms of Release of Biogenic Amines. Pergamon Press, London, 1968.
6. Zaimis, E.: Vasopressor drugs and catecholamines. Anesthesiology *29:* 732, 1968.
7. Cotzias, G. C., Papavasiliou, P. S., Aronson, R. B., Mena, I.: Third Symposium on Parkinson's Disease. E. & S. Livingston, Edinburgh, 1968.
8. Duvoisin, R., Barrett, R., Schear, M., Hoehn, M. M., Yahr, M.: Third Symposium on Parkinson's Disease. E. & S. Livingston, Edinburgh, 1968.
9. Calne, D. B., Karoum, F., Ruthven, C. R. J., Sandler, M.: Brit. J. Pharmacol. *37:* 57, 1969.
10. Harrison, W. H., Levitt, M., Udenfriend, S.: J. Pharmacol. Exp. Ther. *142:* 157, 1963.

GENERAL DISCUSSION

Dr. Sandler: I should like to draw attention once more to the possible role of 6-hydroxydopamine in parkinsonism. A pretty good *prima facie* case has been made out for some degree of sympathetic nervous system deficiency: if 6-hydroxydopamine did not exist, we would almost be forced to invent it. Now that gas chromatographic-mass spectrometric techniques are available to tackle this type of problem,[1] we would be unwise to neglect looking for this compound or its metabolites in the urine of affected subjects.

I should like to make another point. Dr. Goldstein was kind enough to draw attention to our work on monoamine oxidase isoenzymes,[2–5] and, indeed, this work may be of possible relevance to future work in Parkinson's disease.

Monoamine oxidases are likely to be conformational in nature [6] and they differ in number and quality in the different tissues of the body. Thus both human and rat livers possess five isoenzymes,[4] whereas human platelets have only one isoenzyme.[7]

In the human brain there are four such isoenzymes, each with its characteristic pH activity curve, heat inactivation curve, substrate specificity and inhibitor specificity.[8] We have quite arbitrarily termed them MAO-1 to 4. I should like to draw attention to one particular enzyme band, MAO-4, which is not only extremely active toward dopamine, as substrate, but shows the greatest activity in the basal ganglia area. This isoenzyme shows comparatively low activity to the classical substrate of this enzyme, tyramine.

I am not quite sure what these data imply. It has been claimed by Metzel and colleagues,[9] although I have not seen their precise data, that there is some increase in monoamine oxidase in the basal ganglia of patients with parkinsonism. I do not know whether this is true or not, but I would certainly like to look at the individual isoenzyme bands in parkinsonian patients to see if there were any specific increase in MAO-4; if there were, specific inhibition of this particular band might be one possible approach to therapy.

Dr. Everett: In reference to the pupil studies, has anyone studied patients who are taking Dopa to see whether their pupils react in the same way, since they might have dopamine substituted for norepinephrine in these neurons?

Dr. Calne: We have not investigated the effect of L-Dopa or dopamine eye-drops in parkinsonian patients receiving oral L-Dopa.

Dr. de Champlain: I would like to ask Dr. Barbeau if he has treated some normal subjects with L-Dopa to see if it changed their renin activity?

Dr. Barbeau: We have not treated control subjects for that specific purpose, and we have no normal subjects. But some patients with non-Parkinson disorders have received Dopa. In these subjects Dopa again produced a decrease in PRA.

Dr. Goldstein: With regard to Dr. Sandler's comments, I would like to mention that the excretion of 6-hydroxydopamine in the urine was previously reported.

The findings of Dr. Sandler's group that monoamine oxidase appears in different forms seems to me to be of great importance. It is conceivable that in the striatum of Parkinson patients a specific monoamine oxidase form is missing. It might be noteworthy that monoamine oxidase activity is susceptible to changes following hypothalamus stimulation (D. J. Reis and M. Goldstein, unpublished data).

Dr. Sandler: We have treated rats with fairly high doses of L-Dopa and measured monoamine oxidase activity; the only place where any change was detected was the uterus.[10] There seems to be one specific isoenzyme band—in the rat uterus at least [11] and we think it is probably present in the human uterus also—which responds to Dopa administration by an increase in activity. Presumably generation of the monoamine oxidase substrate dopamine is responsible for this. Enzyme activity is also potentiated by progesterone and inhibited by estrogen.[10] Has anybody treated parkinsonian women, while still in the childbearing period of life, with L-Dopa? Do they show any menstrual irregularities?

Dr. Hornykiewicz: This is not an answer to the question, but I want to ask another one. What is the relative percentage of occurrence of MAO-4 in the human caudate? Small? It might mean then that it is not too significant.

Dr. Sandler: I do not know.

Dr. Ansel: I am following a 48-year-old female with a thirty-year history of postencephalitic Parkinson's disease. One of the distressing side effects of L-Dopa therapy in her case has been the development of a bloody vaginal discharge. This persists with no apparent gynecological explanation.

Dr. Kopin: I would like to ask a question of Dr. Sandler, regarding formation of 6-hydroxydopamine. Would the implication here be that treatment with Dopa results in formation of sufficient 6-hydroxydopamine to destroy the adrenergic neurons? Is it the implication that treatment with Dopa over a long interval might be more harmful than good?

Dr. Sandler: The implications are two-fold: there might be an error of metabolism responsible for some degree of endogenous overproduction of 6-hydroxydopamine in idiopathic parkinsonism, and it is also possible, even if administered L-Dopa is completely unadulterated by 6-hydroxydopamine, that such an error of metabolism would result in an increased production of the toxic compound when large doses of L-Dopa are given. It may well be that some side actions of the drug, especially hypotension, arise by this mechanism.

Dr. Goldstein: I would like to mention that in some preliminary experiments we have injected 6-hydroxydopamine intracerebrally into monkeys. Some of the monkeys have shown some postural changes but as yet we have not been able to demonstrate tremor. Further studies are now in progress.

REFERENCES

1. Dalgliesh, C. E., Horning, E. C., Horning, M. G., Knox, K. L., and Yarger, K.: A gas-liquid-chromatographic procedure for separating a wide range of metabolites occurring in urine or tissue extracts. Biochem. J. *101:* 792, 1966.

2. YOUDIM, M. B. H., and SANDLER, M.: Isoenzymes of soluble monoamine oxidase from human placental and rat liver mitochondria. Biochem J. *105:* 43, 1967.
3. COLLINS, G. G. S., YOUDIM, M. B. H., and SANDLER, M.: Isoenzymes of human and rat liver monoamine oxidase. FEBS Letters *1:* 215, 1968.
4. YOUDIM, M. B. H., COLLINS, G. G. S., and SANDLER, M.: Isoenzymes of monoamine oxidase. *In* Enzymes and Isoenzymes: Structure, Properties and Function. Proc. FEBS Symposium (D. Shugar, ed.). Academic Press, London, 1970.
5. YOUDIM, M. B. H., COLLINS, G. G. S., and SANDLER, M.: Multiple forms of rat brain monoamine oxidase. Nature *223:* 626, 1969.
6. YOUDIM, M. B. H., and COLLINS, G. G. S.: Dissociation and reassociation of mitochondrial monoamine oxidase. Biochem. J. (in press.).
7. COLLINS, G. G. S.: In preparation.
8. COLLINS, G. G. S., SANDLER, M., WILLIAMS, E. D., and YOUDIM, M. B. H.: Multiple forms of human brain mitochondrial monoamine oxidase. Nature, in press.
9. METZEL, E., WEINMANN, D., and RIECHERT, T.: A study of the enzymes of DOPA metabolism in parkinsonism from biopsies of the basal ganglia. *In* Third Symposium on Parkinson's Disease (F. J. Gillingham and I. M. L. Donaldson, eds.). E. & S. Livingston, Edinburgh, 1969.
10. COLLINS, G. G. S., PRYSE-DAVIES, J., SANDLER, M., and SOUTHGATE, J.: Effect of pre-treatment with oestradiol, progesterone and Dopa on monoamine oxidase in the rat. Nature, in press.
11. COLLINS, G. G. S., and SOUTHGATE, J.: The effect of progesterone and oestradiol on rat uterus monoamine oxidase activity. Biochem. J., in press.

PART 4

SIGNIFICANCE OF BEHAVIOR AND MENTATION CHANGES DURING THERAPY WITH L-DOPA

CHAPTER 10

Clinical Effects of L-Dopa on Behavior and Mentation

CHANGES IN BEHAVIOR AND MENTATION

F. H. McDowell

This discussion concerns changes in behavior and mentation seen in patients with Parkinson's disease receiving L-Dopa therapy. It might be pertinent to begin by reminding ourselves what Parkinson patients are like in regard to behavior, mentation and mood before they undergo treatment. Most patients come to us with some degree of depression, usually a reactive depression. In our series of 100 patients, we have only one with a cyclic depression which preceded Parkinson's disease for many years. Patients with parkinsonism generally, if asked, will report that they are depressed, sometimes severely so.

As for behavior, most of them, I think, suffer from a certain degree of akinesia of motivation. They have rigidity of intellect and, on examination before treatment, half of them or more show gross evidence of dementia manifested by deficiencies in recent memory, poor calculation, and sometimes disorientation.

How do they appear after treatment? I am sure all of you who have treated a number of patients have had similar experiences. After a brief period of L-Dopa treatment, patients begin to look better. Family and friends often notice this change first and comment about it. Patients frequently report a new sense of well-being. We've not been sure whether this is simply a relief of their depression resulting from the fact that something promising is being done for them, or whether it accompanies the decrease in physical disability that frequently occurs at this time. Nevertheless, there is usually a relief of their depression and some improvement in outlook.

In analyzing mentation, we have found that patients tested with the Weschler Adult Intelligence test before starting treatment, and again six months after starting Dopa, show an improvement in score varying from 10 to 20 percent. This change in score is outside the usual variation that one would expect for these patients. It is not entirely due to improvement in motor function, on which the performance part of the test depends. It seems that there is also some loosening of intellectual capacities.

The mental side reactions which occur in patients treated with L-Dopa need some background. Most of the patients who come to us are elderly; our average age is 65. Often they are taking a large number of other drugs, including anticholinergic medications, sedatives and phenothiazines in an attempt to control their Parkinson's disease. These drugs in combination with dementia, I think, lay the substrate for some of the abnormal mental reactions we see. Table 1 shows the kinds of mental reaction found in our patients.

Fourteen percent showed a disturbance in behavior or mentation. The most common reaction was a confused, delirious state. Patients with this condition were usually disoriented, had hallucinations, and sometimes were delusional, anxious, and confused. Their symptoms tended to be worse in the evening with the decrease in environmental lighting. This state closely resembles an acute drug intoxication or a toxic delirium.

Two of our patients became depressed while on Dopa. One of them attempted suicide at the beginning of his treatment, but since then has shown no further marked depression and has not been suicidal. The other depressed patient had cycles of overactivity and depression for many years before beginning L-Dopa.

Two patients on very high doses of L-Dopa (8 to 9 grams) became stuporous.

We've tried to correlate these phenomena with other clinical findings. Table 2 shows that most of the patients who have abnormal mental reactions are female, but the numbers of patients are too small to consider this significant. Most of the patients are over 60. Mental reactions are much more common in patients over 70. Nine of the 10 patients who showed either delirium or stupor had marked evidence of dementia, and 8 of the 10 had severe Parkinson's disease. We believe that the elderly patient with advanced Parkinson's disease who has some evidence of dementia is in a precarious

TABLE 1. Disturbances in Behavior and Mentation

Disturbances occurred in		14%
Confused delirious		10%
Depressed		2%
Suicidal	1	
Stuporous		2%

TABLE 2. Patients with Delirium or Stupor (12 Patients)

	Male 5	*Female* 7
Age		
50-59	1	0
60–69	1	3
70-79	3	4
Dementia		9
Severe disease		8

balance, and that anything which depresses central nervous system function further may produce a drug-induced delirium.

Table 3 shows the correlation of mental reactions with the L-Dopa dose. Mental reactions are not particularly dose related. The Table shows the number of patients at each dose, and the number at that dose with mental disturbances. Although there is a peak around an intake of 5 to 6.5 grams, the incidence does not rise with increasing doses of L-Dopa.

In an effort to enlarge this experience and to see if ours was similar to that with larger groups of patients, I have been able, through the courtesy of the Hoffmann-La Roche Company, to look at the data from their combined, retrospective study. This represents the experience from our patients, those of Dr. Melvin Yahr and those of Dr. Charles Markham.

In this group of 365 patients, 178 psychiatric side reactions occurred in 127 patients. The variety of reactions observed can be seen in Table 4. There were 19 patients who reported depression. There is no evidence, however, from another analysis of this study to suggest that L-Dopa produces depression. In fact, the reverse seems to be the case.

The most common reaction encountered was a manifestation of delirium. Some patients became confused and had hallucinations, and I think these could easily be combined with the patients who were restless and agitated. If this is done, it makes delirium the most common mental reaction.

TABLE 3. Mental Disturbances with L-Dopa

Dose	*Patients at that Dose*	*Mental Disturbances*
0.5	100	0
1.0	100	0
1.5	100	1
2.0	100	1
2.5	99	4
3.0	98	6
3.5	89	3
4.0	86	9
4.5	74	9
5.0	64	6
5.5	52	8
6.0	43	12
6.5	26	4
7.0	20	0
7.5	9	0
8.0	5	0

TABLE 4. Psychiatric Side Effects (178 Reactions)

Depression	19
Delirium, confusion	
Hallucinations	25
Restless agitated	16
Insomnia	66
Nightmares	13
Paranoia	2
Lethargy	8
Mental reactions	
Type unspecified	29
	178

Insomnia was reported by a sizable number of patients. I am not sure exactly what one can make of this. In our experience, if one asks a leading question about sleep, such as "Do you sleep badly?" or "Do you sleep well?" the answer is usually yes. The patient's family frequently tells us, however, that the patient sleeps quite well. A patient occasionally reports having nightmares.

We attempted to correlate the occurrence of mental reactions with age of the patient and stage of his disease (Tables 5 and 6). Mental reactions were

TABLE 5. Relation of Psychiatric Side Effects To Sex and Stage of Disease

Males	76	60%	
Females	51	40%	
Stage 1	1		
Stage 2	8		
Stage 3	53		
Stage 4	32		92%
		43%	
Stage 5	16		

more common in this group among the males, and among those with the more advanced stages of Parkinson's disease. Ninety-two percent of the patients with psychiatric reactions were in Stage 3 of the disease or higher, and 43 percent were in Stages 4 or 5.

Table 7 correlates mental reactions with intake of other drugs. Many patients who are treated for parkinsonism with anticholinergic medications often report toxic delirium. Their thinking is frequently clouded and slowed by these agents. We thought that perhaps the addition of L-Dopa to their drug

TABLE 6. Relation of Age To Psychiatric Side Effects (117 Patients)

Age	*Patients*
39-49	8
50-60	35
61-71	53
72-82	20
82+	1
	117

TABLE 7. Cooperative Retrospective Study (365 Patients)

	Patients	%
Psychiatric side effects (178 reactions)	127	34
With drugs other than L-Dopa	107	84
Anti-Parkinson agents	82	64
Antihistamines	19	11
Psychotropic agents	24	22
Sedatives	9	8

regimen might be sufficient to push them into delirium. In the whole group there is a high incidence of intake of other drugs. There is not, however, a higher percentage of patients taking other drugs among those who have psychiatric side reactions when compared to the whole group.

In summary, the most common mental reaction observed during L-Dopa treatment is an acute toxic delirium. It is most often seen in the elderly, demented patient with more advanced evidence of disease of the nervous system. There is no evidence that L-Dopa causes depression.

The occurrence of migraine headache was added to this topic. We have one patient who for many years had daily migraine attacks and who now has not had a migraine headache during the year and a half he has received L-Dopa. This is, I think, experience without much meaning but nevertheless intriguing.

PSYCHIATRIC ASPECTS OF L-DOPA TREATMENT IN PARKINSON'S DISEASE

Robert S. Schwab

In managing patients with Parkinson's disease it is clearly evident that psychiatric symptoms occur quite frequently. With Fabing and Prichard in 1951 [1] we reported on the incidence and types of such complications in 200 consecutive patients. In brief we found there were four groups:

1. Preexisting mental disorders, such as manic depressive personality, unrelated to the disease or its treatment.
2. Reactive states (usually depression) directly related to the burdens of the disease.
3. States directly related to drugs used in therapy. These are toxic states that reverse as the drugs are reduced or eliminated.
4. Conditions, often paroxysmal in their setting, that seem to be part of the many facets of the disease itself. Confusion during occulogyric crisis is an example.

Since this paper was published, a fifth group became manifest. This fifth group consists of gradual mental deterioration due to a degenerating process concomitant with the progressive Parkinson's state and primarily due to cerebral arteriosclerosis.

In discussing the psychiatric complications or side effects from L-Dopa treatment of Parkinson's disease, it is highly important to keep these five groups in mind. We are now obviously concerned with the third group and the possible mechanisms whereby L-Dopa can produce clear-cut behavioral, confusional-delusional, or hallucinatory symptoms. We will first report briefly on the prevalence of such side effects in various series of patients receiving L-Dopa.

McDowell states that 10 percent of his group had psychiatric symptoms. Yahr in the October *Archives of Neurology* cites 16 percent of his 60 patients. Rikland in Cooper's series reports 10 percent in 40 suitable selected cases with intact mentation; in unsuitable cases with obvious signs of mental impairment, there was a much higher prevalence of psychiatric complications, namely, 25 percent. Finally Timberlake, in his hospital series studied under double-blind conditions, found between 5 and 10 percent showing psychiatric states.

Among the cases reported to us by other physicians, the following three of Dr. Cawein, from a total of 23 treated patients, are of interest because of the appearance of hallucinations:

1. A. K., an 80-year-old man on 6 gm per day of L-Dopa, developed nocturnal delusions, thinking that people were parking cars in his room. The dose was decreased to 4.8 gm per day and his symptoms have not returned. Dose was subsequently reduced to 4 gm per day because of anorexia.
2. G. V., a 64-year-old man, became confused on 3.6 gm L-Dopa per day, but the confusion subsided with no reduction in dosage. On advancing the dose to 4.8 gm per day the patient had hallucinations which consisted of seeing ants crawling in a circle around him. The dose was decreased to 4 gm per day and he became asymptomatic.
3. F. O., a 66-year-old man, began seeing chickens and collecting eggs while on 4.8 gm of L-Dopa per day. When the dose was reduced to 4 gm these hallucinations subsided.

All three patients have been continued on the doses noted above without further difficulty. There was no associated movement disorder and no hypotension. All have had an excellent response to the drug. The patients were taking Artane (6 mg per day) at the time they developed hallucinations, and the Artane was not stopped. I doubt that this is significant since most of the other patients were also taking Artane.

England, from nearly 100 office patients receiving L-Dopa—a highly favorable group—reported that less than 5 percent showed any mental symptoms attributable to L-Dopa. In my own series of 42 private patients, 4 (10 percent) had psychiatric problems. Thus, in summary, between 5 and 15 percent of any group may show this complication.

Grouping of patients, into the highly selected favorable ones vs. unfavorable with concomitant mental difficulty, is logical and practical because of the tremendous number of patients seeking help from L-Dopa in a comparatively few centers in the United States and Canada. Rikland stated that the suitable patients leave the hospital in 10 to 20 days, whereas poor L-Dopa candidates often require twice this period.

In assessing the cause of psychiatric symptoms with L-Dopa, we must remember that the mildly to moderately involved patient responds best to any treatment—surgical, medical or physiotherapeutic. This is also true of psychotherapy, transplant surgery, and other branches in medicine. Therefore an L-Dopa project with restrictions against the severely ill, bedridden, or demented patient should have a lower incidence of psychiatric complications than a project accepting all comers in an effort to be totally fair to the whole Parkinson's population.

We will now briefly describe the types of complications encountered and described by others:

1. Agitation, insomnia, anxiety, helplessness, depression due to unwise abrupt withdrawal of previous medicine while the L-Dopa dosage is wisely slowly increased.
2. Anxiety with some agitation and discouragement largely due to misunderstanding of the slow rate of drug increment.
3. Insomnia and anxiety usually added to nausea and anorexia because of too rapid drug increase or taking of drug on an empty stomach.
4. a. Development of some confusion directly related to the L-Dopa schedule (no previous confusion).
   b. Increase in slight confusion previously present, related to L-Dopa.
   c. Increase in preexisting confusion from combination of L-Dopa and other drugs.
   d. Paranoid reactions related to L-Dopa or increase in preexisting paranoid symptoms from L-Dopa.
   e. Hallucinations (usually nocturnal) related to L-Dopa or increase of inconsequential hallucinations preexisting treatment.
   f. Combinations of a, d, and e.
5. Reactive depression from total failure of L-Dopa or lack of benefit to a particular symptom, such as tremor or speech defects.

Examples

(N.K.): A 74-year-old female with eight years of akinetic bilateral Parkinson's disease with mild periods of forgetfulness and confusion (brief in time and thought to be due to lack of attention and interest) entered our program in May, 1969. As the dose was increased from 100 mg four times a day to 250 mg four times a day, these symptoms increased. They were accompanied by increasing unsteadiness despite reduction in rigidity. Reducing both the rate of increase and total dose failed to help. The patient left the hospital taking 750 mg a day, gave it all up in a week, gradually improved, and was even better, taking no medicine, when seen three months later.

(R.B.): A 54-year-old male department store owner with ten years of severe rigidity bilaterally had some relief from left thalamotomy five years before. He had been subject to periodic depressions all of his life. He entered the L-Dopa program in January, 1969. He became depressed while on the placebo series and began to lose weight. L-Dopa increased both his weight loss, by 35 pounds, and his depression. A movement disorder developed on the left (face and hand) at a dosage of 6 grams, and akinesia and rigidity became more severe. When psychotic delusions developed, he was transferred to a psychiatric hospital where he recovered from depression, gained 30 pounds, and went back to work. A retrial of L-Dopa three months later, with his weight and mood normal, produced moderate degrees of both subjective and objective benefit, at 4.5 grams per day, without any side effect.

Conclusions

Psychiatric side effects from L-Dopa treatment are reviewed. About 10 percent of patients show these complications, and nearly all are reversible. About half of the psychiatric side effects are either preexisting or are due to other causes. In a highly selected group such complications should be much rarer.

REFERENCE

1. Schwab, R. S., Fabing, H. D., and Prichard, J. S.: Psychiatric symptoms and syndromes in Parkinson's disease. Amer. J. Psychiat. *197:* 12, 1951.

PSYCHIATRIC ASPECTS OF L-DOPA TREATMENT

M. D. Yahr

I am delighted that Dr. Schwab has accepted the concept of double-blind procedures in evaluating therapeutic response. This has been sorely lacking in many previous studies in parkinsonism, and it is my feeling that it is essential in screening drugs for therapeutic efficacy.

Dr. McDowell has well covered the psychiatric aspects of the problem. The data he reviewed include all the cases we initially studied, and I have very little to add. I would like, however, to indicate some recent experiences not encountered in the earlier case material. We have recently had two attempted suicides. Fortunately, both were unsuccessful and the patients survived. Both patients had had minimal benefit from the administration of Dopa and, I believe, became rather depressed when they realized that failure was at hand. This certainly fits in with Dr. Schwab's comment about the impractical expectations patients may have and the need for the physician to set realistic goals.

It is not unusual to encounter a variety of psychological and behavioral alterations in patients receiving Dopa. The most frequent have been a general sense of anxiety and heightened nervous tension, and an increased tendency to dreaming—occasionally so vivid that patients may feel they are experiencing hallucinatory episodes. Readjustment of dosage has usually alleviated these manifestations, but it is important to caution patients about them.

In view of Lance's work on the relationship of migraine attacks to serotonin, we have been interested in the occurrence of headache phenomena in parkinsonism. To date, we have had only one patient who exhibited the combination of migraine and parkinsonism. The migrainous attacks have occurred erratically, a situation which makes evaluation of therapeutic results difficult. However, to date, while on Dopa she has not experienced a migraine episode. It will be of some interest to follow this case over the years to come.

Major psychiatric disturbances have not been a problem to us. We have had one toxic psychosis which we felt was precipitated by Dopa. Dopa was discontinued for a period of time and subsequent retreatment did not produce a recurrence. The entire problem of affective disturbances and alterations in mentation in this group of patients, however, needs further study. It has been our general experience that many of the behavioral difficulties attributed to the Parkinson process are related to the use of drugs, particularly the anticholinergic agents, in its treatment. In our experience about 30 percent of the patients taking standard anti-Parkinson drugs experience some degree of the so-called organic mental syndrome. One of the salutary effects of Dopa has been the elimination or decrease in dosage of these agents with clearing of the mental abnormalities in many patients.

Regarding affective changes, particularly depression, we have recently undertaken a combined study with our Department of Psychiatry. Preliminary indications are that Dopa may alter depression episodes in some psychiatric conditions. Whether it also can induce them is something we have no data on as yet.

BEHAVIOR AND MENTATION CHANGES DURING THERAPY

J. Arbit, B. Boshes, and R. Blonsky

We are reporting results on 36 patients of an initial group of 47 who began treatment with L-Dopa. Because of various incapacities, 9 of the original group could not complete the intellectual testing and 2 who were able to complete the intellectual testing could not complete the tremor studies. Of the 36 patients, 23 were men and 13 were women. Ages ranged from 45 to 78 years, with a mean of 65.9 (standard deviation of 8.0) and a median age of 67.

The research design employed consisted of measurements obtained prior to L-Dopa treatment and post-treatment measures obtained just prior to the patient's discharge from the hospital, a point usually characterized by the patient's having attained maximum therapeutic benefits. We realize the need for results with control groups to ascertain the level of improvement that would occur upon simple retesting, and are in the process of obtaining these.

When we make comparisons relative to patterns of change, the individual serves as his own control. All of the statistics presented are based upon t-tests and the criterion of maximum drug improvement used in establishing the time of retesting is determined by improvement in the tasks we are measuring, whether or not improvement has occurred.

Figure 1 presents the results of the four performance subtests of the Wechsler Adult Intelligence Scale that were administered (Digit Symbol, Picture Completion, Block Design and Picture Arrangement). The ordinate axis is the age-corrected scaled score. This score may range from 0 to 19, with the data for patients of different ages made comparable through a correction for normal aging effects. The mean scaled score would be 10.

Though all four performance subtests improve from the pre-Dopa period to the post-drug period, only one shows a statistically significant improvement. Picture Completion is a purely perceptual task; Block Design involves fairly crude motor functions and a large perceptual component; Picture Arrangement, like Block Design, involves only gross motor functions and a perceptual-conceptual quality. Only in the Digit Symbol test, which shows a significant improvement, do we find, in addition to the perceptual aspects, a need for fine motor coordination and a task that is essentially also a measure of "power" (productivity in a given time interval).

If the improvement is merely the result of previous exposure, we would expect that there would be a significant improvement on all four tasks. That there was not seems attributable to improvement not in the perceptual aspects of the task, not in the gross motor demands required of the patient (else the

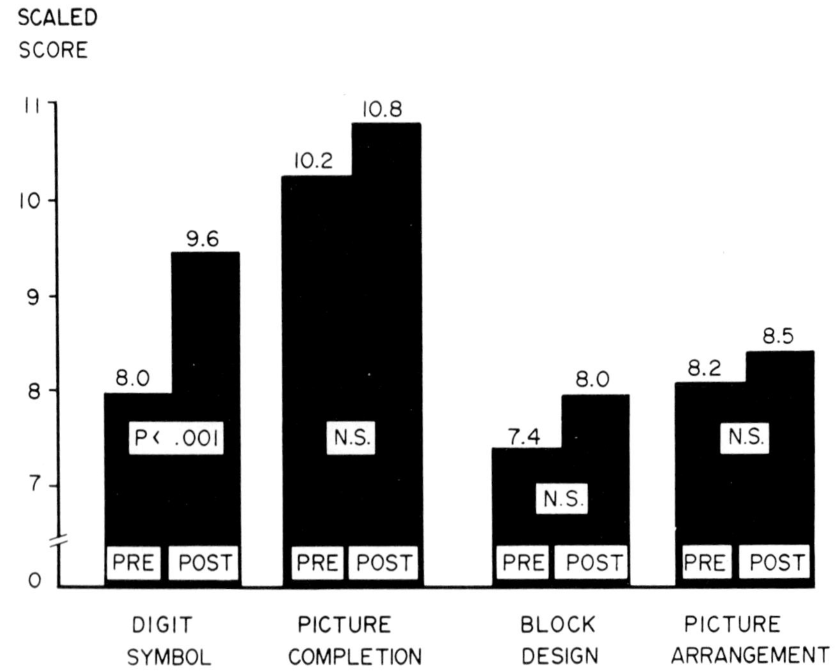

Fig. 1. WAIS performance subtest scores (N=36).

Block Design and Picture Arrangement might also show an improvement), but in speed, vigor, productivity and/or fine motor movements. It may be as important to note that three of the performance subtests did *not* show a significant improvement as to indicate the one that did and its particular characteristics. There are certainly functions that drug therapy did not effect.

Figure 2 presents the results of the four verbal subtests of the WAIS (Information, Comprehension, Similarities and Digit Span) also using an ordinate of 0 to 19 age-corrected scaled score points. As on the performance series, all of the verbal subtests show some improvement after drug treatment (at the point of hospital discharge) as opposed to the test score obtained prior to L-Dopa treatment. However, only two of the four tasks show statistically significant improvement. Information and Comprehension, of the four verbal tasks administered, are likely to show the least sensitivity to the presence of underlying disturbances in the neurological substrate. These tasks are relatively insensitive to any but the most extreme neurological disturbances. Improvement on these tasks may be the result of increased motivation, greater zest and vigor, possibly a greater desire to do well.

The two verbal subtests that show no improvement are Similarities and Digit Span, the former a measure of concept formation, the latter of immediate memory. Of the four verbal tasks administered, these two tend to be most sensitive to the presence of an organic brain involvement; we find that the drug produces no improvement in these abilities. The drug not only does not significantly affect concept formation and immediate memory functions, but

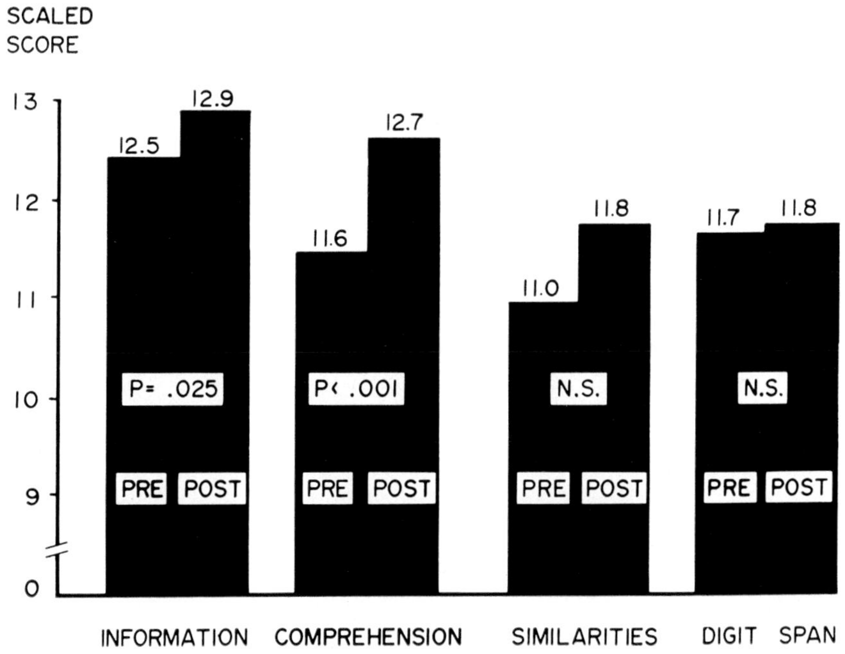

Fig. 2. WAIS verbal subtest scores (N=36).

precludes an expected improvement from the repetition of this test material. We shall emphasize that the most sensitive indicators of organic brain disturbances in terms of the verbal tasks show no change.

In attempting to interpret the pattern of results on the verbal and performance series of the WAIS, it seems that one must consider not only the L-Dopa effect directly upon the ability to perform fine motor functions but also the increased motivation that results in improved performance. We are exploring this possibility through the establishment of a "conditioning laboratory" wherein Parkinson patients may be evaluated relative to drug effects in terms of various magnitudes of both aversive and appetitive motivational needs.

Figure 3 presents the results, pre- and post-drug, relative to the verbal IQ score, the performance IQ score and the full-scale IQ score. Though only one of the four performance tasks and two of the four verbal tasks showed a significant improvement, it is apparent that when we combine all four subtests on each series there is an overall significant improvement of approximately four IQ points on each of the two series. This could be expected in the average population, where no possible drug effect is interpolated (merely a test-retest paradigm), and we can make no inferences from this material.

Figure 4 presents the results obtained with the Wechsler Memory Scale administered prior to L-Dopa treatment and at the point at which maximum hospital benefits occurred. The Wechsler Memory Scale consists of seven

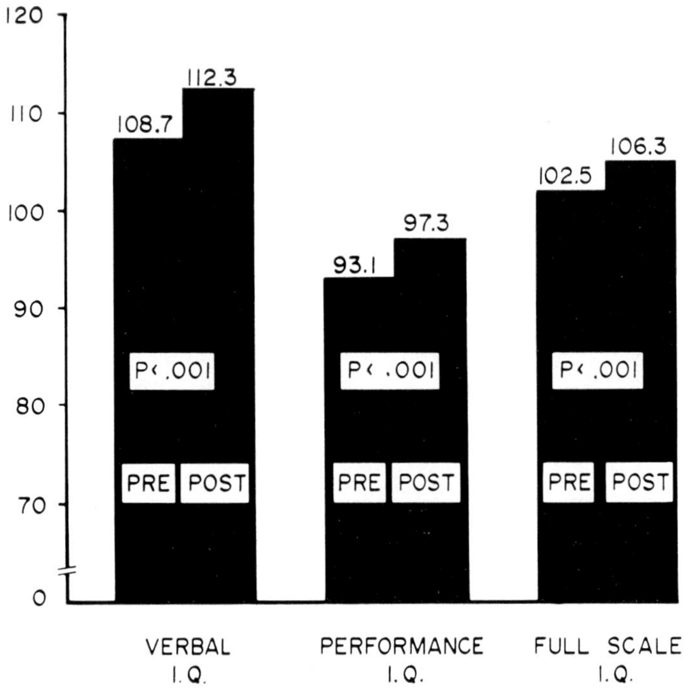

Fig. 3. Wechsler Adult Intelligence Scale (N=36).

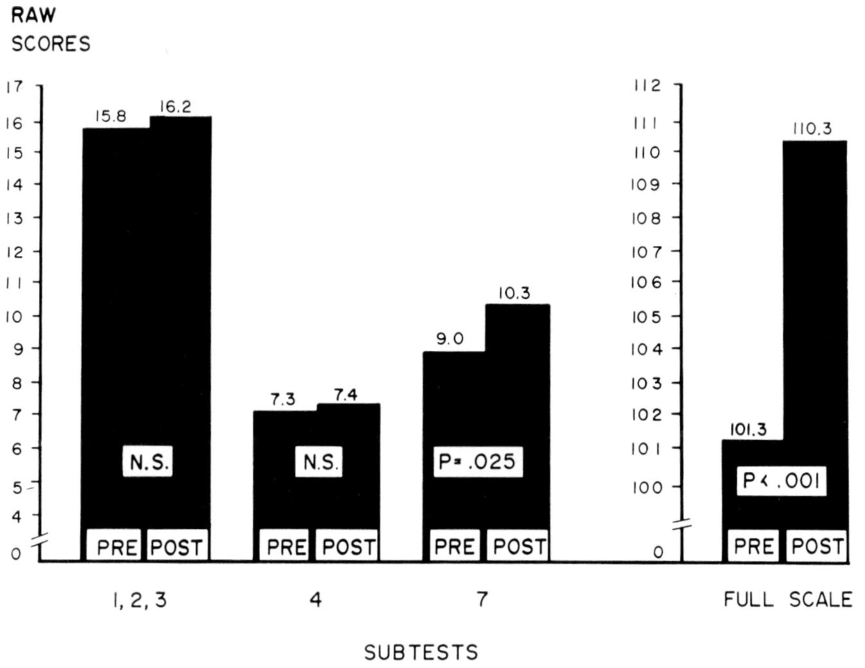

Fig. 4. Wechsler Memory Scale.

subtests with the first three (Personal and Current Information, Orientation and Mental Control) grouped together into a single score; the fourth subtest, Logical Memory, provides a single score, and the seventh subtest (Associate Learning) also provides a single score. Subtest five is the Digit Span subtest of the WAIS already discussed. Subtest six is a visual reproduction task essentially duplicating the Bender Visual-Motor Gestalt Designs Test and the Memory for Designs Test. The vertical axis represents raw scores, i.e., number correct, since there is no age correction for specific subtests. However, since the patients were essentially the same age at the pre-Dopa and post-drug testing, the pre/post comparison can be evaluated despite the lack of age correction. We cannot, however, compare the performance on one subtest or group of subtests with other subtests or groups of subtests. The only improvement appears in the Associate Learning subtest. With the maximum score obtainable on the first group of three subtests being 20, it seems that our group before drug administration functions at such a high level that any additional improvement is most difficult due to a "ceiling effect." Subtest four may well be too difficult to show any improvement, utilizing as it does a complex paragraph recall procedure. The Associate Learning task shows a significant improvement after drug treatment as compared to the pre-drug measure, and this may be because the total score reflects a partial score obtained in learning easy associations and a partial score obtained in learning difficult associations. There is room for improvement on the easy tasks not notably affected by the "ceiling" relative to the lack of improvement on the difficult part.

The right-hand side of this same figure shows the Memory Quotient as a function of test period. The MQ is age corrected and we see a significant post-drug improvement, an improvement in part affected by the patient's past experience with this task. The full-scale Memory Quotient score is based not on the total sample of 36 that was used for the other comparisons but upon the 13 patients under 65 years, for the correction factor extends only to this age level.

Figure 5 shows the pattern of change in patients who have had a 6-month follow-up evaluation. The three tests indicated were obtained immediately prior to L-Dopa treatment, approximately one month after treatment was begun, and six months later.

The pattern of change seems to have some consistent qualities. Patients who do significantly less well after initial L-Dopa treatment show a significant improvement in functioning with a change in medication.

Figure 6 presents the results from ratings of the designs reproduction material. The Bender figures and the Memory for Designs figures were rated separately in terms of the degree of gestalt form distortions and the occurrence of rotations; a rating of 3 indicates a maximum amount of gestalt form distortion and a maximum amount of figure rotation. The Bender figures are administered twice: first, to test memory, each stimulus is exposed for 5 sec-

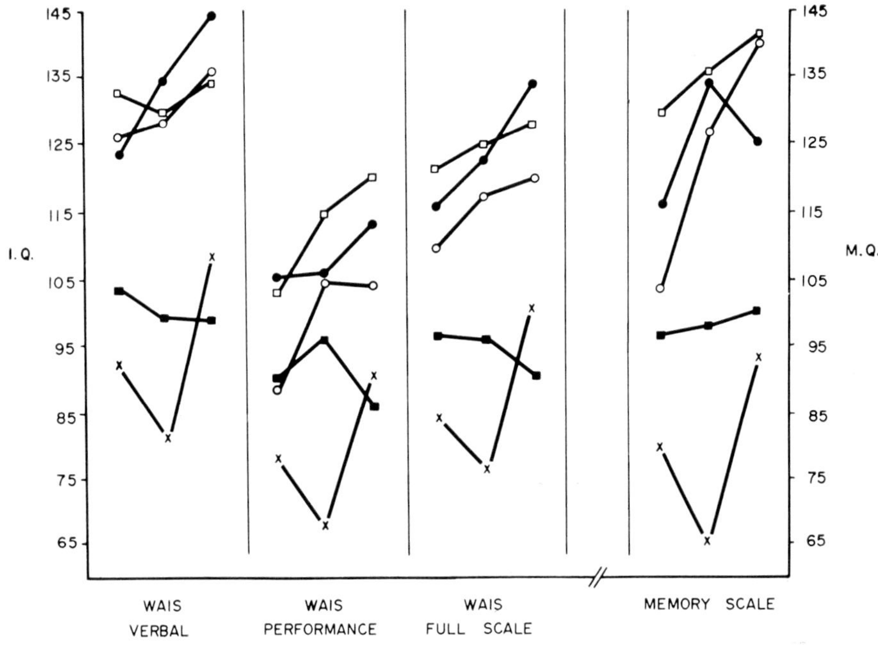

Fig. 5. Selected cases with 6-month follow-up.

○—○ Male, 65, professor, 9-year history
●—● Female, 70, housewife, 8-year history
□—□ Male, 57, electrical engineer, 3-year history
×—× Male, 67, retired industrial machine sales, 12-year history
■—■ Male, 66, retired company manager, 6-year history

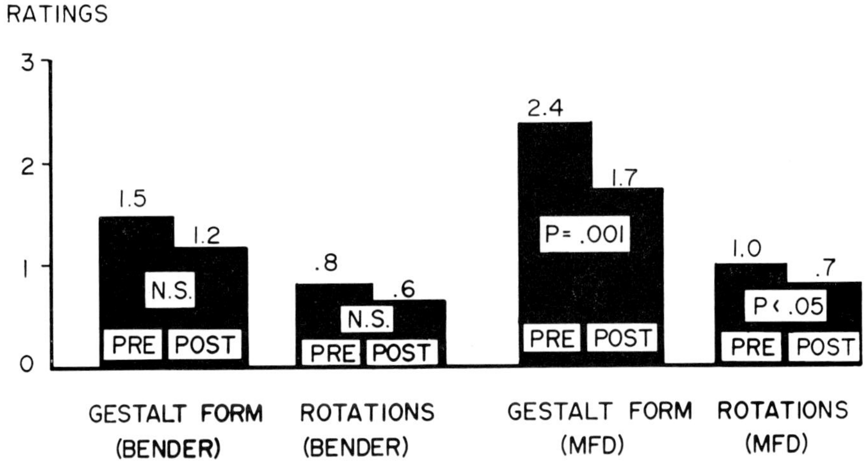

Fig. 6. Bender Visual-Motor Gestalt Test and Memory for Designs Test.

onds, removed, and then drawn from memory by the patient; secondly, the patient is asked to copy each of the figures directly from the stimulus cards. The rating is the average of these two test administrations. While the scoring for the Memory for Designs test is the same, the latter task entails only memory administration.

Though the ratings of the Bender figures both indicate improvement, as do the ratings of the Memory for Designs test, only the improvement in the Memory for Designs test is significant. This may be due to the fact that the Memory for Designs, since it involves a larger number of figures, would be expected to produce a more reliable score. Also, it may be that the improvement during the memory administration of the Bender figures is obscured by the relatively unchanged score on the "easy" direct-copy administration. Considering all four measures on the Memory for Designs, we believe that the Bender Visual-Motor Gestalt is an "easier" task. It may well be that room for improvement is less than on the Memory for Designs, where the figures are more complex. In any case, there seems to be reliable evidence of improvement in the visual-motor integrative ability revealed by this test, though the effects of improved motivation, predicated upon improved attention span and greater concentration, must be considered.

Figure 7 contains our findings from patient performance on the Rorschach Psychodiagnostic Test. While constriction, impotence and impoverishment are characteristics of older-age groups generally, Parkinson patients on the whole show an even greater paucity of productivity than a comparably aged, non-Parkinson group. It may be that this is the psychic equivalent of the anergic, akinetic physical state, it may be the lack of motivation about which we have spoken, or it may be an association between the Parkinson's disease process and a depressive psychological state. Whatever the case, objective analysis of more formal scoring categories seemed inappropriate because of data of low reliability and validity. We note from this Figure that 1.6 fewer

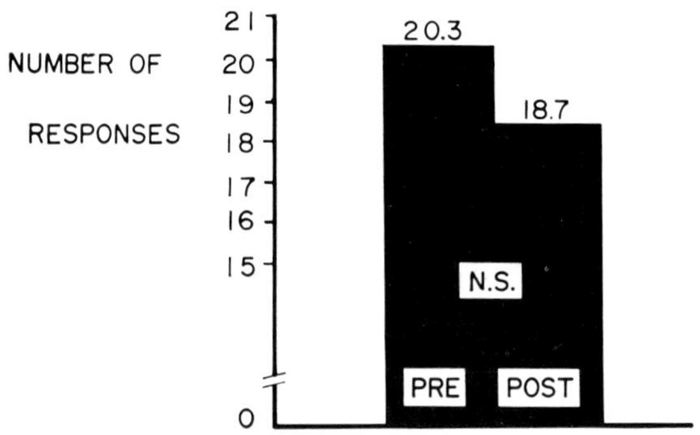

Fig. 7. Rorschach Psychodiagnostic Test.

responses are produced after than before the Dopa treatment. There is certainly not a psychological "loosening" at all comparable to the physical.

In summary, we see some improvement that might well be the result of improvement in motivation and drive and energy relative to L-Dopa treatment. We note that this change does not affect those test activities that are sensitive to organic brain involvement unless they also involve a significant improvement in fine motor coordination. Tests that are too easy show no improvement with drug treatment; tests that are too difficult also show no improvement. In the "in-between" area, increased motility and motivation provide a basis for improvement after drug treatment, though the increase is seen in motor functions and not necessarily at a psychological level.

THE L-DOPA EFFECT IN DYSTONIA AS A MODEL OF COUNTERBALANCING DOPAMINE-SEROTONIN SYSTEMS IN THE CENTRAL NERVOUS SYSTEM

Mary P. Coleman

We will consider the effects of L-Dopa on blood pressure, affect, sleep and tone seen in patients with dystonia. The patient with Parkinson's disease has certain limitations as an experimental model of the effect of massive doses of L-Dopa in human beings. This is mainly because he is an aged adult, usually over 60 years of age, who often has mutiple vascular and blood pressure problems, insomnia, and the chronic depression frequently encountered in older people. Our patients with dystonia musculorum deformans or torsion dystonia occurring in late childhood and early adult life may be better models for understanding the physiological effects of L-Dopa.

Blood Pressure

Patients with dystonia, if they do have side effects from L-Dopa, show hypertension rather than hypotension. This effect, seen in the early stages of therapy, can be overcome by adjusting the dose downward for a day or so.

Sleep

In patients with parkinsonism, it has been reported by Cotzias[1] and by many participants in this symposium, there is an awakening effect from L-Dopa administration. A child of eight years with dystonia before L-Dopa is wide awake, alert, bright and stimulated by lights. This child is Jewish, with the recessive form of dystonia musculorum deformans. After administration of intravenous L-Dopa, the child has to be stimulated by testing in order to remain awake. At this point she has received about 10 mg per kg. The patient at the end of the experiment actually fell asleep during the intravenous injection despite the large number of lights and people in the room. This effect of sleep induction and drowsiness after intravenous Dopa has been seen in other dystonia patients. It has also been observed in two 30- and 40-year-old normal people, our two pharmacists, who have complained of being sleepy when they are preparing a large number of L-Dopa capsules.

In dystonia patients at least, and perhaps in younger people, L-Dopa, rather than being an alerting, awakening agent, may in fact be a relaxing, sleep-inducing agent.

In studying sleep patterns, we found that dystonia patients increased their night's sleep after being placed on L-Dopa (Fig. 1). These patients tend to sleep less than normal for their age before treatment and L-Dopa makes them sleep longer. Figure 1 illustrates a study of a child on a maximum dose of L-Dopa to see how much she slept; the time slept after taking L-Dopa is normal for her age.

Figure 2 shows an electroencephalographic sleep analysis done by Dr. Ann Barnet. Although sleep is actually increased in length, the proportions of sleep remain much the same; that is, the same proportions of slow-wave sleep and rapid-eye-movement (REM) sleep are present. Since almost all drugs decrease REM sleep, it is interesting that Dopa increases such sleep only slightly.

Affect

Dr. Barbeau has mentioned that, although the majority of parkinsonian patients showed an alerting effect, five treatment failures occurred in patients already depressed who became more depressed on L-Dopa. I would like to suggest that perhaps some of the confusing theories as to whether L-Dopa is a euphoriant or a depressant may be explained by a recent study of Bunney et al.[2] These authors suggested that there may well be two types of depression.

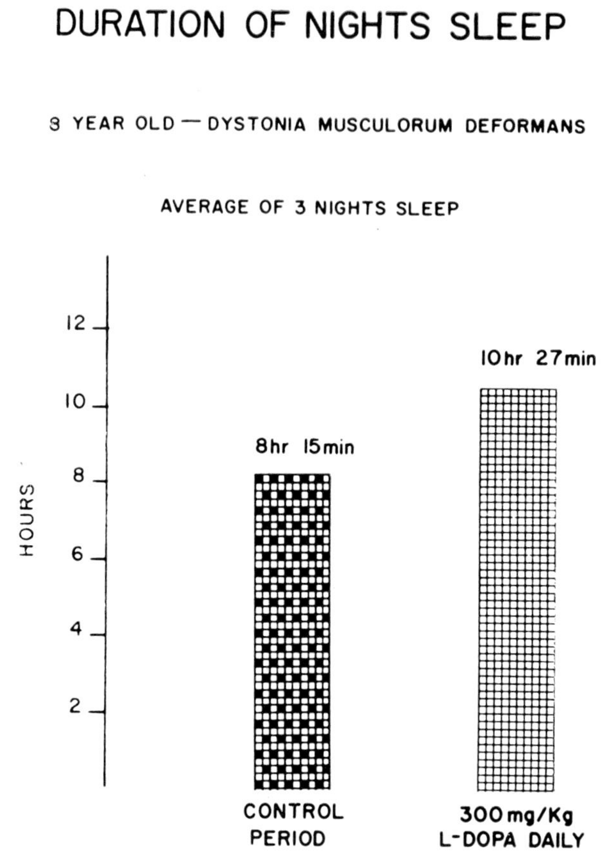

Fig. 1. Sleeping time before and after treatment with L-Dopa in an 8-year-old with dystonia musculorum deformans.

The first type is a "retarded" depression, in a clinically slowed-down patient. This type of depression may be related to low catecholamine levels, and may in fact respond to L-Dopa as reported in the paper. The authors also suggest that the second type of depression, an "agitated" depression, may be biochemically different because the patient may have a low serotonin level and may, in fact, respond to agents which raise serotonin. (Our work with hyperactive children suggests that these children may fall into the classification of agitated depression with low serotonin, and that this may explain, at least partially, the fact that they respond to antidepressant agents such as amphetamines and monoamine oxidase inhibitors.) Because they share many transport and enzyme systems, catecholamines and serotonin are maintained in close balance with one another, and it is this relationship, rather than absolute levels, that may determine affect.

Generally, in patients with dystonia, we notice no lasting effect on mood. The patients who tended to be cheerful when they came remained so through-

EEG SLEEP STUDIES

DYSTONIA -- 8 year old

SLEEP STAGE	CONTROL	% Total sleep time	L-DOPA
1.	14.4		10.6
2.	50.3		50.7
3. & 4.	14.0		14.0
REM	21.2		23.8
TST (hours)	7.9		9.6
Dose of L-DOPA	0		150 mg/kg

Fig. 2. EEG analysis of sleep stages in an 8-year-old patient with dystonia musculorum deformans before and after receiving L-Dopa.

out the treatment period. We were very fortunate to treat a set of identical female twins with dystonia and were able to carry out a double-blind, cross-over study on these twins. At the point of the crossover, we withdrew very gradually the unknown capsule (which was later shown to be L-Dopa) from the patient who had been receiving it for months. The twin became tearful and depressed and had suicidal thoughts, becoming more marked each day of the crossover. We are not sure that this is completely due to withdrawal of the drug, however, because the patient was clinically deteriorating quite rapidly and had good reason to be discouraged. When she came to us, she had marked truncal torsion and was crawling. With L-Dopa she began walking and her torsion dystonia was relieved. By the final day of the crossover, when we decided, because of her severe depression and clinical deterioration, to complete the crossover portion of the study, she was back to crawling although her truncal torsion still remained somewhat improved. After the study we had the impression that it was probably the withdrawal of L-Dopa that caused her depression.

There are many examples in the literature of conflicting information about the effect of L-Dopa on depression. Some early papers report no effect at low doses.[3] Wagshul and Daroff felt that in older patients the too-rapid increase of increments of L-Dopa was responsible for the depression observed.[4] It may be that the effects on mood may be a factor of rate of change either up or down, rather than the absolute levels. We have had several suicidal patients mentioned in these discussions, so an understanding of these factors is of great importance.

Tone

The last point I want to mention is the effect of L-Dopa on muscle tone. Figure 3 shows a patient with Down's syndrome, that is, a patient who is serotonin-deprived, not catecholamine-deprived. In such a patient it is now clear that muscle tone can be increased by administration of 5-hydroxytryptophan (5-HTP), the precursor of serotonin. The patient shown is being tested in the Landau posture at four days of age before administration of 5-HTP and again at five days of age after 5-HTP. The improvement of tone that occurs within 24 hours can be seen. In the serotonin-deprived patient, there is hypotonia to start with; this then moves gradually into normal tone as the patient's serotonin level is raised. I offer this as an interesting contrast to the effects seen with L-Dopa. In a patient with the hypertonia of the type seen in Parkinson's disease or in dystonia, the increased tone is modified with L-Dopa. Thus, serotonin-deprived patients have hypotonia relieved by 5-HPT; catecholamine-deprived patients have hypertonia relieved by L-Dopa. This contrast suggests that one of the balances in the human central nervous system is the balance between serotonin and catecholamines (probably dopamine).

Finally, our results in the children suggest a theoretical formulation that is in marked contrast to many of the results observed in Parkinson's patients and in many studies with experimental animals. We are suggesting that in children dopamine is essentially an ergotrophic agent and that serotonin is a trophotrophic agent, to use the Hess distinctions.[5] This is the opposite of the Brodie hypothesis.[6] Our results suggest that when young human beings with amine-deprivation syndromes are moved gradually into physiologically normal ranges as judged by clinical examination, it appears that dopamine is a muscle-relaxing and sleep-inducing agent and that serotonin is an activity-increasing and muscular tonus-stimulating agent. An explanation of this fact is found in thinking of these agents as "neurotransmitters," perhaps such as acetylcholine. As we know from studies at the myoneural junction, if acetylcholine is either too high or too low it can cause similar blocking effects. Only at the correct

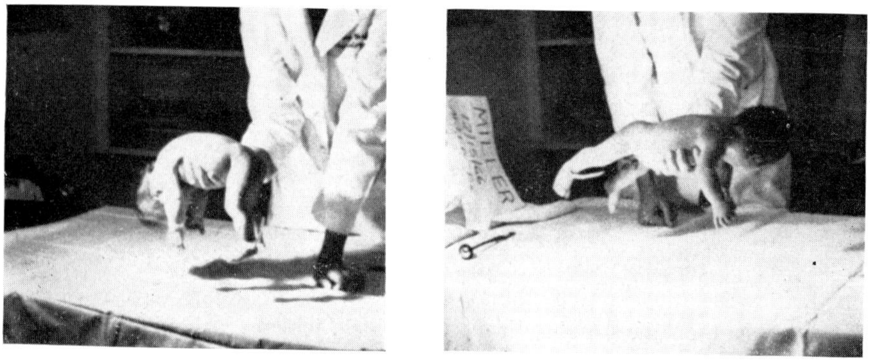

Fig. 3. A 4-day-old patient with Down's syndrome in the Landau posture before and 24 hours after the administration of 5-HTP, the precursor of 5-hydroxytryptamine.

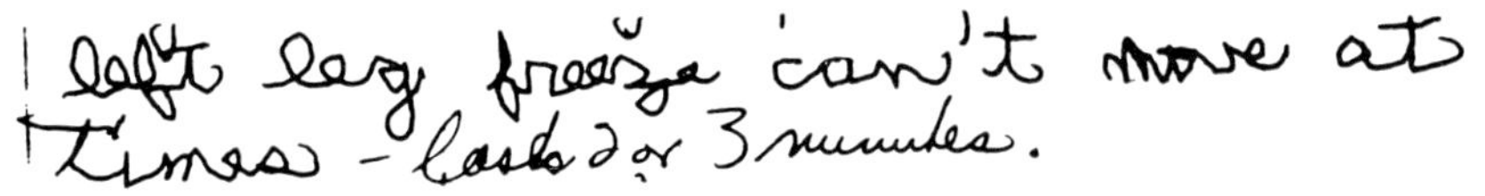

Fig. 4. A 10-year-old dystonia patient on L-Dopa reported her leg was temporarily "frozen," a word usually used by Parkinson patients.

physiological level does the transmitter effect of acetylcholine function well. It may very well be that the effects of dopamine and serotonin may be related to dosage effects. Our amine-deprived children who receive the precursor amino acids and improve may be the most accurate models we have to study the effects of normal physiological levels of these amines in human beings.

Figure 4 illustrates another contradiction to an already complex problem. Most physicians who have either operated on Parkinson's patients or treated them with L-Dopa have occasionally seen dystonia as a side effect. The statement on Figure 4 was written by an 8-year-old girl with dystonia receiving L-Dopa treatment; she called me up one day in complete amazement to report that a strange thing had happened to her foot. She uses the same word—frozen—as the parkinsonian patient. This Parkinson-like symptom was relieved by dropping the dose to 5 mg per kg for the next 48 hours.

REFERENCES

1. Cotzias, G. C., Papavasiliou, R. S., Gellene, R.: Modification of parkinsonianism—chronic treatment with L-Dopa. New Eng. J. Med. *280:* 337, 1969.
2. Bunney, W. E., Janowsky, D. S., Goodwin, F. K., Davis, J. M., Brodie, H. K. H., Murphy, D. L., Chase, T. N.: Effect of L-Dopa on depression. Lancet *1:* 885, 1969.
3. Klerman, G. L., Schildkraut, J. J., Hasenbush, L. L.: Clinical experiences with dihydroxyphenylalanine (Dopa) in depression. J. Psychiat. Res. *1:* 289, 1963.
4. Wagshul, A. M., and Daroff, R. B.: Depression during L-Dopa treatment. Lancet *2:* 592, 1969.
5. Hess, W. R.: Das Zwischenhian. Benno Schwabe, Basel, 1945.
6. Brodie, B. B., and Shore, P. A.: A concept for the role of serotonin and norepinephrine as chemical mediators in the brain. Ann. N.Y. Acad. Sci. *66:* 631, 1957.

BIOCHEMISTRY OF DEPRESSION IN PARKINSON'S DISEASE

W. Birkmayer

In a series of 100 postencephalitic parkinsonian patients verified at autopsy, 25 percent had one or more depressive episodes, while none was present in

90 patients with idiopathic or vascular causes of the illness. Previous studies of these patients from our group [1, 2] showed decreases in brain serotonin and norepinephrine in addition to the well-known deficit in dopamine. However, administrations of 5-hydroxytryptophan (5-HTP) and/or 3,4-hydroxyphenylserine (DOPS) were clinically without effect. Chronic administration of monoamine oxidase inhibitors [2] produced increases in the concentration of both serotonin and norepinephrine, but not of dopamine. It was our opinion that deficiencies in norepinephrine or serotonin could be the cause of some transient effects such as hyperthermia or depressive crises. The studies of Schildkraut [3] and of Ashcroft [4] covering catecholamine and tryptophan metabolism in depressive disorders led to our own investigations. Since both the above authors had found a decrease in normetanephrine and 5-hydroxyindoleacetic acid in the cerebrospinal fluid of depressed patients, it seemed unlikely that an accelerated turnover of the metabolism of these amines was present. For that reason we studied the effect of the precursors tyrosine and tryptophan in 14 patients with endogenous depression and in 4 patients with depressive episodes in Parkinson's disease. In the depressed patients the values for tyrosine (at 8:00 a.m.) were found to be significantly lower than in 9 control subjects. The picture for tryptophan is not as clear. In half the cases the values were lower than normal; in the other half they were higher. In depressed patients, kynurenine excretion in the urine was found to be lower than normal.

A further finding has been the inversion in depressed patients of the normal circadian rhythms in the levels of tyrosine and tryptophan (first demonstrated by Wurtman and Coburn). Normal persons show a peak for both substances around noontime and a low at night. Depressed patients, however, show the highest levels of both substances during the night and the lowest in the morning.[5] We feel that deficiencies in the precursors may be responsible for the symptoms of depression, even in Parkinson's disease.

REFERENCES

1. Ehringer, N., and Hornykiewicz, O.: Verteilung von Noradrenalin und Dopamin im Gehirn des Menschen und ihr Verhalten bei Erkrankungen des extrapyramidalen Systems. Klin. Wschr. *38:* 1236, 1960.
2. Bernheimer, H., Birkmayer, W., and Hornykiewicz, O.: Zur Biochemie des Parkinson-Syndroms des Menschen. Klin. Wschr. *41:* 465, 1963.
3. Schildkraut, J., Gordon, A., and Durell, J.: Catecholamine metabolism in affective disorders. J. Psychiat. Res. *3:* 213, 1965.
4. Ashcroft, G. W., Crawford, T. R. R., Eccleston, D., Sharman, D. F., and Dougall, E. J.: 5-Hydroxyindole compounds in the cerebrospinal fluid of patients with psychiatric or neurological diseases. Lancet *2:* 1049, 1966.
5. Birkmayer, W., Neumayer, E., Stockl, W., and Weilir, G.: Defizit essentieller Aminosäuren bei endogener Depression. Wien. Klin. Wschr. *45:* 832, 1968.

THE EFFECTS OF DOPA ON MOOD AND BEHAVIOR

Jonathan O. Cole

Dopa can currently be accused of threatening to bring psychiatry and neurology back together again. If one assumes that L-Dopa is converted to dopamine and norepinephrine in the brain and that it is an effective treatment for parkinsonism, L-Dopa ties itself nicely into the two major problems of psychiatry: schizophrenia and depression.

The only clearly efficacious treatments for schizophrenia are neuroleptic antipsychotic drugs such as the phenothiazines, the butyrophenones and, less clearly, reserpine. All these agents cause parkinsonism as a side effect which cannot be clearly separated from their main clinical action, and are suspected of blocking the effects of dopamine at the synapse.[1]

The abnormal movements seen when L-Dopa is used in the treatment of parkinsonism also have a parallel in schizophrenia. Grossly similar abnormal movements of the mouth, tongue and face, and the extremities also occur in psychiatric patients [2] as a presumed late effect of phenothiazine treatments. These are generally called "tardive dyskinesias." Their basic etiology is unclear. However, they can be suppressed by phenothiazines and persist for months or longer after phenothiazine treatment is stopped. It has been suggested that they are caused by a phenothiazine-induced increase in the sensitivity of dopaminergic synapses to dopamine, since they are, in some ways, the opposite of drug-induced parkinsonism.

Current biological theories about severe depression postulate a lack of biogenic amines (presumably norepinephrine but probably also dopamine) in the brain.[3, 4] Both the tricyclic antidepressants and the monoamine oxidase (MAO) inhibitors act to increase the availability of norepinephrine at the synapse. With the tricyclics there is reason to believe that dopamine levels are not affected [3] but with the MAOIs dopamine levels are also raised. However, since dopamine is a precursor of norepinephrine, Dopa administration should raise both dopamine and norepinephrine levels in the brain.

Having thus done something of which I generally disapprove—that is, engaged in naive pharmacological theorizing which I like to call metapsychopharmacology—I will proceed to make predictions from the above theories: (1) Dopa should alleviate depression. (2) Dopa should aggravate schizophrenia. What evidence is there in the literature to back either hypothesis?

Dopa has been studied in depression for the obvious reasons noted above. Unfortunately most of the work was done before L-Dopa became relatively available and before it was realized that large doses were needed to affect the symptoms of parkinsonism.

Klerman et al.[5] used DL-Dopa in doses of 400 to 1200 mg daily for two to four weeks in severely depressed women. Seven carefully studied patients were found to show no improvement as compared to a group receiving a placebo. There was a trend toward aggravation of psychopathology. In the same study, 120 to 800 mg of DL-Dopa were added to 45 to 60 mg phenelzine a day for 8 to 91 days in five patients, again with no improvement. Elevation of blood pressure did occur.

Pare [6] gave DL-Dopa, intravenously in doses of 25 mg going up to 275 mg over a 48-hour period, with or without iproniazide. He observed no improvement in depressed patients. Ingvarsson,[7] however, obtained marked improvement in depressed patients on doses of 25 to 40 mg of L-Dopa a day, intravenously; improvement lasted 24 to 48 hours. Matussek,[8] using intravenous doses of 25 to 100 mg L-Dopa, also observed stimulation, increased spontaneity, improved facial expression, and increased verbal activity in six retarded depressions. Turner and Merlis [9] gave Dopa (presumably DL-Dopa in doses of 25 to 50 mg t.i.d. orally in combination with an MAO inhibitor, pargyline 25 or 50 mg t.i.d.) postoperatively to 13 chronically hospitalized psychotic patients with depressive symptoms. Those with clear depression showed some improvement. Others developed increased schizophrenic disorganization or delirium, plus sleeplessness. The investigators felt that the mood elevation was more lasting than that sometimes obtained with amphetamines, even though the most severe effects might be said to resemble those of amphetamine.

More recently Bunney [10] has reported limited success with L-Dopa in depressed patients, although success has not been obtained in all patients treated. The single published successful case received up to 7 gm a day orally and appeared to maintain improvement only in doses higher than 3 gm a day. This patient had a retarded depression. Bunney et al. tried L-Dopa on three agitated depressions without success.

I have found only one real study of L-Dopa in schizophrenia, although the Merlis study could be taken as very weak evidence for aggravation of schizophrenia. Since his Dopa effect was confounded by concomitant use of an MAO inhibitor, it makes any firm conclusion impossible; MAO inhibitors themselves can cause exacerbations in schizophrenia, as can tricyclic antidepressants.

Pollin et al.[11] did administer up to 108 mg of L-Dopa intravenously to a small number of chronic schizophrenics but saw no change in psychopathology.

L-Dopa in Parkinsonism

Based on the above thinking, one should see alleviation of depression and occasional precipitation of schizophrenic-like states when L-Dopa is used to treat parkinsonism. Unfortunately, the available literature reports psychic reactions in limited detail. Yahr et al.[12] report increased vigor, feeling of well-being, increased freedom and spontaneity of movements and greater alertness, and go on to add that L-Dopa "unmasked" a dementia not previously apparent and caused marked agitation in seven patients. Two patients had agitation, nocturnal hallucinations and insomnia and were called toxic deliria.

Cotzias [13] reports some "awakening effect" occurring sometimes before improvement in motor symptoms. One patient showed "irritability, anger, hostility and paranoid misinterpretation" which were dose related. Insomnia appears to be a common side effect. Duvoisin [14] also describes a case of paranoid psychosis.

Wagshul and Daroff [15] describe clear depression occurring in 2 ot 125 patients treated with L-Dopa, one at 3 gm a day, the other at 6 gm. In both cases, with the addition of amitriptyline, the depression cleared.

In an informal conversation about his experiences with L-Dopa, Timberlake [16] reported to me a 10 to 15 percent rate of significant psychological side effects including:

1. Severe anxiety or apprehension, possibly dose related, occurring just before or just after abnormal movements also appear.
2. Depression.
3. Overactive playfulness with poor judgment, possibly hypomania.
4. Paranoid states.
5. Deliria.

From even more informal talks with others who have attended meetings on L-Dopa and parkinsonism, I gather such variegated effects are not common.*

Having gone through this very limited review, I can conclude little. L-Dopa probably relieves depression and causes mild euphoria or even hypomania, but it also can probably produce depression. It can, rarely, produce a schizophrenic-like picture, but it also presumably causes anxiety and organic delirium in some patients.

Although the ongoing collaborative study, which has a placebo control, may help elucidate matters, it is hard to disentangle real L-Dopa effects from those attributable to the anticipation of a magic treatment by patients or to the major changes in mood, thinking and interpersonal relationships which must occur when the symptoms of a highly incapacitating disease are removed.

I am strongly reminded of the psychological sequelae seen with the use of ACTH and cortisone in the treatment of chronic arthritis. Here, too, chronic disabling symptoms were removed and a variety of psychological sequelae were observed. Here also an "alerting" mood elevating effect was observed even in some patients who showed no change in physical symptoms,[17] but a smaller proportion of patients showed increased anxiety, depression or transient toxic deliria or a paranoid psychosis or hypomanic reactions. Malitz et al.[18] described two studies of ACTH or cortisone in patients with less severe illness (frost bite and hepatitis) in whom almost no psychological effects were noted.

Rome and Braceland [19] classify reactions to cortisone and ACTH into four groups:

I. Buoyant well-being with some acceleration of thinking.

* From the presentations and informal discussions held during the conference at Val David, Quebec, I learned two further points of importance. One was that organic confusional states were perhaps the commonest form of serious behavioral toxicity—a fact perhaps attributable to the high average age (ca. 60 to 65) of the patients treated. The second was the occurrence of increased sex drive in some patients.

II. Effusive and expansive statements of well-being, restlessness, insomnia, with poor judgment, hypomania.
III. Inappropriate anxiety and/or depression, regression in behavior.
IV. Gross psychotic reactions with a variety of sorts, usually of brief duration.

They report that about 60 percent of patients on cortisone or ACTH show responses I or II, with 25 to 30 percent showing response III and about 10 percent response IV, basing these figures on their total experience at the Mayo Clinic. They believe patients showing responses III or IV had abnormal personalities before therapy and were therefore predisposed to adverse effects.

All the above rather gross reports seem to me quite similar to results seen with L-Dopa. If one could be sure that cortisone and ACTH had only peripheral effects, one could blandly assume that patients with chronic severe illness show abnormal psychological responses when the illness is alleviated. However, this is presumably not the case [20] and cortisone, like L-Dopa, is entwined in depression in a complex way, since severely depressed patients show very high adrenocortical activity.[21]

One must assume for both cortisone and L-Dopa some undeterminate mix of drug effects and of the response of the individual to either the drug effects or the psychosocial situation in which the drug is given. The rather varied psychological responses to both drugs must, similarly, be attributed to an unknown mix of differences in brain structure and function, underlying personality structure and, quite possibly at least for L-Dopa, quantitative differences in the effects of exogenous L-Dopa in the brain.

It seems possible, but not highly likely, that some combination of a more detailed study of the pre-illness and pretreatment mental status and psychodynamics of parkinsonism patients and of the biochemical sequelae of L-Dopa might enable one to predict and/or explain the type of psychological response patients show to L-Dopa. Better description and documentation of such effects are worthwhile first steps.

REFERENCES

1. Pletscher, A.: Pharmacological changes of the dopamine metabolism in the basal ganglia. *In* Psychotropic Drugs and Dysfunctions of the Basal Ganglia, (Crane, G. E. and Gardner, R., eds.). Public Health Service Publication No. 1938, Government Printing Office, Washington, 1969.
2. Crane, G. E.: Tardive dyskinesia in patients treated with major neuroleptics: A review of the literature. Amer. J. Psychiat. *124:* 40, 1968.
3. Schildkraut, J. J.: Neuropsychopharmacology and the affective disorders. New Eng. J. Med. *281:* 197; 248; 302, 1969.
4. Bunney, W. E., Jr., and Davis, J. M.: Norepinephrine in depressive reactions: review. Arch. Gen. Psychiat. *13:* 483, 1965.
5. Klerman, G. L., et al.: Clinical experience with dihydroxyphenylalanine (Dopa) in depression. J. Psychiat. Res. *1:* 289, 1963.
6. Pare, C. M. B., and Sandler, M.: Clinical and biochemical study of trial of iproniazid in treatment of depression. J. Neurol. Neurosurg. Psychiat. *22:* 247, 1959.
7. Ingvarsson, C. G.: Orientierende klinische Versuche zur Wirkung des

Dioxyphenylalanins (l-dopa) bei endogener Depression. Arzneimittel-Forsch. *15:* 849, 1965.
8. MATUSSEK, N., POHLMEIER, H., and RUTHER, E.: Die Wirkung von Dopa auf gehemmte Depressionen. Klin. Wschr. *44:* 727, 1966.
9. TURNER, W. J., and MERLIS, S.: Clinical trial of pargyline and Dopa in psychotic subjects. Dis. Nerv. Syst. *25:* 538, 1964.
10. BUNNY, W. E., ET AL.: Effect of L-Dopa on depression. Lancet *1:* 885, 1969.
11. POLLIN, W., CARDON, P. V., JR., and KETY, S. S.: Effects of amino acid feedings in schizophrenic patients treated with iproniazid. Science *133:* 104, 1961.
12. YAHR, M., ET AL.: Treatment of parkinsonism with levodopa. Arch. Neurol. *21:* 343, 1969.
13. COTZIAS, G. C., ET AL.: Modification of parkinsonism: Chronic treatment with l-dopa. New Eng. J. Med. *280:* 337, 1969.
14. DUVOISIN, R. C.: Pharmacological treatment of basal ganglia disorders. *In* Psychotropic Drugs and Dysfunction of the Basal Ganglia (Crane, G. E. and Gardner, R., eds.). Public Health Service Publication No. 1938, Government Printing Office, Washington, 1969.
15. WAGSHUL, A. M., and DAROFF, R. B.: Depression during L-Dopa treatment. Lancet *2:* 592, 1969.
16. TIMBERLAKE, W. H.: Personal communication.
17. REES, L.: Psychological concomitants of cortisone and ACTH therapy. J. Ment. Sci. *99:* 497, 1953.
18. MALITZ, S., HAMBURG, D. A., and MODELL, S.: Effects of ACTH on mental function. J. Nerv. Ment. Dis. *118:* 315, 1953.
19. ROME, H. P., and BRACELAND, F. J.: The psychological response to ACTH, cortisone, hydrocortisone and related steroid substances. Amer. J. Psychiat. *108:* 641, 1952.
20. FAWCETT, J. A., and BUNNEY, W. E.: Pituitary adrenal function and depression: an outline for research. Arch. Gen. Psychiat. *16:* 517, 1967.
21. BUNNEY, W. E., FAWCETT, J. A., DAVIS, J. M., and GIFFORD, S.: Further evaluation of urinary 17-hydroxycorticosteroids in suicidal patients. Arch. Gen. Psychiat. *21:* 138, 1969.

CHAPTER 11

Biochemical Implications of Effects of L-Dopa on Behavior and Mentation

SOME ASPECTS OF THE PHARMACOLOGY OF THE PSYCHIC AND BEHAVIORAL EFFECTS OF L-3,4-DIHYDROXY-PHENYLALANINE (L-DOPA)

William G. Clark

Psychic Effects of L-Dopa

During treatment with L-Dopa in Parkinson's disease, the occurrence of pronounced psychic and behavioral effects, both bad and good, are not uncommon. This has been discussed by others. A scan of the world's literature shows that the following effects have been reported: severe hyperactive confusional states, delusions, hallucinations, nocturnal illusions, delirium, vivid dreaming, insomnia, fear, apprehension, agitation, irritability, anger and hostility which is paranoid at times, drowsiness and depression, increased affective tonus and attentiveness (alerting, arousal), increased contact and emotional participation when it was previously lacking, antidepressant action, euphoria, and subjective experiences centered around kinesthetic mental images which appear to be psychiatrically correlated with the hyperkinetic effect of Dopa. In normal subjects, as discussed by Ansel elsewhere in this text, there were no psychological effects seen in four female individuals given up to 5 gm daily for up to a month, which may suggest that the psychic effects reported in patients with Parkinson's disease, other dyskinesias and mental depression might require brain damage as a substrate. This will not be settled, however, until either controls or patients in which brain damage has been ruled out have received Dopa for several months.

Behavioral-EEG Arousal

The behavioral and cortical and subcortical electroencephalographic (EEG) arousal effects of L-Dopa, as well as its neurophysiological effects on reflex and neuronal excitation in the spinal cord, single unit neurons of the brain, etc., have been well reviewed by Hornykiewicz [1] and considered at this symposium by others, especially by Cordeau, Andén, Krnjevic, and Jasper (in discussion of Krnjevic).

L-Dopa induces neocortical and limbic activation and hippocampal EEG arousal in cats. Pretreatment with monoamine oxidase inhibitor (MAOI) lowers the threshold and prolongs this effect. Large lesions of the hypothalamus, leaving the septum intact, abolish it, but destruction of the caudate nucleus does not. In cats with the caudal midbrain transected, bilateral caudate lesions block the neocortical activation effect. Chlorpromazine (CPZ) also blocks it but does not block the responses to electrical stimulation of the hypothalamus or the EEG activation caused by cholinergic or anticholinesterase agents. Disulfiram has no effect (but see Thut and Rech [2]). This evidence suggests that there are dopaminergic links in the hypothalamus to the hippocampus and in the caudate to the neocortex which are blocked by CPZ.[3]

The Dopa activation of the EEG in rabbits is also observed to be blocked by CPZ, as well as by morphine, but not by haloperidol or reserpine. There is no correlation between the blocking doses and those which inhibit conditioned reflexes.[4]

L-Dopa given after tetrabenazine causes a sharp fall in the threshold voltage for EEG arousal and a concomitant marked rise in brain dopamine (3, 4-dihydroxyphenethylamine, DA), both of which show a close time correlation. This is additional evidence that DA has a facilitating effect on the EEG brain-stem reticular-activating (arousal) system. This work differs from earlier studies in that the latter concerned EEG changes due to supranormal amine levels, since no amine-depleting agent was given.[5]

Further support for a correlation between EEG activation by Dopa (and thus increased brain levels of DA) and behavior was shown by Thut and Rech,[2] who measured EEG and DA brain levels and depression of rotarod performance following treatment with L-Dopa and diethyldithiocarbamate, which blocks DA-β-oxidase and thus the formation of norepinephrine.

In chicks, which lack a blood-brain barrier, Dopa, DA and amphetamine given peripherally produce EEG arousal patterns but also akinetic behavioral responses, while epinephrine (E) and norepinephrine (NE) cause EEG and behavioral patterns that are sleep-like, unlike those seen in adult animals. The arousal EEG seen in the latter are probably related to cerebrovascular effects, but the doses used in the chicks caused no blood pressure-related EEG and behavioral changes.[6] This is supported by the observation that intravenous administration of 3, 4-dihydroxyphenylserine (DOPS), which is decarboxylated directly to NE, produces EEG and behavioral sleep in adult rats.[7] Blaschko and Cruściel [8] had shown earlier that Dopa but not DOPS causes arousal in reserpinized mice. Although reports on a soporific effect of E and NE on the brain prevail, the question is by no means settled; this may depend upon which areas of the brain are involved. For example, if NE is injected into the anterior hypothalamus of rats a marked hyperactivity occurs; this is not seen when the injection is made intracisternally.[9]

Microphoretically applied to single units, DA inhibits discharge activity of most neurons in the corpus striatum.[10]

Physostigmine and tetrabenazine given by tail vein in rats cause rigidity, increase α-reflex discharges from ventral roots of the spinal cord, shorten their latency, and decrease γ-reflex discharges. Monoaminergic agents, including Dopa, and cholinolytic agents inhibit rigidity and effect on α-motor activity. This supports the concept that the Parkinson-like rigidity in the rat results from hyperactivity of the α-motor system due to an imbalance between monoaminergic and cholinergic mechanisms in the brain.[11]

Dopa and 5-hydroxytryptophan (5-HTP), the precursor of serotonin (5-HT), modify specifically the rhythmic efferent bursts from two antagonistic hind limb nerves (flexor and extensor) of rabbits, which occur spontaneously or after electrical stimulation of the legs, and simultaneously in both legs, as would be expected from the locomotion pattern. Dopa increases extensor activity while 5-HTP increases flexor activity. It is possible that activation of flexors and extensors in *stereotyped locomotion* (see below) depends on different neurochemical systems, closely controlled by supraspinal mechanisms in the brain.[12, 13]

Psychomotor Stimulating Effects

This subject has been well reviewed by Hornykiewicz[1] through 1965, including the extensive literature on the antireserpine effect of Dopa, and the reader is referred to this and to the review by Carlsson.[14] A marked stimulation of locomotor activity has been related to brain DA, and is characteristically different in different species. Dopa and DA alone, however, depress motor activity in rats probably because of the marked peripheral autonomic effects. When a peripherally acting Dopa decarboxylase is given before Dopa, these autonomic effects are blocked and motor activity is increased due to the central action of Dopa and DA formed from it.[15, 16]

Conditioned Avoidance Response (CAR)

Butcher and Engel[15, 16] have reviewed the effects of Dopa on the conditioned avoidance response (CAR) in several species of animals. In their own experiments they found that 150 mg per kg L-Dopa intraperitoneally was lethal to rats maintained on a CAR schedule, but that if given in even higher doses after pretreatment of a peripherally acting Dopa-decarboxylase inhibitor Dopa caused an increase in the rate of avoidance responding, without the autonomic responses seen in the absence of the inhibitor. This was correlated with a marked increase (7 to 8x) in brain DA.

Butcher and Andén [17] extended this by observing that tetrabenazine abolished lever-pressing in rats maintained on a variable interval schedule of food reinforcement which was correlated with a decrease in brain DA and, to a less extent, decreases in NE and 5-HT levels. Apomorphine, which stimulates DA receptors, decreased the response rate when given alone, partially restored the rate after tetrabenazine, and amphetamine was even more effective. Apomorphine also decreased the rate of decrease of brain DA, but not of NE, after pretreatment with α-methyltyrosine-N-methyl ester, which inhibits CA biosynthesis by blocking the enzyme system tyrosine hydroxylation, the rate-limiting step in CA biosynthesis. This suggests that apomorphine acts by stimulating DA receptors. Their collaborators, Carlsson and Lindquist,[18] already had emphasized the importance of DA for maintaining normal exploratory activity and motor activity. If, however, dopamine-β-hydroxylase, the enzyme which converts DA to NE, is blocked by disulfiram, the effectiveness of L-Dopa in reversing reserpine-induced suppression of CAR in mice is reduced.[19] This suggests that both DA and NE play roles in maintaining CAR. Fuxe and Hanson [20] drew the same conclusion from their finding that a tyrosine hydroxylase inhibitor decreased the rate of CAR in rats maintained on a discriminated avoidance schedule, and this was correlated with a decrease in brain NE.

Taken together with Butcher and Andén's [17] findings, it is indicated that both DA and NE are necessary for the effective performance of some types of conditioned behavior, and that there are possible differences in the relative importance of DA and NE for unconditioned motor activity and conditioned lever-pressing behavior.

It has been observed by Scheckel, Boff and Pazery,[21] confirmed and extended by Butcher et al., that, in rats trained to press a lever to avoid foot shock, high doses of Dopa had a marked depressant effect both on CAR and on motor activity and that lower doses without behavioral effects became depressant by pretreatment with MAOI. If the increase in peripheral DA was blocked by Dopa-decarboxylase inhibitors (Dopa-D-I), the Dopa became stimulating in several species, including monkeys. Nonstimulating or depressant doses of Dopa plus the inhibitor became stimulating on CAR when DOPS was added. Inhibition of DA-β-hydroxylase prevented the stimulation by Dopa+inhibitor. Methionine, a methyl donor, given after Dopa+Dopa-D-I caused marked stimulation, indicating the involvement of a methylated catecholamine (CA). Pyrogallol, a potent CA-O-methyl transferase (COMT) inhibitor, blocked the stimulation by Dopa+Dopa-D-I. All these data suggest that the stimulant could be normetanephrine. When Dopa decarboxylation is blocked, Dopa is converted to 3,4-dihydroxyphenylpyruvic acid (DOPPA) by a reversible transaminase enzyme system which might be involved in the stimulation. This is supported by blocking the transamination with cycloserine, which inhibits stimulation. Injection of DOPPA now restores stimulation. That the transamination can occur was established by us some years ago.

Scheckel et al.[21] went on to propose that Dopa and DOPPA condense to form a Schiff base peripherally; this cyclizes and then enters the brain where it is decarboxylated, O-methylated and hydroxylated. Since the CA releaser, tetrabenazine, causes stimulation of CAR in rats treated with Dopa-D-I plus doses of Dopa which otherwise are ineffective, it also was proposed that the hypothetical stimulatory tetrahydropapaveroline-like complex is stored in neuronal CA vesicles.

That such cyclizations can occur was shown by us [22] when we demonstrated that pyridoxal-5′-phosphate (and other aldehydes, thus presumably DOPPA) cyclized nonenzymatically with Dopa and some other amino acids and amines to form tetrahydroisoquinolines.*

Scheckel et al.[21] showed that no stimulation occurred in Dopa + Dopa-D-I treated rats depleted of 5-HT by *p*-chlorophenylalanine, which inhibits tryptophan hydroxylase, but did if Dopa then is given, suggesting that 5-HT must be present. Dopa + Dopa-D-I cause a marked decrease in brain 5-HT; thus it was further suggested that the CA formed by Dopa + Dopa-D-I releases 5-HT, and stimulation of behavior is dependent upon this released 5-HT.

The objection to this provocative scheme is that the proposed cyclized Schiff base probably is not a substrate—or a very poor one—of Dopa-D-I. Furthermore, in vitro studies must be made to see if its decarboxylated form also is a substrate of DA-β-oxidase and COMT.

Of considerable interest are some recent theoretical considerations resulting from a study of structure-activity relationships of one-ring hallucinogenic (psychotomimetic) phenethylamine analogues of amphetamine, DA and mescaline.[23] This was an extensive study of newly synthesized substituted psychotomimetic phenethylamines tested in human subjects. Including results from previous studies, 44 structures were tabulated with activity expressed in "mescaline units" (M.U.) as the reference compound. From structural-activity considerations and known metabolic pathways and enzyme systems, and by following an argument analogous to that applied to 3,4-dimethoxyphenethylamine (DMPEA, "pink spot," discussed by Barbeau and others), Shulgin et al. proposed that 2-hydroxy-4,5-dimethoxyphenethanolamine could be a possible endogenous psychotogenic endotoxin which might be present in the brain and involved in psychotic states. Since this and related compounds are DA and amphetamine analogues, they also could be involved in behavioral effects, including stereotypy, rage and other bizarre behavioral states, seen after giving exogenous Dopa, amphetamine, and related agents.

Brain Self-Stimulation

Eiduson and Geller [24] studied effects of Dopa after MAOI in rats self-stimulating through electrodes chronically implanted in various parts of the hypothalamus and septum. Low doses of L-Dopa, 25 mg per kg given intraperitoneally, killed the rats *when* they then self-stimulated. The lethal symptoms included hemorrhagic pulmonary edema. The cause is not known,

---

* It might be possible that some such mechanism, if it occurs in vivo, might explain why pyridoxine administration decreased the effectiveness of Dopa therapy in Parkinson's disease, as shown by Duvoisin, Yahr, and Cote in 1969.

but may be associated with increased excitation of the central nervous system (CNS) since it is well known that CNS excitation induced in several ways, including brain concussion which causes pulmonary edema, increases 5-HT toxicity and the increased brain DA may have potentiated this. It would be of interest to study the effect after peripherally acting Dopa-D-I, DOPS, and CNS-stimulants such as amphetamine and pentylenetetrazol, as well as to give these agents intraventricularly and to train the animals to self-administer the drugs intracerebrally, etc.

Wise and Stein [25] showed that self-stimulation in rats with electrodes chronically implanted in the median forebrain bundle was markedly reduced by disulfiram or diethyldithiocarbamate, which are DA-β-oxidase inhibitors. This blockade was removed by intraventricular injection of NE, but not DA and 5-HT; thus, the rewarding effect of self-stimulation of the median forebrain bundle may require NE as a transmitter but not DA or 5-HT. DA has not, however, been ruled out, because only the intraventricular route was used and no other pharmacological manipulations were made.

"Stereotyped" Behavior, Rage, and Aggression

Randrup and Munkvad [26] summarized the literature to 1967–68, including their own publication on these phenomena. Stereotypy (tendency toward bizarre, frequent, prolonged, almost mechanical or "organized" aimless repetition or perseveration of the same movements, postures, acts or individual or social behaviors) is a qualitative phenomenon elicited in animals (and man) by central excitation by drugs, pain, fear, sex, electrical, auditory and other types of stimulation, by isolation stress, and by lesions or local injections of drugs in various brain areas. In some respects "rage," fighting, killing and aggressive behavior also elicited by similar manipulations, can be separately treated because different brain areas seem involved. Everett and Wiegand [27] emphasized that such behavior should be distinguished from motor activity alone, which can be quantitated by various recording devices. These types of behavior are characteristically and specifically different in different species. Stereotypies caused by amphetamine occur in all mammals and birds tested. "Rage" or aggressive reactions are seen after large doses of stimulant drugs such as amphetamine and other analeptic amines, MAOI followed by Dopa in male rats, especially after a peripherally acting Dopa-D-I followed by Dopa, and a noise stimulus.[28] In the latter case the rats rise to a mouth-to-mouth, immobile fighting position, synchronously pushing each other's front paws up and down, rhythmically moving their tongues and vocalizing but not biting. They may remain thus for 15 minutes or more until fatigued, after which they fall apart and assume a catatonic position. Amphetamines can produce the same effect but only briefly. A similar but briefer social interaction was observed in reserpinized male rats given amphetamine by Morpurgo and Theobald.[29] Females exhibited masculine sexual behavior.

MAOI followed by tricyclic antidepressants produces a vocalization rage reaction but not stereotypy in rats and mice and, in the latter, fighting also occurs (see Randrup and Munkvad [26]).

Amphetamine stereotypies, but not all rage reactions, are specifically inhibited by antipsychotic agents but not by barbiturates or antianxiety agents. Stereotypies produced by MAOI plus Dopa, or imipramine, amphetamine and apomorphine, Dopa plus peripherally acting DOPA-D-I, etc., are related to brain DA and its turnover, while the drug-produced rage reactions may be related to brain NE. The effects of brain DA and NE are probably in homeostasis with effects of other amines since, for example, some stereotypies can be enhanced by anticholinergics and weakly antagonized by cholinergics; the addition of acetylcholine, 5-HT and tryptamine also influenced these stereotypies (for literature see Randrup and Munkvad [26]).

Both rage and stereotypies may play important roles in the actions of antipsychotic, antidepressant, antianxiety agents and lithium. Reference should be made to the review of Randrup and Munkvad for their possible relations to normal and psychotic behaviors in man. Other important reviews include a book on abnormal animal behavior,[30] a review on drugs and aggressiveness,[31] and the first international symposium on aggressive behavior, held in Milan.[32]

Sexual Behavior

The participation of biogenic amines, especially DA, in sexual behavior in both sexes and in maternal behavior has only recently gained attention. For reviews see Butcher, Butcher and Larsson [33] and Tagliamonte et al.[34] This interest was stimulated because of the accumulating evidence of the mediation of dopaminergic neurons in the tuberoinfundibular part of the hypothalamus on estrus, ovulation, pregnancy and lactation as well as in male sexual behavior. It is in this region of the brain, the median eminence, that DA nerve terminals are especially rich. Interest in these mechanisms was pioneered by Sawyer et al. who observed that reserpine lowers the levels of pituitary prolactin and induces lactation in ovariectomized, estrogen-primed rabbits, and blocks ovulation in the rat if given before the critical period of luteinizing hormone (LH) release. MAOI plus Dopa overcome the reserpine block in the non-reserpinized animals, cause luteinizing hormone release and increase uterine weight.[35, 36]

Ovarian hormones such as estradiol and progesterone, when given simultaneously, decrease NE and increase DA in the anterior hypothalamus of ovariectomized rats to levels found in normal estrous rats in which the NE is low.[37, 38]

Coppola et al.[39] found that depletors of CA, such as reserpine, α-methyldopa and tetrabenazine, induce pseudo-pregnancy in rats on the first day of diestrus, as confirmed, and fluorescence intensity of the small cells in the median eminence and certain fluorescent cells of the anterior hypophysis. Lichtensteiger,[40] who measured the fluorescence intensity of these cells by a microfluorimetric analytical technique, found increased fluorescence intensity during estrus.

Fuxe et al.[41, 42] extended these observations to include the determination of the turnover of DA in these neurons in the rat brain as well as their fluorescence, and found that small doses of testosterone and estrogen caused marked and selective increases of the DA neuronal terminals of the median eminence of both castrated males and females over that found at any stage of the estrus cycle. The amine turnover was unaffected. Castration had no such effect in males but blocked the changes in DA turnover during the estrus cycle.

Kamberi et al.,[43] using their elegant and sensitive method, found that pituitary halves incubated in pituitary stalk plasma release more LH than their opposite halves incubated in peripheral blood plasma. Several substances tried were inactive when added directly to the incubates in vitro.[44] Much more was released from stalk plasma from DA-treated rats than from control rats. The authors postulated that the secretion of luteinizing release factor (LRF) may be under control of median eminence neuronal DA.

Kordon and Glowinski [45] showed that depletion of brain CA by α-methyltyrosine blocks ovulation in gonadotrophin-treated immature rats, but only when given during a critical time of the day of proestrus, and that restoration of brain DA by Dopa treatment but not NE by *threo*-dihydroxyphenylserine treatment. This again suggests that a DA mechanism is involved in LH regulation. Lichtensteiger [40] demonstrated a functional relationship between different states of fluorescence intensity of these fibers and control of secretion of gonadotrophins. By a combination of histochemical fluorescence, biochemical analyses and isotope techniques, Fuxe and Hökfelt [41] obtained evidence to support their earlier suggestion that DA is released locally in the median eminence, from axon terminals which make axoaxonic contracts with neurons containing LRF, and *inhibits* its synthesis and/or release by way of presynaptic inhibition during pregnancy. In the same report, Fuxe and Hökfelt,[41] Lichtensteiger,[40] Glowinski and Kordon,[46] and also Van Loon and Ganong [47] and Rinne and Hyyppä [48] suggest that these DA (and possibly NE) neurons are involved in the control of other pituitary hormones (prolactin, growth hormone, and corticotrophic hormone [ACTH]).

Butcher, Butcher and Larsson [33] found that tetrabenazine abolishes male sexual behavior and that apomorphine, a DA receptor stimulator, *d*-amphetamine and nialanide antagonize this. They discuss this and implicate the central DA *receptors* in the median eminence.

Finally, it might be mentioned in passing, with respect to male sexual behavior, that DA is the predominant catecholamine in the sympathetic nerves of bovine penile erectile tissues and may play a role in erection.[49]

One might therefore look for effects of Dopa therapy on male and female sexual behavior.

In conclusion, no sexual behavioral effects were seen by Ansel (see elsewhere in this text) in normal human subjects, but this was no doubt due to their ages and to the fact that their spouses were undergoing Dopa therapy for Parkinson's disease.

Effects on Brain Carbohydrate Metabolism and Their Behavioral Implications

Tyce, elsewhere in this text, has reviewed this subject. Kakimoto et al. (cf. Tyce) previously have also studied it, and their work is reviewed by Tyce.

In considering the mechanisms and implications of the effects of Dopa on brain carbohydrate metabolism, Kakimoto et al., knowing that the concentration of glycogen is controlled by phosphorylase, measured it and found that the ratio of phosphorylase A to total phosphorylase was about normal. This probably was due to the rapid postmortem conversion of phosphorylase A to B. Since adenylcyclase is activated by CAs, they postulated that brain CAs formed from Dopa activate phosphorylase B kinase by the formation of adenyl-3′-5′-phosphate, thus explaining the effect of Dopa on glycogenolysis. Also, it is known that reserpine increases brain glycogen levels and decreases phosphorylase activity. Various chemical and physical stimuli, e.g., convulsants and electrical stimulation, stimulate central neurons, which may affect glycolysis by shifts of ions through neuronal membranes or by some change in metabolism; it is known that increased nerve activity increases adenyl-5-phosphate to levels which activate inactive phosphorylase B. The CAs could do the same thing, and thus explain the glycolysis possibly by feedback reactions caused by stimulation of nerve function by monoamines. The possible implications of this as a role of DA on the CNS and behavior are profound.

REFERENCES

1. Hornykiewicz, O.: Dopamine (3-hydroxytyramine) and brain function. Pharmacol. Rev. *18:* 925, 1966.
2. Thut, P. D., and Rech, R. H.: Electrocorticographic (EC_0G) and behavioral effects of diethyldithiocarbomate (DDC) and L-Dopa. Pharmacologist *11:* 254, 1969.
3. Kawamura, H., Hitzemann, R., and Domino, E. F.: Observations on the mechanism of neocortical and limbic EEG activation induced by Dopa. Pharmacologist *10:* 160, 1969.
4. Burov, Y. V., and Racvskii, K. S.: Effect of "major tranquilizers" on the EEG activation in awake rabbits, induced by DL-3,4-dihydroxyphenylalanine. Farmakol. Toksikol. *31:* 387, 1968. (In Russian with English summary.)
5. Takagi, H., Satol, M., Yamatsu, K., Kimura, K., and Nakama, M.: Central effects of 3,4-dihydroxyphenylalanine and 5-hydroxytryptophan on tetrabenazine pretreated rabbits, with special reference to the possible role of catecholamine and serotonin in the brain. Int. J. Neuropharmacol. *7:* 265, 1968.
6. Spooner, C. E., and Winters, W. D.: The influence of centrally active amine induced blood pressure changes on the electroencephalogram and behavior. Int. J. Neuropharmacol. *6:* 109, 1967.
7. Havliček, V.: The effect of dl-3,4-dihydroxyphenylalanine (precursor of noradrenaline) on the EC_0G of unrestrained rats. Int. J. Neuropharmacol. *6:* 83, 1967.
8. Blaschko, H., and Chruściel, T. L.: The decarboxylation of amino acids related to tyrosine and their awakening action in reserpine-treated mice. J. Physiol. *151:* 272, 1960.
9. Bankert, O.: Measurement of hyperactivity in rats in a dose-response curve after hypothalamic norepinephrine injection. Life Sci. *8:* 943, 1969.
10. Hertz, A., and Zieglgansberger, W.: The influence of microphoretically applied biogenic amines, cholinomimetics and procaine on synaptic excitation in the corpus striatum. Int. J. Neuropharmacol. *7:* 221, 1968.
11. Jurna, I., Theres, C., and Bachman, T.: The effect of physostigmine and tetrabenazine on spinal motor control and its inhibition by drugs which influ-

ence reserpine rigidity. Naunyn-Schmiedebergs Arch. Pharmakol. Exp. Pathol. *263:* 427, 1969.

12. VIALA, D., and BUSER, P.: Décharges efférentes rythmiques dans les pattes postérieures chez le lapin et leur mécanisme. J. Physiol. *7:* 287, 1965.
13. VIALA, D., and BUSER, P.: The effects of Dopa and 5HTP on rhythmic efferent discharges in hind limb nerves. Brain Res. *12:* 437, 1969.
14. CARLSSON, A.: 5-Hydroxytryptamine and related compounds. Handbook Exp. Pharmacol. *9:* 529, 1965.
15. BUTCHER, L. L., and ENGEL, J.: Behavioral and biochemical effects of L-Dopa after peripheral decarboxylase inhibition. Brain Res. *15:* 233, 1969.
16. BUTCHER, L. L., and ENGEL, J.: Peripheral factors in the mediation of the effects of L-Dopa on locomotor activity. J. Pharm. Pharmacol. *21:* 614, 1969.
17. BUTCHER, L. L., and ANDÉN, N. E.: Effects of apomorphine and amphetamine on schedule-controlled behavior: Reversal of tetrabenazine suppression and dopaminergic correlates. Europ. J. Pharmacol. *6:* 255, 1969.
18. CARLSSON, A., and LINDQVIST, M.: Metatyrosine as a tool for selective protection of catecholamine stores against reserpine. Europ. J. Pharmacol. *2:* 187, 1967.
19. SEIDEN, L. S., and PETERSON, D. D.: Blockade of L-Dopa reversal of reserpine-induced conditioned avoidance response suppression by disulfiram. J. Pharmacol. Exp. Ther. *163:* 84, 1968.
20. FUXE, K., and HANSON, L. C. F.: Central catecholamine neurons and conditioned avoidance behavior. Psychopharmacologia *11:* 439, 1967.
21. SCHECKEL, C. L., BOFF, E., and PAZERY, L. M.: Hyperactive states related to the metabolism of norepinephrine and similar biochemicals. Ann. N. Y. Acad. Sci. *159:* 939, 1969.
22. SCHOTT, H. F., and CLARK, W. G.: Dopa decarboxylase inhibition through the interaction of coenzyme and substrate. J. Biol. Chem. *196:* 449, 1952.
23. SHULGIN, A. T., SARGENT, T., and NARANJO, C.: Structure-activity relationships of one-ring psychotomimetics. Nature *221:* 537, 1969.
24. EIDUSON, S., and GELLER, E.: Personal communication.
25. WISE, C. D., and STEIN, L.: Facilitation of brain self-stimulation by central administration of norepinephrine. Science *163:* 299, 1969.
26. RANDRUP, A., and MUNKVAD, I.: Pharmacological studies on the brain mechanisms underlying two forms of behavioral excitation: stereotyped hyperactivity and "rage." Ann. N. Y. Acad. Sci. *159:* 928, 1969.
27. EVERETT, G. M., and WIEGAND, R. G.: Central amines and behavioral states: A critique and new data. *In* Pharmacological Analysis of Central Nervous Action (Paton, W. D. M., and Lindgren, P., eds.). Proc. First Int. Pharmacol. Meet. *8:* 85, Pergamon Press, London, and Macmillan Company, New York, 1962.
28. LAMMERS, A. J. J. C., and VAN ROSSUM, J. M.: Bizarre social behavior in rats induced by a combination of a peripheral decarboxylase inhibitor and Dopa. Europ. J. Pharmacol. *5:* 103, 1968.
29. MORPURGO, C., and THEOBOLD, W.: Behavioral reactions to amphetamine in reserpinized rats. Int. J. Neuropharmacol. *5:* 375, 1966.
30. FOX, M. W., ed.: Abnormal Behavior in Animals. W. B. Saunders, Philadelphia, 1968.
31. VALZELLI, L.: Drugs and aggressiveness. Adv. Pharmacol. *5:* 79, 1967.
32. LÉONARDELLI, J.: Modifications des fibres et cellules monoaminergiques de la région infundibulo-tuberienne du cobaye au cours du cycle oestral. Compt. Rend. Soc. Biol. *162:* 1937, 1968.
33. BUTCHER, L. L., BUTCHER, S. G., and LARSSON, K.: Effects of apomorphine, (+)-amphetamine, and nialamide on tetrabenazine-induced suppression of sexual behavior in the male rat. Europ. J. Pharmacol. *7:* 283, 1969.

34. TAGLIAMONTE, A., TAGLIAMONTE, P., GESSA, G. L., and BRODIE, B. B.: Compulsive sexual behavior induced by p-chlorophenalanine in normal and pinealectomized male rats. Science *166:* 1433, 1969.
35. KANEMATSU, S., HILLIARD, J., and SAWYER, C. H.: Effect of reserpine on pituitary prolactin content and its hypothalamic site of action in the rabbit. Acta Endocrinol. *44:* 467, 1963.
36. MEYERSON, B. J., and SAWYER, C. H.: Monoamines and ovulation in the rat. Endocrinology *83:* 170, 1968.
37. DONOSO, A. O., and STEFANO, J. E.: Sex hormones and concentration of noradrenaline and dopamine in the anterior hypothalamus of castrated rats. Experientia *23:* 665, 1967.
38. DONOSO, A. O., and CUKIER, J. O.: Oestrogen as depressor of noradrenaline concentration in the anterior hypothalamus. Nature *218:* 269, 1968.
39. COPPOLA, J. A., LEONARDI, R. G., LIPPMANN, W., PERRINE, J. W., and RINGLER, I.: Induction of pseudopregnancy in rats by depletors of endogenous catecholamines. Endocrinology *77:* 485, 1965.
40. LICHTENSTEIGER, W.: Effects of endocrine manipulations on the metabolism of hypothalamic neurotransmitters. Abstr. 2nd Int. Meet. Int. Neurochem. Soc., Milan, September 1969.
41. FUXE, K., and HÖKFELT, T.: Participation of central monoaminergic mechanisms in the regulation of the hormone secretion from the pituitary gland. Abstr. 2nd Int. Meet. Int. Neurochem. Soc., Milan, September 1969.
42. FUXE, K., HÖKFELT, T., and NILSSON, O.: Castration, sex hormones, and tubero-infundibular dopamine neurons. Neuroendocrinology *5:* 107, 1969.
43. KAMBERI, I. A., MICAL, R. S., and PORTER, J. C.: Luteinizing hormone-releasing activity in hypophysial stalk blood and elevation by dopamine. Science *166:* 388, 1969; A possible role for dopamine in the stimulation of the release of the LH-releasing factor. Abstr. 2nd Meet. Soc. Study of Reproduction, Univ. Calif. at Davis, September 1969.
44. KAMBERI, I. A., and MCCANN, S. M.: Effect of biogenic amines, FSH-releasing factor (FRF) and other substances on the release of FSH by pituitaries incubated in vitro. Endocrinology *85:* 815, 1969.
45. KORDON, C., and GLOWINSKI, J.: Selective inhibition of superovulation by blockade of dopamine synthesis during the "critical period" in the immature rat. Endocrinology *85:* 924, 1969.
46. GLOWINSKI, J., and KORDON, C.: Role of catecholamines in the control of anterior pituitary function. Abstr. 2nd Int. Meet. Int. Neurochem. Soc., Milan, September 1969.
47. VAN LOON, G. R., and GANONG, W. F.: Effect of drugs which alter catecholamine metabolism on the inhibition of stress induced ACTH secretion produced by L-Dopa. Physiologist *12:* 381, 1969.
48. RINNE, U. K., and HYYPPÄ, M.: Effect of adrenalectomy on the ultrastructure and catecholamine fluorescence of the nerve ending in the median eminence of the rat. Abstr. 2nd Int. Meet. Int. Neurochem. Soc., Milan, September 1969.
49. PENTTILÄ, O., and VARTIANEN, A.: Acetylcholine, histamine, 5-hydroxytryptamine and catecholamine contents of mammalian penile erectile and urethral tissue. Acta Pharmacol. Toxicol. *21:* 145, 1964.

L-DOPA AND ABNORMAL METHYLATION *

André Barbeau

There is no doubt that during therapeutic trials with L-Dopa definite changes in behavior occur in a certain number of patients. These changes range from slight modifications of the personality (usually in the direction of more aggressiveness), to severe depressions and even to organic confusional syndromes with hallucinations and paranoid overtones. The presence of depressive episodes is in contrast to the more frequent alerting and "mood elevating" effects of L-Dopa. The question can certainly be raised as to whether this is a direct result of Dopa toxicity or a simple unmasking of an underlying psychosis.

It is difficult to find a proper approach to this problem on biochemical terms, because much is said while little is known about the biochemistry of neuroses and psychoses. One handle we may try to grasp is an old approach. For many years Smythies and co-workers have postulated a defect in transmethylation in schizophrenia.[1] This led to a search for abnormal methylated products, such as mescaline and, more recently, 3,4-dimethoxyphenylethylamine (DMPEA). Friedhoff and Van Winkle in 1962 renewed the interest in this aspect of the problem when they claimed to have isolated DMPEA in the urine of schizophrenic subjects.[2, 3] In 1963, a similar compound was found in Parkinson's disease.[4] For the next five years the controversy raged about these findings, marred by the unfortunate choice of the term "pink spot" to describe a chromatographic spot of specific Rf. By 1967 and the Second International Congress of Neuro-Genetics in Montreal,[5] it had become evident that at least 17 substances, ranging from cadaverine to drug metabolites, masqueraded as a so-called "pink spot" of the same Rf.[6, 7] The subject was almost forgotten, but in 1967 Creveling and Daly[8] produced mass spectrographic and other evidence for the actual presence of some DMPEA in the urine of schizophrenics. DMPAA, the acid metabolite of DMPEA, was also clearly identified in normal urine.[9] Thus there is no doubt that the substance is present in normal, parkinsonian and schizophrenic patients; however, some authors have claimed that it comes from a component of the diet (tea). Even if the quantities involved are not much different from normal, in some individuals DMPEA or a similar compound may remain in that state sufficiently long to produce abnormal symptoms. Another possibility, of course, is that the sensitivity or reactivity of some subjects to such substances may be increased.

There is now clear-cut evidence, from our own studies[10, 11] and those of others,[12, 13, 14] that DMPEA and mescaline can produce extrapyramidal akinesia or catatonia-like symptoms, as well as occasional tremors, when injected into animals.

* Supported by Grants MA-1967, MA-2530, and SP-1 from the Medical Research Council of Canada, and a grant from the Huntington's Chorea Foundation.

Previous reports from our laboratory [15, 16] demonstrated that DMPEA interferes with the metabolism of dopamine. This substance, which theoretically could itself derive from dopamine, would tend to produce a more rapid turnover of dopamine, thereby probably accounting for akinesia or catatonia. Our studies [17] showed also that DMPEA blocked the pressor action of norepinephrine to some degree but completely blocked the smaller pressor effect of dopamine. Similarly, DMPEA only partially blocked the re-uptake of tritiated norepinephrine in the isolated cat superior mesenteric area. With tritiated dopamine, however, the uptake was almost completely inhibited.[17]

Thus DMPEA, if it is present in abnormal quantities or in normal quantities for long periods, can cause direct interference with dopamine metabolism resulting in rapid turnover of dopamine to homovanillic acid (HVA). All this, of course, is theoretically important only if it can be demonstrated that a DMPEA-like substance is in abnormal quantities in specific brain areas. No one has yet produced this evidence but a series of recent observations may indicate that the DMPEA story is not closed. Dill and co-workers [18] have shown that direct injections of DMPEA or mescaline into the striatum of rats produce marked, sustained tremors which they have clearly recorded on films and electromyographic tracings. Thus DMPEA can produce, depending on the dose and site of action, either tremor or akinesia. Moreover, large quantities of 3-methoxytyramine, a normal metabolite, can also produce similar symptoms.[19]

When L-Dopa is given to human patients there is a considerable increase in the excretion of methylated derivatives: O-methyldopa, 3-methoxytyramine and HVA are found in very high concentrations. There is then considerable mobilization of methyl groups by the body. This may be sufficient to produce metabolic changes resulting in psychiatric manifestations, directly or through an effect upon various other brain amines. It is conceivable that in this massive mobilization of methyl groups some abnormal, albeit transient, formation of di- or trimethoxy derivatives of dopamine may happen. A study of transmethylases in patients receiving L-Dopa is now being undertaken to clarify this point.

The other side of the coin may also result in damage to the parkinsonian patient. Such a considerable loss in methyl groups will eventually become a strain to general metabolic cycles if the available methyl donors (cysteine, S-adenosyl-methionine) reach insufficient levels to carry out their other duties. This phenomenon may in part explain the loss of weight seen in some of our patients under long-term Dopa therapy.

In conclusion, I would like to ask a purely philosophical question which may eventually have a bearing on our problem: With long-term L-Dopa treatment, are we risking the transformation of some of our akinetic-rigid parkinsonian patients into hyperkinetic, hypercatabolic, occasionally demented choreics of the Huntington type?

Such a question is not meant as a scarecrow, but as an honest evaluation of some of the phenomena we are now observing. It is hoped that we will have the courage to face up to it and to try to give an equally honest answer.

REFERENCES

1. SMYTHIES, J. R.: Schizophrenia: Chemistry, Metabolism and Treatment. Charles C Thomas, Springfield, 1963.
2. FRIEDHOFF, A. J., and VANWINKLE, E.: The characteristics of an amine found in the urine of schizophrenic patients. J. Nerv. Ment. Dis. *135:* 550, 1962.
3. FRIEDHOFF, A. J., and VANWINKLE, E.: Isolation and characterization of a compound from the urine of schizophrenics. Nature *194:* 897, 1962.
4. BARBEAU, A., DEGROOT, J. A., JOLY, J. G., RAYMOND-TREMBLAY, D., and DONALDSON, J.: Urinary excretion of a 3,4-dimethoxyphenylethylamine-like substance in Parkinson's disease. Rev. Canad. Biol. *22:* 469, 1963.
5. BARBEAU, A.: The significance of the "pink spot": A panel. *In* Progress in Neuro-Genetics (A. Barbeau, and J. R. Brunette, eds.). Excerpta Medica Foundation, Amsterdam, 1969.
6. PUSHPATHADAM, J. J., and BARBEAU, A.: The "pink spot" and Parkinson's disease. *In* Progress in Neuro-Genetics (A. Barbeau, and J. R. Burnette, eds.). Excerpta Medica Foundation, Amsterdam, 1969.
7. BARBEAU, A.: The "pink spot," 3,4-dimethoxyphenylethylamine and dopamine. Relationship to Parkinson's disease and to schizophrenia. Rev. Canad. Biol. *26:* 55, 1967.
8. CREVELING, C. R., and DALY, J. W.: Mass spectrographic identification of biogenic amines as their 'dansyl' derivatives. Pharmacologist *9:* 185, 1967.
9. KUEHL, F. A., HICHENS, M., ORMOND, R. E., MEISINGER, M. A. P., GALE, P. H., CIRILLO, V. J., and BRINK, N. G.: Para-O-methylation of dopamine in schizophrenic and normal individuals. Nature *203:* 154, 1964.
10. BARBEAU, A., TETREAULT, L., OLIVA, L., MORAZAIN, L., and CARDIN, L.: The pharmacology of akinesia. Studies on 3,4-dimethoxyphenylethylamine. Nature *209:* 719, 1966.
11. BARBEAU, A., SINGH, P., and JOUBERT, M.: Effect of 3,4-dimethoxyphenylethylamine injections on catecholamine metabolism in rats and monkeys. Life Sci. *5:* 757, 1966.
12. ERNST, A. M.: Phenomena of the hypokinetic rigid type caused by O-methylation of dopamine in the para-position. Nature *193:* 178, 1962.
13. CARLINI, G. R. S., and CARLINI, E. A.: Inibicao do poder histaminolitico da diamino oxidase (DAO) pela papaverina, mescalina e homoveratrilimina. Ciencia Cult. *17:* 274, 1965.
14. BAKER, W. W., CONNOR, J. D., ROSSI, G. V., and LALLEY, P. M.: Production of tremor by intracaudate cholinergic agents and its suppression by locally administered catecholamines. *In* Progress in Neuro-Genetics (A. Barbeau, and J. R. Brunette, eds.). Excerpta Medica Foundation, Amsterdam, 1969.
15. BARBEAU, A., LESCOP, J., DUPLESSIS, P., and ELIE, R.: Effect of 3,4-dimethoxyphenylethylamine injections upon dopamine metabolism in rats and dogs. Experientia *23:* 536, 1967.
16. BARBEAU, A., SINGH, P., GAUDREAU, P., and JOUBERT, M.: Effect of 3,4-dimethoxyphenylethylamine injections on the concentration of catecholamines in the rat brain. Rev. Canad. Biol. *24:* 229, 1965.
17. ELIE, R., BARBEAU, A., BOURDOIS, P., and PANISSET, J. C.: Interaction of 3,4-dimethoxyphenylethylamine and dopamine at adrenergic nerve endings. *In* Progress in Neuro-Genetics (A. Barbeau, and J. R. Brunette, eds.). Excerpta Medica Foundation, Amsterdam, 1969.
18. LITTLE, M. D., and DILL, E.: Mescaline and other O-methylated β-phenylethylamines: intrastriatal induction of tremor in rats. Brain Res. *13:* 360, 1969.
19. DILL, R. E.: Personal communication, 1969.

PSYCHIATRIC ASPECTS OF L-DOPA TREATMENT

Jonathan O. Cole

In both the formal and informal discussion sections of this meeting, I have been intrigued by several aspects of the treatment of parkinsonian patients by L-Dopa from my psychiatric viewpoint.

First, I gather that a substantial number of parkinsonian patients show a good deal of neurological improvement with L-Dopa but fail to show any increase in socially affective behavior. They continue to be chronic invalids, and stay at home without enlarging their meaningful activities appreciably.

Two causes for this have been suggested by others, and I would like to add a third.

First, many patients with parkinsonism have, in addition, other neurological defects in their brains. Relieving the parkinsonism does not result in the patient's having a normal central nervous system. Chronic organic brain disease may interfere with his showing any real improvement in interpersonal or socially affective behaviors.

Second, the marital relationship in patients with parkinsonism can become highly rigid and highly specialized with the unaffected spouse taking over almost all functions for the ill spouse. When the neurological handicap is relieved with L-Dopa, the roles of the two spouses do not appear to shift appreciably; wives continue to spoon-feed husbands who are now quite capable of feeding themselves. If the patient begins to show more evidence of behavioral and social competence, the well spouse may become very anxious and upset. This is particularly intriguing from a psychiatric standpoint since one would assume that parkinsonism was not present when the couple were married. This pathological marital relationship must therefore be secondary to disease, rather than predictable at the time of marriage. Wives of alcoholic men show a similar disruption in their personalities when their alcoholic husbands become sober and well adjusted. In this case, however, psychiatrists usually assume that the wife married the man because she knew he was an alcoholic. This presumption of psychopathology in the well spouse may have to be reexamined.

My third point is that, from the standpoint of an institutional psychiatrist, some of the behavioral lacks one sees in parkinsonian patients whose neurological symptoms have improved look very much like the effects of institutionalization. One sees a similar thing in chronic schizophrenics who have been confined to a monotonous, boring and restricting environment for many years. The schizophrenic symptomatology has subsided substantially but the patient is left with many social defects resulting from institutionalization. Perhaps some of the social and behavioral lacks seen in parkinsonian patients are also a result of chronic institutionalization, although the institution may have been a single bedroom at home. A similar serious restriction of environmental stimuli and meaningful activities may have occurred. If this is true, a program of intensive resocialization using group techniques may be quite valuable.

This has worked well in chronic inert schizophrenics and may work equally well in chronic parkinsonian patients who are now relatively symptom free neurologically. A group from Northwestern University did in fact report that such a program worked well.

EVIDENCE FOR DOPAMINE AS A CENTRAL NEUROMODULATOR: NEUROLOGICAL AND BEHAVIORAL IMPLICATIONS

G. M. Everett

Although evidence for dopamine as a neuromodulator has been accumulating over a period of ten years, it is only now receiving wide recognition as an important neurohumor in the regulation of behavior and motor output.

Studies of the neurological and behavioral deficits resulting from the repletion of various neurohumors in the brain after treatment with reserpine have been the basis for the important developments in this area.

In the replacement studies of Carlsson et al. in rats and rabbits [1] and our work in mice and monkeys,[2, 3, 4] it was shown that Dopa can bring about a remarkable reversal of the neurological and behavioral deficits produced by reserpine. Carlsson first pointed out that increased brain levels of dopamine were better correlated with behavioral changes than were levels of norepinephrine. The latter is apparently more slowly formed and in smaller amounts.

The investigations of Hornykiewicz et al.[5] established the biochemical deficit of dopamine in the basal ganglia of Parkinson patients. Although some preliminary clinical tests with DL-Dopa in parkinsonism were conducted, the doses were generally inadequate and the therapeutic results equivocal.

The need for large doses of Dopa to produce changes in brain levels of dopamine and behavioral effects is clearly shown in many pharmacological studies. Despite early attempts made to rouse interest in Dopa as a replacement therapy, the neurologists were generally indifferent or skeptical.

The studies of Cotzias [6] using large and adequate doses of Dopa in the treatment of parkinsonism reopened this area to clinical utility and effective clinical response, especially on akinesia and rigidity.

The relative importance of norepinephrine versus dopamine in various behavioral and neurological states is at present a major area of controversy that cannot be completely resolved at this time.

For this discussion we offer one point of view. It will be highly controversial. The studies supporting these ideas come from many sources and seem particularly compelling to us, but they are certainly open to revision and obviously need further clarification.

Dopamine, the "Cinderella" of the biogenic amines, was at first considered only a precursor to norepinephrine. However, dopamine's presence in large amounts in the brain and its special distribution in the basal ganglia suggested a possible and unique function. In mice, L-Dopa given alone produces a large increase in brain dopamine levels, while there may actually be decreases in norepinephrine and serotonin levels. In various animals the administration of Dopa and the resulting increases in dopamine produce marked behavioral effects including alerting, increased irritability, aggressiveness and motor responsiveness. Monkeys show a striking Parkinson-like state and behavioral deficits resulting from reserpine depletion of biogenic amines. The administration of Dopa causes a dramatic and remarkable reversal of the Parkinson-like signs, with a return of normal neuromuscular organization, increased motor activity, irritability and aggressive behavior. These changes have a gradual onset, revealing full effects within one hour. This represents the period of formation of dopamine from Dopa and its distribution in the brain (see Fig. 1).

The widespread distribution of the dopamine formed in the brain from Dopa is a complicating factor in interpreting these studies. Nevertheless, we can postulate that the dopamine is made available at receptor sites and produces the effects seen. Although it is possible that small amounts of norepinephrine are formed from the increased dopamine pool, it seems unlikely that this is of much importance in the effects observed.

The differentiation of the role of dopamine and norepinephrine could be very difficult if they both acted essentially in the same direction on neurological and behavioral systems. However, recent studies suggest they may be quite different in their functional roles. It has long been known that intraventricular injection of norepinephrine induces motor depression and sleep in animals. Further studies in chicks showed that norepinephrine given intravenously caused sedation while dopamine caused alerting and increased activity. A number of investigators have reported that dihydroxyphenylserine (DOPS), a precursor that forms norepinephrine directly, produces only sedative effects. More behavioral work with determinations of brain amine levels after giving DOPS is needed to establish this sedative action. Several studies have shown that the neurological and behavioral deficits from reserpine could not be restored by giving DOPS but were rapidly reversed by Dopa. Further, the blocking of the enzyme involved in the formation of norepinephrine from dopamine did not prevent this reversal.[7] Thus, the reversal of the behavioral and neurological deficits is due primarily to the replacement of dopamine in the brain. If these lines of evidence are confirmed, then norepinephrine may be stripped of many functions attributed to it and become primarily a sedative neuromodulator. On dopamine would then fall the mantle as the neuromodulator concerned with alerting, motor mobilization and aggressive behavior.

In the complex organization of the nervous system it seems probable that norepinephrine and dopamine may be synergistic in some areas and functional antagonists in others. Both dopamine and norepinephrine are primarily inhibitory when applied to individual neurons. Further, increasing levels of

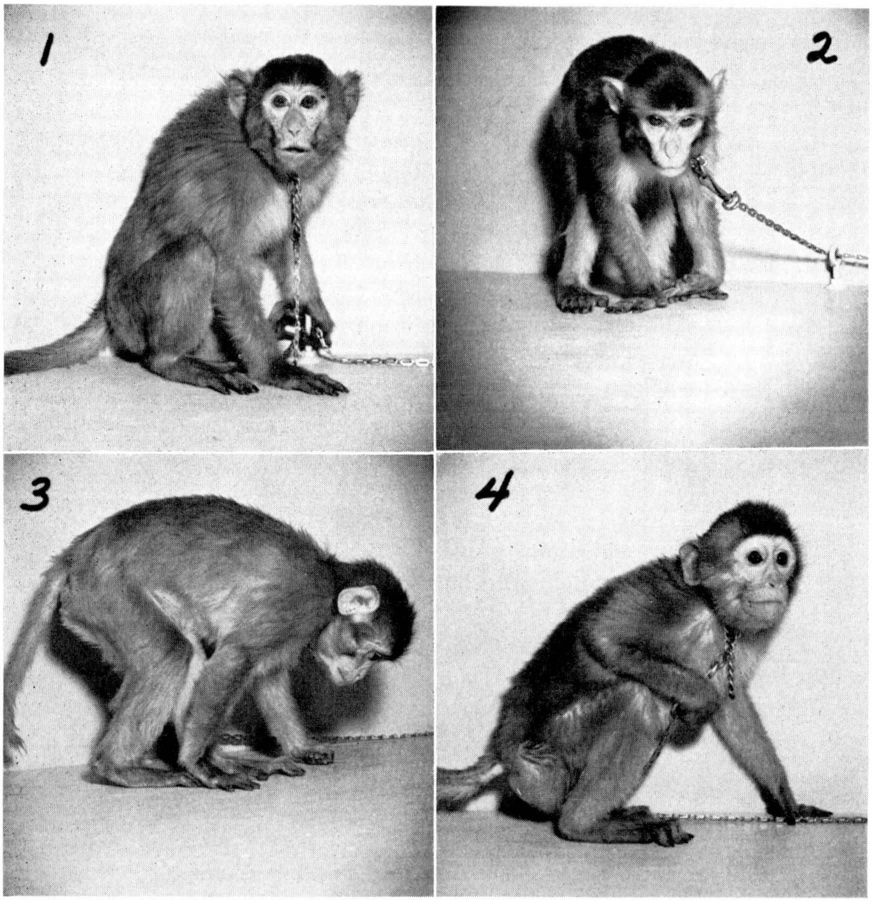

Fig. 1. Dopa reversal of parkinson-like syndrome in monkeys (reserpine-induced). *1.* Normal monkey. *2.* Monkey five hours after 10 mg per kg reserpine given intraperitoneally. Note increased rigidity and extensor in forelimbs, hunched back, abnormal posture, ptosis, masked fasces. Marked behavioral akinesia and loss of aggressiveness observed. *3.* Same monkey. Side view. *4.* One hour after giving 50 mg per kg pargyline + 50 mg per kg DL-Dopa intraperitoneally. Note return of normal posture, piloerection. Alert aggressive behavior and normal motor organization observed.

either amine result in marked increases in thresholds to electroshock seizures. Thus both are important in decreasing brain excitability.

If, indeed, norepinephrine is an antagonist of dopamine (and vice versa) the clinical implications will require some revised thinking. Better methods for manipulating brain levels of these amines individually are much to be desired in resolving these problems.

The relation of cholinergic systems to the biogenic amine system is an important consideration in the organization of the extrapyramidal system. Tremorine and oxotremorine which increase acetylcholine levels in brain [8] also cause decreases in dopamine [9] and norepinephrine [10] and markedly increase the turnover rates of dopamine in the basal ganglia.* Further, we have shown a considerable antagonism of oxotremorine effects by L-Dopa [11] and postulated that this is due to rebalancing the cholinergic and biogenic amine systems. The work of Steg [12] has shown that the alpha and gamma fiber systems in the spinal cord are organized along these lines, the alpha being cholinergic and the gamma a biogenic amine system. Loss of the latter amines (after reserpine) results in an imbalance with cholinergic dominance and increased rigidity is observed.

The relation of the central catecholamines to serotonin systems must also be considered. In mice, doses of 200 and 400 mg per kg of L-Dopa cause a decrease in brain serotonin of up to 50 percent. Metabolic studies in man also suggest a decrease in serotonin and/or its metabolites. The importance of this effect on brain serotonin levels is not yet clear. Also germane to this problem are the effects of Dopa and 5-HTP on rhythmic efferent discharges in hind limb nerves. Dopa increases extensor nerve activity while 5-HTP affects increased firing in the flexor nerves.[13]

Finally, we wish to place dopamine back into the brain along with the other "functional juices" and complex anatomical organizations that make up the nervous system. Certainly the cholinergic systems and serotonergic systems are also very much involved, and any final picture must include the integration of all these neurohumors to bring about the final behavioral and motor output.[14]

Thus, the signs of parkinsonism such as tremor, akinesia and rigidity should be thought of as due not to the loss of dopamine but to the resulting imbalance of the neurological systems involved. The concept of rebalancing neurohumoral systems is basic to the present and future progress in the understanding and treatment of neurological and psychiatric illness.

REFERENCES

1. Carlsson, A., Lindqvist, M., and Magnusson, T.: 3,4-Dihydroxyphenylalanine and 5-hydroxytryptophan as reserpine antagonists. Nature *180:* 1200, 1957.
2. Everett, G. M., and Toman, J. E. P.: Mode of action of rauwolfia alkaloids and motor activity. *In* Biological Psychiatry. Grune and Stratton, New York, 1959.
3. Everett, G. M., and Wiegand, R. G.: Central amines and behavioral states: A critique and new data. Proc. 1st Int. Pharmacol. Meeting, *8:* 85, 1962.
4. Everett, G. M., Wiegand, R. G., and Rinaldi, F. U.: Pharmacologic studies of some nonhydrazine MAO inhibitors. Ann. N. Y. Acad. Sci. *107:* 1068, 1963.
5. Ehringer, H., and Hornykiewicz, O.: Verteilung von noradrenalin und dopamine (3-hydroxytyramin) im Gehirn der Menschen und ihr Verhalten bei Erkrankungen des Extrapyramidalen Systems. Klin. Wschr. *38:* 1236, 1960.
6. Cotzias, G. C., Van Woert, M. H., and Schiffer, L. M.: Aromatic amino acids and modification of parkinsonism. New Eng. J. Med. *276:* 374, 1967.
7. Creveling, C. R., Daly, J., Tokuyama, T., and Witkop, B.: The combined

* See Dr. Gessa's paper in this volume.

use of α-methyltyrosine and threo-dihydroxyphenylserine: Selective reduction of dopamine levels in the central nervous system. Biochem. Pharm. *17:* 65, 1968.

8. Hornykiewicz, O.: Metabolism of brain dopamine in human parkinsonism: Neurochemical and clinical aspects. *In* Biochemistry and Pharmacology of the Basal Ganglia (Costa, E., Cote, L. J., and Yahr, M. D., eds.). Raven Press, Hewlett, New York, 1966.
9. Friedman, A. H., and Everett, G. M.: Pharmacological aspects of parkinsonism. Adv. Pharmacol. *3:* 83, 1964.
10. Everett, G. M.: Pharmacological studies on tremorine. *In* Biochemical and Neurophysiological Correlation of Centrally Acting Drugs. Proc. 2nd Int. Pharmacol. Meeting, Prague, August 1963.
11. Everett, G. M.: Effects of oxotremorine on the biogenic amine and cholinergic systems in rat and mouse CNS. Fed. Proc. *26(2):* 764, 1967.
12. Steg, G.: Efferent muscle innervation and rigidity. Acta Physiol. Scand. *61:* Suppl. 225, 1964.
13. Viala, D., and Buser, P.: The effects of Dopa and 5-HTP on rhythmic efferent discharges in hind limb nerves in the rabbit. Brain Res. *12:* 437, 1969.
14. Everett, G. M.: Discussion of Studies on tremorine antagonism by structurally related compounds, by D. Jenden. *In* Biochemistry and Pharmacology of the Basal Ganglia (Costa, E., Cote, L. J., and Yahr, M. D., eds.). Raven Press, Hewlett, New York, 1966.

CHAPTER 12

Physiology of L-Dopa and Dopamine

MONOAMINES AND THE PHYSIOLOGY OF SLEEP AND WAKING

J. Pierre Cordeau

Doctor Barbeau has asked me to summarize briefly the physiological literature on the role of monoamines in the regulation of the sleep-wakefulness cycle. This purpose might best be served by a series of more or less dogmatic statements on this topic together with some of the experimental observations which have led to them. Each of these observations, or at least the conclusions drawn from them, could be challenged in one way or another and often have been. As a matter of fact, from the mass of recent scientific literature on this subject—depending on which papers one chooses to retain and which to reject—one could argue for several apparently contradictory and mutually exclusive hypotheses. Moreover, this review will, of necessity, be incomplete and sketchy and will undoubtedly be colored by the author's prejudices and preferences. Finally, after listening to the reports on the clinical use of L-Dopa and its side effects in man, I am not at all sure that the review will be of use to the purpose of this symposium. We seem to be dealing with two different time scales altogether. The phenomena that I shall deal with take place within minutes, hours or at most a few days after the experimental procedure, while most of those that have been reported thus far occur weeks or months after continuous administration of high doses of L-Dopa. With these reservations in mind I would like to take each of the monoamines in turn and with quotations from current literature try to assign to each a role in sleep and in wakefulness.

Norepinephrine

Role of norepinephrine in waking. From animal experiments it would appear that norepinephrine has a role to play in the waking process and in the maintenance of wakefulness. This statement is based on the following observations:

1. When injected intravenously in acute preparations (usually "encéphale isolé" cats), whose electrocortical activity is synchronized and whose brain is therefore in a state resembling sleep, norepinephrine will desynchronize the cortex and dilate the pupils, a state resembling wakefulness.[1, 2, 3] Under these conditions, its principal site of action (it may not be the only one, vide infra) appears to be neuronal aggregates situated in the upper pons and lower mesencephalon, since lesions placed in these regions, and gradually advanced more and more rostrally, will decrease and finally abolish the effectiveness of this substance to desynchronize the electrocorticogram. These effects do not appear to be secondary to an increase in blood pressure since they can occur in its absence or precede it in time.
2. Norepinephrine and amphetamine when deposited directly into the brain substance of acute or more often of intact chronic cats, either in the form of crystals or in solution, produce behavioral and electrocortical waking sometimes accompanied by excitation and increased motor activity.[4, 5, 6, 7, 8] Several brain loci appear to be responsive, including the pontine and mesencephalic reticular formation, the preoptic region, some regions of the hypothalamus and the nucleus centralis medialis of the thalamus. To this list we can add the effect of L-Dopa when injected directly into the ergotropic zone of the hypothalamus as reported by Spiegel.
3. Lesions of the brain stem reticular formation in cats, when selectively placed to destroy norepinephrine-containing neurons in the upper pons and lower mesencephalon, i.e., neuronal aggregates corresponding to groups A6 and A8 of Dahlström and Fuxe,[9, 10] are followed by a considerable increase in the duration of "cortical" sleep, i.e., electrocortical synchronization, in the 24-hour sleep-wakefulness cycle. The increase will last several days. The long-lasting cortical synchronization may or may not be accompanied by behavioral sleep and this dissociation between cortical and behavioral sleep is not seen if the lesion simultaneously reaches groups A9 and A10, i.e., dopamine-containing neurons of the ventral tegmental mesencephalon. In such cases the cats are comatose behaviorally when cortical synchronization is present (vide infra). There appears to be a definite correlation between the number of norepinephrine-containing neurons destroyed, the increase in the daily duration of electrocortical synchronization and the lowered content of norepinephrine in telencephalic and diencephalic structures. Rapid-eye movement (REM) sleep duration is not necessarily changed appreciably by such lesions; it may at times show a slight reduction.[11]
4. Dopa given intraperitoneally in cats in doses of 30 to 50 mg per kg is followed, after a latency of 15 to 20 minutes, by a state of quiet wakefulness, with mydriasis and cortical desynchronization, which lasts some 3 to 4 hours.[12]
5. The effects on wakefulness of MAO inhibitors (MAOI), of reserpine, of α-methyldopa, and other drugs which affect monoamine stores in one way or another are not as clear as one might wish, probably because they affect all monoamines indiscriminately but to varying degrees. These drugs have been used extensively in animals for the study of their effects on REM sleep (which is usually deeply affected, as we shall see below). However, it has

been said that such MAOI as harmaline (20 mg per kg) and tranylcypromine (2 to 5 mg per kg) have a strong amphetamine-like effect in cats and induce a state of restless wakefulness, polypnea and mydriasis which lasts from 6 to 10 hours. Reserpine injected in cats in doses of 0.5 mg per kg suppresses slow wave sleep (SWS) for 12 to 14 hours and produces a stage which could be described as depressed or quiet wakefulness, probably because it depletes all monoamine stores.[12] Alpha-methyldopa, which seems to be metabolized into a false transmitter, α-methylnorepinephrine,[13, 14] does not affect waking and slow wave sleep but suppresses REM sleep.[12] The excitation or raised level of arousal and decrease in total sleep time produced by amphetamine would appear to be related to its effect on the liberation of norepinephrine, since this excitation is prevented by a blockade of norepinephrine synthesis such as that obtained by inhibition of tyrosine hydroxylase with α-methyltyrosine.[12, 15]

In clinical fields it seems to have become something of a truism that in man any procedure or drug which increases the amount of readily available norepinephrine will raise the arousal or excitation level while, conversely, decreasing the amount of available norepinephrine has a tranquilizing effect.[16] It has also been suggested that more marked changes in norepinephrine available at functional synaptic sites could also lead to such pathological states as mania or depression.[17, 18, 19] I shall not dwell on this particular aspect, leaving it to others who are more familiar with this field of investigation.

These statements as to the role of norepinephrine in waking, or at least some of the observations on which they are based, could and have been challenged in the following way.

1. Norepinephrine injected intravenously does not produce wakefulness through a direct action on brain stem neurons:

   Epinephrine and norepinephrine do not cross the blood-brain barrier in any "appreciable" amounts except perhaps at the hypothalamic level.[20, 21] Of course the keyword here is "appreciable." The method used for evaluating the effectiveness of the blood-barrier consists in measuring the content of norepinephrine and sometime its metabolites in rather large chunks of brain tissue such as the medulla, pons or mesencephalon, following the intravenous injection of the radioactive tracer, while at the same time applying a "correction factor" which tries to take into account the estimated blood content of these regions. It could be suggested that with such a technique physiologically significant increases in norepinephrine, but increases localized to small and restricted regions of the brain stem, could be missed. Such regions which appear to escape the blood-barrier are known to exist (area postrema, locus ceruleus, etc.; see ref. 22 for review). Furthermore, it has been reported that following intravenous epinephrine infusion in cats, appreciable amounts of the drug can be found in cerebrospinal fluid.[23]

   Another argument used against the direct action of intravenous norepinephrine on brain stem neurons mentions that it is impossible to dissociate the arousal produced by intravenous norepinephrine from the concomitant rise in blood pressure. When such blood pressure changes are prevented there is no arousal. Furthermore, when a rise in blood pressure is obtained through mechanical or other means (vasopressin), this is accompanied by electrocortical desynchronization.[24] Finally, norepinephrine injected into the carotid artery does not always produce cortical desynchronization or is followed by electrocortical arousal only after long latencies (several seconds), long enough for the drug to distribute itself throughout the whole circulation, and is accompanied by a rise of blood pressure.[25, 26] There is no definitive answer to this question except to note that the literature contains contradictory statements concerning the need for a rise in blood pressure for the cortical desynchronization following intravenous norepinephrine on the one hand, and that the carotid and vertebral arteries have different irrigation territories in the brain stem, the frontier between the two being somewhere in the neighborhood of the mesencephalon. I am unaware of any experiments which have studied the effects of norepinephrine injected into the vertebral artery.

2. Other objections concern the wakefulness or arousal obtained following the injection of norepinephrine directly into the brain substance. It has been said that this effect cannot, with confidence, be attributed to the action of the drug on neurons with noradrenoceptive sites. Such an arousal may be the result of local changes in pH or osmotic pressure, or of mechanical stimulation of neurons, or of changes in the local vasculature with resulting CO_2 accumulation, etc.[27] Furthermore, it is said that the sites from which this effect can be obtained are too widespread in the brain and therefore this procedure has no localizing value for an arousal system.[28] (This last argument is similar in kind to that used by those who reject the sleep-inducing properties of low frequency electrical stimulation of the brain which can also be obtained from widespread areas.[28])

   These arguments are probably valid and are difficult to ignore. One can only stress again that in some experiments the pH of the injected solution was carefully controlled and the injection made so as to minimize any local mechanical disturbance of the nervous tissue.[4] It should also be mentioned as a counter-argument that very often the local injection of acetylcholine, in the same regions and using identical techniques, is usually followed by electrocortical synchronization and behavioral sleep.[4, 5, 6] This would seem to eliminate pH, osmolarity and mechanical disturbances as causal factors (see ref. 29 for critical review). I shall attempt to deal with the question of the too-great spread in the brain of noradrenoceptive sites producing arousal later in this paper.

3. Jouvet's results quoted above [11] where destruction of norepinephrine-containing neurons in the upper pons and lower mesencephalon leads to decreased wakefulness, a change correlated to a decreased norepinephrine content of the brain, can also be questioned, since the regions concerned can be shown to contain high concentrations of cholinesterase and have been said to be at the origin of an ascending cholinergic system.[30, 31] The possible role of the destruction of this system in Jouvet's results is not taken into account. This difficulty of course is present in all lesion-making experiments in the reticular core of the brain stem since it is, in practice, impossible to limit the lesion to a single functionally homogeneous neuronal population.

4. One final argument which runs counter to the possible role of norepinephrine in arousal is the frequently quoted observation that when this substance is injected directly into the ventricular system of man, dog or cat it produces within a few minutes a sleep-like state which has been variously described as similar to light anesthesia, very lethargic or comatose behavior, total lack of responsiveness, etc., a state which can last 2 to 3 hours.[32, 33, 34, 35, 36] The apparent contradiction of these results with those obtained from other means of administering norepinephrine is difficult to resolve. What regions of the brain are reached by intraventricular injections? How deep can the drug penetrate from the surface of the ventricular walls within the time necessary for the sleep-like effect to manifest itself? It has been suggested that only a rather superficial layer of neurons can thus be reached. Strangely enough,

most of these observations are limited to clinical descriptions and to my knowledge completely ignore the effects of intraventricular norepinephrine on the electroencephalogram. Recent publications by Matsuda [37, 38] attempt to bridge this gap and describe the effect of norepinephrine injected in the ventricular system of rabbits on their electrocortical activity. Apparently, the initial effect here is one of desynchronization followed after some minutes (10 to 40 depending partly on the dose) by synchronization of the electroencephalogram. With the use of alpha- and beta-adrenergic blocking agents Matsuda suggests that the electrocortical desynchronization of intraventricular norepinephrine is mediated by beta receptors while the synchronizing effect is due to alpha receptors. Unfortunately in this instance observations on the concomitant behavioral effects are missing or so sketchy as to be of no use to the reader. It may not be too farfetched to suggest that perhaps intraventricular norepinephrine produces something resembling REM sleep (vide infra) which would explain both the anesthesia-like behavior and the electrocortical desynchronization. This, of course, could be checked experimentally if it has not already been done.

In summary, it would seem that the bulk of the evidence is in favor of norepinephrine's having a role in setting up, in the brain, some of the patterns of activity which accompany the waking state, and this in spite of criticisms aimed at some of the experimental procedures which have led to this concept. From a teleological point of view, if I may use such an argument, concepts linking norepinephrine with waking and arousal would appear at least plausible if norepinephrine and its sister hormone epinephrine are in fact involved at the periphery in Cannon's "fright, fight or flight" mechanisms and Hess' "ergotropic" class of activity. If these substances can "arouse" the somatic and vegetative components of the organism, one would think they should also arouse its brain. But of course this is not a final argument. Biological phenomena have a way of defying apparently logical deductions which should render us cautious in their use.

Role of norepinephrine in REM sleep. Apart from its role in wakefulness, norepinephrine appears also to be involved in the production of REM sleep. Briefly the evidence for this statement is as follows:

1. Bilateral destruction, in the pons, of the locus ceruleus, a nucleus which is known to be rich in norepinephrine-containing neurons,[9, 10] is followed according to Jouvet [39, 40] by a total or subtotal disappearance of both tonic and phasic events of REM sleep. Here also there appears to be a correlation between the extent of the lesion, in terms of norepinephrine-containing neurons destroyed, the decrease in REM sleep time and the amount of norepinephrine remaining in rostral structures. However, the same criticism mentioned earlier concerning lesion-making experiments also applies here. The region of the locus ceruleus is rich in cholinesterase [30, 31] and no account is taken of the possibly cholinergic system originating there in the syndrome produced following its destruction.
2. In animals, drugs which interfere with norepinephrine synthesis in one way or another may selectively suppress REM sleep. These include α-methyl-*p*-tyrosine[39, 41] which would act at the tyrosine hydroxylase level [42] and disulfiram [43] which appears to impair the synthesis of norepinephrine from dopamine as a probable inhibitor of dopamine β-hydroxylase.[44] Similarly, α-methyldopa, which is probably metabolized to become a false transmitter

and thereby reduces the norepinephrine content of the brain,[45] decreases or suppresses REM sleep[12, 46] as well as the rebound phenomenon which normally follows REM sleep deprivation.

3. Reserpine given in large doses in cats will suppress most manifestations of REM sleep for several hours. During this time REM sleep reappears following injections of Dopa [47] which presumably replenishes norepinephrine stores. The PGO spikes of REM sleep which persist after reserpine will be increased in frequency by injection of Dopa.[12]
4. Most MAO inhibitors, such as harmaline, tranylcypromine, pargyline, nialamide, will decrease or suppress REM sleep for long periods of time (up to 4 days), a suppression which is not followed by the usual rebound phenomenon.[12] This has led to the suggestion that transition from SWS to REM sleep necessitates the intervention of MAO.[28]
5. Finally, rats which have been deprived of REM sleep for several days show an increased turnover of cerebral norepinephrine during the well-known rebound phenomenon which follows deprivation.[48] This, however, may be a nonspecific effect due to the stress involved in REM sleep deprivation.
6. In man the evidence for the role of norepinephrine in REM sleep is less clear and sometimes in conflict with results obtained in animals. Of course, the administration of drugs is usually made at much lower dosages. As in animals, most MAOI in man will produce a decrease in REM sleep time as well as an increase in REM sleep latency (time of onset of first REM sleep period after going to sleep). Reserpine, however, will increase REM sleep time and decrease REM sleep latency [16] but this is with a dose which, according to body weight, is much smaller than that reported to abolish REM sleep in cats. I was unaware of any study on the effects of L-Dopa on the sleep-wakefulness cycle in man until the paper presented at this symposium by Dr. Ansel.

5-Hydroxytryptamine (5-HT)

Evidence is rapidly accumulating, in experimental animals and in man, that 5-hydroxytryptamine is probably involved in some way in the induction and probably in the maintenance of slow wave sleep (SWS). Since SWS appears to be a prerequisite for REM sleep under normal conditions, procedures which affect 5-HT contents of the brain will also affect REM sleep in most cases. The main observations which have led to this suggestion are as follows:

1. Destruction of the raphe nuclei in cats, known to be made up in great part of 5-HT-containing neurons,[9, 10] leads to a state of more or less permanent cortical desynchronization with behavioral arousal, i.e., to a definite shift toward waking in the sleep-wakefulness cycle. There is also a complete suppression of REM sleep if SWS is decreased to less than 15 percent of recording time. It appears also that there is a correlation between the extent of the raphe lesion, the decrease in sleep time and the loss of serotonin in diencephalic and telencephalic structures.[39, 49]

   Another explanation has been offered [50] concerning the results of such midline lesions. It suggests that the increase in waking time is due to the interruption of ascending fibers originating in a "synchronizing" system of the caudal brain stem [51, 52, 53] which would cross the midline through the raphe nuclei. This possibility exists but seems unlikely as the sole explanation in view of the results obtained with other means of depleting the brain of its content in 5-hydroxytryptamine.

2. Injection in cats,[49, 54, 55] rats,[56, 57] rabbits,[58] and monkeys [59] of p-chlorophenylalanine (PCPA), which seems selectively to deplete the brain of its content of 5-HT by inhibition of the enzyme tryptophan hydroxylase,[60, 61] is followed in the next two to three days by a gradual decrease in sleep time (both slow wave and REM sleep) and produces a more or less permanent state of wakefulness both in the electroencephalogram and behaviorally. This effect lasts a few days, following which the animal gradually returns to a normal sleep-wakefulness pattern. During the height of the effect, when the animal shows very little sleep, the injection of low doses of 5-hydroxytryptophan (5-HTP) the serotonin precursor bypassing the enzyme tryptophan hydroxylase, will trigger off, in a few minutes, a state of SWS followed by REM sleep which lasts a few hours before the animal returns to its sleepless condition.[39, 55] PCPA has also been reported to decrease REM sleep time in man.[62]
3. 5-HTP, the precursor of serotonin, when injected intravenously in doses of 15 to 25 mg in cats, is followed within a few minutes by a state resembling slow wave sleep which lasts 5 to 6 hours.[12] During this time REM sleep is suppressed and shows a rebound when the effect of the drug has passed. Rats on tryptophan-free or tryptophan-rich diets show a significant change in their SWS/REM sleep cycles, the former showing longer and the latter shorter cycles.[63] Similar results have been reported in man.[64, 65] Also in man tryptophan and 5-hydroxytryptophan have been reported to decrease sleep latency and increase REM sleep time.[64, 65, 66, 67, 68]
4. Reserpine is known to suppress SWS and REM sleep totally when given in a single dose of 0.5 mg per kg in cats, a suppression which lasts from 10 to 12 hours. 5-HTP given 2 to 3 hours after reserpine will restore slow wave sleep for periods of 4 to 6 hours but not REM sleep; the latter requires Dopa to reappear.[12]

Therefore, on the basis of this evidence, it would seem that the presence of 5-HT in the brain is a necessary condition for the patterns of sleep, both slow wave and REM, to become manifest. Its mode of action is not known. From iontophoretic studies showing that 5-HT is overwhelmingly inhibitory [69] the 5-HT system could be thought of as a "brake" on the neuronal system or systems promoting wakefulness. However, its effect is undoubtedly not limited to brain stem reticular neurons since 5-HT containing terminals are widespread and present in most if not all diencephalic and telencephalic loci, as well as in the spinal cord—a distribution not unlike that of norepinephrine containing terminals.[70]

Dopamine

Very little has been said in the past on the possible role of dopamine in the regulation of the sleep-wakefulness cycle except as a precursor of norepinephrine. Recently, experiments from Jouvet's laboratory [11] in which destruction of the ventral tegmental area of the mesencephalon, lesions which include the pars compacta of the substantia nigra (dopamine-containing neurons, Group A9 and A10 of Dahlström and Fuxe [9, 10]), and lesions similar to those described by Doctor Moore, have produced more or less completely akinetic and behaviorally comatose animals while leaving relatively intact the sleep-wakefulness cycle as measured by the electroencephalogram. Whenever the cortex is "awake" the animal is incapable of manifesting his state of wakefulness behaviorally. The extent of the akinesia, or depth of coma, seems to be related to the decrease in dopamine content of the forebrain. Can one speak of body sleep in this instance, while the brain is awake? I somehow doubt it. More parameters in the somatic and vegetative spheres would have to be measured before such a suggestion could be made. More probable is the suggestion that the decrease in brain dopamine manifests itself as a more or less complete deficit in motor activity, leaving intact the basic mechanisms involved in the sleep-wakefulness cycle. This explanation would fit the clinical data relating dopamine deficiency with akinesia and the improvement obtained following L-Dopa treatment. It should again be noted that the lesions described by Jouvet cover to a large extent the ventral tegmental area of Tsai which is known to contain large amounts of cholinesterase.[30, 31] The extent to which the destruction of this possible cholinergic system influences experimental observations is not known.

Acetylcholine

I would be remiss if I were to leave out completely all references to cholinergic mechanisms, even though this neurotransmitter is outside the main topic of discussion of this conference. On the one hand, several lines of evidence suggest that acetylcholine is somehow involved in all three states that we have been concerned with: wakefulness, slow wave sleep and REM sleep. On the other hand, cholinesterase-containing neurons are present in several brain stem loci where they overlap with monoamine-containing neurons. This overlap is such that thus far it has been impossible to destroy one system while leaving the other intact. The rostral projections of the cholinesterase-containing neurons are said to cover large territories of the forebrain and overlap with regions where monoamine terminals can be found.[30, 31, 70]

For the sake of brevity, what I will say concerning cholinergic mechanisms can be summed up in a few brief sentences:

1. Acetylcholine when injected directly in scattered brain loci in chronic animals is followed by SWS and sometime by REM sleep episodes.[4, 5, 6, 29]
2. The cortical desynchronization of wakefulness, whether occurring spontaneously or provoked by reticular formation activation or sensory stimulation, is accompanied by an increased amount of acetylcholine released by the cortex.[71, 72, 73]
3. Anticholinergic drugs such as atropine will produce a state of cortical synchronization resembling slow wave sleep while leaving the animal behaviorally awake and sometimes overexcited. Atropine will also abolish REM sleep.[12]
4. Anticholinesterases (example eserine) have been said to have a mild effect on REM sleep, i.e., increasing REM sleep time and possibly decreasing REM sleep latency in animals and man.[16, 74]

Therefore, as mentioned previously, acetylcholine probably acting at different levels of the neuraxis seems to be somehow involved in all three main stages that we have been concerned with, i.e., wakefulness, slow wave sleep, and REM sleep.

Summary and Discussion

In summary, norepinephrine seems to be involved in setting up brain patterns of both wakefulness and REM sleep. The presence of 5-hydroxytryptamine in the brain appears to be a necessary condition for slow wave sleep to appear and this is probably also true for REM sleep. Dopamine, though possibly not directly involved in the sleep-wakefulness cycle except as a precursor of norepinephrine, is however necessary as such for the waking brain to manifest its wakefulness behaviorally in the sphere of motor activity. Cholinergic and cholinoceptive mechanisms at the brain stem, hypothalamic and basal forebrain levels seem to promote slow wave and possibly REM sleep, while at the cortical level they seem to be involved in the electrocortical desynchronization of wakefulness. It remains to be seen if they play a similar role in the desynchronization patterns of REM sleep.

I think you realize as well as I do that this summary is not particularly helpful. In great part, what it says in essence is that, in any given species of mammals, for an animal to show a normal sleep-wakefulness cycle it needs a normal brain, at least insofar as monoaminergic and cholinergic mechanisms are concerned, and that there are undoubtedly other, as yet unknown, substances involved. Nothing has been said here of the possible role of GABA and other amino acids in sleep and waking, yet some experimental evidence exists that they are also involved. We are still very far from a complete understanding of the complex repatterning of brain activity which occurs when an animal wakes up or goes to sleep and of the role of monoamines and acetylcholine in this reorganization.

I might end this rather sketchy presentation with a few random thoughts which do not necessarily follow from what I have already said. The first is that—in the same way that we had to abandon some years ago the idea that during sleep the brain was "at rest" in terms of neuronal activity or more recently the proposition that sleep was a "passive" phenomenon, i.e., the absence of wakefulness—we should stop thinking, if we have not already done so, of wakefulness, slow wave sleep and REM sleep as completely different states, each with its own system anatomically and biochemically well defined, with precise points of origin in the brain stem and traceable projections both rostrally and caudally. This search for a "localization of function" may not be applicable to sleep and waking processes, as indeed it is not for several other brain functions. Perhaps we should think instead of sleep and wakefulness as "mixed" states in which all of the systems we have mentioned, and probably others, are involved to greater or lesser degree. We might consider an animal as being either awake or asleep depending on the patterning of the activity in each and every one of these systems. I personally believe, and there is some evidence for this which I cannot go into, that the brain is never, or very rarely, wholly awake or wholly asleep (i.e., that wakefulness and sleep do not exist as "pure" states) but that it is always oscillating in a continuum between these two extremes, being more or less awake or more or less asleep. Even REM sleep, which in some respects seems to be at one end of the spectrum, appears as a queer mixture of sleep and wakefulness. What is a dream if not some form of consciousness with the brain generating its own sensory input? On the other hand, during REM sleep, thalamocortical circuits seem to be cut off from peripheral sensory input through active inhibitory processes and the somatic and vegetative spheres show their most marked sleep patterns. (This reminds us of the old distinction proposed by Von Economo of brain sleep or waking as opposed to body sleep or waking, it being possible to dissociate the two under abnormal or pathological conditions.) The suggestion that both cholinergic and noradrenergic mechanisms may be involved in wakefulness and REM sleep and the observation that very often regions of the brain stem with norepinephrine-containing neurons also show large amounts of cholinesterase may be somehow related, and the anatomical overlap not a mere coincidence.

Secondly, the so-called reticular systems and their projections have been called diffuse systems since they seem to affect the whole brain. They should possibly be thought of as diffuse in another sense, i.e., in the sense that, even if their highest concentration or density is in the core of the brain stem where they may derive their tonic drive in normal conditions, they may also be omnipresent in the whole nervous system. What is being suggested is that neuronal aggregates responsible for patterns of sleep or waking may be present at all levels of the neuraxis in a multitiered type of organization with several substations projecting both rostrally and caudally, each substation reciprocally linked and providing positive or, more probably, negative feedback loops responsible for the so-called reticular homeostasis, a concept developed some years ago by Dell.[75] This might explain, among other things, why electrical and chemical stimulation in so many brain loci can affect the states of sleep and waking in one way or another, or why telencephalic structures chronically separated from the reticular core of the brain stem may eventually show patterns approaching those of wakefulness.[76, 77]

Thus far we have implicitly assumed that monoamines, in the regulation of sleep and wakefulness, act as true neurotransmitters in the sense that acetylcholine is the transmitter at the neuromuscular junction, producing localized changes in membrane permeability to various ions due to their combination to specific receptor sites, and thereby causing local depolarization or hyperpolarization. This might well be the case, and results from iontophoretic studies would seem to support such a concept. However, we could also think of monoamines as "modulators" of neuronal activity, affecting the overall level or pattern of excitability of neuronal aggregates which they bathe, in the same way for example that some hormones may affect behavior through their "modulation" of some groups of neurons. Studies on circadian and ultradian changes in the local concentration of brain monoamines,[78, 79, 80, 87] even though far from a conclusive argument, might still be thought of as supporting this concept. If we were to couple to this the suggestion already made for the differential release of acetylcholine or norepinephrine in the peripheral autonomic nervous system depending on the frequency of stimulation of efferent sympathetic nerves,[82] we may have the beginning of an explanation why low and high frequency stimulation of so many brain loci result in sleep in one case and wakefulness in the other, as has been already suggested.[29]

I might end this rather schematic review of current literature with one last question. To this point we have been concerned uniquely with the "how" of sleep and waking, leaving out altogether the question of the "why": Why do we sleep, why do we need to sleep, why, particularly, do we need REM sleep so badly that deprivation is invariably followed by a rebound phenomenon where REM sleep episodes are more frequent and of longer duration? The answer still eludes us. The more prevalent hypothesis suggests that some "substance" accumulates in the brain during waking (and possibly slow wave sleep) and needs REM sleep to be disposed of in one way or another. What this substance could be, we do not know. As far as I am aware, it has been impossible to show that sleep deprivation, whether it be slow wave sleep and/or REM sleep, results either in a measurable increase in monoamines in the brain as a whole or in a change in their regional distribution.

REFERENCES

1. Bonvallet, M., Dell P. et Hiebel, G. Tonus sympathique et activité électrique corticale. Electroenceph. Clin. Neurophysiol. *6:* 119, 1954.
2. Rothballer, A. B.: Studies on the adrenaline-sensitive component of the reticular activating system. Electroenceph. Clin. Neurophysiol. *8:* 603, 1956.
3. Rothballer, A. B.: The effects of catecholamines on the central nervous system. Pharmacol. Rev. *11:* 494, 1959.
4. Cordeau, J. P., Moreau, A., Beaulnes, A., and Laurin, C.: EEG and behavioural changes following micro-injections of acetylcholine and adrenaline in the brain stem of cats. Arch. Ital. Biol. *101:* 30, 1963.
5. Hernandez-Peon, R.: Sleep induced by localized electrical or chemical stimulation of the forebrain. Electroenceph. Clin. Neurophysiol. Suppl. *24:* 188, 1963.
6. Hernandez-Peon, R., Chavez-Ibarra, G., Morgane, P. J., and Timo-Iaria, C.: Limbic cholinergic pathways involved in sleep and emotional behavior. Exp. Neurol. *8:* 93, 1963.
7. Rothballer, A. B.: EEG activation from micro-injections of adrenaline into the brain stem of cats. Anat. Rec. *127:* 39, 1957.
8. Yamaguchi, N., Ling, G. M., and Marczynski, T. J.: The effects of chemical stimulation of the preoptic region, nucleus centralis medialis or brain stem reticular formation with regard to sleep and wakefulness. Recent Advances Biol. Psychiat. *6:* 9, 1964.
9. Dahlström, A., and Fuxe, K.: Evidence for the existence of monoamine-containing neurons in the central nervous system. Acta Physiol. Scand. *62:* Suppl. 232, 1964.
10. Pin, C., Jones, B., et Jouvet, M.: Topographie des neurones monoaminergiques du tronc cérébral du chat: étude par histofluorescence. Comp. Rend. Soc. Biol. *162:* 2136, 1968.
11. Jones, B. E., Bobillier, P., et Jouvet, M.: Effets de la destruction des neurones contenant des catécholamines du mésencéphale sur le cycle veille-sommeil du chat. Comp. Rend. Soc. Biol. *163:* 176, 1969.

12. JOUVET, M.: Mechanisms of the states of sleep: A neuropharmacological approach. *In* Sleep and Altered States of Consciousness. ARNMD Vol. 45, 1967.
13. GESSA, G. L., COSTA, E., KUNTZMAN, R., and BRODIE, B. B.: On the mechanism of norepinephrine release by α-methyl-metatyrosine. Life Sci. *1:* 353, 1962.
14. UDENFRIEND, S., and ZALTZMAN-NIRENBERG, P.: On the mechanism of the norepinephrine release produced by α-methyl-meta-tyrosine. J. Pharmacol. Exp. Ther. *138:* 194, 1962.
15. BLOOM, F. E., and GIARMAN, J.: Physiologic and pharmacologic considerations of biogenic amines in the nervous system. Ann. Rev. Pharmacol. *8:* 229, 1968.
16. HARTMAN, E.: Pharmacological studies of sleep and dreaming: chemical and clinical relationships. Recent Advances Biol. Psychiat., in press.
17. BUNNEY, W., DAVIS, J., WEIL-MALHERBE, H., and SMITH, E.: Biochemical changes in psychotic depression. Arch. Gen. Psychiat. *16:* 448, 1967.
18. SCHILDKRAUT, J. J., and KETY, S. S.: Biogenic amines and emotion. Science *156:* 21, 1967.
19. SCHILDKRAUT, J. J., SCHANBERG, S., BREESE, G., and KOPIN, I.: Norepinephrine metabolism and drugs used in affective disorder: A possible mechanism of action. Amer. J. Psychiat. *124:* 600, 1967.
20. WEIL-MALHERBE, H., AXELROD, J., and TOMCHICK, R.: Blood-brain barrier for adrenaline. Science *129:* 1226, 1959.
21. WEIL-MALHERBE, H., WHITBY, L. G., and AXELROD, J.: The uptake of circulating (H^3) norepinephrine by the pituitary gland and various areas of the brain. J. Neurochem. *8:* 55, 1961.
22. KOELLA, W. P., and SUTIN, J.: Extra blood-brain barrier brain structures. Int. Rev. Neurobiol. *10:* 31, 1967.
23. DRASKOCI, M., FELDBERG, W., and HARANATH, P. S. R. K.: Passage of circulating adrenaline into perfused cerebral ventricles and subarachnoid space. J. Physiol. *150:* 34, 1960.
24. BAUST, W., NIEMCZYK, H., and VIETH, J.: The action of blood pressure on the ascending reticular activating system with special reference to adrenaline-induced EEG arousal. Electroenceph. Clin. Neurophysiol. *15:* 63, 1963.
25. CAPON, A.: Nouvelles recherches sur l'effet d'éveil de l'adrénaline. J. Physiol. *51:* 424, 1959.
26. MANTEGAZZINI, P., POECK, K., and SANTIBAÑEZ, H.: The action of adrenaline and noradrenaline on the cortical electrical activity of the "encéphale isolé" cat. Arch. Ital. Biol. *97:* 222, 1959.
27. RECH, R. H., and DOMINO, E. F.: Observations on injections of drugs into the brain substance. Arch. Int. Pharmacodyn. *121:* 429, 1959.
28. JOUVET, M.: Neurophysiology of the states of sleep. Physiol. Rev. *47:* 117, 1967.
29. MARCZYNSKI, T. J.: Topical application of drugs to subcortical brain structures and selected aspects of electrical stimulation. Ergeb. Physiol. *59:* 86, 1967.
30. LEWIS, P. R., and SHUTE, C. C. D.: The cholinergic limbic system: Projections to hippocampal formation, medial cortex, nuclei of the ascending cholinergic reticular system, and the subfornical organ and supra-optic crest. Brain *90:* 521, 1967.
31. SHUTE, C. C. D., and LEWIS, P. R.: The ascending cholinergic reticular system: Neocortical, olfactory and subcortical projections. Brain *90:* 497, 1967.
32. BASS, A.: Uber eine Wirkung des Adrenalins auf das Gehirn. Neurol. Psychiat. *26:* 600, 1914.

33. Feldberg, N., and Sherwood, S. L.: Injection of drugs into the lateral ventricle of the cat. J. Physiol. *123:* 148, 1954.
34. Leimdorfer, A.: The action of sympathomimetic amines on the central nervous system and the blood sugar. J. Pharmacol. Exp. Ther. *98:* 62, 1950.
35. Leimdorfer, A., Arana, R., and Hack, M. A.: Hyperglycemia induced by the action of adrenaline on the central nervous system. Amer. J. Physiol. *150:* 588, 1947.
36. Leimdorfer, A., and Metzner, W. R. T.: Analgesia and anesthesia induced by epinephrine. Amer. J. Physiol. *151:* 116, 1949.
37. Matsuda, Y.: Effects of intraventricularly administered adrenaline on rabbit's EEG and their modifications by adrenergic blocking agents. Jap. J. Pharmacol. *18:* 139, 1968.
38. Matsuda, Y.: Effects of some sympathomimetic amines administered intraventricularly on rabbit's EEG. Jap. J. Pharmacol. *19:* 102, 1969.
39. Jouvet, M.: Biogenic amines and the states of sleep. Science *163:* 32, 1969.
40. Roussel, B., Buguet, A., Bobillier, P., et Jouvet, M.: Locus coeruleus, sommeil paradoxal et noradrénaline cérébrale. Comp. Rend. Soc. Biol. *161:* 2537, 1967.
41. Weitzman, E. D., McGregor, P., Moore, C., and Jacoby, J.: The effect of alpha-methyl-para-tyrosine on sleep patterns of the monkey. Life Sci. *8:* 751, 1969.
42. Spector, S., Sjoerdsma, A., and Udenfriend, S.: Blockade of endogenous norepinephrine synthesis by α-methyl-tyrosine, an inhibitor of tyrosine hydroxylase. J. Pharmacol. Exp. Ther. *147:* 86, 1965.
43. Peyrethon-Dusan, D., et Froment, J. L.: Effets du disulfiram sur les états de sommeil chez le chat. Comp. Rend. Soc. Biol. *162:* 2141, 1968.
44. Goldstein, M., Anagnoste, B., Lauber, E., and McKereghan, M. R.: Inhibition of dopamine-β-hydroxylase by disulfiram. Life Sci. *3:* 763, 1964.
45. Carlsson, A., and Lindquist, M.: In vivo decarboxylation of α-methyl dopa and α-methyl metatyrosine. Acta Physiol. Scand. *54:* 87, 1962.
46. Dusan-Peyrethon, D., Peyrethon, J., et Jouvet, M.: Suppression élective du sommeil paradoxal chez le chat par α-methyl dopa. Comp. Rend. Soc. Biol. *162:* 116, 1968.
47. Matsumoto, J., et Jouvet, M.: Effets de réserpine, dopa et 5-HTP sur les deux états de sommeil. Comp. Rend. Soc. Biol. *158:* 2137, 1964.
48. Pujol, J. F., Mouret, J., Jouvet, M., and Glowinski, J.: Increased turnover of cerebral norepinephrine during rebound of paradoxical sleep in the rat. Science *159:* 112, 1968.
49. Pujol, J. F., Bobillier, P., Buguet, A., Jones, B., et Jouvet, M.: Biosynthèse de la sérotonine cérébrale: étude neurophysiologique et biochimique après para-chlorophenylalanine et destruction du système du Raphé. Comp. Rend. Acad. Sci. *268:* 100, 1969.
50. Mancia, M., Desiraju, T., and China, G. S.: The monkey split brain-stem: Effects on the sleep-wakefulness cycle. Electroenceph. Clin. Neurophysiol. *24:* 409, 1968.
51. Bonvallet, M., et Allen, M. B.: Localisation des formations bulbaires intervenant dans le contrôle de différentes manifestations de l'activation réticulaire. Comp. Rend. Soc. Biol. *156:* 597, 1962.
52. Cordeau, J. P., and Mancia, M.: Evidence for the existence of an electroencephalographic synchronization mechanism originating in the lower brain stem. Electroenceph. Clin. Neurophysiol. *11:* 551, 1959.
53. Dell, P., Bonvallet, M., and Hugelin, A.: Mechanisms of reticular deactivation. *In* The Nature of Sleep, (G. E. Wolstenholme and M. O'Connor, eds.). Ciba Symposium 1961.
54. Delorme, F., Froment, L., et Jouvet, M.: Supression du sommeil par la

p-chlorométhamphétamine et la p-chlorophenylalanine. Comp. Rend. Soc. Biol. *160:* 2347, 1966.
55. Koella, W. P., Feldstein, A., and Cziman, J. S.: The effect of para-chlorophenylalanine on the sleep of cats. Electroenceph. Clin. Neurophysiol. *25:* 481, 1968.
56. Mouret, J., Bobillier, P., et Jouvet, M.: Effets de la para-chlorophenylalanine sur le sommeil du rat. Comp. Rend. Soc. Biol. *161:* 1600, 1967.
57. Torda, C.: Effect of brain serotonin depletion on sleep in rats. Brain Res. *6:* 375, 1967.
58. Florio, V., Scotti de Carolis, A., and Longo, V. G.: Observations on the effect of DL-parachlorophenylalanine on the electroencephalogram. Physiol. Behav. *3:* 861, 1968.
59. Weitzman, E. D., Rapport, M. M., McGregor, P., and Jacoby, J.: Sleep patterns of the monkey and brain serotonin concentration: effect of p-chlorophenylalanine. Science *160:* 1361, 1968.
60. Jequier, E., Lovenberg, W., and Sjoerdsma, A.: Tryptophan hydroxylase inhibition: The mechanisms by which p-chlorophenylalanine depletes rat brain serotonin. Molec. Pharmacol. *3:* 274, 1967.
61. Koe, K. B., and Weisman, A.: p-Chlorophenylalanine: A specific depletor of brain serotonin. J. Pharmacol. Exp. Ther. *154:* 499, 1966.
62. Wyatt, R., Engleman, K., Kupfer, D., Scott, J., Sjoerdsma, A., and Snyder, F.: Effects of MAO inhibitors and para-chlorophenylalanine in longitudinal sleep studies. Report to the Association for the Psychophysiological Study of Sleep, 1968.
63. Hartman, E.: The sleep-dream cycle and brain serotonin. Psychon. Sci. *8:* 295, 1967.
64. Williams, H. L., Lester, B. K., and Coulter, J. D.: Monoamines and the EEG stages of sleep. Report to the Association for the Psychophysiological Study of Sleep, 1968.
65. Williams, H. L., Lester, B. K., and Coulter, J. D.: Monoamines and the EEG stages of sleep. Activ. Nerv. Sup. *11:* 188, 1969.
66. Evans, J., and Oswald, I.: Some experiments in the chemistry of narcoleptic sleep. Brit. J. Psychiat. *112:* 401, 1966.
67. Hartman, E.: The effect of tryptophane on the sleep dream cycle in man. Psychon. Sci. *8:* 479, 1967.
68. Mandell, M., Mandell, A., and Jacobson, A.: Biochemical and neurophysiological studies of paradoxical sleep. Recent Advances Biol. Psychiat. *7:* 115, 1964.
69. Stefanis, C.: Discussion of serotonin effects in central neurons. Adv. Pharmacol. *6A:* 414, 1968.
70. Fuxe, K.: The distribution of monoamine terminals in the central nervous system. Acta Physiol. Scand. *64:* Suppl. 247, 37, 1965.
71. Celesia, G., and Jasper, H. H.: Acetylcholine released from cerebral cortex in relation to state of activation. Neurology *16:* 1053, 1966.
72. Mitchell, J. P.: Acetylcholine released from brain. *In* Mechanisms of Release of Biogenic Amines, Vol. 5. Pergamon Press, Oxford, 1966.
73. Sie, G., Jasper, H. H., and Wolfe, L.: Role of ACh release from cortical surface in "encéphale" and "cerveau isolé", cat preparations in relation to arousal and epileptic activation of the ECoG. Electroenceph. Clin. Neurophysiol. *18:* 206, 1965.
74. Jouvet, M.: Recherches sur les structures nerveuses et les mécanismes responsables des différentes phases du sommeil physiologique. Arch. Ital. Biol. *100:* 125, 1962.
75. Dell, P.: Reticular homeostasis and critical reactivity. *In* Progress in Brain Research, Vol. I. Elsevier, Amsterdam, 1963.

76. Cordeau, J. P., and Mancia, M.: Effect of unilateral chronic lesions of the midbrain on the electrocortical activity of the cat. Arch. Ital. Biol. *96:* 374, 1958.
77. Villablanca, J.: Electroencephalogram in the permanently isolated forebrain of the cat. Science *138:* 44, 1962.
78. Friedman, A. H., and Walker, C. A.: Circadian rhythms in rat midbrain and caudate nucleus biogenic amine levels. J. Physiol. *197:* 77, 1968.
79. Quay, W. B.: Differences in circadian rhythm in 5-hydroxytryptamine according to brain region. Amer. J. Physiol. *215:* 1448, 1968.
80. Reis, J. D., Carvelli, A., and Connors, J.: Circadian and ultradian rhythms of serotonin regionally in cat brain. J. Pharmacol. Exp. Ther. *167:* 328, 1969.
81. Scheving, L. E., Harrisson, W. H., Gordon, P., and Pauly, J. E.: Daily fluctuation (circadian and ultradian) in biogenic amines of the rat brain. Amer. J. Physiol. *214:* 166, 1968.
82. Burn, J. H., Dromey, J. J., and Large, B. J.: The release of acetylcholine by sympathetic nerve stimulation at different frequences. Brit. J. Pharmacol. *21:* 97, 1963.

THE PHYSIOLOGY OF DOPAMINE

H. McLennan

My consideration of this subject will be confined to those areas most germane to the topic of this symposium: the possible role of dopamine as a transmitter of synaptic action and the sites where it may act in this capacity. One of the criteria for the identification of a synaptic transmitter is the presence of the substance throughout the presynaptic neuron whose activity is to be transmitted. Histofluorescent techniques have demonstrated such neurons in the zona compacta of the substantia nigra, whose axons are believed to run in the medial forebrain bundle to reach the neostriatum (Fig. 1).[1, 2] Within the striatal nuclei abundant terminals may be visualized which contain the highest concentrations of dopamine within the central nervous system.[3] Other dopamine terminals have been reported in the tuberculum olfactorium, nucleus accumbens, median eminence, nucleus centralis of the amygdala and nucleus interstitialis striae terminalis, all of which probably also derive from fibers running in the medial forebrain bundle.[4]

A second criterion of a synaptic role for a particular chemical entity [5] is that the substance can be liberated from activated synaptic terminals and collected if the region of the terminals is artificially perfused. Attempts to do this have been made by myself [6, 7] and by McKenzie and Szerb [8] in the caudate and putamen, using a push-pull cannula which irrigates a very small volume of tissue near the tip, and by Portig and Vogt [9, 10] who bathed the dorsal and lateral surfaces of the head of the caudate. The results obtained are summarized in Table 1, and indicate that small amounts of dopamine can be

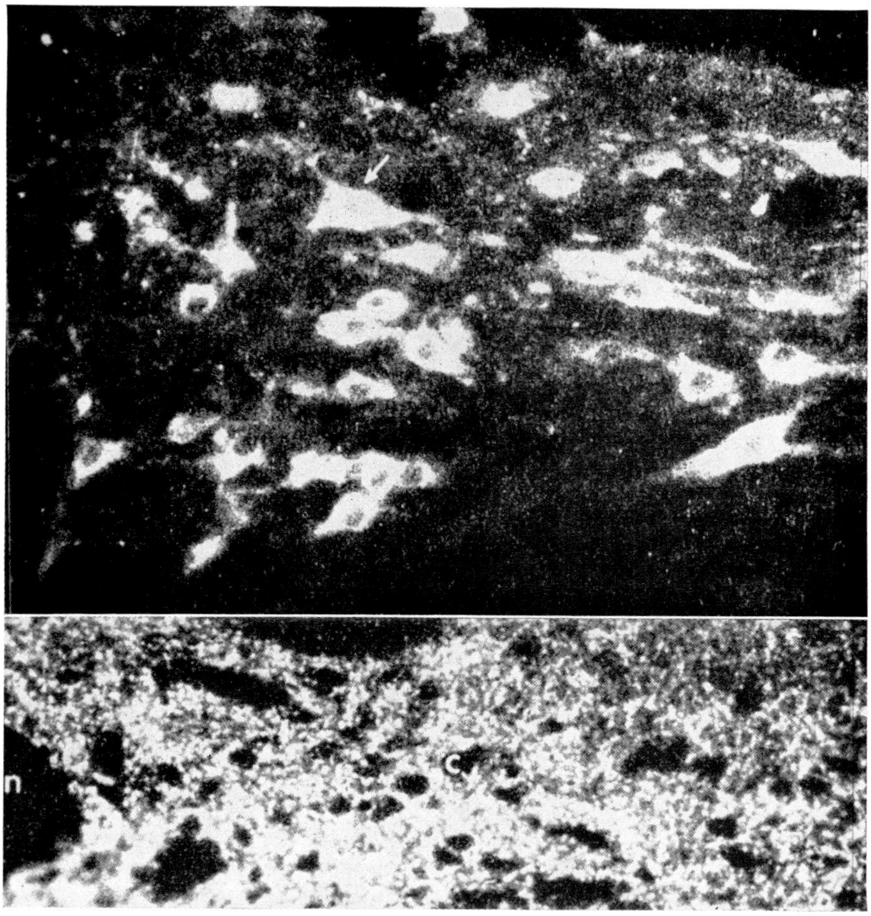

Fig. 1. *Above:* Neuronal somata in the zona compacta of the substantia nigra of rat, showing a fluorescent reaction for dopamine (from Dahlström and Fuxe [25]). *Below:* Histofluorescent appearance of dopamine-containing terminals in the caudate nucleus of rat. *c* indicates a nonfluorescent cell body with a number of terminals apparently in close association with it; *n* myelinated nerve bundles (from Andén, Fuxe, Hamberger and Hökfelt [4]).

detected in fluids allowed to irrigate the caudate or putamen. The concentrations in these fluids of the metabolic derivatives of dopamine (dihydroxyphenylacetic acid—DOPAC—and homovanillic acid) are considerably greater than elsewhere. Appropriate stimulation appears to result in an increased output of dopamine and of its major metabolite in this region.

The results obtained by various investigators when dopamine has been applied to neostriatal neurons by the iontophoretic technique are summarized in Table 2. These results seem clearly to indicate that more than half of the cells tested in the caudate are depressed or inhibited by dopamine and that relatively few are excited, but that the same situation may not be true for the putamen.

If these results are compared with those which have been obtained following

TABLE 1. The Release of Dopamine and its Metabolites from the Neostriatum in Response to Nigral and Thalamic Stimulation. (SN: substantia nigra; CM: nucl. centromedianus thalami)

Source	*Stimulation*	*Dopamine*	*DOPAC*	*Homovanillic Acid*	*Reference*
Caudate	Control	970 pg/min	—	—	McLennan [6]
	SN	No change	—	—	
	CM	1640 pg/min *	—	—	
Caudate	Control	<20 pg/min	0.4-0.5 ng/min	—	McKenzie and Szerb [8]
Caudate	Control	Small	—	3-6 ng/min	Portiz and Vogt [9, 10]
	SN	—	—	Increased	
Putamen	Control	380 pg/min	—	—	McLennan [7]
	SN	720 pg/min **	—	—	
	CM	560 pg/min ***	—	—	

* $P = 0.05$
** $P = 0.02$
*** $P >> 0.1$

electrical stimulation of the substantia nigra, the situation becomes more complicated. McLennan and York [11] described the action of dopamine upon caudate cells excited by nigral stimulation as either depressant (Fig. 2*A, B*) or without effect. They could never locate a neuron which was excited both by stimulation and by applied dopamine. Feltz [12] failed to find a depressant action of dopamine on nigrally excited cells, but agrees with McLennan and York that this substance could depress neurons which were not influenced by nigral excitation. The preliminary experiments of York [13] suggested that the situation in the putamen may not be identical, since he found that both nigral- and dopamine-induced excitations could frequently be detected in the same neurons.

The effects of stimulation of the intralaminar and centromedian nuclei of the thalamus upon events in the caudate have also been examined. McLennan [6] (Table 1) showed that low-frequency repetitive stimulation of the centromedian thalamic nucleus evoked a release of dopamine from the caudate.

TABLE 2. The Actions of Dopamine Applied Iontophoretically to Neurons of the Neostriatum

Structure	*Anesthesia*	*No. of Cells Examined*	*% Excited*	*% Depressed*	*Reference*
Caudate	None	86	14	50	Bloom, Costa and Salmoiraghi [24]
Caudate	Barbiturate	80	2.5	60	Bloom, Costa and Salmoiraghi [24]
Caudate	None	189	9	60	McLennan and York [11]
Caudate	None	213	10	63	York [14]
Caudate	None	100	(small)	70	Herz and Zieglgängsberger [16]
Putamen	?	120	39	30	York [13]

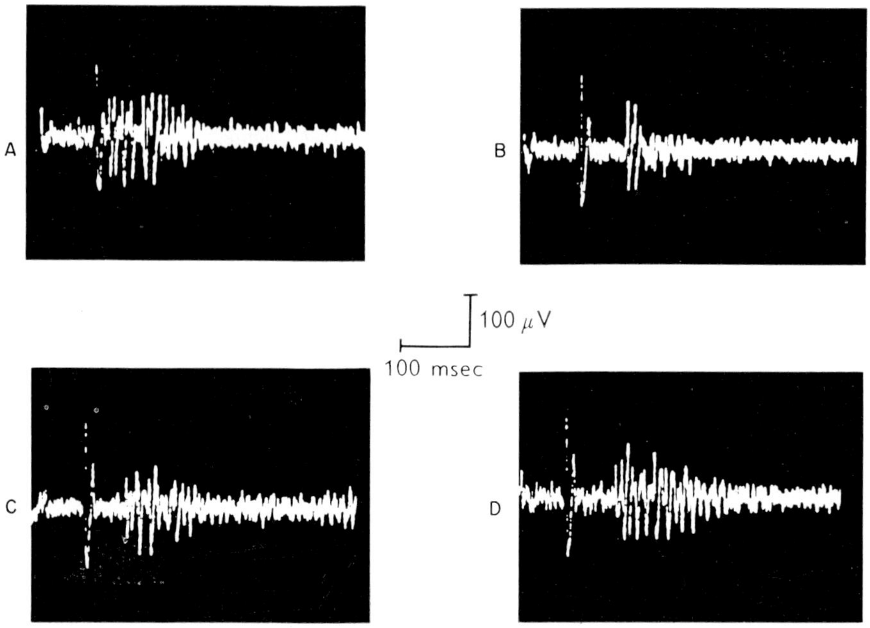

Fig. 2. Recordings obtained from a neuron of the caudate nucleus of cat during stimulation of the substantia nigra. *A:* Control; *B:* 20 sec after start of iontophoretic application of dopamine to the cell; *C:* 5 sec after cessation of phenoxybenzamine; *D:* 20 sec after start of dopamine application, immediately after *C*. The depressant action of dopamine has been prevented by phenoxybenzamine which had a negligible effect upon the nigral excitation (from McLennan and York [11]).

McLennan and York [11] and York [14] found that such stimulation would depress the firing of individual caudate neurons whether these were activated by nigral stimulation or by the iontophoretic application of an excitatory amino acid. Herz and Zieglgängsberger,[15, 16] stimulating the same region of the thalamus with single shocks rather than repetitively, reported that the effect upon caudate neurons took the form of a brief initial excitation followed by an inhibition of firing lasting ca. 200 msec, succeeded by a period of enhanced excitation. This response was reported to be suppressed by the application of dopamine.

These two sets of observations are not fully in accord since, although the repetitive stimuli used by York and myself might be expected to interact so that the inhibitory pauses observed by Herz and Zieglgängsberger [15, 16] might give an overall effect of continuous depression during the period of stimulation, the actions produced by dopamine do not appear to be entirely compatible. Whether a species difference between cats and rabbits can fully explain this discrepancy is not known.

Leaving aside this problem, the compendium of evidence suggesting that dopamine is involved as a synaptic transmitter in the caudate, specifically in a pathway leading from the medial thalamus, is rather good. The salient points are: (1) Stimulation of nucleus centromedianus caused the liberation of dopamine in the caudate; (2) Stimuli delivered repetitively depressed the activity of caudate neurons whether their activity was initiated by amino acid application or by nigral stimulation; (3) These effects were mimicked by the exogenous application of dopamine to the cells; (4) The effects of dopamine [11] (Fig. 2) and of centromedian stimulation [14] were prevented by the iontophoretic administration of phenoxybenzamine to the caudate neurons. This drug was without action on nigral excitation.

In the putamen the brief preliminary communications of York [13, 17] to the effect that nigral stimulation and dopamine application excited neurons, and the finding of McLennan [7] that dopamine release from this structure was increased by nigral stimulation, are more in accord with the histofluorescent data on the existence of nigroneostriatal dopamine-containing fibers. The further report that α-adrenergic blocking agents were effective in preventing both electrically and chemically induced excitation of neurons in the putamen is further evidence in favor of the view that dopamine here is acting as a synaptic excitatory transmitter. Further work is required both on this point and on the significance of the depressant responses which have also been reported to follow iontophoretic application of this material to other neighboring cells.

In the caudate nucleus however the evidence appears at present somewhat contradictory. On the one hand the histofluorescent results and the abundant biochemical data, of which much has been heard during this conference, are indicative of a functional relationship involving dopamine between the nigra and both the caudate and putamen. The iontophoretic data do not support the concept that the excitation of caudate neurons following nigral stimulation is a dopaminergic process.

It is true that nigral stimulation can also lead to a depression of firing of many caudate neurons (65 percent reported by Albe-Fessard, Raieva and Santiago,[18] and 44 percent by Connor [19]). These figures are much in accord with the proportions of cells depressed by iontophoretic dopamine (Table 2). The temptation to state that there is a direct nigrocaudate action which is dopaminergic and inhibitory is therefore strong, but many of the critical experiments have not yet been performed. The iontophoretic and pharmacological evidence for a similar link between the medial thalamus and caudate is better, as described earlier. It is worth noting that while direct connections between the zona compacta of the substantia nigra and the caudate have not been demonstrated by the classical methods of neuroanatomy [20, 21] or by electron microscopy [22, 23] they certainly exist between the substantia nigra and the nucleus centromedianus thalami.

Summary

Evidence concerning the possible role of dopamine as a synaptic transmitter in certain basal ganglial interconnections has been considered.

Histofluorescent and biochemical studies imply that axonal terminals stemming from the substantia nigra exist in the neostriatum.

Iontophoretic and pharmacological investigations do not support the hypothesis that dopamine is the transmitter of nigrocaudate excitation.

Dopamine may function as an inhibitory transmitter in the caudate via pathways in which the medial thalamus may be involved.

Preliminary evidence suggests that dopamine may be an excitatory transmitter in the nigroputamen pathway.

REFERENCES

1. Andén, N.-E., Dahlström, A., Fuxe, K., Larsson, K., Olson, L., and Ungerstedt, U.: Ascending monoamine neurons to the telencephalon and diencephalon. Acta Physiol. Scand. *67:* 313, 1966.
2. Parent, A., and Poirier, L. J.: The medial forebrain bundle (MFB) and ascending monoaminergic pathways in the cat. Canad. J. Physiol. Pharmacol. *47:* 781, 1969.
3. Bertler, Å., and Rosengren, E.: Occurrence and distribution of catechol amines in brain. Acta Physiol. Scand. *47:* 350, 1959.
4. Andén, N.-E., Fuxe, K., Hamberger, B., and Hökfelt, T.: A quantitative study on the nigro-neostriatal dopamine neuron system in the rat. Acta Physiol. Scand. *67:* 306, 1966.
5. Werman, R.: Criteria for identification of a central nervous system transmitter. Comp. Biochem. Physiol. *18:* 745, 1966.
6. McLennan, H.: The release of acetylcholine and of 3-hydroxytyramine from the caudate nucleus. J. Physiol. *174:* 152, 1964.
7. McLennan, H.: The release of dopamine from the putamen. Experientia *21:* 725, 1965.
8. McKenzie, G. M., and Szerb, J. C.: The effect of dihydroxyphenylalanine, pheniprazine and dextroamphetamine on the *in vivo* release of dopamine from the caudate nucleus. J. Pharmacol. *162:* 302, 1968.
9. Portig, P. J., and Vogt, M.: Release of homovanillic acid from the caudate nucleus. J. Physiol. *194:* 65, 1968.
10. Portig, P. J., and Vogt, M.: Activation of a dopaminergic nigro-striatal pathway. J. Physiol. *197:* 20, 1968.
11. McLennan, H., and York, D. H.: The action of dopamine on neurones of the caudate nucleus. J. Physiol. *189:* 393, 1967.
12. Feltz, P.: Dopamine, amino acids and caudate unitary responses to nigral stimulation. J. Physiol., in press.
13. York, D. H.: Is dopamine the transmitter of the nigro-striatal pathway? Proc. 1st. Winter Meet. Canad. Physiol. Soc., 1969.
14. York, D. H.: The inhibitory action of dopamine on neurones of the caudate nucleus. Brain Res. *5:* 263, 1967.
15. Herz, A., and Zieglgängsberger, W.: Synaptic excitation in the corpus striatum inhibited by microelectrophoretically administered dopamine. Experientia *22:* 839, 1966.
16. Herz, A., and Zieglgängsberger, W.: The influence of microelectrophoretically applied biogenic amines, cholinomimetics and procaine on synaptic excitation in the corpus striatum. Int. J. Neuropharmacol. *7:* 221, 1968.
17. York, D. H.: A microiontophoretic study of neurones in the globus pallidus-putamen complex. Aust. J. Exp. Biol. Med. Sci. *46:* P3, 1968.
18. Albe-Fessard, D., Raieva, S., and Santiago, W.: Sur les relations entre substance noire et noyau caudé. J. Physiol. *59:* 324, 1967.
19. Connor, J. D.: Caudate unit response to nigral stimuli: evidence for a possible nigro-neostriatal pathway. Science *160:* 899, 1968.

20. METTLER, F. A.: Fiber connections of the corpus striatum of the monkey and baboon. J. Comp. Neurol. *82:* 169, 1945.
21. AFIFI, A., and KAELBER, W. W.: Efferent connections of the substantia nigra in the cat. Exp. Neurol. *11:* 474, 1965.
22. FRIGYESI, T. L., and PURPURA, D. P.: Electrophysiological analysis of reciprocal caudato-nigral relations. Brain Res. *6:* 440, 1967.
23. HÖKFELT, T., and UNGERSTEDT, U.: Electron and fluorescent microscopical studies on the nucleus caudatus putamen of the rat after unilateral lesions of ascending nigro-neostriatal dopamine neurons. Acta Physiol. Scand. *76:* 415, 1969.
24. BLOOM, F. E., COSTA, E., and SALMOIRAGHI, G. C.: Anesthesia and the responsiveness of individual neurons of the caudate nucleus to acetylcholine, norepinephrine and dopamine administered by microelectrophoresis. J. Pharmacol. *150:* 244, 1965.
25. DAHLSTRÖM, A., and FUXE, K.: Evidence for the existence of monoamine-containing neurons in the central nervous system. I. Demonstration of monoamines in the cell bodies of brain stem neurons. Acta Physiol. Scand. *62:* Suppl. 232, 1964.

PART 5

HOW DOES L-DOPA WORK IN PARKINSONISM?

CHAPTER 13

Mechanisms of Action

HOW DOES L-DOPA WORK IN PARKINSONISM?

O. Hornykiewicz

When upon my suggestion Dr. Birkmayer early in 1961 gave small doses of L-Dopa (50 to 150 mg) intravenously * to parkinsonian patients and observed that this compound was capable of reversing, in some cases quite dramatically, the parkinsonian akinesia,[1] we had no doubts that L-Dopa did so because it was converted to dopamine in the brains of the patients. We based our assumption on the finding published a few months earlier by Ehringer and myself [2] showing a severe dopamine deficiency in the caudate nucleus and the putamen of parkinsonian patients. By that time the use of Dopa in animals as a precursor of brain dopamine and as an antireserpine agent was well established by Carlsson and his group [3, 4] and in 1959 Degkwitz and his colleagues [5] had used Dopa to antagonize the reserpine syndrome in man. (Everett and his colleagues [6] showed that the chemically related metamphetamine had also a strong antireserpine action.) We also had no doubts that the anti-parkinsonian action of L-Dopa was a real one: In the same year, that is to say in 1961, and quite independently from us, Barbeau, Sourkes and Murphy [7] described the antirigidity effect of oral Dopa. Later the issue became a little confused because a few neurologists either denied that Dopa had an anti-parkinsonian activity or found that the drug was inferior to the conventional atropine-like drugs. Such a controversy about the therapeutic value of a new drug is nothing unusual, and in the case of L-Dopa

* Since only about 10 percent of orally administered L-Dopa reaches the general circulation unchanged, the above doses correspond to 0.5 to 1.5 gm oral L-Dopa.

quite nicely shows how difficult it is to assess clinically the activity of new drugs in a disease as complex as the parkinsonian syndrome. Finally time decided in favor of L-Dopa, particularly in the light of the decisive report in 1967 by Cotzias and his team [8] on excellent results with high oral doses of DL-Dopa, given in increasing amounts over prolonged periods of time.

Although the connection between the anti-parkinsonian action of L-Dopa and the dopamine deficiency in the basal ganglia seemed very logical and direct, we soon realized that this was far from being the only possibility. Theoretically, either the amino acid itself or any of its numerous metabolites could be the active compound, and the action could be either peripheral or central in origin. In the central nervous system, the site of action could be the striatum or some other area.

The first question which arose was, therefore: What is the active compound responsible for the anti-parkinsonian actions of L-Dopa? Apart from L-Dopa, the following compounds for L-Dopa in the body are quantitatively of major importance, and therefore have to be considered here: 3-methoxy-4-hydroxy-phenylalanine (3-methoxydopa), dopamine, 3-methoxydopamine, de-aminated dopamine metabolites (DOPAC, HVA etc.), norepinephrine and its methoxylated and/or deaminated metabolites. Dr. Sandler mentioned a multitude of other, probably minor, metabolites of L-Dopa.

Most of my discussion of the above problem will be based on findings in man, that is to say, parkinsonian patients. Clinical-pharmacological studies were performed together with Dr. Birkmayer in Vienna on patients in whom the beneficial effect of intravenous L-Dopa was both excellent and reproducible.

The anti-parkinsonian action of L-Dopa could be due (1) to L-Dopa itself or (2) to metabolites formed from the amino acid. One Dopa metabolite can be easily excluded as being responsible for the anti-parkinsonian action of L-Dopa: this compound is 3-methoxydopa which is formed in the body, as animal experiments show, in fair amounts after injection of L-Dopa. In our clinical experiments [9] injection of up to 100 mg of 3-methoxydopa to parkinsonian patients did not have any L-Dopa-like anti-akinesia effect. This finding is more significant as it seems that 3-methoxydopa is not attacked by the mammalian aromatic amino acid decarboxylase [10]; therefore we can expect this compound to be more stable in the body than L-Dopa itself.

The next major group of compounds formed from L-Dopa are the amines dopamine and norepinephrine and their methoxylated and/or deaminated metabolites. Useful pharmacological tools stand at our disposal to decide whether the amines formed from L-Dopa participate in its anti-parkinsonian (and other pharmacological) actions; these are the monoamine oxidase (MAO) inhibitors. From animal experiments it has long been known that inhibition of MAO greatly potentiates the pharmacological actions of L-Dopa, causing an increased accumulation of the amines formed from the amino acid in different tissues. Experiments in parkinsonian patients showed that in fact chronic pretreatment of the patients with a MAO inhibitor greatly potentiated the anti-parkinsonian activity of L-Dopa; the effective dose could be reduced from 150 mg to 25 mg intravenously.[9, see also 7]

The above observations using 3-methoxydopa and MAO inhibitors allow the conclusion that probably one of the amines formed from L-Dopa is the compound responsible for the anti-parkinsonian activity of the amino acid. Neither L-Dopa itself nor its metabolite 3-methoxydopa in these experiments qualified for this role.

The anti-parkinsonian action of L-Dopa could be (1) peripheral or (2) central in origin. This question is comparatively easy to answer, particularly if we keep in mind the conclusion reached above, namely, that the anti-parkinsonian action of L-Dopa is due not to the amino acid itself but apparently to an amine formed from it. From the amines formed from L-Dopa, neither dopamine [9] nor norepinephrine, when given systemically, has any Dopa-like activity in parkinsonian patients. On the contrary, norepinephrine, when infused peripherally, has been shown to increase parkinsonian tremor.[11] These observations already suggest that the site of the L-Dopa action might be the central nervous system. This conclusion is supported by recent findings on the combined treatment of parkinsonian patients with L-Dopa and one of the peripheral decarboxylase inhibitors. Substances belonging to the latter group prevent the decarboxylation of L-Dopa in the periphery only, and by doing so they increase the amount of L-Dopa going into the brain; Dr. Pletscher presented the pertinent data. Peripheral decarboxylase inhibitors have been shown to potentiate the anti-parkinsonian actions of L-Dopa (cf. this volume: Birkmayer; Siegfried; Tissot). From these observations we can conclude that there is evidence for the view that the site of action of L-Dopa in parkinsonism is the central nervous system.

The anti-parkinsonian action of L-Dopa is probably due to (1) amine(s) or (2) amine metabolite(s) formed from it in the body. The question of which of the amines formed from L-Dopa in the central nervous system is responsible for the anti-parkinsonian actions of the amino acid is certainly the most difficult one to answer. However, it seems safe to assume that none of the deaminated metabolites of dopamine or norepinephrine can play a role. This is clearly shown by the observation that MAO inhibitors potentiate L-Dopa actions in parkinsonian patients; MAO inhibitors effectively decrease the concentration of deaminated metabolites. It is more difficult to decide which of the amines, dopamine or norepinephrine, is the decisive one. Experiments in parkinsonian patients showed that threo-dihydroxyphenylserine (threo-DOPS, 200 mg i.v.) had no Dopa-like activity.[9] Since threo-DOPS is converted in the body to norepinephrine without the intermediate formation of dopamine, the conclusion that norepinephrine is of no importance for the anti-parkinsonian effects of L-Dopa suggests itself. This is in agreement with similar findings in animals.[12, 13] However, one must remember that the decarboxylation of threo-DOPS to norepinephrine is a slow process (as judged from animal experiments); furthermore, nothing is known about the ability of the human brain to convert threo-DOPS to norepinephrine.

In spite of several questions which will remain open (e.g., role of 3-methoxydopamine), it seems that at the moment the anti-parkinsonian action of L-Dopa is best explained by assuming that it is the *dopamine formed in the brain* which is the active compound, and that there is a close relationship between striatal dopamine deficiency and parkinsonian akinesia and rigidity. This conclusion is in keeping with a large body of fairly solid evidence obtained in animal experiments, especially in respect to striatal dopamine metabolism in reserpinized animals and Dopa-reserpine antagonism. Therefore it would appear that, for the time being, we are justified to say that in principle we have a rational basis for the use of L-Dopa in parkinsonism.

If L-Dopa is to act in parkinsonism by dopamine formed from it, obviously the most important prerequisite is that the enzyme which forms dopamine be present in the parkinsonian brain. Simple as this question may seem, we still do not have a final answer to it. The major difficulty is undoubtedly the fact that, despite several attempts, it has been thus far impossible to detect in postmortem material of control patients a reasonable level of Dopa decarboxylase activity; the values reported by several research groups are very low indeed.[14, 15, 16] This is the reason why results on Dopa decarboxylase activity in parkinsonian brains have to be interpreted with caution. A study by Bernheimer and myself[14] showed no significant difference in enzyme activity in homogenates of the striatum between normal and parkinsonian patients. Interestingly enough, when we estimated the Dopa decarboxylase activity in striatal *slices* (instead of in homogenates) the values were in general higher than in homogenates, and there was some indication that in the putamen of two parkinsonian patients the enzyme activity was lower than in the putamen of control patients; however, the scatter in the control group was rather high.[17] These findings could be interpreted as indicating that in parkinsonism there is no major impairment of the enzymatic conversion of L-Dopa to dopamine, but it is clear that this question should be carefully reexamined. Animal experiments seem to prove that after making lesions in the nigrostriatal dopamine pathway in the monkey there is a decrease of striatal Dopa decarboxylase activity[18, 19] (see also Dr. Sourkes' paper, this volume).

The question is sometimes raised as to how dopamine can act in parkinsonism, where often nearly 100 percent of the nigrostriatal dopamine neurons have degenerated. To answer this question, one has to keep in mind two important points: (1) Degeneration of the dopamine neurons does not mean degeneration of the dopamine-sensitive striatal (receptor) cells. Therefore the problem comes down to the question just discussed—whether the parkinsonian striatum has enough decarboxylase activity left to form dopamine from the administered L-Dopa. (2) Clinical experience seems in fact to indicate that severe cases of parkinsonism show a smaller reaction to L-Dopa than mild cases; however, this issue, like many other clinical questions, needs further clarification.

In my attempt to indicate that L-Dopa acts in parkinsonism through the formation of dopamine in the brain, all my conclusions are based on the therapeutic effects of low intravenous doses of L-Dopa as used routinely by Dr. Birkmayer in Vienna. However, we have to realize that the situation may be profoundly different when the high oral doses, as introduced in 1967 by Cotzias,[8] are used. It is obvious that low doses do not exert some of the therapeutic effects which the high doses seem to have (such as antitremor activity), and they also do not produce hyperkinetic side effects. High doses of L-Dopa undoubtedly give rise to an accumulation of very large amounts of L-Dopa and its metabolites in the periphery and in the central nervous system. The possibility must, therefore, be kept in mind that under these conditions the drug's action might be a combination of specific, anti-parkinsonian, dopamine effects and more or less unspecific effects produced by some other L-Dopa metabolite or even L-Dopa itself. This possibility also implies that, with high doses of L-Dopa, the site of action could be, besides the striatum, nearly any other brain region. Still unpublished work in my laboratory on the regional distribution and metabolism of high L-Dopa doses in the brain of the mouse shows that the amounts of L-Dopa, 3-methoxydopa, dopamine and homovanillic acid which accumulate after an intraperitoneal injection of 400 mg per kg L-Dopa in the cortex, "caudate nucleus," hypothalamus and the midbrain area are rather high. Therefore one cannot dismiss the possibility that these compounds, when accumulating in brain areas from which they normally are absent, may interfere with metabolic or neurotransmission processes of the corresponding brain regions; they also may interfere with the action of the active compound formed from L-Dopa, i.e., dopamine. A good example for such a possibility was given by Dr. Krnjevic during this symposium. In 1963 Krnjevic and Phillis [20] observed in the cat that Dopa increased the rate of firing of single cortical neurons when applied locally to them. The excitatory effect of Dopa on cortical neurons was in contrast to the action of dopamine which inhibited single unit activity in the cat's cortex.

The question of how L-Dopa works in parkinsonism not only includes the question as to the active compound mediating the L-Dopa effect but also the neurophysiological and neuropharmacological modes of action. It is my opinion that these questions are intimately connected with the question of whether or not there does exist a dopaminergic nigrostriatal pathway in the human brain. Animal experiments seem to support the existence of such a pathway,[21, 22] and Dr. Moore has presented beautiful work showing that, using the appropriate staining techniques, degeneration of nerve terminals can be observed in the striatum (cat) after tegmental lesions associated with cell loss in the substantia nigra. In the human brain, a nigrostriatal pathway containing dopamine can be assumed from the characteristic pattern of homovanillic acid distribution within the internal capsule.[23] The concentration of this dopamine metabolite increased in the caudorostral direction; lowest concentrations can be found in the caudal part of the internal capsule next to the substantia nigra, and highest amounts in the rostral parts close to the caudate nucleus and putamen. To date, however, the evidence in favor of a dopaminergic nigrostriatal pathway in man is certainly not complete.

On the whole the neurophysiological activity of dopamine in the brain seems to be mainly inhibitory. It would thus appear that in parkinsonism the removal of inhibitory influences normally exerted by dopamine in the striatum and the globus pallidus leads to a dysfunction of these centers, resulting in the symptoms of akinesia and rigidity. This hypothesis was put forward more than three years ago.[24] Dr. Poirier, Dr. Krnjevic and Dr. McLennan have already touched on this problem, and I would like to hear some additional provocative views on this question.

The possibility has been also raised that L-Dopa might act in parkinsonism not so much by replenishing the dopamine in the striatum but by exerting its (pharmacological) influence upon some other brain areas and/or mechanisms. Although, as far as I can see, most of the neurochemical and pharmacological evidence at present available is in favor of the "striatal dopamine hypothesis," it would be wise to be alert to other possibilities. The uncertainty in this field of research shows that we need more basic neuropharmacological research on the extrapyramidal motor system.

In this context I should also like to remind you of Denny-Brown's observations on degeneration of the substantia nigra without any parkinsonian symptomatology. One might disagree with Denny-Brown as to how to interpret these findings, but there is no reason to doubt the validity of his observations. It seems to me that it would be rewarding indeed to examine such cases of nigral degeneration without parkinsonism in respect to the striatal monoamine metabolism.

Before ending my remarks, I should like to say a few words about L-Dopa as a drug. If it is really true that the striatal dopamine deficiency is causally connected to some of the symptoms in parkinsonism, and, furthermore, if L-Dopa works in parkinsonism by replenishing the missing dopamine in the striatum, then L-Dopa is quite obviously the most natural substance we can have for treating what I should like to call *"the striatal dopamine deficiency syndrome."* However, it is quite clear to me as a pharmacologist that, whatever the mode and site of its action may be, L-Dopa is far from being perfect as a drug. Some of the side effects, especially of high doses of L-Dopa as used at the present time, would have been classified by classical pharmacological standards as being already in the "subtoxic" range. This shows that we need new and safer compounds which act like L-Dopa, that is, compounds which have a dopamine-like anti-parkinsonian activity.

REFERENCES

1. BIRKMAYER, W., and HORNYKIEWICZ, O.: Der L-Dioxyphenylalanin (= DOPA)-Effekt bei der Parkinson-Akinese. Wien. Klin. Wschr. *73:* 787, 1961.
2. EHRINGER, H., and HORNYKIEWICZ, O.: Verteilung von Noradrenalin und Dopamin (3-Hydroxytyramin) im Gehirn des Menschen und ihr Verhalten bei Erkrankungen des extrapyramidalen Systems. Klin. Wschr. *38:* 1236, 1960.
3. CARLSSON, A., LINDQVIST, M., MAGNUSSON, T., and WALDECK, B.: On the presence of 3-hydroxytyramine in brain. Science *127:* 471, 1958.
4. CARLSSON, A., LINDQVIST, M., and MAGNUSSON, T.: 3,4-Dihydroxyphenylalanine and 5-hydroxytryptophan as reserpine antagonists. Nature *180:* 2100, 1957.

5. DEGKWITZ, R., FROWEIN, R., KULENKAMPFF, C., and MOHS, U.: Über die Wirkungen des L-DOPA beim Menschen und deren Beeinflussung durch Reserpin, Chlorpromazin, Iproniazid und Vitamin B_6. Klin. Wschr. *38:* 120, 1960.
6. EVERETT, G. M., and TOMAN, J. E. P.: Mode of action of Rauwolfia alkaloids and motor activity. *In* Biological Psychiatry (H. J. Masserman, ed.). Grune and Stratton, New York, 1959.
7. BARBEAU, A., SOURKES, T. L., and MURPHY, G. F.: Les catécholamines dans la maladie de Parkinson. *In* Monoamines et Systeme Nerveux Centrale (J. de Ajuriaguerra, ed.). Georg, Genève, and Masson, Paris, 1962.
8. COTZIAS, G. C., VAN WOERT, M. H., and SCHIFFER, L. M.: Aromatic amino acids and modification of parkinsonism. New Eng. J. Med. *276:* 374, 1967.
9. BIRKMAYER, W., and HORNYKIEWICZ, O.: Der L-Dioxyphenylalanin (= DOPA)-Effekt beim Parkinson-Syndrom des Menschen: Zur Pathogenese und Behandlung der Parkinson-Akinese. Arch. Psychiat. Nervenkr. *203:* 560, 1962.
10. SOURKES, T., HENEAGUE, P., and TRANO, Y.: Enzymatic decarboxylation of isomers and derivatives of dihydroxyphenylalanine. Arch. Biochem. *40:* 185, 1952.
11. BARCROFT, H., PETERSON, E., and SCHWAB, R. S.: Action of adrenaline and noradrenaline on the tremor in Parkinson's disease. Neurology *2:* 154, 1952.
12. SCHEEL-KRÜGER, J., and RANDRUP, A.: Stereotyped hyperactive behavior produced by dopamine in the absence of noradrenaline. Life Sci. *6:* 1389, 1967.
13. CREVELING, C. R., DALY, J., TOKUYAMA, T., and WITKOP, B.: The combined use of α-methyltyrosine and threo-dihydroxyphenylserine—selective reduction of dopamine levels in the central nervous system. Biochem. Pharmacol. *17:* 65, 1968.
14. BERNHEIMER, H., and HORNYKIEWICZ, O.: Das Verhalten einiger Enzyme im Gehirn normaler und parkinsonkranker Menschen. Arch. Exp. Path. Pharmak. *243:* 295, 1962.
15. LANGEMANN, H., and ACKERMANN, H.: Über die Aktivität der Aminosäuren-Decarboxylasen im Gehirn des Menschen. Helv. Physiol. Acta *19:* 399, 1961.
16. ROBINS, E., ROBINS, J. M., CRONINGER, A. B., MOSES, S. G., SPENCER, S. J., and HUDGENS, R. W.: The low level of 5-hydroxytryptophan decarboxylase in human brain. Biochem. Med. *1:* 240, 1967.
17. BERNHEIMER, H., and HORNYKIEWICZ, O.: Unpublished data.
18. ANDÉN, N.-E., DAHLSTRÖM, A., FUXE, F., LARSSON, K., OLSON, L., and UNGERSTEDT, U.: Ascending monoamine neurons to the telencephalon and diencephalon. Acta Physiol. Scand. *67:* 313, 1966.
19. GOLDSTEIN, M., ANAGNOSTE, B., BATTISTA, A. F., OWEN, W. S., and NAKATANI, S.: Studies of amines in the striatum in monkeys with nigral lesions. J. Neurochem. *16:* 645, 1969.
20. KRNJEVIĆ, K., and PHILLIS, J. W.: Iontophoretic studies of neurons in the mammalian cerebral cortex. J. Physiol. *165:* 274, 1963.
21. ANDÉN, N.-E., CARLSSON, A., DAHLSTRÖM, A., FUXE, K., HILLARP, N.-Å., and LARSSON, K.: Demonstration and mapping out of nigro-neostriatal dopamine neurons. Life Sci. *3:* 523, 1964.
22. POIRIER, L. J., and SOURKES, T. L.: Influence of the substantia nigra on the catecholamine content of the striatum. Brain *88:* 181, 1965.
23. HORNYKIEWICZ, O., LISCH, H.-J., and SPRINGER, A.: Homovanillic acid in different regions of the human brain: Attempt at localizing central dopamine fibres. Brain Res. *11:* 662, 1968.
24. HORNYKIEWICZ, O.: Dopamine (3-hydroxytyramine) and brain function. Pharmacol. Rev. *18:* 925, 1966.

LESSONS FROM CLINICAL EXPERIENCE *

André Barbeau

How does L-Dopa work? How does it improve and permit improvement in akinesia, rigidity and even tremor? What is the mechanism through which L-Dopa causes psychotic symptoms or abnormal involuntary movements? Many hypotheses have been proposed in the last few years, and a few more during this symposium, but the final understanding is not yet at hand.

Often the answer to many basic problems comes from careful clinical observations of patients. Drawing from our own experience and that of others over these last nine years, we would like to summarize our thoughts and offer a few hypotheses to be tested in the laboratory.

Beneficial Results

There is little doubt that L-Dopa causes its beneficial effects through its transformation, within the brain, into dopamine, which in turn probably acts upon dopaminoceptive receptors in the striatum. The evidence for this argument has been throughly reviewed by Hornykiewicz in 1966 [1] and again in this volume. At low oral or intravenous doses (i.e., 50 to 150 mg intravenously and up to 2 grams daily postoperatively), such as were used until 1967, this was evident from the time course of the measurable effects: within 30 minutes after an oral dose of 200 mg of L-Dopa, rigidity and reaction times were significantly reduced.

With the progressively higher doses now being used, the problem is more complex. Large quantities of dopamine probably flood not only the striatum but also other brain areas. However, it is our contention that the improvement in motility is still due to stimulation of striatal dopaminoceptive receptors. This impression is based on observation slightly at variance to those reported by others at this symposium. We do not think that success or failure of L-Dopa depends on the absolute number of functioning nigrostriatal "dopaminergic" neurons, or the state of pigmentation within the substantia nigra, both factors otherwise important in determining the extent, topography and severity of the illness.

We cannot relate failure to the duration or severity of the illness as long as the symptoms are purely extrapyramidal. We have observed a few patients who had been bedridden for more than three years, totally immobile because of rigidity and akinesia (Stage 5 of Yahr), but who were entirely clear mentally and had no evidence of pyramidal tract damage. These patients responded very well to L-Dopa treatment and some were among our most complete reversals. As long as the extrapyramidal illness is "pure," it is entirely amenable to therapy. Failures start to appear when there is concomitant involvement of other systems, and particularly with generalized

* Supported by Grant SP-1 from the Medical Research Council of Canada.

arteriosclerosis clinically evident through dementia and Babinski signs. We feel that the important factor determining the success or failure rate is the *amount of damage to the dopaminoceptive receptor cells in the striatum.* These cells *must* be intact for dopamine, from endogenous or exogenous origin, to exert any kinetic effect. In Parkinson's disease, probably as a consequence of the damage to the nigrostriatal pathways and as a necessary compensation, these cells may already be hypersensitive (denervation hypersensitivity), thus permitting a measurable kinetic response with a single oral dose of 200 mg or an intravenous dose of 50 mg.[2, 3] The long-term effect of high, repeated doses of Dopa (i.e., of dopamine) is probably to reduce this hypersensitivity, thus to return the "set" of responsiveness of these cells to more normal levels. Such an hypothesis is compatible with the theories of Andén [4] and of Poirier [5] assuming direct dopamine release upon these cells from nigrostriatal fibers, or with the theory of McLennan in this volume whereas the nigrostriatal system is not dopaminergic but influences dopamine-releasing cells in the striatum. In our hypothesis the all-important factors in determining the kinetic effects are the absolute number of dopaminoceptive receptor cells in he striatum and their "set" of responsiveness.

Stimulation of the dopaminoceptive cells results in a direct kinetic effect which reverses the akinesia produced by a long deprivation of dopamine. This symptom (and all its components of decreased associated movements, bradypsychia, apraxia, loss of postural control [6]) is the only true "negative symptom," [7] and therefore the only one *directly* helped by L-Dopa. The other manifestation of parkinsonism, such as rigidity and tremor, are *release phenomena.* They are probably due to the increased activity of other servomechanisms and loops in various areas of the brain, brain stem or even spinal cord, liberated from the inhibitory influence of the striatum. Such a servomechanism, the rubro-olivo-cerebello-rubral loop, has been demonstrated by Poirier and collaborators [8] as necessary for tremor production. It is likely that these other mechanisms are modulated by a variety of neurohormones, which may or may not be involved in Parkinson's disease but are dependent on the state of activity of the striatal dopaminoceptive system. Any *functional* alteration in this striatal system (such as with harmaline, oxotremorine, α-methyl-p-tyrosine, phenothiazines or haloperidol) will result in release *symptoms* if there is an accompanying *structural* lesion in one or more of the other loops. Otherwise bradykinesia is the only symptom elicited. L-Dopa, by normalizing the unsettled balance between these servomechanisms and neurohormones, eventually decreases the hyperactivity of the other systems and corrects still other symptoms. The time course of tremor reversal in L-Dopa therapy, as observed by most authors,[19] is compatible with this hypothesis.

Adverse Effects

It is obvious that L-Dopa in high doses is being converted into large quantities of dopamine, even within the brain. This is more evident still with the concomitant use of a peripheral Dopa decarboxylase inhibitor such as

Ro 4-4602.[10] Side effects have been observed as previously noted [11] with such long-term, high-dose therapy. The most troublesome are undoubtedly abnormal involuntary movements and mental changes and we will limit our remarks to these two. If it is true that the kinetic effect of Dopa is due to direct stimulation of dopaminoceptive receptors in the striatum, the side effects may be due to any of the following possibilities:

1. Exaggerated stimulation of striatal dopaminoceptive receptors by excess dopamine.
2. Stimulation of the same receptors by derivatives, normal or abnormal, of dopamine or Dopa.
3. Displacement of other neurohormones in other sites by excess dopamine or dopamine derivatives.
4. Modification of the balance between various neurohormonal systems in the striatum or elsewhere.

A few clinical observations may guide us in our evaluation of the above possibilities:

1. Abnormal involuntary movements were never observed before 1967 on doses of L-Dopa not exceeding 2 grams: there is thus a *factor of total dose*.
2. Abnormal involuntary movements occur usually after weeks or months of continuous treatment: thus a *factor of time* (duration of administration) must also be considered.
3. Abnormal involuntary movements are not observed in normal subjects receiving Dopa,* in nonextrapyramidal neurological patients also treated with the drug,[9, 11] but they can occur in dystonia musculorum deformans; [12] and finally, Dopa certainly aggravates choreic symptoms in classical Huntington's chorea.[13] Thus there is the *necessity for preexisting damage* to the nigrostriatal system, before Dopa can produce abnormal movements.
4. Abnormal movements involving more than the cephalic area are almost always accompanied by a reversal of the reactivity of the deep tendon reflexes. From hyperactive, usual in Parkinson's disease, the reflexes become markedly hypoactive or even absent. Hypotonia invariably accompanies the appearance of the involuntary movements in other parts of the body.
5. The main factors controlling the appearance of toxic psychotic manifestations (vivid dreams, hallucinations, paranoid delusions, anxiety reactions, depression) are the preexisting terrain (subclinical psychiatric illness?), the total daily dose of Dopa and the rate of administration. The larger and the faster the increase, the higher the incidence of these toxic manifestations.

From such observations we would like to postulate the following:

1. The stereotyped orobuccofacial grimaces of choreic type (see classification earlier in this symposium) are due to overstimulation of the striatal dopaminoceptive receptors by excess dopamine. The presence of a previously damaged nigrostriatal system is necessary for these effects, probably by creating a state of hypersensitivity of the receptors. This hypersensitivity can be reduced by very low increments of L-Dopa.
2. Increases in slow, large-amplitude tremor and dystonic peripheral manifestations, with generalized hypotonia, result from displacement of other neurotransmitters elsewhere in the brain (most likely the brain stem) by dopamine or some of its methylated derivatives (3-methoxytyramine or epinine?). This is akin to a false transmitter effect.

* See Ansel and Markham, this volume.

3. The toxic psychotic symptoms are due to displacement of other neurotransmitters in the brain stem and/or limbic system. Serotonin appears to be the most likely candidate. As an example we refer to the pseudo-aphrodisiac effects of L-Dopa and of *p*-chloro-phenylalanine.

Experiments exploring these hypotheses are under way in our department and will be reported shortly. For example, we have been able to produce dyskinesias associated with generalized hypotonia with large doses of 3-methoxytyramine, a normal dopamine derivative.[14]

Conclusions

L-Dopa does not exert its beneficial and adverse effects through similar mechanisms. Because of this, it will certainly prove to be a most useful tool in the study of extrapyramidal disorders and even of psychoses.

REFERENCES

1. HORNYKIEWICZ, O.: Dopamine (3-hydroxytyramine) and brain function. Pharmacol. Rev. *18:* 925, 1966.
2. BARBEAU, A.: Biochemistry of Parkinson's disease. *In* Proceedings of the Seventh International Congress of Neurology, Rome, September 1961. Societa Grafica Romana, Rome, 1961.
3. BIRKAMYER, W., and HORNYKIEWICZ, O.: Der L-3,4-dioxyphenylalanin (Dopa)-Effek bei der Parkinson Akinese. Wien. Klin. Wschr. *73:* 787, 1961.
4. ANDÉN, N. E., CARLSSON, A., DAHLSTRÖM, A., FUXE, K., HILLARP, N. Å., and LARSSON, K.: Demonstration and mapping out of nigro-neostriatal dopamine neurons. Life Sci. *3:* 523, 1964.
5. POIRIER, L. J., and SOURKES, T. L.: Influence of the substantia nigra on the catecholamine content of the striatum. Brain *88:* 191, 1965.
6. JOUBERT, M., and BARBEAU, A.: Akinesia in Parkinson's disease. An experimental study. *In* Progress in Neuro-Genetics (A. Barbeau and J. R. Brunette, eds.). Excerpta Medica Foundation, Amsterdam, 1969.
7. MARTIN, J. P.: The Basal Ganglia and Posture. J. P. Lippincott Co., Philadelphia, 1967.
8. POIRIER, L. J., BOUVIER, G., BEDARD, P., BOUCHER, R., LAROCHELLE, J., OLIVIER, A., and SINGH, P.: Essai sur les circuits neuronaux impliqués dans la tremblement postural et l'hypokinésie. Rev. Neurol. *120:* 15, 1969.
9. BARBEAU, A.: L-Dopa therapy in Parkinson's disease: a critical review of nine years' experience. Canad. Med. Assoc. J. *101:* 791, 1969.
10. BARBEAU, A., and GILLO-JOFFROY, L.: Treatment of Parkinson's disease with L-Dopa and Ro 4-4602. Excerpta Medica I.C.S. *193:* 171, 1969.
11. COTZIAS, G. C., PAPAVASILIOU, P. S., and GELLENE, R.: Modification of Parkinsonism: chronic treatment with L-Dopa. New Eng. J. Med. *280:* 337, 1969.
12. ELDRIDGE, R., ed.: Symposium on the torsion dystonias. Neurology, in press.
13. BARBEAU, A.: L-Dopa and juvenile Huntington's disease. Lancet *2:* 1066, 1969.
14. BARBEAU, A.: Importance and pathogenesis of abnormal movements during L-Dopa therapy of Parkinson's disease. Neurology, in press.

L-DOPA IN PARKINSON'S DISEASE: POSSIBLE MODES OF ACTION *

T. L. Sourkes

Godwin-Austen et al.[1] have recently stated clearly the theoretical basis for the use of L-Dopa in Parkinson's disease, pointing to the depletion of striatal dopamine in patients with the disease,[2] presumably due to degeneration of nigrostriatal fibers, and to the mounting evidence that dopamine acts as a neurotransmitter at certain striatal synapses. Administration of Dopa would increase the concentration of dopamine and favor restoration of lost function. This was the basis for the earliest trials of Dopa in Parkinson's disease [3,4] and for some of the more recent ones. Cotzias was led to test Dopa for other reasons.[5] The concept of replacement therapy using L-Dopa may well prove to be correct, but the treatment deserves sober assessment in terms of intermediary metabolism of Dopa and dopamine and in relation to the clinical results reported thus far.

The major path of metabolism of L-Dopa seems to be its decarboxylation to dopamine, catalyzed by a pyridoxal phosphate-linked enzyme. This reaction takes place in nervous tissue (and elsewhere) and, according to the above view, would permit Dopa to replace the missing dopamine, for dopamine does not pass the blood-brain barrier readily but L-Dopa does. In this way dopamine would become available at receptors in the striatum and perhaps elsewhere in the brain, where it would act as a neurohumoral agent.[6,7] In order to overcome losses of Dopa by decarboxylation outside the brain, large doses must be given; the limited accessibility of certain antidecarboxylases such as Ro 4-4602 (N-trihydroxybenzy-N′-serylhydrazide) or MK-485 (DL-alpha-hydrazino-beta-3,4-dihydroxy-phenylpropionic acid monohydrate) to the brain may aid in preserving the orally administered Dopa from excessive decarboxylation peripherally.[8,9]

If the only function of L-Dopa were to replace the missing dopamine, one should expect to find cases that fail to respond: extensive pathology would entail loss of the normal cellular machinery for decarboxylation of Dopa for temporary storage of the dopamine that is formed, and for its release by physiological stimuli. In fact, there is experimental evidence in monkeys with brain lesions for just such a loss,[10] and the results of some clinical studies [1] provide a correlation of efficacy of the treatment with severity of the disease, i.e., greatest improvement in the mildest cases. However, the clinical investigations are not consistent, for Cotzias et al.[9] obtained good results with L-Dopa in severe as well as mild cases, whereas Calne et al.[11] could not predict therapeutic response from a review of clinical details. This uncertainty necessarily generates scepticism about the replacement-therapy hypothesis.

Is some substance other than dopamine formed which serves as an anti-Parkinson agent? There is little information about the metabolism of Dopa

* Supported by grants from the Medical Research Council of Canada.

in brain with respect to other pathways such as catechol-O-methylation and transamination. These reactions have been invoked by Cotzias et al.[9] in a general way. They would lead to a number of products of as yet unknown biological activity or chemical reactivity, e.g., the oxo-acids corresponding to Dopa and 3-O-methyldopa (3-methoxytyrosine). These oxo-acids undergo oxidative decarboxylation to dihydroxyphenylacetic and homovanillic acids, respectively, but they may follow other routes of which we are now ignorant. Even dimethoxyphenethylamine (i.e., dimethyldopamine), the original "pink spot," has been proposed for a role in Parkinson's disease, although the evidence for this is as uncertain as in schizophrenia.[12]

Many amines and amino acids in the catechol series can react with aldehydes, forming Schiff bases.[13, 14] The ubiquitous coenzyme pyridoxal phosphate is one such aldehyde giving rise to an isoquinoline product.[13] Pyridoxamine phosphate, present to a small extent in some tissues, could form an analogous product with the aldehyde resulting from the action of monoamine oxidase on dopamine. Since 1940 there have been sporadic reports of beneficial effects in Parkinson's disease obtained with very high doses of vitamin B-6.[15] If these observations are correct, then excess pyridoxine entering the brain may be reacting in that organ with an endogenous metabolite to form an anti-Parkinson substance. Endogenous pyridoxine might also do this. A discovery by Holtz et al.[16] makes this possibility more realistic. The product of the action of monoamine oxidase on dopamine, 3,4-dihydroxyphenylacetaldehyde, can condense with unchanged dopamine to yield tetrahydropapaveroline, a hypotensive drug. Thus far this reaction has been demonstrated only in vitro; [16] perhaps the same or an analogous reaction occurs in vivo.

Thus, the property of L-Dopa rendering it useful in the treatment of Parkinson's disease is its penetrability of the blood-brain barrier; in parenchymal cells it would undergo conversion to a specific anti-Parkinson chemical by some minor pathway of metabolism. The long-lasting effect (ca. three weeks) after withdrawal of L-Dopa [1, 9] speaks for the formation of some metabolite with a greater length of life in the brain than dopamine possesses. If such a metabolite could be identified, one ought to be able to synthesize it in the laboratory and test it directly in Parkinson's disease, using smaller dosages than necessary for L-Dopa. An obvious experiment would be the chronic treatment of animals with labeled L-Dopa, followed by examination of the brain for radioactive products. Drugs that would sustain high levels of Dopa or dopamine in the brain, to allow for formation of a putative therapeutic derivative, would be useful in such experimental studies and perhaps even in clinical ones. This means the use not only of antidecarboxylases but also of inhibitors of catechol-O-methyl transferase and dopamine-beta-hydroxylase. Disulfiram, which inhibits the latter enzyme, is already under clinical trial in Parkinson's disease.[17] If a minor metabolite or product of Dopa is ultimately active, these basic experimental studies may prove to be the shortest route to the discovery of a new series of therapeutic agents for the treatment of Parkinson's disease.

Finally, L-Dopa itself may have a direct action on certain neurons.

Recent success with the antiviral agent amantadine in Parkinson's disease [18] and evidence for increased turnover of brain biogenic amines in animals infected with herpes simplex virus [19] emphasize that, with our present state of knowledge of Parkinson's disease and the metabolism of L-Dopa in the brain, we must be open-minded about the mechanism of action of this compound.

REFERENCES

1. Godwin-Austen, R. B., Tomlinson, E. B., Frears, C. C., and Kok, H. W. L.: Effect of L-Dopa in Parkinson's disease. Lancet *2:* 165, 1969.
2. Ehringer, H., and Hornykiewicz, O.: Verteilung von Noradrenalin und Dopamine (3-Hydroxytyramin) im Gehirn des Menschen und ihr Verhalten bei Erkrankungen des extrapyramidalen Systems. Klin. Wschr. *38:* 1236, 1960.
3. Barbeau, A., Sourkes, T. L., and Murphy, G. F.: Les catécholamines dans la maladie de Parkinson. *In* Monoamines et Système Nerveux Central (de Ajuriaguerra, J., ed.). Paris, Masson, 1962.
4. Birkmayer, W., and Hornykiewicz, O.: Der L-dioxyphenylalanine (=DOPA) Effekt beim Parkinson-syndrom des Menschen: Zur Pathogenese und Behandlung der Parkinson-akinese. Arch. Psychiat. Nerv. *203:* 560, 1962.
5. Cotzias, G. C., Van Woert, M. H., and Schiffer, L. M.: Aromatic amino acids and modification of parkinsonism. New Eng. J. Med. *276:* 374, 1967.
6. Carlsson, A.: The occurrence, distribution and physiological role of catecholamines in the nervous system. Pharmacol. Rev. *11:* 490, 1959.
7. Sourkes, T. L.: Formation of dopamine *in vivo:* relation to the function of the basal ganglia. Rev. Canad. Biol. *20:* 187, 1961.
8. Tissot, R., Gaillard, J. M., Guggisberg, M., Gauthier, G., de Ajuriaguerra, J.: Thérapeutique du syndrome de Parkinson par la L-Dopa "per os" associée à un inhibiteur de la décarboxylase. Presse Mèd. *77:* 619, 1969.
9. Cotzias, G. C., Papavasiliou, P. S., and Gellene, R.: Modification of parkinsonism: chronic treatment with L-Dopa. New Eng. J. Med. *280:* 337, 1969.
10. Lancaster, G., Larochelle, L., Bédard, P., Missala, K., Sourkes, T. L., and Poirier, L. J.: Effect of brain lesions and of harmaline on the dopa decarboxylase activity of striatum of the cat. In preparation.
11. Calne, D. B., Stern, G. M., Laurence, D. R., Sharkey, J., and Armitage, P.: L-Dopa in postencephalitic parkinsonism. Lancet *1:* 744, 1969.
12. Anonymous: Lessons of the "pink spot." Lancet *1:* 382, 1969.
13. Schott, H. F., and Clark, W. G.: Dopa carboxylase inhibition through the interaction of coenzyme and substrate. J. Biol. Chem. *196:* 449, 1952.
14. Sourkes, T. L.: Inhibition of dihydroxyphenylalanine decarboxylase by derivatives of phenylalanine. Arch. Biochem. Biophys. *51:* 444, 1954.
15. Holtz, P., and Palm, D.: Pharmacological aspects of vitamin B_6. Pharmacol. Rev. *16:* 147, 1964.
16. Holtz, P., Stock, K., and Westerman, E.: Pharmakologie des Tetrahydropapaverolins und seine Enstehung aus Dopamin. Arch. Exp. Path. Pharmakol. *248:* 387, 1964.
17. Serrano, P.: Personal communication, August, 1969.
18. Schwab, R. S., England, A. C., Poskanzer, D. C., and Young, R. F.: Use of amantadine hydrochloride in Parkinson's disease. J.A.M.A. *208:* 1168, 1969.

19. LYCKE, E., and ROOS, B. E.: Some biological and biochemical aspects of the pathogenesis of Parkinson's disease. *In* Third Symposium on Parkinson's Disease (Gillingham, J. F., and Donaldson, I. M. L., eds.). E. & S. Livingston, Edinburgh, 1969.

ENZYME CHEMISTRY AND L-DOPA

A. Meister

The accumulated data seem to show that the function of the extrapyramidal system involves, at least in part, a balanced system of complex interactions between certain highly active compounds: dopamine, serotonin, norepinephrine, histamine, glutamate, and γ-aminobutyrate. It occurs to me that the synthesis of all of these requires the action of vitamin B-6 enzymes. To my knowledge, none of these has yet been isolated from mammalian brain in homogeneous form. Yet the answers to many of the detailed questions that have been considered at this meeting relating to inhibitions, competition, "teeter-totter" interactions, etc., will undoubtedly require isolation of these (and related brain enzymes) in pure form. Such matters as the existence and types of isoenzymes, allosteric phenomena, mechanisms of inactivation, and control mechanisms need to be studied.

It seems clear that the recent dramatic clinical achievements with Dopa have developed directly from basic biochemical studies which have dealt largely with *metabolism.* I would propose that a new phase of understanding and clinical application would closely follow basic developments at the *enzyme level.* By this I mean isolation of pure enzymes from brain and their examination and study as chemical entities. Clearly Dopa decarboxylase is one of the crucial enzymes in this important group. Although brain enzymes are not always easy to isolate, progress has been made in this area—even in the isolation of enzymes from human brain obtained at autopsy. In our own laboratory we have been able to obtain in homogeneous form glutamine synthetase and two isoenzymes of γ-glutamylcyclotransferase from human brain. Once one has the pure enzyme in hand, a number of new experimental possibilities arise. For example, although it has been known for many years that brain glutamine synthetase is inhibited by the convulsant agent methionine sulfoximine, the true nature of the inhibition was elucidated only when experiments could be done on homogeneous enzyme; thus, it is now known that methionine sulfoximine is phosphorylated on the enzyme to yield methionine sulfoximine phosphate, which binds tightly to the enzyme and inhibits it irreversibly.

In relation to this, I wonder whether Dopa might be inhibiting certain of the vitamin B-6 enzymes of the intestinal flora of patients receiving large doses of this compound. Dr. Sourkes has already cited earlier work which indicates the formation of a stable cyclic derivative of Dopa and pyridoxal-5′-phosphate. Formation of such an inactive complex might lead to vitamin B-6 deficient bacterial flora, which would tend to destroy less of the administered Dopa. When patients are given pyridoxine—even in very low doses—the ability of the enzymes of the bacteria to destroy Dopa might be increased. This would explain the clinical finding that pyridoxine administration seems to decrease both the beneficial and the untoward effects of Dopa. As a practical matter, this idea might be worth checking in relation to oral Dopa therapy.

NEUROPHYSIOLOGICAL MECHANISMS IN PARKINSONISM

Herbert H. Jasper

I will confine myself to a few general comments concerning the interpretation of some of the data which has been presented and the overall problem of neurophysiological mechanisms which may be involved in the syndrome of parkinsonism.

Microiontophoretic Studies

The local application of minute quantities of chemical substances in the near vicinity of central neurons while recording electrical activity from a separate extracellular and even intracellular micropipette has provided a wealth of precise data concerning the local action of possible neurotransmitter substances. However, the interpretation of such data must be made with considerable caution, and always together with evidence from other sources using quite different neurochemical or electrophysiological techniques combined with behavioral studies. For example, acetylcholine (ACh), the most thoroughly established central transmitter substance, appears to be an inhibitory substance when viewed from effects upon anterior horn cells, but excitatory for Renshaw cells which exert their inhibitory action on anterior horn cells by means of another substance, probably glycine. Furthermore, Renshaw cells appear to have both rapidly acting "nicotinic" ACh receptors and delayed long-lasting "muscarinic" type receptors.

In the cerebral cortex, some (perhaps about 25 percent) deep-lying pyramidal cells appear to be activated in a slow "muscarinic" manner by ACh while about the same percentage of more superficial cells are inhibited.

In the lateral geniculate and sensory nuclei of the thalamus many cells appear to be activated by ACh, but conduction of afferent impulses through these nuclei to the cerebral cortex is not blocked by specific ACh blocking agents; therefore, it cannot be assumed that ACh is the mediator of rapid synaptic transmission in sensory pathways. In fact, the mediator involved in rapid transmission across synaptic stations in either the principal sensory or motor pathways of the brain has yet to be established with assurance. It seems unlikely to me that it could be the universally excitatory amino acid glutamate, as suggested by Krnjevic. Certainly the biogenic monoamines, NE, 5-HT, or dopamine, cannot be considered as likely candidates for direct synaptic transmission in the principal sensory and motor pathways, though there is some evidence to suggest that NE and 5-HT may act indirectly as modulator substance. They may play a major rôle in the regulation of general excitatory or inhibitory states associated with states of consciousness and arousal, the evidence for which has been clearly outlined by Cordeau in this volume. What then is the specific rôle of Dopa and dopamine?

Iontophoretic evidence would suggest that dopamine has an inhibitory action upon cells of the caudate nucleus, although the precise nature of this action has yet to be clarified. It may have an excitatory action upon interneurons, which in turn inhibit other neurons, as described for the action of ACh upon Renshaw cells and motor neurons of the cord. Such a mechanism would be consistent with the observation that unit discharge can be recorded from caudate cells in response to electrical stimulation of the substantia nigra. Since there is a growing body of evidence to suggest that ACh is an excitatory transmitter in the caudate, dopamine deficiency in Parkinson patients would be consistent with the theory that disinhibition of the striatum may be a most important mechanism in at least a part of the syndrome, especially since anticholinergic drugs have also been effective in its treatment. This hypothesis implies that excessive unregulated discharge of the caudate may cause bradykinesia, tremor, and loss of associated movements. In addition one must assume that the administration of very large amounts of Dopa serves only to replace the deficient dopamine to restore striatal inhibitory-excitatory balance. This appears to me to be a gross distortion and oversimplification of the physiological action of the striatum in relation to other components of the extrapyramidal motor system which have shown to be involved in the generation of the Parkinson syndrome, and also a gross oversimplification of the action of high concentrations of Dopa, as has been pointed out by Carlsson in this volume. It may well be, as Ward has suggested, that the effectiveness of Dopa may even be unrelated to striatal dopamine deficiency ("the right treatment for the wrong reason").

The Conception of Alpha/Gamma Imbalance

Ward has suggested that the principal defect in parkinsonism is a disturbance in balance between alpha and gamma motoneurons, a balance which tends to be reestablished by thalamotomy and by high concentrations of Dopa, but not necessarily by the reestablishment of dopamine balance in the striatum. This seems to me to be a reasonable proposition, since it is highly unlikely that the Parkinson syndrome can be due only to striatal dopamine deficiency and just as unlikely that all of the beneficial effects, as well as the remarkable dyskinetic side effects, with prolonged administration of Dopa can be due to a presumed restoration of dopamine balance in the striatum. The potent activating influence of Dopa upon the central nervous system as a whole is probably due to the stimulation of noradrenergic as well as dopaminergic receptors throughout the neuraxis, including noradrenergic systems of the spinal cord itself where spinal motor systems could be directly affected.

Imbalance of Phasic vs. Tonic Motor Systems

I would like now to suggest a modification of Ward's hypothesis of alpha/gamma imbalance for your consideration. It is based upon two premises for which there is considerable experimental support: (1) that the syndrome of parkinsonism is primarily a disturbance in tonic-postural as opposed to phasic-dynamic components of the motor system, and (2) that there may be a neurochemical basis for reciprocal antagonism between tonic-postural and phasic-dynamic components of motor systems throughout the neuraxis. This is merely an extension of Sherrington's demonstration that movement must involve a change in posture; tone or rigidity must be arrested for movement to occur. Defective negative bias in a feedback control system, allowing oscillation, could explain the tremor. An additional defect in a generalized motor arousal system may be necessary to explain some cases in which bradykinesia appears in excess of that which might be explained by rigidity alone.

It has been clearly demonstrated that both alpha and gamma motor systems can be divided into phasic and tonic or "dynamic and static" components with quite different functional properties and anatomical structure, including different kinds of anterior horn cells, different kinds of muscle spindles and probably different kinds of striatal muscle fibers. Tonic and phasic motoneurons can also be distinguished among pyramidal tract cells of the motor cortex. Differences in the physical properties, conduction velocity, and anatomical connections of tonic and phasic pyramidal tract cells have recently been studied in considerable detail by Oshima. He concludes that tonic pyramidal tract (pt) cells, in addition to subserving muscle tone or "static" motor functions, also serve to provide a positive excitatory bias in preparation for the execution of rapid phasic movements, a bias which seems deficient in bradykinetic Parkinson patients. In addition, he has shown that tonic pt cells have axonal terminals in the cuneate nucleus to provide presynaptic inhibitory control on somatic afferent pathways, while a similar postsynaptic inhibitory feedback control is exerted by the more rapid phasic pt cells. This is just another important aspect of the "sensory" functions of motor systems which must be kept constantly in mind when attempting to understand neurophysiological mechanisms of the dyskinesias. Is it not possible that it is an imbalance between tonic and phasic aspects of motor systems throughout the neuraxis which may have distinctive neurochemical mediators, and that this, rather than a simple restoration of dopamine balance in the striatum, may explain the beneficial effects of Dopa in Parkinson patients?

Importance of Combined Neuroanatomical and Neurochemical Mechanisms

I would like to put on record an impression, or perhaps I should say a growing conviction, which has developed over many years of studying Parkinson patients with Gilles Bertrand of the Montreal Neurological Institute, and experimental monkeys with parkinsonian tremor with Drs. Poirier, Cordeau and Gybels many years ago. I am becoming convinced that neither an exclusively neuroanatomical nor neurochemical approach will ever provide a complete understanding of the physiopathological mechanisms involved. It seems to me that both must be considered together. For example, it is not easy to produce chronic spontaneous tremor in monkeys with local tegmental lesions near the substantia nigra. There are many failures. Following such lesions, however, a typical Parkinson-like tremor can be induced by means of certain pharmacological agents such as harmaline or thioproperazine in doses which do not produce tremor in intact animals. Recently, in collaboration with de Champlain, we induced a typical sustained Parkinson tremor in a monkey with a tegmental lesion and only rare brief episodes of spontaneous tremor by a single injection of thioproperazine in the amount of 1 mg per kg. The tremor then persisted actively for one month without further injection.

This is just one of many observations which could be cited in favor of the conception that we must search for interacting neuroanatomical and neurochemical factors if we are to succeed in our attempts to provide a comprehensive and accurate description of physiopathological mechanisms involved in the Parkinson syndrome and in the action of Dopa on the various components of that syndrome.

CHAPTER 14

Summation of the Conference

SUMMATION OF THE CONFERENCE AND CONCLUSIONS

Arvid Carlsson

This conference, which I have the honor and privilege to conclude, has been centered around one single compound, L-Dopa. We have witnessed how L-Dopa has enormously fertilized basic research in the fields of neuroanatomy, neurobiochemistry, and neurophysiology and how it has been introduced as a new and important therapeutic principle in Parkinson's disease.

The general theme of the conference has been the exploration of the basic mechanisms underlying the therapeutic results with L-Dopa and the production of side effects. The structure has been somewhat unusual, at least according to my experience: clinical observations have formed a base from which the basic scientists have been invited to explain—or at least try to explain—these observations. I should like to congratulate the organizing committee on this interesting approach to the problem of interdisciplinary communication, for this problem is one of the most serious in science today, and I think that the intentions of the committee did in fact materialize.

One cannot demand from any conference that all aspects of the problems it serves to illustrate are covered equally well. One aspect which I think important for understanding the particular profile of the research field is the historical one. This has been treated in a rather fragmentary manner, and this is to a certain extent a shortcoming. I feel that it is almost as important critically to describe and analyze the historical development of our research field as our actual scientific observations.

I have therefore decided to review very briefly the history of L-Dopa from its synthesis to its introduction in routine clinical therapy (Table 1), and have divided this historical development into three phases:

The first phase I call the *early phase,* simply because it was before my time as an active scientist. Immediately, we come across a very famous name, Casimir Funk, the man who coined the term "vitamin" because he believed that all the essential food constituents of this category were amines. Funk took the trouble of synthesizing Dopa because he thought it might be the precursor of epinephrine, the only biogenic monoamine known at the time. In this case he was indeed right; this was a brilliant idea.

TABLE 1. History of L-Dopa

Early Phase

1911—Synthesis of DL-Dopa (Casimir Funk).
1913—Isolation and identification of L-Dopa in Vicia faba.
Emetic action in self experiment (2.5 g orally) (Guggenheim).
1923—Synthesis of dopamine (Waser and Sommer).
1938—Detection of Dopa decarboxylase in mammalian tissue (Holt, Heise and Lüdtke).
1942—Blood pressure increased by L-Dopa.
—Dopamine identified as normal constituent of human urine (Holtz, Credner and Koepp).

L-Dopa and Parkinsonism; Preclinical Phase

1954/
1955—Reserpine-induced parkinsonism described by several clinical groups.
1955—Reserpine causes release of 5-HT (Shore, Silver and Brodie).
1956—Reserpine depletes also catecholamine stores (Carlsson and Hillarp; Bertler, Carlsson and Rosengren; Holzbauer and Vogt) and causes adrenergic transmission failure (Bertler, Carlsson and Rosengren).
1957—Reserpine-induced akinesia and sedation reversed by DL-Dopa (Carlsson, Lindqvist and Magnusson).
—Presence of dopamine in brain suggested (Montagu).
1958—Dopamine identified and quantitatively measured in brain. Depletion of dopamine by reserpine demonstrated; restoration of dopamine levels by Dopa. Independent role of brain dopamine proposed (Carlsson, Lindqvist, Magnusson and Waldeck).
—High dopamine levels in caudate nucleus-putamen. Dopamine suggested to be involved in parkinsonism and chorea (Bertler and Rosengren; Carlsson).
1959—Independent demonstration of high dopamine level in caudate nucleus-putamen (Sano, Gamo, Takimoto, Taniguchi, Takesada and Nishinuma).

L-Dopa and Parkinsonism; Clinical Phase

1960—Low levels of dopamine, norepinephrine and 5-HT in the brains of parkinsonian patients (Ehringer and Hornykiewicz).
1961—Low urinary dopamine in parkinsonian patients (Barbeau, Murphy and Sourkes).
—Intravenous L-Dopa (50-150 mg) has a beneficial effect on the akinesia of parkinsonian patients (Birkmayer and Hornykiewicz).
—Oral L-Dopa (200 mg; single dose) reduces rigidity and tremor in parkinsonian patients (Barbeau, Sourkes and Murphy).
1965—". . . the effect of L-Dopa on akinesia and rigidity is undeniable. On the other hand the question as to the therapeutic value of L-Dopa in parkinsonism seems still to be unsettled. This, in my opinion is mainly due to the fact that, because of the unpleasant side effects of L-Dopa, it is practically impossible to inject doses of the drug which are high enough to increase the dopamine level in the striatum" (Hornykiewiz).
1967—Oral DL-Dopa (up to 16 g/day) induces striking and sustained improvement in parkinsonian patients (Cotzias, Van Woert and Schiffer).
—Oral L-Dopa (up to 8 g/day) also active (Cotzias and Papavasiliou).
1967—The observations of Cotzias et al. confirmed by several groups.
1968—Introduction of L-Dopa in routine treatment of parkinsonism.

Guggenheim, the author of the comprehensive handbook of biogenic amines, isolated and identified L-Dopa from the bean Vicia faba. He took 2.5 grams of it orally and was lucky enough to demonstrate its emetic action. In animals he did not observe any effects because the doses he used were too small. If he had saved some of the material he took himself for the animal experiments, he might have speeded the development considerably.

Another grand old man in pharmacology, Peter Holtz, is still an active scientist. There is a series of important papers by Holtz and his co-workers starting with the discovery of Dopa decarboxylase in mammalian tissue in 1938. He then injected Dopa and observed an increase in blood pressure, and finally he identified dopamine as a normal constituent of human urine.

We now come to what I would like to call the *preclinical phase.* It starts out with a very important clinical observation, however: it was observed by several groups that reserpine could induce the picture of parkinsonism; it is difficult to single out any particular group responsible for this observation. Then came, of course, a very great breakthrough, the discovery that reserpine releases 5-HT. It is to be regretted that Dr. Bernard Brodie, because of illness, has not been able to be with us; his presence would have added to this conference.

I had the privilege of being with Dr. Brodie shortly after his discovery and, after returning home, together with Hillarp and other co-workers found that reserpine also depletes catecholamine stores. This was also reported independently by Martha Vogt. We also found that reserpine causes transmission failures in the adrenergic system, and this made us believe that, in contrast to the concept of Brodie, we might be dealing with a deficiency of amine rather than an excess. Thus we injected Dopa into animals and found that the reserpine-induced akinesia and sedation could be reversed by Dopa.

The same year (1957) Montagu suggested that dopamine was present in the brain, and in 1958 we independently identified and quantitatively measured it in the brain. We demonstrated a depletion of dopamine with reserpine and its restoration with Dopa; we thus proposed that dopamine might play an independent role in the brain. Finally, we found high dopamine levels in caudate nucleus and putamen and suggested that dopamine may be involved in parkinsonism and chorea. Japanese workers led by Sano independently demonstrated high dopamine levels in the caudate nucleus and putamen.

We now arrive at the *clinical phase.* This began with the paper of Ehringer and Hornykiewicz demonstrating low levels of dopamine, norepinephrine and 5-HT in the brains of parkinsonian patients. It was followed by the observation of Barbeau, Murphy and Sourkes that there were low urinary dopamine levels in parkinsonian patients. Here we should note that both in Vienna and in Canada the important step from laboratory to clinic took place through the cooperation of two people, one laboratory man and one clinician (Hornykiewicz and Birkmayer in Austria, Sourkes and Barbeau in Canada); this collaboration of course was vitally necessary.

The activity of intravenous Dopa was demonstrated by Birkmayer and Hornykiewicz, and of oral L-Dopa by Barbeau, Sourkes and Murphy. Their results were repeated in many different laboratories, and after several years Hornykiewicz concluded that the situation was as shown in Table 1.

A different kind of thinking was required to proceed further, and this was provided by Dr. Cotzias and his co-workers who began the oral therapy we now know so well. Cotzias himself has defined his contribution as the slow titration of each individual patient—and this was a great contribution indeed. His observations have been confirmed by several groups, and Dopa is being introduced into routine therapy.

To make a real summation of a conference such as this in a short time would be impossible. Let me make just two points.

First, some of the confusion about treatment failures will probably be clarified after we have learned more about the pharmacokinetics of L-Dopa and convenient methods for measuring L-Dopa levels in plasma have become generally available. We learned a bit about the fate of L-Dopa in the body during this conference, but our knowledge is still incomplete.

Second, in order to master some of the side effects, we will evidently have to modify the treatment. Peripheral Dopa decarboxylase inhibitors seem to have something to offer against emetic side effects, and possibly also against hypotension. Dopamine β-oxidase inhibitors might perhaps be useful for preventing norepinephrine formation and thus possibly mental side effects or orthostatic hypotension.

With respect to abnormal movements, we need a deeper analysis of the underlying mechanism before it is possible to suggest a way to eliminate this problem. Perhaps we can hope for results using Dopa in combination with various drugs known to modify response to L-Dopa. We had, in fact, a few observations indicating that this is a promising pathway.

There are still discoveries to make in this field, and I would now like to suggest that we go home and make them. Before that, however, I have the great honor and privilege to speak for all the participants and to thank Dr. Barbeau and his staff once again for a magnificent conference and a most generous hospitality. Our thanks also go to the National Institutes of Health, The Medical Research Council of Canada, and the W. Garfield Weston Charitable Foundation, for having supported this conference.

REFERENCES

1. Barbeau, A., Murphy, G. F., and Sourkes, T. L.: Excretion of dopamine in diseases of basal ganglia. Science *133:* 1706, 1961.
2. Barbeau, A., Sourkes, T. L., and Murphy, G. F.: Les catecholamines dans la maladie de Parkinson. *In* Monoamines et Système Nerveux Central (de Ajuriaguerra, J., ed.). Georg, Genève and Masson, Paris, 1962.
3. Bertler, A., Carlsson, A., and Rosengren, E.: Release by reserpine of catecholamines from rabbits' hearts. Naturwissenschaften *22:* 521, 1956.
4. Bertler, A., and Rosengren, E.: Occurrence and distribution of dopamine in brain and other tissues. Experientia *15:* 10, 1959.
5. Birkmayer, W., and Hornykiewicz, O.: Der L-Dioxyphenylalanin (=Dopa)-Effekt bei der Parkinson-Akinese. Wien. Klin. Wschr. *73:* 787, 1961.

6. CARLSSON, A.: The occurrence distribution and physiological role of catecholamines in the nervous system. Pharmacol. Rev. *11:* 490, 1959.
7. CARLSSON, A., and HILLARP, N.-A.: Release of adrenaline from the adrenal medulla of rabbits produced by reserpine. Kgl. Fysiogr. Sällsk. Lund Förh. *26:* 8, 1956.
8. CARLSSON, A., LINDQVIST, M., and MAGNUSSON, T.: 3,4-Dihydroxyphenylalanine and 5-hydroxytryptophan as reserpine antagonists. Nature *180:* 1200, 1957.
9. CARLSSON, A., LINDQVIST, M., MAGNUSSON, T., and WALDECK, B.: On the presence of 3-hydroxytyramine in brain. Science *127:* 471, 1958.
10. COTZIAS, G. C., and PAPAVASILIOU, P. S.: Therapeutic studies of parkinsonian patients. Long-term effects of DL- and L-Dopa. *In* Progress in Neuro-Genetics (Barbeau, A., and Brunette, J.-R., eds.). Excerpta Medical Foundation, Amsterdam, 1969.
11. COTZIAS, G. C., VAN WOERT, M. H., and SCHIFFER, L. M.: Aromatic amino acids and modification of parkinsonism. New Eng. J. Med. *276:* 374, 1967.
12. EHRINGER, H., and HORNYKIEWICZ, O.: Verteilung von Noradrenalin und Dopamin (3-Hydroxytyramin) im Gehirn des Menschen und ihr Verhalten bei Erkrankungen des extrapyramidalen Systems. Klin. Wschr. *38:* 1236, 1960.
13. FUNK, C.: Synthesis of dl-3:4-dihydroxyphenylalanine. J. Amer. Chem. Soc. *99:* 554, 1911.
14. GUGGENKEIM, M.: Dioxyphenylalanin, eine neue Aminosäure aus Vicia faba. Z. Physiol. Chem. *88:* 276, 1913.
15. HOLTZ, P., CREDNER, K., and KOEPP, W.: Die enzymatische Entstehung von Oxytyramin im Organismus und die Physiologische Bedeutung der Dopa decarboxylase. Arch. Exp. Path. Pharmakol. *200:* 356, 1942.
16. HOLTZ, P., HEISE, R., and LÜDTKE, K.: Fermentativer Abbau von 1-Dioxyphenylalanin durch die Niere. Arch. Exp. Path. Pharmakol. *191:* 87, 1938.
17. HOLZBAUER, M., and VOGT, M.: Depression by reserpine of the noradrenaline concentration in the hypothalamus of the cat. J. Neurochem. *1:* 8, 1956.
18. HORNYKIEWICZ, O.: Metabolism of brain dopamine in human parkinsonism: neurochemical and clinical aspects. *In* Biochemistry and Pharmacology of the Basal Ganglia (Costa, E., Coté, L. Y., and Yahr, M. D., eds.). Raven Press, New York, 1966.
19. MONTAGU, K. A.: Catechol compounds in rat tissues and in brains of different animals. Nature *180:* 244, 1957.
20. SANO, I., GAMO, T., TANKIMOTO, Y., TANIGUCHI, K., TAKESADA, M., and NISHINUMA, K.: Distribution of catechol compounds in human brain. Biochim. Biophys. Acta *32:* 586, 1959.
21. SHORE, P. A., SILVER, S. L., and BRODIE, B. B.: Interaction of reserpine, serotonin and lysergic acid diethylamide in brain. Science *122:* 284, 1955.

PART 6

SELECTED BIBLIOGRAPHY ON L-DOPA, DOPAMINE AND PARKINSONISM

(Prepared by André Barbeau and Nicole Guay)

I. DOPAMINE AND PARKINSONISM 1954-1969

1954

1. VOGT, M.: The concentration of sympathin in different parts of the central nervous system under normal conditions and after administration of drugs. J. Physiol. *123:* 451, 1954.

1955

2. SHORE, P. A., SILVER, S. L., and BRODIE, B. B.: Interaction of reserpine, serotonin and lysergic acid diethylamide in brain. Science *122:* 284, 1955.

1956

3. CARLSSON, A., and HILLARP, N. Å.: Release of adrenaline from the adrenal medulla of rabbits produced by reserpine. Kgl. Fysiogr. Sällsk. Lund Förh. *26:* 8, 1956.

1957

4. CARLSSON, A., LINDQVIST, M., and MAGNUSSON, T.: 3,4-dihydroxyphenylalanine and 5-hydroxytryptophan as reserpine antagonists. Nature *180:* 1200, 1957.
5. MONTAGU, K. A.: Catechol compounds in rat tissues and in brains of different animals. Nature *180:* 244, 1957.

1958

6. CARLSSON, A., LINDQVIST, M., MAGNUSSON, T., and WALDECK, B.: On the presence of 3-hydroxytyramine in brain. Science *127:* 471, 1958.
7. HORNYKIEWICZ, O.: The action of dopamine on the arterial blood pressure of the guinea pig. Brit. J. Pharmacol. *13:* 91, 1958.

1959

8. BERTLER, A., and ROSENGREN, E.: Occurrence and distribution of dopamine in brain and other tissues. Experientia *15:* 10, 1959.
9. CARLSSON, A.: The occurrence, distribution and physiological role of catecholamines in the nervous system. Pharmacol. Rev. *11:* 490, 1959.
10. SANO, I., GAMO, T., KAKIMOTO, Y., TAMIGUCHI, K., TAKEGADA, M., and NISHINUMA, K.: The occurrence, distribution and physiological role of catecholamines in the nervous system. Biochim. Biophys. Acta *32:* 586, 1959.

1960

11. Barbeau, A.: Preliminary observations on abnormal catecholamine metabolism in basal ganglia diseases. Neurology *10:* 446, 1960.
12. Blaschko, H., and Chrusciel, T. L.: The decarboxylation of amino acids related to tyrosine and their awakening action in reserpine-treated mice. J. Physiol. *111:* 272, 1960.
13. Ehringer, H., and Hornykiewicz, O.: Verteilung von noradrenalin und dopamin (3-hydroxytyramin) im Gehirn des Menschen und ihr Verhalten bei Erkrankungen des extrapyramidalen Systems. Klin. Wschr. *38:* 1236, 1960.

1961

14. Barbeau, A.: Dopamine and basal ganglia diseases. Arch. Neurol. *4:* 97, 1961.
15. Barbeau, A., Murphy, G. F., and Sourkes, T. L.: Excretion of dopamine in diseases of basal ganglia. Science *133:* 1706, 1961.
16. Bernheimer, H., Birkmayer, W., and Hornykiewicz, O.: Verteilung des 5-hydroxytryptamins (serotonin) im Gehirndes menschen und seim Verhalten bei Patienten mit Parkinson-Syndrom. Klin. Wschr. *39:* 1056, 1961.
17. Bertler, A.: Occurrence and localization of catecholamines in the human brain. Acta Physiol. Scand. *51:* 97, 1961.
18. Cotzias, G. C., Borg, D. C., Hughes, F. R., Bertinchamps, A. J., and Papavasiliou, P. S.: Phenothiazines: curative or causative in regard to parkinsonism? Rev. Canad. Biol. *20:* 289, 1961.
19. Sourkes, T. L.: Formation of dopamine in vivo: relation to function of basal ganglia. Rev. Canad. Biol. *20:* 2, 1961.

1962

20. Barbeau, A.: The pathogenesis of Parkinson's disease: a new hypothesis. Canad. Med. Assoc. J. *87:* 802, 1962.
21. Bernheimer, H., and Hornykiewicz, O.: Das Verhalten einiger Enzyme im Gehirn normales und Parkinson Kranker Menschen. Arch. Exp. Path. Pharmakol. *243:* 295, 1962.
22. Everett, G. M., and Wiegand, R. G.: Central amines and behavioral states: a critique and new data. *In* Proc. First Int. Pharmacol. Meeting, *8:* 85, 1962.
23. Hornykiewicz, O.: The occurrence of dopamine (3-hydroxytyramine) in the central nervous system: its relationship to parkinsonism. German Med. Month. *7:* 344, 1962.

1963

24. Carlsson, A.: Evidence for a role of dopamine in extrapyramidal functions. Acta Neuroveg. *26:* 484, 1963.
25. McGeer, P. L.: Central amines and extrapyramidal functions. J. Neuropsychiat. *4:* 247, 1963.

1964

26. Andén, N. E., Carlsson, A., Dahlström, A., Fuxe, K., Hillarp, N. Å., and Larsson, K.: Demonstration and mapping out of nigro-neostriatal dopamine neurons. Life Sci. *3:* 523, 1964.

27. Bernheimer, H.: Distribution of homovanillic acid in the human brain. Nature *204:* 587, 1964.
28. Cotzias, G. C., Papavasiliou, P. S., and Miller, S. T.: Manganese in melanin. Nature *201:* 1228, 1964.
29. Cotzias, G. C., Papavasiliou, P. S., Van Woert, M. H., and Sakamoto, A.: Melanogenesis and extrapyramidal diseases. Fed. Proc. *23:* 713, 1964.

1965

30. Bruno, A., and Cumer-Bruno, S.: Dopamine e sistema nervoso centrale. Riv. di Neurobiol. *11:* 813, 1965.
31. Heymans, C., and DeSchaepdryver, A.: Dopamine and central neurotransmission. Pontif. Acad. Sci. Acta *761,* 1965.
32. Poirier, L. J., and Sourkes, T. L.: Influence of the substantia nigra on the catecholamine content of the striatum. Brain *88:* 181, 1965.

1966

33. Cotzias, G. C.: Manganese, melanins and the extrapyramidal system. J. Neurosurg. *24:* 170, 1966.
34. Goldstein, M., Anagnoste, B., Owen, W. S., and Battista, A. F.: The effects of ventro-medial lesions on the biosynthesis of catecholamines in the striatum. Life Sci. *5:* 2171, 1966.
35. Hornykiewicz, O.: Dopamine (3-hydroxytyramine) and brain function. Pharmacol. Rev. *18:* 925, 1966.

1967

36. Barbeau, A.: The "pink spot," 3,4-dimethoxyphenylethylamine and dopamine: relationship to Parkinson's disease and to schizophrenia. Rev. Canad. Biol. *26:* 55, 1967.
37. Van Praag, H. M.: The possible significance of cerebral dopamine for neurology and psychiatry. Psych. Neurol. Neurochir. *70:* 361, 1967.

1968

38. Barbeau, A.: Effect of phenothiazines on dopamine metabolism and biochemistry of Parkinson's disease. Agressologie *9:* 195, 1968.
39. Barbeau, A.: Dopamine and dopamine metabolites in Parkinson's disease—A review. Proc. Aust. Assoc. Neurol. *5:* 95, 1968.
40. Hornykiewicz, O.: Gegenwärtiger Stand der biochemisch-pharmakologischen Erforschung des extrapyramidal-motorischen Systems. Pharmakopsychia. Neuro-Psychopharmak. *1:* 6, 1968.
41. Klawans, H. C., Jr.: The pharmacology of parkinsonism. Dis. Nerv. Syst. *29:* 805, 1968.
42. Papavasiliou, P. S., Miller, S. T., and Cotzias, G. C.: Functional interactions between biogenic amines, 3′,5′-cyclic AMP and manganese. Nature *220:* 74, 1968.

1969

43. Aranda, L. C., and Asenjo, A.: Tremor induced by iron overloading. J. Neurosurg. *30:* 35, 1969.
44. Barbeau, A.: Parkinson's disease as a systemic disorder. *In* Third Sym-

posium on Parkinson's Disease (Gillingham, F. J., and Donaldson, I. M. C., eds.). E. & S. Livingston, Edinburgh, 1969.
45. COYLE, J. T., and SNYDER, S. H.: Antiparkinsonian drugs: inhibition of dopamine uptake in the corpus striatum as a possible mechanism of action. Science *166:* 899, 1969.
46. GOODALL, McC, and ALTON, H.: Dopamine (3-hydroxytyramine) metabolism in parkinsonism. J. Clin. Invest. *48:* 2300, 1969.
47. NEFF, N. H., BARRETT, R. E., and COSTA, E.: Selective depletion of caudate nucleus dopamine and serotonin during chronic manganese dioxide administration to squirrel monkeys. Experientia *25:* 1140, 1969.
48. POIRIER, L. J., BOUVIER, G., BEDARD, P., BOUCHER, R., LAROCHELLE, L., OLIVIER, A., and SINGH, P.: Essai sur les circuits neuronaux impliqués dans le tremblement postural et l'hypokinésie. Rev. Neurol. *120:* 15, 1969.

II. CLINICAL USES OF L-DOPA 1960-1969

1960

1. DEGKWITZ, R., FROWEIN, R., KULENKAMPFF, C., and MOHS, V.: Über die Wirkungen der L-Dopa beim Menschen und deren Beinglussung deerch Reserpin, Chlorpromazin, Iproniazid und Vitamin B_6. Klin. Wschr. *38:* 120, 1960.

1961

2. BARBEAU, A.: Biochemistry of Parkinson's Disease. *In* Proceedings of the Seventh International Congress of Neurology, Rome, September 1961. Societa Grafica Romana, Rome, 1961.
3. BARBEAU, A.: Biochemistry and treatment of Parkinson's disease. Un. Med. Canad. *90:* 1000, 1961.
4. BIRKMAYER, W., and HORNYKIEWICZ, O.: Der L-3,4-dioxyphenylalanin (Dopa)-Effekt bei der Parkinson Akinese. Wien. Klin. Wschr. *73:* 787, 1961.
5. McGEER, P. L., BOULDING, J. E., GIBSON, W. C., and FOULKES, R. G.: Drug-induced extrapyramidal reactions: treatment with diphenhydramine hydrochloride and dehydroxyphenylalanine. J.A.M.A. *177:* 665, 1961.
6. POLLIN, W., CARDON, P. V., JR., and KETY, S. S.: Effects of amino acid feedings in schizophrenic patients treated with iproniazid. Science *133:* 104, 1961.

1962

7. BARBEAU, A., SOURKES, T. L., and MURPHY, G. F.: Les catécholamines dans la maladie de Parkinson. *In* Monoamines et Système Nerveux Central (de Ajuriaguerra, J., ed.). Masson & Cie, Paris, 1962.
8. BARBEAU, A.: The pathogenesis of Parkinson's disease: a new hypothesis. Canad. Med. Assoc. J. *87:* 802, 1962.

9. Birkmayer, W., and Hornykiewicz, O.: Der L-dioxyphenylalanin (=L-Dopa) Effekt beim Parkinson-Syndrom des Menschen: zur Pathogenese und Behandlung der Parkinson-Akinese. Arch. Psychiat. Nervenkr. *203:* 560, 1962.
10. Birkmayer, W.: Features of medical therapy of Parkinson's syndrome. Wien. Klin. Wschr. *74:* 700, 1962.
11. Carlsson, A., and Hillarp, N. A.: Formation of phenolic acids in brain after administration of 3,4-dihydroxyphenylalanine. Acta Physiol. Scand. *55*: 95, 1962.
12. Gerstenbrand, F., and Pateisky, K.: Uber die Wirkung von L-Dopa auf die Motorischen Störungen beim Parkinson Syndrom. Wien. Z. Nervenheilk. *20:* 90, 1962.

1963

13. Barbeau, A.: Etudes récente sur les catécholamines. Un. Med. Canad. *92:* 42, 1963.
14. Bernheimer, H., Birkmayer, W., and Hornykiewicz, O.: Zur Biochemie des Parkinson-Syndroms des Menschen. Einfluss der mono-amino-oxydase-Hemmer-Therapie auf die Konzentration des Dopamins, noradrenalins und 5-hydroxytryptamins im Gehirn. Klin. Wschr. *41:* 465, 1963.
15. Friedhoff, A. J., Hekiman, L., Alpert, M., and Tobach, E.: Dihydroxyphenylalanine in extrapyramidal disease. J.A.M.A. *184:* 255, 1963.
16. Gerstenbrand, F., Pateisky, K., and Prosenz, P.: Erfahrungen mit L-Dopa in Therapie des Parkinsonismus. Psychiat. Neurol. *146:* 246, 1963.
17. Greer, M., and Williams, C. M.: Dopamine metabolism in Parkinson's disease. Neurology *13:* 73, 1963.
18. Hornykiewicz, O.: Die topische Lokalisation und das Verhalten von Noradrenalin und Dopamin (3-Hydroxytyramin) in der Substantia nigra des normalen und Parkinson-Kranken Menschen. Wien. Klin. Wschr. *75:* 309, 1963.
19. Klerman, G. L., Schildkraut, J. J., Hasenbush, L. L., Greenblatt, M., and Friend, D. G.: Clinical experience with dihydroxyphenylalanine (Dopa) in depression. J. Psychiat. Res. *1:* 289, 1963.
20. Schildkraut, J. J., Klerman, G. L., Friend, D. G., and Greenblatt, M.: Biochemical and pressor effects of oral D, L-dihydroxyphenylalanine in patients pretreated with antidepressant drugs. Ann. N. Y. Acad. Sci. *107:* 1005, 1963.

1964

21. Barolin, G. S., Bernheimer, H., and Hornykiewicz, O.: Seitenberschiedenes Verhalten des Dopamins (3-hydroxytyramin) im Gehirn eines Falles von Hemiparkinsonismus. Schweiz. Arch. Neurol. Psychiat. *94:* 241, 1964.
22. Bernheimer, H., and Hornykiewicz, O.: Das Verhalten der Dopamine-Metaboliten Honovanillinsaure im Gehirn von normalen und Parkinson-Kranken Menschen. Arch. Exp. Path. Pharmakol. *247:* 305, 1964.
23. Birkmayer, W., and Hornykiewicz, O.: Weitere experimentelle Untersuchungen uber L-Dopa beim Parkinson-Syndrom und Reserpine-Parkinsonismus. Arch. Psychiat. Nervenkr. *206:* 367, 1964.
24. Bruck, H., Gerstenbrand, F., Grundig, E., and Teuflmayr, R.: Uber Ergebnisse von Liquoranalysen beim Parkinson-Syndrom. Acta Neuropathol. *3:* 638, 1964.
25. Hirschmann, J., and Mayer, K.: Zur Beeinflussung der Akinese und

anderer extrapyramidal motorescher Störungen mit Dopa. Deutsch. Med. Wschr. *89:* 1877, 1964.
26. HORNYKIEWICZ, O.: Zur frage des Verlaufs dopaminerger Neurons im Gehirn des Menschen. Wien. Klin. Wschr. *76:* 834, 1964.
27. LEREBOULLET, J.: Nouveaux médicaments antiparkinsoniens. Rev. Pratic. *14:* 178, 1964.
28. MCGEER, P. L., and ZELDOWICZ, L. R.: Administration of dihydroxyphenylalanine to parkinsonian patients. Canad. Med. Assoc. J. *90:* 463, 1964.
29. TURNER, W. J., and MARLIS, S.: Clinical trial of pargyline and Dopa in psychotic subjects. Dis. Nerv. Syst. *25:* 538, 1964.
30. UMBACH, W., and BAUMANN, D.: Die Wirksamkeit von L-Dopa bein Parkinson-Patienten mit und ohne stereotaktischen Hirneingriff. Arch. Psychiat. Nervenkr. *205:* 281, 1964.

1965

31. BERNHEIMER, H., and HORNYKIEWICZ, O.: Herabgesetzte Konzentration der Homovanillinsaure im Gehirn von Parkinsonkranken Menschen als Ausdruck der Strorung des Zentralen Dopaminstoffwechsels. Klin. Wschr. *43:* 711, 1965.
32. BRIGIDA, E., BRUNO, A., and CUMER-BRUNO, S.: Modification by L-Dopa of chlorpromazine-induced parkinsonism. Lavoro Neuropsich. *37:* 123, 1965.
33. BRUNO, A., and BRIGIDA, E.: Azione della L-Dopa sulla sintomatologia extrapiramidale da haloperidol. Riv. di Neurobiol. *11:* 646, 1965.
34. BRUNO, A., and CUMER-BRUNO, S.: Dopamina e sistema nervoso centrale. Riv. di Neurobiol. *11:* 813, 1965.
35. BRUNO, A., and CUMER-BRUNO, S.: Effetti della L-3,4-diidrossifenilalanina (L-Dopa nei pazienti parkinsoniani). Riv. Sper. Fren. *89:* 5, 1965.
36. BRUCK, J., GERSTENBRAND, F., GRUNDIG, E., and PROSENZ, P.: Stoffwechselveränderungen bei extrapyramidalen Syndrome und vorlaüfige Therapeutische Konsequenzen. Fortschr. Neurol. Psychiat. *33:* 677, 1965.
37. DIEMATH, H. E., HEPPNER, F., and ENGE, S.: Hirnstrombilder während der verabreichung von L-Dopa vor, bei (Tiefenableitungen und Reizeffekte) und nach stereotaktischen Hirnoperationen bei Parkinson-Patienten. Acta Neuroveg. *27:* 137, 1965.
38. DUVOISIN, R. C.: A review of drug therapy in parkinsonism. Bull. N. Y. Acad. Med. *41:* 898, 1965.
39. FRIEND, D. G., BELL, W. R., and KLINE, N. S.: The action of L-dihydroxyphenylalanine in patients receiving nialamide. Clin. Pharmacol. Ther. *6:* 362, 1965.
40. GERSTENBRAND, F., and PROSENZ, P.: Über die behandlung der Parkinsonsyndroms mit monoaminoxydasehemmern allein und in Kombination mit L-Dopa. Praxis *54:* 1373, 1965.
41. INGVARSSON, C. G.: Orientierende Klinische Versuche zeur Wirkung des Dioxyphenylalanins (L-Dopa) bei endogenes Depression. Arzneimittelforsch. *15:* 849, 1965.
42. INGVARSSON, G.: L-Dopa och astmatisk bronkit. Nord. Med. *18:* 1166, 1965.
43. JOHANSSON, B., and ROOS, B. E.: Acid monoamine metabolites in the cerebrospinal fluid of patients with Parkinson's disease. *In* Proc. 8th Int. Congr. Neurol. Vienna, 1965.
44. METZEL, E.: On disturbances in space perception caused by Parkinson's syndrome. Deutsch. Med. Wschr. *90:* 1955, 1965.

45. O'Reilly, S., Loncin, M., and Cooksey, B.: Dopamine and basal ganglia disorders. Neurology *15:* 980, 1965.
46. Rinaldi, F., Margherita, G., and De Divitiis, E.: Effetti della somministrazione di Dopa a pazienti parkinsoniani pretrattati con inibitore delle monoaminossidasi. Ann. Fren. Sc. Aff. *78:* 105, 1965.
47. Sourkes, T. L., Pivnicki, D., Brown, W. T., Wiseman-Distler, M. H., Sankoff, I., and Saint-Cyr, S.: A clinical and metabolic study of Dopa (3,4-dihydroxyphenylalanine) and methyldopa in Huntington's chorea. Psychiat. Neurol. *149:* 7, 1965.
48. Umbach, W., and Tzavellas, O.: Zur behandlung akinetischer Begleitsymptome beim Parkinson-Syndrom. Deutsch. Med. Wschr. *90:* 1941, 1965.

1966

49. Bente, D., Stoerger, R., and Tautz, N. A.: Klinische elektroencephalographische und biochemische Aspekte des L-Dopa-Effektes. Arzneimittelforschung *16:* 265, 1966.
50. Bernheimer, H., Birkmayer, W., and Hornykiewicz, O.: Homovanillinsaure im Liquor cerebrospinalis: Untersuchungen beim Parkinson-Syndrom und anderen Erkrankungen des ZNS. Wien. Klin. Wschr. *78:* 417, 1966.
51. Bernheimer, H., and Hornykiewicz, O.: Dopaminumsatz im Gehirn von normalen und parkinsonkranken menschen. Wien. Z. Nervenheilk. *23:* 110, 1966.
52. Birkmayer, W.: Experimental findings and new aspects in extrapyramidal system disorders. Wien. Z. Nervenheilk. *23:* 128, 1966.
53. Bruno, A., and Cumer-Bruno, S.: Effect of L-Dopa on pharmacological parkinsonism. Acta Psychiat. Scand. *42:* 264, 1966.
54. Fehling, C.: Treatment of Parkinson's syndrome with L-Dopa. A double blind study. Acta Neurol. Scand. *42:* 367, 1966.
55. Hornykiewicz, O.: Dopamine (3-hydroxytyramine) and brain function. Pharmacol. Rev. *18:* 925, 1966.
56. Hornykiewicz, O.: Neuere Aspekte der biochemischen Pharmakologie des Parkinson-Syndroms. Wien. Z. Nervenheilk. *23:* 103, 1966.
57. Matussek, N., Pohlmeier, H., and Ruther, E.: Die Wirkung von Dopa auf gehemmte Depressionen. Klin. Wschr. *44:* 727, 1966.
58. Oehme, P., and Schwartz, K.: Ättologie und Therapie des Parkinsonsyndroms und parmakologische Modell vorstellungen. Deutsch. Gesundh *21:* 792, 1966.
59. Pazzagli, A., and Amaducci, L.: La sperimentazione clinica del Dopa nelle sindromi Parkinsoniane. Riv. di Neurobiol. *12:* 138, 1966.
60. Prosenz, P.: Changes in amino acid levels in the cerebrospinal fluid following drug administration. Wien. Z. Nervenheilk. *1:* 136, 1966.
61. Travenec, I. A.: Therapy of parkinsonian akinesia. Zh. Nevropat. i. Psik. *66:* 703, 1966.
62. Travenets, I. K.: Dopamine metabolism in the brain of parkinsonism patients. Zh. Nevropat. Psikhiat. Korsakov. *66:* 1505, 1966.
63. Tzavellas, O.: New therapeutic possibilities in the akinetic forms of Parkinson's disease. Nosokomiaka Chronica *28:* 597, 1966.
64. Umbach, W.: Different effects of drugs before and after stereotaxic operation. Confin. Neurol. *27:* 258, 1966.

1967

65. Aebert, K.: What does L-Dopa accomplish in the treatment of parkinsonian akinesia? Deutsch. Med. Wschr. *92:* 483, 1967.

66. AEBERT, K.: What does L-Dopa accomplish in the treatment of parkinsonian akinesia? Conclusions. Deutsch. Med. Wschr. *92:* 1242, 1967.
67. BARTHOLINI, G., BURKARD, W. P., and PLETSCHER, A.: Increase of cerebral catecholamines caused by 3,4-dihydroxyphenylalanine after inhibitor of peripheral decarboxylase. Nature *215:* 852, 1967.
68. BIRKMAYER, W.: Die Bedeutung des Monoaminmetabolismus für die Pathologie des Extrapyramidalen Systems. *In* Ass Medici Verlag Lüdin A G Liestal. Schweiz, 1967.
69. BIRKMAYER, W., and MENTASTI, M.: Weitere experimentelle Untersuchungen über Catecholaminstoffwechsel bei extrapyramidalen Erkrankungen (Parkinson- und Chorea-Syndrom). Arch. Psychiat. Nervenkr. *210:* 29, 1967.
70. BISCHOFF, A.: Wirkung von L-Dopa auf die Akinese beim Parkinson-Syndrom. Schweiz. Med. Wschr. *97:* 1158, 1967.
71. Comment: First steps in metabolic control of parkinsonism. Med. World News *8:* 36, 1967.
72. COTZIAS, G. C., VAN WOERT, M. H., and SCHIFFER, L. M.: Aromatic amino acids and modification of parkinsonism. New Eng. J. Med. *276:* 374, 1967.
73. COTZIAS, G. C.: Dopa and parkinsonism. Brit. Med. J. *3*(5563): 497, 1967.
74. Editorial: Dopa and parkinsonism. Brit. Med. J. *2*(5555): 783, 1967.
75. OELSSNER, W.: Pharmacology of antiparkinson drugs. Psychiat. Neurol. Med. Psychol. *19:* 236, 1967.
76. PERRINO, I.: Amino acid precursors of catecholamines and their use in Parkinson's syndrome. Sttim. Med. *55:* 699, 1967.
77. UMBACH, W.: What does L-Dopa accomplish in the treatment of parkinsonian akinesia? Deutsch. Med. Wschr. *92:* 1241, 1967.
78. WATT, J.: Dopa and parkinsonism. Brit. Med. J. *3*(5562): 437, 1967.

1968

79. ARRAS, M. J., and BAILEY, G. W. H.: Metabolism of Dopa in parkinsonism. New Eng. J. Med. *278:* 280, 1968.
80. BARBEAU, A.: The current therapy of Parkinson's disease. *In* Current Therapy (H. F. Conn, ed.). W. B. Saunders, Philadelphia, 1968.
81. BOUYGES, P.: Traitement médical des tremblements. Gaz. Med. France *75:* 4207, 1968.
82. CALNE, D. B.: The pharmacology of parkinsonism. *In* Symposium on Advanced Medicine (O. Wong, ed.). Williams & Wilkins, Baltimore, 1968.
83. Comment: Animating victims of Parkinson's disease. Med. World News *9:* 20, 1968.
84. Comment: New and old drugs for parkinsonism. Med. Letter Drugs *10:* 69, 1968.
85. COTZIAS, G. C.: Dihydroxyphenylalanine treatment of parkinsonism. J.A.M.A. *205:* 541, 1968.
86. COTZIAS, G. C., PAPAVASILIOU, P. S., GELLENE, R., and ARONSON, R. B.: Parkinsonism and Dopa. Trans. Assoc. Amer. Physiol. *81:* 171, 1968.
87. KLAWANS, H. L., JR.: Recent advances in the pharmacology of Parkinson's disease. New Physician *17:* 373, 1968.
88. KLAWANS, H. L., JR.: Recent advances in the pharmacology of Parkinson's disease. Presbyt. St. Luke's Hosp. Med. Bull. *7:* 121, 1968.
89. HORNYKIEWICZ, O., LISCH, H. J., and SPRINGER, A.: Homovanillic acid in different regions of the human brain: attempt at localizing central dopamine fibres. Brain Res. *11:* 662, 1968.
90. MERRILL, A. J.: Treatment of parkinsonism with L-Dopa. J. Med. Assoc. Georgia *57:* 350, 1968.

91. RINNE, V. K., and SONNINEN, V.: A double blind study of L-Dopa treatment in Parkinson's disease. Europ. Neurol. *1:* 180, 1968.
92. URCIVOLI, R., and NIVOLI, G. C.: Prime osservazioni siell'uso di L-Dopa nel trattamento della sintomatologia acinetica del morbo di Parkinson. Minerva Med. *59:* 2516, 1968.
93. VOELLER, G. W.: Ein Beitrag zur biochemischen Therapie akinetischer formen des Parkinson-Syndroms. Med. Welt. *19:* 338, 1968.
94. YAHR, M. D., and DUVOISIN, R. C.: Medical therapy of parkinsonism. Mod. Treatm. *5:* 283, 1968.
95. YAHR, M. D., DUVOISIN, R. C., HOEHN, M. M., SCHEAR, M. J., and BARRETT, R. E.: L-Dopa (L-3,4-dihydroxyphenylalanine). Its clinical effects in parkinsonism. Trans. Amer. Neurol. Assoc. *93:* 56, 1968.

1969

96. Annotations: Supply of L-Dopa for treatment of parkinsonism. Lancet *2:* 1288, 1969.
97. Annotation: Dopa in Parkinson's disease. Lancet *1:* 871, 1969.
98. BARBEAU, A.: Traitement de la maladie de Parkinson par la L-Dopa. Un. Med. Canad. *98:* 183, 1969.
99. BARBEAU, A.: L-Dopa and juvenile Huntington's disease. Lancet *2:* 1066, 1969.
100. BARBEAU, A.: L-Dopa therapy in Parkinson's disease: a critical review of nine years' experience. Canad. Med. Assoc. J. *101:* 791, 1969.
101. BARBEAU, A.: Dopamine and blood pressure regulation. Clin. Res. *17:* 634, 1969.
102. BARBEAU, A., and GILLO-JOFFROY, L.: Eight years of experience with L-Dopa therapy in Parkinson's disease. Neurology *19:* 284, 1969.
103. BARBEAU, A., and GILLO-JOFFROY, L.: Treatment of Parkinson's disease with L-Dopa and Ro 4-4602. *In* Excerpta Medica, International Congress Series 193, Amsterdam, 1969.
104. BARBEAU, A., GILLO-JOFFROY, L., BOUCHER, R., NOWACZYNSKI, W., and GENEST, J.: Renin-adolesterone system in Parkinson's disease. Science *165:* 291, 1969.
105. BARTHOLINI, G., TISSOT, R., and PLETSCHER, A.: Selective increase of cerebral dopamine: a therapeutic possibility of parkinsonism. *In* Third Symposium on Parkinson's Disease (F. J. Gillingham and I. M. L. Donaldson, eds.). E. & S. Livingstone, Edinburgh, 1969.
106. BETTAG, W., and HOLBACH, K. M.: Effect of L-Dopa on different symptoms of Parkinson's disease. *In* Third Symposium on Parkinson's disease (F. J. Gillingham and I. M. L. Donaldson, eds). E. & S. Livingston, Edinburgh, 1969.
107. BIRKMAYER, W.: Experimentelle Ergenknisse über die Kombinationsbehandlung des Parkinson-Syndroms mit L-Dopa und einem Decarboxylasehemmer (Ro 4-4602). Wien. Klin. Wschr. *81:* 677, 1969.
108. BOSHES, B., and BRUMLIK, J.: Recent advances in the treatment of parkinsonism. Illinois Med. J. *135:* 253, 1969.
109. BRAHAM, J., SAROVA-PINHAS, I., CRISPIN, M., GOLAN, R., LEVIN, N., and SZEINBERG, A.: Oral phenylalanine and tyrosine tolerance tests in parkinsonian patients. Brit. Med. J. *2:* 552, 1969.
110. BRAHAM, J., and SZEINBERG, A.: Metabolism in parkinsonian patients. Brit. Med. J. *4:* 432, 1969.
111. BUNNEY, W. E., JANOWSKY, D. S., GOODWIN, G. K., DAVIS, J. M., BRODIE, H. K. H., MURPHY, D. L., and CHASE T. N.: Effect of L-Dopa on depression. Lancet *1:* 885, 1969.

112. Butcher, L. L., and Engel, J.: Peripheral factors in the mediation of the effects of L-Dopa on locomotor activity. J. Pharm. Pharmacol. *21:* 614, 1969.
113. Calne, D. B.: L-Dopa in parkinsonism. Prescriber's J. *9:* 97, 1969.
114. Calne, D. B., Karoum, F., Ruthven, C. R. J., and Sandler, M.: The metabolism of orally administered L-Dopa in parkinsonism. Brit. J. Pharmacol. *37:* 57, 1969.
115. Calne, D. B., Spiers, A. S. D., Stern, G. M., and Laurence, D. R.: L-Dopa in idiopathic parkinsonism. Lancet *2:* 973, 1969.
116. Calne, D. B., Stern, G. M., Laurence, D. R., Sharkey, J., and Armitage, P.: L-Dopa in postencephalitic parkinsonism. Lancet *1:* 744, 1969.
117. Cawein, M. J., and Hewins, J.: False rise in serum uric acid after L-Dopa. New Eng. J. Med. *281:* 1489, 1969.
118. Comment: A second report on levodopa. Med. Letter *11:* 73, 1969.
119. Comment: Living with Parkinson's disease. Nature *223:* 889, 1969.
120. Comment: Tyrosine and Parkinson's disease. Nature *222:* 1231, 1969.
121. Comment: Progress in Parkinson's disease. Science *9:* 20, 1969.
122. Comment: Drugs used in manganese poisoning. J.A.M.A. *208:* 956, 1969.
123. Comment: Devices to evaluate therapy in Parkinson's disease. J.A.M.A. *209:* 1621, 1969.
124. Comment: La dopamina y el parkinsonismo. Dia Medico *41:* 1961, 1969.
125. Comment: Parkinsonism bows to levodopa, usually. J.A.M.A. *210:* 434, 1969.
126. Comment: Orthopedic surgery may complement levodopa. J.A.M.A. *210:* 436, 1969.
127. Coleman, M. P., and Barnet, A.: L-Dopa reversal of muscular spasm, vomiting and insomnia in a patient with an atypical form of familial dystonia. Trans. Amer. Neurol. Assoc., in press.
128. Cotzias, G. C.: Levodopa (L-Dopa) treatment of parkinsonism. J.A.M.A. *207:* 1522, 1969.
129. Cotzias, G. C.: Metabolic modification of some neurologic disorders. J.A.M.A. *210:* 1255, 1969.
130. Cotzias, G. C.: Parkinsonism and Dopa. J. Chron. Dis. *22:* 297, 1969.
131. Cotzias, G. C., Mena, I., and Papavasiliou, P. S.: Amelioration of the dystonia of chronic manganese poisoning by L-Dopa. Neurology *19:* 284, 1969.
132. Cotzias, G. C., and Papavasiliou, P. S.: Auto-immunity in patients treated with levodopa. J.A.M.A. *207:* 1353, 1969.
133. Cotzias, G. C., and Papavasiliou, P. S.: Long term effects of D-L and L-Dopa on parkinsonian patients. *In* Progress in Neuro-Genetics (A. Barbeau, and J. R. Brunette, eds.). Excerpta Medica Foundation, Amsterdam, 1969.
134. Cotzias, G. C., Papavasiliou, P. S., and Gellene, R.: L-Dopa in Parkinson's syndrome. New Eng. J. Med. *281:* 272, 1969.
135. Cotzias, G. C., Papavasiliou, P. S., and Gellene, R.: Modification of parkinsonism: chronic treatment with L-Dopa. New Eng. J. Med. *280:* 337, 1969.
136. Cotzias, G. C., Papavasiliou, P. S., Gellene, R., Aronson, J. B., and Mena, I.: Long term effects of Dopa on parkinsonism: a proposal. *In* Third Symposium on Parkinson's Disease (F. J. Gillingham and I. M. L. Donaldson, eds.). E. & S. Livingston, Edinburgh, 1969.
137. Cotzias, G. C., Papavasiliou, P. S., and Mena, I.: Parkinsonism, Dopa and chronic manganese poisoning. Excerpta Medica, I.C.S. 193, Amsterdam, 1969.
138. Crawford, J. P.: Dopa in Parkinson's disease. Lancet *2:* 1014, 1969.

139. Crawford, J. P.: Dopa in parkinsonism. Lancet *1:* 984, 1969.
140. Degkwitz, R.: L-Dopa-Behandlung des Parkinson-Syndroms und Wirkungsweise der Neuroleptika. Med. Klin. *64:* 1451, 1969.
141. Duvoisin, R. C.: Pharmacological treatment of basal ganglia disorders. *In* Psychotropic Drugs and Dysfunctions of the Basal Ganglia: a Multidisciplinary Workshop (G. E. Crane and R. Gardner, Jr., eds.). USPHS Publication 1938, Washington, 1969.
142. Duvoisin, R., Barrett, R., Schear, M., Hoehn, M., and Yahr, M. D.: The use of L-Dopa in parkinsonism. *In* Third Symposium on Parkinson's disease (F. J. Gillingham and I. M. L. Donaldson, eds.). E. & S. Livingston, Edinburgh, 1969.
143. Duvoisin, R., Yahr, M. D., Schear, M., Hoehn, M. M., and Barrett, R. B.: The present status of L-Dopa in the treatment of parkinsonism. Excerpta Medica, I.C.S. 193, Amsterdam, 1969.
144. Editorial: New approaches in the treatment of parkinsonism. J.A.M.A. *209:* 1358, 1969.
145. Fangman, A., and O'Malley, W. E.: L-Dopa and the patient with Parkinson's disease. Amer. J. Nurs. *69:* 1455, 1969.
146. Fasano, V. A.: Long term results in 250 patients with cryosurgery and L-Dopa. *In* Third Symposium on Parkinson's disease (F. J. Gillingham and I. M. L. Donaldson, eds.) E. & S. Livingston, Edinburgh, 1969.
147. Fenton, L. M.: L-Dopa decision. New Scientist *44:* 427, 1969.
148. Gajdos, A.: Acides aromatiques et maladie de Parkinson. Presse Med. *77:* 916, 1969.
149. Gilbert, J. J., and Feldman, R. G.: L-Dopa for progressive supra-nuclear palsy. Lancet *2:* 494, 1969.
150. Godwin-Austen, R. B., Lind, N. A., and Turner, P.: Mydriatic responses to sympathomimetic amines in patients treated with L-Dopa. Lancet *2:* 1043, 1969.
151. Godwin-Austen, R. B., Tomlinson, E. B., Frears, C. C., and Kok, H. W. L.: Effects of L-Dopa in Parkinson's disease. Lancet *2:* 165, 1969.
152. Greenberg, J.: L-Dopa treatment too successful. New Eng. J. Med. *281:* 621, 1969.
153. Gross. M.: L-Dopa for progressive supra-nuclear palsy. Lancet *2:* 1359, 1969.
154. Grundig, E., Gerstenbrand, F., Bruck, J., Gnad, H., and Prosenz, P.: Der Einfluss der Verabreichung von Aminosäuren speziell von L-Popa und α-methyldopa, auf die Zusammensetzung des Liquor cerebrospinalis bei extrapyramidalen Syndromen. I. Veräderungen der Liquorzusammensetzung nach L-Dopa-Gaben bei Parkinson-Patienten und Gesunden. Deutsch. Z. Nervenheilk. *196:* 236, 1969. II. Veränderungen der Liquorzusammensetzung bei Parkinson und Chorea-Patienten nach Verabreichung von o-methyldopa. Deutsch. Z. Nervenheilk. *196:* 256, 1969.
155. Hoehn, M. M., and Yahr, M. D.: Evaluation of the long term results of surgical therapy. *In* Third Symposium on Parkinson's Disease (F. J. Gillingham and I. M. L. Donaldson, eds.) E. & S. Livingston, Edinburgh, 1969.
156. Kaeser, H. E.: L-Dopa-Behandlung des Parkinson-Syndroms. Praxis *44:* 1401, 1969.
157. Klawans, H. L., Jr.: L-Dopa for progressive supra-nuclear palsy. Lancet *2:* 1359, 1969.
158. Klawans, H. L., Jr., and Erlich, M.: L-Dopa and the negative symptoms of parkinsonism. J. Neurol. Neurosurg. Psychiat. *32:* 460, 1959.
159. Klawans, H. L., Jr., Garwin, J. S., and Shekelle, R. B.: The effect of previous anti-parkinson surgery on L-Dopa treatment. Presbyt. St. Luke's Hosp. Med. Bull. *8:* 115, 1969.

160. Klawans, H. L., Jr., and Goodwin, J. A.: Reversal of the glabellar reflex in parkinsonism by L-Dopa. J. Neurol. Neurosurg. Psychiat. *32:* 423, 1969.
161. Lavy, S.: Mahalat Parkinson: aspektim atiulugüm utifulüm. Harefuah *77:* 103, 1969.
162. Leading Article: L-Dopa in parkinsonism. Brit. Med. J. *2:* 202, 1969.
163. McDowell, F. H., Lee, J. E., Sweet, R., Swift, T., Ogsbury, J., and Kessler, J.: The treatment of Parkinson's disease with dihydroxyphenylalanine. Excerpta Medica, I.C.S. 193, Amsterdam, 1969.
164. Mena, I., Court, J., Fuenglalida, S., Papavasiliou, P. S., and Cotzias, G. C.: L-Dopa en el control del manganismo cronico. Rev. Med. Chile *97:* 343, 1969.
165. Mones, R. J., and Elizan, T. S.: Short term evaluation of L-Dopa therapy in 34 patients with parkinsonism. J. Mt. Sinai Hosp. *36:* 503, 1969.
166. Natelson, B. H.: Beans—A source of L-Dopa. Lancet *2:* 640, 1969.
167. O'Reilly, S.: Dopa in Parkinson's disease. Lancet *1:* 984, 1969.
168. Pakkenberg, H.: L-Dopa og Parkinsonisme. Ugeskr. Laeg. *131:* 959, 1969.
169. Papavasiliou, P. S., and Cotzias, G. C.: Experiments with therapy of parkinsonism. Amer. Chem. Soc., Meeting 157, 37 M, 1969.
170. Papavasiliou, P. S., Gellene, R., and Cotzias, G. C.: Modification of parkinsonism: dyskinesias accompanying treatment with Dopa. *In* Psychotropic Drugs and Dysfunctions of the Basal Ganglia: a Multidisciplinary Workshop (G. E. Crane and R. Gardner, J., eds.). USPHS Publication 1938, Washington, 1969.
171. Pletscher, A.: Pharmacological changes of the dopamine metabolism in the basal ganglia. *In* Psychotropic Drugs and Dysfunctions of the Basal Ganglia: Aa Multidisciplinary Workshop (G. E. Crane and R. Gardner, Jr., eds.). USPHS Publication 1938, Washington, 1969.
172. Poskanzer, D. C.: L-Dopa in Parkinson's syndrome. New Eng. J. Med. *280:* 382, 1969.
173. Refsum, S.: L-Dopa-behandling ved Parkinsonisme. Nord. Med. *81:* 570, 1969.
174. Renzi, E. de, and Vignolo, L. A.: L-Dopa for progressive supra-nuclear palsy. Lancet *2:* 1360, 1969.
175. Rosenberg, H.: Dopa in Parkinson's disease. New Eng. J. Med. *281:* 730, 1969.
176. Sacks, O. W.: L-Dopa for progressive supra-nuclear palsy. Lancet *2:* 591, 1969.
177. Sadwin, A.: Watch that tremor. New Eng. J. Med. *281:* 962, 1969.
178. Sandler, M., Karoum, F., and Ruthven, C. R. J.: Metabolism of Dopa in parkinsonism. New Eng. J. Med. *281:* 1429, 1969.
179. Sandler, M., Karoum, F., Ruthven, C. R. J., and Calne, D. B.: m-Hydroxyphenylacetic acid formation from L-Dopa in man: suppression by neomycin. Science *166:* 1417, 1969.
180. Schwab, R. S.: Limitations in treatment of Parkinson's disease. New Eng. J. Med. *279:* 943, 1969.
181. Schwab, R. S.: L-Dopa vs. surgery in Parkinson's disease. Excerpta Medica, I.C.S. 193, Amsterdam, 1969.
182. Sedgwick, E. M.: Parkinson's disease and L-Dopa. New Scientist *48:* 234, 1969.
183. Siegfried, J.: Traitement du parkinsonisme avec la L-Dopa associée à un inhibiteur de la décarboxylase. Médecine et Hygiène *27:* 543, 1969.
184. Siegfried, J., Klaiber, R., Perret, E., and Ziegler, W. H.: The treatment of Parkinson's disease with L-Dopa combined with a decarboxylase inhibitor. Excerpta Medica, I.C.S. 193, Amsterdam, 1969.
185. Siegfried, J., Ziegler, W. H., Regli, F., Fischer, C., Kaufmann, W., and

PERRET, E.: Treatment of parkinsonism with L-Dopa in association with a decarboxylase inhibitor: first objective results. Pharmacol. Clin. *2:* 23, 1969.

186. SPIEGEL-ADOLF, M., SPIEGEL, E. A., FABIANI, F. R., and CALISE, J.: Urinary excretion of catecholamines after ingestion of fava bean extracts. Fed. Proc. *28:* 898, 1969.

187. SPIEGEL, E. A., SPIEGEL-ADOLF, M., WYCIS, H. T., SZEKELY, E. G., FABIANI, F. R., and SCHWARTZ, H. A.: Restoration of speed in parkinsonian and experimental bradykinesia and akinesia. Excerpta Medica, I.C.S. 193, Amsterdam, 1969.

188. SPIERS, A. S. D.: L-Dopa in parkinsonism. Brit. Med. J. *2:* 448, 1969.

189. SPIERS, A. S. D.: Mydriatic responses to sympathomimetic amines in patients treated with L-Dopa. Lancet *2:* 1301, 1969.

190. SPIERS, A. S. D., and CALNE, D. B.: Action of dopamine on the human iris. Brit. Med. J. *2:* 333, 1969.

191. STEG, G.: Side effects during L-Dopa treatment in parkinsonism. Excerpta Medica, I.C.S. 193, Amsterdam, 1969.

192. STERN, G.: Parkinsonism. Brit. Med. J. *4:* 541, 1969.

193. SUAREZ, M. M. V.: Instilacion intracerebral de Dopa. Neurol. Neurocir. Psiquiat. *10:* 39, 1969.

194. TIMBERLAKE, W. H., ZIEPER, I., and SCHWAB, R. S.: Double blind study of L-Dopa treatment of parkinsonism. Excerpta Medica, I.C.S. 193, Amsterdam, 1969.

195. TISSOT, R., GAILLARD, J. M., GUGGISBERG, M., GAUTHIER, G., and AJURIAGUERRA, J. DE: Terapeutica del sindrome de Parkinson per la L-Dopa per os asociada a un inhibidor de la descarboxilasa (Ro 4-4602). Neurol. Neurocir. Psiquiat. *10:* 1, 1969.

196. TISSOT, R., GAILLARD, J. M., GUGGISBERG, M., GAUTHIER, G., and AJURIAGUERRA, J. DE: Thérapeutique du syndrome de Parkinson par la L-Dopa "per os" associée à un inhibiteur de la décarboxylase (Ro 4-4602). Presse Méd. *77:* 619, 1969.

197. WAGSHUL, A., and DAROFF, R. B.: L-Dopa for progressive supra-nuclear palsy. Lancet *2:* 105, 1969.

198. WAGSHUL, A. M., and DAROFF, R. B.: Depression during L-Dopa treatment. Lancet *2:* 592, 1969.

199. WEINER, W., HARRISON, W., and KLAWANS, H. L., JR.: L-Dopa and cerebrospinal fluid homovanillic acid in parkinsonism. Life Sci. *8:* 971, 1969.

200. WOLCOTT, G. J., and HACKETT, T. N., JR.: Urine test for L-Dopa consumption. Lancet *2:* 1201, 1969.

201. YAHR, M. D., DUVOISIN, R. C., SCHEAR, M. J., BARRETT, R. E., and HOEHN, M. M.: Treatment of parkinsonism with levodopa. Arch. Neurol. *21:* 343, 1969.

Joss Whedon's *Dollhouse*

Science Fiction Television
Series Editor: A. Bowdoin Van Riper

From Starship Captains to Galactic Rebels: Leaders in Science Fiction Television, by Kimberly Yost, 2014

Joss Whedon's Dollhouse*: Confounding Purpose, Confusing Identity,* edited by Sherry Ginn, Alyson R. Buckman, and Heather M. Porter, 2014

Joss Whedon's *Dollhouse*

Confounding Purpose, Confusing Identity

Edited by

Sherry Ginn
Alyson R. Buckman
Heather M. Porter

ROWMAN & LITTLEFIELD
Lanham • Boulder • New York • Toronto • Plymouth, UK

Published by Rowman & Littlefield
4501 Forbes Boulevard, Suite 200, Lanham, Maryland 20706
www.rowman.com

10 Thornbury Road, Plymouth PL6 7PP, United Kingdom

British Library Cataloguing in Publication Information Available

Library of Congress Cataloging-in-Publication Data

Joss Whedon's Dollhouse : confounding purpose, confusing identity / edited by Sherry Ginn, Alyson R. Buckman, and Heather M. Porter.
pages cm. – (Science fiction television)
Includes bibliographical references and index.
ISBN 978-1-4422-3312-6 (cloth : alk. paper) – ISBN 978-1-4422-3313-3 (ebook)
1. Dollhouse (Television program) 2. Whedon, Joss–Criticism and interpretation. I. Ginn, Sherry, editor of compilation. II. Buckman, Alyson R., editor of compilation. III. Porter, Heather M., editor of compilation.
PN1992.77.D6277J68 2014
791.45'72–dc23
2013044999

∞™ The paper used in this publication meets the minimum requirements of American National Standard for Information Sciences Permanence of Paper for Printed Library Materials, ANSI/NISO Z39.48-1992.

Printed in the United States of America

Sherry says, “Once again, for Larry.”

For Al, Ben, and Sam, who watch and re-watch with me. ~ Alyson

For my Mom, who encourages me, and for Chris, who watches with me over and over. ~ Heather

Contents

Acknowledgments

From the moment I thought, "Why not just do it myself?" when no one at Slayage 5 (summer 2012) was contemplating such a project, to this moment in summer 2013 when we are putting the finishing touches on the manuscript, everyone involved has put forth great effort to finish their chapters in a timely fashion. This book would never have been possible without the hard work from all involved. A big thank-you goes to Alyson and Heather for agreeing that a book on *Dollhouse* was overdue and that they would work on it with me. Heather has encyclopedic knowledge of all things Whedon as well as an intimate knowledge of Hollywood. Her assistance was invaluable, and so is her friendship. Alyson managed to find time during a difficult period to edit these chapters in such a way as to draw out the best in each and every one. I really learned a lot just by watching her in action. Shiny!

—Sherry Ginn

I would like to thank Sherry Ginn for asking me to coedit this volume with her and Heather M. Porter. As my first major editing assignment, it turned out to be quite an eye-opener! I am grateful for the hard work put forth by everyone involved, especially Sherry, who made everything coalesce. Each of us brought different strengths to this work, and we worked well as an editing team. My thanks go to Sherry and Heather for their patience and support and to our authors for their excellent work. I would also like to express my appreciation to Heather for her wide-ranging knowledge of all things Whedon.

—Alyson R. Buckman

From the moment this fantastic editing team came together at Slayage 5 to this moment as we finish up the manuscript, I still cannot believe how amazing this process has been. Sherry and Alyson are both supportive and wonderful to work with, and the authors have worked hard to produce excellent chapters: thank you all. My involvement with this project would not have been possible without the encouragement and support that I received along the way from people such as Professor Jeremy Murray-Brown, with his encouragement to do something with the *Buffy* paper I wrote, to Rhonda Wilcox, who reached out and made me feel welcome at the first Slayage Conference, to Tanya Cochran, who convinced me to start attending Popular Culture Association meetings, and Sherry, who mentored me and helped me get my first paper published. All of these people, whom I now call friends, helped me get to today, and for all of them and all of you, "I try to be my best."

—Heather M. Porter

Fantasy Is His Business, but It Is Not His Purpose

An Introduction to Joss Whedon and His Storytelling

Alyson R. Buckman

Since the inception of Joss Whedon's career, he repeatedly has interrogated, deconstructed, and played with the process of telling stories. Although television and film direction is the skill for which he is best known, having received his BA in film studies from Wesleyan University in 1987, he asserts his primary identity as that of a writer in his 2013 British Academy of Film and Television Arts interview with Francine Stock.[1]

He learned the craft of writing for television, in part, in the writer's room on the set of *Roseanne*, a series revolutionary at the time for its construction of a smart, funny, strong working-class woman; as one of several writers, he was credited on four of the scripts produced in the 1989 season. After leaving *Roseanne,* Whedon not only penned the screenplay for the *Buffy: the Vampire Slayer* film (1992) but was an oft-sought script doctor for years, working on *Speed* (1994), *Toy Story* (1995), and *Twister* (1996), among others. *Toy Story* broke boundaries not only in computer animation but also in its construction of plot and smart, witty dialogue, especially for an animated feature. About the experience, Whedon says, "I was sort of teaching [the directors] about comic structure, and they were teaching me about visual style."[2] The *Toy Story* screenplay won him an Academy Award nomination. He also wrote the screenplays for *Alien: Resurrection* (1997) and the animated film *Titan A.E.* (2000), the former of which he has designated one of the lowest points in his career.[3] His first major on-screen effort, *Buffy the Vampire Slayer*, was specifically designed like the television show that followed

(1997–2003): to confront the trope of young woman as victim and to turn it on its head.[4]

The multiple productions that ensued—*Angel, Firefly, Serenity* (2005), *Dr. Horrible's Sing-Along Blog* (2008), *Dollhouse* (2009–2010), *The Cabin in the Woods* (2012), *The Avengers* (2012), and *Much Ado about Nothing* (2013)—exhibit, to varying degrees and among other elements, a concern with the form and function of storytelling, in specific episodes and overall arcs. In a 2002 interview with James Longworth, Whedon admitted, "I just want to keep trying, because I think of narrative and everything I do, whatever form it takes, as this sort of great, vast library that can be explored, that you'll never, ever see all of, but that you can just try and see as much of it as you can."[5] At what well may be the pinnacle of Whedon's career—with the third-highest-grossing film ever (*The Avengers*, released in 2012), two other films (*The Cabin in the Woods* and *Much Ado about Nothing*, released in 2012 and 2013 to critical acclaim), and a contract for *The Avengers 2* (to be released in 2015)—it seems appropriate to give a brief critical overview before proceeding to an introduction to the subject of this anthology, *Dollhouse*.

The world of the imagination is not a world of rainbows, unicorns, and sparkles to Joss Whedon; where there *are* unicorns, they gore people—mercilessly, like any other horror movie monster (*Cabin*). The imagination is more likely to be represented by a demonic force, such as a triple-threat demon ironically named "Sweet" facilitating one's own death by dancing while attempting to take one's sister to a hell dimension (*Buffy*, "Once More, with Feeling," 6.7). The enactment of imaginative power may bring havoc or temporary control over one's life, but it is fraught with dangerous consequences, even if not enacted by the literal demon of the imagination of "Once More," the minor character trying to write himself into the hero role of "Superstar" (*Buffy*, 4.17), or the self-proclaimed (and misleading) biographer of the hero in "Storyteller" (*Buffy*, 7.16). Casualties inevitably occur, and Whedon presents this as at least bittersweet and often tragic ("Once More," "Superstar," *Cabin*). Even when the storyteller is not center stage, the remnants of his or her imagination—prophecies, for example, as in *Buffy* and *Angel*—can wreak havoc. Stories have power, especially those that seek to change the world, and this includes historical myths (*Angel*, *Buffy*, *Firefly/Serenity*, *Dr. Horrible's Sing-Along Blog*, *Dollhouse*, *The Avengers*, *Cabin*). One of the more disturbing images from *Serenity* is of River Tam seemingly remembering, from childhood, a teacher's story of the "heroic" Alliance's civilizing impulse along with River's rejection of it melded with both a Reaver attack and a needle being stuck in River's head by Alliance scientists, both of which ironically comment on this story of civilization.

However, when reimagined by those concerned, stories can become empowering, as in Buffy's near death in "Prophecy Girl" (1.12): the Master, flummoxed by Buffy's lively reappearance, sputters, "You were destined to die! It was written!" Buffy retorts, "What can I say? I flunked the written," and proceeds to demolish the given story.[6] Whedon has deconstructed and reimagined many of our cultural tropes, ideologies, and systems: the blonde, female victim (*Buffy*); the lustful, destructive vampire modeled on Dracula and designed to punish and contain women who express their own sexuality (*Buffy* and *Angel*); the "comfort and enlightenment of true civilization" (*Serenity*); the simply evil mad scientist (*Dr. Horrible's Sing-Along Blog*); the woman concerned only with her looks and true love (throughout the Whedonverses); the power of the puppet-master studio (*Angel*, "Smile Time," 5.14; *Dr. Horrible*; *Cabin*); the gaze of the audience, whose demands must be satisfied lest the world be destroyed (*Cabin*). Such tensions between the commercial and the aesthetic as suggested here are implied in Whedon's oft-repeated signature statement: "Give the audience what they need, not what they want." The changes that he provides as a storyteller are their own destructive force. In an interview with James Longworth, Whedon explains the destructive potential of stories:

> [Stories] come from violence, they come from sex. They come from death. They come from the dark places that everybody has to go to, kind of wants to, or doesn't, but needs to deal with. . . . And, ultimately, to access these base emotions, to go to these strange places, to deal with sexuality, to deal with horror and death, is what people need and it's the reason that we tell these stories. . . . Stories are sacred.[7]

While they may be sacred, change is necessary to keep them vital and relevant.[8] Unfortunately for fans, such upheaval often includes the death of beloved characters, who are generally untouchable in genre fiction. Xander Bennett alludes to this in discussing screenwriting and what he calls "the Whedon Gambit":

> Joss Whedon is the master of this particular trick. If you've seen the first or last season of *Angel*, or if you've ever sat down to enjoy *Serenity* immediately after watching *Firefly*, you know exactly what I'm talking about. Hey, I still haven't forgiven that callous bastard for the ending of *Buffy*'s sixth season. It's gotten so bad that fans of Whedon's work now go into automatic paroxysms of fear if two of his characters start exhibiting signs of happiness together, or worse still, profess love for each other. The fans know that means their chances of horrible, horrible death or disfigurement just increased exponentially.[9]

As a result, the viewer cannot mindlessly consume Whedon's fictions, secure in the knowledge that the main characters will remain safe.

Form also is in flux in the Whedonverse, as indicated by the discussion of language in one of the first book-length academic studies published within

Whedon Studies, Michael Adams's *Slayer Slang*; Adams argues that the language of *Buffy* relied on changes in usage, coinage, and experimentation.[10] "Once More, with Feeling," "The Body," "Restless," and "Hush," to name only four examples from *Buffy*, are all experiments in the connection between form and content: the famous musical episode, which neatly makes all characters aware of the various emotions that have been simmering within the Scooby gang; the episode that features the death of an important character, an absence of nondiegetic music, and an emphasis on real time; an episode that takes place in the dreams of the main characters, which Whedon refers to as "basically a forty-minute poem";[11] and the allegedly "silent" episode, which is not silent at all but removes the voices of the characters for much of the episode to focus on communication. "The great thing about ["Hush"]," Whedon has asserted, "was I knew I wanted to do it silent, because I was starting to devolve into sort of a hack TV director, and I wanted to push myself visually. But what it became about was the way language interferes with communication."[12] Regarding *Buffy*, Matthew Pateman writes that the series

> asks its characters, and its viewers, to assume a longer vision, to track questions of knowledge, of ethnicity, of history and of ethics (among others) and to participate in the awkward, politically contradictory, socially messy, culturally fraught, personally painful effects of these questions. And it does this by choosing, carefully, critically, knowingly, aesthetic styles that complement or stand in juxtaposition to these ideas. . . . The show never fails to be complex, to demand of its audience a reflective as well as sensuous response.[13]

The aesthetic power of Whedon's work is illustrated by the quick rise of critical interest in his work. Roz Kaveney, David Lavery, and Rhonda Wilcox published the first two books in Whedon Studies in 2002, *Reading the Vampire Slayer: The Unofficial Critical Companion to "Buffy" and "Angel"* (Kaveney) and *Fighting the Forces: What's at Stake in "Buffy the Vampire Slayer"* (Lavery and Wilcox).[14] *Reading*, *Fighting*, and *Why Buffy Matters* (Wilcox's second book on the series) assert the aesthetic power of Whedon's storytelling.[15] For example, *Why Buffy Matters* considers, in part, why Whedon profits from a comparison with Shakespeare and Dickens, in that there are similarities in both context and reception of the works.[16]

Whereas *Buffy* played with the tropes of horror, teen romance, and supernatural fiction, *Angel* combined the supernatural with film noir, placing a vampire detective and his pursuit of redemption at the center of this fiction. While engaging with the same universe as *Buffy*, Whedon respun the parameters of that universe, and the stories changed as a result.[17] *Angel* is darker in form and function, and the treatment of women often is more in keeping with its film noir aesthetics than with that of *Buffy*.[18] However, Whedon and company continued to privilege innovation in their storytelling in *Angel*. For example, Tammy Kinsey provides an excellent discussion of the use of experimental technique in *Angel* and argues that the series represents "five years of risk taking in prime time."[19] Stacey Abbott concurs, writing, "What stands out about *Angel* is its commitment to exploring dark and morally ambiguous themes, enhanced by the show's irreverent humor, the flouting of generic conventions, and an emphasis on a cinematic visual style."[20]

The next series developed by Whedon (in association with Tim Minear), *Firefly,* functioned as Whedon's exploration of the interplay between science fiction and the Western; as with *Buffy*, linguistic play was foregrounded as well, with Whedon combining Chinese and English languages and no subtitles.[21] In large part an homage to *Stagecoach* and its stock characters, *Firefly* also upended *Star Trek*'s Federation ideology.[22] Whedon has stated several times that *Firefly* is the flip side of the *Star Trek* universe, representing those people who saw *Star Trek*'s Federation not as the iconic great white hope but as conquerors imposing their ideology on others. As he did with *Buffy*, Whedon revised the trope of the victimized female character and invested her with power, this time in relation to the Western genre rather than the horror: victimized by the centralized government (instead of racial minorities, as in the standard Western), she became, as some joked after *Serenity*, "River the Reaver Slayer."[23] She reclaimed her own story, much like Buffy, and became more powerful as a result.[24] *Firefly* continues many of Whedon's trademark emphases: created family, strong female characters, humor interlaced with pathos, a wisecracking central character, a character who speaks uncomfortable truths, resistance to centralized patriarchal authority, mixing of genres, and visual and narrative innovation. For example, "Objects in Space" (*Firefly*, 1.14), directed and written by Whedon, is a beautifully filmed exploration of existentialist philosophy.[25] Form and content again reinforce each other as multiple aspects of the narrative are captured in the visuals, including the instability of both River and Jubal Early's minds, River's tactile and emotional connection to the ship, the ship's isolation as they attempt to avoid Alliance ships, and River's vulnerability. It is little wonder that Whedon, the cast, and fans lost their hearts to this series and demanded a feature-length film: *Serenity*, a film that was darker than the television series and tied up the loose ends left by the sudden cancellation of the series.

Whedon did not stop with film and television, either, as he made multiple forays into the comic form, one of his first loves. In *Fray* (2001–2003)—and the still-running Buffy, Angel, Angel and Faith, and Spike comics, which launched after *Buffy*'s end in 2003—he further explored his vampire-centric universe and the mythology of the Slayer in comic form, although Whedon-penned issues of those comics have become less frequent, and Whedon has not penned any of the Spike comics.[26] From 2004 to 2008, Whedon also wrote Marvel's *Astonishing X-Men* series, and in 2006 he authored *Runaways*, already established by Brian Vaughan.[27] He launched the short series

Sugarshock in 2007 on the Dark Horse MySpace page; it was released in hard copy in 2009.[28] He directed episodes of *The Office* ("Business School," 3.16; "Branch Wars," 4.6) and *Glee* ("Dream On," 1.19) as well. The latter are humorously intertextual: *The Office*'s "Business School" episode features one character pretending to turn into a vampire after another finds a bat (in a *Buffy* homage), and *Glee* features Neil Patrick Harris, who also starred (and sang) in *Dr. Horrible*, as a competitive high school glee club has-been who is now on the school board and attempting to cut the club's funding.

Aired in three parts over the course of a week in 2008, *Dr. Horrible's Sing-Along Blog* played not only with traditional tropes—making the evil scientist into the protagonist more sympathetic than the traditional hero—but with form and platform. Dr. Horrible, for instance, addresses the camera in a breaking down of the fourth wall; features of the commentary and DVD format take this further.[29] As an independently produced Internet sensation[30] specifically designed to challenge the prevailing Hollywood model in the face of the 2007–2008 writer's strike, *Dr. Horrible* was recognized by such publications as *Forbes* as a successful new model of film and television production, even as it played with aesthetic form and commented ironically on political, social, and cultural arenas.[31] The production earned Whedon the 2010 Vanguard Award, given by the Producers Guild of America to acknowledge "achievements in new media and technology": David Friendly and Laurence Mark, cochairs of the Guild, stated that "Joss Whedon has mastered the art of melding the newest technology with inspired storytelling, truly exemplifying the spirit of the Vanguard Award."[32]

The Cabin in the Woods—directed by Drew Goddard and cowritten by Goddard and Whedon—is yet another playful retelling of the horror genre. From its title to the use of Sigourney Weaver—the so-called final girl par excellence—as a menacing government agent who encourages murder, Whedon and Goddard wrote a film that revels in its metatextual deconstruction of the horror film.[33] The control room scene—in which technicians are partying in the foreground while Dana agonizingly crawls away from her death on the large control room screen in the background—works as a marvelous deconstruction of the tropes of the horror genre and the gaze and implicates the audience in the torture and death of the film's characters. This is only one example of the metatextuality; the use of cast members from other Whedon fictions serves as well to highlight the constructed nature of the text.[34]

While *The Avengers* offers little room for playful deconstruction due to its position within the Marvel universe, Whedon again provided a powerful female character (in the person of Natasha Romanov, aka Black Widow) within an often patriarchal genre. In two scenes in which she seems defeated, she masterfully shows herself to have been in control the entire time. He also wrote the characters of Captain America, Iron Man, and the Hulk not as towering behemoths but as vulnerable and nearly broken real men. Much of the dialogue reveals Whedon's classic deflation through humor as well; even silent moments, such as the postcredit shawarma scene, convey humorous character elements. Whedon has worked on some of the associated film scripts in this Marvel reboot, including *Thor* (2011), *Captain America: The First Avenger* (2011), and *Iron Man 3* (2013). Additionally, he is developing and cowriting *Agents of S.H.I.E.L.D.* for television.

Much Ado about Nothing, in a different, nonblockbuster approach, is Whedon's interpretation of Shakespeare's play of the same title, set within contemporary America yet maintaining the original dialogue. The film was converted to black and white in postproduction. Shot over the course of twelve days at Whedon's home in Santa Monica, *Much Ado* offers eager fans and cinephiles an independent labor of love. Philip French posits the common worry that directors of blockbusters may not return to simpler filmmaking but asserts that Whedon, for one, has shown that he skillfully can traverse both modes of storytelling; Killian Fox argues that Whedon's use of black-and-white cinematography—a daring choice within a filmic economy that emphasizes new gimmicks—was perfect for the glamorous tone that he wanted to set.[35]

Dollhouse also was a daring production. Actively interrogating fantasy, its audiences, and its purveyors, its concept alone was risky: "It was so clear to me what was interesting about this show—the idea of identity, and the idea that everybody is compromised. But knowing going in that the premise was going to be offensive to some people scared the shit out of me."[36] The conceit of the show is that a living Dollhouse, peopled not only by Dolls but also by management, security, and medical and technical support, exists underneath the streets of Los Angeles (and in cities around the globe). Such Dollhouses contain people who allegedly have volunteered for a five-year stint as a Doll: a person who has had his or her memory completely wiped and replaced with a rudimentary self. Eating, sleeping, creating art, receiving massages, and exercising in a daily regimen designed so that they can "be [their] best," these Dolls are regularly sent out on assignments as Actives. An Active is a Doll with a specific imprint constructed by technology; the parameters of the imprint are generally provided by a client willing to pay large sums of money for a programmed person. These assignments range from the humanitarian (e.g., saving a child) to the criminal (e.g., stealing parts of the Parthenon from a vault) to the sexual. Meanwhile, FBI agent Paul Ballard searches for one particular woman, Caroline, whom he is certain has been made into a Doll. Echo, played by Eliza Dushku, is the Doll in question and the lead of the series. A genetic predisposition and some manipulation of events, as the audience later finds out, enable Echo to gradually resist the mind wipes and thus strengthen the gene's expression and her value to Rossum Corporation.[37] Gradually, she builds a composite personality that incorporates all her imprints. Two episodes, "Epitaph One" (1.13) and "Epitaph Two: Return" (2.13), detail the dystopic results of Dollhouse imprinting technology: wiping the brains of most humanity so that it becomes either "dumbshows" (people whose minds are remotely wiped and without the ability for speech or complex thought) or "butchers" (people imprinted with a bloodthirsty desire to kill dumbshows and "actuals," those who have managed not to be wiped).[38] Those with the money, power, and influence to do so—especially the heads of Rossum Corporation—claim new bodies upon which to imprint themselves, thereby gaining beauty, health, and youth. Meanwhile, Echo and most other main cast members fight against the tech and attempt to save the world.

While Whedon had planned a six-season arc for the series, the lack of investment by Fox quickly became evident.[39] After meeting with Fox executives, Whedon scrapped the already-filmed pilot—bits and pieces would find their way into other episodes—and revised his plan to meet Fox's demands that the show be less about relationships and character development and more about explosions and stand-alone episodes.[40] In the surprising second-season renewal, Whedon lost money from his budget and consequently switched from 35-mm film to HD digital film as well as handheld cameras.[41]

Whedon has cited several elements in regard to Fox's difficulty with the series: a prurient attitude toward sexuality in mainstream television, a complex concept irreducible to two or three sentences, a focus on intimacy and potentially disturbing ideas, a plotline that focused on slow revelation, and a desire for stand-alone episodes rather than complex mythology. Whedon had his own disturbing and thought-provoking anxieties about the show as well, ones that he wanted to work through on-screen.[42]

Whedon seems pressed to create, seized by creative necessity: In 2002–2003, he had three series airing simultaneously; in two weeks on vacation from *Buffy,* he began writing *Firefly*; he developed *Dr. Horrible* during the 2007 writer's strike and produced it with his own money; and during the two-week downtime that he was contractually obligated to take after the filming wrapped on *The Avengers*, he made *Much Ado about Nothing* in twelve days at his home in Santa Monica. The need to create slightly twisted versions of our world and use them to interrogate the morality and humanity of our own world is an essential aspect of Whedon's oeuvre. Like many science fiction and fantasy writers, he utilizes a strategy of displacement in which alternative worlds are created that signify back on our own world, enabling a subtextual discussion of our society through seemingly unrelated characters and events. Such a perspective calls for active participation from its audience, although that call sometimes goes unheard.

In the case of *Dollhouse*, the strategy of displacement relates the world of the Dolls, one might argue, to that of the entertainment industry.[43] Don Tresca argues that *Dollhouse* is "nothing less than a treatise on the art and science of making a television series."[44] The parallels are multiple: Rossum as the uncaring network with shady morality; Adelle and Topher as the directors and writers of television, actively selling fantasy and sometimes selling what they find repugnant; the Dolls as the actors "pimped" by the industry; the self-justification through the argument that they are providing what people "need" (and, thus, even Whedon becomes implicated). The commentary and main plot in "Man on the Street" (1.6) is one of the clearest examples of this subtext. Joel Mynor, for instance, details Agent Ballard's clichéd fantasy: the knight in shining armor riding to rescue the damsel in distress, whom he then will sweep off her feet and kiss. Mynor asserts that Ballard's fantasy is "pathetic" and clichéd. One person on the street makes the subtext fully textual: "You think it's not happening? You think they're not controlling you? Don't worry about it. Just sit back and wait for 'em to tell you what to buy." However, the series is about far more than the relationship among commercialism, the audience, and network television, as our contributors illustrate.

This collection adds to the critical academic interest in Joss Whedon and, specifically, *Dollhouse*. More than thirty books relating to Whedon's work have been published,[45] and, at this count, thirty-two issues of *Slayage: The Journal of the Whedon Studies Association* have been posted as well as eight editions of *Watcher Junior*, the undergraduate Whedon Studies journal. These figures do not include the multiple conference presentations or essays published in non–Whedon Studies Association journals.[46] While the Association has produced a special issue on *Dollhouse*, this is only the second book to specifically focus on *Dollhouse*, and it is the first book with a sustained scholarly lens.

The anthology is divided into three parts: "Self and Identity," "Ethics," and "Structure and Form." In the first section, Michael Starr, Meg Saint Clair Pearson, and Margo Collins consider the postmodernist and poststructuralist aspects of *Dollhouse*: what makes us human and how far that category can be stretched certainly is a clear consideration of the series. If one considers that Donna Haraway wrote in 1991 that simply the acts of having dental work or wearing sneakers constitute modifications of the natural human form and make us cyborgs, then certainly having one's personality wiped and another inserted qualifies as cyborg identity. All three of our authors in this section also emphasize the importance of human connection in their chapters.

"'I've Watched You Build Yourself from Scratch': The Assemblage of Echo" is Michael Starr's exploration of Echo's character in relation to Gilles Deleuze's theories of self and identity. Although Echo initially functions as a model of the Deleuzian assemblage, ultimately her composite identity illustrates the liberatory potential within Gilles Deleuze's theory of the schizophrenic.

Meg Saint Clair Pearson explores posthumanism and its consequences in "'We Are Not Just Human Anymore': Accepting the Posthuman Future." Pearson takes the provocative stance that Topher, rather than the Dolls, presents the most alarming construction of the perils of posthumanism within the series: his estrangement from humanity indicates that he has become an inhuman object rather than subject. *Dollhouse* does not simply excoriate posthumanism, however: in their connection to humanity and an ethics of care, the Dolls present a more positive version of posthumanism than Topher.

Margo Collins interrogates the distinctions between memories located within the body and those that are artificial, in "Anamnesis, Hypomnesis, and the Failure of the Posthuman in Whedon's *Dollhouse*," finding that the former enable human connection far better than the latter. This argument clearly jibes with Whedon's other productions, which also privilege connection in their emphasis on chosen family and embodiment.

Tom Garbett, Ananya Mukherjea, Samira Nadkarni, and Heather M. Porter and Sherry Ginn's chapters constitute the section on what we are broadly calling "ethics." The ethical treatment of women of color, one's fellow human beings, and oneself, and the most vulnerable human population—children—is at the heart of these chapters.

In "'What about the Laws?' Regulation and the Celebration of Resistance," Tom Garbett frames his argument with Adelle DeWitt's assertion that the laws of humanity must be obeyed. Garbett asserts that characters who assert their personal will, such as Agent Paul Ballard and Echo, personalize regulatory functions and thus strengthen them.

"Somebody's Asian on TV: Sierra/Priya and the Politics of Representation" is Ananya Mukherjea's exploration of the weight carried by characters of color, who must signify in multiple ways that white characters do not. While both Echo/Caroline and Sierra/Priya are female, Sierra/Priya's body performs in ways very different from that of Echo/Caroline; this may be attributed to her Asian, Australian, and transnational migrant identities. Like Hélène Frohard-Dourlent in her Slayage conference essay on the construction of Asian female identity in the commentary for *Dr. Horrible*, Mukherjea contributes valuable criticism of racial embodiment in the Whedonverse.

Samira Nadkarni tackles a difficult subject in her chapter, "'In My House and Therefore in My Care': Transgressive Mothering, Abuse, and Embodiment," as she discusses the construction of motherhood within the world of *Dollhouse* and carefully lays out an argument for reading the Dolls as abused children. Giving examples from other works in the Whedonverse, Nadkarni argues that abuse is explored throughout Whedon's oeuvre and without any pat answers provided.

In "'I Possess the Means to Satisfy My Vagaries': What Motivates the Dollhouse Clients?" Heather M. Porter and Sherry Ginn find that while there are two main categories of clients of the Dollhouses—those who use the Dolls for sexual fantasies, wish fulfillment, or romance and those who use the Dolls for nonsexual purposes, such as undercover operative or thief—the chief customer of the Dollhouses was actually its parent company, the Rossum Corporation. As a result, fears about government control, big business, and loss of the self are linked to more common representations of prostitution and crime.

The structure and form of *Dollhouse* is considered by Bronwen Calvert, Lorna Jowett, Joel Hawkes, Derrick King, David Kociemba, and Ian G. Klein. Our authors consider the relation between pleasure and horror, the allure of Sierra/Priya and Victor/Tony's "true love" narrative, the ways in which *Dollhouse* functions as a theater of the self, the tension between fear and hope within the *Dollhouse* dystopia, the importance of the title sequences in building narrative and authorship (among other elements), and the pervasive water symbolism of the series.

"'Who Did They Make Me This Time?' Viewing Pleasure and Horror" considers the overlap between pleasure and horror in the series, using Julia Kristeva's theory of the abject. Bronwen Calvert manages to skillfully show that the abject is inclusive of both pleasure and horror; the interplay creates productive tension for the viewer, due to the viewer's engagement in the horror at the heart of the Dollhouse.

Lorna Jowett explores the power of cultural constructions of romance in "'I Love Him . . . Is That Real?' Interrogating Romance through Victor and Sierra." Although she has extensively analyzed and critiqued the social and cultural construction of gender and sexuality, even Jowett finds the trope of true love hard to resist in Sierra/Priya and Victor/Tony's relationship.

Joel Hawkes utilizes a creative format to illustrate his points about fragmentation, theater, and the self in "The Theatre of the Self: Repetitious and Reflective Practices of Person and Place." Just as the episodes begin with an explosion of fragments mirroring the fragmented identities of Echo, fragments and mirrors abound throughout episodes and seasons. *Dollhouse*, Hawkes argues in part, functions as a double for our society such that we might analyze and better understand ourselves.

While Derrick King emphasizes dystopia in "'We're Lost. We Are Not Gone': Critical Dystopia and the Politics of Radical Hope," he focuses on the utopic impulse as well in *Dollhouse*'s assertion of a radical politics of hope within its narrative. While the two "Epitaph" episodes and the second season in general are firmly embedded within a dystopic world, Echo's narrative and her character provide what King calls "stubborn hopefulness" in the face of the hegemonic reach of global capitalism.

David Kociemba decodes the *Dollhouse* title sequence in "Welcome to the *Dollhouse*: Reading Its Opening Title Sequences." As with his reading of these sequences in, for example, *Buffy the Vampire Slayer*, Kociemba provides excellent reasons to pay attention to them in his interesting exploration of their signifying power. Narrative promises, assertions of authorship and creative control, and thematic cues are only some of the elements of title sequences discussed in Kociemba's chapter.

"Ritual, Rebirth, and the Rising Tide: Water and the Transcendent Self" meditates on water symbolism throughout *Dollhouse*. Ian G. Klein traces the many ways in which water functions in the series as a transformative agent, especially in Echo's journey. Greek mythology, Japanese design elements, and psychology all come together in Klein's contribution to illustrate how important water is to *Dollhouse.*

Nonacademics—and even many academics outside of cultural studies—often question the worth of studying media. Our authors illustrate the worth of such examinations and why critical thinking about popular culture is necessary. In reading these chapters, I came to appreciate *Dollhouse* and understand facets of it that had not occurred to me before. The authors herein have accomplished the goal of criticism: not just explaining the text but opening it up to new understandings of the material, to fully excavate it and explore its multiple meanings. Just as art shows us what it means to be human, criticism of that art can illustrate the richness of the endeavor and our humanity.

In his recent British Academy of Film and Television Arts interview, Whedon was asked what he thought about the critical attention that his work has received from academics, starting with *Buffy*. He responded,

> The other thing I learned early on was not just that people were getting what it was I was doing. In some cases, they were getting it either better or differently than I was but just as legitimately. And everything we did, we did with enormous intent. . . . Something is worthy of study even when it isn't doing any of that. I mean, you could study any show that has a cultural impact to find out why and to find out what it is, what the message is. . . . We didn't think that they would *get* that we were working so hard on it, but we absolutely were.[47]

The critical response that has continued from *Buffy* through *Much Ado about Nothing* suggests that Whedon and cadre continue to approach their subject matter thoughtfully and artistically, even as they de- and reconstruct. The chapters in this volume certainly argue for continued critical analysis of Whedon's work and demonstrate the value of popular culture scholarship.[48]

BIBLIOGRAPHY

Abbott, Stacey. *Angel*. Detroit, MI: Wayne State University Press, 2009.

Adams, Michael. *Slayer Slang: A "Buffy the Vampire Slayer" Lexicon.* New York: Oxford University Press, 2003.

Angel: After the Fall (Season 6). Created by Joss Whedon. November 2007–April 2011. IDW.

Angel & Faith. Created by Joss Whedon. August 2011–present. Dark Horse.

Astonishing X-Men. Written by Joss Whedon. Penned by John Cassaday. July 2004–July 2008. Marvel. Print.
Bennett, Xander. "Screenwriting Tip #103: The Joss Whedon Gambit." MasteringFilm.com.
Brown, Rebecca. "Orientalism in *Firefly* and *Serenity*." *Slayage: The Journal of the Whedon Studies Association* 7, no. 1 (2008).
Buckman, Alyson R. "'Didn't Get the Memo? Hero of the People Now': Hat Tricks, and the Complication of Viewer Responses in the Works of Joss Whedon." Presented at the Slayage Conference on the Whedonverses 5, Vancouver, BC, July 2012.
———. "'Go Ahead, Run Away! Say It Was Horrible!' *Dr. Horrible's Sing-Along Blog* as Resistant Text." *Slayage*: *The Journal of the Whedon Studies Association* 8, no. 1 (2010).
———. "'Much Madness Is Divinest Sense': *Firefly*'s 'Big Damn Heroes' and Little Witches." In *Investigating "Firefly" and "Serenity": Science Fiction on the Frontier*, edited by Rhonda V. Wilcox and Tanya R. Cochran, 41–49. New York: Tauris, 2008.
Buffy the Vampire Slayer (Season 8–Season 9). Created by Joss Whedon. March 2007–present. Dark Horse.
Clover, Carol. *Men, Women, and Chain Saws: Gender in Modern Horror Film.* Princeton, NJ: Princeton University Press, 1993.
Comeford, AmiJo. "Cordelia Chase as Failed Feminist Gesture." In *Buffy Meets the Academy: Essays on the Episodes and Scripts as Texts*, edited by Kevin K. Durand, 150–60. Jefferson, NC: McFarland, 2009.
———. "Helping the Helpless: Medieval Romance in *Angel*." In *The Literary Angel: Essays on Influences and Traditions Reflected in the Joss Whedon Series*, edited by AmiJo Comeford and Tamy Burnett, 175–86. Jefferson, NC: McFarland, 2010.
da Silveira, Luciana Hiromi Yamada. "How Cancellation Told the Story of the Dollhouse." In *Inside Joss' "Dollhouse": From Alpha to Rossum*, edited by Jane Espenson, 147–59. Dallas, TX: Smart Pop, 2010.
Fox, Killian. "How Black-and-White Movies Made a Comeback." *The Observer*. June 29, 2013.
Fray. Created by Joss Whedon. Written by Joss Whedon. Penned by Karl Moline. 2001–2003. Dark Horse.
French, Philip. "*Much Ado about Nothing*—Review." *The Observer*. June 15, 2013.
Frohard-Dourlent, Hélène. "Somebody's Asian in *Dr. Horrible*: Humor and Racial Representations in the Whedonverse." Presented at the Slayage Conference on the Whedonverses 5, Vancouver, BC, July 2012.
Haraway, Donna. "A Cyborg Manifesto: Science, Technology, and Socialist-Feminism in the Late Twentieth Century." In *Simians, Cyborgs and Women: The Reinvention of Nature*, 149–81. New York: Routledge, 1991.
Harrison, Janine. "Gender Politics in *Angel*: Traditional vs. Non-traditional Corporate Climates." In *The Literary Angel: Essays on Influences and Traditions Reflected in the Joss Whedon Series*, edited by AmiJo Comeford and Tamy Burnett, 117–31. Jefferson, NC: McFarland, 2010.
Hornick, Alysa. "Whedonology: An Academic Whedon Studies Bibliography." http://www.alysa316.com/Whedonology/. June 30, 2013.
"Joss Whedon to Be Honored with 2010 Vanguard Award." Producers Guild of America. http://www.producersguild.org/news/34895/. November 23, 2009.
Kaveney, Roz, ed. *Reading the Vampire Slayer: The Unofficial Critical Companion to "Buffy" and "Angel."* New York: Tauris, 2002.
Kinsey, Tammy. "Transitions and Time: The Cinematic Language of *Angel*." In *Reading "Angel": The TV Spin-Off with a Soul*, edited by Stacey Abbott, 44–56. New York: Tauris, 2005.
Knowledge@Wharton. "A New Vision for Online Entertainment." *Forbes*. February 5, 2009. See also: Whedon, Joss. "Joss Whedon's Plan to Monetize Internet Content (Watch Out, Hollywood)." By Knowledge@Wharton. In Lavery and Burkhead, *Joss Whedon*, 175–83.
Lavery, David, and Cynthia Burkhead, eds. *Joss Whedon: Conversations.* Jackson: University Press of Mississippi, 2011.
Longworth, James. "Joss Whedon: Feminist." In Lavery and Burkhead, *Joss Whedon*, 42–63.

Mandala, Susan. "Representing the Future: Chinese and Codeswitching in *Firefly*." In *Investigating "Firefly" and "Serenity": Science Fiction on the Frontier*, edited by Rhonda V. Wilcox and Tanya R. Cochran, 31–40. New York: Tauris, 2008.

Murray, Noel. "How *Dollhouse* Toyed with the Idea of How People and Institutions Are Formed." AV Club. http://www.avclub.com/article/how-idollhousei-toyed-with-the-idea-of-how-people--93682. March 14, 2013.

Pateman, Matthew. *The Aesthetics of Culture in "Buffy the Vampire Slayer."* Jefferson, NC: McFarland, 2006.

Press, Joy. "Joss Whedon Just Wants to be Loved." In Lavery and Burkhead, *Joss Whedon*, 184–89.

Rosen, Lisa. "New Media Guru: Meet Joss Whedon the Web Slayer." In Lavery and Burkhead, *Joss Whedon*, 170–74.

Runaways. Created by Brian Vaughan. Written by Joss Whedon. Penned by Michael Ryan. June 2007–August 2008. Marvel.

Ryan, Maureen. "Sex, Secrets and *Dollhouse*: Joss Whedon Talks about the End of His Fox Show." *Chicago Tribune*. December 3, 2009.

Schwartz, Terri. "Joss Whedon Reflects on the Mistakes of *Alien: Resurrection*." IFC. http://www.ifc.com/fix/2013/04/joss-whedon-reflects-on-the-mistakes-of-alien-resurrection. April 18, 2013.

Spike. Created by Joss Whedon. Written by Brian Lynch. Penned by Franco Urru. October 2010–May 2011. IDW.

Stock, Francine. "Joss Whedon: A Life in Pictures." British Academy of Film and Television Arts. http://www.bafta.org/film/features/joss-whedon-a-life-in-pictures,3816,BA.html. June 17, 2013.

Sugarshock. Created by Joss Whedon. Written by Joss Whedon. Penned by Fabio Moon. Portland, OR: Dark Horse Comics, 2009.

Tresca, Don. "'Fantasy Is Their Business, but It Is Not Their Purpose': The Metaphor of *Dollhouse*." In *Joss Whedon: The Complete Companion*, edited by Mary Alice Money, 411–25. London: Titan, 2012.

Whedon, Joss. "Objects in Space." Commentary. *Firefly*. Season 1, episode 14. Twentieth Century Fox Home Entertainment, 2003.

———. "Vows." Commentary. *Dollhouse*. Season 2, episode 1. Twentieth Century Fox Home Entertainment, 2010.

Whedon, Joss, and George Jeanty. "The Long Way Home, Part I." *Buffy the Vampire Slayer*, 8.1. March 14, 2007. Dark Horse.

Whedon, Joss, Karl Moline, and Andy Owens. *Fray*, 1–8. November 2003. Dark Horse.

Wilcox, Rhonda V. "Breaking the Ninth Wall with *Dr. Horrible's Sing-Along Blog*: Internet Creation." *Critical Studies in Television*. http://cstonline.tv/breaking-ninth-wall. November 22, 2012.

———. "Echoes of Complicity: Reflexivity and Identity in Joss Whedon's *Dollhouse*." *Slayage: The Journal of the Whedon Studies Association* 8, nos. 2–3 (2010).

———. *Why Buffy Matters: The Art of "Buffy the Vampire Slayer."* New York: Tauris, 2006.

Wilcox, Rhonda V., and David Lavery. "Afterword: The Depths of *Angel* and the Birth of *Angel* Studies." In *Reading Angel: The TV Spin-Off with a Soul*, edited by Stacey Abbott, 221–29. New York: Tauris, 2005.

———. *Fighting the Forces: What's at Stake in "Buffy the Vampire Slayer."* Lanham, MD: Rowman & Littlefield, 2002.

FILMOGRAPHY

Alien: Resurrection. Written by Joss Whedon. Directed by Jean-Pierre Jeunet. Twentieth Century Fox, 1997.

The Avengers. Written and directed by Joss Whedon. Walt Disney Studios Home Entertainment, 2012.

"The Body." *Buffy the Vampire Slayer.* Season 5, episode 16. Written and directed by Joss Whedon. First broadcast February 27, 2001.
"Branch Wars." *The Office*. Season 4, episode 6. Written by Mindy Kaling. Directed by Joss Whedon. First broadcast November 1, 2007.
Buffy the Vampire Slayer. Written by Joss Whedon. Directed by Fran Kuzui. Twentieth Century Fox, 1992.
"Business School." *The Office*. Season 3, episode 16. Written by Brent Forrester. Directed by Joss Whedon. First broadcast February 15, 2007.
The Cabin in the Woods. Written by Joss Whedon and Drew Goddard. Directed by Drew Goddard. Lionsgate, 2012.
Dr. Horrible's Sing-Along Blog. Written by Joss Whedon et al. Directed by Joss Whedon. New Video Group, 2009.
"Dream On." *Glee*. Season 1, episode 19. Written by Brad Falchuk. Directed by Joss Whedon. First broadcast May 18, 2010.
"Hush." *Buffy the Vampire Slayer*. Season 4, episode 10. Written and directed by Joss Whedon. First broadcast December 14, 1999.
Iron Man 3. Written by Drew Pearce and Shane Black. Directed by Shane Black. Walt Disney Studios Home Entertainment, 2013.
Much Ado about Nothing. Screenplay by Joss Whedon, play by William Shakepeare. Directed by Joss Whedon. Bellwether, 2012.
"Objects in Space." *Firefly*. Season 1, episode 14. Written and directed by Joss Whedon. First broadcast December 13, 2002.
"Once More, with Feeling." *Buffy the Vampire Slayer.* Season 6, episode 7. Written and directed by Joss Whedon. First broadcast November 6, 2001.
"Prophecy Girl." *Buffy the Vampire Slayer.* Season 1, episode 12. Written and directed by Joss Whedon. First broadcast June 2, 1997.
"Restless." *Buffy the Vampire Slayer.* Season 4, episode 22. Written and directed by Joss Whedon. First broadcast May 23, 2000.
Serenity. Written and directed by Joss Whedon. Universal Studios Home Entertainment, 2005.
"Smile Time." *Angel.* Season 5, episode 14. Written by Ben Edlund and Joss Whedon. Directed by Ben Edlund. First broadcast February 18, 2004.
Speed. Written by Graham Yost. Directed by Jan de Bont. Twentieth Century Fox, 1994.
Stagecoach. Screenplay by Dudley Nichols. Directed by John Ford. United Artists, 1939.
"Storyteller." *Buffy the Vampire Slayer.* Season 7, episode 16. Written by Jane Espenson. Directed by Marita Grabiak. First broadcast February 25, 2003
"Superstar." *Buffy the Vampire Slayer*. Season 4, episode 17. Written by Jane Espenson. Directed by David Grossman. First broadcast April 4, 2000.
Thor. Written by Ashley Edward Miller, Zack Stentz, and Don Payne. Directed by Kenneth Branagh. Paramount, 2011.
Titan AE. Screenplay by Ben Edlund, John August, and Joss Whedon. Directed by Don Bluth, Gary Goldman, and Art Vitello. Twentieth Century Fox, 2000.
Toy Story. Written by Joss Whedon et al. Directed by John Lasseter. Pixar, 1995.
Twister. Written by Michael Crichton and Anne-Marie Martin. Directed by Jan de Bont. Warner Brothers, 1996.

NOTES

The title to this introduction was inspired by Don Tresca's contribution to *Joss Whedon: The Complete Companion*, "'Fantasy Is Their Business, but It Is Not Their Purpose': The Metaphor of Dollhouse."

1. Stock, "Joss Whedon: A Life in Pictures."
2. Stock, "Joss Whedon: A Life in Pictures."
3. Schwartz, "Joss Whedon Reflects."

4. Whedon wrote but did not direct the film. The television series was sometimes directed and/or written by Whedon, who, during the last two seasons of *Buffy*, had three projects airing simultaneously, including *Angel* (1999–2004) and *Firefly* (2002–2003).

5. Longworth, "Joss Whedon: Feminist," 62–63.

6. In "Storyteller," Andrew reimagines his role, purposely distorting facts so that he can create a grand, heroic story in which he is not culpable for his villainous deeds; this, too, suggests the empowerment of stories. However, this is not the sort of reimagining that can remain in the Whedonverse; the power in this story ultimately comes from Andrew taking responsibility for what he has done. In "Superstar," "Storyteller," and *Serenity*, lasting empowerment comes from truthful storytelling.

7. Longworth, "Joss Whedon: Feminist," 57.

8. I am not arguing here that Whedon's work is above clichés of storytelling. Some of the less well-received episodes of his work do just that, such as *Dollhouse*'s "The Target," 1.2, which is a retelling of Richard Connell's 1924 short story entitled "The Most Dangerous Game." Steven S. DeKnight wrote the episode, but in keeping with general Whedon Studies practice, I am including works created by Whedon and written and/or directed by others as within his oeuvre.

9. Bennett, "Screenwriting Tip #103."

10. Adams, *Slayer Slang*.

11. Longworth, "Joss Whedon: Feminist," 63.

12. Longworth, "Joss Whedon: Feminist," 63.

13. Pateman, *Aesthetics of Culture*, 209–10.

14. Kaveney, *Reading*; Wilcox and Lavery, *Fighting the Forces*.

15. Wilcox, *Why*.

16. Wilcox, *Why*.

17. Rhonda Wilcox and David Lavery argue that the richness of *Angel* is better appreciated in concert with *Buffy* ("Afterword," 222).

18. See, for example, AmiJo Comeford's discussions of Angel as a medieval hero (*The Literary Angel*) and Cordelia as a "failed feminist gesture" (*Buffy Meets the Academy*). Conversely, Janine Harrison finds that Cordelia and Fred "actualize their potential" (130) within the corporate culture of Angel Investigations.

19. Kinsey, "Transitions and Time," 56.

20. Abbot, *Angel*, 8.

21. For more on Whedon's use of Chinese, see Mandala, "Representing the Future"; Brown, "Orientalism."

22. Gene Rodenberry promoted *Star Trek* as "*Wagon Train* to the stars."

23. In the case of traditional Westerns, the woman generally has been kidnapped by Indians, with the subtext that she has been raped/penetrated. River has been penetrated as well, albeit through her brain. Dr. Albert Cinelli is the first person who I heard call River "the Reaver Slayer." For more on the empowerment of River, see, for instance, Buckman, "'Much Madness.'"

24. Buckman, "'Much Madness.'"

25. Whedon, "Objects in Space" (commentary).

26. *Fray*, *Angel: After the Fall* (season 6), *Angel & Faith*, *Buffy the Vampire Slayer* (seasons 8 and 9), *Spike*.

27. *Astonishing X-Men*, *Runaways*.

28. *Sugarshock*.

29. Wilcox, "Breaking."

30. Whedon and fellow creators Jed Whedon, Zach Whedon, and Maurissa Tancharoen joked about how they "broke the Internet" with the production's huge fan response.

31. Knowledge@Wharton, "A New"; Buckman, "Go."

32. "Joss Whedon to Be Honored."

33. Clover, *Men, Women, and Chain Saws*.

34. Buckman, "'Didn't Get.'"

35. French, "*Much Ado*"; Fox, "How Black-and-White."

36. Press, "Joss Whedon Just," 188.

37. Her poisoning by Richard Connell ("The Target") may be one of these manipulations: she first shows the ability to remember Caroline after drinking water tainted by Connell. Her kidnapping and manipulation by Alpha in "Omega" (1.12) also affected her ability to develop a composite personality.

38. As Sherry Ginn stated in editorial comments, it does seem odd that dumbshows would be killed, considering that those manipulating the technology want to have plenty of spare bodies to utilize as they body hop.

39. Press, "Joss Whedon Just," 186.

40. Rosen, "New," 173; Murray, "How *Dollhouse*."

41. Whedon, "Vows" (commentary).

42. Press, "Joss Whedon Just," 186–88; da Silveira, "How Cancellation," 151; Ryan, "Sex."

43. Tresca, "Fantasy Is Their Business"; Wilcox, "Echoes."

44. Tresca, "Fantasy Is Their Business," 412.

45. Lavery and Burkhead, *Joss Whedon: Conversations*, viii.

46. Citations for most of these can be found at Whedonology, thanks to Alysa Hornick.

47. Stock, "Joss Whedon: A Life in Pictures."

48. Thanks go to Erin Giannini as well as my fellow editors for their careful reading of this introduction and their useful suggestions for revision.

Part I

Self and Identity

Chapter One

"I've Watched You Build Yourself from Scratch"

The Assemblage of Echo

Michael Starr

With its central themes of identity and personhood, *Dollhouse* presents a variety of conflicting philosophical conceptions of body, mind, and soul via the medium of its science fiction setting. Such concepts are principally espoused via the cipher of Echo. Although Echo is initially devoid of both identity and memory, we witness her gradual construction of "self" over the course of the narrative arc as she is imprinted with a large number of personalities, eventually resulting in her becoming a "person" in her own right.

Dollhouse initially foregrounds a variety of philosophical perspectives pertaining to the nature of the self in relatively unambiguous terms, via the subjective convictions of various characters. For example, the notion of the Dolls as "tabula rasa," as espoused by Adelle in "The Target" (1.2), is a cornerstone of Enlightenment philosophy, while dialogue in "Omega" (1.12) between Ballard and Topher presupposes other conceptions. The former accuses the latter of "stealing [their] souls," an indictment that Topher refutes: "Yeah. And then we put 'em in a glass jar with our fireflies." In the same episode, Ballard is incredulous when Alpha's psychopathic condition is described merely as a "technological anomaly"; hence, with a belief in a soul (or at least some form of intangible human self) that cannot be manipulated or imprinted, Ballard is essentially espousing classical Aristotelian philosophy, while Topher, driven by scientific rationalism, espouses Cartesian metaphysics, assuming that the mind is the operating principle of a person. However, in effect, the two are merely arguing opposite sides of the same coin in terms of conceptions of the dualism of body and mind.[1] Regardless of the initial convictions of both characters and the perceived scientific rationalism of the series's central concept, as the story arc progresses, the notion that mind and body can be regarded as separate entities is consistently challenged and complicated;[2] indeed, such classical ontologies (Platonic or Cartesian, e.g.) may regard the "self" as a stable, bounded entity, but the emergence of Echo as a person in her own right is an altogether more esoteric proposition. Jawetz touches on this issue when stating that "it is not clear where this personality came from, or how Echo manages to assemble it," concluding that "our identities shift from one moment to the next, depending on our circumstances."[3]

What is significant is that *Dollhouse* negotiates these concepts by its portrayal of identity and personhood not existing in isolation but inherently tied into wider discourse: Vinci writes that "through his depiction of the dolls, Whedon explores not only the ineffability of identity but also the myriad possible relationships between body, mind, soul, culture, and technology."[4] The issue that Vinci touches on here and that this chapter reconceptualizes is that, as Echo builds "herself from scratch" (as described by Boyd in "Getting Closer," 2.11), she becomes Echo, not despite the multitude of personality imprints, but because of them. Furthermore, it is this construction of self as multiplicity that makes possible her eventual resistance to both the Dollhouse and the Rossum Corporation.

Dollhouse's complex theoretical underpinnings—incorporating all the aforementioned issues of body, mind, soul, and technology—can be negotiated and extrapolated on to great effect via the concepts of philosopher Gilles Deleuze[5] (1925–1995), whose process of "political anthropology" provides tools to give us unprecedented ways to map the "immanence" of body and mind, providing an in-depth and radical poststructural interpretation of how *Dollhouse* both confronts and negotiates issues of self and identity. Just as *Dollhouse* uses its science fiction premise to explore sophisticated philosophical themes, Deleuze himself proclaimed that philosophy itself must be a kind of science fiction in that the strange rhetoric and "monstrous" slang of science fiction estrange one from the historical inertia of the "now" and allow a leap into "untimely" futures with singular self-consistency.[6] As Fancy states, "it is clear that science fiction, with its core concerns around future assemblages of technologies, humans, other planets and species, is a series of assemblages that can provide the 'lines of flight' and 'bodies without organs' Deleuze felt would be necessary to help undo the various stratifications he saw gathering."[7] Hence, vital to Deleuze is the sheer fact of conceptual invention: the creation of new concepts (and, hence, the undoing of established stratifications) means that we see the world in a new way, one not available to us before: "a perception as it was before men (or after) . . . released from human coordinates."[8] Inherent to this is the idea of creation as rhizomatic (as opposed to arboreal) thinking; Deleuze describes the standard epistemology of Western thought as functioning on the principle of an arboreal structure in which a trunk grows vertically and produces branches that subdivide into smaller and lesser categories—hence, an insistence on linear, hierarchic, and totalizing principles. To challenge this, the image of the rhizome is utilized, in which the stem of a plant sends out interconnecting roots and shoots as it spreads underground, to describe the way in which ideas should be conceived as multiple, interconnected, and self-replicating. In this manner, rhizomatic thinking advocates the formation of intricate paths of free-flowing associations, as opposed to arboreal, flat trains of thought. Hence, life for Deleuze is an "open, creative whole of proliferating connections" of which the human subject is only the "effect of one particular series of experiential connections."[9] *Dollhouse* demonstrates just such a Deleuzian aesthetic, most overtly via its depiction of Echo's construction of self as multiplicity, which (being predicated on change, transformation, and mutation) serves well as an exemplification of rhizomatic identity, the "release from human coordinates" that Deleuze advocates.[10] In keeping with this rhizomatic remit, this chapter explores Echo's gradual construction of "self" via Deleuzian conceits such as the "assemblage," the "body without organs," and "schizoanalysis," in the process demonstrating how *Dollhouse*'s notions of identity rhizomatically intersect with meditations on wider discourses such as technology, capitalism, and patriarchy.

THE ECHO ASSEMBLAGE

In specific terms of the conceptualization of identity and personhood, a rhizomatic approach acknowledges that a body has multiple potentials, each of which depends on other "bodies." Such rhizomatic connections are key to understanding Echo's construction of self, which can be conceptualized via one of the most prominent terms in Deleuze's oeuvre: the concept of the "assemblage." In basic terms, an assemblage is any number of "things" or pieces of "things" gathered into a single context. These bodies are not stable entities, instead being assemblages of forces, and are undergoing constant change dependent on the encounters they have with other entities with whom they enter into relationships. In this manner, Deleuzian theory has significant ramifications for traditional philosophical concepts that emphasize being, originality, and essence, such as the aforementioned Aristotelian or Cartesian ontologies. For example, we can no longer conceive of a human being as a stable individual experiencing change but essentially remaining the same person. For Deleuze, the self must instead be envisaged as an assemblage of different forces in a state of constant flux:

> A human being . . . is a multitude of physiological, neurological, genetic, hormonal, cognitive and conative assemblages, a phenomenon that arises from

> the organization of the assemblages. Its identity, personality and self-consciousness, similarly, are molar properties that arise from these assemblages' organization. A person, consequently, is an integral, organized assemblage of multiplicities.[11]

Assemblages bring together entities that are organic and nonorganic, material and abstract, technological and natural; "the connections that occur between the most disparate and the most similar of objects, places and people: any network of things brought into contact with one another, functioning as an assemblage machine for new affects, concepts, bodies, thought."[12] The relevance of this approach to *Dollhouse* is clearly evident due to its portrayal of the alteration of human identity via speculative technology, such as "mind wipes," "memory wedges," and the "imprinting chair."

According to Deleuze and his concept of difference, there is no identity; there is only difference: change arises from repetition; copies are something new; and reality is a becoming, not a being.[13] In the context of personhood then, this is not a concept of a person as an individual but the idea that "the power of any given body resides not in the power of the body itself but in its relations to other bodies."[14] He argues that our brains are embodied in the flux of the material world; hence, the subject only has meaning dependent on its relationship with other assemblages—be they, for example, biological, chemical, social, political, or technological—resulting in transformations and the potential for the development of new formations. Such ideas are useful in conceptualizing *Dollhouse*, in which Echo is not portrayed as a stable ontological entity but instead the result of multiple personality imprints, a conglomeration of forces undergoing constant change dependent on the encounters that she has with other entities with which she enters into relationships. As these various personalities form collective elements of Echo's "self," we are witness to a particularly visceral representation of the assemblage concept, where "self" is malleable and dependent on external forces, with Echo comprising an amalgamation of many forces and personalities.

The body conceived of as a machinic[15] assemblage becomes a body that is multiple, since its function or meaning no longer depends on an interior truth or identity but on the particular assemblages that it forms with other bodies.[16] The significance of this approach in regard to *Dollhouse* is that the assemblage offers a theoretical means to elaborate the multiple and collective dimensions of Echo's unconscious and hence comprehend how she is inherently tied into wider discourse: "not ego as subject," Deleuze explains, "but rather these peoples who are in us and who make us speak, and who are the source of our statements."[17] As the narrative progresses, Echo forms just such an assemblage, as various imprinted personalities remain within her, in one form or another, ranging from complete imprints to mere traits that the personality had or had witnessed in others—in Deleuzian terms, a rhizomorphic flow, leading to new formations.

Of course, the assemblage can contain any number of disparate elements; Echo's status as such does not prevent her from containing assemblages within herself or entering into new assemblages. The often-used narrative device of Echo's "imprint of the week" gives us explicit reference to specific imprints, which are often linguistically marked—for example, the expression "blue skies" to indicate the Taffy personality ("Gray Hour," 1.4) and "goodness gracious" for Terry Karrens ("Belle Chose," 2.3). In this way, we are witness to "these people who are in us and make us speak, and who are the source of our statements."[18] Hence, the *Dollhouse* narrative clearly presents us with the multitude of elements that constitute the Echo assemblage, in keeping with Deleuzian theory, in which "this functioning of an assemblage can be explained only if one takes it apart to examine both the elements that make it up and the nature of its linkages."[19] Of course, not all imprints may be desirable, or "good," for Echo (serial killer Terry Karrens, for example), but Deleuze does warn that we cannot decide in advance whether the elements of different assemblages will be "healthy" to us. Indeed, as assemblages are in flux and not restrained by expectations about "appropriate" behaviors and connections, they can be associated with contesting notions of, for example, fixed subjectivity and gender identity. *Dollhouse* engages with such contestations on several occasions: Victor is accidentally imprinted with the female Kiki personality in "Belle Chose," and Echo admits in "Meet Jane Doe" (2.7) that, as an Active, her romantic encounters have been "at least seven times gay."

The idea of a "healthy" or "good" assemblage raises issues of ethics and morality, in which regard *Dollhouse* is very problematic. Ostensibly, the Dolls are volunteers who have surrendered their minds and bodies to the Rossum Corporation for the duration of their contracts, but of course, this is complicated by a great many issues: that Sierra has been made a Doll against her will, for example[20] ("Belonging," 2.4). In terms of the ethics of the assemblage, Deleuze emphasizes that we must "come to know what a body can do, its potential for interacting with other bodies, whether or not that interaction will bring harm to either body and whether there is the potential for exchange or a joining together to form a still stronger body."[21]

Drawing on the work of Spinoza, Deleuze distinguishes between good and bad effects on the body according to whether a particular assemblage enhances or harms each body's "life force"—in other words, whether it increases or reduces each body's power to act and its potential to go on forming new relations.[22] This distinction is complex, especially given that a body can form multiple relations with other bodies: "we have many constituent relations, so that one and the same object can agree with us in one respect and disagree with us in another."[23] In essence, an ethical event for Deleuze is one in which bodies emerge with a strengthened potentiality or, at least, a potentiality that is undiminished. In these terms, although frequently depicted as traumatic, the imprints performed on the Dolls may not be intrinsically "bad" in themselves; what is important is that bodies are able to go on connecting with other bodies, creating new flows of desire and undertaking new becomings. The Dolls are controlled by virtue of their "treatments," when they are wiped of personality at the end of every assignment, which, in these terms, equates to restriction of potentiality of their assemblage (as is subsequently explored in terms of Deleuze's "body without organs").

Deleuze's embodied ethics aims to reduce unethical assemblages (which reduce bodily potentials) and increase ethical, life-enhancing assemblages: that is, assemblages that increase a body's power to form creative, productive relations and that increase its capacity for life.[24] The vital issue here is that, ideally, assemblages should be consensual; if they are enforced, with the existential rights of the "singularities" concerned denied, suffering can result.[25] It is unsurprising, therefore, that wipes are frequently portrayed as traumatic experiences—particularly, the initial extensive and painful "tissue mapping" as seen at Sierra's introduction in "Ghost," in which she is wiped of all bodily potential, and Bennett's uses of the imprinting chair (a direct technological source of the assemblage process) as a device with which to torture Echo ("The Left Hand," 2.6). Ultimately, in "Epitaph One" (1.13) and "Epitaph Two: Return" (2.13), the *Dollhouse* narrative takes such ramifications to their ultimate conclusion. The majority of the world's population has been remotely mind wiped, thus demonstrating the ultimate result of the enforcement of unethical assemblages: the removal of human potentiality on a global scale.

Before each engagement, the Doll is imprinted with a personality most suited for the client's needs, and, significantly, the finished personality is an amalgam of different personae assembled into a custom design. As an assemblage, the more Echo is imprinted with "personalities" (or at least amalgamations of mental states), the more she has access to potentialities (or, in Deleuzian terms, "lines of flight"), and the more she becomes a "person." Early in the narrative arc, she is unable to conceive of herself in terms of either personhood or potentiality; "they hollowed me out, there's no me. I'm just a container," she states in "Omega." However, the assemblage of potentialities via multiple imprints subsequently enables her to perceive herself as a person in her own right, not merely as a vessel reserved for Caroline; by season 2's "Meet Jane Doe," she states, "I've been saving this body for her [Caroline], but I'm not her. My name is Echo." In conceptual terms of the assemblage, this is because "once machines are assembled, they have an identity and a life of their own."[26] Boyd also validates this conception when he states, "I know you. I've watched you grow. I've watched you build yourself from scratch. . . . You're stronger than everyone here, Echo" ("Getting Closer").

In these terms, the character of Boyd is interesting in that (contrary to Topher and Ballard), he appears to conceive of Echo in terms of her potentialities.[27] However, it is subsequently revealed that he has his own agenda. In "The Hollow Men" (2.12), Boyd claims that Echo's uniqueness is grounded in the biological; she has a specific gene in her spinal fluid that allowed her to evolve from her tabula rasa state, resist wipes, and hence grow into Echo. As discussed previously, assemblages bring together entities that are organic and nonorganic, material and abstract, technological and natural; hence, the potentiality of Echo's spinal fluid (in that Boyd plans to harvest it to produce a vaccine that provides immunity to imprinting) is merely another element in the wider assemblage, a body of pure intensities. However, although Echo may be the only known person with this genetic anomaly, it appears that she is not the only one with a resistance to the imprinting process; we see signs early on in the series of individuality creeping into the Actives when they are in a blank Doll state.[28] Although Echo is unique in her ability to function as a coherent amalgamation of personalities, the other Dolls do display some degree of potentiality[29] but do not develop a "self" in the same manner, despite being subject to the same processes and hence also possessing potentiality of assemblage. This can be further explored in terms of the processes that these assemblages take in regard to their flow of potentialities.

ECHO'S BODY WITHOUT ORGANS

In their *tabula rasa* state, the Dolls can be conceptualized as what Deleuze calls the "body without organs," as defined in *Anti-Oedipus*[30] and *A Thousand Plateaus*,[31] the two volumes that constitute *Capitalism and Schizophrenia*. The term is not literal[32] but instead an assemblage or body with no underlying organizational principle: "the body that will become identity, waiting to be created."[33] The body without organs is hence "a body that breaks free from its socially articulated, disciplined, semioticized, and subjectified state (as an 'organism'), to become disarticulated, dismantled, and deterritorialized, and hence able to be reconstituted in new ways."[34] This state is experienced when an organization breaks down or is revealed to be arbitrary or culturally determined, as is the case with the Dolls; their original personalities are quite literally "dismantled" through the initial mind wipe process when they first enter the Dollhouse: in their Doll state, they are disarticulated and no longer connected to potentialities. Without such links, a body becomes incapable of forming new assemblages and creating new "lines of flight."[35]

> You have to keep enough of the organism for it to reform each dawn; and you have to keep small supplies of significance and subjectification, if only to turn

> them against their own systems when the circumstances demand it, when things, persons, even situations force you to; and you have to keep small rations of subjectivity in sufficient quality to enable you to respond to the dominant reality.[36]

However, the body without organs can be reterritorialized via the process of the assemblage. In this potential "reconstitution," the body without organs has multiple potential destinations, which Deleuze defines as "empty," "full," or "cancerous." The empty body without organs is a nonproductive and desireless body, with which the blank Dolls (as opposed to the Actives) in their *tabula rasa* state can be equated. They are bodies of pure intensities, but the Dollhouse prevents (or at least attempts to restrict) the potential of these intensities; they are specifically conditioned for assignments and wiped clean of the new personality as soon as it is over. This clearly shows potentiality but also a stagnant cycle: bodies are empty, then written upon for "engagements" (and hence given potentiality), but then given a "treatment" and reduced to empty again. Hence, the Dolls have no potential to move beyond the state of the empty body without organs. At best, they become "cancerous."

The inherently political nature of a Deleuzian reading results in no separation between the personal and political, the individual and the collective, and this bears implications not only for individuals but also for the society and culture to which they are machinically attached. In this way, the body without organs can be related to capitalism and capital; indeed, *Anti-Oedipus* defines the body without organs mainly in these terms.[37] Deleuze equates the empty, stagnant cycle of the cancerous body without organs to capitalism in that it is endlessly reproducing but never changing. In accordance with this conception, the Rossum Corporation—and, hence, the Dollhouse itself—can be positioned as a cancerous body because of its position as a capitalistic corporation, exploiting its workers while (literally) seeking world domination. The Dollhouse itself as a metaphor for capitalism is a constant theme through the series, as explored by Connelly and Rees: "wage labor under capitalism can only ever be forced labor, always a kind of slavery, and while in a technical sense we all 'volunteer' to enter the work force, our choices are as compromised as Caroline's because we are every bit as coerced."[38] Additionally, the "sign system of economics—contracts and remuneration" is used to convince prospective Actives that the Dollhouse can provide them with "escape from responsibility for their labor and their selves without any consequences for their humanity."[39]

This is further compounded by the terms and conditions to which the Dolls are bound. As demonstrated by Victor in "Stop-Loss" (2.9), the idea that a Doll can leave the Dollhouse for a "normal" life after the completion of its five-year contract is called into question;[40] Dolls still have "Active architecture" installed (the permanent alteration to the chemical and physical makeup of the brain to allow imprinting), are still tracked by surgically implanted GPS systems, and hence are still, for all intents and purposes, property of the Rossum Corporation. In these terms, the Rossum Corporation literally owns the material underpinnings of the Active's subjectivity, and the latter's sense of personhood cannot be thought of apart from its bureaucratic organizational structure; the virtual bodies of Actives are part of the assemblage of potentialities, but this assemblage is tied to the assemblage of the Dollhouse and Rossum Corporation. Due to the capitalistic nature of this system, these parts are prevented from being "free flows" of Deleuzian potentiality. It is indeed apt that Topher describes the wiping and imprinting process as "not unlike being born," while Ballard equates the process with death ("Gray Hour," 1.4). Both events are equatable with Deleuze's conception of the capitalistic state apparatus, which needs its "people to be born that way, crippled and zombie-like. The myth of the zombie, of the living dead, is a work myth and not a war myth. . . . The State apparatus needs . . . predisabled people, pre-existing amputees, the still-born, the congenitally infirm."[41] The Dolls, reduced to a state of childlike dependence, certainly qualify as literal manifestations of the zombified, living dead required by the State such that it might function correctly. This analogy is further compounded via the depiction of the Dollhouse "Attic," a place of physical and mental torture where all captives within function as elements of a computer system, forced to be passive conduits for information ("The Attic," 2.10). Laurence Dominic describes how the Attic's residents are kept in a "constant, repeating nightmare," and this is a particularly visceral depiction of the cancerous body without organs, characterized by "assembly-line personalities . . . endlessly repeating homogenized individuals in a runaway process of conformity."[42] The fact that the Dollhouse Attic takes the form of a computer system is significant: from the Deleuzian perspective, capitalism and technology have a mutual relationship; along with the multitude of assemblages that influenced the emergence of capitalism, technology can be seen as its symbiotic partner, with the interaction between the two serving to accelerate said symbiosis.[43]

However, despite these restrictions upon her potentiality, by the end of the narrative arc, Echo has become a full body without organs: a productive body "populated by multiplicities."[44] In the episode "Meet Jane Doe," she describes herself to Ballard very much in these terms: "There's a lot of noise from the chorus girls, but they're not me. There is a me." By the time of the events in "Epitaph Two," Echo has over one hundred imprints, over which she has total control. Philosophically far removed from the simplistic, ontological concepts of self proffered by Topher and Ballard, she has successfully experimented with her potentiality and is able to view herself as a Deleuzian construct of a constantly changing assemblage of forces, an epiphenomenon arising from a confluence of personalities. In keeping with Deleuze's remit in *A Thousand Plateaus* (in the plateau titled "How do you make yourself a body without organs?"), Echo has orchestrated enough intensities upon the body without organs that new connections have become possible and a new "plateau" is achieved.[45] Deleuze describes this process thus: "This is how it should be done . . . experiment with the opportunities it offers, find an advantageous place on it, find potential movements of deterritorialization, possible lines of flight, experience them, produce flow conjunctions here and there, try out continua of intensities segment by segment, have a small plot of new land at all times."[46] Hence, in her assimilation and control over multiple imprints, Echo has performed these flows of potentiality, filling up her body without organs in the process; she is able to view herself as a Deleuzian construct of a constantly changing assemblage of forces, an epiphenomenon arising from chance confluences of personalities. By virtue of this process, Deleuze maintains that "you have constructed your own little machine, ready when needed to be plugged into other collective machines."[47] These other machinic potentialities quite literally flow in and out of Echo and in and out of one another; as Deleuze would put it, she is full of "free intensities and nomadic singularities."[48] She does suffer from "glitches" ("Vows," 2.1) and painful headaches ("Meet Jane Doe") during the process of assemblage, but this is unsurprising as "no identity category is ever entirely stable; no subject totally unified and consistent; and no mode of organisation fully sedimented."[49]

Echo is, however, the only Doll to have harnessed the flow of forces successfully, while the others remained either trapped in a stagnant cycle (e.g., November and Sierra) or unable to control multiple imprints (as is the case for Alpha). In terms of the latter, Deleuze is keen to emphasize that the process of making oneself a body without organs is not only fraught with problems but can also be inherently dangerous: "Deleuze recognizes . . . that there are great risks and perils in the pursuit of a body without organs and thus in *A Thousand Plateaus* [he] speaks of both a suicidal and a cancerous body without organs, each with their own specific dangers."[50] In "Omega," flashbacks from "The Target" (1.2) suggest that Alpha's instability and psychopathic behavior are a result of an incident that Topher dubs a "composite event"; all his previous imprinted personalities (some of which are multiple personalities themselves) resurfaced simultaneously and uncontrollably. Such failings of the empty or cancerous body without organs can "engender the complete collapse of subjectivity," expressed in *A Thousand Plateaus* thus: if you free the body without organs "with too violent an action, if you blow apart the strata without taking precautions, then instead of drawing the plane [of consistency] you will be killed, plunged into a black hole, or even dragged towards catastrophe."[51] Deleuze was very aware of such failed becomings, whose ultimate consequences are often worse than the conventional forms of subjugation that they seek to escape, describing these "catastrophic" failings as "vitreous bodies, cancerous bodies, totalitarian and fascist."[52] Alpha's composite event is a sudden event of violence, resulting not in potentiality and flows of assemblage but in a cancerous body without organs, "a fascist version of the Overman who has not really ascended to anything."[53] Echo's response to the composite event forced upon her is very different from Alpha's experience. While Alpha appeared to be "constantly pulled in different directions by his simultaneous selves," Echo's multiple personalities have "made her more secure in who she was on her own."[54]

SCHIZOANALYZING ECHO

In this sense, Echo's harnessing of these virtual potentials can be related to the body without organs being an element of Deleuzian "schizoanalysis" (or "schizophrenic process"). The Deleuzian schizophrenic does not correspond to the medical use of the term;[55] indeed, Deleuze "makes a very clear distinction between schizophrenia as a pathological disease and schizophrenia as a process (strategy)," the latter being his primary focus.[56] Schizoanalysis is instead a subversion of psychoanalysis that Deleuze performs within the two volumes of *Capitalism and Schizophrenia.* In this context, schizophrenia is not an illness or biological state but a "potentially liberatory psychic condition produced within capitalistic social conditions, a product of absolute decoding"; it is a state of activity in which the full potential of the unconscious can be realized.[57] In other words, the Deleuzian schizophrenic refuses to be defined within a Freudian framework, which "insists on the individual, Oedipus and the family," instead showing "what the unconscious really is, namely a factory full of machinic connections."[58] Significantly, toward the end of the series's narrative arc, Echo resists Ivy's attempt to define her in terms of a monadic psychoanalytic model that privileges the individual. When discussing the implications of imprinting Caroline into Echo, Ivy states, "You're carrying multiple imprints and something in you has allowed you to fuse them into a kind of superego, but Caroline isn't just one more personality build you'd be adding to the mix—she's real." Topher concurs with this diagnosis, concerned that Caroline "might fight back," but Echo rejects this model of the psyche in no uncertain terms: "She'd lose!" ("Getting Closer"). As a multiplicity, schizoanalytic process posits that an exploration of identity as rhizome can be performed, employing a multitude of character traits that make use of the unconscious, repressed memories and desires in a way that the nonschizophrenic cannot.[59] Echo can be conceptualized in these terms as she explores, assumes, and utilizes the many identities of her composite personality. This process takes place throughout the *Doll-*

house narrative, with overt instances occurring; for example, in "Stop-Loss," when Echo states that the imprints of Roma Klar and Eleanor Penn are both contributing their specific skills to a crime scene analysis, and in "Meet Jane Doe," when she uses Rebecca Mynor's medical knowledge to free Galena from prison; it is in these actions that she is harnessing the "liberatory power" of the Deleuzian schizophrenic.

In *Anti-Oedipus*, Deleuze claims the schizophrenic is an anticapitalist metaphor, someone "free to roam outside of predictable paths" who refuses to be sublimated, hence offering a "schizoanalysis" of capitalism designed to expose and help eliminate the arbitrariness of its constraints, thereby "freeing the creative power of schizophrenia."[60] In accordance, "the schizophrenic deliberately seeks out the very limit of capitalism: he is . . . its proletariat, and its exterminating angel."[61] As previously discussed, Echo is a product of absolute decoding by the capitalistic Rossum Corporation. Boyd maintains that this has been a deliberate act to harness the potentiality of her spinal fluid: "Who do you think allowed you to grow as an individual, as Echo . . . making sure you had the space to become your own person. . . . You're the key. The key to everything. You will save us all" ("The Hollow Men"). However, events do not work in accordance with Boyd's plans, as Echo resists and defeats him; in conceptual terms, the Dollhouse has assembled a Deleuzian schizophrenic that possesses liberatory power, an "exterminating angel" of capitalism. Significant here, and tied into Deleuze's critique of capitalism and psychoanalysis, is the fact that Echo's ability to challenge Boyd is inherently tied into the way that he defines and attempts to control her via a patriarchal framework. In these terms, St. Louis and Riggs observe that the Dollhouse functions as a family unit, citing Boyd's history of sleeping at the L.A. Dollhouse ("The Attic") and the fact that Topher also lives full-time at the 'House ("Vows"); "the Dollhouse is thus home, making the people in its walls family."[62] This family unit is highly controlled and stratified, with the Dolls named in accordance to a phonetic alphabet, encouraged to "be their best" and correspondingly ranked hierarchically in terms of their popularity with clients. The concept of "being number one" is a stratified, linear (or, as Deleuze puts it, "arboreal") way of thinking and hence is incompatible with the rhizomorphic associations advocated via such concepts as the assemblage.[63] Significantly, Echo resists such definitions, stating that "no one is their best in here [the Dollhouse]" ("Vows") and rebuking Priya (Sierra's "original" personality) for referring to the Actives as a family unit: "Call us what you want, just not family" ("The Hollow Men").

In the same episode, Boyd refers to Echo precisely in these terms: "Don't you get it? I brought you here because you're my family." As the self-proclaimed head of the "Dollhouse family," Boyd is a symbolic representative of patriarchy, states Wilcox:

> Boyd is presented as an ex-cop, an authority figure; as the director of Rossum, he stands for the corporate world. He is a father figure. . . . All of these identities point towards a symbolic representative of the patriarchy—and the patriarchy in this story is highly complicit in the technological/corporate choices that lead to the destruction of civilization.[64]

Deleuze maintains that the nuclear family is actually a means of psychological repression where the desires of the child are perverted and, as a result, docile and unquestioning individuals are created; an analogy can clearly be seen here in terms of the subjugation and childlike behavior of the Dolls. Just such an Oedipal family structure is "one of the primary modes of restricting desire in capitalist societies," and the strict hierarchical structure embodied by the family is a primary source of the Oedipal oppressive power that the schizophrenic resists.[65] Indeed, the schizophrenic "rejects any attempt to impose on it any sort of triangulation implying that it was produced by parents."[66] In other words, Echo has gained resistance to being defined by an arboreal framework of Freudian Oedipalization: "I have thirty-eight brains; not one of them thinks you can sign a contract to be a slave" ("Omega"). In keeping with the Deleuzian schizophrenic, she may be the "proletariat," a creation, product, and slave of Rossum and the Dollhouse, but ultimately, by the end of the narrative, she has transformed into its "exterminating angel": Boyd is "wiped" (hence reduced to an empty body without organs); the hierarchical Oedipalization process is thus rejected; and the Rossum laboratory is destroyed ("The Hollow Men"), thus demonstrating Echo's radical potential. As the apocalyptic events of the "Epitaph" episodes depict, cancerous cycles are not broken easily but potentially always remain, as "a living, desiring body will always form new assemblages that have the potential to transform it and its territories,"[67] as demonstrated in the final imprint of the narrative, Echo's assimilation of Ballard in "Epitaph Two."[68]

Deleuze states that fictional characters strike him with the same force as philosophical concepts: "just as a philosopher creates concepts, within fiction a concept adopts the form of a character, and as a result the character takes on the various dimensions of the concept that was their genesis."[69] Hence, in positioning Echo via the rhizomatic conceptions of the assemblage, the body without organs and schizoanalysis, we are witness to the manner in which *Dollhouse*, as a work of science fiction, performs a radical negotiation of issues of self and identity, via the assemblage of body, mind, soul, technology, and culture. In the act of "building herself from scratch," Echo is "released from human coordinates"[70] and, in performing this process, is a visceral manifestation of what Deleuze describes as the possibility of a new postmodern mode of existence, where individuals overcome repressive arboreal forms of identity and stratification, instead embracing a constant process of becoming and transformation.

BIBLIOGRAPHY

Ansell-Pearson, Keith. *Germinal Life: The Difference and Repetition of Deleuze*. London: Routledge, 1999.

Ballantyne, Andrew. *Deleuze and Guattari*. Oxford, England: Routledge, 2007.

Best, Stephen, and Douglas Kellner. *Postmodern Theory: Critical Interrogations*. New York: Guilford Press, 1991.

Bogue, Ronald. *Deleuze and Guattari*. London: Routledge, 1993.

———. "Violence in Three Shades of Metal." In *Deleuze and Music*, edited by Ian Buchanan and Marcel Swiboda, 95–118. Edinburgh, Scotland: Edinburgh University Press, 2004.

Brown, Lori. "Becoming-Animal in the Flesh: Expanding the Ethical Reach of Deleuze and Guattari's Tenth Plateau." *PhaenEx: Journal of Existential and Phenomenological Theory and Culture* 2, no. 2 (2007): 260–78.

Calvert, Bronwen. "Mind, Body, Imprint: Cyberpunk Echoes in the *Dollhouse*." *Slayage: The Journal of the Whedon Studies Association* 8, nos. 2–3 (2010).

Colebrook, Claire. *Gilles Deleuze*. Oxford, England: Routledge, 2002.

Colman, Felicity J. "Rhizome." In *The Deleuze Dictionary*, edited by Adrian Parr, 231–33. Edinburgh, Scotland: Edinburgh University Press, 2005.

Connelly, Tom, and Shelley S. Rees. "Alienation and the Dialectics of History in Joss Whedon's *Dollhouse*. *Slayage: The Journal of the Whedon Studies Association* 8, nos. 2–3 (2010).

Deleuze, Gilles. *Cinema 1: The Movement-Image*. Minneapolis: University of Minnesota Press, 1986.

———. *Difference and Repetition*. London: Continuum, 2004.

———. *Spinoza: Practical Philosophy*. San Francisco: City Lights Books, 1988.

Deleuze, Gilles, and Félix Guattari. *Anti-Oedipus: Capitalism and Schizophrenia*. London: Continuum, 2004.

———. *Kafka: Towards a Minor Literature*. Minneapolis: University of Minnesota Press, 2003.

———. *A Thousand Plateaus: Capitalism and Schizophrenia*. London: Continuum, 2004.

Deleuze, Gilles, and Claire Parnet. *Dialogues II*. New York: Columbia University Press, 2007.

Fancy, David. "Difference, Bodies, Desire: The Collaborative Thought of Gilles Deleuze and Félix Guattari." *Science Fiction Film and Television* 3, no. 1 (2010): 93–106.

Ginn, Sherry. "Memory, Mind, and Mayhem: Neurological Tampering and Manipulation in *Dollhouse*." *Slayage: The Journal of the Whedon Studies Association* 8, nos. 2–3 (2010).

Gutting, Garry. *French Philosophy in the Twentieth Century*. Cambridge, England: Cambridge University Press, 2001.

Heba, Gary, and Robin Murphy. "Heroes and Villains: Morality, the Will to Power and the Overman in the Works of Joss Whedon." In *The Philosophy of Joss Whedon*, edited by Dean A. Kowalski and S. Evan Kreider, 133–48. Louisville: University Press of Kentucky, 2011.

Jawetz, Ryan. "Identity and Memory in *Dollhouse*." In *Joss Whedon: The Complete Companion*, 393–97. London: Titan Books, 2012.

Malins, Peta. "Machinic Assemblages: Deleuze, Guattari and Ethico-Aesthetics." *Janus Head* 7, no. 1 (2004): 84–104.

Muntersbjorn, Madeline. "Disgust, Difference, and Displacement in the Dollhouse." *Slayage: The Journal of the Whedon Studies Association* 8, nos. 2–3 (2010).

Pisters, Patricia. "Delirium Cinema or Machines of the Invisible?" In *Deleuze and the Schizoanalysis of Cinema*, edited by Ian Buchanan and Patricia MacCormack, 102–15. London: Continuum, 2008.

Powell, Anna. *Deleuze: Altered States and Film*. Edinburgh, Scotland: Edinburgh University Press, 2007.

———. *Deleuze and Horror Film*. Edinburgh, Scotland: Edinburgh University Press, 2006.

Protevi. John. "A Problem of Pure Matter: Fascist Nihilism in *A Thousand Plateaus*." In *Nihilism Now! Monsters of Energy*, edited by Keith Ansell-Pearson and Diane Morgan, 167–87. New York: St. Martin's Press, 2000.

Quilty, Susan. "Negative Space in the 'House." In *Inside Joss' "Dollhouse": From Alpha to Rossum*, edited by Jane Espenson, 133–46. Dallas, TX: Smart Pop, 2010.

Schatzki, Theodore R. *The Site of the Social: A Philosophical Account of the Constitution of Social Life and Change*. University Park, PA: Penn State University Press, 2002.
Starr, Michael. "Untimely Mutations: Deterritorializing H. G. Wells's Scientific Romances." PhD diss., University of Northampton, 2011.
St. Louis, Renee, and Miriam Riggs. "A Painful, Bleeding Sleep: Sleeping Beauty in the *Dollhouse*." *Slayage: The Journal of the Whedon Studies Association* 8, nos. 2–3 (2010).
Sutton, Damian, and David Martin-Jones. *Deleuze Reframed*. London: Tauris, 2008.
Vinci, Tony M. "Not *an* Apocalypse, *the* Apocalypse: Existential Proletarisation and the Possibility of Soul in Joss Whedon's *Dollhouse*." *Science Fiction Film & Television* 4, no. 2 (2011): 224–47.
Wilcox, Rhonda V. "Echoes of Complicity: Reflexivity and Identity in Joss Whedon's *Dollhouse*." *Slayage: The Journal of the Whedon Studies Association* 8, nos. 2–3 (2010).

NOTES

1. For an in-depth analysis of how *Dollhouse* negotiates the mind-body problem, see Muntersbjorn's "Disgust, Difference, and Displacement in the Dollhouse."
2. Calvert, "Mind, Body, Imprint."
3. Jawetz, "Identity and Memory," 397.
4. Vinci, "Not *an* Apocalypse," 227.
5. Though several of the texts and theories utilized in this chapter were coauthored by Deleuze in conjunction with Félix Guattari (1930–1992), the term *Deleuzian* is used here as opposed to *Deleuzoguattarian*. In keeping with the rhizomatic nature of his theoretical approach, when Deleuze collaborated with Guattari (and while their work developed its own unique concepts), he was, in fact, already in collaboration with Friedrich Nietzsche, Henri Bergson, Michel Foucault, and others; hence, Deleuze is given primacy.
6. Deleuze, *Difference*, xix–xx.
7. Fancy, "Difference, Bodies," 105.
8. Deleuze, *Cinema*, 122.
9. Colebrook, *Gilles Deleuze*, 5.
10. Deleuze, *Cinema*, 122.
11. Schatzki, *Site of the Social*, 204.
12. Colman, "Rhizome," 231–32.
13. Deleuze's concept of difference interprets Nietzsche's "Eternal Return." Rather than the rotation of a cycle, Deleuze's eternal return is the repetition of that which differs from itself: "The subject of the eternal return is not the same but the different, not the similar but the dissimilar, not the one but the many" (Deleuze, *Difference*, 126). Hence, change arises from repetition, as *Dollhouse* demonstrates in broad terms; Echo's growth into personhood is a result of the cyclical process of wiping and imprinting.
14. Deleuze and Guattari, *Thousand*, 9.
15. In regard to the use of the term *machinic* used here and elsewhere, Deleuze specifically distinguishes between this and the mechanical. As Powell states, "machinic is distinguished sharply from mechanism, which . . . is used negatively for repetition that reduces energy, causes damage and eventually stops working. Machinic connections are potentially energising" (*Altered States*, 78). An assemblage itself is not opposed to either mechanical machines or organic bodies but encompasses both.
16. Malins, "Machinic Assemblages," 84.
17. Deleuze and Guattari, *Thousand*, 256.
18. Deleuze and Guattari, *Thousand*, 256.
19. Deleuze and Guattari, *Kafka*, 53.
20. Such ethical issues are further compounded in "Omega" (1.12), where it is revealed that the Dollhouse's first subjects were criminals who were given the option of spending five years as a Doll in lieu of serving a protracted jail sentence.
21. Brown, "Becoming-Animal," 268.
22. Malins, "Machinic Assemblages," 97.

23. Deleuze, *Spinoza*, 33.
24. Malins, "Machinic Assemblages," 98.
25. Powell, *Horror*, 74.
26. Ballantyne, *Deleuze and Guattari*, 23.
27. Adelle also appears to adopt this way of thinking in "A Spy in the House of Love" (1.9), when she orders Topher not to "scrub" Echo, so she can see her grow to her full potential.
28. Quilty, "Negative Space," 139.
29. Examples of such potentialities include Sierra and Echo's friendship (initially established in "Stage Fright," 1.3) and Sierra and Victor's romantic relationship (which commences in "True Believer," 1.5), both of which develop throughout *Dollhouse*'s narrative arc.
30. Deleuze and Guattari, *Anti-Oedipus*.
31. Deleuze and Guattari, *Thousand*.
32. The body without organs is not an actual organless body, and a body can be anything, not just a human body (though the term can be used to describe a relationship to one's literal body).
33. Sutton and Martin-Jones, *Deleuze Reframed*, 111.
34. Best and Kellner, *Postmodern Theory*, 90–91.
35. Malins, "Machinic Assemblages," 90.
36. Deleuze and Guattari, *Thousand*, 160.
37. Deleuze and Guattari, *Anti-Oedipus*.
38. Connelly and Rees, "Alienation and the Dialectics."
39. Connelly and Rees, "Alienation and the Dialectics."
40. Calvert, "Mind, Body, Imprint."
41. Deleuze and Guattari, *Thousand*, 470.
42. Protevi, "Problem of Pure Matter," 171–72.
43. Starr, *Untimely Mutations*, 115.
44. Deleuze and Guattari, *Thousand*, 30.
45. Deleuze and Guattari, *Thousand*.
46. Deleuze and Guattari, *Thousand*, 178.
47. Deleuze and Guattari, *Thousand*, 179.
48. Ansell-Pearson, *Germinal Life*, 153.
49. Malins, "Machinic Assemblages," 87.
50. Bogue, "Violence," 111.
51. Malins, "Machinic Assemblages," 90; Deleuze and Guattari, *Thousand*, 178.
52. Deleuze and Guattari, *Thousand*, 165.
53. Heba and Murphy, "Heroes and Villains," 145.
54. Quilty, "Negative Space," 139.
55. For an in-depth analysis regarding how *Dollhouse* (incorrectly) approaches schizophrenia as a medical and neurological condition, see Ginn's "Memory, Mind, and Mayhem: Neurological Tampering and Manipulation in *Dollhouse*."
56. Pisters, "Delirium Cinema," 103.
57. Best and Kellner, *Postmodern Theory*, 90.
58. Pisters, "Delirium Cinema," 108.
59. Sutton and Martin-Jones, *Deleuze Reframed*, 144.
60. Gutting, *French Philosophy*, 340.
61. Deleuze and Guattari, *Anti-Oedipus*, 38.
62. St. Louis and Riggs, "A Painful, Bleeding Sleep."
63. Alpha also appears to advocate such stratification, mutilating Whiskey to favor Echo: "Why don't you let Echo be number one?" ("Omega"). His insistence on arboreal processes, rather than rhizomatic, may explain why he is an unstable assemblage.
64. Wilcox, "Echoes of Complicity."
65. Bogue, *Deleuze and Guattari*, 88.
66. Deleuze and Guattari, *Anti-Oedipus*, 16.
67. Malins, "Machinic Assemblages," 88.
68. Interestingly, in the same episode, Alpha's personality appears to have stabilized significantly, possibly as a result of imprinting Ballard upon himself in "A Love Supreme" (2.8).

69. Deleuze and Parnet, *Dialogues II*, 113–15.
70. Deleuze, *Cinema*, 122.

Chapter Two

"We Are Not Just Human Anymore"

Accepting the Posthuman Future

Meg Saint Clair Pearson

> You can't program people. We're not frickin' computers.
> Not that different. Our brains are natural motherboards. Everything we think, feel, do—electrical impulses fired by the brain. I just hack the system.
> —"Needs" (1.8)

Throughout the course of the two-year series, *Dollhouse* pushed its viewers to contemplate varying levels of humanity. With each new episode, viewers were challenged to develop and cultivate a relationship with a character that, for all intents and purposes, was a manipulated version of what had always been considered human. In *Dollhouse*, Whedon subtly poses the idea that perhaps civilization is undergoing a profound shift in what constitutes integral aspects of humanity. In a similar fashion, theorist Elaine Graham speculates that the entire structure of human nature is currently being challenged: "the digital and biotechnological age engenders renewed scrutiny of the basic assumptions on which matters such as personal identity, the constitution of community, the grounds of human uniqueness and the relationship between body and mind are founded."[1]

Whedon's fictional series parallels issues being dealt with in our own world. Whether the show's examples of humanity are viewed through a negative or positive lens, *Dollhouse* depicts a believable futuristic approach to traditional humanity by using elements of contemporary posthuman theory. Although *Dollhouse* is a work of science fiction, the idea of technological implantation is not far-fetched. From pacemakers and laser eye correction or even something as common as contact lenses, technology and science are becoming more and more a part of our daily human experience and a part of our physical selves. In a world where technology is moving at an impossible rate, humanity can no longer ignore its direct impact. As the show progresses, viewers witness two differing reactions to the proposed posthuman existence: one specifically seen within the characters of the Dolls and the other viewed in the character of Topher Brink. In the fictitious world of *Dollhouse*, the argument is not whether posthumanism exists but how society might handle living in the world of the posthuman. Ultimately, the show suggests that the posthuman future has happened, whether we admit it or not, and only those who are open and accepting of the change will survive.

THIS WORLD IS FOR PEOPLE WHO CAN EVOLVE . . . DOES THAT INCLUDE US?

The audience's first glimpse into the 'House is during Adelle's explanation of the process involved in the construction of an Active ("Ghost," 1.1). Under threat of possible incarceration, Caroline enters into a contract with the Dollhouse, seeing this as her only acceptable alternative. Within the confines of the *Dollhouse* universe, technology and science seem to have advanced to the point to allow this type of extensive biological and neurological procedure to believably occur, yet Caroline immediately questions the concept of erasing a person's identity. Clinging to the idea that some form of her original self will remain in the body with her, Caroline argues with Adelle, who holds fast to the claim that each Active is simply wiped clean like a chalkboard; Caroline points out that residue still remains even after the board has been erased ("Ghost"). When reassured that she will be unaltered upon completion of the contract, Caroline still questions how it would be possible for her original identity to remain unchanged after her body and mind have undergone such drastic changes. In other words, can Caroline still be Caroline if she will be "filled" with multiple other identities?

Caroline Walker Bynum posed a similar question in "Why All the Fuss about the Body?" Bynum chose to use film and television as the clearest examples of a lived identity, noting the parallels between the two media.[2] In her investigation, Bynum specifically suggested that film and television "explore the problem of identity and personal survival through asking whether 'I' will still be 'I' if transplanted into a body clearly marked by the personal characteristics . . . of 'someone else.'"[3] While Bynum wonders if an original identity still exists when placed into an alternate container, *Dollhouse* poses a similar question: is the original person or human still there when the container of that person's body has been filled with someone else? Kate Rennebohm states that the unique human body is an intrinsic part of someone's identity, unable to be lost, forgotten, or ignored.[4] On the surface, Rennebohm's concept seems definitively accurate; however, within the confines of the Dollhouse technology, many of the prescribed alterations do seem to affect the physical body container. While never witnessing an outward transformation in the physical body shape or appearance of the Dolls, in one episode the 'House is able to trigger lactation within a Doll, implying that physical changes are possible ("Instinct," 2.2). Taking into account that the Active architecture is able to affect physical aspects, the idea of the body as a constant in the preservation of a person's identity can be thrown out of the metaphorical window. Our identity is tied to the physical changes that take place in the body; once the body has been changed, even if seemingly returned to its original state, identity can no longer remain the same.

In his introduction to the book collection *Posthumanism*, Neil Badmington tries to explain when the shift beyond traditional humanity occurs. While his argument is limited specifically to that of the human and the machine, Badmington imposes his own distinctions separating the two and explains that when the two begin merging, an "other" is created.[5] This "other" that Badmington introduces—whether the human affected directly by machine or, in the case of *Dollhouse*, the human affected by any sort of advancement beyond the body's original capability—characterizes what is meant by the term "posthuman." Taking Badmington's argument to a new level, Dominique Janicaud looks at what these changes would do to the definition of human. In his book *On the Human Condition*, Janicaud specifically targets the influence of technology and its observable effect on humanity's construction.[6] Concerned with where humanity's existence is headed in light of advances in electronics and science, he suggests that our future will be directly affected by genetic and biotechnological alterations, and he argues this has the possibility "to transform 'the human race' to the point of rendering it unrecognizable, biologically, technically, culturally."[7] With the posthuman evolution, this societal shift will spawn a separation among humans into two divergent roads—the superhuman and the inhuman.[8] Using Badmington's and Janicaud's theories, we are provided a baseline for examining the posthuman characteristics of the characters in *Dollhouse*, making us aware when characters start on the path of the superhuman, diverge into the path of the inhuman, or tightrope walk in between the two.

In the case of Echo and the other Dolls, the technology brought out the most beneficial aspects of humanity and revealed a glimpse into Janicaud's concept of the superhuman.[9] Echo learned how to use the Dollhouse's technology, not as a foreign tool, but fluidly, as an aspect of herself and her identity; she felt an innate desire to help others, utilizing the gift that the 'House had given her. In the episode "A Spy in the House of Love" (1.9), the wiped, assumed incognizant, Echo brings herself to the programmer without prior instruction to do so. Echo recognizes that she could be of assistance to the 'House if she is imprinted with the correct programming. Sitting down in the programming chair, she asks to be imprinted to assist the 'House. Echo understands that the chair can make people different, and while she is unsure of how the technology works, she knows that she can become a tool to help ("Spy"). Later in the series, as Echo embodies the amalgamated identity of various Active personalities, her ambition shifts to an internal need to help the Dolls and, as in the flash-forward episodes, assist all of humanity through the transition into the world of the posthuman ("Epitaph One," 1.13; "Epitaph Two: Return," 2.13).

However, not all characters playing a role in the radical posthuman revolution are portrayed in such a heroic light. Along with the superhuman, Janicaud predicted that the effects of such technological advancements could bring out the worst in someone, making a human appear almost monstrous, thus symbolizing the inhuman.[10] While often endearing to fans, Topher clearly had a personality that was harsh and often unsettling to other characters in the show. This is set up initially in Boyd's resistance to Topher's attitude concerning the Dolls, as he urges Boyd to understand that the Dolls are similar to a herd of bison ("Gray Hour," 1.4). As the show progresses, more characters tend to distance themselves from Topher, recognizing that, in some way, he should be feared. The posthumanism of Topher sprouts from his deeply rooted need to be connected with technology. While not implanted into his body, the tech was essential for Topher just as much as it was for a Doll.

Two theorists currently discussing posthuman theories at length are Donna Haraway and Jean Baudrillard. Haraway leans toward the idea that the posthuman is best represented as a cyborg, a form of human-machine hybrid, containing materials both organic and nonorganic.[11] An alternative take on the posthuman can be found in the writings of Jean Baudrillard, who views the posthuman as a being who only exists sustained by machines, much like the theory of NASA's "Boy in the Bubble."[12] Both of these conceptualized theories are relevant to the posthuman experiences occurring in *Dollhouse*—the work of Haraway best resembles that of the superhuman Echo and the other Actives, whereas the studies of Baudrillard relate more closely to the character of Topher Brink as inhuman.

ASCENDING, EVOLVING, BECOMING SOMETHING MORE

In her essay "A Cyborg Manifesto," Haraway deconstructs many aspects of this posthuman body.[13] As part of her feminist critique, Haraway explicates the female struggle to maintain equality in today's world, finding a possible solution to inequality in the posthuman vision of a cyborg human able to escape many of the confines of normal humanity, such as gender. Although the Actives in the Dollhouse are biologically men and women, the actual physical gender of a Doll was not significant to the process of imprinting.

The programming that Actives undergo is said to give them a life's worth of memories, feelings, and connections. Topher explains that the experiences given to an Active are taken from an assortment of people: "I can create amalgams of those personalities, pieces from here or there, but it's not a greatest hits; it's a whole person" ("Ghost"). The 'House is responsible for programming every aspect and mannerism that makes someone an authentic human being.

Haraway believed that the posthuman would also allow a place for the postgender world, a world where gender played no role in identity as well.[14] In *Dollhouse*, the postgender is most evident in "Belle Chose" (2.3). This episode makes clear that the Active imprint technically could fill a body of any given gender, making it unnecessary for the physical sex of the Doll to correspond to that of the Active identity. For instance, Kiki—the imprint of a fun-loving female college student—is imprinted on Echo, but due to an unexpected glitch in the system, this imprint is suddenly transferred to Victor. Victor, now believing that he is Kiki, appears unaware of his personality switch or the fact that his body shape no longer corresponds with the memories suddenly prescribed him. Unaware of his outward appearance, Victor is able to seamlessly switch as his identity is separated from the physical body. Nonetheless, those who are able to see Victor's appearance are instantly aware that the identity and body no longer match correctly. Opposing the postgendered belief of Haraway, Caroline Bynum says that there is no escaping the physical form of gendered reproductive organs, claiming that "for better or worse, one could not shed gender or appearance."[15] Without having seen himself in a mirror, Victor wholeheartedly believes that he is Kiki, yet as Bynum tries to explain, one cannot merely let go of gender and appearance. Victor's unexpected switch from socially proscribed masculine attributes to feminine attributes causes a problem when others in the club notice his characteristics and mannerisms, which appear different from what the other partygoers expect. Victor's physical sex severed from his personality causes conflict, and he is unable to anticipate how those around him will perceive his actions.

The cyborg mechanics of Victor are technically without sex; he is a metaphorical "Ken doll," as in Haraway's explanation: "The cyborg is a creature in a post-gender world."[16] In *Dollhouse,* the cyborg Dolls, while they may be posthuman, are far from living in a postgender world. Gender is still supposed to be an anchored aspect of a person's identity, even in Whedon's futuristic setting. Victor's complete ignorance of his gender-switched personality represents a scenario really only possible within the body of the cyborg. His posthuman identity relies entirely on programming and is completely severed from his biological sex. Whether or not it correlates with the outward appearance of the physical body, any gendered identity is able to fill it as if it were an empty container.

Another key component of Haraway's theory concerning the cyborg is the absence of an origin story, which seems to suggest that the cyborg body can gain a freedom normally unavailable to human identity.[17] The character of a cyborg, created without a mother, father, or form of sexual reproduction, eliminates many subconscious needs found within most humans, such as the need to separate oneself from the phallic mother and to individuate oneself from the parents.[18] In *Dollhouse*, the primary example of this is found in the character of Alpha/Karl William Kraft. Alpha develops an obsession with destroying his original self, his origin story. As soon as Alpha becomes self-aware, he proceeds to attack Whiskey and the others in the Dollhouse, but his first real victim is himself. Removing those in his way, Alpha goes into the storage room and destroys the data storage device containing his original, unaltered identity as Karl William Kraft ("Omega," 1.12). This is an incredibly vital part of Alpha's transition from human to cyborg. As Alpha destroys the identity device, he effectively destroys any origin story that he had, showing his deeply rooted need to distinguish himself as cyborg and separate himself from humanity. Alpha is psychotic and unstable, and despite their similarities, Echo is never viewed in such a negative light. Echo is willing to share the body with Caroline, her original self. Echo wants to retain this human attribute, while Alpha wants to throw it away. Echo's desire to keep her origin story shows the audience that she does not wish to claim the rights to her own creation; despite the huge changes occurring within her, it is still crucial to her existence to remember and be proud of where she came from. While other factors contribute to Alpha's assumed psychopathology, this lack of identity and urge to divorce himself from his remaining humanity cannot be ignored.

While relatively similar in concept, Alpha and Echo were different in the way that they developed their unique attributes. Whereas Echo had the ability to retain memories from each implantation, Alpha was created by an anomaly where he was unexpectedly imprinted with dozens of personalities, one of which had multiple personality disorder ("Omega"). Relishing this experience and recognizing its "reality" for him, Alpha refused to allow himself to be wiped again for fear of losing this newly found identity. This is where Echo's similarity to Alpha differs: she is able to withstand the memory wipes and hold on to her previous imprints. Living with the knowledge that he is a composite of many people, Alpha recognizes that none of them is singularly capable of representing his identity and that each of the almost fifty imprints somehow creates the being that is Alpha. His pride in his ability to be creator and creation simultaneously contributes to his convoluted thinking that he has become a new, superior type of human, or posthuman. As such, out of fear, he knows that he must destroy anything that possibly could make him revert to the lesser being, that of Kraft.

In terms reminiscent of Haraway's definition of cyborg, Alpha argues with Echo, "We are not just humans anymore. . . . [We are] perfected, objective, something new" ("Omega"). Although we assume that Kraft had no part in choosing his Active name, that Whedon gave him the name "Alpha" seems, in itself, incredibly intentional. The word "Alpha" is riddled with underlying meaning, specifically in relation to the biblical use of Alpha and Omega.[19] Alpha sees himself as a god, perhaps even *the* God, yet recognizes that he needs his other half, his Omega. When trying to convince Echo of the importance of being transformed, Alpha pleads that she is the same as him and has the ability to ascend and evolve, illustrating that he fully believes that he and Echo are new types of human beings. Yet, despite his insistence, Alpha does not appear human anymore; his actions and mannerisms are unsettling, monstrous, and evocative of Janicaud's futuristic inhuman, and in the end, Echo is unable to see herself mirrored in the body of Alpha despite their apparent similarities.[20] Alpha is clearly the villain of the first season, illuminating Janicaud's monstrous version of the posthuman that uses technology to work for a character, as opposed to working with a character.[21]

Illuminating another representation of the posthuman role in *Dollhouse*, Topher Brink has no foreign object implanted into his brain, yet something about him is different than the other employees in the 'House. How can he be considered posthuman? Topher's role in *Dollhouse* is more similar to the posthuman approach of another theorist, Jean Baudrillard, who takes to referencing the image of NASA's "Boy in the Bubble," an attempt to raise a human without influence from any outside elements.[22] In the case of Topher, he is living in his own type of bubble, contained almost entirely within the technological bubble provided by the 'House. Through practically the entire first season of *Dollhouse*, Topher is isolated from the outside world, limiting his interactions to characters within the 'House. Topher's affinity for the operation of the biotechnological systems used on the Actives is nothing short of stunning; even the staff that hired him is astonished by his ability to substantially decrease the time of imprinting. Be that as it may, the others view Topher's profound connection to, or his relationship with, the machines and programming as abnormal. Topher's admission that he sees all of humanity as motherboards that he can control and predict also comes across as dehumanizing. In the unaired pilot episode, "Echo" (1.0), Topher is shocked by Boyd's inability to grasp this concept. He explains the rationale behind his thought processes, providing a multitude of examples of everyday social conventions and claiming that everybody has been programmed in one way or another ("Echo"). Topher forces himself into this protective dome of technology and knowledge, making himself a prime example of Baudrillard's theory.[23] Baudrillard asserts that a human being who no longer interacts with other humans is unable to be considered fully human and has begun to merge into an entirely new type of creation.[24]

Topher's intense relationship with technology strictly limits the type of human connections that he is capable of making. Despite their kindness and caring attitudes, others in the 'House attempt to remain at a distance from him. He never seems to appear in a sincere relationship with any other human being; this is made candidly clear in the episode "Haunted" (1.10). Unable to maintain friendships in the natural world, Topher imprints a Doll to be his friend. Upon seeing this, Boyd jokes that he is surprised that Topher actually is able to make friends, oblivious to the fact that Topher did not "make" a friend, as in building a relationship with another person, but literally "created" a friend, as one would create a robot ("Haunted"). Topher is lonely and seems to lack the skills that it takes to build a relationship with another human being; instead, he skips the relationship-building step and creates a friend preprogrammed to like him. Upon closer examination of the episode, the Doll's imprint is never even called by a name: even young children give names to their dolls, toys, and imaginary friends. The nameless Doll suggests many things in regard to the person of Topher. For instance, the simple act of naming the personality was unimportant to Topher, as Topher is unable to make a human connection that even young children know is necessary for cultivating relationships. Alternatively, the lack of a name emphasizes the machinelike state of the Dolls, as Topher sees them merely as tools to be manipulated, not people with their own personalities and characteristics. Sadly, in an even more depressing realization, the end of the episode reveals that it is Topher's birthday, and he celebrates it without family, friends, or colleagues, with only his programmed, nameless buddy by his side ("Haunted").

Topher's relational ineptitude is only more obvious when he attempts to venture outside his protective bubble of the 'House. Baudrillard asserts that in the world of the posthuman, someone's dependency on a world absent of disease would make one more susceptible to new, developing illnesses; in addition, a life without passions and thought-provoking discussions would make a person more vulnerable to varied psychological complications.[25] Topher's self-enforced, strict limitations on human contact make him vulnerable in situations requiring him to communicate effectively with others. This is no more evident than in the episode "Belonging" (2.4), revolving around Topher's complicated interaction with Sierra, the Doll who originally lived the life of an artist named Priya. From even the first introduction to her file, the character of Priya struck a chord with Topher but not because of who she was but what she was: a paranoid schizophrenic. Topher considered her merely as his next project, one needing to be repaired and rebuilt; in fact, Priya's broken "hardware" fascinated Topher, and he could not wait to get his metaphorical hands on her brain. Topher's obsession with the technological world and his detachment from the human world prevented him from thoroughly investigating what was right in front of him: Priya had been drugged by her doctor and was, in reality, falsely hallucinating. Later realizing his crucial mistake, he felt an obligation to fix things for Priya, but his ineptitude for building relationships ultimately made this impossible.

At this point, Topher is forced to confront that he has become the impaired, flawed creation that Adelle recognized from the very beginning: "The cold reality is that everyone here was chosen because their morals have been compromised in some way. Everyone except you, you Topher were chosen because you have no morals. You had always thought of people as playthings" ("Belonging," 2.4). Topher comes to a point of internal confrontation: has his posthuman form traveled down the path of the superhuman, or does he more closely resemble the grotesque inhuman? Topher's last-ditch effort to further a friendship and perform an act of human compassion ultimately leads Priya into murdering the doctor who misled the 'House into admitting her in the first place. Conscious of the fact that Priya cannot live with herself because of what she has done, she requests that Topher imprint her back into the dreamlike state of Sierra, the Doll. Knowing full well that living as Sierra, she will be unable to remember what Priya has done, she asks Topher how he will be able to cope with their secret. Topher does not have the advantage of a memory wipe, and although he can keep the secret, he admits that he does not know if he will be able to live with that secret suppressed inside of him ("Belonging"). Finally recognizing the destructive path upon which he has placed himself and where it will lead if he continues down it, Topher realizes that he must make the choice to allow a part of himself to metaphorically die to survive in this posthuman world; he must extinguish his inhuman self to find his way toward the superhuman.

"WHEN HUMANS BECOME OBSOLETE . . . LIKE BLOCKBUSTER AND OLD PEOPLE"[26]

Literary theorist Katherine Hayles explains that the success of these types of futuristic narratives is due to their representation of the posthuman future as already in full swing, with a culture fully immersed in intellectual discussions and everyday actions redolent of the posthuman experience.[27] In *Dollhouse*, how the characters deal with living in the world of the posthuman creates a clear, obvious conflict between acceptance and rejection of the posthuman reality. As a result, the most unsettling characters are not the Actives as they attempt to become the cyborg form of the posthuman but, rather, the characters who attempt to place a barrier between their identity and the technology around them, desiring to live within their isolated bubble and drifting into the realm of inhuman or monstrous.

Likewise, within the confines of the *Dollhouse* universe, Whedon creates an intricately woven balance of assumed fictional technology with nonfictional quotidian situations. Every character in the series is directly or indirectly affected by the role of the posthuman through the effects of the Active architecture, although most bystanders carry on with their lives as if nothing has changed. In the second season, characters in the 'House discover the Rossum Corporation's political takeover plot using Senator—and Active—Daniel Perrin; the fictional civilization is being thrown headfirst into the realm of the posthuman whether they like it or not, whether they even realize it or not ("The Public Eye," 2.5). The biggest dilemma for the characters in the imaginary world of *Dollhouse* is how to fight for the posthuman, as opposed to fighting against it.

Early in the narrative of *Dollhouse*, it is clear that Caroline is not the same as everyone else. It seems implied that Caroline's behaviors are different from those of Actives who have come before her, even different from Karl William Kraft's. Her recognition and critique of the processes that will be performed on her create such an inherent curiosity that it carries through into her despite the extreme procedure of memory wiping. Opposed to the other Dolls, who seem to accept whatever they are told, Echo expresses concern over the fact that her knee has been injured, yet she lacks memory of how it happened. She also worries over the injuries to Dr. Saunders and is concerned whether someone will take care of the doctor. Echo, intrigued by the bright lights upstairs, finds herself face to face with the reality of Topher's chair as she witnesses Priya's first visit to the 'House ("Ghost"). Echo's inquisitive nature should have been stripped away as easily as writing erased from a slate, yet it remains. Dolls are expected to desire one thing only—to try to be their best—and Echo's behaviors are clearly a departure from that specific behavior. Echo's innate ability to locate pain and suffering in others, Priya and Victor's ability to retain love for each other, and November's grief over her lost daughter are all emotions nonexistent to Topher. These remaining human emotions allow for the Dolls to maintain their humanity through the transition into their cyborg bodies.

Before "Belonging," Topher's extremely apathetic attitude toward the needs and emotions of the Dolls in the 'House throws him out of balance in his internal struggle between human and machine and, as such, makes Topher incapable of connecting to the humans around him. Topher's apathy made him appear inhuman. His lack of empathy and sympathy—and, even more so, his inability to recognize that these Actives were people, as opposed to just walking computers—appalled many original audience members.[28] Perhaps the writers intentionally chose to create Topher as monstrous, as a way of almost forcing the viewers into creating a stronger bond with a human-machine hybrid than a seemingly human character.

Janicaud points out from the story of *Frankenstein* that there is both psychological and moral risk when humanity is pushed beyond its normal limits and emerges as a monstrous creation.[29] Topher, like Victor Frankenstein, was so obsessed with the future of technology that he blinded himself to what was around him, unconcerned with the consequences that might result because of it. In a twist of his very own words, Topher seems to be running incredibly fast, but he is only running toward something; he had nothing from which to escape ("Ghost"). This lasts until the end of the episode "Belonging," as Topher finally recognizes that what he is afraid of, what he is running away from, is himself. His apathy, up until the end of this episode, made him seem like he was incapable of fear and made him appear inhuman; however, as he finally comes to his own shocking realization of his principal role in the formation of Priya's character, he recognizes how absent was his humanity. Topher sees himself for the inhuman monster that he has become and recognizes, just as Frankenstein's creature does in the end of Shelley's story, that he must destroy himself, or at least this version of himself, to save those around him.[30]

In respect to other works in the "Whedonverse," Maggie Burns argues that the worst crime is allowing objects to have mannerisms and characteristics of humans.[31] Up until his drastic character changes, Topher exists as no more than an inhuman object, a piece of property owned by the 'House. The audience's uncomfortable feeling when watching Topher would have been directly affected by this confusion of object pretending to be human. Unlike the Actives in the Dollhouse, there is no need to install the architecture into his brain as Topher willingly allows the technology of the 'House to dictate his actions, desires, and daily routine. Before this pivotal episode, Topher has no freedom or independence, living a stationary life in the confines of his laboratory and the 'House; hypothetically, he becomes just another machine, and whatever humanity he had been born with seems extinct. Ironically, it is after the episode titled "Belonging" that Topher is finally able to make an effort to start "belonging" to the other people who surround him daily.

As I explain earlier, both Topher and the Actives are versions of the posthuman future that may await us, or, if *Dollhouse* is an accurate portrayal of our society, the posthuman future that we are already living. An attempt to control and overpower this new beneficial technology will force society to exist as an Alpha or as the original Topher trapped in his bubble. Yet, by utilizing and allowing the tech to become a part of society, we will ultimately be successful. Society has the chance to follow in the footsteps of the Dolls once they are in full possession of their biotechnological abilities as well as in the footsteps of the changed character of Topher after his allegorical death in "Belonging."

In considering *Dollhouse* as a metaphor for humanity's evolution into the posthuman, it is the way that characters negotiate through these changes around them that brings about the "thought-pocolypse"[32] within the audience. Whereas the beginning of the narrative often depicts Echo as helpless and unsure, as in the episode "Gray Hour" and even in "The Target" (1.2), her confidence level changes by leaps and bounds as she becomes more familiar with the new, superior version of herself. Given the show's intended focus on the impact of technology on humanity, the audience was drawn to this confrontation between the two most similar Dolls: Echo and Alpha. What Alpha thought was pride in himself gave way to insecurity and embarrassment of the origin story found within Karl William Kraft; this led to the greatest difference between the two, as Echo wanted to accept who she was as a part of who she had become.

In a similar fashion, Topher is faced with choosing his pride and reliance on machinery or choosing to acknowledge and accept his humanity. While it may not be an easy choice, the acceptance of the posthuman is necessary to society's survival. The end of *Dollhouse* sets up the final challenge for Topher as he is forced to show his true humanity once again, by choosing whether or not to sacrifice himself for the rest of society. Adelle offers to take on the challenge, but Topher rejects it. He is cognizant of the challenge that humans are facing: that of having to completely redefine themselves, since he has undergone the process himself. As Topher willingly accepts the explosive that will save the world yet destroy him, he attains his metaphorical seat in the realm of the superhuman, throwing away any conflict with his previously destroyed inhuman self. Topher Brink illustrates that all individuals have the ability to change their future and restore their humanity. In the words of another Active, the world was created for people to evolve, and only those with the ability to do so will continue to exist ("The Hollow Men," 2.12).

BIBLIOGRAPHY

Badmington, Neil. "Introduction: Approaching Posthumanism." In *Posthumanism*, edited by Neil Badmington, 1–10. Houndmills, England: Palgrave, 2000.

Baudrillard, Jean. "Prophylaxis and Virulence." In *Posthumanism*, edited by Neil Badmington, 34–41. Houndmills, England: Palgrave, 2000.

Burns, Maggie. "Mars Needs Women: How a Dress, a Cake and a Goofy Hat Will Save Science Fiction." In *Serenity Found: More Unauthorized Essays on Joss Whedon's Firefly Universe*, edited by Jane Espenson, 15–24. Dallas, TX: Smart Pop, 2007.

Bynum, Caroline Walker. "Why All the Fuss about the Body? A Medievalist's Perspective." *Critical Inquiry* 22, no. 1 (1995): 1–33.

Graham, Elaine L. *Representations of the Post/Human: Monsters, Aliens, and Others in Popular Culture*. New Brunswick, NJ: Rutgers University Press, 2002.

Haraway, Donna J. "A Cyborg Manifesto: Science, Technology, and Socialist-Feminism in the Late Twentieth Century." In *Posthumanism*, edited by Neil Badmington, 69–84. Houndmills, England: Palgrave, 2000.

Hayles, N. Katherine. *How We Became Posthuman: Virtual Bodies in Cybernetics, Literature, and Informatics*. Chicago: University of Chicago Press, 1999.

Holy Bible: King James Version. Peabody, MA: Hendrickson, 2004.

Janicaud, Dominique. *On the Human Condition.* London: Routledge, 2005.

Jones, Andrew Z. "The Redemption of Topher Brink." In *Inside Joss' "Dollhouse": From Alpha to Rossum*, edited by Jane Espenson, 78–93. Dallas, TX: Smart Pop, 2010.

Rennebohm, Kate. "'The Mind Doesn't Matter, It's the Body We Want': Identity and the Body in *Dollhouse*." In *Inside Joss' "Dollhouse": From Alpha to Rossum*, edited by Jane Espenson, 4–19. Dallas, TX: Smart Pop, 2010.

Shelley, Mary Wollstonecraft. *Frankenstein; or, The Modern Prometheus*, edited by Maurice Hindle. London: Penguin, 2003.

NOTES

1. Graham, *Representations*, 1–2.
2. Bynum, "Why All the Fuss," 8.
3. Bynum, "Why All the Fuss," 9.
4. Rennebohm, "The Mind Doesn't," 8.
5. Badmington, "Introduction," 1.
6. Janicaud, *Human Condition.*
7. Janicaud, *Human Condition*, 1.
8. Janicaud, *Human Condition*, 4.
9. Janicaud, *Human Condition.*
10. Janicaud, *Human Condition.*
11. Haraway, "A Cyborg Manifesto," 69.
12. Baudrillard, "Prophylaxis and Virulence," 34.
13. Haraway, "A Cyborg Manifesto."
14. Haraway, "A Cyborg Manifesto," 71–84.
15. Bynum, "Why All the Fuss," 32.
16. Haraway, "A Cyborg Manifesto," 71.
17. Haraway, "A Cyborg Manifesto," 71.
18. Haraway, "A Cyborg Manifesto," 71.
19. *King James Bible*, Revelation 22.13.
20. Janicaud, *Human Condition.*
21. Janicaud, *Human Condition.*
22. Baudrillard, "Prophylaxis and Virulence," 34.
23. Baudrillard, "Prophylaxis and Virulence."
24. Baudrillard, "Prophylaxis and Virulence," 34.
25. Baudrillard, "Prophylaxis and Virulence," 34–35.
26. "A Love Supreme," 2.8.
27. Hayles, *How We Became Posthuman*, 39.
28. Jones, "Redemption," 80.
29. Janicaud, *Human Condition*, 26.
30. Shelley, *Frankenstein.*
31. Burns, "Mars Needs Women," 23.
32. The thought-pocalypse is what Topher calls Rossum's attempted control of civilization ("The Hollow Men," 2.12).

Chapter Three

Anamnesis, Hypomnesis, and the Failure of the Posthuman in Whedon's *Dollhouse*

Margo Collins

Memory plays an important role in the works of Joss Whedon, from the season 5 arc of *Buffy the Vampire Slayer* involving the insertion of Dawn into both the narrative and the characters' memories, to the removal of Connor from the memory of most of the characters in *Angel*, to the nonlinear (and therefore problematic) memory of River Tam in *Firefly* and *Serenity*. In *Serenity*, River says, "People don't like to be meddled with. We tell them what to do, what to think, don't run, don't walk. We're in their homes and in their heads, and we haven't the right. We're meddlesome." In River's case, that meddling takes the form of invasive brain surgery that leads to her problematic memory. In the Whedonverse, "meddling" in minds, particularly in memory, is dangerous, often leading to apocalyptic scenarios.[1]

Nowhere is this as obvious as in the short-lived series *Dollhouse*, in which characters' memories are erased, stored in "wedges," and replaced with entirely new memories, enabling them to become new people, ready to be rented out to play parts in the fantasies of the rich and powerful. More than any other work in Whedon's oeuvre, this series examines the connections between the Platonic ideas of anamnesis and hypomnesis—technically translated as memory and forgetting but employed here in their more common definitions as somatic, bodily memory, as opposed to artificial memory. I take as my starting point the idea of anamnesis as both personal, individual memory and somatic memory—memory predicated on and mediated through the body. In opposition to this kind of memory is hypomnesis, or technological memory. For Plato, hypomnesis meant textual memory; more recently, in the works of philosopher and critic Michel Foucault, for example, hypomnesis has come to mean any kind of recorded, reproducible memory, from notebook to computer.[2] These two categories of memory are often placed in opposition to one another, but in *Dollhouse*, they are inextricably intertwined. Throughout *Dollhouse*, reliance on hypomnesis alone inevitably leads to social chaos and personal monstrosity, which can be remediated only through anamnesis, indicating that it is personal remembrance that enables community, connection, and a move toward becoming fully human. Ultimately, Whedon uses hypomnesis and anamnesis to illustrate the impossibility of a viable posthuman existence based on a split between body and mind. Personal remembrance serves as the antidote to the social ills arising from a disconnected, disaffected, and hypertechnologized culture, and hypomnesis functions only when mediated by anamnesis. Even those memories that come to us via hypomnesis, the show suggests, must be integrated via anamnesis, through living with them in our own bodies, before they can influence our identities. Only those memories that function anamnestically can lead us to connection and community.

MEMORY IN THE DOLLHOUSE

A number of critics have touched on the role of embodiment and memory in *Dollhouse*.[3] Indeed, it is virtually impossible to discuss *Dollhouse* without also discussing memory, given the show's fascination with the role of memory in identity formation. Much of the first season is taken up with depictions of the various "engagements" of the Active Echo—engagements that are meant to be controlled by the personas serially imprinted upon her. The implication is that she is often sent out on "romantic engagements," sexual trysts with rich clients, but the series more often shows other kinds of engagements: for instance, she is a friend turned bodyguard in "Stage Fright" (1.3), the safe-cracking leader of an art heist in "Gray Hour" (1.4), a blind religious cult member in "True Believer" (1.5), a rich woman solving her own murder in "Haunted" (1.10), and an adult survivor of abuse now volunteering at an orphanage in "Briar Rose" (1.11).[4] In all these cases, her identity is given to her through a session in scientist Topher Brink's chair, which ostensibly changes her identity by installing a complete set of memories, then removing them again at the end of the engagement, returning her to her Doll state. However, as Caroline, the woman who becomes Echo, notes in "Ghost" (1.1), a clean wipe is virtually impossible. Throughout season 1, Echo begins to remember parts of her past as the imprinting and wiping technology begins to break down for her in a variety of ways. In "The Target" (1.2), she begins having flashes of herself as Caroline; in "Gray Hour," an anonymous cell phone transmission eliminates Echo's imprint as Taffy midengagement; in "Stage Fright," Echo goes beyond the programmed parameters of the imprint; in "Echoes" (1.7), a trip to a college campus reminds her of her past as Caroline; and in "Omega" (1.12), Echo whispers the word "Caroline" to herself after a postengagement memory wipe at the end of the originally aired season.[5] Much of the rest of the series is devoted to Echo's use of composite memories from all her imprints to create a coherent self that opposes the Rossum Corporation's determination to put memory-wiping and personality-imprinting technology to use against the general population.

In her article "Memory, Mind, and Mayhem: Neurological Tampering and Manipulation in *Dollhouse*," Sherry Ginn discusses the plausibility of the memory wipes and imprints in the series by discussing various types of memory and the psychological and physiological storage of memories.[6] In *Power and Control in the Television Worlds of Joss Whedon*, Ginn expands on this discussion by examining, among other issues, the ways in which neural plasticity and pharmacological tampering with memory function in *Dollhouse*.[7] While Ginn's discussions are important, their primary focus on the various neurological realities of memory leaves room for continued discussion of the contingencies of Whedon's fictional depiction of memory, as Ginn acknowledges when she writes that *Dollhouse* "can be read as an intellectual exercise in how memories might be manipulated for both nefarious (e.g., stealing artworks) and altruistic (e.g., rescuing kidnapped children) purposes, and how such manipulations might affect those involved."[8] Indeed, Whedon himself has suggested that the memory wipes and imprints should be read metaphorically, stating that the series is about how we "incorporate other people in ourselves; . . . how we project ourselves onto people and how everybody relates to everyone in their lives through the filter of their own beliefs, experiences and memories."[9] Other critics have argued that the series's focus on memory is indicative of the importance of identity and that memory and the body are ultimately inseparable, as does Lisa K. Perdigao when she writes that "the body's experiences, including pain, are written onto the body and the self. In the story of forgetting and remembering pain is the story of constructing (or sustaining) one's identity."[10] Ultimately, as Oluwafemi Morohunfola notes, "although Rossum's technology appeared to create a world where bodies were interchangeable and personalities could live forever, on closer inspection, the lives of the Actives demonstrate that there is more to people than their personalities, and what's underneath can never be erased."[11] This inability to eliminate "what's underneath" is a function of the distinction between hypomnesis and anamnesis.

HYPOMNESIS VERSUS ANAMNESIS

At first, *Dollhouse* seems most concerned with hypomnesis in its strict denotation of "forgetting" and in its Platonic definition of memory stored outside the self—material, technological memory in the form of text or computer disk or the Dollhouse wedges, memory that is "repeatable and transmissible" and "becomes knowledge."[12] For the Actives of the Dollhouse, this form of memory is the only form of memory to which they are supposed to have access; the body forgets while technology remembers. Thus, in the Dollhouse, the self, as stored in the wedge, becomes a form of Foucauldian hypomnémata—a "technology of the self" that, rather than acting as an aid to identity, is instead designed to replace physical, bodily memory: the hypomnémata replaces anamnestic memory.[13] Theoretically, this makes the Dolls endlessly interchangeable. The Actives' memories are repeatedly wiped and replaced with information from the wedges to create new people. Information is combined and recombined to create desired personalities in a manner that situates the Dolls as texts to be written and rewritten. In this, *Dollhouse* relies on a definition of posthumanity in which "a person—a being with intellect, self-awareness, and self-interest—might come to exist in a nonbiological substance."[14]

The series intimates that these computerized wedges hold complete personalities. In "Epitaph One" (1.13), Adelle tells the newly hired Topher (in a flashback), "We have access to over 100,000 brain models. Every scan Rossum makes sends a full 3D map to our mainframe. In three years, we'll have a million." And in "Ghost," Topher asserts, "These personality imprints, they come from scans of real people. I can create amalgams of those personalities, pieces from here or there, but it's not a greatest hits; it's a whole person. Achievement is balanced by fault, by a lack." In "Man on the Street" (1.6), Topher says that he wants "precision," "a *soupçon* of rage, not a whole tasting menu," in the personality that he is creating. Thus these new personalities, as created by Topher, also become beings with "intellect, self-awareness, and self-interest,"[15] existing only as hypomnémata unless and until imprinted onto the Dolls.

Hypomnesis functions in the early episodes as a means to the clients'—and the Dollhouse's—ends, which Adelle and Topher both characterize as inherently benign. In a conversation with Echo's Handler, Boyd Langton, Topher claims, "We're great humanitarians. . . . We're all so misunderstood, which great humanitarians often are," justifying his stance by quoting Shakespeare's *Hamlet*: "There's nothing good or bad but thinking makes it so" ("Ghost"). Similarly, Adelle rationalizes her participation in the Dollhouse as meeting the needs of clients by providing the "heart's desire made flesh" ("The Target," 1.2). Hypomnesis is presented as a way of acknowledging and meeting those needs that otherwise cannot be met in the real world. Of course, the problem with this idea is that hypomnesis alone cannot meet those needs—the hypomnémata, the technology of the self, is merely software and cannot function in the world without a platform on which to run. In the Dollhouse, that platform is the Active architecture installed in the Dolls' brains. In this series, hypomnesis cannot exist without anamnesis; there is no disembodied self, and throughout *Dollhouse*, selfhood implies embodiment. The wedges are inactive, inanimate objects until the Doll is imprinted. It is this imprinting that leads to the horror inherent in the idea of the Dollhouse and also to the apocalyptic scenario of the "Epitaph" episodes that end each season.[16]

Similarly, one of the most salient moments of horror in the show comes with the realization that the "Attic" to which unruly Dolls and Rossum employees are routinely banished is actually the site of a computer run on human brains. When Echo is sent by Adelle to discover the secret that Rossum holds in the Attic, Clyde Randolph, one of the original two creators of the technology that powers the Dollhouse, tells her in cyberspace that the Attic was founded when the other creator (later revealed to be Boyd Langton) and his accomplices "cleared my head out, hooked me up, and I became a human computer—first one in history, I suppose" ("The Attic," 2.10). The people linked in the Attic function as one giant supercomputer, turning anamnestic memory—the physical, somatic memory capability and capacity of the human brain—into hypomnestic memory, using people as mere components in a computing conglomerate: "minds engaged, integrated" ("The Attic"). Everyone in the attic is kept in a nightmare loop, inducing fear to create the adrenaline that powers the supercomputer, and this—along with the fact that in the Attic people's bodies are wrapped in plastic and inanimate (much like those of real dolls)—highlights Rossum's desire to limit people to their hypomnestic uses; the Corporation has no use for the Attic bodies beyond their ability to create and maintain computing power. The horror implicit here is in Rossum's continued willingness to invalidate the significance of individual identity and memory, of anamnesis as the basis of self.

But Echo is able to overcome the connections in the Attic by allowing herself to "die" in the Attic's virtual reality, thereby causing her body to die in the real world and be automatically disengaged from the human-mind supercomputer to "avoid any corrupted data" ("Attic"). To overcome the hypomnesis of the Attic's supercomputer, Echo uses anamnesis, drawing on what the series has set up as an inevitable mind-body connection. Engaging somatic, anamnestic memory, Echo tricks the hypomnestic system into believing that she is physically dead—an act that can occur only when she mentally dies in the Attic. And although the series claims that Echo is genetically unique in her ability to remember events despite mind wipes, Echo's ability is shown to be perhaps less singular than previously implied when the Dolls Sierra and Victor also are able to remove themselves from the Attic.

Like Echo (though perhaps to a lesser degree), Sierra and Victor are also able to overcome hypomnesis via the use of anamnesis.

Indeed, *Dollhouse* consistently illustrates the failure of hypomnesis, imposed memory, in the face of anamnesis, somatic memory, thereby reinforcing the importance of both individuality and community in the face of a hypertechnologized society. Echo, rather than simply existing as a rewritten text of the Dollhouse, instead becomes a palimpsest of the individuals with whom she has been imprinted. She is that slate that never can be erased fully. In "Vows" (2.1), she tells former FBI agent tuned Doll Handler Paul Ballard: "I remember everything. Sometimes I'm someone else and then I come back. But I still feel them. All of them. I've been many people. I can hear them, sometimes suddenly. I'm all of them, but none of them is me. Do you know who's real?" Paul responds, "Caroline," but Paul is wrong; Caroline is no longer any more "real" than Echo, and this has happened through anamnesis. By holding and retaining the memories of others in her body (despite the mind wipes), Echo has created her own identity, comprising all those personalities and mediated through the controlling figure of Echo.

This need for anamnesis to mediate hypomnesis is developed gradually throughout the first season. In "The Target," Echo is sent on what is meant to be a romantic engagement, but she instead becomes the client's prey in a scenario reminiscent of Richard Connell's 1924 short story "The Most Dangerous Game."[17] Drugged and stumbling through the woods, Echo reaches out to a retreating figure who spins around and reveals herself to be Caroline—but rather than being any sort of "real" Caroline, she is the Caroline of the college yearbook video that Paul Ballard watches obsessively as he works to track down the Dollhouse. In this video, Caroline says, "Oh, no, no. Get that thing out of my face," a statement that refers, in the video, to the original camera being wielded by another student. Here, though, it serves another purpose: it reveals the artificiality of Echo and her situation. She, like the video version of Caroline, is simply a recording, and in her drugged state, she realizes it on some level. When Boyd tracks her down, she tells him, "There was something in the water . . . I've been seeing things . . . a girl that looks like me but isn't, dead bodies in a shower, and a man standing over me holding a knife, but I can't see his face. I don't think he has one." The fact that the drug in her system interferes with the imprint is an early indication that the hypomnestic imprinting is not a replacement for anamnestic memory but rather an overlay; despite repeated wipes, Echo's mind still holds memories. As Kate Rennebohm notes about the series, "the unique, specific human body is an integral aspect of identity, not to be forgotten or left behind."[18] Echo might not be able to access the details of her memories, but those memories still exist.

In "Stage Fright," Echo's decision to move beyond the programmed parameters of her engagement to save pop star Rayna from a crazed fan and her own suicidal tendencies concerns Dollhouse head of security Laurence Dominic and interests a number of other Dollhouse employees. Dominic's concern is that Echo "went off mission," whereas Adelle and Boyd believe that she improved on those parameters. Dr. Saunders worries that Echo's ability to "create a new approach to the problem" ultimately will be detrimental to Echo's safety in the Dollhouse. Thus, even those who run the Dollhouse and its security and who care for the Actives are uncertain of the degree to which an imprinted Doll should be able to function beyond the programming of the hypomnémata instilled into them. Bronwen Calvert notes, "In contrast to the language used by those in control of the Dollhouse, it seems clear that the 'imprinting' process carried out upon dolls is also physical, and has embodiment at its core."[19] I argue that it is not merely the "language" of those in control of the Dollhouse that indicates their belief in the ascendency of imprint over body, of hypomnesis over anamnesis; their actions, too, illustrate that belief. Adelle DeWitt seems particularly able to interact with the Actives as the people they become, despite seeing them every day in their Doll state—generally, she responds to hypomnesis as if it were anamnesis. For example, when her friend Margaret dies and is imprinted onto Echo, Adelle interacts with Echo as if she truly were Margaret—although she is aware of the distinction, commenting that she is both ten years younger than her friend and "oddly, fifteen years older" ("Haunted"); Adelle gives primacy to the imprint, indicating, at least initially, her own determination to value hypomnesis over anamnesis. This ability is also what allows her to run the Dollhouse. She is concerned not with the body but with the consciousness within that body. However, that stance is ultimately unsustainable, as Adelle eventually discovers.

The distinctions among an Active's ability to "think outside the pieces," the hypomnestic imprints, and anamnestic somatic memory become increasingly blurred as the first season progresses. By the episode "Omega," the fact that the escaped Active Alpha imprints Echo with almost forty identities simultaneously seems almost secondary; she already is able to function beyond the parameters that she is given, and she is moving toward developing an identity beyond that with which she is imprinted. By giving her multiple imprints, Alpha merely hastens a process that has already begun. Despite Alpha's intentions—he plans to have her join him in becoming "a superior creature, an ascended being"—Echo's assimilation of the multiple imprints leads, not to a desire to kill her original personality (as Alpha did) and rule over the rest of humankind, but to a desire to protect that personality despite the fact that, as she says to Caroline (currently residing in the body of the kidnapped Wendy), "you walked away from me. You left me alone in that place." When Caroline asks, "Who are you?" Echo replies, "I'm Echo," but follows it with, "She's nobody. I'm just the porch light, waiting for you." However, when Alpha threatens to shoot the wedge (after killing Wendy and thereby destroying the only other copy of the original Caroline personality), Echo says, "I said shoot it! I don't care! . . . She won't know. She's me! And we're both coming to get you!" In doing so, Echo indicates that for her new, composite personality, identity resides in both anamnesis and hypomnesis—she is both Caroline and Echo. Moreover, the fact that she remembers Alpha attacking Whiskey with a knife indicates that the imprints interact with, rather than overlay, her physical memory, as does the fact that Alpha's original personality influenced his tendency to attack women with knives. Hypomnesis, the show insists, cannot overcome anamnesis. Rather, the two forms of memory work together.

The second season furthers that claim as Echo's autonomous memory and character develop more fully. However, lest the viewer think that Echo's genetic profile—which makes her particularly capable of maintaining a number of imprints at once—differentiates her from the other Actives, the characters of Sierra and Victor also develop through anamnesis, despite hypomnestic imprints. The first indication that the Actives might develop without imprints comes when Sierra, in her Doll state, screams when Victor touches her. Though she cannot articulate it, she carries the anamnestic memory of her Handler repeatedly raping her ("Man on the Street"). In spite of that visceral reaction, Sierra and Victor bond. Topher calls it "grouping," indicating that this kind of bonding is not uncommon in Actives, but he denies its significance to their humanity, claiming that they are "a little bit bison" ("Gray Hour"). However, the fact that the connection between the two persists through imprinting indicates that, as with Echo, anamnestic memory at least troubles hypomnestic memory, even when it cannot overcome it. When confronting the man who placed her in the Dollhouse after she rejected him, Priya (the "original" of the Active Sierra) asserts that there is "someone I trust. He helps me. He thrills me. . . . I managed to fall for someone else. It's absurd. I don't remember meeting him or even spending a moment with him, but I can feel it stronger than anything. I'm crazy about him. I love him" ("Belonging," 2.4). Similarly, Victor, in his persona of Roger, tells "Miss Lonelyhearts" (Adelle, who is secretly a client), "It's as if I met this woman in a previous life, as if our love transcends our very being" ("Stop-Loss," 2.9). In both of these cases, anamnesis interacts with the hypomnesis of the imprints.

This interaction persists even after Victor and Sierra are both reincorporated into the Dollhouse. Their love for one another remains strong—so much so that when Victor (in his actual identity of Anthony) is released and joins the group mind of Rossum's military division, he is able to break free through the intervention of Priya/Sierra. While still part of the group mind, Anthony holds Priya at gunpoint:

Anthony: "Anthony's gone."

> Priya: "Anthony's gone, but what about Victor? I know he's in there."
>
> Anthony: "You're trying to confuse us. You're not going to confuse us." ("Stop-Loss")

His use of "us" here could be read as either the "us" of the group mind or the "us" of Victor and Anthony. In any case, Priya calls out to Victor, not Anthony: "Victor. Victor! Victor remembers me! He couldn't forget. I didn't forget." Anthony/Victor replies, "They don't want me to remember. They want me to kill you." His use of "they" separates him from the group mind. By appealing to Victor, Priya taps into the same anamnestic memory that forced Roger/Victor to break off his relationship with Adelle. When he says, "I remember," he identifies himself as Victor rather than part of the group mind; later, he introduces himself as Anthony—a move indicating that he has incorporated the lived experiences of both Anthony and Victor into his identity, since he has responded to one name and used the other. Through anamnesis, he remembers and incorporates multiple identities into his self-identity, and as Ginn notes, "memories are what make us who we are; we remember information about ourselves, and these memories serve as our selves"—here, those memories as self are both somatic, drawing as they do on those lived experiences, and hypomnestic, since they incorporate at least some of the memories created during imprints and wipes.[20]

THE PROBLEMS OF POSTHUMANISM

More than simply an examination of the core of identity, though, *Dollhouse* also functions as an exploration of the inevitable failure of any concept of posthumanity based solely on hypomnesis. Both Stephen Asma and Daniel Dinello have argued that fictional scenarios of robots or other artificial intelligence systems rising up against their human masters (as in the *Terminator* films or the reimagined *Battlestar Galactica* television series) tie directly to fears of the ascendancy of the technological over the biological.[21] These fears often connect directly to posthumanism—the philosophical stance that humans are developing (generally via technology) toward a future in which the body and mind are no longer inherently connected.[22] *Dollhouse* evokes many of those fears. Donna Haraway's ruminations on the potentially liberating connections between women and cyborgs are, the show suggests, naïve fantasies—the ability to move the consciousness out of the body leads not to female empowerment but to female exploitation.[23] Female (and, to be fair, male) bodies are, in the Dollhouse, not freed from the needs of physicality but enslaved to them, their bodies becoming the highly sexualized playthings of the extremely rich while their individual minds stay locked inside their wedges, unaware of the world around them until they are "imprinted" onto a brain. Minds, the series suggests, are useless until they are embodied.

Moreover, the ability to move minds via hypomnesis is depicted as potentially dangerous. In the two "Epitaph" episodes of the series, we learn that Topher Brink's invention of a portable wipe-and-imprint device has allowed the members of the Rossum Corporation to use that hypomnémata, the technological artifact itself, to take over other human beings. In fact, virtually all humans have become "dumbshows" (those whose identities have been erased), "butchers" (those imprinted to kill others), or imprinted with apparently random identities. Only a few "Actuals" (people with their own identities) remain.

This apocalyptic vision relies in part on the clear focus that the Rossum Corporation has maintained on the value of hypomnesis. When preparing to send Echo to the Attic, Adelle DeWitt says, "There's only one place for a Doll as irreparably damaged as you" ("Stop-Loss"). Despite the fact that Adelle is playing a part to help sabotage the corporation, the representation of Echo here as a broken Doll indicates that, for Rossum, anamnesis indicates damage; the Corporation is interested only in hypomnesis, in the ability of humans to act as repositories of recorded (and controlled) memory rather than as creators of embodied, physical memories. That Boyd, who turns out to be the mastermind behind Rossum, says of the Dolls, "They're all broken" ("Man on the Street"), further indicates the corporation's belief that any need for embodiment is an indication of damage.

Nonetheless, Rossum executives need bodies for their consciousnesses to survive the abuse that they inflict upon their original bodies, and they do not hesitate to imprint dumbshows with their own hypomnestic identities. *Dollhouse* suggests that posthumanism can lead to the erasure of the individual, as revealed by the dumbshows—the "group mind" soldiers who try to recruit Tony, the Attic's computer created from the brains of former Actives (and anyone else who crosses the Corporation)—and even by the identities formed by Echo and eventually Alpha—identities that, though encompassing some part of the body's former occupants, still replace it. Julie L. Hawk argues that Echo "becomes able to use the technology subversively and to create a posthuman subjectivity that is both at odds with and in debt to the technological tampering that at first rendered her powerless."[24] However, Echo's characterization throughout the series as particularly adept at integrating multiple imprints eliminates any possibility that she might act as a model of viable posthuman existence. Similarly, Alpha's comment that it is "harder for me than most" ("Epitaph Two: Return," 2.13) to find peace indicates that a personality in which multiple imprints function at once is, at best, less than ideal, particularly since the original personality can break through imprinting (albeit as only one of many, as in Alpha's case).

Similarly, Anthony's post-Dollhouse attempt at an upgrade—hardware that allows him to upload skills, as long as he eliminates other elements of his memory—draws him away from his connection to other people; even his interactions with the other "tech heads" is at least agonistic, if not antagonistic. Indeed, nowhere in the postapocalyptic world of the "Epitaph" episodes is there a fully viable posthuman model. Tom Connelly and Shelley S. Rees argue that *Dollhouse*'s apparent technophobia departs from Whedon's usual character-driven television but "still addresses recognizably Whedonesque concerns about the place of the Self in a larger context."[25] Although I certainly agree that the series is indeed about placing the self in a larger context, I do not believe that *Dollhouse* is a departure from Whedon's earlier works. Repeatedly, as I have argued elsewhere, Whedon's work addresses the need for personal remembrance to serve as the basis for social connection.[26] Despite its focus on technology as the means of replicating hypomnesis and attempting to divorce individual memory from identity, *Dollhouse* illustrates this continuing theme in Whedon's oeuvre. Ultimately, in *Dollhouse* as elsewhere, only through anamnesis can the individual ever become part of a viable community, and it is only through community that humanity can survive.[27]

Indeed, many survivors of the Dollhouse and its technology find redemption through a focus on anamnesis and humanity and a move away from the kind of technological posthumanity seen in the postapocalyptic "Epitaph" episodes, including Echo, Adelle DeWitt, Anthony, Priya, and Alpha. In "The Attic," Rossum cofounder Clyde says, "My loop has been to run statistical probability scenarios to where the technology might lead. All but 3 percent of them include the end of civilization"—and, clearly, none of the scenarios in that 3 percent played out, as the "Epitaph" episodes find the characters in the midst of an apocalypse very like the ones that Clyde envisions. Both Adelle and Topher feel responsible for the breakdown of society, and eventually, both seek redemption. Topher, having realized the enormity of what he has done, moves through insanity and toward humanity, working to create a device that will eliminate all imprints. In his final scene, Topher creates and detonates a bomb carrying the pulse that will "fix what we did to their heads" ("Epitaph Two"). However, his very last action is to take a step toward a wall covered with photos from the Dollhouse and labeled with the heading "To Remember." In doing so, he moves once again toward a representation of memory, toward a form of hypomnesis. Despite everything, he still is drawn to the technological records of memory, to hypomnesis over anamnesis, and therefore his destruction is virtually assured—he must be killed in the explosion that restores the population of the world to its preapocalyptic memory state because he is the representative of hypomnesis, the one remaining character who might still on some level value technologically recorded memory over memory recorded in the body.

That Adelle, however, has come to value anamnesis is made clear when she leads the remaining "dumbshows" out of the remains of the Dollhouse. Her job is to "fix what we did to the rest of the world" (a job that Topher characterizes as "way harder" in "Epitaph Two"), and she can do so best by working with the humans whose memories will be reset—the people who will find themselves back in their own bodies with no memories of the actions of those bodies, people who will perhaps not remember any part of the apocalypse or how they came to be surrounded by the bombed-out buildings around them. In this, Adelle, as a representation of everyone who has never been imprinted, must use her own personal memory to explain the past. Kirsten Strayer notes that in the first "Epitaph" episode,

> the chair became the vehicle for the telling of history, and it was Doll bodies (and, through the bodies, the not-quite-human Actives' minds) that narrated key events of the past ten years, a decade that we the viewers had not seen and knew nothing about. Even though the survivors argued over the relevance of the memory wedges to their situation, they had no doubt that the chair (and whoever was in it) would explain the Dollhouse's role in making the present dystopian world.[28]

However, by the end of the second "Epitaph" episode, the bodies of those who remember the past decade serve to explain those same events. As she leaves to lead the dumbshows and imprinted people out of the Dollhouse, Adelle tells Echo that Alpha said to "dismantle all the tech in the building. He said you should start with the chair" ("Epitaph Two"). Any history to be shared, any explanation of the events of the past ten years, will now come out of anamnesis, not hypomnesis.[29] Adelle and the other survivors with their original memories intact will be the ones to remember, without the aid of hypomnestic memory devices. Even Tony has allowed his identity as Victor—the technologically enhanced rebel leader who carries a variety of memories in miniaturized wedges on a chain around his neck—to be destroyed in favor of becoming the father of T (Tony Jr.), his son with Priya. In doing so, Tony chooses biological reproduction—the ultimate form of anamnesis, of remembrance through the body, through biology—over technological reproduction. Priya herself eschewed "the tech" many years before in favor of caring for her son, and she and Tony are now moving back into the Dollhouse to protect their memories of their child (as Topher's bomb would reset their memories back to the time before their Dollhouse experiences and thus eliminate their memories of T). Priya says, "You'd have to kill [her]" ("Epitaph Two") to take the memories of her son, highlighting the inherent physicality of those memories.

But, as ever in the series, it is Echo who best illustrates the need to combine anamnesis with hypomnesis. Like Victor and Sierra, Echo must return to the Dollhouse to protect her new identity—the one created anamnestically over a number of imprints and years—from Topher's bomb. Paul is dead, but in what is apparently the last imprint that the chair will ever make, Echo integrates Paul into herself. By incorporating Paul into her own memory, Echo becomes complete, illustrating the fact that a combination of anamnesis and hypomnesis allows for connection. In their separate forms, Echo and Paul are both incomplete—Paul tells her in "Epitaph Two," "You've got a hundred people living inside your head and you're the loneliest person I know." Echo's response, "That's kind of sweet," is countered by Paul's comment, "Not for the person who's with you," but, in their separately embodied state, they never can be truly "with" one another, never truly together. Only after Echo incorporates Paul's imprint into her own physical body can she understand him. Paul's claim that he has been "knocking ten years; you still won't let me in" is answered by the ultimate form of letting him in. When he becomes a real part of her, understood anamnestically as part of her own body, Echo is finally complete. That this incorporation takes place via the chair's hypomnesis is secondary to the fact that only anamnesis will eventually integrate them entirely:

> Paul: "Am I. . . ? Are we . . . ?"
>
> Echo: "You wanted me to let you in."
>
> Paul: "You sure you got room? I got a lot of baggage. Childhood stuff."
>
> Echo: "We'll work through it. We've got time." ("Epitaph Two")

Echo's comment that they will "work through it" indicates her willingness to use this new anamnestic bond to Paul to finally connect fully. Christopher Souza claims, "Echo's final act of imprinting herself with Paul's memories after his death . . . further demonstrated the desperate need we feel to confide in other human beings."[30] That this "desperate need" can be assuaged through connection and community as mediated through anamnesis and aided by hypomnesis is ultimately, perhaps, the clearest idea the Dollhouse can imprint upon anyone.

BIBLIOGRAPHY

"Anamnesis and Hypomnesis." Ars Industrialis. http://arsindustrialis.org/anamnesis-and-hypomnesis.

Asma, Stephen. *On Monsters: An Unnatural History of Our Worst Fears*. Oxford, England: Oxford University Press, 2009.

Bennett, Eve. "Deconstructing the Dream Factory: Personal Fantasy and Corporate Manipulation in Joss Whedon's *Dollhouse.* " *Slayage: The Journal of the Whedon Studies Association* 9, no. 1 (2011).

Buckman, Alyson R. "'Much Madness Is Divinest Sense': Firefly's 'Big Damn Heroes' and Little Witches." In *Investigating Firefly and Serenity: Science Fiction on the Frontier*, edited by Rhonda V. Wilcox and Tanya R. Cochran, 41–49. New York: Tauris, 2008.
Calvert, Bronwen. "Mind, Body, Imprint: Cyberpunk Echoes in the Dollhouse." *Slayage: The Journal of the Whedon Studies Association* 8, nos. 2–3 (2010).
Campbell, Norah, and Mike Saren. "The Primitive, Technology and Horror: A Posthuman Biology." *Ephemera: Theory & Politics in Organization* 10, no. 2 (2010): 152–76.
Collins, Margo. "Memory and Monstrosity in *Buffy the Vampire Slayer*." In *Creatures of the Night*, edited by Nadine Farghaly. Newcastle upon Tyne, England: Cambridge Scholars Press, forthcoming.
Connell, Richard. "The Most Dangerous Game." *Collier's Weekly*. January 20, 1924. Reprinted in *Feedbooks*.
Connelly, Tom, and Shelley S. Rees. "Alienation and the Dialectics of History in Joss Whedon's *Dollhouse*." *Slayage: The Journal of the Whedon Studies Association* 8, nos. 2–3 (2010).
Dinello, Daniel. *Technophobia! Science Fiction Visions of Posthuman Technology*. Austin: University of Texas Press, 2005.
Espenson, Jane, ed. *Inside Joss' "Dollhouse": From Alpha to Rossum*. Dallas, TX: Smart Pop, 2010.
Foucault, Michel. "Self Writing." In *Ethics: Subjectivity and Truth*, 207–21. New York: New Press, 1997.
Ginn, Sherry. "Memory, Mind, and Mayhem: Neurological Tampering and Manipulation in *Dollhouse*." *Slayage: The Journal of the Whedon Studies Association* 8, nos. 2–3 (2010).
———. *Power and Control in the Television Worlds of Joss Whedon*. Jefferson, NC: McFarland, 2012.
Haraway, Donna. "A Cyborg Manifesto: Science, Technology, and Socialist-Feminism in the Late Twentieth Century." In *Simians, Cyborgs and Women: The Reinvention of Nature*, 149–81. New York: Routledge, 1991.
Hawk, Julie L. "Hacking the Read-Only File: Collaborative Narrative as Ontological Construction in *Dollhouse*." *Slayage: The Journal of the Whedon Studies Association* 8, nos. 2–3 (2010).
Hayles, N. Katherine. *How We Became Posthuman: Virtual Bodies in Cybernetics, Literature and Informatics*. Chicago: University of Chicago Press, 1999.
MacCormack, Patricia. "Posthuman Teratology." In *The Ashgate Research Companion to Monsters and the Monstrous*, edited by Asa Simon Mittman and Peter J. Dendle, 293–310. Farnham, England: Ashgate, 2012.
Morohunfola, Oluwafemi. "What Echo's Journey to Self-Awareness Tells Us about the Human Soul." In Espenson, *Inside Joss' "Dollhouse,"* 218–31.
Perdigao, Lisa K. "'This One's Broken': Rebuilding Whedonbots and Reprogramming the Whedonverse." *Slayage: The Journal of the Whedon Studies Association* 8, nos. 2–3 (2010).
Quilty, Susan. "Negative Space in the 'House: How Caroline Is the Vase." In Espenson, *Inside Joss' "Dollhouse,"* 132–45.
Rennebohm, Kate. "'The Mind Doesn't Matter, It's the Body We Want': Identity and the Body in *Dollhouse*." In Espenson, *Inside Joss' "Dollhouse,"* 4–19.
Ryan, Maureen. "Sex, Secrets and *Dollhouse*: Joss Whedon Talks about the End of His Fox Show." *Chicago Tribune*. December 3, 2009.
Shuster, Martin. "What It Means to Mourn." In Espenson, *Inside Joss' "Dollhouse,"* 232–45.
Souza, Christopher. "Boyd Langton and the Fantasy of Trust." In Espenson, *Inside Joss' "Dollhouse,"* 204–17.
Strayer, Kirsten. "*Dollhouse*'s Future History Machines." In Espenson, *Inside Joss' "Dollhouse,"* 174–87.

FILMOGRAPHY

"Tabula Rasa." *Buffy the Vampire Slayer*. Season 6, episode 8. Written by Rebecca Rand Kirshner. Directed by David Grossman. First broadcast November 13, 2001.
The Terminator. Written by James Cameron and Gale Anne Hurd. Directed by James Cameron. Hemdale Film, 1984.

NOTES

1. Editors' note: Sometimes it's just personal. Tara, for instance, is twice the victim of memory meddling, once by Glory and once by Willow. Tara breaks up with Willow as a result ("Tabula Rasa").
2. Plato develops the idea of anamnesis as memory in the dialogues *Meno* and *Phaedo*. The Platonic version of anamnesis implies memory as carried by the soul from life to life. In this chapter, I am using anamnesis in the sense of personal memory. The Platonic idea of hypomnesis as textual memory appears in *Phaedrus*. For Foucault's view, see "Self Writing."
3. See, for example, Martin Shuster's statement: "At every step, the show highlighted our continual dependence on our body, but also our all-too-human ability not only to lose sight of this dependence, but to reach continually beyond it; indeed, *Dollhouse*'s story arc neatly illustrated a continual drive to forget our human embodiment through hopes of reaching beyond that embodiment. Ultimately, then, we saw that this tension between the body and mind (or spirit) was precisely what drove Echo's development, her coming of age. This tension was also formally on display in the strange state of affairs where a show about the transference and manipulation of personalities and minds ended up thematizing to such a large degree the body" (239). See also Bennett, Calvert, Ginn, and Quilty.
4. See Porter and Ginn in this collection for a discussion of *Dollhouse* clients.
5. "Epitaph One" (episode 1.13) was never aired by the Fox network in the United States (though it was aired in other countries); it is available only via DVD.
6. Ginn, "Memory."
7. Ginn, *Power*.
8. Ginn, "Memory."
9. Ryan, "Sex."
10. Perdigao, "This One's Broken."
11. Morohunfola, "What Echo's Journey," 221.
12. "Anamnesis and Hypomnesis."
13. Foucault, "Self Writing."
14. Asma, *On Monsters*, 267.
15. Asma, *On Monsters*, 267.
16. For more on horror in *Dollhouse,* see Calvert, this volume.
17. Connell, "Most Dangerous."
18. Rennebohm, "Mind Doesn't Matter," 8.
19. Calvert, "Mind, Body, Imprint."
20. Ginn, *Power*, 98.
21. Asma, *On Monsters*; Dinello, *Technophobia!*
22. For more detailed discussions of posthumanity (simplified here to the idea of human identity contained in technological forms), see Haraway, Hayles, MacCormack, and Campbell and Saren.
23. Haraway, "A Cyborg Manifesto."
24. Hawk, "Hacking the Read-Only File."
25. Connelly and Rees, "Alienation and the Dialectics."
26. Collins, "Memory and Monstrosity."
27. Buckman makes a similar argument in her discussion of River in Whedon's series *Firefly*, suggesting that community is vital for her, much as it is for Buffy in *Buffy the Vampire Slayer*.

28. Strayer, "*Dollhouse*'s Future," 182.

29. The fact that Adelle passes this as a verbal message from the reformed Alpha is also significant in that it highlights a distrust of any form of recording—the message is passed verbally, bodily, from Alpha to Adelle to Echo, thus moving through anamnesis rather than hypomnesis.

30. Souza, "Boyd Langton," 39.

Part II

Ethics

Chapter Four

"What about the Laws?"

Regulation and the Celebration of Resistance

Tom Garbett

As the first season reaches its climax, Adelle DeWitt, previously cold and reserved, reacts passionately to the news that the bodies of Actives are now to be sold permanently to ultra-high bidders. "What about the laws?" she asks, not rhetorically but hopelessly in the face of overwhelming opposition. After a pause, she adds, "The laws of humanity?" ("Epitaph One," 1.13). The distinction that she makes between "law" and "the laws of humanity" is necessary, given that the Dollhouse exists in a shadow world, literally underground and seemingly beyond the reach of criminal law (especially given the influence bought by the Rossum Corporation in government). Adelle appeals to something that one hopes is inherent in the human condition—a feeling that to sell the Actives would be repugnant and against some fundamental rule, one that defines humanity and, once broken, makes us less than human. Yet this suggestion is misguided—that the legal system, in its widest sense, has no role to play in the *Dollhouse* story (other than as a net that Rossum and others must avoid or control). In fact, as this chapter demonstrates, it is a driving force for the *Dollhouse* narrative.

ADELLE DEWITT: REGULATION[1]

The Dollhouse is a paradoxical entity. It exists in the world's peripheral vision, at once known and unknown. It seemingly is beyond the reach of the police and yet is subject to extensive surveillance; it seemingly is ungoverned by the political powers that be, yet it is run to strict rules. However, *regulation*, "in the sense of the intentional activity of attempting to control, order, or influence the behavior of others," is evident throughout the 'House, both inside and out.[2] Actives are disciplined in an increasingly draconian fashion, from being repeatedly wiped,[3] to being segregated, to being sent to the Attic (if nothing else works). For non-Actives, regulation is more direct. FBI special agent Paul Ballard is warned off the Dollhouse case, then dismissed, and finally ignored by his former colleagues; DeWitt is reprimanded on the phone and in person; Laurence Dominic is castigated and then condemned to the Attic after being unmasked as a mole.

Threatening Actives as a means of regulation, one might argue, will have no effect on future behavior since, as a result of their lack of memory, they have no sense of consequences; still, the sense of a threat is ever present, albeit subtextual. They are regulated by the fact that not only have their personalities been excised but, in an underlying sense, they also have been captured by Foucault's theory of discipline—the regulation of bodies by subliminal means—and its relative, "biopower," which is applied to the management and promotion of people and populations.[4] The Dollhouse is a make-believe world that makes explicit to the viewer the illusion of free will: the Dolls imagine that they think for themselves—"Shall I go now?" / "If you like," as the postwipe ritual so memorably expresses it—but, of course, they do not.

Disciplinary power, for Foucault, is apparent only when one investigates the use of power from the bottom up (instead of top down).[5] Power should be investigated "at its extremities, in its ultimate destinations, with those points where it becomes capillary . . . in its more regional and local forms and institutions."[6] It is the formative, localized exercise of power that is important: the idea is not to focus on the self-conscious use of power but the fundamental "level of on-going subjugation, at the level of those continuous and uninterrupted processes which subject our bodies, govern our gestures, dictate our behaviours."[7] So the disciplines are the basic, inherent, and unnoticed methods of control that inform the development of society by conditioning individuals from birth. Rather than looking at power from the center and working outward (i.e., identifying the seat of power and following where it leads), we must look at "the manner in which the phenomena, the techniques and the procedures of power enter into play at the most basic levels."[8] This is an ascending analysis of power, starting from small mechanisms which have their own history, their own trajectory. We can point to this as a summary of treatment of the Actives in the Dollhouse: they are recruited at their most vulnerable and encouraged to "be [their] best" while at their most psychologically weak. The manipulation of their bodies and (residual) minds is conducted through the simple exercise and craft regimes propagated in the Dollhouse—the promise of betterment is entirely illusory. Even when we think that the Actives have circumvented the system and found a way to escape Rossum's clutches in, perhaps, the most poignant moment of calculated deception, we see Echo and other Dolls emerge into the sunlight of freedom before collapsing around a subtly victorious Adelle DeWitt in "Needs" (1.8) as their programming, there all along, stops the escape in its tracks.

It is not just the Actives, however, who are subject to this unknown conditioning. Adelle's rather pathetic assertion, "I run this House" ("Echoes," 1.7)—blurted out after and as a result of her exposure to an inhibition-limiting drug—is manifestly untrue. She is betrayed and manipulated by subordinates and superiors—for example, in relation to the proposed sale of the Actives, as noted, and ultimately by employing the duplicitous head of Rossum, Boyd Langton. As we learn later, there are many 'Houses, many "Adelles," and a host of shadowy figures running the enterprise behind the scenes (both in the organization itself and in the realms of politics and law enforcement)—a community of duped operatives. Arguably, the Dollhouse emerges from a society that anticipates such a place to exist (as we see from the "street cam" comments that pepper "The Man in the Street" [1.6]) and wishes it into being: as one "interviewee" puts it, "You think it's not happening? You think they're not controlling you? Don't worry about it. Just sit back and wait for them to tell you what to buy." In fact, as Canavan has suggested, one might argue that "the Dollhouse [is] the ultimate synthesis between biopower and market logic, [and] . . . thus the culminating moment of neoliberal hegemony Foucault identifies in *The Birth of Biopolitics*."[9] In other words, the final expression of a society subject to (or in thrall to?) commodification will be the sale of the human body itself. This sale is both a product of such a society and its downfall:

> For Foucault, this paradoxical double action of the biopolitical state—what might be called the fundamental insincerity of biopower—can be seen clearest in its furthest extremity, the atom bomb: biopower develops a capacity to kill that is in essence a power to destroy life altogether, which is completely incompatible with the supposed aims of the biopolitical nation state.[10]

The apocalyptic "Epitaph" episodes, with which each of the first and second seasons culminate, suggest that Whedon, at least, sees a very real risk that such commodification could result in the potential for the destruction of life, just as Foucault argues.

Of course, there are both embedded and explicit methods of regulation in *Dollhouse*. For overt regulation, such as DeWitt's punishment of Laurence Dominic ("A Spy in the House of Love," 1.9), Ballard's dismissal ("Briar Rose," 1.11), or Hearn's murder by November (engineered by DeWitt as punishment for his abuse of Sierra ["Man on the Street"]), we can look toward John Braithwaite's theory of the regulatory pyramid for illumination.[11] Braithwaite's theory is intended to suggest ways in which those who regulate should be responsive to the conduct of those whose behavior they seek to control in deciding whether a more or less interventionist response is needed; lapses from regulatory requirements are dealt with (relatively) lightly at the bottom of the pyramid and meet with more serious consequences as one moves up, depending on the severity of the breach or its repetition.[12] In this form of regulatory control, one can (indeed must) make a choice about whether or not to comply with what the regulator asks in light of the threatened consequences and, crucially, what has gone before.

The pyramid is a conceptual way of attempting "to solve the puzzle of when to punish and when to persuade."[13] At the bottom of the pyramid there is a "restorative dialogue-based approach," which focuses on the "justice of the law." As one moves up the pyramid, responses become more punitive and more interventionist, but the presumption should remain that one starts at the base. This is a dynamic model that always starts with dialogue—unless there is a compelling reason not to—and then escalates the response and moves up the pyramid only if the regulatee refuses to engage or the dialogue is needlessly prorogued. It also is possible, once compliance or agreement is achieved, to "de-escalate down the pyramid," rewarding the regulatee's compliant response.[14] This way of considering regulation, Braithwaite says, recognizes not only that persuasion works but also that sometimes it does not.[15] In cases where it fails, escalation up the pyramid treats the regulatee as a rational actor, ramping up the pressure until it becomes "rational to comply," weighing up "the likely costs of law enforcement [or any other regulatory command] compared with the gains from breaking the law."[16]

Of course, the actors in the Dollhouse are rarely rational. Ballard, for example, has what we can assume is a stable job with the FBI—excitement aplenty but a solid job contributing to society. However, his response to an order to drop the Dollhouse investigation is not to toe the line but to commit more wholeheartedly to the project. He is a classic example of a movement through the pyramid's stages: spoken to in the first instance, suspended, then fired. The viewer knows that Ballard is right, that the Dollhouse does exist (and, we learn later, there is a high-level conspiracy to keep its existence hidden ["The Public Eye," 2.5]). The pyramid works, and it appears as though regulation has both failed and succeeded: failed in the sense that Ballard has not been turned from his path and in the sense that the authorities—the personification of law—cannot or will not do anything to end the abuse of the Dollhouse; succeeded in the sense of the Foucauldian disciplines—it is not just Ballard's law enforcement colleagues but society in general that is busy ignoring something that could (and almost does) destroy it.[17] Interestingly, Ballard is accused by a Dollhouse client of reacting in this irrational manner—risking everything—because of his obsession with Caroline, something that will become important shortly:

> I mean, she, she changed things for you. So you're the head of this FBI task force to uncover the Dollhouse, and you're working hard, you're chasing leads, you're cracking skulls, but it's just work. And then you meet this girl or you . . . you see her somewhere, huh? Caroline? And suddenly . . . it gets personal. ("The Man on the Street")

So we have two methods of determining behavior in the Dollhouse: first, the Foucauldian disciplines, which ensure that the governed are conditioned to compliance, a method of regulation that takes place in the manipulation of the Dolls but also goes to the heart of a society brought up with the (potential) excesses of rampant consumerism; second, an overt effort at ensuring compliance, escalating from censure of the individual to one's committal to the Attic.[18] This is of interest to us because these methods of regulation all fail but, in their failure, drive the narrative. Tony Vinci argues that "there is an authentic identity residing at the centre of each individual that is obfuscated and trapped by ideology, though this identity can be 'freed,'" and it is indeed in rejection of these attempts at control—at the imposition of an ideology of compliance—that characters find that identity.[19] It is discovered through acts of resistance.

ECHO: RESISTANCE

It is intriguing that resistance is seen in both those who are coerced to conform without knowing it—the Dolls (and, for Foucault, society at large)—and those who are faced with a choice of obeying demands or not.[20] Ballard remains committed to his cause despite its destroying his life; Echo develops awareness against all the odds. There are other examples, including, most obvious, Dr. Claire Saunders/Whiskey and her desire to remain in the Dollhouse as the world she knows crumbles around her. This resistance takes the form of querying "what is right": for the Dolls, things are frequently "wrong," triggering a memory or a feeling of uncertainty in the definitive world of the Dollhouse. For a nonimprinted character, resistance is a moral choice: hence, DeWitt's "What about the laws?" ("Epitaph One").

The source of this resistance is found in Robert Cover's conception of a *nomos*, where legality is constructed from myth—and I extrapolate from this to account for regulation, which is in many ways a more pervasive form of control.[21] Cover's revolutionary suggestion is that it is the continuities constructed from a community's stories that give (their) law substance and shape.[22] In *Dollhouse*, this chapter argues, it is the personal myth of each character that drives their increasing awareness; it is the past of both Actives and non-Actives that colors their future. For Caroline, "although [she] has acted with good intentions, she has been caught breaking the law for political reasons," and in return for her becoming a Doll, the Rossum Corporation "will make her legal problems go away and make her very rich."[23] This directly influences her future attitude as Echo: she is driven to save the other Dolls in the same way that she was driven to attempt an infiltration of Rossum headquarters, and, similarly, she attempts to save Ballard to assuage memories of her lost, former lover ("Getting Closer," 2.11). Yet this is more than simply mirroring, more than a reiteration of personal history or originating identity. It is a demonstration of a personal mythos, the view of oneself as "rescuer," "soldier," "mother," and, in Echo's case, "savior." Ultimately, this leads directly to resistance, an assertion of personal choice over circumstance, because it provides a framework for behavior—a framework within which certain things are wholly off-limits—akin to Cover's system of *nomos* or narrative law.[24]

Cover's assertion may be summarized in three parts: that the way in which we make sense of the world is by reference to the narrative of the past, combined with its ongoing importance in the present, eventually creating an approach to the future.[25] His discussion of *nomos* is an attempt to "grant all collective behavior entailing systematic understandings of our commitments to future worlds equal claim to the word 'law.'"[26] Such a position results in a corresponding weakening of the state's monopoly on the setting and maintaining of the norms of a society: the norms of a state are "not denied the dignity of 'law'" but "must share the dignity with thousands of other social understandings."[27] The Dollhouse, then, is not beyond the law but of the law: the commitment by those who are in command to its existence and ethos (whether or not this is sincere, and there is certainly a disconnect between the overarching intentions of Rossum and the managers of the various Houses) may grant the collective behavior of the Dollhouse's members a kind of internal law. The collective belief by the Actives in the very existence of the Dollhouse and the regularized set of basic rules to which they are subject might suggest that a Cover-style "law" is in effect. But this is unsatisfactory given the lack of a cohesive system and the simmering tensions and moral compromises that are lurking just under the surface.

What we learn from Cover is that a society—of more than *de minimis* congregation—that commits to and lives by a set of rules should be seen as having a system of "laws."[28] This results in a variegation of law, meaning that Cover's view of regulatory frameworks is therefore dramatically different from Foucauldian homogenization; it allows for a plurality of regulatory viewpoints that, importantly, defeats a "one-size-fits-all" type of control.[29] Might we then claim not that the Dollhouse itself constitutes a legal microclimate but that the resistance seen in and from the individual central characters in *Dollhouse*, particularly Ballard and Echo, who are figureheads for an embryonic resistance movement, is best described as a manifestation of Cover's law? Their collective action then would not be characterized as a breach of regulatory control, as it might first appear, but instead a personalization (within their microsociety) of it: "I work to my own laws, my own sense of regulation," they seem to say.

One example of this, to illustrate the point, is Claire Saunders's confrontation with Topher, voicing simultaneously her hatred toward him and her suffering over that hatred. She wants to know why he made her hate him ("Vows," 2.1). The answer is that Topher wanted to be challenged, and, by creating a different view from his own, he creates a conscience.[30] Here is not only a personalization of regulation—Topher eventually rejects his regulatory surroundings by working against the Dollhouse ("The Attic," 2.10), making the ultimate sacrifice in the final episode ("Epitaph Two: Return," 2.13)—but the creation of his own regulator in the form of an external conscience (another microsociety capable of having "law").

An even more compelling example is Paul Ballard. He struggles to pursue the Dollhouse case in the face of apathetic Bureau colleagues and faces the full force of regulatory censure in his refusal to walk away. Yet there is something more complicated going on. As Vinci puts it,

> Ballard's quest to unveil and destroy the Dollhouse reveals the complex exchanges between identity and one's perception of self. On the surface, and according to his own perception, Ballard is performing an altruistic cultural function: he is attempting to stop the murder of innocents. However, as "Man on the Street" and future episodes in season one indicate, once the notion of a distressed damsel is introduced, Paul is revealed to be living out a homogenized heroic narrative that can be traced back to a patriarchal fairy tale. He is the prince whose sense of identity and purpose come from his desire to save the girl.[31]

Once again, then, we see that one of the driving narrative forces in the show is the form of persona with which characters self-identify: the form of their personal mythology. It is not so much that Ballard is wholly aware of the motivation behind his actions—the point is inscrutable—but that Ballard is spurred on to effect resistance by a vision of how he should act.

At this point in our discussion the separation between Active and non-Active is beginning to collapse. The Actives—Echo and Alpha aside—possess (for narrative purposes) at least three personalities at any one time, although not in the same body: their original self, their Doll state, and their imprint. Similarly, we see that the nonimprinted characters create their own alternative personalities or, more precisely, they project a separate personality to fulfill their personal narrative. There seems little doubt that Ballard wants to be the knight in shining armor, but he tells himself that he is a public functionary, acting for the general good. Topher is not the amoral technician that he pretends to be: he does not need the external "conscience" that he alleges to have created in Dr Saunders; rather, she is his masochistic spur.

The point here is that although "the Dollhouse . . . coercively preys on the disenfranchised and the desperate to acquire its dolls," this subaltern caste retains a personal mythos that drives the direction of their lives.[32] And this is both a collective and a personal mythology because the Actives are amalgams. As Echo begins to piece together the amalgam personality that she slowly develops, we see the ghost of Caroline as a crucial part of her and her self-appointed mission: "It's gotta be me," Echo says as the team considers bringing Caroline back to ask her who the head of Rossum really is. "I need to keep her close. She can lie to the rest of you—she can't lie to me" ("Getting Closer," 2.11). Where Vinci suggests that "Whedon envisions a postmodern apocalypse created by embracing and embodying ideology as a way imaginarily to resolve the problems of identity inherent in a post-industrial, post-postmodern world,"[33] I am arguing that he is presenting something much simpler but much more fundamental. Rather than a move toward the collective, I think Whedon wants us to consider the irreducible power of the individual and, furthermore, the power of an individual to self-regulate.

TOPHER: NARRATIVE

This returns us to Cover's vision of the role of regulation, of law. For Cover, law is constitutive of the world, and it is the "language and . . . mythos" of law that combine to create "narratives in which the corpus juris[34] is located by those whose wills act upon it."[35] Cover's use of *corpus* is telling here: the "body" of law is created from a combination of language and myth.[36] The body is crucial in *Dollhouse*, given a cast of "characters who ostensibly have no fixed identity or personality" but exist through their physicality.[37] Relating the body of law to the body itself is more than mere semantics: the body of law, for Cover, is the narrative that a community tells itself to establish continuity among past, present, and future; the continuity in the Dollhouse is the body. Despite the repeated wipes the Actives' bodies remain constant, and they retain a body memory, contributing to a behavioral consistency.[38]

It is myth that is important here: "myths establish paradigms for behavior . . . [and] build . . . relations between the normative and the material universe, between the constraints of reality and the demands of an ethic."[39] Furthermore, myths "establish . . . a lexicon of normative action . . . that may be combined into meaningful patterns culled . . . from . . . the past."[40] In other words, the way in which we, as people, as humans, move from the bare world of existence to the creation of the world as we wish it to be is through narrative. An accepted myth is a forward-urging paradigm,[41] and the frame of reference and sense of self-expression provided by the past are found in everything that we take to be law, including "the history of ordinary legal doctrine . . . in utopian and messianic yearnings . . . in the critiques that may be leveled at . . . law."[42] "Law" is therefore a continuum because it is a response to (and a direct consequence of) what has come before: it is "a connective between two states of affairs, both of which can be represented in their normative significance only through the devices of narrative. . . . Law may be viewed as a system of tension or a bridge linking a concept of reality to an imagined alternative."[43]

Bringing this together, it is my position that law—at least as characterized by Cover—is an exciting lens through which to view *Dollhouse*. Its contribution is twofold. First, it is the sense of a past and the retained sense of self in the multiple personalities of the Actives and the collective commitment to the Dollhouse by all characters (leading to creation of a personal and group mythology) that drives both Active and non-Active. An Active enters the program to escape his or her past, yet, as viewers, we are left in little doubt that some essence, some core of their "real" self remains: Priya and Victor's enduring love is one example. The same is true for non-Actives: to choose but one example, Bennett's past rejection (at high school) and the passing kindness of Caroline lay the groundwork for the ill-fated attack on the Rossum Corporation. Bennett is eager to be used, sacrificing all, just like an Active. Caroline asks Bennett, "You're not pissed because I used you?" and Bennett replies, "I'm hurt because you won't" ("Getting Closer"). If we accept Cover's proposition that law provides a bridge or framework to allow us to move from where we are to where we want to be, we can suggest that a personalized "law" exists to drive all the Dollhouse protagonists.

So far, all this emphasizes is that the characters have a past, which, of course, we all possess. I think the unique contribution that law makes when used as a viewfinder is that it explains why the characters behave as they do. They react against or in line with their personal law, and this reaction provides them with direction: they act to create a world different from that of the present. We see this in the "Epitaph" episodes where, given the breakdown in law and order, our characters yearn for a world of structure and seek to achieve it. This is true whether the resulting behavior is "good" or "bad"—Alpha is just as much a product of his personal law as Echo as they act out the narrative of their lives, reacting against their pasts, each setting standards that they accept or reject. In fact, Cover suggests, "Law gives a vision [of the future—of how things should be] depth of field, by placing one part of it in the highlight of insistent and immediate demand while casting another part in the shadow."[44] This is something more than saying that the characters of *Dollhouse* make moral choices; it is "law . . . which holds our reality apart from our visions and rescues us from the eschatology that is the collision in this material social world of the constructions of our minds."[45] Personal law is a manageable combination of moral, practical, and teleological rules that one follows because it leads to a certain desirable position. These rules may be wholly separate from those externally imposed; resistance, then, is inherent in both Active and non-Active.

The reference to eschatology is also appropriate. When the potential end of the world arrives, Whedon is at pains to show the success of resistance to indoctrination and the potential for successful adoption of personal law. Topher and Ballard die and, in the process, combine with their projected personalities: Topher destroying the tech and Ballard saving the girl, fulfilling their self-imposed narrative. Even Alpha's "home for the Actives" ("Epitaph Two") is an indication that self-regulation can be effective. After all, building a safe haven in the face of an apocalypse suggests not contrariness or caprice but a reckoning or balancing of what has gone before. And this is, in my opinion, the very definition of law or regulation: the attempt to maintain balance between past and present and between regulation and utter freedom.

The fascinating result, if one accepts this proposition, is that legal and regulatory mechanisms become more important in *Dollhouse*, not less so, for although the very mechanism of law—which we might have characterized previously as something consistently applied that acts to regulate all behavior—seems to fracture as a result of the multiplicity of differing positions that logically result from having a personalized lore, it is this intersection of external regulatory stimulus and personal myth/lore/law that is a catalyst for action (whether it be resistance or compliance). *Dollhouse*, as seen through the lens of law, is therefore a celebration of resistance rather than an exploitative or superficial commentary of passivity: there is always a point at which a person says "I will" or "I won't" based on their past and their vision of a future.

BIBLIOGRAPHY

Braithwaite, John. "Responsive Regulation and Developing Economies." *World Development* 34, no. 5 (2006): 884–98.

Canavan, Gerry. "Fighting a War You've Already Lost: Zombies and Zombis in *Firefly/Serenity* and *Dollhouse*." *Science Fiction Film and Television* 4, no. 2 (2011): 173–203.

Cover, Robert. "The Folktales of Justice: Tales of Jurisdiction." In *Narrative, Violence and the Law: The Essays of Robert Cover*, edited by Martha Minow, Michael Ryan, and Austin Sarat, 173–202. Ann Arbor: University of Michigan Press, 2004.

———. "Nomos and Narrative." *Harvard Law Review* 97, no. 4 (1983): 4–68.

Foucault, Michel. "Governmentality." In *The Foucault Effect: Studies in Governmentality*, edited by Graham Burchell, Colin Gordon, and Peter Miller, 87–104. London: Harvester Wheatsheaf, 1991.

Ginn, Sherry. *Power and Control in the Television Worlds of Joss Whedon*. Jefferson, NC: McFarland, 2012.

Parker, Christine, and John Braithwaite, eds. *Regulating Law*. Oxford, England: Oxford University Press, 2004.

Randell-Moon, Holly. "'I'm Nobody': The Somatechnical Construction of Bodies and Identity in Joss Whedon's *Dollhouse*." *Feminist Media Studies* 12, no. 2 (2012): 265–80.

Vinci, Tony. "'Not an Apocalypse, the Apocalypse': Existential Proletarisation and the Possibility of Soul in Joss Whedon's *Dollhouse*." *Science Fiction Film and Television* 4, no. 2 (2011): 225–48.
Wilcox, Rhonda. "Echoes of Complicity: Reflexivity and Identity in Joss Whedon's *Dollhouse*." *Slayage: The Journal of the Whedon Studies Association* 8, nos. 2–3 (2010).

NOTES

1. In this chapter, I link concepts with characters to give my argument the kind of dramatic tension the "Whedonverse" is known for: for those unfamiliar with some of the works used, the personality flavor should help to provide a dramatic tone. The argument in this chapter is that both the collective and the personal myths of each character inform action, and the association of themes with characters is intended to embed this.
2. Parker and Braithwaite, *Regulating Law*, 1.
3. "Wiped" is a shorthand used to refer to the process of deleting the personality and memories of an Active and subsequent personality "imprints."
4. "Biopower" is the Foucauldian term for his description of power over populations rather than individuals (who are regulated by what he calls the "disciplines" and which take hold within the sphere of biopower). See Foucault, "Governmentality."
5. Foucault, "Governmentality."
6. Foucault, "Governmentality," 96.
7. Foucault, "Governmentality," 97.
8. Foucault, "Governmentality," 99.
9. Canavan, "Fighting a War," 196.
10. Canavan, "Fighting a War," 174.
11. Braithwaite, "Responsive Regulation."
12. Braithwaite, "Responsive Regulation."
13. Braithwaite, "Responsive Regulation," 30.
14. Braithwaite, "Responsive Regulation," 31.
15. Braithwaite, "Responsive Regulation."
16. Braithwaite, "Responsive Regulation," 32.
17. Foucault, "Governmentality."
18. Foucault, "Governmentality."
19. Vinci, "Not an Apocalypse," 243.
20. Foucault, "Governmentality."
21. Cover, "Nomos."
22. Cover, "Nomos."
23. Wilcox, "Echoes of Complicity."
24. Cover, "Nomos."
25. Cover, "Nomos."
26. Cover, "Folktales," 176.
27. Cover, "Folktales," 176.
28. Cover, "Folktales," 176.
29. Cover; Foucault, "Governmentality."
30. Wilcox, "Echoes of Complicity."
31. Vinci, "Not an Apocalypse," 240.
32. Canavan, "Fighting a War," 191.
33. Vinci, "Not an Apocalypse," 227.
34. The body of law, meaning the collection of laws in a given jurisdiction.
35. Cover, "Nomos," 9.
36. Cover, "Nomos," 9.
37. Randell-Moon, "I'm Nobody," 267.
38. Ginn, *Power*.
39. Cover, "Nomos," 9.
40. Cover, "Nomos," 9.

41. "Urging" because it encourages an evaluation of the present, of reality, to see what might be changed in the future to attain or move toward the paradigm (hence, forward-looking).

42. Cover, "Nomos," 9.

43. Cover, "Nomos," 9.

44. Cover, "Nomos," 9.

45. Cover, "Nomos," 10.

Chapter Five

Somebody's Asian on TV

Sierra/Priya and the Politics of Representation

Ananya Mukherjea

Sierra/Priya—by virtue of her narrative arc and her ethnic, racial, and gendered representations—is a fascinating character, a problematic one, and unusual in the Whedonverse.[1] This chapter focuses on how her story reflects not only on Sierra/Priya as an individual (albeit, an individual who is a kind of hybrid of multiple selves) but on some of the demographic categories that she more obviously, or visibly, inhabits—as a woman, as an Asian woman, and as a migrant who crosses all sorts of borders, including national and ethnic ones. These demographic categories matter, especially because there are still so few regular characters of color on television and in the Whedonverse. So while it is tempting to let Sierra/Priya's story be just *her* story and nothing more, that is not yet possible at this historical moment. When there are relatively so few minority characters on television, those that are present bear a representational burden that might be out of proportion to their intended role as a character. Sierra/Priya exemplifies this phenomenon because her character is central to *Dollhouse*'s plot and the relationships that drive the show and because her biracial international status means that the actress and character both straddle multiple demographic categories.

In a featured talk at the Fifth Slayage Conference on the Whedonverses, Hélène Frohard-Dourlent considered the role that institutionalized racism plays in the production of television in general and Joss Whedon's shows in particular.[2] Institutionalized racism, as Frohard-Dourlent discussed, is a better way to understand the complicated political significance of a character such as Sierra/Priya than to focus on individual intentions or interpretations.[3] While much discussion of racism in the United States focuses on discrete incidences and the provability of individual "malintent," the greatest material consequences of racial (or other types of) prejudice are wrought through systemic and institutional bias, which can go unidentified and unchallenged. Because television still falls short of accurately representing the full diversity of our society, Sierra/Priya—and Dichen Lachmann, the biracial Tibetan Australian actress who portrays her—necessarily signifies whole groups of people who remain underrepresented on-screen and in Whedon's productions. Whedon seems to acknowledge this as Sierra/Priya becomes pivotal to a secondary context of the series as well as its primary narrative. This secondary context requires viewers to remember that the backdrop to the science fiction of *Dollhouse* is the reality of economies reliant on sexual commodification and exploitation, so it is important to consider how an Asian female character's arc is especially implicated by that reality.

THE "COLOR" IN COLOR TELEVISION: DIVERSITY ON THE SMALL SCREEN

In June 2008, Jennifer Armstrong and Margeaux Watson wrote a special report for *Entertainment Weekly* called "Diversity in Entertainment: Why Is TV So White?"[4] In it, they examined the continued dearth of characters and writers of color on television as of the spring 2008 season. At that time, Cleveland Brown—an African American character on the cartoon show *Family Guy*, voiced by white actor Mike Henry—was the only minority character to be the focus of a new series on major network television. Armstrong and Watson reported, "According to an *Entertainment Weekly* study of scripted-programming casts for the upcoming fall 2008 season, each of the five major broadcast networks is whiter than the Caucasian percentage (66.2 percent) of the United States population, as per the 2007 census estimate."[5] They go on to cite an NAACP study reporting that one of every three people in the United States is a racial minority, therefore suggesting that one-third of those actors working in television and films should be minority as well, although the actual proportion is far smaller.[6] The presence of characters of color has grown significantly, and the number of writers of color has also risen somewhat over the past five years, but there is still a large gap between how the U.S. population actually looks and how it is represented on-screen.[7] A further issue is giving those characters of color who do appear rich storylines that keep them from being mere sidekicks or ethnic or racial stereotypes. Producer Tim Hobert is quoted as saying, "For me, the goal would be to tell interesting stories [for minority characters]."[8]

Per this criticism, some critics have faulted Joss Whedon for failing to offer characters of color many roles or long runs on his shows, although Whedon does quite rightly win praise for consistently and consciously creating opportunities for white female characters and writers.[9] In speaking to the

Los Angeles Times about casting Eliza Dushku as the lead in *Dollhouse,* Whedon says, "She's a feminist. She's an activist."[10] In regard to hiring writers Liz Craft and Sarah Fain, he says, in the same article, "I never set out to hire women. I set out to hire good writers. The exception being that I did want women running the show."[11] Whedon explains that women writers were necessary to prevent the sexual aspects of the show becoming an "exploitation fest."[12] He does therefore think politically and progressively in his casting, hiring, and writing decisions, at least with respect to gender and sexuality.

Dollhouse is distinctive, both in the Whedon oeuvre and on television more generally, for offering a cast that is diverse in multiple respects. It features two major minority characters—Harry Lennix's Boyd Langton, who is African American, and Dichen Lachmann's Sierra/Priya—and several minor characters of color, including Topher's beleaguered assistant, Ivy. It also features Miracle Laurie as Mellie/November/Madeline, who, though still slender, is noticeably larger than the extremely thin standard for women in Hollywood, a fact that Whedon apparently sought out, as Laurie herself reported in the season 1 featurette "Making Dollhouse."[13] At least one online critic suggests that Miracle Laurie is a person of color, although I was not able to verify this, and she is not, at any rate, a clearly, visually recognizable racial minority, which points to how multilayered the issue of racial representation is.[14] The question of Laurie's racial background, though, and of the complexity of demographic diversity on the *Dollhouse* set points to the need to study race and racism in the media without doing more harm than good through concretizing already existing but constantly shifting borders.[15] Boyd Langton, like all the major characters on *Dollhouse,* deserves thorough consideration himself, but this chapter focuses on Sierra/Priya because *Dollhouse* lavishes so much story on her and imbricates her narrative arc with those of Victor/Tony and Echo/Caroline.[16] More than Boyd, Ivy, Sam (from the Rossum labs at the university), or the other characters of color on *Dollhouse,* Sierra/Priya is a central character with an independent storyline that develops with those of Echo/Caroline and Victor/Tony.

Sierra/Priya is a central character for this series, and her narrative is marked by the depth and richness that Armstrong and Watson argue that characters of color very much need on television, but her characterization, like every character's, is wrapped up in history, politics, and cultural expectations.[17] Again, when there are relatively few minority characters onscreen—especially, perhaps, Asian female characters who get complex storylines—they are always loaded with a representational burden, a representational load that demands unpacking. Sierra/Priya's character, as one that is central to the plot of this show and also a multilayered minority, holds a heavier-than-usual representational load that bears serious critical investigation.

Dollhouse received much online criticism,[18] some of it obviously warranted, for giving in to the sexploitation of its actors and presenting some misogynistic themes; for instance, watching these nonsuperhero women be beaten repeatedly by multiple men throughout the series was intensely jarring. However, I believe that its overall story arc is a basically feminist and laudatory one, and I would like to proceed with that as a given. Despite that overall success, the show deals in problematic stereotypes and doublespeak on some important issues, particularly violence against women and the justification of answering violence with violence. As sociologist Patricia Hill Collins explains, using the concept of the matrix of domination throughout her book *Black Feminist Thought*, different systems of privilege and oppression always interact and work together.[19] The racial significance of the characters on *Dollhouse* is markedly gendered and sexualized through their body-conscious costuming and filming because of the overtly sexual scenes in which they are repeatedly placed and because Whedon's intentions for the show are explicitly feminist.

Sierra/Priya must be considered simultaneously as a female character, as a character of color, and as a heterosexual character who has the sexual fantasies of straight male clients projected onto and played out via her body (one assumes that female clients also hired her for romantic engagements, as Echo comments that seven of her imprints are lesbian, and male clients must have hired Victor and other male Dolls, but viewers of the series do not see this happen on-screen). The ways in which Sierra/Priya is shown to be strong or weak, to be objectified or to have agency, then, reflect on her as a female character and as an Asian character. Sierra's imprints are sometimes shown to be dangerous and battle capable, as in "A Spy in the House of Love" (1.9), but this is not usually the case for her in the televised episodes. In the featurette "Making Dollhouse," Whedon says that he cast Dichen Lachmann because she "was most convincingly tough," but much of that toughness she portrays on the show is an *inner* strength.[20] Whereas Echo is imprinted with multiple skill sets until she feels "locked and loaded" in "Stop-Loss" (2.9), Priya is sent into the field to rescue Tony, armed only with the love they share. Her strength in that episode and through much of the series is caring, creating, and reciprocating more than fighting. Her very name, "Priya," means "beloved" in Sanskrit.

Sierra is, however, repeatedly physically harmed. In the unaired pilot episode "Echo" (1.0), as the camera pans through the Dollhouse, Sierra stumbles in from an assignment, clearly worse for wear and shaky, dressed in an iconic pink Jackie Kennedy–type suit, with pearls around her neck and a bloody wound on her head. Early in the first season, she is on assignment as an Australian pop music fan and is kidnapped by a psychotic stalker. Later that season, it is revealed that she has been raped four times by her Handler, Hearn, whom Adelle quickly dispatches to his own death when she learns what Hearn has done. The episodes "Needs" (1.8) and "Belonging" (2.4) reveal Sierra's tragic backstory: Priya, an Australian free spirit and artist, was abducted, drugged, and forced into being a Doll so that she could become the imprinted plaything of the execrable Nolan Kinnard, whose advances she had rebuffed in real life. In "The Attic" (2.10), Sierra's nightmare consists of an endless loop in which her passionate lovemaking with Victor turns into a vicious sexual assault by the dead and decaying Kinnard. While Victor/Tony is traumatized by combat, Sierra/Priya is traumatized by repeated victimization. Victor/Tony *acts*, even if it is in the service of a military institution that is larger than and dominant to him; thus, he plays an unwitting part in the production of his own trauma as a classically masculine agent of action. Sierra/Priya is *acted upon* and, in a more classically feminine manner, is horribly punished for what she refuses to do, which is to be a willing lover to Kinnard. Hearn coldly comments to Adelle that Sierra's inability to resist his assaults was less a turn-on than simply a matter of convenience for him, and Adelle lures him to his death at November's hands by playing to his predatory nature in promising him that this victim will put up a fight. The image of the innocent Asian woman mistreated, abused, or forced into sex work is a repeated motif in Western media, from *Madama Butterfly* (1904) to *Miss Saigon* (1991) to *Slumdog Millionaire* (2008), and the reference to that image stings both for its stereotyped nature and for its roots in the historical, structural vulnerability of poor women, perhaps especially poor women in colonized or war-torn places. Seeing this theme played out on a Whedon series is unnerving and, I argue, meant to trouble the viewer.

When Priya does stab Kinnard, satisfyingly, to death, she needs Topher and Boyd to come help her clean up the situation and huddles, frightened, with bloody hands until Topher arrives. He finds the strength to dismember and disintegrate Kinnard's body *and* keep that secret. Priya, however, finds that continuing with her awareness of the events of the day is too hard, and she chooses the peace of being Sierra, with her memory of that day wiped, as well as to return to the Dollhouse to be with Victor. The show makes clear that she is a lover more than a fighter, but this initially seems like weakness. Sierra metaphorically depicts Kinnard as a looming black blotch overshadowing the lighthearted birds that she paints in her art classes; but, after Priya kills Kinnard, she stands slowly up from his body and comes into the frame, her silhouette becoming the black blotch against the backdrop of her beautiful bird painting hanging on his wall. As a killer, therefore, Priya is out of character, not herself. Because Priya, despite this, always does what must be done to protect herself and the ones she loves, her nonviolence is more than simple weakness. That nonviolence, in fact, is antithetical to and a balance to the misuses of power that surround her.

Sierra/Priya is one of at least three recurring Asian female characters on *Dollhouse*; the other two are Ivy and the Active Kilo, who is played by show writer Maurissa Tancharoen and who eventually becomes one of the tech heads fighting with Victor. All three are multilayered characters who initially seem to fulfill, in some ways, the traditional role expected of an Asian woman in mainstream media—to be submissive and victimized or hypersexualized—and then belie the limits of that role as they expand beyond it.[21] Whereas television currently does boast more female Asian characters acting outside the stereotype, with women such as Lucy Liu's Dr. Watson on *Elementary* and Grace Park's Sharon Valerii on *Battlestar Galactica*, the inclusion of three recurring Asian female characters on one show is noteworthy. It is still more significant because Maurissa Tancharoen is married to Joss Whedon's brother Jed, and the three worked on *Dollhouse* and *Dr. Horrible* together. Tancharoen, therefore, is integrated into Joss Whedon's repertoire of writers, actors, and producers who work on multiple Whedon productions. She plays a creative role in the larger Whedonverse, and I argue that a question she raises elsewhere in that 'verse is, in part, answered on *Dollhouse*. Tancharoen sings (and also cowrote) the song "Nobody's Asian in the Movies" for *Dr. Horrible's Commentary! The Musical*, which Frohard-Dourlent used to frame her presentation on institutionalized racism at Slayage 5.[22] In the song, Tancharoen opines in verse that Asians rarely get cast on television except to be typecast or marginalized.[23] This metacritique, with which Joss Whedon must be familiar, carries through to *Dollhouse* in its casting of multiple Asian actors.

The Active Kilo's small size, a Western stereotype of Asian women, is repeatedly referenced. When she is introduced in "The Public Eye" (2.5), Topher feels the need to joke about how appropriate her designation is: "Isn't she so cute and tiny? . . . She weighs one kilo!" She seems tiny, adorable, and tractable but next appears in "Epitaph Two" (2.13), where Mag shocks Zone by telling him that she thinks that "the little Asian" is cute. That little Asian turns out to be a tough, power-hungry soldier who later intimidates Zone with her weapons expertise and hard demeanor. *Dollhouse* offers up the type, has Topher and Mag name-check it, and then turns the type around and alters it, all with humor and a lesbian subtext.

Similarly, Ivy is introduced as Topher's frustrated assistant who initially fetches his lunch and juice boxes instead of doing the sophisticated technical work that she was trained and hired to do. In "Vows" (2.1), Sierra—imprinted as the same woman in the pink suit with pearls whom we first meet in "Echo"—comes in for a treatment, a task that Topher dismissively relegates to Ivy as he goes in search of caffeine. The Sierra imprint is unhappy having Ivy provide her treatment because, as she tells her, "I'm not comfortable with Orientals," leaving Ivy wide-eyed. She goes on to explain that it is not a "racist issue," as her discomfort is simply owing to the fact that "[Asian] culture's not really the thing, is it?" Dressed in the vintage costume of her imprint, Sierra speaks in a posh, vaguely British accent, her clothes emphasizing the outdated, colonial nature of her rhetoric. She then shifts course and suggests, confidentially, that Ivy might want to restrain and spank her, which, along with the head wound that Sierra earlier sported in this same costume, hints at the convoluted dynamics of her likely assignment. This scene is both funny and discomfiting, and it is another instance of drawing explicit attention to the Asianness of the actors. Later in the series, Ivy demonstrates her expertise and authority and eventually becomes a member of the inner circle working to take down Rossum. She also plays a crucial role in the *Dollhouse, Volume 1: Epitaphs* comics.[24] Again, this is an expansion and jostling, with humorous self-awareness, of how she is first presented. While Kilo and Ivy are recurring characters who play important roles in the series, Sierra/Priya's role is still more central and full of representational significance.

THE SHADOW TO THE LIGHT: THEMES OF SEX TRAFFICKING AND PROSTITUTION

If the serene spa-like set, *couture* costumes, and pretty Actives are the glossy exterior of *Dollhouse*, then the dark context that underlies all that surface flash is the question of whether an Active's contract really amounts to sexual slavery by another name. In the real world, human trafficking and sexual exploitation on a large scale are even trickier, more nefarious problems than they are usually depicted to be on-screen. Trafficking is not always so clear-cut and easily distinguished by sweaty middlemen with bad teeth as they can appear on *Law and Order: SVU*.[25] Whereas such depictions do portray a type of crime that is all too prevalent, much of what falls under the classification of "trafficking" is more blurry. The underground economy that buys and sells the young and helpless into slavery of all kinds does not stand on its own and cannot be explained simply in moral terms. Rather, it is thoroughly enmeshed with the global economy, with the inequities of extreme wealth and widespread poverty and with local economies; for example, southern Asian countries such as Thailand now rely on a lively tourist trade that depends, in part, on a flourishing sex industry. For example, a 2010 CNN report on EMPOWER, a sex workers' advocacy organization in Thailand, quoted its director as saying, "Much of the tourist industry is dependent on sex workers and [sex work] makes up around 7 percent of the [Thai] GDP—more than rice exports. The International Labour Organisation found sex workers send home US$300 million a year to rural areas, which is more than [that of] any government development project."[26]

EMPOWER speaks to the need that most sex workers have for decent working conditions and basic resources such as child care, yet sex trade–based economies also give way to the worst types of opportunism and exploitation, such as child prostitution and human trafficking. In southern Asia, much migration for sex work—voluntary and coerced—happens across the borders of neighboring nations. A 2012 report from the Tibetan Women's Association writes about "a growing problem of Tibetan women forced into marriage, domestic slavery, or prostitution in China."[27] In Nepal, financial hardship and limited work opportunities contrive to move thousands of children, women, and men across the open border to India to become sex workers, although they do not all go voluntarily and they do all make sacrifices.[28] The economies of southern Asia and the nations of the former Soviet Union have proved fertile grounds for an industry of sex work and migration, and the typical image of a twenty-first-century "mail-order bride" or sex tourism companion is of a woman or child from one of these regions. Fictional portrayals of characters from these regions who are involved in the global sex trades refer to this reality. On *Dollhouse*, one such portrayal is Victor's imprint as a very minor member of the Russian mob (Enver Gkokaj, himself, is half Albanian) who may or may not have information to offer Paul about trafficked girls. And, of course, another such portrayal is Sierra/Priya, a character who is kidnapped, raped, and forced into service in a brothel of sorts, albeit a very expensive and high-tech one.

The complicated politics of prostitution and human trafficking provide a series of not-so-subtle metaphors for the moral quandary at the heart of *Dollhouse*. In the unaired pilot "Echo," Paul first meets Echo when he thinks that he is busting a trafficking job to which Victor's mob underling imprint, Lubov, leads him. Although Paul's mission to find and save Caroline is ostensibly driven by his belief that she is being held against her will to service rich men, it is Sierra's story that more closely corresponds to the current narrative of sex trafficking, which usually involves more financially and legally vulnerable women than what most college students like Caroline are likely to be. By virtue of being half Asian, Dichen Lachmann physically resembles the large numbers of women moving through sex work networks in southern Asia. As well, Sierra/Priya's story is full of trafficking and assault. Priya is herself a migrant, though one guesses a voluntary one, as her Australian accent is clear in the flashback to her earlier encounter with Kinnard, and she acknowledges her vulnerable legal status to him, admitting that she does not have a work permit. Kinnard subsequently kidnaps, drugs, and imprisons her in Adelle DeWitt's house of sin, attempting, finally, to move her permanently into his own home, albeit without the knowledge of those at the L.A. Dollhouse itself.

Dollhouse evokes the sex trades in several ways, highlighting how morally and politically murky and multilayered the sex industry is, and it is not surprising that some viewers responded to Whedon's highly complex portrayal as antifeminist for not condemning the Dollhouse's sex engagements more harshly and immediately. I argue, however, that Whedon offers a more honest depiction of the spectrum of scenarios that compose the international sex industry, giving the viewer stories that illustrate a variety of trajectories into the Dollhouse. Echo agrees to sign her five-year contract under great duress but not exactly with a gun pointed to her head. The acutely shell-shocked Tony is simply assigned to be an Active and, as a good soldier, never questions his orders. The show does not make clear how much Tony understands about how, or even if, being an Active will cure his posttraumatic stress disorder before he signs his contract. Madeline—whose last name, appropriately, is Costley—is persuaded by Adelle that Active architecture will ease her overwhelming grief at the death of her daughter. Sam, from the episode "Echoes" (1.7), enters the Dollhouse to secure a good financial future for his mother and himself and as a consequence for his legal misconduct while working at the Rossum lab.

Sierra, however, is pressured by economic need (selling paintings on the boardwalk is seldom lucrative), duped by Kinnard's elaborate reception for her painting, coerced when she refuses his advances, and, finally, drugged and physically forced into a psychiatric hospital, from whence she is recruited into Active status by Topher and Adelle (although Kinnard certainly seems to engineer this transfer from behind the scenes). These five stories are very different in terms of the path that each character takes into the Dollhouse, but Sierra's story puts the lie to all the others. Adelle seems to believe her own line when she persuades Echo, Madeline, and Sam that she is offering them a chance to start over. Topher believes that he is helping Tony and Priya, curing their brains of posttraumatic stress disorder and schizophrenia, respectively. In contrast, Paul Ballard just assumes that the capabilities of the Dollhouse must inevitably lead to the exploitation of the vulnerable. In the end, Paul is proven right, and, as with those trafficked in real life, the Actives find that they cannot "seamlessly" return to their pre-Dollhouse lives as all of Rossum's recruitment rhetoric would have them believe. Each is fundamentally changed by his or her tenure as an Active.

For example, during his five years in the Dollhouse, Victor/Tony has been hired out for countless sexual "romance" engagements (only a couple of which the audience witnesses), has had his face horrendously slashed, and, unknown to Tony, has also fallen in love. While his mind has been tucked away within the Active architecture, Victor's laboring body has made enormous sums of money for Rossum's executives and shareholders and served their political purposes. The laboring sex worker, the laboring Active, and the laboring retail salesperson are all working to generate profits for a larger organization, profits of which the workers themselves will see only a small to negligible portion. While these systems are very different, they have important similarities, a point that *Dollhouse* presents through metaphor. For example, with respect to the laboring body and her or his sense of self, Tom Connelly and Shelley Rees suggest analyzing the exploitation of Actives through Karl Marx's critique of capitalism's diminishment of the lives of workers.[29] In their argument, the Actives are alienated from their labor in a hyperbolic instantiation of what Marx meant by the worker's estrangement from her or his own effort and its products, in that they cannot remember having worked at all, let alone the fruits of that labor. Connelly and Rees write, "Whedon incorporates representations of dialectical history to dramatize what Karl Marx identifies as 'alienation,' experienced by the Dolls/Actives through their compromised identities and confronted by all the show's characters through their relationship to historical time."[30] Understanding dialectical history and Marx's notion of alienation can help us understand, first, the particular kind of estrangement that the Actives experience—with their multiplied, sometimes disembodied consciousness—and, second, the hops in the *Dollhouse* timeline from present to postapocalyptic future to the pre-tech world that Topher's pulse yields.

About the significance of the "Self" with respect to time, they write,

> The diachronic, dialectical version [of the Self] places the Self on a continuum that runs from one's ancestors through her past and her present, pointing on toward descendants, *with all those experiences, influences, and moments contributing to a multivalent, dynamic conversation*. . . . The synchronic Self, on the other hand, is static; it's the "soul" that Paul Ballard references throughout the series. It stands as a constant, fixed snapshot in Time.[31]

Until Alpha's and Echo's composite events and their eventual "evolution" (as Echo calls it in "Epitaph Two" [2.13]) into identities that can incorporate multiple selves and the memories of each one, along with the original, "synchronic" self, the diachronic and synchronic selves are held resolutely apart by Dollhouse technology. There is a parallel between someone who agrees to be an Active, whose body is reunited with her or his consciousnesses at the end of the contract, and a prostitute whose mind might be strategically focused elsewhere, thinking through matters far away from the present circumstance, while her or his body goes through the motions of sex. In both situations, the labor that the individual does may well be of a piece with the experiences of her or his body but not necessarily with her or his identity. The client, however, might not know or care about this fragmentation.[32] Kinnard, therefore, is shocked to learn that Sierra/Priya—as a whole consciousness that has some continuity from Priya's pre-Dollhouse self, through all of Sierra's imprints, to Priya's post-Dollhouse self—is falling in love with Victor/Tony while only one of her imprints believes herself to love Kinnard.

Let me, for this discussion, leave aside the large sector of the sex industry comprising workers who enjoy the support of labor unions or advocacy organizations, who balance their time "on the clock" with families and other pursuits, integrating their histories with their present work lives and plans for the future. Let me focus, instead, on those sex workers who are coerced, deceived, and held hostage—those whose plights are meant to be referenced in the moral questions that *Dollhouse* raises—who experience a vacuum-like existence, concentrated around their present labor, cut off from their own pasts, and denied much sense of the future. Whiskey, for example, does not know who she was before she entered the Dollhouse, but Topher has rigged her thinking so that she is afraid to leave it. Her entire identity revolves around Rossum's goals for her, either within the Dollhouse or while living in Boyd's apartment and serving his strategic, emotional, and physical needs. Similarly, Victor wants to forget Tony's past at war in Afghanistan, and he is afraid of a future self without Active technology. In contrast, Sierra/Priya—who was unambiguously trafficked, having never signed consent documents and having no expectations of lavish compensation at the end of her contract—actually resolves Priya's terrible past by confronting Kinnard twice and then killing him in self-defense. Furthermore, she yearns for a future beyond the Dollhouse, taking, however, the alliances that she built while there with her, demonstrating the integrity of herself through time, as Connelly and Rees discuss in their article.[33] She strives for and achieves a continuity of "Self," recapturing something of the sunny warmth of the old Priya and creating a future with her family. Priya wants to move forward and agrees to return to the Dollhouse in "Epitaph Two" only to retain her memories of the past, especially those of her son. Her motif is the bird because she is aloft in motion, above the fray, and refuses to be a still, pretty thing in a cage, as others would have her be. As I argue about Kilo's character, *Dollhouse* purposely allows all the political and cultural associations that accompany an Asian female character whose narrative is full of sexual exploitation and trafficking, but the series also uses those associations to underscore the complex moral questions that drive its plot. Furthermore, the show ultimately resolves some of that moral uncertainty by not letting the type define who Priya becomes or how her arc ends on the series. The tragic figures who die in the Dollhouse—Whiskey and Topher—are both introduced, counterintuitively, as white professionals who are influential figures in the running of the L.A. 'House, not merely coglike, childlike workers in its machine.

According to Sherryl Vint, Marx primarily criticized capitalism for its diminishment of personhood and its degradation of social relations.[34] It is unexpected and uplifting that a character who is forced horribly into the Dollhouse, who is repeatedly sexually victimized, and who physically resembles some of those in our real world who are most likely to be trafficked is the one who emerges as not diminished but instead enhanced by how she makes sense of her time in the Dollhouse. Priya is not defined by the roles forced on her by Kinnard and Rossum. At the end of the series, Priya is ready, willing, and able to embrace and defend her relationships. As the tech heads' truck drives away from Safe Haven, Priya leans out to wave sadly to those she is leaving, having clearly put down roots and built alliances in this new home. Back on Venice Beach, too, Priya had a mutually helpful, affectionate friendship with her neighbor vendor. Priya is quick to reach out to others, and she forms lasting connections with them. Kinnard tells her in "Belonging" that he opted to have her be a regular Active rather than his exclusive Doll so that she would be reduced to servicing other clients. He means to debase her, to make her into a fully estranged body with no agentic being, a thing that Kinnard believes that he can actually own. Far from succumbing to that fate, Sierra falls in love with another Active. In so doing and in going on to prioritize her family—chosen and biological—she asserts her own desires, makes lasting and personal investments, and chooses and defends her own alliances.

Dr. Saunders/Whiskey turns out to the one Active whose life is entirely given over to being a Doll. More than simply not remembering who she was before arriving at the Dollhouse, she does not *want to know*; she is afraid that she will "die" if her body returns to the woman she was before signing Rossum's contract. Unlike Sierra, when Whiskey falls in "love," it is through Boyd's design, to serve his goals, and she forgets that love as soon as he imprints her with another personality, also to serve his goals. Her fate parallels Topher's in many ways. Through the episode "Vows," Dr. Saunders/Whiskey struggles to come to terms with the revelation that she is a Doll and a damaged one at that. In her struggle, she focuses on Topher, who programmed her to disagree with him, to even dislike the way that he smells, so that she could balance him in their assessments of the Actives. The audience sees Topher asleep twice in this episode, once napping on a cot just outside his control room and another time sleeping on a mattress in what seems to be a semipermanent space for him. This suggests that he rarely leaves the Dollhouse himself. This episode also reveals that Topher has programmed Dr. Saunders with agoraphobia to keep her from leaving the Dollhouse of her own accord. She does leave at the end of the episode, at Boyd's urging, but it turns out to be for his purposes. Whiskey and Topher both return to the Dollhouse to die, suggesting that they do not really have lives outside of it, that they have become defined by the Dollhouse, its economy, and its logic. Topher is even dependent on the Dollhouse for his personal life, programming Sierra as a fellow gamer for recreation and falling in love with his counterpart at the D.C. 'House. Whereas the reptilian (no offense meant to reptiles) Rossum higher-up Harding may ruin each new body with gluttony before transferring his mind into a new fit one in "Epitaph Two," it is Topher's brilliant and evolving mind that is irreparably damaged by the apocalypse that he helps Rossum manifest. Until the end of the television series, Topher is not so much an independent agent as a duped means to the ends designated by others. It is a role belied by the arrogance and privilege that he seems to embody through most of the show, casually condescending to Boyd, ordering Ivy around, and pettily protecting the domain of his work space and his minifridge while Rossum plays on his huge ego to trick him into building their weapons.

Sierra/Priya, however, turns out to be resilient, with a solid sense of self and a drive to set the terms for her own existence. Topher's initially arrogant, highly educated white male character is slowly destroyed by Rossum's scheming, but the trafficked and victimized Sierra/Priya ultimately survives the apocalypse, totally intact. She refutes the tech outright after "The Attic," sustains her close relationships, and, finally, "wins"—finding her happy ending with her family. When Victor disappears into his need to continue as a soldier, Priya remains at Safe Haven, raising their son, cultivating her garden, and creating a real home. The series ends with Priya still and again stuck in the Dollhouse, but this stay is meant to be voluntary and temporary, making way for a brighter future. While her Dollhouse history and the time immediately before it are dark, Whedon brings her character on a true arc and a dialectical one, in which she adjusts and develops through each experience and each significant interaction with another person, always growing. *Dollhouse* allows her to be supple and strong, to survive as an intact individual, and to move well beyond her abuse. This proposes a more potentially hopeful, empowering telling of a trafficked person's full story than to simply cast her as a victim whose life is destroyed by others who lust only after her body and the price that it can bring. We first see Sierra/Priya in the unaired pilot as the unsteady, wounded, pink-suited imprint with so little self-consciousness that she does not realize that her very own body is that of an Asian (the imprint would say "Oriental") person. We last see Priya returned to the Dollhouse to keep her identity and memories intact, nestled securely with her family, with her own terms for her life met. This is Whedonesque optimism.

CONCLUSION

Whedon takes the questions that drive the narrative of *Dollhouse* beyond the simple matters of good and evil or even the muddier issue of making good moral choices in difficult circumstances. *Dollhouse* tackles structural inequity and the abuses it spawns. In the season 1 featurette "Designing the Perfect Dollhouse," Whedon describes seeking a set that evoked "kind of an Eastern feel."[35] The generic term "Eastern" often holds the wide, nonspecific space that "Oriental" used to, connoting anywhere in Asia or the Middle East, anything exotic, spicy, and possibly from a former British, French, or Dutch colony. Whedon's use of the term suggests a linguistic sloppiness that might not bode well for Sierra's story. But maybe the harshness of Sierra/Priya's arc is a nod to the fact that women, specifically vulnerable women, still have more obstacles to overcome and that, with social support to balance out those structural injustices—and Sierra/Priya does receive support from a number of people, including Victor/Tony, Echo/Caroline, Adelle, Boyd, and Topher—they sometimes can surmount them.

Dollhouse was a hard sell for me initially. I found the way that it put actors' gym-defined bodies on display to be problematic. For example, the back shot of Echo in her dominatrix costume in "A Spy in the House of Love" is nothing short of gratuitous. I was troubled by the stereotyped presentation of Sierra and Ivy as Asian women who were constantly subject to the whims and demands of others, particularly white men. It irked me that a central couple of the show should be so seemingly heteronormative, with a physically strong and protective man who has a hard time settling into the domestic sphere that his highly feminine, whisper-thin female partner lovingly cultivates. But repeated viewings of *Dollhouse* revealed layers, nuances, and connections that thoroughly changed my perspective on the show, and I came to love the series for its complex portrayals.

I have tried to offer, here, that very skeptical first viewing along with the subsequent ones that more impressed and moved me. *Dollhouse* is a series that ended just as it was really finding its feet, but it made important contributions to the still small field of feminist television, to richer stories for characters of color or mixed race or ethnicity on the screen, and to the philosophical consideration of the politics of individual liberties, of social attachments, the misuse of power, and the commodification of bodies. These are not easy topics, and *Dollhouse* ultimately most impressed me in its refusal to offer easy answers to questions about them. What it did do is extend the scope for intelligent shows interrogating gender, sexuality, the family, and capitalism, and it added a great deal to Whedon's body of work presenting progressive, prodiversity, and emancipatory stories.

BIBLIOGRAPHY

American Anthropological Association. *Statement on "Race."* Arlington, VA: American Anthropological Association, 1998.

American Sociological Association. *The Importance of Collecting Data and Doing Social Scientific Research on Race*. Washington, D.C.: American Sociological Association, 2003.

Armstrong, Jennifer, and Margeaux Watson. "Diversity in Entertainment: Why Is TV So White?" *Entertainment Weekly*. June 13, 2008.

Buckman, Alyson. "'Didn't Get the Memo? Hero of the People Now': Hat Tricks and the Complication of Viewer Responses in the Works of Joss Whedon." Presented at the Slayage Conference on the Whedonverses 5, Vancouver, BC, July 2012.

Bussolini, Jeffrey. "Television Intertextuality after *Buffy*: Intertextuality of Casting and Constitutive Intertextuality." *Slayage: The Journal of the Whedon Studies Association* 10, no. 1 (2013).

Chambliss, Andrew, et al. *Dollhouse, Volume 1: Epitaphs*, #0-5. Created by Joss Whedon. Written by Andrew Chambliss, Jed Whedon, and Maurissa Tancharoen. Penned by Cliff Richards. Illustrated by Andy Owens. Milwaukee, WI: Dark Horse Comics, 2012.

Chan, Rebecca. *The Hyphen Project: Asian Americans and Alternative Media*. Wordpress, n.d.

Cheng, Joy, et al. *Media Representations of Asians*. University of Michigan, n.d.

Collins, Patricia Hill. *Black Feminist Thought: Knowledge, Consciousness, and the Politics of Empowerment.* New York: Routledge, 2000.

Connelly, Tom, and Shelley Rees. "Alienation and the Dialectics of History in Joss Whedon's *Dollhouse.*" *Slayage: The Journal of the Whedon Studies Association* 8, nos. 2–3 (2010).

"Countdown to *Dollhouse*: Joss Whedon on Women (and Sex!)." *Los Angeles Times.* February 6, 2009.

"*Dollhouse* and Joss Whedon's Commitment to Garbage." MrDestructo.com. February 17, 2009.

"The Expired Feminism of Joss Whedon." Overthinkingit.com. December 23, 2009.

Frohard-Dourlent, Hélène. "Somebody's Asian in *Dr. Horrible*: Humor and Racial Representations in the Whedonverse." Presented at the Slayage Conference on the Whedonverses 5, Vancouver, BC, July 14, 2012.

Garcia, Arturo R. "Revisiting *Dollhouse.*" Racialicious.com. April 7, 2009.

"Increase in Tibetan Women Trafficked to China." TibetanWomen.org. September 17, 2012.

Kirkland, Ewan. "The Caucasian Persuasion of *Buffy the Vampire Slayer.*" *Slayage: The Journal of the Whedon Studies Association* 5, no. 1 (2005).

Lim, Thea. "Joss Whedon and the Blurry Line between Homage and Appreciation." Racialicious.com. February 16, 2009.

Perdigao, Lisa. "'This One's Broken': Rebuilding Whedonbots and Reprogramming the Whedonverse." *Slayage: The Journal of the Whedon Studies Association* 8, nos. 2–3 (2010).

Short, Philippa. "25 Years in Thailand's Sex Industry." CNN.com. August 3, 2010.

FILMOGRAPHY

"Debt." *Law and Order: Special Victims Unit.* Season 6, episode 2. Written by Dick Wolf and Amanda Green. Directed by David Platt. Universal Studios, 2004.

"Designing the Perfect Dollhouse." *Dollhouse: Season 1.* Directed by Joss Whedon. Featurette. Twentieth Century Fox, 2009.

"Making Dollhouse." *Dollhouse: Season 1.* Directed by Joss Whedon. Featurette. Twentieth Century Fox, 2009.

The Making of Miss Saigon. Directed by David Wright. A&E Home Video, 2006.

"An Open Secret: Sex Trafficking in Nepal." GlobalPressInitiative.org. March 10, 2010.

Puccini: Madama Butterfly. Directed by Anthony Minghella and Carolyn Choa, 1904. Sony Classics, 2011.

Slumdog Millionaire. Written by Simon Beaufoy. Directed by Danny Boyle. Twentieth Century Fox, 2009.

Tancharoen, Maurissa, perf. "Nobody's Asian in the Movies." *Dr. Horrible's Sing-Along Blog. Commentary! The Musical.* Written by Maurissa Tancharoen and Jed Whedon. Mutant Enemy, 2008.

NOTES

1. In her talk "'Didn't Get the Memo? Hero of the People Now': Hat Tricks and the Complication of Viewer Responses in the Works of Joss Whedon," Alyson Buckman suggested that the collective Whedonverses might compose a singular, unified "Whedonverse," which features shared messages and themes. In her talk, she cited Jeffrey Bussolini's work on "intertextuality of casting" (2013), whereby actors such as Eliza Dushku, Fran Kranz, and Enver Gjokaj appear in multiple Whedon productions. I use Whedonvers*es* to indicate the different worlds that Whedon builds and the unitary Whedonverse to consider especially Whedonian themes.

2. Frohard-Dourlent, "Somebody's Asian."

3. Frohard-Dourlent, "Somebody's Asian."

4. Armstrong and Watson, "Diversity in Entertainment."

5. Armstrong and Watson, "Diversity in Entertainment."
6. Armstrong and Watson, "Diversity in Entertainment."
7. Armstrong and Watson reported that as of 2008, it was Gen Y shows aimed at teenage audiences that were most likely to feature diverse casts and interracial friendships or relationships in a casual, unproblematized way. As they explain this, younger viewers grew up in a world already so racially diverse as to expect that reality to be reflected on shows that purport to represent some version of their lives.
8. Armstrong and Watson, "Diversity in Entertainment."
9. See Thea Lim and Ewan Kirkland for more discussion of this issue.
10. "Countdown to *Dollhouse*."
11. "Countdown to *Dollhouse*."
12. "Countdown to *Dollhouse*."
13. "Making Dollhouse."
14. Garcia, "Revisiting *Dollhouse*."
15. It is important to note that race and ethnicity are not biological but lived realities and therefore deeply contextual to people's experiences and social spheres. They determine people's perceptions of groups to and with which they belong, do not belong, have allegiance, and/or have conflicts, and they affect people's access to resources and opportunities. Race and ethnicity are tremendously important but defy definition, and a primary challenge for anyone studying them is how to meaningfully address these issues without reifying them into something that they do not mean or do not *always* mean. What race and ethnicity mean is made still more complex by politics. So the sovereign right of Tibet to resist its unlawful occupation by China could be associated metaphorically with the sovereign right of the character played by the half-Tibetan actress Dichen Lachmann to resist her unlawful captivity by Kinnard and by Rossum. Of course, really delving into that issue is a much larger task than what space here allows. For fuller discussion of my assertion that race is social and political rather than biological, see the American Anthropological Association's statement on their official position on "race" (1998) or the American Sociological Association's statement on the need to study race as a social and political category and an official organizational position that it should not be considered a biological category (2003).
16. This list is not exhaustive, but here are some more examples of experiences or identities that contribute to diversity in the cast: Enver Gjokaj and Eliza Dushku are both half-Albanian, and Dushku is an Albanian, as well as a U.S., citizen; Tahmoh Penikett is Canadian, and his mother is a member of the White River First Nation; and Olivia Williams is English, married to an African American man, and the mother of two biracial children. My point is only that racial, ethnic, and national diversity is often subtle and ever present but that it is also necessary to simply represent more—and more complex—characters of color on-screen.
17. Armstrong and Watson, "Diversity in Entertainment."
18. See, for example, "*Dollhouse* and Joss" and "Expired."
19. Collins, *Black Feminist Thought.*
20. "Making Dollhouse."
21. For fuller discussions of this topic, see Chan and Cheng et al.
22. Tancharoen, "Nobody's Asian"; Frohard-Dourlent, "Somebody's Asian."
23. Tancharoen, "Nobody's Asian."
24. Chambliss, *Dollhouse, Volume 1: Epitaphs.*
25. See, for example, "Debt," 6.2.
26. Short, "25 Years."
27. "Increase in Tibetan Women."
28. See "Open Secret."
29. Connelly and Rees, "Alienation and the Dialectics."
30. Connelly and Rees, "Alienation and the Dialectics."
31. Connelly and Rees, "Alienation and the Dialectics" (emphasis mine).
32. See Perdigao for further discussion of fragmentation on *Dollhouse.*
33. Connelly and Rees, "Alienation and the Dialectics."
34. Cited in Connelly and Rees, "Alienation and the Dialectics."
35. "Designing the Perfect Dollhouse.

Chapter Six

"In My House and Therefore in My Care"

Transgressive Mothering, Abuse, and Embodiment

Samira Nadkarni

Dollhouse's basic premise of Cartesian dualism introduces a postmodern world in which the links between individual minds and bodies are simultaneously sidelined and underlined. The show plays with the notion of memory, constantly stripping and reaffirming its viability as a source of "authentic" identity or experience. Bronwen Calvert notes that "it is possible to trace two distinct, contradictory viewpoints in the narrative of *Dollhouse:* mind over body versus body over mind," using multiple examples to emphasize the fact that, despite the supposed wiping process, some core memory or identity remains with the body.[1] Indeed, this is not the first time that Joss Whedon's work has led the viewer to question memory as a basis for identity; as Sherry Ginn notes, this theme of (re)working identity can be traced back to his work on *Buffy the Vampire Slayer*. Whedon's canon offers numerous examples of wiping certain memories and (re)inserting others to play with identity, such as the *Buffy* episodes "Halloween" (2.6) and "Tabula Rasa" (6.8), as well as Dawn's magical emergence in season 5; Cordelia's visions being inserted into her mind by the Powers That Be, Gunn's magical mental transition into the perfect lawyer, and the replacement of Fred's consciousness with that of demon Illyria on the series *Angel*; and River's fractured/augmented mental state in *Firefly* and its role in producing her nonnative memory of Miranda in *Serenity*. This play with memories and personas reaches its peak in *Dollhouse*, where every episode involves at least one Active being imprinted and/or wiped, taking any assertion of a fluidity of identity politics to its extreme.

However, Calvert and Ginn both note that, despite this exaggerated performance, the show emphasizes the presence of some form of core identity that remains embodied.[2] The Actives in their resting state are able to function, though they are stripped of any sense of personality and self-awareness that might suggest maturity or independence. Topher, in a typical rejection of sentiment, views the Actives as beings reduced to instinct; his presumptions are based on the fact that the Dolls exhibit largely submissive behavior that speaks of herd mentality and ingrained habit rather than any genuine understanding ("Gray Hour," 1.4). Adelle DeWitt's pitch to a client in "The Target" (1.2) suggests another means by which to read this return to a more instinctual self, associating the Actives with the innocence and vulnerability of children. Between this speech and the general behavior of the Actives in their wiped states, it seems clear that the Actives—despite their mature physical attributes clearly aimed to appeal to a variety of tastes—are emotionally immature: innocent, vulnerable, and unable to defend themselves against a real or perceived threat. Adelle's comparison of the Actives to children who, for the most part, evolve toward independence in the course of the series thus enters *Dollhouse* into a conversation that explores issues of child abuse, albeit with physically mature characters.

In doing so, the show presents itself as part of an ongoing narrative within the Whedonverse that draws links between a certain type of mothering and sexual abuse, what I term "transgressive mothering." The use of this term draws on Sara Ruddick's theory of mothering, defined in her book *Maternal Thinking* as encompassing the three demands of children upon their caregivers: preservation, growth, and social responsibility. Ruddick states,

> In this sense of demand, children "demand" that their lives be preserved and their growth fostered. In addition, the primary social groups with which a mother is identified, whether by force, kinship, or choice, demand that she raise her children in a manner acceptable to them. These three demands—for *preservation, growth,* and *social acceptability*—constitute maternal work; to be a mother is to be committed to meeting these demands by works of preservative love, nurturance, and training.[3]

Building on this, I use the term "transgressive mothering" to denote a particular gender-neutral portrayal of maternal work that, while outwardly displayed as socially acceptable, often transgresses these boundaries in ways that connote an abuse of the authority granted by the act of mothering itself. *Dollhouse*'s engagement with the issue of child abuse thus opens onto a larger narrative of transgressive mothering that has begun to reveal itself within the Whedonverse, wherein the act of mothering is transgressed by the child figure's sexual abuse by the maternal figure.

Arguably, *Dollhouse*'s employment of physically mature characters that appear regressed is one of the only ways in which the issue of child abuse and transgressive mothering could be approached on mainstream network television. It is easy to foresee issues regarding the use of child actors in the scenarios that *Dollhouse* chooses to explore, particularly given its emphasis on the explicitly sexual nature of the Dolls for the majority of the series.[4] There is no doubt that if *Dollhouse* were to depict child actors in these scenes, Fox would in all likelihood have suffered large losses in viewership and drawn heavy fire from the censorship boards. Indeed, even with the use of mature actors, Fox had numerous issues with the content under discussion and, as Whedon states in his interview with Maureen Ryan after the show's cancellation, "pretty much wanted to back away from the concept five minutes after they bought it."[5]

The volatility of *Dollhouse*'s quite explicitly themed content is therefore already played down to a large extent throughout its narrative.[6] Child actors in the show are almost always depicted as removed from the situations wherein they might have been sexually harmed; for example, the viewer is informed that the young girl Susan in the episode "Briar Rose" (1.11) was molested but is introduced to her in the aftermath of the abuse, when Echo is sent to provide a form of counseling. The viewer is presented only with a verbal summary of the events, rather than an infinitely more inflammatory visual flashback. Similarly, in "Ghost" (1.1), Echo's composite persona of hostage negotiator Eleanor Penn contains a personality who was molested as a child and is forced to face her molester (in a manner of speaking).[7] The viewer is never directly confronted with this persona's visual recollection of her abuse, only a version of the events through Topher's recounting and Eleanor/Echo's reactions.

The problem of child actors aside, it still is made clear at the outset of the show that the Dollhouse is largely a playground for wealthy or influential predators. Whedon himself acknowledges this element in his commentary for "Man on the Street" (1.6), stating that Sierra's abuse demonstrates everything that is wrong with a system of this nature,[8] leading Eve Bennett to note that "the system *in practice* is unacceptable . . . because the Dolls are being exploited."[9] In the episode, the viewer watches helplessly as Sierra is molested by her Handler, Hearn. Sierra's terrified reactions and childlike incomprehension make this incident particularly hard to watch as all the conventional signs of child abuse play themselves out on-screen. Sierra's choice to isolate herself, her screaming reaction to Victor's unexpected touch, her bleak acceptance of this ongoing abuse at the hands of her primary authority figure, and Echo's observation that she hears Sierra crying alone in her bed at night all underlie the latter's trauma and current inability to extract herself from the situation. The fact that it is Sierra's Handler who is abusing her—someone whom the Dollhouse's bonding process has forced her to trust with her safety and well-being—paints an all-too-familiar (some might say incestuous) picture. In fact, Hearn's use of the bonding process in his prequel to the rape itself simply underlies his status as de facto authority figure and her enforced compliance: Sierra is made to repeat her trust of Hearn (per the bonding process), and her abuse is positioned as a silent game in which she is forced to partake despite her wishes to the contrary. Hearn's description of the abuse as a "game" they play together further highlights his awareness of her childlike mental status and his choice to take advantage of this disparity.

The characters on-screen (and presumably the viewer as well) react with anger and repulsion when they discover the extent of Hearn's betrayal. Not only does Boyd punch Hearn through a glass wall upon discovering this hidden abuse, but Adelle goes so far as to send him to his death at November's hands, essentially executing him for his crime. The reactions engendered by Sierra's abuse among the administrators of the Los Angeles Dollhouse provide some form of catharsis for the viewer in the aftermath of these events. The viewer seemingly is assured implicitly that any Doll, specifically Sierra, in her or his wiped state will be safe, cared for, and protected as a child would be. She or he will not be placed in further positions wherein she or he will be unable to refuse unwanted advances.

Yet this implicit promise is never fulfilled; it remains simply another fantasy within the Dollhouse. The fact remains that this is what the Dollhouse sells—the simulacrum of perpetual consent. Sierra's rape in her childlike state is repulsive in the extreme because she is incapable of processing what happens to her; she is unable to say no. But the viewer remains aware that this is also what occurs in her imprinted state. She is not imprinted with the ability to make a choice and refuse—that is not the fantasy (unless it is). Despite the Handlers overseeing the Dolls' safety and preservation during their engagements, it is made clear that their aim is to minimize damage rather than prevent the abusive scenarios from occurring altogether. This aspect is introduced early on and reinforced in the course of the series; Echo is repeatedly harmed during the course of her engagements, this aspect occasionally being paired with a Handler presence shown to be ineffective (e.g., Boyd's presence in "The Target" and the anonymous Handler who misreads her brain patterns in "Vows" [2.1]). The onus of Echo's safety during an engagement is thus largely on her own shoulders, with her Handlers providing limited protection. This aspect of allowable damage is repeated with Victor, who, in his guise as Lubov ("Ghost"), is placed in a situation in which he is physically threatened by Paul Ballard without any action taken by his Handler, and Sierra, whom we see having sustained a wound to her head in the aftermath of an engagement in "Vows"; the latter's wound is never explained in the episode, suggesting that such wounds are commonplace and unremarkable. This constant reiteration of seeming protection and its transgression in terms of physical abuse echoes the Dollhouse's stance on sex: that, within specific parameters, everything is possible. In essence, this is also Hearn's defense to DeWitt and Mr. Dominic: that theirs is a business of "using people," and so neither DeWitt nor Dominic can claim the moral high ground because they are also exploiting the Actives' inability to comprehend or act against the situation. Hearn's response suggests that the abuse was inevitable because the Dolls lack both memory and power; an ability to fight back would not necessarily preclude abuse, but its lack did make the occurrence easier.

The viewer, horrified by Hearn, is reassured by Adelle's claim that he fails to understand the extent of the Dollhouse's purposes (reminding the viewer of her repeated claim that the Dollhouse helps people). The events are set aside with the implicit assumption that taking advantage of an imprinted Active with the illusion of free will is in some way less reprehensible that doing so with a Doll in his or her wiped state, even if this fails to stand up under a closer inspection. The viewer's complicity in the ideology that allows for the Dollhouse to exist is simply another layer of its metanarrative, since the show's commentary extends far beyond the bounds of the entertainment industry to the society that demands and consumes its products. In the episodes that follow after "Man on the Street," such as "Echoes" (1.7) and "Needs" (1.8), we see the Actives relive past traumas and then seek and receive what might be considered a form of resolution. DeWitt's theorizing that what the Dollhouse caters to goes beyond simple wants to what is needed—by both clients and Dolls—is emphasized here, and the show returns to its baseline, ambiguous moral territory. The viewer is now conditioned to distrust Hearn and, thanks to her extreme reaction at the abuse of an Active in her care, to trust in Adelle's protection. Yet this is abruptly undermined as the viewer watches another set of events play themselves out in a distinct parallel to Hearn's abuse of Sierra, except this time between Adelle—the Dolls' supposed protector—and Victor.

The episode "A Spy in the House of Love" (1.9) reveals Adelle to be the "Miss Lonelyhearts" of Victor's early romantic engagements. Interestingly, Martin Shuster and Eve Bennett (who build on Shuster's interpretation of these events) both suggest that DeWitt's sexual encounters with the imprinted Victor in this episode can be viewed alongside Topher's playful birthday celebration with the imprinted Sierra in "Haunted" (1.10) as touching events that add depth to otherwise emotionally isolated characters.[10] In fact, Bennett goes further to infer that these events fold into Whedon's larger ideas for *Dollhouse* itself before Fox's interference. She argues that these events (and others like these) were intended to complicate any strict moral guidelines by which to judge the narrative; not only would the Dolls be presented as objects of exploitation, but this exploitative aspect would be complicated by the Dolls being used to provide otherwise unavailable emotional assistance (much like the events of "Briar Rose"). Thus, the Dolls would be making available the emotional connection that certain clients might need, rather than the transient physical satisfaction that they might want, and this would occur irrespective of the Actives' own ability to consent to the events in question.[11]

While Shuster and Bennett's point about these events adding depth to these characters could be argued, their interpretation of these events as touching is incredibly problematic, particularly in DeWitt's case. Given (1) that Adelle chooses to define the Actives as children in her care, underlining her own supposedly protective stance; (2) that she takes upon herself an authoritarian role, referring to the Los Angeles Dollhouse as "her House" and the Actives there as charges in her care ("Omega," 1.12; "Belonging," 2.4); and (3) that her role in the series is almost consistently portrayed as that of a matriarch, her choice to program Victor, an Active in her care, for a sexual encounter borders on the unconscionable. This is perhaps most notably underlined when, during their tryst in "A Spy in the House of Love," Adelle tells the imprinted Victor/Roger that he has to trust her, and he responds that he does, completely, this statement evoking images of the Doll-Handler relationship and further paralleling Hearn's abuse of Sierra. Adelle's growing discomfort with her relationship with Victor/Roger does not stop her from using him for the length of their engagement, and, despite her eventual choice to cease this liaison at the episode's conclusion, the point remains that the engagement was one among many, as is evidenced by the crude familiarity of the Handlers with the name "Miss Lonelyhearts." The fact that an abuse of authority of this nature is a common occurrence is emphasized in the episode "Belonging," wherein DeWitt's superior, Matthew Harding, reveals that he has been aware of her liaisons with Victor all along, stating, "We don't care; everyone likes to take a little something home from the office once in a while." Her attempt to resume this liaison in the later episode "Stop Loss" (2.9), despite her continued reservations, shows that her conduct is part of a larger pattern of predatory behavior: she abuses her authority to control Victor whenever she feels out of control elsewhere.[12]

Victor eventually is able to free himself from her abuse in the episode "Stop-Loss" due to the embodied love that he feels for Sierra/Priya, a love that transcends his lack of memory and fluid identity. DeWitt's disbelief at Victor/Roger's refusal to renew their relationship is extreme and displays what would occur should the Actives ever choose to exercise their rights to consent: she reduces their interaction to its basic transactional nature, insisting that he is simply an imprint and cannot refuse despite his having just done so. She also intimates that he is more animal than human being since he is to be "released into the wild" the next day upon completion of his contract, implying that her consideration for the Actives in her care, her view of them as children to be protected, extends only so far as the length of their contract and outside of their engagements. Her actions display not only an abuse of her authority in the use of Victor's body for these liaisons but also a lack of respect for the imprinted personality of Roger, her supposed equal, and his supposed right to consent, as well as a lack of respect for Tony, the body's original inhabitant.[13]

Whereas some might suggest that DeWitt's actions are somehow mitigated by the fact that Victor was imprinted at the time with a distinctly adult persona while Sierra was still in her wiped, childlike state during her repeated sexual and psychological abuse by Hearn, the underlying similarities cannot be ignored—an authority figure who was supposed to protect someone in her or his care has taken advantage of that person. Any sympathy being elicited in these moments is sympathy for someone who has taken advantage of his or her position to create and play out a fantasy that is not about the other person at all and is distinctly paraphilic.[14] If the viewer feels any sympathy for Adelle in these moments—feels in any manner that these events can be condoned by paralleling the situation with Topher's choice to imprint Sierra to celebrate his birthday with the platonic comfort of a friend—then he or she must accept that these events also parallel Hearn's rape of Sierra in "Man on the Street" and Sierra's continued abuse at the hands of Dr. Nolan Kinnard in "Belonging."[15] In fact, these events are strongly linked in the narrative, for it is Harding's knowledge of DeWitt's transgressions that allows him to force her to comply with his demand that Sierra be released to "a raping scumbag one tick shy of a murderer" in spite of her own reservations.

Adelle's role as a matriarch is presented not simply by her own assertions—that the Los Angeles Dollhouse is her House or that the Actives are in her care—but also due to the fact that she outwardly appears to follow Ruddick's outline of a mother figure: she ensures the preservation, growth, and social acceptability of the Actives while they are placed within her care.[16] Although her role in allowing the Actives' engagements (physically and emotionally taxing as they are, whether sexual or otherwise) might be morally questionable, it is possible to suggest that this still falls within Ruddick's parameters, as DeWitt's behavior would conform to the expectations of the primary social groups with which she identifies—that is, primarily figures already involved in the Rossum Corporation and the clients who request engagements with the Dolls. The Dolls are, to her knowledge, assured of economic, physical, and psychological well-being in the aftermath of her care. As such, Adelle's transgressive maternity presents itself due to the fact that her actions in public appear to conform to expectations regarding the preservation, growth, and social acceptance of the Actives, while allowing for her hidden transgressions as Catherine/Miss Lonelyhearts, which are kept physically and nominally separate from her role as the matriarch of the Los Angeles Dollhouse.

Problematically, Adelle is never confronted with her abuse in the course of the series, other than by Harding, who excuses it as commonplace within a Dollhouse while using his knowledge of it to manipulate her, and Topher, who uses this knowledge to mock her briefly but also seems to excuse this action as mundane, if unlikely, with regard to DeWitt. Thus, although she is forced to atone for her actions as a bastion of the Los Angeles Dollhouse in the course of the show, her transgressive maternity appears to have no repercussions at all; even Harding's attempts to blackmail her into giving Sierra/Priya to Kinnard are foiled. More disturbingly, like a number of Whedon's darker heroes, she is provided with an element of redemption that is specifically linked to her role as a maternal figure and without the mitigating circumstances that we have come in some way to expect: Angelus did what he did because he was a vampire without a soul; Faith accidentally stakes and kills someone and then denies her own accountability in this action before she is, in part, manipulated by the Mayor;[17] Willow goes dark because Tara is killed;[18] and so on. In each of these cases, redemption was consistently depicted as one where the abuser was repeatedly confronted by his or her past transgressions and made to atone for them over and over. This is not the case when it comes to DeWitt's choice to abuse Victor. In *Dollhouse*, DeWitt's actions occur because she was lonely, because she needed to exert a sense of control, and/or perhaps simply because she could. Her actions are not provided with any sort of extenuating circumstances, possibly because no form of mitigation could make this behavior palatable. Her matriarchal stance in the show's two-part finale remains unopposed by any knowledge of her transgression, and her behavior is even lauded as a "class act" by Zone. Her mothering at this time encompasses not only Topher and the remaining Actuals but also the actual children, underscoring not simply her own maternal side but its current depiction as redemptive, suggesting that her role as maternal figure matures alongside the evolving Actives and rebellious Actuals in her care to allow for independence in her charges. At the close of the series, she is shown standing tall while the world wakes around her and Zone, gently touching Iris's face in a clearly maternal gesture, playing shepherd to this new flock (as per her earlier conversation with Echo in "Epitaph Two: Return" [2.13]).

The choice to emphasize Adelle's maternity in the show's finale avoids addressing any sense of real resolution with regard to her transgressions: Adelle was and still is a maternal figure, albeit one that appears more sympathetic; she was surrounded by people in her care, then and now; and she was lonely then and, with the death of Topher—her primary charge and one of the few people who seemed to understand her—is now more alone than ever. The viewer is not given to believe that she is in a relationship. If the reasons for her predatory behavior were based on these original conditions, as the series would have us believe, then nothing has changed. If anything, her role as a shepherd "leading them into the light" simply places her in another position wherein an abuse of authority is a possibility; the symptoms of her various transgressions may no longer be present (i.e., she no longer works for

Rossum and does not have the assumption of absolute authority, nor does she seek emotional refuge from her actions in either drink or encounters with "Roger"), and the cause itself remains unexplored and, in all likelihood, unresolved. And this lack of resolution then feeds into a larger issue within the Whedonverse—namely, confrontation with and independence from particular forms of authority.

As previously stated, Adelle is not the only character in the Whedonverse to enact a transgressive mothering that remains unresolved on the part of the abuser, with the victims achieving a form of resolution through some form of embodied memory. For example, Maggie Walsh's character in season 4 of *Buffy the Vampire Slayer* is also concerned primarily with the initiative soldiers in her care, with her caregiving finding a special focus in Riley.[19] Although she does not depict the same paraphilic traits as Adelle, her mothering is similarly transgressive in its emphasis on authority and control of both Riley and her lab at the Initiative (i.e., "her House"). In fact, her role as a maternal figure is explicitly stated in "Goodbye Iowa" (4.14) by Adam, who hails Riley as a brother since both of them had been created with specific focus by Dr. Walsh: to be Initiative supersoldiers through body modification (the repercussions of which are evident in "Goodbye Iowa" and "Out of My Mind" [5.4]). Maggie Walsh's transgressive maternity is tied to her role within the Initiative: her monitoring of Riley's home appears to be part of the surveillance afforded by the Initiative (as seen in "The I in Team" [4.13]); his cover identity when on campus is that of a teaching assistant for her class; and she controls the changes to his body chemistry as the chief scientist at the Initiative laboratory. Her maternal work of preservation, growth, and social responsibility is carried out within this overlap of caregiver and scientist, and, much like Adelle, the overlap of these roles also allows for her to abuse this authority while keeping this abuse hidden. Her decision to use the Initiative's surveillance to spy on Riley and Buffy's sexual intimacy, as well as her use of its military excursions as a pretext to eliminate Buffy from Riley's life, shows an abuse of authority linked to her dual roles as Initiative scientist and de facto caregiver. Riley's independence eventually emerges as he removes himself from the dual arenas under Dr. Walsh's purview: his unexpected viewing of Buffy's one-sided confrontation with Dr. Walsh fractures the trust that Dr. Walsh had engendered in him; Dr. Walsh's eventual death and body modification by Adam end her ability to emotionally manipulate him or play voyeur; and, finally, perhaps most important, his rejection of and rehabilitation from the Initiative's medication all play a hand in gradually removing him from her systemic control. However, the effects of her abuse persist far past her death and the destruction of the Initiative facility: Riley's ingrained notion of himself as primarily a soldier, his subsequent reckless behavior caused by the shutdown of his pain receptors and withdrawal from the drugs that Dr. Walsh was secretly feeding him, and his paranoia regarding the military can all be traced back, at least in part, to her abuse and the breach of his trust. His abrupt departure from the show to resume his role as a soldier fails to provide the viewer with much resolution with regard to this abuse,[20] and the single episode in which he appears following this departure reveals him to be happily married,[21] suggesting his attempts to deal with the aftermath of this abuse ended off-screen, albeit happily. The issue of Dr. Walsh's transgressive mothering is thus presented to the viewer, and its traumatic aftermath is made visible in Riley's actions, but the subsequent actions taken by Riley to come to terms with the memory of his abuse and abuser remain unresolved on the part of the viewer, sidelined and soon forgotten.

The transgressive maternity displayed by Maggie Walsh and Adelle DeWitt is further paralleled in the character of Cordelia Chase in season 4 of *Buffy*'s spin-off series *Angel.* Cordelia acts in a manner that is distinctly maternal with regard to Angel's son Connor[22] and, despite the moral delineations of this role, has sex with him supposedly to give him "something that's real."[23] This act then both underlines her role as maternal figure (emphasized by her choice to call him "baby" and speak of his short life) and transgresses what would be appropriate to this role. Her choice to have sex with Connor while he is in her care and under her authority is a clear breach of the trust engendered by the act of mothering but also only comes about through this relation. Connor himself does not initiate the act; in fact, he questions it, and Cordelia's response then distinctly emphasizes her authority as a figure of mothering, tying this to his wants and needs rather than her own. The Cordelia that enacts this is supposedly no longer the Cordelia that the viewer knew and is retrospectively recognized as Jasmine, an ancient being that manipulated events to take control of Cordelia's body and ensure her own birth into a new vessel, that of Cordelia's child. Jasmine's control of Cordelia's body while Cordelia remained aware yet unable to intercede allows her to seduce Connor to ensure that Cordelia is impregnated. Cordelia (while controlled by Jasmine) thus gives birth to Jasmine in a new form before she subsequently falls into a magical coma.

Cordelia's transgressive maternity is thus far more complex than that displayed by Maggie Walsh and Adelle DeWitt since she is simultaneously both victim and abuser. Her own persona is tampered with (like that of the Actives) to allow for the potentiality of her monstrous motherhood,[24] and freedom from this abuse arrives in the form of a final out-of-body experience wherein she might make reparations and reclaim her own identity as a guiding force within Angel Investigations. Cordelia herself suggests that her presence within her own body, despite her inability to intervene in the events that unfolded, leaves her feeling culpable and with a wish to atone. However, while her final return in "You're Welcome" (5.12) resolves a number of issues with regard to her actions while under Jasmine's control (regardless of the level of her own culpability in these events)—her apology to Wesley for the death of Lilah being one of them—she is never confronted with Connor or with the results of her transgressive maternity despite her wish to encounter him.

Connor experiences numerous betrayals by the authority figures in his life: he is largely manipulated and emotionally abused by Daniel Holtz, is taken advantage of by Cordelia/Jasmine in her role as maternal figure, and has his memories wiped and his reality remade (supposedly for his own good) by Angel. None of these events allow for Connor's own control or independence. Indeed, he does not achieve resolution or independence until after Cyrus Vail's spell is shattered by Wesley and his own memories begin to reemerge.[25] Connor's identity then resolves itself as somewhere between his two personas: the angry, demon-killing son of Angel and the "normal" son of the Reilly family. Much like Echo, this composite personality finally allows him the ability to make independent decisions outside these previous abusive authorities. He then freely chooses both to return to his previous life with the Reillys and to take a stand in the final battle against Wolfram and Hart. However, no mention is made of his conflicted memories regarding his time with Cordelia or the pseudo-incestuous relationship that gave rise to Jasmine, though there is the underlying assumption that Connor has come to terms with the events themselves.

The narrative of survival traced within the Whedonverse is thus one that lays emphasis on the reclamation of aspects of one's identity as a means by which to free oneself from abuse. For example, Riley's eventual return in *Buffy the Vampire Slayer* suggests that he has once again come to terms with being a soldier despite Maggie Walsh betraying him and attempting to turn him into a supersoldier without his knowledge or consent. Similarly, in *Angel*, Connor returns to reclaim his role as a warrior while divesting it of its associations with regard to his family and the abuse inflicted on him—that is, his paternity, his affair with Cordelia/Jasmine and Jasmine herself. *Dollhouse* continues this trend as Sierra, Victor, and Echo are each provided with an episode in the narrative arc wherein they remove themselves from situations of long-standing abuse. In "Belonging," the focus is on Sierra, who signals through her dark, moody drawings that something is terribly wrong. Both Sierra's deeply embedded memory of her repeated abuse at the hands of Dr. Kinnard and Echo's growing ability and self-awareness produce an unexpected result—Topher disobeys his official orders and sends the reimprinted Priya to confront and deal with her rapist. As previously discussed, Victor's embodied memories, specifically his love for Sierra, allow him to remove himself from his exploitative liaison with DeWitt, and it is this same embodied love that allows Priya to save him when he is recruited into project "Mind Whisper" later in the same episode ("Stop-Loss"). Finally, as the series shows, Echo's growing self-awareness makes issues of consent a notably sticky point; she appears to reach a stage at which she is capable of making choices, even if these choices are largely impacted by the situations that she is in at the time.

It appears that Adelle DeWitt's transgressive mothering lies somewhere between Cordelia's being overtaken by forces beyond her control and Maggie Walsh's deliberate misuse of Initiative authority, but her redemption occurs without any of the punishments or attempts to atone that the other narratives showed. What does this suggest for the ongoing and sustained narrative within the Whedonverse that not only appears to link transgressive mothering to a growing lack of confrontation between victims and abusers but increasingly links these abusers with the possibility of a redemption that fails to acknowledge this abuse or whitewashes it on the basis of its being a contextual situation? It remains that in each of the cases discussed, the authority provided by mothering sustained the relation, engendering affection and gratitude, while providing the conditions that allowed for the abuse to occur. The resolution or independence of the survivors of this abuse through the persistence of embodied memories does not suggest the abuser's own redemption, particularly given the fact that they continue, in most cases, to retain their authority as providers of mothering. Perhaps this is part of Whedon's greater narrative, depressing as it might seem: the assumption that systemic abuse can and possibly will continue under the apparently resolved façade that our systems of thought almost make it inevitable. The narrative being traced here is not simply one of transgressive mothering in the Whedonverse but also one of our own complicity in refusing to recognize its presence and choosing to ignore these tricky gray areas. As Rhonda Wilcox notes, "to reach forgiveness, we must first know our guilt."[26] The length of this narrative suggests that these events can no longer be viewed as isolated instances, cannot be chalked up merely to the assumption of context. This pattern of behavior cannot be left unquestioned or unmentioned anymore, despite our discomfort with it.

BIBLIOGRAPHY

Bennett, Eve. "Deconstructing the Dream Factory: Personal Fantasy and Corporate Manipulation in Joss Whedon's *Dollhouse*." *Slayage: The Journal of the Whedon Studies Association* 8, nos. 2–3 (2010).

Calvert, Bronwen. "Mind, Body, Imprint: Cyberpunk Echoes in the Dollhouse." *Slayage: The Journal of the Whedon Studies Association* 8, nos. 2–3 (2010).

Giannini, Erin. E-mail. December 28, 2012.

Ginn, Sherry. "Memory, Mind, and Mayhem: Neurological Tampering and Manipulation in *Dollhouse*." *Slayage: The Journal of the Whedon Studies Association* 8, nos. 2–3 (2010).

Richardson, J. Michael. Conversation at Southwest Texas Popular/American Culture Association. February 10, 2012.

Ruddick, Sara. *Maternal Thinking: Towards a Politics of Peace*. Boston: Beacon Press, 1995.

Ryan, Maureen. "Sex, Secrets and *Dollhouse*: Joss Whedon Talks about the End of His Fox Show." *Chicago Tribune*. December 3, 2009.

Shuster, Martin. "'What It Means to Mourn': *Dollhouse* and Aporia." In *Inside Joss' "Dollhouse": From Alpha to Rossum*, edited by Jane Espenson, 233–46. Dallas, TX: Smart Pop, 2010.

Tresca, Don. "Images of Paraphilia in the Whedonverse." In *Sexual Rhetoric in the Works of Joss Whedon*, edited by Erin B. Waggoner, 146–72. Jefferson, NC: McFarland, 2010.

Whedon, Joss. "Man on the Street." Commentary. Season 1, episode 6. *Joss Whedon's Dollhouse: The Complete First Season*, disc 2. Los Angeles: Twentieth Century Fox Home Entertainment, 2009.

Wilcox, Rhonda. "Echoes of Complicity: Reflexivity and Identity in Joss Whedon's *Dollhouse*." *Slayage: The Journal of the Whedon Studies Association* 8, nos. 2–3 (2010).

FILMOGRAPHY

"Apocalypse Nowish." *Angel*. Season 4, episode 7. Written by Steven S. DeKnight. Directed by Vern Gillum. First broadcast November 17, 2002.

"As You Were." *Buffy the Vampire Slayer*. Season 6, episode 15. Written and directed by Doug Petrie. First broadcast February 26, 2002.

"Bad Girls." *Buffy the Vampire Slayer*. Season 3, episode 14. Written by Douglas Petrie. Directed by Michael Lange. First broadcast February 9, 1999.

"Expecting." *Angel*. Season 1, episode 12. Written by Howard Gordon. Directed by David Semel. First broadcast January 25, 2000.

"Goodbye Iowa." *Buffy the Vampire Slayer*. Season 4, episode 14. Written by Marti Noxon. Directed by David Solomon. First broadcast February 15, 2000.

"Halloween." *Buffy the Vampire Slayer*. Season 2, episode 6. Written by Carl Ellsworth. Directed by Bruce Seth Green. First broadcast October 27, 1997.

"The I in Team." *Buffy the Vampire Slayer*. Season 4, episode 13. Written by David Fury. Directed by James A. Contner. First broadcast February 8, 2000.

"Into the Woods." *Buffy the Vampire Slayer*. Season 5, episode 10. Written by Marti Noxon. Directed by Marti Noxon. First broadcast December 19, 2000.

"Origin." *Angel*. Season 4, episode 18. Written by Drew Goddard. Directed by Terrence O'Hara. First broadcast April 21, 2004.

"Out of My Mind." *Buffy the Vampire Slayer*. Season 5, episode 4. Written by Rebecca Rand Kirshner. Directed by David Grossman. First broadcast October 17, 2000.

Serenity. Director Joss Whedon. Universal Pictures, 2005. DVD.

"Tabula Rasa." *Buffy the Vampire Slayer*. Season 6, episode 8. Written by Rebecca Rand Kirshner. Directed by David Grossman. First broadcast November 13, 2001.

"Villains." *Buffy the Vampire Slayer*. Season 6, episode 20. Written by Marti Noxon. Directed by David Solomon. First broadcast May 14, 2002.

"You're Welcome." *Angel*. Season 5, episode 12. Written by David Fury. Directed by David Fury. First broadcast February 4, 2004.

NOTES

1. Calvert, "Mind, Body, Imprint."

2. Calvert, "Mind, Body, Imprint"; Ginn, "Memory, Mind, and Mayhem."

3. Ruddick, *Maternal Thinking*, 17.

4. Moreover, protection laws for child actors may have prevented their involvement in some of the show's often conflict-ridden sexual scenarios.

5. Ryan, "Sex, Secrets."

6. The only child actor who is potentially sexualized on-screen is Iris, from "Epitaph One" (1.13) and "Epitaph Two: Return" (2.13). The use of "potentially sexualized" indicates the conflict of Iris's childlike body being imprinted with Caroline's memories. Iris therefore plays the opposite of the usual Doll figures by showcasing a mature mind in an immature body. Her

verbal interjection into the sexual relationship between Echo and Paul—informing Paul that Echo does want him in a sexual manner—positions her in a liminal space wherein she both is and is not sexualized; her mind is mature, and she interjects in a manner that suggests sexual awareness and knowledge disjunctive from her physical form. The awkwardness of this liminality is undercut not simply by Echo's position as Paul's sexual counterpart but also eventually dispersed by the realignment of her "real" body and mind at the close of "Epitaph Two." I am extremely grateful to Fiona Noble from the University of Aberdeen for drawing my attention to the liminality expressed by the body of the child figure.

7. Although the composite persona of Eleanor Penn contains an imprint that confronts her childhood attacker in "Ghost," she does so as an adult and part of an identity amalgam in Echo's body. Eleanor is brought to resolution through Echo/Topher's intervention, much like Susan, who is provided with the same chance for some form of resolution in "Briar Rose."

8. Whedon, "Man on the Street," commentary.

9. Bennett, "Deconstructing the Dream."

10. Shuster, "'What It Means"; Bennett, "Deconstructing the Dream."

11. Bennett, "Deconstructing the Dream."

12. An early draft of this chapter was presented at the Southwest Texas Popular Culture and American Culture Association Conference, Albuquerque, New Mexico, 2012. During the question-and-answer session, I was informed that numerous people read the interaction between Adelle and Roger as her attempt to recapture a past love, suggesting that a version of him may actually have existed. Therefore, they argue that DeWitt's engagements with Roger do in fact suggest an emotional entanglement that would go beyond simple abuse. I disagree with this position for two reasons. First, Adelle notes the irony in Roger occasionally seeming like the most real person whom she has ever met, distinctly identifying her awareness of him as a fictional persona ("A Spy in the House of Love"). Second, in the episode "Vows" (2.1), we distinctly see Adelle caress Victor's face once he has undergone reconstructive surgery, clearly producing a tender and sexual moment outside their usual constructed scenarios and while Victor is distinctly still in his childlike Active state. She is clearly aware of this transgression, immediately removing her hand and choosing to ignore his questioning discomfort. The episode "Stop-Loss" then simply underscores her objectification of Victor, since Adelle's disbelief is focused around the fact that she was rejected by an Active that was specifically programmed to love her. Her formulation of this event never suggests her own feelings for Roger or for Victor; the only focus within this relationship is Adelle. These events suggest Adelle's objectification of and fixation on Victor, and Roger is simply the persona she feels to be her mental equal placed into Victor's body.

13. I am extremely grateful to Namarata Nadkarni for her close reading of this chapter and her analysis of what she terms "the love debacle" in "Stop-Loss." I have repurposed part of her argument here with her permission.

14. I use the definition of paraphilia per Don Tresca in his chapter "Images of Paraphilia in the Whedonverse," wherein he defines it as "sexual arousal in response to objects or situations not part of normal arousal activity patterns" and an activity in which the paraphile's personal sexual pleasure is paramount.

15. This is not to suggest that Topher's choice to imprint Sierra as a platonic birthday buddy is any less problematic than these other scenarios; the lack of sexual pleasure derived from his choice to do so does not reduce the abusive act of using Sierra's body to facilitate a fantasy that is specific to Topher and his needs without care or concern for Priya/Sierra. Topher's choice to use Sierra stemmed from the fact that she had gone the longest without being imprinted among the Dolls, something any Active would have done in this scenario. Schuster and Bennett's argument that this scene provides emotional depth for Topher is therefore dangerously double-edged because, whereas it intimates an emotional depth to Topher by suggesting that he might long for the intimacy of friendship, it also presents his absolute disregard for the selfhood of the person he chooses to achieve it with, as the scene underlines just how much the Dolls are simply interchangeable bodies to him.

16. Ruddick, *Maternal Thinking*.

17. "Bad Girls," *Buffy the Vampire Slayer*.

18. "Seeing Red," *Buffy the Vampire Slayer*.

19. I am extremely grateful to J. Michael Richardson for this insightful comment. I am also grateful to Erin Giannini, Alyson Buckman, Heather M. Porter, and Sherry Ginn for their suggestions and thoughtful commentary on this chapter.

20. "Into the Woods," *Buffy the Vampire Slayer.*

21. "As You Were," *Buffy the Vampire Slayer.*

22. While this maternal role is briefly hard to define due to Cordelia's memory loss following her return from the higher plane, it remains the most consistent role in defining her relationship with Connor.

23. "Apocalypse Nowish," *Angel.*

24. Cordelia's persona is overtaken twice in the course of the series in favor of a monstrous child that she is forced to carry, first in the episode "Expecting" (1.12) and then during the Jasmine storyline.

25. "Origin," *Angel.*

26. Wilcox, "Echoes of Complicity."

Chapter Seven

"I Possess the Means to Satisfy My Vagaries"

What Motivates the Dollhouse Clients?

Heather M. Porter and Sherry Ginn

> A brothel? You mean to tell me that the tech that punk-kicked the ass of mankind was originally designed to create more believable hookers?
> —Zone ("Epitaph One," 1.13)

The television medium shows many depictions of sex workers and their clients. Police procedurals, such as *Law and Order: Special Victims Unit* (1999–present), generally depict sex workers as the victims of crimes, rarely as perpetrators, and focus on the police investigation and prosecution of the crime. Other programs about law enforcement, such as *CSI* (2000–present), focus on the forensic investigation of a crime scene and offhandedly reference prostitution when they discuss the difficulty of working a crime scene in a hotel room. Television series such as Lifetime's *The Client List* (2012–2013) and HBO's *Hung* (2009–2011) focus on sex workers and their attempts to maintain a normal life while working in the sex industry. In both series, the main characters turn to sex work to support their children when they have no other options. Almost all depictions of sex workers are of females; the interesting twist in *Hung* is that the main character is a male prostitute. Prostitutes in television and film are most often shown as being forced into the situation through desperation, hoping to be rescued by that one person who will look beyond their job and see them for who they really are. This idealized prostitute and client are encapsulated in the movie *Pretty Woman*, in which Julia Roberts's character solicits customers (slang: turns tricks) on Hollywood Boulevard, and Richard Gere's character buys her for a week to pretend to be his girlfriend. Instead, he falls in love with her and saves her from her life.[1]

Dollhouse (2009–2010), created by self-professed feminist Joss Whedon, depicts the sex trade in a particularly provocative way and ignited controversy and dismay in his longtime fans. In this series, wealthy clients are able to hire men and women for (generally) sexual engagements designed to satisfy their every desire. These men and women have supposedly entered into the various worldwide Dollhouses of their own volition to escape from personal problems that rendered their lives unbearable. For example, the character Madeline was unable to bear the death of her child and became the Active November. Tony, Active-appellation Victor, suffered from posttraumatic stress disorder as a result of his service in Iraq and Afghanistan. Caroline, Active-name Echo, hoped to escape prosecution for her terrorist activities against the Rossum Corporation, owner of the Dollhouses, one of which had also resulted in the death of her lover. In exchange for a five-year contract, each Active was guaranteed that her or his memories of unpleasant events or the unpleasant events themselves would be cleansed and her or his bank accounts strengthened considerably. All that the person had to do was sign the contract, allowing her or his memories to be "wiped," removed, and placed on a data storage unit, and wait until the five-year term had expired. During those five years, the Active would live in a Dollhouse and have her or his every need taken care of, and, of course, her or his bodies would be imprinted with new personas as appropriate to clients who could rent her or his services, usually sexual in nature. Adelle DeWitt, director of the Los Angeles Dollhouse, describing the process to a client, tells him, "In their resting state our Actives are as innocent and vulnerable as children. We call it the *tabula rasa*, the blank slate. Now imagine the imprint processes filling it, creating a new personality, a friend, a lover, a confidant in a sea of enemies, your heart's desire made flesh. . . . Everything you want, everything you need, she will be honestly and completely" ("The Target," 1.2). The idea of an Active being made into whomever a client desires opens the door for many types of examination, including why a client chooses to pay for the use of an Active. Unlike other shows that focus on the sex worker (with the clients as secondary characters) or shows that use the sex worker as a secondary character (and the client absent), the role of Actives in *Dollhouse* puts the desires of the client front and center. This aspect of the series allows us to examine the clients of the Dollhouse and attempt to answer the question of who buys these Actives and why.

Scholarly research on the series has examined various themes and characters depicted in the twenty-six episodes filmed. Nicholas Greco, for example, suggests that we, the audience, want the Dolls; we desire them just as their clients do, because they are exotic and "other."[2] Such a reading of the series can be compared to comments made by Joss Whedon and others[3] who have analyzed the series in terms of its similarity to the way in which a television series works in general: an author who programs actors to be exactly what she or he wants them to be, all at the behest of a "shadowy" corporation (i.e., the network).[4] Catherine Coker compares the Dollhouse to society, noting that we are all programmed to perform in various roles, including those proscribed for gender, from the time that we are small children, just as the Dolls enact childlike behaviors before programming by Topher (a very disturbing father image).[5] Sharon Sutherland and Sarah Swan have discussed the issue of consent with respect to those five-year contracts, noting that the clients are complicit in the sex trade, being perfectly willing to purchase these men and women and "blithely [accepting] the programming as equivalent to genuine consent."[6] To date, no one, to our knowledge, has actually examined the Dollhouse clients and their motivations for utilizing the Actives. Season 1 of the series led the viewer to believe that the Dollhouses were nothing more than high-class—very high-class—brothels, and season 2 revealed that such a "reading" of the series was incorrect, finally revealing the truth behind the statement "Fantasy is their business, but that is not their purpose" ("Man on the Street," 1.6). Many of Whedon's fans were dismayed at the subject matter of the series, especially from a man so upset by exploitation of women that he has gone on record many times to discuss it. How then, these fans wondered, could Whedon produce a series that appeared to glamorize sexual exploitation and abuse? It is not our intention to discuss that issue here, as it would take more space than we have herein; however, we believe that Whedon's purpose in producing *Dollhouse* was to address the issue and create a dialogue about the sex trade and its abuse of men, women, and children throughout the world.

Recent years have seen a substantial increase in literature examining the sex trade and the men and women who work therein. An overwhelming number of these studies focus on the workers, especially their reasons for entering and remaining in the trade. Only a few have studied the (predominately) men who frequent prostitutes and other sex workers. Martin A. Monto, in his "Prostitutes' Customers," notes that "men are attracted to paid sex because they desire sexual acts they cannot receive from their partners; they are able to have sex with a larger number of partners; they are attracted to specific physical characteristics; they like the limited emotional involvement; and they are excited by the illicit nature of the act."[7] These ideas on the use of sex workers are mirrored by psychotherapist and arousal researcher Michael Bader when asked why men, specifically politicians with so much to lose, use prostitutes.[8] He tells us that "the appeal of hookers lies in the temporary psychic relief they supply to men struggling with conflicts about guilt and responsibility." He further explains, "For these men, a prostitute is sought as a relationship in which the man can 'let go' and freely express his most selfish desires without feeling guilty and worried about the effect of these desires on his partner." Nevertheless, a careful viewing of the series will reveal that not all the clients were motivated by the reasons suggested by Monto and Bader. Some of the clients actually entertained ideas about love and romance, even within the confines of a preprogrammed engagement.

While examining the differences inherent in indoor and outdoor sex workers, Ronald Weitzer, in his "Sex Work: Paradigms and Policies," describes the "girlfriend experience," whereby the client is seeking counsel and friendship as much as sex.[9] He states, "The customer buys a kind of 'relationship' with an escort rather than just sex. Some customers who become 'regulars' have long-term relationships with providers and develop a real emotional connection, albeit one that is paid for."[10] In their "Call Girls and Street Prostitutes: Selling Sex and Intimacy," Janet Lever and Deanne Dolnick, in interviewing both call girls and street prostitutes, found that call girls provide greater emotional support in their services.[11] They state that "in particular, clients expect call girls to create at least the illusion of intimacy in the exchange."[12] Their study extended research conducted by Arlie Hochschild on "emotional labor," such as the "listening occupations" of a hairdresser or bartender.[13] Lever and Dolnick expanded such occupations to include sex workers, explaining, "Perhaps what many call girls and escorts are really selling is the illusion of intimacy. Hochschild states that, 'Once an illusion is clearly defined as an illusion, it becomes a lie.' But part of the emotional labor performed by the call girl is to keep the illusion alive; she defines her behavior as a playful 'fantasy' rather than an illusion."[14] Indeed, the very notion that a prostitute is "turning tricks" undermines his or her attempt to create the illusion. Hochschild's comment is interesting when compared to what Adelle tells one prospective client about the Dollhouse: "Tricks are illusions, Mr. Connell, created to deceive. What we offer is truth" ("Target"). Although a client in the "real world" knows that the girlfriend experience is not real, a client in the world of Dollhouse could experience just such a fantasy, knowing that at least the Active believes it to be real.

Although, undoubtedly, patrons of the Dollhouse used Actives for sexual and other purposes, some of them utilized Actives for love and romance. In this chapter, we first examine one theoretical orientation designed to explain different types of love and how various relationships within the Dollhouse illustrate its tenets. These relationships occurred among Actives and between Actives and clients. In addition, a number of patrons utilized Actives for various business enterprises, some of which were illegal. However, the patron utilizing Actives most often was in fact the Rossum Corporation itself, in what we refer to as special-operation assignments, or special-ops. Some of these assignments involved love, romance, and sexual relations, although not all did. We therefore analyzed the engagements depicted on-screen to determine the various reasons that clients give for choosing to engage an Active. Our hypothesis was that clients actually chose to engage an Active for a romantic engagement more often than for a purely sexual one or one contracted for another purpose.

> I feel love and pain, fear. It's not pretend for me.
> —Echo ("Instinct," 2.2)

One lens through which to analyze the romantic relationships between characters on Dollhouse is via the triangular theory of love, as proposed by psychologist Robert Sternberg.[15] He proposes that love consists of three components that combine to produce seven types of love. The first component is *intimacy*, which "refers to those feelings in a relationship that promote closeness, bondedness, and connectedness." Intimacy is the foundation of love, and it develops slowly, over time. It can be hard to achieve and even harder to maintain. As one becomes more involved with someone, a fear of losing the self may occur. The task is to maintain the sense of intimacy while maintaining a sense of autonomy. The second component of love is *passion*, which involves "a state of intense longing for union with the other." Sternberg states that passion consists of the expression of our needs and desires, one of which is sexual fulfillment, but that is not the only need that can be fulfilled passionately. The final component of love is *commitment*, which consists of both a short- and long-term aspect. Sternberg states that the "short-term aspect is the decision to love a certain other, whereas the long-term one is the commitment to maintain that love." These two decisions do not necessarily occur at the same time or within any given relationship. It is commitment that keeps a relationship going.

Excluding nonlove, seven types of love can be combined using the aforementioned components. *Liking* consists only of intimacy. This is the type of love that one feels for close friends and family members. Sternberg uses the term "liking" to refer to the type of feeling that one finds in friendships: feeling close to the friend but feeling no passion toward him or her and not expecting or desiring a long-term commitment. *Infatuated love* consists only of passion. This is what most would consider "love at first sight." It consists of intense physical arousal, but it can dissipate as quickly as it arises. *Empty love* consists only of commitment. Love may sometimes devolve into this type of love, as in the case of couples who have been together for many years. In cultures where marriages are arranged, empty love may mark the beginning of a relationship. *Romantic love* consists of passion plus intimacy. Liking in combination with the arousal of physical attraction characterizes this type of love. Romantic love includes the feeling that one has met the right person, coupled with the desire to fuse spirits. *Companionate love* consists of intimacy and commitment. One could think of this type of love in terms of a committed friendship. Many relationships devolve into this type of love once physical attraction, which is a major source of passion, has waned.

Fatuous love consists of passion plus commitment. We often read about this type of love in the tabloids, when two people meet, fall in love, and marry after a whirlwind romance. Since intimacy takes some time to develop, the people who fall "head over heels in love" and rush to the altar may wake one morning to realize that they do not even like their partner. As Sternberg notes, "The partners commit . . . to one another on the basis of passion without the stabilizing element of intimate involvement." Finally, a combination of all three components yields *consummate love*. Many people would consider this to be a complete love. Sternberg notes, however, that consummate love is often like meeting your goal in a weight-loss program: it is easier to achieve than to maintain: "like other things of value, [it] must be guarded carefully."

> I love him. Is that real?
> —Sierra ("Belonging," 2.4)

Although we can analyze the Dollhouse patrons and their reasons for engaging an Active, we can examine the relationships held by those who work within the Dollhouse as well. However, the relationships held by the Actives on *Dollhouse* are harder to evaluate given the fact that the entire series is about fantasy and the distortion of memory.[16] Nevertheless, several relationships within the Dollhouse itself might fit the taxonomy proposed by Sternberg.[17] Victor and Sierra begin their relationship with passion: his erections are easily observed and are exclusive to her. The two continue to bond with each other; eventually, they escape the Dollhouse, have a child together, separate, and reunite. Quite interestingly, we do not, at least during season 1, observe Sierra engaging in sexual encounters with clients. All her televised encounters reinforce Echo's assignments; usually, Sierra is brought in to complete a (nonsexual) assignment that Echo has been unable to finish. Unfortunately, while we are not privy to most of her engagements (since they occur off-screen), Sierra's sojourn in the Dollhouse is fraught with danger. We learn very early in season 1 that Sierra has been raped several times by her Handler, Hearn ("Man on the Street," 1.6). Although she does not consciously remember the encounter, she reacts violently to touch afterward. This example of procedural memory, the unconscious memory of actions, leads to the discovery of the rape. Later we learn that Sierra did not volunteer to become an Active; rather, she was forced into servitude by a rejected lover named Nolan, who repeatedly uses her afterward ("Belonging"). We can consider that Nolan's use and abuse of Sierra speaks to all of Monto's suggestions why men chose to engage with sex workers, including some reasons that Monto did not mention, such as the ability to become physically aggressive with the sex worker.[18]

Given that the majority of Dollhouse clients are male, it would stand to reason that Monto's prediction that men want to have sex with as many women as possible is supported by their actions within the Dollhouses.[19] One question raised by this is, if men desire variety, why is Echo the most requested Active? Dr. Saunders refers to Echo as being "number one" in several instances (once she herself is injured by Alpha and no longer serves as "number one" ["Omega," 1.12]). Over the course of season 1, we do learn that two men have requested Echo more than once: Matt Cargill and Joel Mynor. Given what we see of Matt during his engagements with Echo, he definitely is seeking sexual adventure, as implied by the fast motorcycles and in the flashback of bondage scenes with Echo. Joel Mynor apparently thinks that Echo is special as well. Although he requests Echo multiple times, apparently he does so only on his wedding anniversary, one that he is unable to spend with his wife because of her death on that date. We do not know if it is possible to "engage" an Active permanently,[20] as it was possible to hire a Companion exclusively on *Firefly*, but Joel is apparently happy enough to engage Echo's services once a year. Echo is extremely beautiful, as are all the Actives, but she is also different. It may be that the something that makes her different appeals to her clients and that is why they request her more often than others.

Echo is attractive to both male and female clients, as is Victor. Although referred to earlier in the series as one of his repeat customers, we do not learn until later in season 1 that "Miss Lonelyhearts" is actually Adelle DeWitt. Adelle has created an elaborate cover for herself to keep anyone from knowing that she is attached to one of the Actives. After all, Miss Lonelyhearts has engaged Victor eight times prior to events in "A Spy in the House of Love" (1.9). The encounter is definitely sexual but also emotional: it is obvious that Adelle loves "Roger," and we are left to wonder about this Roger and their relationship. That she is emotional is obvious from her remarks, but that she knows she is "cheating" is clear as well. She tells him that she does not have to hide anything from him. In a "normal" relationship, one that contains intimacy, this would mean that she trusts him enough to be able to tell him anything. In this fake relationship, she does not have to hide anything from him, because he will never remember it. Sutherland and Swan as well as Samira Nadkarni point out that Adelle's usage of Victor/Roger is really no different than Hearn's use of Sierra.[21] Both Actives are unable to give consent to the acts committed against their bodies.

> Your heart's desire made flesh.
> —Adelle DeWitt ("The Target," 1.2)

Robert Sternberg's triangular theory of love can be helpful in elucidating different types of love and providing a means of classifying relationships.[22]

Using Sternberg's taxonomy, we analyzed eighty-two engagements occurring in twenty-six filmed episodes of *Dollhouse* to determine the various reasons that clients give for choosing to engage an Active. Each engagement was examined, and the following was noted: client (including gender), who paid for the engagement, the imprint, the Doll used, and the type of engagement. These eighty-two engagements depicted one or more types of love as described by Sternberg's theory: specifically, five of the seven types of love are shown in the series.[23] A summary of these engagements can be seen in Table 7.1. Surprisingly, about 41 percent of incidents code as romantic, something not expected of a brothel—even a high-class brothel—if the clients were primarily interested in anonymous sex. Note, however, that the majority of incidents (64 percent) were not sexual at all but rather were what we are labeling special-ops, engagements that may or may not involve sexuality (e.g., Mellie and Paul Ballard; Senator Daniel Perrin and his wife). Such engagements reflect the underlying mission of the Dollhouses as subsidiaries of the Rossum Corporation, a mission reflected in the oft-repeated "Fantasy is their business, but that is not their purpose."

The one incident of *liking* involved Topher on his birthday when he created a friend for himself using the Active Sierra. This interaction for Topher was about platonic friendship, as the gender of the Active was not a factor. Topher simply requested an Active for his "test," allowing Boyd to choose any Active for him. Sierra just happened to be "free." She was programmed to be exactly the type of friend Topher desired; he could not help but "like" her.

Three incidents of *infatuation* occurred. In this type of love, no emotional connection between partners is made; it involves sexual acts only. The clients in these incidents were Mr. Caviezel and his desire for a submissive ("Meet Jane Doe," 2.7), Echo as a dominatrix (to an unknown client; "Spy," 1.9), and Tempura Joe, who liked to be battered in eggs and flipped ("Vows," 2.1).

Thirty-four romantic engagements were coded. *Romantic love* combines passion and intimacy, a connection between two people but without the commitment. Though the Active, through his or her programming, may be-

Table 7.1. Types of Love Observed in Two Seasons of Joss Whedon's *Dollhouse*

Type of Love	Incidents	Active/Client
Liking	1	Sierra/Topher
Infatuation	3	Echo/Mr. Caviezel and others
Romantic love	34	Echo/Matt; Echo/Richard and others
Companionate love	1	Echo/Joel Mynor
Nonlove/special ops	52	Mellie/Paul Ballard; Daniel Perrin

lieve this relationship to be more, the client knows the truth about the engagement. This is illustrated in the first-season episode "The Target," when Adelle tells Richard Connell that the Active will be "your heart's desire made flesh. And when the engagement has been completed all memory of you and your time together wiped clean." Examples of this type of love include Matt and Alice/Echo ("Ghost," 1.1), Richard and Jenny/Echo ("Target," 1.2), Professor Gossan and Kiki/Echo ("Belle Chose," 2.3), Nolan and Sierra/Priya ("Belonging," 2.4), Adelle/Miss Lonelyhearts and Roger/Victor ("Spy," 1.8; "Stop-Loss," 2.9), and Megan Crantz and Echo, who are briefly married in San Francisco ("A Love Supreme," 2.8). It is these engagements that are indicative of Weitzer's explanation of the "real" girlfriend experience: clients who are seeking romance, dating, friendship, and companionship but not a lifelong (or even a nightlong) commitment.[24] Such people may be reflective of more traditional moral values suggesting that sex without love is evil, wrong, or just "not right" and consider sex within a relationship okay as long as the relationship lasts. Clients who engage the "girlfriend experience" can thus delude themselves into thinking that they are not just using people for their sexual services. Rather, they really like the person they hire, which will be especially true when engaging Actives that have been created to the clients' specifications and who sincerely believe that they are in love with the person with whom they are interacting.

However, Sternberg notes that most people want it all—intimacy, passion, and commitment—and this is considered to be *consummate love*, depicted on *Dollhouse* (perhaps) in the case of Joel Mynor and his wife, Rebecca, as imprinted on Echo.[25] Though Joel knows that Echo is not his wife in the engagements that he books each year for the anniversary of her death, he still has a commitment to her and his memory of her. That the fantasy he lives involves showing her the house that she never got to see because she was killed in a car accident on the way there shows that there is more to this engagement than sex. Echo's desire in her Doll state—to finish this interrupted engagement—shows a connection between the two that had not been seen before in the Active-client relationships ("Man," 1.6). The moment when this relationship shows all aspects of love on the part of both Joel and the imprint occurs in the episode "A Love Supreme" (2.8) when Echo, in control of all her imprints and thus Rebecca's "persona," tells Joel to move on. This moment gives both Joel and Rebecca the closure they were seeking in their relationship, a relationship so complete that it needed proper closure on both sides.

Unlike the aforementioned, the majority of the overall engagements fell into the category of nonlove; there was no intimacy, no commitment, and most often no sexual desire: these were the special-ops engagements. Such engagements included Eleanor Penn/Echo as a negotiator specializing in kidnappings ("Ghost"), Jordon/Echo as a background singer who is also a bodyguard ("Stage Fright," 1.3), Esther Carpenter/Echo as a blind cult member ("True Believer," 1.5), and Anton Lubov/Victor as a Russian mobster programmed to throw special agent Paul Ballard off the track of the Dollhouse ("Ghost," "Stage Fright"). Although, at times, special-ops engagements included sex and, in some cases, even romantic love, the primary purpose was to complete a set of mission parameters, and the sexual relationship was just a part of the mission. One example is when Active Taffy pretends to be a call girl. She is really a thief using her guise as a hurt prostitute to get past security to help her team break into a vault and steal what appears to be a piece of the Parthenon ("Gray Hour," 1.4). Another example is when Adelle imprints Echo as Bree, the prostitute trying to blackmail Senator Perrin in an attempt to find out if the senator's wife is an Active ("The Public Eye," 1.5). In these missions, the sexual aspect is only a means to an end, and thus we do not consider them to fall under the aegis of "love."[26]

An interesting trend can be observed when the actual clients of the Dollhouses are examined more closely. Of the eighty-two engagements that we observed or heard discussed, 51 percent were ordered or paid for by the Dollhouse and/or Rossum itself, indicating that over half their business is for the exclusive use of the company for its own means and toward its own goals. When the engagement types are divided between in-house engagements (paid for by Rossum) and out-of-house engagements (paid for by actual clients), the following results are found. The engagements booked by paying clients were romantic engagements 65 percent of the time, special-ops 30 percent of the time, and sexual only 7 percent of the time.[27] These results show that fantasy for paying customers was their business at least 72 percent of the time. We also see with respect to the paying customers what Monto tells us we should expect: 77 percent of the clients were identified as male, 14 percent as female, and 9 percent unidentifiable, as we only saw the Active, not the client.[28] The "real" clients illustrate the trend that we would expect in a brothel: a heavily male clientele wanting sex and the girlfriend experience in the most realistic way possible. Such clients get the fantasy that Adelle's sales pitch promises: "It isn't about what you want. It is about what you need. An Active doesn't judge, doesn't pretend. This will be the purest, most genuine human encounter in your life and hers" ("Epitaph One").

Though the Active may not pretend, the Rossum Corporation does. It pretends that it is in the business of fulfilling fantasies and giving people what they need. However, when these engagements are examined, the majority of the time Rossum is fulfilling its own needs. As noted earlier, 51 percent of the engagements in the series are contracted by the Dollhouse itself or Rossum. Of these, 93 percent are special-ops,[29] some of which are short-term, as when Sierra and November are imprinted as bounty hunters to find Echo after Alpha kidnaps her ("Omega," 1.12) or when Echo is imprinted as Margaret Bashford to discover her own murderer ("Haunted," 1.10). Others are long-term engagements, such as the programming of Daniel Perrin to influence the way that laws are passed—and to perhaps even be elected president one day—or the double programming of Mellie: first, to influence special agent Paul Ballard's investigations of the Dollhouse and, second, as a sleeper agent in case romance does not work and she needs to kill him ("The Hollow Men," 2.12). Overall, it appears that Rossum is its own best client, as long as it suits its needs.

> Isn't that why we're in business? To give people what they want?
> —Adelle ("Stop-Loss," 2.7)

Our original hypotheses were supported with respect to the reasons why clients engaged Actives. Our results show that the clients actually chose to engage an Active for a romantic engagement more often than for a purely sexual one. Nevertheless, the majority of overall encounters were actually special-ops engagements. Although *Dollhouse* was ostensibly about an expensive brothel with questionable ethical standards, the reality was much darker and much more dangerous.

Many longtime fans of Joss Whedon were unable to understand his creation of a show about sexual exploitation, and apparently neither could the Fox network, whose executives balked at the idea that viewers would think that they were airing a program about the "p" word or the "s" word.[30] Fans who continued to watch the show eventually "got it" and understood that the program was not really about sex at all. The sex was a "red herring" designed to mask what was really happening: rich and powerful people finding a way to be rich and powerful forever, to live beyond death, to be immortal. *Dollhouse* proposed that this would be possible via "mind/soul"[31] transplants rather than the usual way in science fiction: via clones or androids. However, Rossum executives needed a valid cover story for their experiments on these mind/soul transplants; thus, the Dollhouses were born. Rich people could pay for sexual as well as other types of favors, and Rossum could use the profits to fund the research as well as use the Dolls as research subjects and power their mainframe in the Attic.

Rossum provided a service for its clients, and those clients used the service in a variety of ways. Although many of these services were sexual in nature, not all were. As noted earlier, the majority were actually operations ordered by Rossum to further its agenda of increasing its power and control over the (supposed national) government of the country as well that country's citizens where the various Dollhouses were located. As long as there were people who needed money, to escape from their problems, to avoid time in jail, or to disappear for awhile, there would be Dolls. When Topher's portable remote wiping device was perfected, Rossum was able to create a world of victims, bodies to be used and abused then discarded ("Epitaph Two: Return," 2.13).

Our attempt in this chapter was to analyze the various clients of the Dollhouse and determine their motivation for utilizing these sex workers. This analysis showed that indeed many clients were engaging Dolls and using them for sexual purposes, albeit for a variety of reasons. Some of these reasons were "the girlfriend experience," sexual variety, sexual fantasy, and kinky sex,[32] all reasons for which customers engage sex workers according to our research. Other reasons were practical: to hire a bodyguard, an investigator, an undercover operative. However, other reasons were more insidious, much more frightening, speaking to human fears of big business out of control, governmental surveillance and the loss of basic human rights, and, even more terrifying, loss of self. Although not as readily apparent in *Buffy*, Whedon's later series *Angel* and *Firefly* also examined these themes, and, as in *Dollhouse*, the outcome was not always positive.

BIBLIOGRAPHY

Bader, Michael. "Why Men Do Stupid Things: The Psychological Appeal of Prostitutes." AlterNet.org. March 12, 2008.

Brown, Scott. "Tale of Two Endings: A Star—and a Hollywood Myth—Was Born When *Pretty Woman* Hit Theaters 11 Years Ago." *Entertainment Weekly*. March 30, 2001.

Coker, Catherine. "Exploitation of Bodies and Minds in Season One of *Dollhouse*." In Waggoner, *Sexual Rhetoric*, 226–38.

Davis, Hugh H. "'The Drama Is in Us': Pirandellian Echoes in Dollhouse." *Slayage: The Journal of the Whedon Studies Association* 8, nos. 2–3 (2010).

Davis, Nanette J., ed. *Prostitution: An International Handbook on Trends, Problems and Politics.* Westport, CT: Greenwood Press, 1993.

Espenson, Jane, ed. *Inside Joss' "Dollhouse": From Alpha to Rossum*. Dallas, TX: Smart Pop, 2010.

Ginn, Sherry. *Power and Control in the Television Worlds of Joss Whedon*. Jefferson, NC: McFarland, 2012.

Greco, Nicholas. "The Companion as a Doll: The Female Enigma in *Firefly* and *Dollhouse*." In Waggoner, *Sexual Rhetoric*, 239–47.

Lever, Janet, and Deanna Dolnick. "Call Girls and Street Prostitutes: Selling Sex and Intimacy." In *Sex for Sale: Prostitution, Pornography, and the Porn Industry*, edited by Ronald Weitzer, 187–204. New York: Routledge, 2009.

Masson, Cynthea, and Rhonda V. Wilcox. "'Fantasy . . . Is Their Business, but That Is Not Their Purpose': Introduction to the *Slayage* Special Issue on *Dollhouse*." *Slayage: The Journal of the Whedon Studies Association* 8, nos. 2–3 (2010).

Monto, Martin A. "Prostitutes' Customers: Motives and Misconceptions." In *Sex for Sale: Prostitution, Pornography, and the Porn Industry,* edited by Ronald Weitzer, 67–84. New York: Routledge, 2000.

Ryan, Maureen. "Sex, Secrets and *Dollhouse*: Joss Whedon Talks about the End of His Fox Show." *Chicago Tribune*. December 3, 2009.

Sternberg, Robert J. *Cupid's Arrow: The Course of Love through Time*. Cambridge: Cambridge University of Press, 1998.

Sutherland, Sharon, and Sarah Swan. "'There Is No Me; I'm Just a Container': Law and the Loss of Personhood in *Dollhouse*." In *Reading Joss Whedon*, edited by Rhonda V. Wilcox,

Cynthea Masson, Tanya R. Cochran, and David Lavery. New York: Syracuse University Press, in press.

Waggoner, Erin, ed. *Sexual Rhetoric in the Works of Joss Whedon*. Jefferson, NC: McFarland, 2010.

Weitzer, Ronald. "Sex Work: Paradigms and Policies." In *Sex for Sale: Prostitution, Pornography, and the Porn Industry*, edited by Ronald Weitzer, 1–45. New York: Routledge, 2000.

Wilcox, Rhonda. "Echoes of Complicity: Reflexivity and Identity in Joss Whedon's *Dollhouse*." *Slayage*: *The Journal of the Whedon Studies Association* 8, nos. 2–3 (2010).

FILMOGRAPHY

Pretty Woman. Written by J. F. Lawton. Directed by Garry Marshall. Touchstone Pictures, 1990.

NOTES

Portions of this chapter were presented at the annual meeting of the Popular Culture Association of the South (Nashville, TN, fall 2012) as well as the Joss in June Conference (Shelby, NC, summer 2013). Portions of this chapter were published previously in Ginn's *Power and Control in the Television Worlds of Joss Whedon*. Used with permission.

1. The original ending of the screenplay for *Pretty Woman* was not a happy ending. In the screenplay titled *Three Thousand*, Vivian, the prostitute (played by Julia Roberts), is dumped by Edward Lewis, the real estate mogul (played by Richard Gere), at the end of the movie and returns to the streets in a drug-fueled rage (Brown). This is a more realistic portrayal of the way this relationship would end.
2. Greco, "Companion as Doll."
3. Wilcox, "Echoes"; Davis, "The Drama Is in Us."
4. One can only speculate as to where Rockne S. O'Bannon's short-lived series *Cult* was headed before its cancellation, as it, too, seemed to mock not only the writer and the network but the fans as well (especially the fans, which may explain why it was canceled).
5. Coker, "Exploitation."
6. Sutherland and Swan, "There Is No Me."
7. Monto, "Prostitutes' Customers," 77.
8. Bader, "Why Men."
9. Weitzer, "Sex Work."
10. Weitzer, "Sex Work," 11.
11. Lever and Dolnick, "Call Girls."
12. Lever and Dolnick, "Call Girls," 199.
13. Lever and Dolnick, "Call Girls," 199.
14. Lever and Dolnick, "Call Girls," 188.
15. All of the information and quotes in this section come from Sternberg. Quotes are taken from pages 6, 9, 11, 20, 21, 22.
16. Masson and Wilcox, "Fantasy."
17. Sternberg, *Cupid's Arrow*.
18. Monto, "Prostitutes' Customers."
19. Monto, "Prostitutes' Customers."
20. The discussion between Harding and DeWitt, when Nolan Kinnard wants to have Sierra permanently imprinted and "sold" to him, suggests that it is not the policy of the Dollhouses to allow this; however, they are apparently willing to do anything for a client or employee if necessary. For more discussion of Sierra and her plight in the Dollhouse, see Mukherjea in this collection.
21. Sutherland and Swan, "There Is No Me"; Nadkarni, this collection.
22. Sternberg, *Cupid's Arrow*.

23. Sternberg, *Cupid's Arrow*.
24. Weitzer, "Sex Work," 11.
25. Sternberg, *Cupid's Arrow*.
26. It must be noted that some special-ops engagements, such as Mellie and Paul as well as Daniel Perrin and his wife/handler, involved romantic love. However, in the case of Perrin, this was one-sided, as his wife knew that the relationship was an assignment. Indeed, in the episode "The Left Hand," "Cindy Perrin" tells Daniel that she cannot stand him; even his smell makes her sick. Neither Mellie nor Paul knew that she was on an assignment for the Dollhouse. These types of relationships were thus coded as both romantic love and special-ops.
27. These numbers are more than 100 percent because some engagements included both special-ops and romance in one engagement.
28. Monto, "Prostitutes' Customers."
29. In examination of the engagements for which Rossum is considered to be the client, there are instances that could be considered "regular," such as Adelle and Roger and Topher and his friend. These engagements are designed by Adelle and Topher for their personal needs, but they do not pay for them; Rossum does. These are the only non-special-ops engagements that fall under Rossum as a client. We have coded Bashford as special-ops, assuming that Dollhouse clients would expect the corporation to expend effort and energy to ensure their safety. Given their friendship, it is possible that Adelle DeWitt requested the investigation into the murder or paid for Echo's services herself. In either case, the investigation would be a special-ops assignment.
30. Ryan, "Sex, Secrets."
31. A common theme within science fiction with respect to immortality is creating either an android that is programmed with one's essence—mind or soul, depending upon one's beliefs—or a clone, which is a twin. *Dollhouse*'s premise is that Rossum created the technology to download one's mind onto a data storage device, which then can be uploaded into any container capable of holding the data. Such a theory proposes that one's mind is noncorporeal and that any old body will do to hold the data. There are data suggesting that this is not possible, that a mind is more than a collection of data and that a body is necessary for its essence. A complete discussion is beyond the scope of this chapter.
32. Kink sexual practices are those that extend beyond what generally is considered to be conventional sexual practices. Kink includes acts such as BDSM (bondage, dominance and submission, and sadomasochism), spanking, cuckoldry, and sexual fetishism. Practitioners of kink consider it to be a means of heightening the intimacy between sexual partners. Echo tells Boyd that it is all about trust when returning from an assignment in her very skimpy dominatrix outfit in the episode "A Spy in the House of Love" (1.9).

Part III

Structure and Form

Chapter Eight

"Who Did They Make Me This Time?"

Viewing Pleasure and Horror

Bronwen Calvert

Taking its inspiration from the different roles that actors are required to adopt, Joss Whedon's "odd and difficult" *Dollhouse* was "created around the concept of 'who do they want me to be?'"[1] The show's complex exploration of identity, memory, and fantasy has divided the reactions of critics and fans. Whereas some commentators express their enjoyment of the performances of actors and characters in a variety of roles in this series, they find disturbing the "uncomfortable discussions about agency and consent"[2] implicit in its themes of slavery, sex trafficking, and the dystopian effects of out-of-control technology. As Eve Bennett notes, "Almost every article, review, blog and discussion forum dealing with the show at least mentions the potential for controversy raised by its themes of sexual exploitation."[3] Bennett and others have pointed out the clear connections between the fictional Dollhouse and the production of the actual television show; developed in response to Eliza Dushku's description of an actor's life, it is recognized by Whedon as a commentary on the process of writing and making television. He comments, "It is sort of about the creation, but also the removable self. And the idea of becoming somebody else for a little while, which the writer gets to do all [the] time if he is lucky."[4]

However, as has been noted, "many of the characters display a high degree of self-consciousness, and the series invites that same self-consciousness from the audience."[5] This self-consciousness extends beyond viewers (who may or may not acknowledge their own complicity in continuing to watch a disturbing television show) to Whedon the writer/director/creator, who has been clear about the problematic implications of the show, and it encompasses the network, the writers, and the actors.[6] All are implicated (and complicit) in this narrative of coercion and exploitation, and the series constantly interrupts viewing pleasure with narrative elements that juxtapose enjoyment with monstrosity and horror. The show invites the viewer to watch, then emphasizes the voyeuristic nature of such viewing, particularly at narrative points that display exploitative, problematic, and even horrific material. In so doing, it forces the viewer to question the uncomplicated enjoyment of watching.

The entire narrative of *Dollhouse* is focused on repeated transformations of a small number of individuals who, episode by episode, take on different traits and characteristics. The transformations enacted by the fictional "imprinting" process allow characters and actors to be seen in new and unexpected ways; they highlight some actors' abilities to mimic; and they may be played for comedy as well as for dramatic potential, all of which feeds into viewing pleasure. However, as this imprinting process dramatizes a destabilization of identity (especially corporeal identity) and control, it allows aspects of horror into the narrative. The imprinting process suggests a separation of body and mind; bodies are presented as empty shells, or "tabula rasa," ready to receive each successive imprint. In narrative terms, the process emphasizes the frailty of bodies and minds and foregrounds the performative artificiality of personality.[7] This is particularly evident in those moments when the illusion of a coherent identity and personality is broken, when the workings of the imprinting process are made apparent, and when an Active's temporary identity is disrupted in various ways. These include instances when the personality is altered or interrupted or when the same personality is present in different bodies, both of which underline the ways in which identity is not fixed or stable but can be altered or reconfigured. The narrative both draws in and repels the viewer with compelling and enjoyable elements that are nevertheless essentially disturbing and horrifying and can be considered monstrous and abject.

The abject, as Julia Kristeva argues, is "what disturbs identity, system, order. What does not respect borders, positions, rules."[8] Kristeva, Barbara Creed, and others discuss the abject in relation to horror, and it is evident that abjection is present in on-screen horror's disruptions and disturbances. However, such disruptions can be connected with other aspects of viewing, including aspects commonly considered pleasurable. Noël Carroll, for example, describes comedy in terms that echo Kristeva, noting that both horror and the kind of pleasure that a viewer encounters in watching comedy is similarly "bound up with the violation, problematization and transgression of our categories, norms and concepts."[9] In her discussion of comic episodes in *Supernatural* (2005–current), Stacey Abbott emphasizes the links between abjection and the presentation of comedy and horror; with both, we see bodies "out of control" and "under attack."[10] "Comedy" may not appear to be a suitable term to describe the experience of viewing *Dollhouse*, but I suggest that it can be included as an aspect of viewing pleasure, and it is evident that different forms of pleasure, including the enjoyment of horrific and comic moments, do operate in this show, as in other Whedon fictions. Viewers of *Dollhouse,* especially those with awareness of Whedon's work as a whole, may find pleasure in being wrong-footed, surprised, or shocked as much as (or more than) in having expectations met. A regular viewer of Whedon productions may enjoy seeing actors who frequently appear in his "ensemble"; in addition to Eliza Dushku, Amy Acker, Summer Glau, and Alexis Denisof appeared in *Dollhouse.* Viewers equally may enjoy the interplay—or what Jeffrey Bussolini calls "intertextuality of casting"—between *Dollhouse* and other television texts, such as *Battlestar Galactica* (2004–2009) and *Terminator: The Sarah Connor Chronicles* (2008–2009), when watching certain actors[11] (e.g., Tahmoh Penikett, whose *Battlestar* role as Helo is frequently commented on by the show's writers as well as by critics and fans).[12] Viewers who respond to such intertextuality engage in what Sue Tjardes calls "sense making," in which "the meaning of a show or episode . . . is a process that includes both the creator and the viewer: a cocreation, negotiation, or even struggle."[13] In the context of *Dollhouse*, part of this "struggle" of "sense making" may include reconciling the tensions between pleasurable and disturbing elements of the narrative. Paradoxically, the disorder of the ever-changing Active personalities may highlight viewing pleasure; the viewer is allowed the pleasurable experience of seeing known actors portraying very different characters on the same show, often within the same episode. These circumstances emphasize an actor's ability to create convincing characters, and this feeds into viewing pleasure: Enver Gjokaj's ability to re-create or mimic his fellow actors' physical and vocal traits is singled out for special praise on several online fan discussion sites; his performance is cited as "totally convincing," "creating [one of] my favorite characters," "he really is the biggest discovery."[14] Yet, because these changing personalities dramatize a destabilization of identity (especially corporeal identity) and control, aspects of horror intrude into the narrative—even in scenes or storylines that play as strongly comic or pleasurable. In such instances, viewers are invited to find pleasure in viewing a monstrous body that transgresses boundaries. Examples of disorder and destabilization reinforce the definition of Active and Doll bodies as abject, and the multiplication of identity renders them uncanny; of particular interest here is the definition of "uncanny" as "those things that relate to the notion of a double: a cyborg; twin; doppelganger; a multiplied object."[15]

While destabilization, abjection, and the uncanny are foregrounded for the Doll characters of this show, there are suggestions that these bodies are not without agency or identity. Here, Elizabeth Grosz's work on corporeal identity is particularly helpful; as she notes, "the body can be approached, not simply as an external object, but from the point of view of its being lived or experienced by the subject."[16] Similarly, Judith Butler's well-known comments on performativity are useful. Butler sees the body as a performative space and examines the processes by which we create "an identity tenuously constituted in time, instituted in an exterior space through a *stylized repetition of act.*"[17] As viewers watch the inhabitants of the Dollhouse in both Doll and Active states, they have numerous opportunities to see "performative" bodies as "external objects." Indeed, the narrative frequently emphasizes these aspects of both Doll and Active, underlining their objectification, especially with regard to their sexual exploitation by Dollhouse clients and by those working in the Dollhouse (both Sierra's Handler Joe Hearn and Adelle DeWitt use Dolls in this way; see "Man on the Street" [1.6] and "A Spy in the House of Love" [1.9]). However, the series narrative does insist on ways in which Doll bodies are still "lived or experienced" from within, especially in relation to Echo's development of a composite identity. Furthermore, the narrative interrogates aspects of performance in various episodes—for instance, in "Gray Hour" (1.4) or "Belle Chose" (2.3), which foreground the very instability of Active identity. Grosz envisions corporeal identity as "an open materiality, a set of (possibly infinite) tendencies and potentialities which may be developed" within or upon bodies that are "neither 'blank' nor programmed."[18] In reference to a television show that repeatedly insists on the "blankness" of certain bodies, this is a useful point, suggesting more positive implications for the "infinite" Active identities that an individual could perform and suggesting, too, that such identities may have the capability to overcome both "blankness" and programming. Such bodies still may become, in Rosi Braidotti's terms, "monstrous": "the monstrous signifies the difficulty in keeping manageable margins of differentiation of the boundaries between self and other."[19] This is particularly relevant to the imprinting process, where the integrity of the body is apparently maintained, while at the same time it is utterly violated, the boundaries of self/other and inside/outside completely transgressed to the extent that, at one point, Echo has to ask, "Who did they make me this time?" ("Vows," 2.1). As is apparent throughout the narrative of the show, the imprinting process does not necessarily mean that the "self" has been erased (though the Dollhouse makes that claim). Although that process suggests a separation of body and mind, with the Doll body presented as a "tabula rasa" ready to receive the imprint, there are suggestions that a Doll may "remember" elements of an engagement or feelings about particular individuals (see Calvert).[20] Holly Randell-Moon's discussion of "somatic history" posits a means by which self and identity are connected with and through the body,[21] and, like Grosz's "lived" or "experienced" body and Butler's "repetition of acts," this is a helpful way to conceptualize the imprinting process—the body is the constant through all the versions of "self" that are overlaid on it, which in turn emphasizes the instability of (imprinted) identity. The imprinting process itself, however, and *Doll-*

house's narrative as a whole still emphasize the frailty of bodies and minds and foreground the performative artificiality of personality, with examples of identity that are disrupted and destabilized.

Through its two seasons, *Dollhouse* presents this destabilization in a number of ways. Identities may be reset or "wiped" so that an Active's imprinted personality is altered or lost. Doubling or duplication likewise complicates identity as a personality appears in different bodies at different times or in more than one body at the same time. The creation of a "composite identity" is itself a destabilizing concept, as it brings together many different aspects of personality in one apparently unified imprint. The imprinting process allows "resurrections" from the Attic (the repository of damaged Dolls and enemies of the Dollhouse) or, literally, from the dead, allowing another chance at embodied life for particular personalities. With the imprinting technology, "ghosts can be made to speak,"[22] and the resurrection of a dead person further emphasizes the Dolls' association with empty bodies, the corpse, and death, bringing abjection to the fore once again.

Instances of resetting or wiping imprinted personalities appear in "Gray Hour," in which Echo's engagement as unstoppable thief Taffy is interrupted by a "remote wipe" apparently set off by rogue Doll Alpha. After a complex setup that has already involved abrupt switching between identities—from exuberant call girl to frightened victim and then to the confident and in-charge Taffy personality—this further switch to the Doll's wiped "tabula rasa" state not only undoes identity but removes it completely. In her Doll state, Echo has no memory, no name at all, and, at first, almost no vocabulary but repeats "Can I go now?" a question that is part of the Doll/Handler "script." For the rest of the episode, other individuals—the two fellow criminals in the vault as well as Adelle and others back in the Dollhouse, including another version of the Taffy identity—attempt to "reprogram" Echo to complete the heist. It is notable that she does not fully comply with any of them but appears to exercise her own will by attacking one of the criminals with a syringe and bringing the other out of the bank vault while insisting, "I'm not broken."[23] Nevertheless, the process of the wipe highlights the performative nature of imprinting; once wiped, Echo finds it impossible to continue any part of the performance.[24] A similar reset can be seen in "Briar Rose" (1.11) when all the L.A. Dolls lose consciousness simultaneously and are reset according to Alpha's programming. Here, not only are their identities destabilized, but also they are co-opted into Alpha's plan and transformed into murderous weapons, a successful reprogramming according to Alpha's aims. Here we see an early example of imprinting "tech" becoming weaponized and, moreover, a further stage of abjection as the Dolls are objectified. The very bodies of the Dolls become weapons; the question is not just into whom but into *what* "they" (in this case, the Rossum Corporation) can make an individual.

"Needs" (1.8) demonstrates a further instance of a reset or wipe, this time with Dolls reverted to identities that are close to their original personalities but with no memory of the Dollhouse or their previous lives. Echo, Sierra, Victor, and November take themselves on quests or, as Adelle puts it, "self-guided journeys," to resolve key moments, "particularly poignant or reoccurring experiences," as Claire Saunders describes them. This storyline poses interesting questions about the formation of identity and the resiliency of memory. Although the Dolls are unable to remember either their names or their lives before or within the Dollhouse, they are nevertheless able to identify those aspects of their lives that are troubling and that need "closure." They are able to dress—or costume—themselves appropriately with clothes that fit their true identities; most obviously, November dresses herself for the role or performance of mother, with a very different choice of costuming than the other three Dolls. The presentation of the "quests" for each Doll and, to some extent, their resolution at the end of the episode allows for a feeling of closure for the viewer. It has been suggested that the possibility of removing painful memories is what persuades some individuals to sign up for the Dollhouse in the first place.[25] However, this episode reveals the disturbing extent to which the Dolls are subject to exploitation and disempowerment, both through the imprinting process and through the consent that each individual has given the Dollhouse to erase identities. For example, because her identity has been removed/erased, Sierra/Priya does not know that she has been kidnapped and forcibly admitted to the Dollhouse and is being sexually exploited—indeed, raped—on a regular basis by Nolan Kinnard. Sierra/Priya's quest in "Needs" is to recognize and confront her past; in "Belonging" (2.4), I might argue that she completes her quest by finding and killing Kinnard, but she disavows this act and her past. Although her memory and identity have been restored, Priya returns to the Dollhouse and specifically asks Topher to "skip this day. . . . I don't ever want this back." Though Priya seems to be appealing to the "good deeds" that the Dollhouse—and the Rossum Corporation as a whole—performs for its clients and makes her claim to Topher on this basis, the episode does acknowledge Topher's troubled response to her request, and it is clear that this kind of selective erasure is a "morally gray area."[26] Its problematic nature is evident, too, if we examine the link between Sierra's identity loss and November's. In "Needs," November/Madeline returns to the site of her daughter's grave; her overriding need is to grieve her child's death. The narrative reveals that Madeline has agreed to become a Doll to be relieved of the pain of bereavement. Like Sierra, the erasing/imprinting process means that she has no memory of her past, but this means that she has no memory of her daughter or the fact that she was a mother. The process therefore disrupts and disavows key moments in her life, moments that have shaped her personality, and she is unable to grieve because she has no recollection of the need to do so. Therefore, a

Doll's identity is destabilized through imprinting or wiping, and it is altered through selective omissions. The suggestion remains that the "moral gray area" applies equally to the exploitation of the Doll's body and mind.

Examples of doubling or duplication introduce uncanny and abject moments to the narrative; however, though these examples show horror, this is frequently overlaid with enjoyment and pleasure in viewing the identities in action. Thus, horror and pleasure are simultaneously present in such moments. Doubling occurs early in the series, with the second Taffy imprint in "Gray Hour" called into action to finish the art heist, this time embodied by Sierra. It is present in "Haunted" (1.10) when Topher performs his annual "diagnostic" with a "super-specific imprint" that turns out to be an ideal best friend. I suggest that this "specific" friend-identity remains constant and is repeated (duplicated) year to year in different bodies; note that while Sierra embodies the friend-identity in this episode, she is selected at random and only because she has been "idle the longest" (i.e., not imprinted and not sent on any engagement). This infers that any Doll could embody this particular identity—that it is not tied to a specific gender or type of embodiment—and this is borne out in Sierra's androgynous performance with clothing, body language, and behavior that are not obviously identifiable as male or female. The uncanny notion that this identity could migrate through different bodies and remain more or less unchanged is muted in these scenes through a focus on apparently harmless activities (Sierra plays computer games and football with Topher and brings him a birthday Twinkie) and is reinforced through Adelle's indulgent eye upon the scenario. Other examples of doubling and duplication resonate more strongly with the uncanny and the monstrous.

The episode "Belle Chose" underlines the performativity of embodiment with a heightened example of imprinting that includes elements of duplication. Indeed, it signals its focus on performativity near the start of the episode, with a sequence in which Ballard accompanies Echo to be costumed for her Active identity. Unlike "Needs," in which the Dolls dress themselves in styles that fit their true personalities, in "Belle Chose" the dressing scene reinforces the way in which an Active imprint overwrites the Doll "tabula rasa." Echo begins this scene in her Active identity but wearing her Doll clothes and finishes in her Active costume, preening before a three-way mirror while Ballard looks on. The emphasis on looking—Echo in the mirror and Ballard at Echo—alerts the viewer to the undercurrents of voyeurism that operate here. In addition, the complex interplay between unease and comedy (e.g., the Handlers are invited to read magazines while they await their charges, as though they were out shopping with their significant others) invites the viewer both to enjoy the scene and to identify its connections with disorder, destabilization, and abjection.

"Belle Chose" begins with Echo imprinted as the ditzy student Kiki (a composite personality), while Victor is imprinted as the serial killer Terry

Karrens (a preexisting identity). The Terry and Kiki imprints are then accidentally switched. While Terry notices, with a "goodness gracious," that his body has changed gender, he does not appear concerned and continues on his murderous mission. The Echo/Terry gender switch is, in fact, downplayed in the remainder of the episode, and Echo does not obviously "perform" a male identity or display distinct contrasts in modes of embodiment, even though two personalities battle for dominance over one (female) body. In contrast, Kiki continues to behave as she did in Echo's body, but she seems to have no idea that anything about her body is different. With Victor "performing" Kiki, the viewer is presented with a heightened disruption of gender norms: an out-of-control *male* body giggling and gyrating on a dance floor and flirting with a group of young men. This scene is funny because of the incongruity of embodied behavior, but it shows an exaggerated transgression of boundaries and so approaches the abject. The collision of pleasure and abjection here may extend to the actor's participation: David Solomon, the episode's director, admitted that he asked Gjokaj for take after take because he found his performance as Kiki so funny.[27] Showing the same imprint in two different bodies certainly presents challenges for the actors involved in creating a consistent performance; the overt gender disruption of Victor/Kiki further overturns boundaries and creates resonances between viewing pleasure and horror.

A "doubled" imprint that is more obviously played for comedy is that of "Topher 2.0" in season 2. The character of Topher occupies a particular place within the narrative. Like Xander in *Buffy*, Fred in *Angel*, or Wash in *Firefly*, Topher is a clever, funny, geeky character, a type often cited as a cipher for Whedon himself (something that Whedon has acknowledged; see Wilcox). In *Buffy*, *Angel*, and *Firefly*, this "geek" character is central to the group, functioning as the heart of the enterprise.[28] Watching *Dollhouse*, a regular viewer of Whedon productions may expect Topher to fulfill the same central, likable role, and although this expectation is played against for much of the series with Topher presented as an antagonist, a deliberately unappealing character with questionable moral values, the character still receives positive reactions from fans. So, the duplication of Topher in Victor's body in "The Left Hand" (2.6) and "The Hollow Men" (2.12) is pleasurable for a variety of reasons, including enjoyment in both Topher's own character and Gjokaj's outstanding performance of that character's signature behavior and speech patterns.[29] The duplication demonstrates Topher's cleverness as he shows that he can be, quite literally, in two places at once ("The Left Hand"). This possibility is hinted at when Topher declares that there is only one person whom he will trust to synchronize the process of retrieving Echo from the Washington, D.C., Dollhouse, but there is still the possibility of surprise for viewers when Topher telephones "himself" to set things in motion. The sight of Victor performing Topher *is* enjoyable; Topher's dress sense and mannerisms are evident, and the banter between the two is entertaining. In the midst of the enjoyment, aspects of the scene still draw attention to the destabilization of identity that is taking place and offer connections with uncanny and abject doubles and doppelgangers.[30] Although Topher 2.0 (i.e., Topher's personality in Victor's body) is, for the most part, aware that he is not "really" in D.C., he appears to periodically forget this, giving rise to statements like "I wish I was there!" or "I see what you see"; he follows this latter statement with a double take and recognition of the situation with a quiet "yeah" that underlines the point. Visually, Topher's doubling is emphasized in the scene—even with the characters in different locations—with fast intercutting between them, which suits their fast-paced, back-and-forth conversation. Later in the episode, when the two characters are both in the L.A. Dollhouse, they mirror each other with hands on their chins and duplicate entire sentences. Once again, the pleasure in watching Topher with his duplicate is complicated by horror; this duplication of a known character is weird and uncanny.[31]

The duplication of identity has a further, more extreme role in the series, with the imprint as a reanimation of the corpse, a means by which the dead can live again. Several episodes focus on imprints that are the fully formed personalities of people who have died. Joel Mynor meets his dead wife, Rebecca, once a year, in a regular, prebooked engagement with the Dollhouse ("Man on the Street," 1.6), presenting a scenario with which a viewer can sympathize and simultaneously find disturbing and uncanny.[32] Adelle's friend Margaret comes back to life to investigate her own murder ("Haunted"), recalling the search for closure that was a key element of "Needs," though, as in that episode, Margaret cannot "change the story itself" and eventually must relinquish her imprinted body.[33] Performance or masquerade is particularly emphasized in these cases; Joel Mynor is fully aware that Rebecca is dead and that all he is experiencing is wish fulfillment. Margaret, meanwhile, must masquerade as a different individual due to her embodiment as Echo while in the midst of her family and friends, but she still gives away her true identity through a "repetition of acts" that her son recognizes.[34] In a more complex example, an imprint allows a modified version of the Dollhouse physician, Dr. Saunders, to live on after Alpha kills him; this modified identity is imprinted onto a former Doll, creating "Claire Saunders" ("Omega," 1.12). A parallel example here is Ballard's transformation into an imprint or Doll of himself ("The Attic," 2.10, and following episodes), which brings him back to life after he is declared brain-dead. This plot device makes apparent another disturbing aspect of the imprinting process. As Randell-Moon notes, Ballard recognizes that Caroline "may as well have been murdered" when her personality was wiped and stored by the Dollhouse ("Ghost," 1.1), and this judgment applies to any and all of the Dolls, "even though they are bodily 'alive.'"[35] The abject, "empty" body of the Doll thus strongly parallels that of the corpse.

Toward the end of season 2, the problematic nature of the imprinting process and the ensuing destabilization of identity is recognized more overtly. Despite the strongly comic overtones of Victor as Topher (and Gjokaj's acclaimed performance),[36] by the conclusion of "The Left Hand," it is evident that Topher is troubled by the presence of another self who tries to argue that he should be allowed to stay in Victor's body; Topher wipes his duplicate quite abruptly and shows his relief when, in response to Victor's "Shall I go now?" he departs from the script and replies, "For the love of God, yes!" And he should be troubled. At the end of season 1, "Epitaph One" (1.13) reveals Executive Vice Chairman Ambrose visiting the L.A. Dollhouse by downloading his personality into the body of Victor.[37] While he gives DeWitt an ultimatum about the new "complete anatomy upgrade" that the company intends to offer and which he is demonstrating, he takes the opportunity to stuff himself with seafood, to which his own body has an allergy. This depiction of the ravenous personality using another body to fulfill its own wishes is chilling, yet it is no different from the behavior of all those throughout the series who hire Dolls to fulfill their pet scenarios. In the case of the "complete anatomy upgrade," it suggests the potential for a kind of immortality whereby a consciousness might travel from body to body, outrunning disease and overturning the boundary between life and death—once again, becoming abject and horrific.

This potential is fulfilled in the dystopian "Epitaph Two: Return" (2.13), in which Ambrose and another Rossum executive, Harding, are shown in Neuropolis. Here, individuals whose minds have been wiped are dehumanized, treated as "dumbshows," "merchandise," or "suits." The "dumbshow" Doll bodies are made available to those who have power over the imprint technology, so Harding can eat and drink as much as he likes and switch to a new "suit" when he feels like it. Ambrose's appearance further underlines the horror present here, since he was killed in an earlier episode ("Getting Closer," 2.11) and, now present in another body, takes on the connotations of the abject, reanimated corpse. According to narrative chronology, the first person to carry out the multibody transfer that Ambrose demonstrates in "Epitaph One" is Topher in "The Left Hand." In a doubling of key narrative moments, the swap that is played as humorous with Topher becomes horrific for the bodies co-opted by Ambrose and Harding, just as the technology that Topher creates becomes militarized and brings about the end of civilization. The remote device used to wipe the Dolls in "Belle Chose" becomes the technology that creates "dumbshows" and "butchers" in the future dystopia of "Epitaph One" and "Epitaph Two." Viewer response to Ambrose and Harding's horrific activities is complicated by the comedy of the Topher 2.0

swap (and it is complicated further if a viewer has not seen "Epitaph One," which never aired in the United States).[38]

These examples show that while audiences may focus on aspects that are enjoyable and pleasurable, it still is possible for a television series to introduce and examine serious, difficult, and even disturbing or distasteful content in the context of mainstream entertainment. That *Dollhouse* frequently suggests that the viewer is not only complicit in watching but also part of a kind of "dollhouse" in reality[39] may make more uncomfortable the series's examination of identity and its destabilization.[40] However, this discomfiting viewing experience, with its echoes of real-world issues of (sexual) exploitation, encourages real discussion. As blogger meloukhia observes, "the question is 'are viewers of *Dollhouse* actually engaging with these issues as a result of seeing them depicted on the show?'"[41] The difficulties that characters have with the "moral gray area" of the imprinting process, the unease around the suggestion that the Dollhouse "can give you life after death" ("Haunted"), and the horror associated with the "complete anatomy upgrades" presented as a service for wealthy clients but enacted as a process of total control and exploitation ("Epitaph One," "Epitaph Two") all contribute to a complex viewing process. Such is the show's insistence on revealing these unpleasant and troubling issues that viewers may find themselves questioning their own enjoyment of a narrative that constantly chides them for their complicity in watching objectified and exploited bodies. In watching *Dollhouse*, the viewing experience itself is destabilized through the interactions of monstrosity, horror, and pleasure.

BIBLIOGRAPHY

Abbott, Stacey. "Rabbits' Feet and Spleen Juice: The Comic Strategies of TV Horror." In *TV Goes to Hell: An Unofficial Road Map of "Supernatural,"* edited by Stacey Abbott and David Lavery, 3–17. Toronto, ON: ECW, 2011.

Bennett, Eve. "Deconstructing the Dream Factory: Personal Fantasy and Corporate Manipulation in Joss Whedon's *Dollhouse*." *Slayage: The Journal of the Whedon Studies Association* 9, no. 1 (2011).

Braidotti, Rosi. "Teratologies." In *Deleuze and Feminist Theory*, edited by Ian Buchanan and Claire Colebrook, 156–72. Edinburgh, Scotland: Edinburgh University Press, 2000.

Buckman, Alyson. "'Didn't Get the Memo? Hero of the People Now': Hat Tricks and the Complication of Viewer Responses in the Works of Joss Whedon." Presented at the Slayage Conference on the Whedonverses 5, Vancouver, BC, July 2012.

Bussolini, Jeffrey. "Television Intertextuality after *Buffy*: Intertextuality of Casting and Constitutive Intertextuality." *Slayage: The Journal of the Whedon Studies Association* 10, no. 1 (2013).

Butler, Judith. *Gender Trouble: Feminism and the Subversion of Identity*. London: Routledge, 1990.

Calvert, Bronwen. "Mind, Body, Imprint: Cyberpunk Echoes in the Dollhouse." *Slayage: The Journal of the Whedon Studies Association* 8, nos. 2–3 (2010).

Canavan, Gerry. "Fighting a War You've Already Lost: Zombies and *Zombis* in *Firefly/Serenity* and *Dollhouse*." *Science Fiction Film and Television* 4, no. 2 (2011): 173–203.

Carroll, Noël. "Horror and Humor." *Journal of Aesthetics and Art Criticism* 57, no. 2 (1999): 145–60.
Chow, Lesley. "A Thousand Blooms inside Joss Whedon's *Dollhouse*." *Bright Lights Film Journal* 68 (2010).
Coker, Catherine. "Exploitation of Bodies and Minds in Season One of *Dollhouse*." In *Sexual Rhetoric in the Works of Joss Whedon: New Essays*, edited by Erin B. Waggoner, 226–38. Jefferson, NC: McFarland, 2010.
Creed, Barbara. *The Monstrous Feminine*. London: Routledge, 1993.
"Discuss the Fifth and Sixth Episodes of *Dollhouse* Season 2." Whedonesque.com. http://whedonesque.com/comments/22556. December 5, 2009.
"Enver Gjokaj Spills Victor's Secrets." Whedonesque.com. http://whedonesque.com/comments/20010. April 24, 2009.
Grosz, Elizabeth. "Notes Toward a Corporeal Feminism." *Australian Feminist Studies* 5 (1987): 1–16.
———. *Volatile Bodies: Toward a Corporeal Feminism*. Bloomington: Indiana University Press, 1994.
Hawk, Julie L. "Hacking the Read-Only File: Collaborative Narrative as Ontological Construction in *Dollhouse*." *Slayage: The Journal of the Whedon Studies Association* 8, nos. 2–3 (2010).
———. "More Than the Sum of Our Imprints." In *Inside Joss' "Dollhouse": From Alpha to Rossum*, edited by Jane Espenson, 247–57. Dallas, TX: Smart Pop, 2010.
Kristeva, Julia. *Powers of Horror*, translated by Louis-Ferdinand Céline. New York: Columbia University Press, 1982.
Lyden, Jacki. "Welcome to the *Dollhouse*: Meet the Anti-Buffy." National Public Radio. February 7, 2009.
meloukhia [S. E. Smith]. "Feminism and Joss Whedon." *This Ain't Livin'* blog. July 17, 2009.
———. "Much Ado about *Dollhouse*." *Deeply Problematic* blog. September 1, 2009.
Randell-Moon, Holly. "'I'm Nobody': The Somatechnical Construction of Bodies and Identity in Joss Whedon's *Dollhouse*." *Feminist Media Studies* 12, no. 2 (2012): 265–80.
Ryan, Maureen. "Sex, Secrets and *Dollhouse*: Interview with Joss Whedon." *Chicago Tribune*. December 3, 2009.
Tjardes, Sue. "'If You're Not Enjoying It, You're Doing Something Wrong': Textual and Viewer Constructions of Faith, the Vampire Slayer." In *Athena's Daughters: Television's New Women Warriors*, edited by Frances Early and Kathleen Kennedy, 66–77. Syracuse, NY: Syracuse University Press, 2003.
Wilcox, Rhonda V. "Echoes of Complicity: Reflexivity and Identity in Joss Whedon's *Dollhouse*." *Slayage: The Journal of the Whedon Studies Association* 8, nos. 2–3 (2010).

FILMOGRAPHY

"Looking Back." *Joss Whedon's Dollhouse: Complete Season 2*. Disc 4. Los Angeles: Twentieth Century Fox Home Entertainment, 2010. DVD.
"Primeval." *Buffy the Vampire Slayer*. Season 4, episode 21. Written by David Fury. Directed by James A. Contner. First broadcast May 16, 2000.

NOTES

1. Quoted in Ryan, "Sex, Secrets and *Dollhouse*." "They" are, of course, the fictional Dollhouse, but equally "they" can stand for the real television network, the viewers, and/or the writers and actors. The comments at the end of Ryan's interview are extremely interesting in what they reveal about the split in reactions to *Dollhouse*: commentators both appreciate and understand what Whedon is attempting with difficult subject matter and are repulsed by what

they see as the "nastier" elements of the show (in particular, the overtones of exploitation and sex trafficking).

2. meloukhia, "Feminism."
3. Bennett, "Deconstructing the Dream."
4. Quoted in Ryan, "Sex, Secrets and *Dollhouse.*"
5. Wilcox, "Echoes of Complicity."
6. See Ryan, "Sex, Secrets and *Dollhouse*"; Lyden, "Welcome to the *Dollhouse.*"
7. As discussed by, among others, Grosz, "Notes Toward"; Butler, *Gender Trouble.*
8. Kristeva, *Powers of Horror*, 4.
9. Carroll, "Horror and Humor," 152.
10. Abbott, "Rabbits' Feet," 5, 11.
11. Bussolini, "Television Intertextuality."
12. For more on hyperdiegetic casting within the Whedonverse, see Buckman, "Didn't Get the Memo?"
13. Tjardes, "Enjoying It," 67.
14. "Enver Gjokaj."
15. Creed, *Monstrous Feminine*, 53.
16. Grosz, "Notes Toward," 9.
17. Butler, *Gender Trouble*, 140 (italics original).
18. Grosz, *Volatile Bodies*, 191, 190.
19. Braidotti, "Teratologies," 167.
20. Sierra and Victor, for instance, though wiped and imprinted repeatedly, still apparently feel love for each other even in their Doll states.
21. Randell-Moon, "I'm Nobody," 275.
22. Chow, "Thousand Blooms."
23. For further discussion of this aspect of the episode, see Coker, "Exploitation of Bodies," 229.
24. This episode emphasizes the *performance* of Echo's Active identities, opening with Echo in the identity of a midwife delivering a baby in a mountain chalet, before showing Echo as "call girl" Taffy, as Taffy, and as her "wiped" Doll-self. While trapped in the vault, Doll-Echo appears to recall aspects of her midwife performance as she stares at a painting of snowy mountains and murmurs, "When I'm there, my name is something else." (Thanks to Alyson Buckman for noting this emphasis.)
25. Canavan, "Fighting a War," 191.
26. Coker, "Exploitation of Bodies," 228.
27. "Looking Back."
28. This is manifested literally when Xander takes on the role of "heart" in a spell to unite the powers of the Scooby Gang in the *Buffy* episode "Primeval" (4.21).
29. "Discuss."
30. Creed, *Monstrous Feminine*, 53.
31. I would go so far as to suggest that it is his experience *as* Topher that makes it possible for Tony to trust Topher at the beginning of "The Hollow Men." Tony decides to activate the wedge that imprints him again with Topher 2.0, a potentially fatal decision. If it is motivated by the "somatic history" of his experience as Topher (Randell-Moon, "'I'm Nobody,'" 275), then it is more comprehensible. This episode reveals that it is Topher who first suggests to Tony the prospect of adding technological "enhancements" to his own identity (Topher lists such useful abilities as "ninja fighting skills . . . cooking"), giving Tony apparently helpful and desirable attributes that, by "Epitaph Two," have become extreme and controversial.
32. As Alyson Buckman notes in her editorial remarks, this episode is a particularly good example of the way *Dollhouse* plays with audience perception over content that can be interpreted as both disturbing and pleasurable.
33. Hawk, "Hacking."
34. Butler, *Gender Trouble*, 140.
35. Randell-Moon, "I'm Nobody," 269.

36. Both actors have been asked questions about their performances; so convincing is Gjokaj's vocal performance that Fran Kranz mentions being asked if he had dubbed both Tophers' voices ("Looking Back").

37. He tells DeWitt that he is simultaneously present in other Dollhouses. Clyde 2.0 reveals a similar, though probably serial, doubling process when he comments, "I use a lot of bodies" ("The Hollow Men").

38. Extremely problematic is the way that viewers are invited to see Paul's "immortality" with Echo as a resolution to their story and therefore pleasurable, especially since, throughout the series, Paul has opposed the Dollhouse's use of any such practices.

39. As Topher insists in the unaired pilot, "everybody's programmed" ("Echo"). Canavan and Bennett note *Dollhouse*'s metafictional tendencies, especially regarding parallels between Rossum and the Fox network.

40. Noted by Wilcox, "Echoes of Complicity"; Bennett, "Hacking"; Hawk, "More Than."

41. meloukhia, "Much Ado."

Chapter Nine

"I Love Him . . . Is That Real?"

Interrogating Romance through Victor and Sierra

Lorna Jowett

While looking for *Dollhouse* images online one day, I came across a screenshot of Doll characters Victor and Sierra walking away from the camera holding hands. The words "Victor and Sierra give me hope that true love exists" had been added across the front of the picture, and the image appeared on a site called Dollhouse Confessions.[1] This one image and its accompanying "confession" capture many of the issues addressed in this chapter, including *Dollhouse*'s interrogation of gender, sexuality, and heterosexual romance as well as how viewer responses to the "true love" between Victor and Sierra might pull against such an interrogation. The chapter derives from my long-term interest in representations of gender, sexuality, and romance, as well as that in the genre of science fiction and how it can offer new representations. It also derives, more subjectively, from the fact that despite years of analyzing, critiquing, and interrogating fictional representations of gender and dismissing romance as a construction designed to perpetuate heteronormative, patriarchal social systems, as I watch *Dollhouse*, Victor and Sierra make me hope that true love exists. They make me root for "true love," something that I otherwise consider part of the myth of heterosexual romance. The word "confessions" in the website title suggests that I am not alone in reluctantly admitting this emotional desire to believe in something that I might intellectually reject.

In part, then, this chapter deals with two conflicting responses in one viewer. This is not quite the same as the split between scholar and fan that Matt Hills and others have analyzed.[2] In places, therefore, this chapter draws on academic analysis of television drama and science fiction. In other places, it draws on my own subjective response, containing warring impulses, to

Victor and Sierra's "true love" as depicted over two seasons of *Dollhouse*. I know, from presenting an earlier version of this chapter at the biennial Whedon Studies Association's Slayage conference to a roomful of Whedon scholar-fans in 2012, that my own responses to Victor and Sierra's relationship are not singular. I cannot promise to offer clear-cut answers to the questions that I pose as part of my interrogation of those responses and *Dollhouse*'s representations. However, I think that even the attempt at analysis raises important issues, not just about this television show and its work with gender, but about larger cultural representations of heterosexuality and romance and how we respond to them.

Because of Joss Whedon's commitment to what he regularly calls feminism in interviews and commentaries, his creations consistently have interrogated the myth of heterosexual romance as part of a broader project of renegotiating traditional gender representation. Long-running television shows such as *Buffy the Vampire Slayer* and *Angel* offered wide scope for examining romance alongside other aspects of gender and sexuality. The mix of genre conventions in these earlier shows also lends itself to negotiating romance from different angles, from *Buffy* drawing on teen drama as it follows characters growing up and changing their own ideas about romantic and sexual relationships, to both series testing what you can "get away with" in a fantasy show about vampires. *Firefly* featured both a happily married couple and a sex worker—neither commonplace in network television drama—allowing that shorter-lived series to move away from obvious conventions of romance. And then there is *Dollhouse*, where almost all the characters are effectively prostitutes or pimps. Melissa Milavec and Sharon Kaye suggest that *Buffy* "owes much of its popularity to making erotic love a dominant theme."[3] *Dollhouse* may owe its lack of popularity to the way it treats much the same theme in a more disturbing fashion.

"Like every good fairy tale, the story grows more intricate, and more divisive, every decade," says a reporter presenting *vox populi* sound bites dealing with the Dollhouse urban myth in "Man on the Street" (1.6). His words are applicable equally to the myth of heterosexual romance as tackled by the Whedonverses on television. The Whedon shows offer a sustained interrogation of gender but are complicated by the demands of mainstream entertainment. The premise of *Dollhouse* apparently continues both trends. In this series, advanced technology controlled by the Rossum Corporation means that the memories of an individual can be wiped somehow and that individual "programmed" with new personalities. People may be persuaded to sign on and become wiped of all conscious memories (and thus become a Doll), conducting "engagements" for a fixed period (though the show eventually reveals that not all Dolls acceded to their stint in the Dollhouse and not all return to their previous lives when that term ends). During this time, clients of the Dollhouses rent romantic and sexual fantasies as embodied by a

Doll, and romance becomes prostitution. As Cynthea Masson points out, this led some viewers to read the show as brutally misogynistic.[4] Yet, since the Dolls have new identities downloaded for each new engagement, *Dollhouse* puts the notion of gender (and sexuality and romance) as social construction and performance front and center.

However, the show's premise means that character development cannot work in the usual ways, since many of the characters change "personalities" from engagement to engagement, sometimes several times during an episode. Thus, the romantic connection between Victor and Sierra is one means of encouraging audience identification with the Dolls and widening engagement with them beyond protagonist Echo. This relationship humanizes the Dolls and creates sympathetic characters. It also suggests that the Dolls retain some form of essential identity, that their "original" personalities are not entirely erased by the "treatments," since this romance manifests as a connection between originals Priya and Anthony as well as Dolls Sierra and Victor. These two are the only *Dollhouse* characters for whom romance remains relatively untarnished, and their relationship also provides hope in the dystopian atmosphere of the show's world. In this chapter, the notion that Sierra and Victor have some form of essential connection, that they are "soul mates," is tackled in the context of genre and how it affects the show's (and the Whedonverse's) negotiation of heterosexual romance, subjectivity, and free will.

Dollhouse's dystopian setting and advanced technology situate it as science fiction, and I argue here that Victor and Sierra fulfill a key function of this genre. Jan Johnson-Smith articulates the way that background in science fiction is often foreground. That is, science fiction generally strives for "an inversion of . . . the 'mundane' . . . through the foregrounding of the background—psychological, physical, or geographical."[5] Thus, features that in other genres might remain background details supporting a story unfolding in the foreground become the stuff in science fiction that tells us that this is indeed science fiction, and those features often become the story. In *Dollhouse*, for instance, the luxurious but bland atmosphere of the LA 'House is created through set design, costume, and actor movement, whereas the chair and its apparatus demonstrate and make concrete more sinister aspects of the "tech" that makes selling Dolls as fantasy companions possible. This is neatly exemplified by the arrangement of sleep pods for Dolls (five pods set into the floor, radiating from a round center), which become a key icon of the show, included in its logo and some promotional material. The pods, recognizable as beds yet disturbingly reminiscent of coffins or storage drawers, make visible in one striking image the commodification and dehumanization that the tech enables. The image thus foregrounds a major theme of the show, and its presentation as an icon in advertising and promotional materials suggests its importance as much more than a background detail.

Language also takes on this function of foregrounding background in science fiction, and novel terms litter the dialogue without being explicitly defined or explained. When their meaning becomes clear, such terminology can tell us much about how a particular science fictional world works: sometimes signaling a world different from our own, more often suggesting extrapolation of existing tendencies in our society. These novel terms might even be familiar words turned to new meanings (tech, engagements, Dolls), another rendering of the genre's ability to estrange the familiar. As Lillian Deritter, Hugh H. Davis, and Tom Connelly and Shelley S. Rees all point out, much of the language adapted in this way for *Dollhouse* is already loaded in terms of gender, forcing us to reappraise what these words usually mean and how they operate in this science fictional scenario. "The use of the word 'Doll' to describe actives," comments Deritter, "immediately charged the universe . . . in terms of gender."[6]

Victor and Sierra's relationship, I suggest, also functions as background/foreground, contributing not only to the show's representation of the tech as dehumanizing but also to its exposure of heterosexual romantic myth. These two characters are often in the background or on the periphery of a scene, yet their actions and their connection serve a key purpose in the structure of the show, as I examine in the rest of this chapter.

The core of much new science fiction drama on television is characterization and emotional realism rather than, or as well as, spectacle, action, and special effects (as seen in shows from the reboot of *Doctor Who* to *Fringe*). The Whedon shows consistently have focused on emotional realism no matter how far-fetched their premise. *Buffy*'s success was ascribed by early commentators as much to its literalization of "high school is hell" as to its writing, acting, directing, or role-reversing premise. Of course, emotional realism is a suspect concept itself in the Dollhouse since Dolls can be programmed to feel as well as perform. As Adelle DeWitt tells a client in "The Target" (1.2), "She won't lie to you, Mr. Connell. Everything you want, everything you need, she will be—honestly and completely."

The genre of science fiction consistently presents emotion as an index of humanity, something that stands in contrast to cold science and technology (as demonstrated in the clash of perspectives between Dr. McCoy and Mr. Spock in *Star Trek*). The demonstration of feeling—physical or emotional—between two "blank slates" such as Victor and Sierra thus demonstrates that Echo is not the only Doll who is still potentially human even in "tabula rasa" state (and consequently aware of being grossly exploited by Rossum). Therefore, on a structural level, as stated earlier, Sierra and Victor's connection is clearly used to make Dolls and other characters sympathetic. One of the problems with having characters who take on different personas each time they come out of the chair is that little opportunity is created for the development of those characters and subsequent engagement with them. Victor and

Sierra are lovers who love against the odds, a trope that audiences are cued to sympathize with, and the two characters even recognize and love each other despite apparently losing all memory and identity.

The couple thus elicit sympathy and empathy, as do several other characters associated with the Dollhouse, including Boyd, Topher, and Adelle. Topher Brink—scientist and tech genius—is, as Madeline Muntersbjorn points out, redeemed by his emotional response to the revelation that Priya has been wrongly diagnosed with schizophrenia by rejected lover Nolan Kinnard, who happens to be a valued Rossum employee ("Belonging," 2.4).[7] Kinnard sends Priya to the Dollhouse, where she becomes Sierra and his favorite plaything. This redemption also encompasses Topher's confirmation that Victor's love for Sierra/Priya is real in the lines of dialogue that I use as the title for this chapter: "I love him," says Priya, "Is that real?" and Topher responds, "Yes, it's real. He loves you back" ("Belonging"). Topher's willingness to help Priya escape Kinnard, as well as his reassurance that her love for Victor/Anthony is genuine, reveals a more caring side to the apparently irresponsible boy wonder.

The Sierra-Victor connection is developed throughout the series, spreading audience engagement with the Dolls beyond protagonist Echo and allowing us further insight into how being a Doll works and why a person might end up becoming one. This is emphasized especially in season 2, where we get backstories for both Priya and then Anthony, reinforcing their importance to the narrative and to the ensemble cast. Although this is not conventional character development (we find out about Tony and Priya, not Victor and Sierra), it serves much the same function and again foregrounds something that might be considered "background" in terms of character. Furthermore, it provides hope in a dystopian world, something emphasized in "Epitaph Two: Return" (2.13), in which we meet their son, we see Tony reconciled with Priya, and we see him introduced to young T as his father after a long estrangement. This reconciliation operates as shorthand for all kinds of resolution, balancing the grim content of the series and its not-so-happy ending. As Julie Hawk points out, "all of this serves to keep a humanist hero in play. What better representation than a walking reminder of that go-to humanist value, sexual reproduction?"[8]

The positioning of Victor and Sierra as Dolls and the development of sympathy for them form an obvious function of their characters. Yet, I suggest that their background/foreground presence also serves to complicate another key element of *Dollhouse*: its interrogation of gender. Our notions of traditional gender and sexuality, including myths of heterosexual romance, are often seen as social constructions. Whereas certain behaviors and characteristics might seem to be "natural," they are produced through a combination of social convention and repetition and designed to perpetuate patriarchal power. The premise of *Dollhouse*, "societal brainwashing" as Whedon calls it,[9] exposes these norms in a pretty scary fashion. Whedon notes, "Some people were made very uncomfortable because we're dealing with issues of sexuality and, by not explaining it, the more it seemed to be the elephant in the room."[10] This element of discomfort presumably accounts for those who saw the show as exploitation rather than about exploitation, as a fantasy of dominance and compliance rather than an exploration of how society perpetuates unequal power relations. Or, as Rhonda Wilcox puts it, "whether we are examining exploitation or participating in it is part of the question Whedon and Dushku put to us."[11]

Similarly, Muntersbjorn sees the show as "[provoking] us to see ourselves, in all of our deviant detail, as unique yet monstrous amalgams" of "slave, master, serial killer, or mad scientist," pointing to Whedon's comment, "'It's supposed to be about the sides of us that we don't want people to see.'"[12] I do not think it is reaching too far to suggest that the latter three roles cited by Muntersbjorn (master, serial killer, mad scientist) traditionally have been gendered male and the conventions that surround their representation tend to align with traditional masculinity. So, in following this line, are the sides of us that we do not want people to see, well, sexist? Or, are they at least complicit in upholding and perpetuating gender inequality and the myth of heterosexual romance? Remember, Whedon also admits to easily being able to identify with more traditional male perspectives: "I understand the motivation of the man with the murderous gaze, of the animal, of the terrible objectifying male, 'cause I'm him."[13] It is not difficult to see *Dollhouse* in this light, as the latest in a series of Whedon creations designed to negotiate the position of "the terrible objectifying male" in contemporary society. The show's exposure of myths of heterosexual romance is central to this negotiation.

Dollhouse poses questions about technology and humanity, fantasy and reality, identity and subjectivity, about corporate society and control, about power of all kinds. My own specific questions, in relation to Victor and Sierra, are as follows: How is Victor's response to Sierra any different from the other male obsessions in the show? Does the apparent innocence of their love suggest that society is the problem with romance? Does the notion that Sierra and Victor are soul mates rule out free will? Much of the existing *Dollhouse* scholarship focuses on Echo and other, more prominent characters, such as Paul, Topher, Adelle, and Boyd, and this seems to be a question nobody thus far has addressed.[14] Looking at the rundown of clients, especially in season 1, there is a preponderance of obsessive heterosexual men. Some, like Joel Mynor in "Man on the Street" (1.6), are presented relatively sympathetically, enabling the show to shed light on Paul's rather more dubious obsession with Caroline. Others, such as rogue Active Alpha, are shown to be capable of change. So, is Victor's "love" for Sierra aligned with or distinguished from the behavior of these other obsessive heterosexual men?

One obvious answer that might distinguish Victor is that his love is innocent, contrasting the corrupted versions of "love supreme" ("A Love Supreme," 2.8) touted by other male characters.

The first evidence of Victor's feelings for Sierra is physical (the "man reaction" Topher notes in "True Believer" [1.5]), and this is open to several interpretations. Given the narrative arc about Sierra being abused in the Dollhouse that continues into the next episode ("Man on the Street"), Victor's sexual interest in her puts him in the frame as potentially her abuser. While the so-called man reaction can be seen simply as a male reaction (male babies get erections too)[15] that possibly implies Victor's innocence, his erection is clearly presented as a response to seeing Sierra naked in the showers. However, it is also presented as being "on a par" with other pleasant physical sensations: Sierra says that the warm water "feels nice," and Victor repeats the same words in response to his erection. The viewer might see this as ironic, yet it can equally be read at face value. This narrative eventually does reinforce the sense that Victor's feelings are innocent and other male characters' are not by showing his concern for Sierra and proving that her Handler is using his position and the Doll's imprinted trust to sexually abuse her. Still, it is notable in reading about these characters that people tend to cite Victor's love for Sierra, not Sierra's love for Victor. It is Victor who is seen seeking out opportunities to spend time with Sierra: walking toward her while she is reading or drawing, coming to the dining table where she is sitting. Like many of the obsessive males depicted across the seasons, this "love" seems rather one-sided. Remember, it is Priya whom we see confess her feelings for him, not Sierra, and this does not come until season 2. The show establishes Victor's physical attraction to Sierra: is this what we mean by "love"? Is obsession merely a physical drive? Commentators such as Deritter note Victor's passivity: is it only this that prevents him from pressing his attention on Sierra?[16] We probably do not want to think so, though arguably the show implies that this might be the case. Posing questions like this, however subtly, indicates how hard it is to tell the difference among "real" love, physical arousal, myths of romance, and abuses of power.

When Echo "marries" in "Vows" (2.1) as part of an engagement designed to catch an illegal arms dealer, the images of a stereotypically perfect wedding day that we are offered might make some people sigh with pleasure and delight. Yet the romance is not real: it is just another engagement (and here the irony of the terminology becomes immediately apparent). The images of flowers, dresses, and the apparently perfect couple with which we are presented and we know are not real are familiar from advertising and other media. They are exactly the type of images that perpetuate the myth of heterosexual romance, and the episode offers them from the start as a performance that quickly breaks down into distrust and violence. Remember, Whedon comments that the whole show demonstrates how "everything that was sweet was kind of creepy."[17] It is worth bearing this in mind when we recall Sierra's words to Dr. Saunders in "Man on the Street": "Victor wants to pretend. He pretends we're married." While a variety of behaviors might constitute "pretending" to be married (from playing house to having sex), the narrative uncertainty here about who is abusing Sierra draws out the more sinister overtones of this statement, aligning marriage with rape, however briefly.

Given the apparent "innocence" of Victor's feelings, something echoed by actor Enver Gjokaj,[18] it is not surprising that we never see Sierra and Victor in a sexual scene, despite his "man reaction" to her. During "The Attic" (2.10), Priya's nightmare transforms sex with Anthony into abuse from Nolan, whom she stabs. This sexual activity is thus distanced from the Dolls, since it takes place in a dream space and it features Anthony and Priya, not Victor and Sierra. In this instance, it is fairly overt that something "sweet" turns into something "creepy," since the Attic sets up loops to imprison its victims. This folding together of creepy/sweet may partially account for the often contradictory ways that the show's writers, creators, and viewers talk about romance in *Dollhouse*. On the DVD commentary to "Belonging," for instance, writer Jed Whedon uses "friendship" to describe Sierra and Victor's relationship, yet a little later cowriter Maurissa Tancharoen calls it "true love."[19]

Furthermore, by virtue of Victor and Sierra being Dolls, Victor's feelings might seem to be more authentic than other emotional or sexual connections because they are not reliant on social context or performance of social roles. That is, if the Dolls are really "tabula rasa," then they have no social conditioning about gender and sexuality that might perpetuate patriarchal standards and inequality. Female Dolls should have no memory of social training that dictates that they remain submissive or try to be sexually attractive; male Dolls should have no memory of role models promoting aggression and dominance as integral to courtship or of imagery presenting women as sexual objects. This starts to answer my second question: is society to blame for the problems attendant on the myth of heterosexual romance? This is not really about social constructions of love in general: memory wipes were integral to plots in both *Buffy* and *Angel*. In those cases, the love that persisted regardless of memory loss was a family bond between sisters Buffy and Dawn or between biological parent Angel and child Connor. *Dollhouse* has a more difficult task in asking whether "love" can be shorn of the myth of heterosexual romance. It may, for example, be easier for an audience to see family love as somehow genetic, despite many real-life examples suggesting that biological connections do not necessarily result in loving relationships. As the cliché goes, you cannot choose your family. But selecting a romantic or sexual partner is presented in contemporary society as a conscious choice made by individuals, and this tends to set it up as something personal rather than social, despite social conditioning around gender or social structures of heteronormativity.

One of the answers is that neither soul nor romance nor even love is a factor; it is just physical. Connelly and Rees note the way that Dolls are compared to animals: "In a sense, they exist as their animal selves, eschewing what Marx describes as 'conscious life activity.'"[20] The authors remind us of Topher describing the Dolls as "a little bit bison" ("Gray Hour," 1.4) and Bennett Halverson's "free range chickens" and "veal" comparison ("Getting Closer," 2.11). Such descriptions are—entirely consciously, I am sure—distasteful, especially by the second season, where we have come to know certain Dolls as individuals. To describe the "tabula rasa" Dolls as animals is an obvious form of dehumanization and has a long history in the language of slavery and oppression. Yet (background) scenes where we see Dolls engaging in artistic activities as well as physical ones offer a counterbalance to this "animal" instinct. Like emotion, aesthetic appreciation is often used in science fiction (for instance in *2001: A Space Odyssey* [1968], *Blade Runner* [1982], or *Star Trek: The Next Generation* [1987–1994]) as a measure of humanity, something that distinguishes us as self-conscious, intelligent life. (There is also historical context for this in terms, for example, of the abolitionist movement in the United States and debates around the authenticity of the work of writers such as Phyllis Wheatley and Frederick Douglass.) At the end of *Blade Runner*, replicant Roy Batty makes a lyrical speech about the wonders that he has witnessed during his travels: "I've seen things you people wouldn't believe. Attack ships on fire off the shoulder of Orion. I watched C-beams glitter in the dark near the Tannhauser Gate. All these moments will be lost in time, like tears in rain."[21] Roy's appreciation and awe at these remembered spectacles, his ability to express them lyrically, and his sense of their transitory nature humanize a character that is basically a created product of the Tyrell Corporation. The film's continuous questioning of difference (or similarity) in relation to the replicants is easily related to historical debates around class or race, which have used similar signifiers to define humanity.

Moreover, the suggestion that, without memories and social identity, we are simply instinctive animals threatens Western society's highly valued notions of individuality and free will. Trailers for *Dollhouse* used the tagline "You can wipe away a memory, but can you wipe away a soul?" Aligning these questions with Echo implies that the key issue is individual identity and whether memories constitute subjectivity. If we are just animals, the fittest survive, and sex is simply a means to reproduction, another biological urge. In the episode "Instinct" (2.2), Echo is engaged to act as the mother of a child by his bereaved father. The title and the narrative of this episode suggest that Echo acquires maternal instinct at a "glandular level" by being imprinted with it, and she is even able to lactate. At this point in season 2, regular viewers already know that Madeline's overwhelming grief for her lost daughter is what causes her to enter the Dollhouse and become November (see "Needs," 1.8). So, is biological essentialism real, or is it just social imprinting? *Dollhouse* keeps its options open. It is not too comfortable; therefore, that Victor and Sierra (albeit as Tony and Priya) have a child by the end of the series . . . is this the sum of their love? That "go-to humanist value, sexual reproduction," as Hawk refers to it? Of course, we are free to interpret the existence of the child T as the culmination of a biological urge on the part of Victor/Anthony and Sierra/Priya, as the imprinting of a heteronormative society, or as the need to repopulate in the face of an apocalyptic event.

Drawing on Elisabeth Bronfen, Bronwen Calvert observes that "the basic situation of *Dollhouse*'s narrative" inevitably "questions . . . whether it is possible to create 'a body devoid of all discursive and cultural delimitations.'"[22] I suppose one reason that viewers may not be too comfortable with the idea of the Dolls as just bodies is that this seems to accede to their positioning as Actives/prostitutes/skinsuits for hire. As Calvert notes, the show asks whether it is possible to create an unacculturated body, rather than answering this question. Even if its answer is yes and we believe that Victor and Sierra can "fall in love" and that love persists when they revert to Priya and Anthony, they still have to live in the world. Notably, when they are Priya and Tony in the postapocalyptic world of "Epitaph One" (1.13) and "Epitaph Two," their relationship is more real and more problematic than it was as Victor and Sierra.

Is this the case because, whereas Tony and Priya have choices to make, Victor and Sierra do not? Many of the Whedon shows have tackled fate, predestination, and prophecy, tending to uphold the value of choice. Buffy kicked against being the Chosen One and survived being "Prophecy Girl" right from the first season ("some prophecies are a bit dodgy," admits Giles);[23] *Angel* consistently overturned and/or played with the idea of destiny and prophecy, often arguing, as Angel says, "Don't believe everything you're foretold."[24] *Firefly* and *Serenity* included stories about mind control (via River Tam) and featured Reavers apparently devoid of anything but violent instinct. All three shows valorized free will and the freedom to make choices. *Dr. Horrible's Sing-Along Blog* takes this to a sinister extreme when Billy chooses to pursue his goal and become supervillain Dr. Horrible rather than win his heterosexual happy ending with Penny. In *Dollhouse*, free will is given up voluntarily by those who sign up to be Dolls.

When the Dolls are temporarily allowed to work through some of their originals' "issues" in "Needs" and appear to escape the Dollhouse, Priya says that they decide for themselves now. Given that this is a plan engineered by those running the LA 'House, it is not exactly free will. While Echo and other Dolls eventually do escape the Dollhouse, at least partially, and start making decisions about how to resist their slavery, this action in the political sphere is not necessarily paralleled in the private, romantic sphere. If Victor and Sierra are "soul mates" to the point that Tony and Priya also fall in love, does this suggest that they are still programmed, not by the Dollhouse or by Rossum, but by social conditioning or biology, acting on some deeper instinct that precludes conscious choice? Is this what love is?

"It's as if I met this woman in a previous life," Victor as Roger tells Adelle in "Stop-Loss" (2.10), "as though our love transcends our very being," and the very fact that he should not remember this love while imprinted as Roger and on engagement with "Miss Lonelyhearts" seems to prove the truth of his words. While the "*Dollhouse* Confessions" image I described at the start of this chapter frames its comment on Victor and Sierra by positioning it as a "confession," other fan products are more wholehearted in their endorsement of the relationship. One montage features images of the pair together sandwiched between separate images of the two characters, positioned so that they seem to be gazing wistfully at each other. "Kindred souls," prominent writing at the top of the montage proclaims.[25] At least one essay in the anthology *Inside Joss' Dollhouse* takes a similar approach. "It was as if their souls had fallen in love with each other in the absence of the complications of their personalities, and that love continued to grow after their personalities returned."[26] Oluwafemi Morohunfola here states that, as well as social pressure and myths of heterosexual romance, personality is irrelevant to Victor and Sierra's connection. This directly contradicts the (perhaps equally idealized) notion that personality is or should be what we fall in love with, rather than, say, an attractive physical appearance, or social status, or economic success, or even the "correct" gender and/or sexual orientation. This falling in love without personality is clearly not intended to be a return to the animal metaphor, since the soul is a factor here. Defining the soul is a rather difficult task, and the ways that it is used in the series suggest some confusion about its referent, as well as its rhetorical power. "Soul" might be defined as the animating principle or the essence of a person, and it generally has a sense of the numinous and mysterious. The word "soul" elicits a different response from, say, "self-consciousness" or "self-awareness" even if all these words attempt to identify what distinguishes us from animals. An Active "is the truest soul among us," says Adelle in "Ghost" (1.1). While much of Adelle's sales pitch seems designed to make us wary and suspicious of the whole Dollhouse setup, readings such as Morohunfola's suggest that viewers come to believe statements like the one just quoted, at least when it applies to Victor and Sierra. "Soul mates" in particular taps into the notion of the soul as something inexplicable, and calling Victor and Sierra soul mates removes romance and sexuality from social context and positions it as something essential or transcendent.

Part of me wants to say that this talk of soul mates is nonsense; it is just a way to describe a feeling that may or may not be an idealization of accumulated factors that make two people "fall in love." The term is implicated in myths of heterosexual romance that present romance as something entirely apart from social structures of power: soul mates are simply destined to be together in the ultimate romantic closure. Another part of me wants to believe it. It would be easier to believe it. It would be nicer. It would remove responsibility for making (good or bad) choices of romantic or sexual partners. It did not work out? The reason is that they were not soul mates, obviously. The treatment of Paul and Echo in *Dollhouse* indicates the failure of heterosexual romance and the damage that social conditioning around gender and sexuality can do. Paul's investment in traditional gender roles and his obsession with a woman whom he has never met make it impossible to have a happy ending, a satisfying romantic closure. When Paul and Echo are united in Echo's mind, this is a disturbing or "creepy" version of the usually "sweet" notion of ultimate togetherness. Victor and Sierra, however, offer a more innocent version of romance, a love that is "true" because it is largely separated from society and social roles.

So do Victor and Sierra give us hope that true love exists? If so, what does that say about how we understand "love" and how invested we are in the myth of heterosexual romance? We want to accept their "love" at face value, and there are pressing narrative reasons to give the characters a form of happy ending to alleviate the dystopian gloom of the *Dollhouse* world. It is easy, watching *Dollhouse*, to see the corporate Big Bad and want to fight the power, to debate Actuals versus Actives, and to condemn the objectification of the Dolls. It is less easy—because it is more in the background—to fight our own programmed desire to see "true love" triumph, to challenge the myths of heterosexual romance, including the notion of soul mates. Not everyone wants to be challenged in that way. But this is what science fiction is for. The estrangement or defamiliarization strategies of science fiction are tools that enable us to see our own society and our own social programming afresh. When Whedon says that *Dollhouse* is about societal brainwashing, this does not just apply to corporate control; it applies equally to heteronormativity and the myths of heterosexual romance. It applies to Victor and Sierra as much as to Echo and Paul.

BIBLIOGRAPHY

Calvert, Bronwen. "Mind, Body, Imprint: Cyberpunk Echoes in the Dollhouse." *Slayage: The Journal of the Whedon Studies Association* 8, nos. 2–3 (2010).

Connelly, Tom, and Shelley S. Rees. "Alienation and the Dialectics of History in Joss Whedon's *Dollhouse*." *Slayage: The Journal of the Whedon Studies Association* 8, nos. 2–3 (2010).

Davis, Hugh H. "'The Drama Is in Us': Pirandellian Echoes in Dollhouse." *Slayage: The Journal of the Whedon Studies Association* 8, nos. 2–3 (2010).

"Defining Moments: A Retrospective with Joss Whedon." *Joss Whedon's Dollhouse: The Complete Second Season.* Disc 4. Los Angeles: Twentieth Century Fox Home Entertainment, 2010.

Deritter, Lillian. "We're Not Men." In *Inside Joss' "Dollhouse": From Alpha to Rossum*, edited by Jane Espenson, 189–203. Dallas, TX: Smart Pop, 2010.

Hawk, Julie L. "Hacking the Read-Only File: Collaborative Narrative as Ontological Construction in *Dollhouse.*" *Slayage: The Journal of the Whedon Studies Association* 8, nos. 2–3 (2010).

Hills, Matt. *Fan Cultures*. London: Routledge, 2002.

Johnson-Smith, Jan. *American Science Fiction TV*. London: Tauris, 2005.

"Kindred Souls." Fan Forum. http://www.fanforum.com/f88/kindred-souls-%7Bvictor%E2%99%A5sierra%7D-1-light-me-will-guide-you-home-all-i-want-your-harbor-62955966/

"Looking Back: Roundtable with Joss Whedon & Cast." *Joss Whedon's "Dollhouse": The Complete Second Season.* Disc 4. Los Angeles: Twentieth Century Fox Home Entertainment, 2010.

Masson, Cynthea. "Who Painted the Lion? A Gloss on *Dollhouse*'s 'Belle Chose.'" *Slayage: The Journal of the Whedon Studies Association* 8, nos. 2–3 (2010).

Milavec, Melissa M., and Sharon M. Kaye. "Buffy in the Buff: A Slayer's Solution to Aristotle's Love Paradox." In *"Buffy the Vampire Slayer" and Philosophy: Fear and Trembling in Sunnydale*, edited by James B. South, 173–84. Peru, IL: Open Court, 2003.

Morohunfola, Oluwafemi. "What Echo's Journey to Self-Awareness Tells Us about the Human Soul." In *"Inside Joss' Dollhouse": From Alpha to Rossum*, edited by Jane Espenson, 219–31. Dallas, TX: Smart Pop, 2010.

Muntersbjorn, Madeline. "Disgust, Difference, and Displacement in the *Dollhouse.*" *Slayage: The Journal of the Whedon Studies Association* 8, nos. 2–3 (2010).

"Victor and Sierra Give Me Hope That True Love Exists." Dollhouse Confessions. http://dollhouse-confessions.tumblr.com/page/2.

Whedon, Jed, and Maurissa Tancharoen. "Belonging." Commentary. Season 2, episode 4. In *Joss Whedon's Dollhouse*: *The Complete Second Season.* Disc 2. Los Angeles: Twentieth Century Fox Home Entertainment, 2009.

Whedon, Joss. "Joss Whedon, Feminist." Interview with James Longworth (2002). In *Joss Whedon Conversations,* edited by David Lavery and Cynthia Burkhead, 42–63. Jackson: University Press of Mississippi, 2011.

———. "Man on the Street." Commentary. Season 1, episode 6. *Joss Whedon's Dollhouse: The Complete First Season.* Disc 2. Los Angeles: Twentieth Century Fox Home Entertainment, 2009.

Wilcox, Rhonda V. "Echoes of Complicity: Reflexivity and Identity in Joss Whedon's *Dollhouse.*" *Slayage: The Journal of the Whedon Studies Association* 8, nos. 2–3 (2010).

FILMOGRAPHY

2001: A Space Odyssey. Written by Stanley Kubrick and Arthur C. Clarke. Directed by Stanley Kubrick. MGM/Stanley Kubrick Productions, 1968.

Blade Runner. Written by Hampton Fancher and David Peoples. Directed by Ridley Scott. Ladd Company/Run Run Shaw/Warner Bros., 1982.

Dr. Horrible's Sing-Along Blog. Written by Joss Whedon et al. Directed by Joss Whedon. 2008.

"Prophecy Girl." *Buffy the Vampire Slayer*. Season 1, episode 12. Written and directed by Joss Whedon. First broadcast June 2, 1997.

Serenity. Written and directed by Joss Whedon. Universal Pictures/Barry Mendel Productions, 2005.

"To Shanshu in LA." *Angel*. Season 1, episode 22. Written and directed by David Greenwalt. First broadcast May 23, 2000.

NOTES

1. "Victor and Sierra."

2. See, for example, Hills, *Fan Cultures*. The scholar and the fan, for example, might have different ways of reading a text, applying different contexts or approaches for extracting information and enjoyment from it. Here, I am really discussing how, as viewers, we can find ourselves divided in terms of how we respond to these particular characters. As a feminist, I might reject the notion of romance and "true love," but as someone who enjoys well-told stories and science fiction, I admire the way that Victor and Sierra are perfectly presented as integral to this narrative.

3. Milavec and Kaye, "Buffy in the Buff," 174.

4. Masson, "Who Painted the Lion?"

5. Johnson-Smith, *American Science Fiction TV*, 4.

6. Deritter, "We're Not Men," 199.

7. Muntersbjorn, "Disgust, Difference, and Displacement," 19.

8. Hawk, "Hacking,"18.

9. Whedon, "Man on the Street," commentary.

10. "Defining Moments."

11. Wilcox, "Echoes of Complicity," 4.

12. Muntersbjorn, "Disgust, Difference, and Displacement," 8.

13. Whedon, "Joss Whedon, Feminist," 58.

14. See Porter and Ginn, this collection.

15. Thanks to Sherry Ginn for pointing this out and to Alyson Buckman for encouraging me to follow this line of thinking further.

16. Deritter, "We're Not Men," 200.

17. "Looking Back."

18. "Looking Back."

19. Whedon and Tancharoen, "Belonging," commentary.

20. Connelly and Rees, "Alienation," 9.

21. *Blade Runner*.

22. Calvert, "Mind, Body, Imprint," 5.

23. "Prophecy Girl," *Buffy the Vampire Slayer*.

24. "To Shanshu in LA," *Angel*.

25. "Kindred Souls."

26. Morohunfola, "Echo's Journey," 225.

Chapter Ten

The Theatre of the Self

Repetitious and Reflective Practices of Person and Place

Joel Hawkes

> The unexamined life is not worth living.
>
> —Plato

FRAGMENTS 1

This is how an episode begins. Not with a single clear picture but with an explosion of tiny images, numerous and streaming toward the viewer on the screen. These images are both a partial index of the show (in that they are perfect copies/recordings of portions of *Dollhouse*) and icons of experience (representations: Dolls' memories and the viewers' memories of episodes already seen). They are the television program and programs for the Dolls. The images are unambiguously fragments and therefore ambiguous in nature. Fragmented and unclear, they are like memories of one's past actions and experiences and, consequently, of the self. An episode is itself a fragment of the whole season, the season half of two. Separate from but still part of the television whole is the graphic novel *Dollhouse, Volume 1: Epitaphs*.[1]

The initial images stream toward us as parts of a tale, the whole unclear. There is meaning to be found in each, if only someone could arrange them all side by side, like the connecting panels of a graphic novel, scenes from a reel of film, or fragmented memories of a Doll. The images are presented without words, without language, without speech or thought balloons, without context other than that which we know already as the viewer of past shows. And then an episode begins, and we no longer see so clearly the Dolls from the House—do not easily see the parts that make the whole. Searching, though, we find that *Dollhouse* is a show of fragments: fragments of Joss Whedon's past productions; of a vast intertextuality referencing myth, literature, music, and film; of memories and experiences lost and found by the Dolls; of short, carefully crafted Whedonesque phrases, scenes, and images; and of key scenes, words, and phrases—muttered, shouted, and written and everywhere repeated. And repeated. These are the practices—physical and linguistic—enacted by writers, actors, and characters; their repetition and reflection act in the creation (in the practice) of Doll, Dollhouse, episode, season, series, *Dollhouse*, and, perhaps most important, us.

These are the fragments, like bricks, that build *Dollhouse*. Each fragment is part of a puzzle, a doll in a series of Russian nesting dolls getting ever bigger, all the players and parts performing on a stage (of the Dollhouse/*Dollhouse*, in performance of the Dollhouse/*Dollhouse*). They mimic the practices of our own society that, according to Pierre Bourdieu's theory of *habitus*, act to structure physical and cultural space—the space in which we live, built from "structuring practices" that have become largely unconscious—practices formed from shared events, culture, history, and myths of its group members.[2] They are the unrecognized configurations of society that result in us objectifying the practices of everyday life.[3] It is easy, also, to objectify the pieces of the show.

FRAGMENTS 2

But perhaps the words, phrases, and actions that build this house, like small philosophical aphorisms, need someone to make sense of them, organize them, like Walter Benjamin's[4] Angel of History:

> His face is turned toward the past. Where a chain of events appears before *us*, *he* sees one simple catastrophe, which keeps piling wreckage upon wreckage and hurls it at his feet. The angel would like to stay, awaken the dead, and make whole what has been smashed. But a storm is blowing from Paradise and has got caught in his wings; it is so strong that the angel can no longer close them. This storm drives him irresistibly into the future, to which his back is turned, while the pile of debris before him grows toward the sky. What we call progress is *this* storm.[5]

The Angel of History organizes by seeing just one catastrophe, pulling fragments of the past together. Perhaps we can position Echo in such a role (is she an angel of sorts, or someone's, everyone's, idea of one?), organizing memories and experiences of the past into a whole, into herself, into a future that she cannot see, one of technological revolution and apocalyptic storm? As Boyd Langton notes, Echo "seems to have the ability to think outside of pieces that we give her and create" ("Stage Fright," 1.3). But is it only her past and future that she is responsible for? Does she also marshal ours? Or is it the viewer's job to make connections? Or suppose that the shape of parts is already there—a theater of fragments.

REPETITION 1

It is said that history is repeated. So is the *Dollhouse*. Repetition gives Dollhouse and *Dollhouse* form, as fragments take on meaning like the repeated sounds of a young child learning to speak, forming the words that will create its world. The repetition of small acts maintains the Dollhouse's shape and order: wiped Dolls do what they are told, eating, painting, receiving a massage, passively filing into the showers, and then going to their pods to sleep.

The repetition of words also gives the location and its occupants form; through the official designations of "Active," "Handler," and "engagement," roles and relationships—and so the Dollhouse—are defined. The repeated, ritually reassuring dialogue between freshly wiped Doll and programmer exemplifies these formative repetitions, first introduced through the exemplary Doll Echo (whose name reflects the importance of such repetitions):

"Hello, Echo."

"Did I fall asleep?"

"For a little while."

"Shall I go now?"

"If you like." ("Ghost," 1.1)

Repeated ad nauseam in the show's twenty-six episodes, this script (perhaps a mirror of the show's script that designates the actors' roles) builds the relationship of Dollhouse to Doll, the script's linguistic illusion of free will highlighting the Doll's lack of choice. They too are actors in someone else's show. The repetition of dialogue in the chair shows us meaning, then takes meaning away, as further repetition makes the dialogue meaningless, until we almost disregard the oppressive and formative power of the words, left just to enjoy and "objectify" the spectacle of the Dollhouse.

Another echo assures the bond between Active and Handler:

"Everything is going to be alright."

"Now that you're here."

"Do you trust me?"

"With my life."

These words are specifically repeated in episode 1 of season 2 to reestablish their significance as a structuring practice of House and show, their meaning compounded in the episode's title, "Vows" (2.1), and in Echo's "marriage" engagement. These "vows" are repeated in variant form between Senator Daniel Perrin and his wife (later exposed as his Handler), Cindy Perrin ("Remind me why I love you so much," "I'm your white knight," "And I'm your beautiful damsel," "Ever after"). In these words, the show's fairytale theme is repeated ("The Public Eye," 2.5), making us question the meanings of the repetitions. We realize that in the banality of repetition, there is also meaning and oppression. But do we ask if this is the same in our lives?

REPETITION 2

Even as repetitions create Doll and Dollhouse, they bring about their demise but also ensure the continued and repetitive creation of *Dollhouse*. Repeat imprints threaten to cause glitches in the program, breaking order but engendering the possibility of new personalities in the Dolls. From the "composite event" of Alpha ("The Target," 1.2) to the development of Echo, the key is repetition of parts. Fragmented survivals of wiped personalities are connected and repeated, helping to create a new whole.

We often know which imprint an Active has been given through a repeated phrase. Echo, as Taffy in "Gray Hour," repeats the words "blue skies"; Sierra utters this line when she is given the same imprint (1.4). However, the repeated practice of these personality-designating phrases eventually breaks the cycle of engagement and wipe. The first fracture of this cycle comes not from words but action (practice): Echo repeating not only the reiterated phrase of her would-be killer, "Shoulder to the wheel," in "The Target," but the accompanying practice of slapping his shoulder as he says this, in profession of his philosophy of earning one's survival, justifying one's life. In the episode's final moments, Echo begins to earn her own life (her personality), repeating his shoulder-slapping action while understanding the act through the repeated phrase. Echo's words are also actions, repeated to create the individual of many pieces, in the same way that we are constructed from our different experiences, words, and practices. As Echo says, "I'm all of them but none of them is me" ("Vows"). This practice of the individual is then reflected in the practice of the episode, in which the same action and phrase create the central theme and so the episode itself.

REPETITION 3

Programmer at the Washington, D.C., Dollhouse Bennett Halverson explains, "Every action affects our neural topography; we literally become what we do" ("Epitaph Two: Return," 2.13). When Echo wants to know how the Dollhouse chair works—in which "civilians" are mapped and Actives are wiped and imprinted—she tells Topher Brink, "Show, don't tell" ("Needs," 1.8): practice is how we learn. Echo's actions rather than programmed memories arguably decide the path of her development. The memories of knowledge and skills allow her to act in particular ways, but she decides how to utilize these memories, adapting them to her situation and her growing understanding of self and world and so helping to create both.

Paul Ballard, too, is created by what he does: searching for the Dollhouse, repeating its name everywhere,[6] laying the pieces of a puzzle side by side—images of Caroline, a college video, and a news station's interviews with the public on the topic of the Dollhouse "myth" ("Man on the Street," 1.6). His apartment wall in "A Spy in the House of Love" (1.9) is a map of fragments connected by colored string (the "deeper" he looks, the more "connections" appear). Linking these, he furthers the story and the viewer's knowledge; he helps create what he sets out to destroy, drawn into the practices of the Dollhouse and his eventual position as Handler, just as the problem-solving viewer is drawn into constructing meaning in *Dollhouse* and, at the same time, is reflected in its practices. As one of the interviewees in the "man on the street" interviews challenges, "you think it's not happening? You think they're not controlling you? Don't worry about it. Just sit back and wait for 'em to tell you what to buy" ("Man on the Street"). This one voice claims, like a paranoid prophet, that all of us are already inside the Dollhouse. Our residence is assured by our actions, as we wait ("just sit back"), acting on command ("tell you what to buy"), our daily routines dictating that we go to work, return home, eat, and bathe. Then, like Dolls, we forget, sitting in a chair in front of the television screen, which tells us what to buy, like, and consume. *Dollhouse* repeats and reflects this litany of fragments (practices), showing us to ourselves.

THEATER

Dollhouse is a theater of fragments, repeated and reflected, where, as anthropologist Kirsten Hastrup suggests, the stage acts as "society's spiritual 'double,'" able to show "what is not" as well as what is, so becoming a "magic space beyond the ordinary."[7] A stage shows the self to the self, exemplifying everyday cultural performances that assert our identity. As Barbara Myerhoff explains, we are "made self aware, conscious of our own consciousness."[8]

Dollhouse remakes us. The moment exemplifies Myerhoff's claim that a stage-induced self-awareness makes us both actor and audience.[9] It is in this moment that our practices of "just [sitting] back and [waiting] for 'em to tell [us] what to buy" ("Man") merge with our practices as viewer and reflect the practices of characters and scriptwriters, to reveal the confusion and illusion of the show, "reality," and the self.

The "theater" of *Dollhouse* reveals what is by showing us what is not: It is science fiction illustrating our society, its beliefs, practices, ideals, and abuses, questioning relationships, religion, politics, love, work, family, the role of women, corporations—in short, all the practices of our society and ourselves. *Dollhouse*, as theatrical double, allows what is normally hidden and inaccessible in society to be brought to "observation and reason."[10] After all, "that's what art's for," an injured thief tells Echo in "Gray Hour," "to show us who we are."

TELEVISION

Dollhouse is also a television show showing us *a* television show, a "treatise on the art and science of making a television series," in which Dolls represent actors, Rossum the television network, Topher the show's creator Joss Whedon, and Adelle DeWitt the producer.[11] Boyd is the director and Ballard the "quasi-psychotic 'obsessed fan.'"[12] The clients (the audience) want "meaningless sex, eye candy, mind-numbing violence, and simplistic story-lines."[13] It might first appear that Whedon gives this type of show to his audience. But then he questions it, reflecting and distorting the pieces of a formulaic television experience and perhaps our expectations of a Whedon production too.

REFLECTIONS

Joseph J. Foy and Dean A. Kowalski, examining "authenticity" in Whedon's work, discuss Martin Heidegger's *Being and Time*:

> Heidegger suggests that in order for a person to begin to see through the banal aspects of the everyday, she must be somehow moved to a state of reflection. Most people, however, are content with an unexamined life, which offers them the comfort of social acceptance and a predefined sense of identity.[14]

We can read this reflection in Echo's development; she sees beyond the pieces, the norms and expectations of her world, and she becomes self-aware. But does *Dollhouse* move us to reflect? Does it make us question our world, our "everyday," our accepted norms and values? Do we see ourselves as Dolls?

"TABULA RASA"

DeWitt tells a client, "In their resting state our Actives are as innocent and as vulnerable as children. We call it the 'tabula rasa': the blank slate" ("The Target"). But, as Caroline tells DeWitt, "You ever tried to clean an actual slate? You always see what was on it before" ("Ghost"). Perhaps we can see the *Buffy* episode "Tabula Rasa" (6.8) on this slate, where a misfiring spell from Willow Rosenberg wipes everyone's memories. In the *Buffy* episode the characters attempt to piece together their identities from clues around them. Spike thinks that his name is Randy because it is stitched on a label in his jacket, while Rupert Giles and Anya Jenkins erroneously believe that they are married because they are joint owners of the magic shop. Similarly, the Dolls search to create meaning and define the self.

VAMPIRES AND OTHER CHARACTERS

If we must search for a beginning of the fragmented, repetitious, and reflective practices of the Dollhouse, then we might look to Whedon's past work. Perhaps, in a sense, we are looking at a marked slate? The quick-witted dialogue, carefully crafted but casual phrases, and strong female lead of his earlier work (*Buffy*, *Angel*, and *Firefly*) remain in place in *Dollhouse*. An echo of the vampire from *Buffy* is found in the empty (soulless?) Dolls. "This is where you steal their souls," accuses Ballard ("Omega," 1.12). Like Whedon's vampire characters Angel and Spike, Echo, too, must earn her soul.

The Dolls sleep in coffin-like pods, while an Active/Handler relationship is not unlike the Watcher/Slayer bond—one of trust that can be abused. "Do you trust me?" a Handler repeatedly asks. "With my life," the Doll replies, the dialogue reminiscent of Willow, in *Buffy*, asking, "Don't you trust me?" Tara Maclay assures her, "With my life" ("Tough Love," 5.19). When Claire Saunders (Amy Acker) shoots Bennett and blood splatters onto Topher's face in "Getting Closer" (2.11), we recall Willow splattered with Tara's blood when she is shot by Warren in *Buffy* ("Seeing Red," 6.19) or Winifred Burkle (Amy Acker) coughing blood into Wesley Wyndam-Pryce's face in *Angel* ("A Hole in the World," 5.15). The sad farewell between Winifred and Wesley in the same *Angel* episode is oddly reminiscent of that between Saunders and Boyd in "Getting Closer." Another reflection sees Caroline tell Bennett (Summer Glau), "I bet you could kill me with your brain" ("Getting Closer"); in *Firefly*, River Tam (Glau, again) tells Jayne Cobb, "I can kill you with my brain" ("Trash," 1.11). Elsewhere, Topher's goofiness occasionally reminds us of *Buffy*'s Xander Harris, particularly when Topher programs Sierra with elements of his own personality in "Haunted" (1.10), their interaction evoking the splitting of Xander in "The Replacement" (5.3). Fragments of other parts of the Whedonverse remain in *Dollhouse*, which itself becomes another part of the Whedonverse.

DOLLHOUSE

Dollhouse. The word evokes a history of meanings, dualities, a Blakean game of innocence and experience, in which we are reminded of other dollhouses: a toy, a place of innocence, a smaller version of the adult ("real") world that also conceals darker parts. Henrik Ibsen's play *A Doll's House* (1879) questions the norms of marriage, the intolerable position of a wife hidden behind the façade of respectable household.[15] Todd Solondz's film *Welcome to the Dollhouse* (1995) traces the unbearable childhood of twelve-year-old Dawn Wiener in a tale of bullying, rape, drugs, and pedophilia.[16] In *Amityville Dollhouse* (1996), the traditional threats of the horror film manifest in a child's toy—a voodoo version of the family home.[17]

A dollhouse is a place of childish dreams—or nightmares, in volume 2 of Neil Gaiman's *The Sandman* graphic novel series, in which a house of dreaming people gives way to night terrors, the machinations of the King of the Dream World, and a convention for serial killers.[18] Alan Moore's *Lost Girls* graphic novels unite Wendy (*Peter Pan*), Dorothy (*The Wonderful Wizard of Oz*), and Alice (*Alice's Adventures in Wonderland*), imagining the grown-up heroines of children's stories discussing sexual experiences together and participating in new ones.[19] The volume 2 cover displays a seemingly innocent dollhouse, but, on the back cover, the dollhouse is opened to reveal dolls in various sexual positions.

"When you hear the words 'doll house' you probably think of little girls playing tea party," says the interviewer in "Man on the Street." Joel Mynor develops the idea: It is "pink, and it opens up, and there's teeny furniture, and you put the boy doll on top of the girl doll, and we learn about urges." Childish play becomes a site of sexual imagery and imagination, reflecting the innocence and experience of the Dolls/Actives in their various states and roles. They work to satisfy client urges but also develop their own.

SEX

Sex is at the heart of the Dollhouse. It comes in two parts. The house stigmatizes and carefully controls sexuality (e.g., concerns over Victor's "man reaction" ["True Believer," 1.5]) yet simultaneously functions as a "brothel" ("Epitaph One," 1.13). Questions of sexual identity, power, enjoyment, and exploitation are reflected everywhere, representing a world that is troubled by such matters, traditionally chastising female sexuality while celebrating the masculine equivalent. Echo, however, is celebrated for being both "virgin and the whore." She can be both: innocent Doll and sensual Active ("A Love Supreme," 2.8). But celebrated by whom? By a world that also desires both in one? Dollhouse technology allows Echo to become "virgin again," but, as she suggests, a slate cannot be completely cleaned ("Ghost"). Like a slate, female Dolls are drawn on by primarily male clients, with expectations (as powerful as the sexual acts themselves that mark the Dolls) of a male gaze, matched by a media gaze (a powerful societal expectation) that similarly sexualizes and wants perfection in the female form. The media, too, demand both virgin and whore, with perfect "teeth and hair." As Echo asks, "Who cares what's underneath?" ("The Target").

Is all sex exploitative? "Man on the Street" is overtly structured around this question, with the story of Sierra's Handler abusing her (a microcosm of the larger Nolan storyline) beginning the storyline, and reflecting the use of the Dolls by the Dollhouse, reflecting Joel Mynor's use of Echo as his wife, reflecting an apparently caring relationship between Ballard and Mellie (November), which is not what it seems. But can we identify an original exploitative act in the series of reflections? It is difficult to do. Has sex always contained an exploitative component? Ballard jokes to Mellie, "I'm not a piece of meat," yet this is what they all appear to be. Still, "nothing is what it appears to be" ("Ghost"); the Dollhouse "deals in fantasy, but that is not their purpose" ("Spy").

DEATH

Is the Dollhouse's purpose death? Sex and death come as a pair. The "little death" of sex might not be shown to America, to a country intolerant of sex on television, but where violence is acceptable and, at times, a substitute. As the scantily clad pop star of "Stage Fright" explains of her audience, "They love to see me die." The Dolls in the chair die over and over again (their lives flashing before their eyes) for their clients' and our pleasure. The technology that offers titillation and fantasy allows Dolls to kill, but it also provides a mastery of death (the ultimate quest), making death a little thing, the resurrection of one client in "Haunted" anticipating Rossum's later use of the technology. But the promise of a return to Eden results in a wasteland.

FAIRYTALES

Dollhouse begins with a fairytale—a child's world (like a dollhouse), concerned with sex and death.[20] Ballard's hunt for the elusive house is mocked as a childish fairytale threatening to disrupt a "real" investigation of people smuggling ("Ghost"). Nevertheless, in the second episode, this ethereal realm begins to manifest, even as Ballard is ridiculed, with coworkers telling him, "We got a call, a couple of kids found a house in the woods all made of candy and gingerbread" ("The Target"). As they laugh at his question as to who left an envelope on his desk ("Granny left it. Man, her teeth looked big"), its content (a photograph of Caroline) begins to give substance to the fable. Yet still the fairytale persists, in which Caroline is lost, and Echo must "get to her carriage before it [turns] into a pumpkin" ("Ghost"), for she, as Joel Mynor points out, is not a "real girl" to Ballard, but a fantasy ("Man").

Repetitions and reflections of "wood" and "wolf" build "The Target" episode, with Ballard's "fairytale" reflected in Echo's assignment, in which a romantic outdoor-pursuits date becomes a deadly pursuit: Echo, like Hansel and Gretel or Red Riding Hood, is lost in the wood ("I hate the woods," bemoans one of Echo's Handlers, anticipating his own death there). Boyd refers to their pursuer as "Robin Hood," another cowled folkloric figure, and another phrase helps construct the fairytale. Another is used when their hunter calls himself the "big bad wolf" ("The Target").

This episode positions itself, and the show, as a dark fairytale, the wolf of this episode reflected in the elusive Alpha who stalks the Dollhouse and is perhaps behind this attack on Echo. Two wolves are reflected in a third when, in the same episode, we are given the backstory of Boyd, who is later revealed as the "big bad" stalking the show.[21]

As the television interviewer in "Man on the Street" says of the Dollhouse, "like every good fairytale the story grows more intricate." The camera pans as the interviewer's voice continues and we see silent video screens playing footage of Caroline at college and Caroline in the religious community of "True Believer," the previous episode. Ballard sits and watches the screens, trying to put the pieces of the "fairytale" together so that he can "[swoop] in on them like Prince Valiant" ("Briar Rose," 1.11). This reference to the Nordic Prince at Camelot from Hal Foster's (2009) comic strip[22] reflects Echo's opening reading of "Sleeping Beauty" to a young, troubled, sexually abused girl named Susan. Repetitions and reflections abound in this episode, in which Echo must "wake," the young girl who is an earlier version of her own imprint. This foreshadows Ballard's entry into the castle (Dollhouse), through the thorns that grow around it (its security), with the help of an agoraphobic (another person trapped inside but with cannabis plants rather than thorns) designer of environmental systems for the Dollhouse, Stephen Kepler. Kepler (the former Active Alpha in disguise) is a wolf in the wrong fairytale, who also doubles as the Prince who awakens Echo. As the pair flee, Ballard is left locked inside the castle (Dollhouse) in Echo's place, his failed rescue foreshadowing the need for Alpha's assistance in the finale, "Epitaph Two," in which Ballard's death frustrates a final rescue. Alpha must intervene, leaving the imprint wedge that finally allows the union of Ballard and Echo to take place, though with Ballard this time locked inside the princess's head.

EDEN

Another fairytale: Eden. The Dolls are innocents, without experience in their wiped state. But Eden is not what it appears to be: Stephen Kepler (Alpha) on breaking into the Dollhouse says, "They told me this would be the new Eden." Ballard replies, "Eden wasn't a prison." To which Kepler replies, "What, are you kidding? The apples were monitored" ("Briar Rose").

Its controlled borders are again highlighted in "Epitaph One" when survivors of the future technological apocalypse break into a still-intact Dollhouse: "Paradise wasn't built to take in strays." Dollhouse and Eden are mirrored in the cult of "True Believer" (1.5). Its leader, Jonas Sparrow, proclaims, "[we] must protect this garden," "if a serpent should enter we must crush it underneath our heel," and "[we] must keep the garden pure." Simultaneously, Victor has his "man reaction" in the shower, and Saunders comments, "Of course, if he'd been a snake. Please pretend I didn't say that" ("True"). The Victor situation highlights the sexual component of the biblical reading of this episode, reminding us that the fall from the garden is often read as a sexual one. Both Houses have a warped sense of purity through oppressive dogmas. Another reflection of us?

TEA

A less obvious act of control is found in tea: a ritual act, calming and reassuring. In Japan the enjoyment of tea often manifests through a carefully scripted ceremony, unifying culture and community. DeWitt pours tea for Caroline when she is offered her Active contract ("Ghost"); the act of tea becomes an origin rite of the Doll's career, bringing the new Doll into the house. When Owen Johnston is presented with a similar proposition at the close of "Echoes," tea is again served, in an episode that traces Echo's entrance (origin) as a Doll through the reflection of Owen's arrival. The episode also marks the first appearance of the Rossum Corporation, the founders (the origin) of the Dollhouse. Tea is additionally a practice of power, exerting DeWitt's control (containing perhaps a hint of Britishness and tea-trading empire). Tea is given to clients as well as new Actives, in a similar act of authority, and to Matthew Harding (DeWitt's superior) in "Meet Jane Doe," when she wrestles back control of the Dollhouse (2.7). Tea creates boundaries. It is an act of command, a structuring action.

THE STAGE

The stage can show us "what is not" as well as what is, thus becoming a "magic space beyond the ordinary."[23] It can show us what is hidden and inaccessible in society.[24] The first lines spoken in *Dollhouse* are DeWitt's to Echo: "Nothing is what it appears to be" ("Ghost"). This dialogue provides another origin, a phrase about which the show is structured. The lines are spoken again in flashback at the beginning of "Echoes," as we retrace Echo's becoming and witness Owen's (1.7). In "Meet Jane Doe," DeWitt tells Ballard, "In the Dollhouse things aren't always what they seem." The same is often true of society.

INTERTEXTUALITY

Fragments of intertextuality illustrate action and implications of an episode and help create layers of meaning in the show. Television and film increasingly do this; it becomes a game. References to Britney Spears, "Doris friggin' Day," and "factory girl" enhance the meaning of "Stage Fright," an episode about an unhinged singer and her obsessive fan. "Bonnie and Clyde" are cited in the vault-robbing episode "Gray Hour," while biblical allusions and references abound when Echo enters a religious cult ("True Believer"). The phrases and words "classic science fiction errors" ("Haunted"), Topher's "disruptor" ("It's not too *Star Trek*, is it?" ["Public"]), Cylons ("Meet Jane Doe"), "Freudenstein" ("A Love"), and "Darth Vader" ("The Attic," 2.10) establish the show's (and Topher's) science fiction/geek credentials as well as add meaning to events and people.

Clyde Randolph 2.0 suggests, "Rossum is just a name, actually. From a play" ("Getting Closer"), failing to note that it is a science fiction play entitled *RUR* (Rossum's Universal Robots) by Czech playwright Karel Câpek; *RUR* is about a synthetic workforce that rebels against its creators.[25] As Clyde says of the name, "although technically you're not robots, it seemed to fit" (2.11). There is dramatic irony in his statement.

The penultimate episode (2.12) takes its title from T. S. Eliot's poem "The Hollow Men," which begins,

> We are the hollow men
> We are the stuffed men
> Leaning together
> Headpiece filled with straw. Alas!
> Our dried voices, when
> We whisper together
> Are quiet and meaningless
> As wind in dry grass.[26]

Like *Dollhouse*, Eliot's poem is fragmentary and repetitious in structure and word use. It is a dark vision of modernity and forms part of Eliot's vision of a modern wasteland (more fully explored in his poem "The Waste Land," which is again a poem of fragments and highly intertextual, containing a plethora of meanings that warn of a wasted land where meaning becomes "a heap of broken images").[27] We see the Hollow Men reflected in the "dead" that walk across London Bridge in "The Waste Land"—the city workers and ourselves.[28] We see them again in the Dolls and in ourselves. The Hollow Men are "stuffed" and filled, forcibly perhaps, by someone else. They are like the Dolls. And they are like us. We, too, are "stuffed" as we "sit back" to be told what to "buy" ("Man"), feel, think, do.

ART

"That's what art's for, to show us who we are" ("Gray Hour"). Trapped inside the vault and episode, Echo and her vault-robbing colleague Scott stare at a Picasso portrait, which not so subtly reflects Echo's own fragmented psyche (and perhaps our own). Art is repeatedly used to reflect something of a Doll or viewer to himself or herself; it is another repetitive and reflective act of creation, a liturgical practice structuring the Doll, episode, and show. Topher claims of his programming, "This is art. It's not an oil change" ("Stage Fright"): it takes time, as does the development of Echo, who begins painting on a blank canvas—the "tabula rasa" that is herself. Her creativity is noted, DeWitt telling Dominic, "Echo took the mission parameter"; the scene switches, leaving the words a fragment of a sentence to be completed by Boyd, who says to Saunders, "And did even better. She seems to have the ability to think outside of pieces that we give her and create." It is left to Saunders to finish the chain of thought that mimics, in structure, its content: "Create a new approach to the problem" ("Stage Fright"). While they construct from fragments, Echo sits below, her discussed creativity reflected in the picture she is painting.

The Elgin Marbles[29] are positioned as fragments in "Gray Hour." Echo must steal one from a private collector. Art is everywhere in this episode, in pieces, reflections, and doubles: Echo dresses as a prostitute in front of a mirror that reflects a work of art behind it; this is then reflected in her makeup and short dress (an art of sorts—doubly so, for she is not a prostitute, rather an art thief playing one). We place this on top of the art that is Topher's program.

Pieces of famous art fill the vault. The Picasso reflects the "broken," fragmented Echo and is then repeated with Topher explaining to Ivy, "I break you down, then I build you back" (his teaching, or artistic, technique)—an art reflected in the remote wipe of Echo (Topher: "It's not even science; it's art";

"Gray Hour") by Alpha, who is another kind of artist using violence and a blade. Seeing the Marble, DeWitt comments, "Michelangelo believed his sculptures already existed inside the marble, waiting to be freed." Echo is not unlike these statues and, similar to the carved statue of Galatea in the myth of Pygmalion, comes to life. The episode is a birth for Echo, the meaning compounded in the first scenes of Echo assisting at a birth and reflected in the final moments with Echo in fetal position, underwater in the pool.

The development of Sierra and Victor is also mapped by art in "Belonging" (2.4). Sierra (as Priya Tsetsang) is an artist. We see Priya take a photograph moments before meeting Nolan Kinnard for the first time; Nolan takes and keeps photographs of every trip Priya makes to him as a Doll. They are an index, a static/broken version of Priya, in contrast to art—an icon of the "real" Priya, a representation that allows interpretation and development. Art here is an expression of the self. It is not Priya caught and molded by someone else. Nolan seeks to keep her trapped and broken in the many photographs that he collects (pieces reflected in Nolan's fake schizophrenia diagnosis and in Dolls more generally).

Art symbolically sets Priya free (the repeated bird image representing this freedom); we begin to see Priya emerge, a sculpture from the stone, as we do the story of the episode. Sierra's repeated bird images in the Dollhouse reflect her work on Nolan's wall; the splash of black paint that she adds to each piece symbolizes the darkness of Nolan and her emptiness as a Doll; this is again reflected in the black, hollow-like sculptured figure sitting in the center of the art room in the Dollhouse and repeated in the black paint that Sierra and Victor put on each other's faces (triggering a dark memory of war for Victor) in "Belonging." The pieces put together (just as Topher looks at pieces of Sierra's art) allow the story to be told.

With Nolan granted the permanent imprint on Sierra, his indexing, his control over her, is almost complete. But, having killed Nolan, Sierra stands as a silhouette in front of the brightness and the bird of her own art, caught in between identities. The death of Nolan is a substitute for the death of Priya, and, where she has been a fragmented figure, Nolan is now cut into pieces by Topher (again the artist), freeing Priya to continue her own development—to paint herself.

The repetitions and reflections of art help illustrate the development of characters and our understanding of the show. While painting an image of the house that Joel Mynor buys for his wife and after the interruption of their engagement, Echo tells DeWitt, "It isn't finished." Like DeWitt, we must ask what isn't finished, "the picture?" ("Man on the Street"), for the repetitions and reflections of art in *Dollhouse* not only illustrate Doll development and engender our understanding but again build the episodes of the show into a picture, in which we might, once more, see something of ourselves.

DOLLHOUSES

There are many dollhouses in *Dollhouse*. They are not Edens. They structure the show, and many of them (organizations and institutions) structure our society. They are both refuges and traps. They might be prisons: the vault and commodity art in "Gray Hour"; religion's dogmatic controls, offering false serenity, in "True Believer"; "playing house" (trapped?) with Joel Mynor in "Man on the Street"; the role-shaping world of the university in "Echoes"; NSA headquarters in "A Spy in the House of Love"; the picture-perfect mansion (but broken family) of "Haunted"; the care home (and princess locked in castle) of "Briar Rose"; marriage in "Vows"; a psychiatric clinic ("Belonging"); the world of government and the L.A. Dollhouse ("The Public"); jail ("Meet Jane Doe"); the military ("Stop-Loss," 2.9); the Attic ("The Attic"); Rossum headquarters at Tucson ("The Hollow Men").

FREE WILL

Later in her evolution, Echo can access past personalities at will, again shown through repetition of key phrases. "Blue skies" tells us she has utilized Taffy's imprint in "Meet Jane Doe." But is this free will? Is Echo still just what others have stuffed into her?

ECHO

Echo's evolution is mapped by repeated key words and actions from past imprints. She slaps her shoulder like her would-be murderer in "The Target" and draws a face on the mirror reminiscent of the Picasso painting that she has seen in "Gray Hour." "I see perfectly," says Echo after being blind in "True Believer," revealing her growing awareness of herself and the Dollhouse. She paints a house after "playing house" with Joel Mynor ("Man on the Street") and mutters "goodness gracious," like the serial killer who inhabited her body in "Belle Chose" (2.3). At the close of the first season, she utters the name "Caroline," a key stage in self awareness ("Omega"). In "The Public Eye," she sees the "gaps," the "other people" that she has been programmed with; she can "see the seams, how it's all constructed." She sees herself, and she sees us. We are the "other people." As Boyd explains, "every person I know is pretty poorly constructed" ("Vows").

FRAGMENTS 3

As the Dollhouse is challenged and Echo evolves, the phrases that give the Dollhouse structure fall apart. When asked if she wants a treatment in

"Echoes," Echo replies, "No," breaching her conditioned state. Later, in "The Left Hand," she tells Senator Perrin, "I don't want to fall asleep, even for a little while." In the same episode, Victor, after being wiped, asks "Shall I go now?" and an exasperated Topher replies, "For the love of God, yes."

THE ATTIC

An attic is where we store "odds and ends." It is also where we might find a mad woman, hidden away from society (Gilbert and Gubar).

The Attic is a loop, a trap, people caught in their own repeating nightmares: Sierra with Nolan the zombie, Victor as a soldier fighting himself, and Matsu from the Tokyo Dollhouse eating small pieces of sushi made from his own severed leg ("The Attic"). They are fragments, parts of the self, or just the self, alone. They connect to create the Rossum mainframe computer, but they do not connect to one another. They remain alone.

WRITING ON THE WALL (GLASS DOOR)

In "Belonging," we see various phrases of identity scratched on the glass of the inside of Echo's pod door, icons of past episodes, experiences in her becoming, solidified in written form: "My son killed me" ("Haunted"); "I am a believer" ("True Believer"); "My husband bought me a house" ("Man"); "Blue skies" ("Gray Hour"); "I love my baby" and "The baby isn't mine" ("Instinct," 2.2); "Women are whores" ("Belle Chose"); "Friends help each other" ("Stage Fright"); "I was blind" ("True Believer"); "Shoulder to the wheel" ("Target"). All entries are a journal that reflects who Echo is.

WATCHING

Cameras and screens fill *Dollhouse*; the practices of watching, viewing, and seeing are integral to the show, mimicking the destructive and constructive male gaze, the similar gaze of the television viewer, the security of the Dollhouse and that of our own CCTV society, and our reliance on electronic screens. In these we watch ourselves. Saunders suggests in "Stage Fright" that "someone else is watching," to which Boyd replies, "Someone always is."

BEST

The word creates a loop, or trap, within the show. "Best" means a judgment of worth/perfection in the Active (and in us?): hard worker, following the rules, high achiever, young, beautiful, athletic. Echo is confronted with the Doll mantra of "I try to be my best" as she hallucinates multiple versions of herself in "The Target." "Echo wasn't always the best," says Saunders ("Stage"). After Alpha slices Victor's face, he asks, "How can I be my best now?" ("Omega").

But the word also offers a way out of the loop when it is questioned, broken into pieces. Echo asks Sierra, "Are you your best?" She answers, "I'm not sure how to know that" ("Briar Rose"), prompting us to question the criteria for judgment in the Dollhouse and society. We may arrive at the same conclusion Echo does: that "no one is their best in here" ("Vows").

RITE OF PASSAGE

Topher tells DeWitt, "Brain science is all about hardware and software. There's no mystical in-between" ("A Love"). And yet each character completes a developmental rite of passage, following, as Victor Turner discusses, the three stages of "separation, transition, and incorporation."[30] It is a performance like that on the stage: to be seen, even if the audience is only the self. Alpha draws our attention to the process, encouraging the new Echo to symbolically kill the old Caroline in a "blood ritual": "new life from death" ("Omega"). Echo, Alpha (from villain to hero), Paul (toward knowledge), Victor (confronting trauma and getting the girl), Sierra (partly shown through her art), and DeWitt all complete such rites, moving from a blank slate into fragments to become a whole, realizing the self, just as we do through life, and, at times, moving through less obvious (personal) rites of passage but also through the prescribed rites that accompany birth, graduation, adulthood, work, marriage, and death.

GROUPING

Topher notes the Dolls sitting together and eating together, "grouping": they are enacting "instinctual survival patterns" ("Gray Hour"), the parts necessarily coming together. We see practices of community, friendship, and love that are stronger than the individual; they evolve beyond the self. No one remains alone. Dolls assist one another's escape, fight alongside one another, and build a community ("Epitaph Two"). We can read this grouping as *communitas*, a term that Turner uses to denote a greater sense of togetherness and solidarity.[31] *Communitas* refers to a community spirit often defined by and experienced through liminality—that period of transition and transformation found within ritual performance and located within the development of the Dolls.

THE FUTURE

The panels of *Dollhouse, Volume 1: Epitaphs* reveal that the future occurs at 10:07, when automated calls divide humanity into "Wielders," "Butchers," and "Dumbshows."[32] These words come to define the world and its practices from the first moments of "Epitaph One": a technological wasteland, everything in ruins, a "heap of broken images,"[33] where "tech heads" ("Epitaph One") stuff new skills into their heads and take old memories out and where Ivy is printed into multiple people, as is Echo.[34] Whedon's *Dollhouse* landscape remains a practice of fragments and repetitions.

MOUNTAINS

Mountains offer a position of height in opposition to the underground Dollhouse. The pure snow-covered mountain is a place of birth when Echo assists a delivery in "Gray Hour," in an episode that helps birth Echo's personality. One of the phrases scrawled on the door to Echo's pod in "Belonging" is "Mountains are safe." Is this a recollection of her engagement in "Gray Hour" or a premonition, or representation, of the end of a spiritual and narratological quest? In *Dollhouse, Volume 1: Epitaphs*, there are stories of Echo helping "survivors sneak into the mountains," with her leaving fragments of cipher on walls and buildings to guide the way.[35] Again, we must connect the pieces to trace a path, this time to the mountains: "If you come to the mountains, we can offer you safe haven."[36]

END FRAGMENTS

The world of Rossum and the Dollhouses falls in two explosions, the first caused by Boyd, wiped in Doll state, detonating a grenade ("The Hollow Men"); Topher creates the second to return people's original personalities. In the last moments before Topher's death and the completion of his own developmental rite of passage from amorality to moral self-sacrifice, he glances at collected photographs of past faces pinned to the wall with the words "To Remember" above them ("Epitaph Two"). We are reminded of the parts, events, people that make *Dollhouse*, as well as the many pieces and practices of our own society, and selves, that the show reflects.

The final moments repeat the structure of an episode as a young girl, her original personality regained, is reassured by Zone that "everything's gonna be alright"; then we see Echo climbing into her sleeping pod one last time, Echo, the episode, and *Dollhouse* complete, its performance finished.

The unexamined life is not worth living (Plato).

BIBLIOGRAPHY

Benjamin, Walter. "On the Concept of History." In *Selected Writings: 1938–1940*. Translated by Edmund Jephcott et al., edited by Howard Eiland and Michael W. Jennings, 4:389–400. Cambridge, MA: Belknap Press, 2006.

Bourdieu, Pierre. *Outline of a Theory of Practice*. Translated by Richard Nice. Cambridge Studies in Social Anthropology 16. Cambridge, England: Cambridge University Press, 1977.

Câpek, Karel. *RUR : Rossum's Universal Robots*. Translated by David Wyllie. Rockville, MD: Wildside Press, 2010.

Chambliss, Andrew, Maurissa Tancharoen, and Jed Whedon. *Dollhouse, Volume 1: Epitaphs*, #0-5. Created by Joss Whedon. Written by Andrew Chambliss, Jed Whedon, and Maurissa Tancharoen. Penned by Cliff Richards. Illustrated by Andy Owens. Milwaukie, OR: Dark Horse Books, 2012.

Connell, Richard. *The Most Dangerous Game*. Rockville, MD: Arc Manor, 2007.

De Certeau, Michel. *The Practice of Everyday Life*. Translated by Steven Rendall. Berkeley: University of California Press, 1988.

Eliot, T. S. "The Hollow Men." In *Collected Poems, 1909–1962*, 77–83. Orlando, FL: Harcourt Brace Jovanovich, 1991.

———. *The Waste Land. Modernism: An Anthology*. Edited by Lawrence Rainey. Oxford, England: Blackwell, 2005.

Foster, Hal. *Prince Valiant*. Vol. 1, *1937–1938*. Seattle, WA: Fantagraphics Books, 2009.

Foy, Joseph J., and Dean A. Kowalski. "Seeking Authenticity in the Whedonverse." In *The Philosophy of Joss Whedon*, edited by Dean A. Kowalski and S. Evan Kreider, 151–67. Lexington: University Press of Kentucky, 2011.

Gaiman, Neil. *The Sandman*. Vol. 2, *The Doll's House*. New York: DC Comics, 2010.

Gilbert, Sandra M., and Susan Gubar. *The Mad Woman in the Attic: The Woman Writer and the Nineteenth-Century Literary Imagination*. New Haven, CT: Yale University Press, 2002.

Hastrup, Kirsten. "Theatre as a Site of Passage: Some Reflections on the Magic of Acting." In *Ritual, Performance, Media*, edited by Felicia Hughes-Freeland, 29–45. London: Routledge, 1998.

Ibsen, Henrik. *A Doll's House*. New York: Dover, 1992.

Moore, Alan. *Lost Girls: Book 2*. Mariette, GA: Top Shelf, 2006.

Myerhoff, Barbara. "Life History among the Elderly: Performance, Visibility, and Re-membering." In *Remembered Lives: The Work of Ritual Storytelling, and Growing Older*, edited by Barbara Myerhoff et al., 231–47. Ann Arbor: University of Michigan Press, 1992.

Tresca, Don. "'Fantasy Is Their Business, but It Is Not Their Purpose': The Metaphor of *Dollhouse*." In *Joss Whedon: The Complete Companion: The TV Series, the Movies, the Comic Books and More*, 411–25. London: Titan Books, 2012.

Turner, Victor. *The Ritual Process: Structure and Anti-structure*. New York: Aldine de Gruyter, 1969.

———. *From Ritual to Theatre: The Human Seriousness of Play*. New York: PAJ, 1982.

FILMOGRAPHY

Amityville Dollhouse. Written by Joshua Michael Stern. Directed by Steve White. Promark Entertainment Group, 1996.

"Bewitched, Bothered and Bewildered." *Buffy the Vampire Slayer*. Season 2, episode 16. Written by Marti Noxon. Directed by James A. Contner. First broadcast February 10, 1998.

"Gingerbread." *Buffy the Vampire Slayer*. Season 3, episode 11. Written by Thania St. John and Jane Espenson. Directed by James Whitmore Jr. First broadcast January 12, 1999.

"A Hole in the World." *Angel*. Season 5, episode 15. Written and directed by Joss Whedon. First broadcast February 25, 2004.

"The Replacement." *Buffy the Vampire Slayer.* Season 5, episode 3. Written by Jane Espenson. Directed by James A. Contner. First broadcast October 10, 2000.
"Seeing Red." *Buffy the Vampire Slayer.* Season 6, episode 19. Written by Steven S. DeKnight. Directed by Michael Gershman. First broadcast May 7, 2002.
"Tabula Rasa." *Buffy the Vampire Slayer.* Season 6, episode 8. Written by Rebecca Rand Kirshner. Directed by David Grossman. First broadcast November 13, 2001.
"Tough Love." *Buffy the Vampire Slayer.* Season 5, episode 19. Written by Rebecca Rand Kirshner. Directed by David Grossman. First broadcast May 1, 2001.
"Trash." *Firefly.* Season 1, episode 11. Written by Ben Edlund and José Molina. Directed by Vern Gillum. First broadcast July 21, 2003.
Welcome to the Dollhouse. Written and directed by Todd Solondz. Sony Pictures Classic, 1995.

NOTES

1. Chambers, Tancharoen, and Whedon, *Dollhouse, Volume 1: Epitaphs.*
2. Pierre Bourdieu, *Outline of a Theory.*
3. Bourdieu, *Outline of a Theory*, 72, 79; De Certeau, *Practice*, 1.
4. The German philosopher and literary and social critic Walter Benjamin (1892–1940) was a writer fascinated with fragments, using them to house his philosophical and sociological ideas.
5. Benjamin, "On the Concept of History," 392.
6. This repetition is epitomized in the first episode, "Ghost," when Ballard makes Victor (imprinted as mobster Anton Lubov) repeat the word "Dollhouse" over and over again.
7. Hastrup, "Theatre as a Site," 34.
8. Myerhoff, "Life History," 234.
9. Myerhoff, "Life History," 234.
10. Turner, "From Ritual to Theatre," 13.
11. Tresca, "Fantasy Is Their Business," 412–13.
12. Tresca, "Fantasy Is Their Business," 413, 415.
13. Tresca, "Fantasy Is Their Business," 414.
14. Foy and Kowalski, "Seeking Authenticity," 152.
15. Ibsen, *A Doll's House.*
16. Solondz, *Welcome to the Dollhouse.*
17. White, *Amityville Dollhouse.*
18. Gaiman, *The Sandman.*
19. Moore, *Lost Girls.*
20. Original versions of classic fairytales are dark tales filled with sex and death. Even the dark nineteenth-century versions of the Brothers Grimm hid the nastier parts of their folkloric antecedents: cannibalism, rape, murder, and bestiality.
21. The term "big bad" originates in *Buffy* and designates a season's main villain. The phrase first was used in "Bewitched, Bothered and Bewildered" when Buffy refers to Angelus as the "big bad thing in the dark" (2.16)—perhaps referring to the big bad wolf of fairytales. The phrase is used as a noun for the first time in "Gingerbread" (3.11), a variation on the Hansel and Gretel story—a fairytale that also shapes *Dollhouse*'s "The Target." "The Target" is more commonly linked by critics to Richard Connell's short story "The Most Dangerous Game," which tells the story of a hunter living on an island who sends shipwrecked sailors out into the jungle as more exciting quarry. See Klein in this volume.
22. Hal Foster created the comic strip *Prince Valiant in the Days of King Arthur* in 1937, tracing the adventures of Valiant at Arthur's Camelot. The comic continues today.
23. Hastrup, "Theatre as a Site," 34.
24. Turner, *From Ritual to Theatre*, 13.
25. Câpek, *RUR.*
26. Eliot, "Hollow Men," 79.
27. Eliot, *Waste Land*, 125
28. Eliot, *Waste Land*, 126–27.

29. The Elgin Marbles are parts of the Greek Parthenon and were removed to Britain by Thomas Bruce, Lord Elgin, between 1801 and 1812. The "occupying" Ottoman Empire gave Lord Elgin permission to remove the marbles, but Greece now demands their return. They continue to be held in London. They are seen as pieces of Greece, "building blocks" of its culture and identity. In "Gray Hour," they help to deconstruct and rebuild Echo.

30. Turner illustrates the stages of the rite of passage, "separation, transition, and incorporation," with the example of an individual going into a temple (a sacred space), where she or he is divided from the rest of society; a ritual act assists her or his spiritual progress/status but also aids in the designation of that space as sacred. She or he emerges from the temple and is reincorporated into society at a more advanced, or higher, level (*From Ritual to Theatre*, 24).

31. Turner, *The Ritual Process*, 96–97.

32. Chambliss, Tancharoen, and Whedon, *Dollhouse, Volume 1: Epitaphs*, 12.

33. Eliot, *The Waste Land*, 125.

34. Chambliss, Tancharoen, and Whedon, *Dollhouse, Volume 1: Epitaphs*.

35. Chambliss, Tancharoen, and Whedon, *Dollhouse, Volume 1: Epitaphs*, 83.

36. Chambliss, Tancharoen, and Whedon, *Dollhouse, Volume 1: Epitaphs*, 145.

Chapter Eleven

"We're Lost. We Are Not Gone"

Critical Dystopia and the Politics of Radical Hope

Derrick King

At the surface level, *Dollhouse* appears to be a rather dispiriting portrait of a near future world in which corporate domination and technological control have advanced to such a point that they literally bring about the apocalypse. Yet, to the show's credit, the bleakness of this setting is actively resisted by the narrative trajectory of Echo, whose unwavering commitment to the struggle for a better world manages to infuse the entire series with a tone of stubborn hopefulness. While the television worlds of Joss Whedon consistently revolve around protagonists struggling for what they believe in despite overwhelming odds, *Dollhouse* is unique in that the villain whom Echo must overcome is not epic in scale like the "First Evil" in *Buffy the Vampire Slayer* or the multidimensional law firm Wolfram and Hart in *Angel* but is rather a more mundane multinational corporation that we could easily imagine existing within our own world. *Dollhouse* is, in this way, a timely and important response to our current world situation, in which economic, environmental, and humanitarian disasters brought about by global capitalism mean that hope itself can sometimes be in short supply. Here I situate *Dollhouse* within the science fiction genre of "critical dystopia" to demonstrate that, despite its gloomy premise, the show has at its core a utopian project of radical hope, or a belief that a systemic transformation of global capitalism is still possible despite the power currently wielded by multinational corporations. Furthermore, I argue that *Dollhouse*'s project of utopian hope not only challenges the hegemony of global capitalism but also unmasks and rejects several antiutopian discourses, including the compensatory function of utopian fantasy itself, which must be distinguished from true utopian praxis. Central to this analysis is a comparison of the two revolutionary acts in the series: the final moments of "The Hollow Men" (2.12), in which the Rossum mainframe is destroyed, and the conclusion of "Epitaph Two: Return" (2.13), a truly utopian moment of historical openness in which the future can be remade in any number of different ways. To demonstrate this utopian horizon, I draw on a Marxist problematic capable of mapping these utopian impulses and situating them within the show's depiction of twenty-first-century global capitalism, which functions as both the historical backdrop to the show's narrative and the central antagonist to Echo's political project.

While previous critics have noted some Marxian ideas in the show—including alienation[1] and Horkheimer and Adorno's vision of the "culture industry"[2]—I propose that we can more comprehensibly understand the connections among labor, disciplinary control, and global capital in *Dollhouse* by situating the show within Marxist theorist Fredric Jameson's description of postmodern "late capitalism," or the third "stage" of capitalism more commonly known as globalization.[3] Briefly, Jameson notes that the term "late capitalism" was first used by the Frankfurt School to describe a "tendential web of bureaucratic control" and "the interpenetration of government and big business" but has since expanded to include not only the emergence of multinational corporations but also "the new international division of labor, a vertiginous new dynamic in international banking and the stock exchange . . . new forms of media interrelationship . . . computers and automation, [and] the flight of production to advanced Third World areas."[4] This description will seem familiar to viewers of *Dollhouse,* of course, as Rossum is indeed an excellent example of the kind of multinational corporation that has developed within the late capitalist system. Other critics have considered some of these issues: Tony M. Vinci, for example, explores how the first season of *Dollhouse* uses its late capitalist setting to negotiate "the problems of identity" in a world plagued by "the naturalization of ideological values."[5] His analysis effectively demonstrates the effect that late capitalism has on subjectivities through the process of what he calls "existential proletarisation," which, he contends, ultimately "eradicates the very notion of authentic identity."[6] Likewise, Jameson notes that one of the features of late capitalism is its increasing commodification of the unconscious itself, which he situates within "the rise of the media and the advertising industry."[7] Eve Bennett extends this analysis in her article on the Dollhouse as a metaphor for the entertainment industry—or Horkheimer and Adorno's "culture industry"—arguing that "by making the Dollhouse part of a larger, multifaceted conglomerate, the series highlights the interconnectedness of the entertainment industry with other branches of the western capitalist infrastructure."[8] The operations of the Dollhouse thereby effectively demonstrate what Jameson describes as the late capitalist "collapse" of the "economic" and "cultural" backing into each other as culture itself becomes a privileged form of economic production in the first world.[9]

Yet, if this entertainment conglomerate is the outside face of the Dollhouse, then perhaps the "Attic" is a figuration of the deeper structures of the late capitalist multinational: not only are the Dollhouses all over the world connected to one another in a massive network, but this network, or the Rossum "mainframe," is quite literally powered by human labor. Moreover, it is powered by a form of postindustrial labor in which, as Clyde explains to Echo in the attic, the brains of those trapped in the attic are being used as "human computers" ("The Attic," 2.10). Jameson has suggested that depictions of complex technological networks in postmodern cultural texts often function as a "representational shorthand for grasping a network of power and control even more difficult for our minds and imaginations to grasp: the whole now decentered global network of the third stage of capital itself."[10] *Dollhouse* effectively makes this connection literal by turning its representation of an immense computer network back into a figuration of capitalist control and domination. In fact, the Attic itself is the ultimate manifestation of the way in which capital deploys its power: during the course of the show, Rossum utilizes its immense networks of control to influence everything from the everyday subjectivity of its employees and customers up to the workings of government through Senator Perrin, who is able to inoculate Rossum from any negative publicity ("The Left Hand," 2.6). The apparent effects of this influence are shown in the flashback portions of "Epitaph One" (1.13), in which Ambrose tells Adelle that while Rossum has always been "above the law," they are now "writing it" themselves. We also glimpse the scope of Rossum's influence in Boyd's speech to Caroline in "The Hollow Men": "We have clinical laboratories in every major city in the world. . . . If you've had your blood drawn . . . Rossum likely tested that blood . . . [and] we like to keep medical records of everyone who crosses our path."

Yet, beyond merely depicting the wide reach of Rossum's influence, these speeches gesture at the complete commodification and control of life itself by multinational capitalism, a process related to what Michael Hardt and Antonio Negri call "biopolitical production."[11] To flesh out this analysis, which is crucial to an understanding of how Rossum operates, it is worth briefly tracing the concept of "biopower" and its relationship to late capitalism's strategies of domination. In his later writings, Foucault begins to theorize the concept of biopower, which he loosely describes as "the set of mechanisms through which the basic biological features of the human species became the object of a political strategy, of a general strategy of power."[12] Biopolitical technologies of power are thus concerned with the administration and management of life itself (or what Giorgio Agamben theorizes as "bare life").[13] With the development of biopower, every part of a person's life is subject to discourses of power and calculation, not merely those parts that take place within a typical sphere of power relations, such as the court system or, as Marx demonstrates in *Capital*, the workplace.[14] Hardt and Negri note that biopower becomes increasingly important for Marxist theory as "productive labor" becomes "increasingly immaterial," leading to a point in which "there is nothing, no naked life, no external standpoint, that can be posed outside the field permeated by money."[15] Not only are the Dolls in the Dollhouse representative of immaterial productive labor or, specifically, "affective labor," which Hardt and Negri describe as immaterial labor "of human contact and interaction," but they also demonstrate the complete erasure of any distinction between "work" time and "personal" time.[16] Their "bare" or "naked" life itself is therefore subject to discourses of power at all times—after all, there is no going home from the Dollhouse at night or on the weekends.

Expanding Hardt and Negri's line of inquiry, Kathi Weeks situates affective labor and the erasure of any distinction between work and personal time within post-Taylorism, or the deindustrialization of the first world that Jameson contends is a crucial element of late capitalism.[17] She argues that in the postindustrial workplace, "many forms of work [require] not just the labor of the hand, but the labors of the head and heart" as employers seek workers willing and able to "enlist their creativity and their relational and affective capabilities."[18] Yet these "affective skills" must be cultivated or, more accurately, produced: Weeks argues that "the complex art of cultural fashioning and emotional engineering" becomes crucial for management, and the injunction to "be professional" is the major discursive mechanism by which this desired subjectivity is constructed.[19] This construction of subjectivity within the field of affective labor reaches its apex in *Dollhouse*: not only do those (such as Caroline) who agree to become "Actives" give up their labor power, but they also agree to give up bodies, minds, and hearts. Every part of their life is completely under the control of the Dollhouse, and their entire subjectivity is programmed by Rossum. *Dollhouse*'s depiction of late capitalism is so potent precisely because it not only reflects the larger political and institutional aspects of late capitalism but also is concerned with its workings at the interpersonal level as the system produces the very subjectivities that it needs to function.

Yet it is when these two levels—the subjective and the system as totality—are examined side by side that the fundamental conceptual problem arises: the global capitalist system appears to be so powerful that we can no longer imagine any form of resistance capable of radical change or any alternative to the status quo. Mark Fisher calls this ideology "capitalist realism," or "the widespread sense that not only is capitalism the only viable political and economic system, but also that it is now impossible even to *imagine* a coherent alternative to it."[20] Jameson sees this imaginative failure as related to postmodernism's ahistoricity, or what he defines as "epochality," which is a kind of "relationship to the present" in "which we defend the historical meaning and significance of the present age against all claims of the past and the future."[21] Epochality is therefore a "blockage of the historical imagination" that prevents us from imagining any kind of alternative to the present by short-circuiting our sense of history itself.[22] This fundamental ideology of late capitalism is reflected in *Dollhouse* through what Tom Connelly and Shelley Rees call the Dolls' alienation from "historical time" as a result of their "treatments."[23] Because their memory is erased on a daily basis, the Dolls quite literally have no past, nor can they imagine a future. Indeed, Sherry Ginn's broader analysis of memory tampering as a form of control in the Whedonverse concludes with a pertinent observation that effectively demonstrates the intense subjective alienation caused by a lack of memory: "memory is what makes us who we are. It allows us to situate ourselves within the world and make connections between our pasts, our presents, and our futures."[24] As long as the Dolls are so alienated from their memories and from history itself that they cannot place themselves within a past or imagine a future, there is no hope of challenging the structures of power responsible for this situation.

In fact, Connelly and Rees's examination of what they call the Dolls' "diachronic" versus "synchronic" selves can be expanded profitably into an exploration of the entire late capitalist problematic within the show as well as the utopian discourse that challenges it.[25] Connelly and Rees argue that, throughout most of the show, the Dolls exist merely as "synchronic" selves since their treatments remove them "from the cause-effect teleology of dialectical history."[26] A postcomposite Echo, however, manages to reject this "regressive, static synchrony and commodification in favor of a dynamic, empowered, progressive participation in the dialectical conversation."[27] Echo manages a return to history itself, a claim that is tremendously important for the argument I am advancing here. Indeed, Echo's historical break is ultimately a stubborn insistence that, despite the ideologies of capitalist realism or epochality, another world is still possible. It is therefore precisely by inserting herself back into the active process of history that Echo is able to formulate a utopian program of radical hope aimed at the production of something radically new. While Echo does not have a specific vision or "blueprint" of an alternative society, her unwavering belief that a radical transformation is possible is itself utopian. Indeed, "utopia," as I use it here, is simply the possibility that a radically different social organization is possible, that Rossum's (and, by extension, late capitalism's) domination does not equate with an end of history as such. Notably, it is also exactly this sort of utopian possibility that Rossum is so concerned with snuffing out: when Echo is trapped in the Attic, she discovers a man who has been sent there simply because he discovered that there was a vulnerability in the Rossum mainframe, even though he does not seem to know exactly what this weakness is ("The Attic"). It is therefore nothing more than the mere *idea* that a weakness exists that is so intolerable to Rossum and necessitates this worker's banishment. Of course, this sequence also reiterates the show's central point: that it is worth struggling for a better future precisely because such weaknesses do exist. Indeed, here we find the fundamental utopian belief that I argue runs throughout the show: while it may be impossible to imagine from our current vantage point, a radical transformation of society is still possible and worth struggling for. Paradoxically enough, it is actually in the more properly dystopian segments of the show—the two "Epitaph" episodes that conclude each season—that this utopian impulse is the strongest. It is thus the genre of "critical dystopia," or a dystopian narrative that remains politically committed to utopian transformation, that places the narrative trajectory of *Dollhouse* firmly within the diachronic procession of history.

Indeed, while "utopia" and "dystopia" appear to be opposites at first glance, this simple distinction has been contested within science fiction studies. While some dystopias, such as Orwell's *1984*, seem to function, in Jameson's words, "to denounce and warn against utopian programs in the political realm," others hold out hope that the systemic transformation of the world is still possible and desirable.[28] I argue that *Dollhouse* belongs to the latter category, or what Tom Moylan calls the "critical dystopia": While the world has tumbled into an apocalyptic hell brought about by Rossum, Echo—and, by extension, the show itself—still holds onto the seemingly impossible hope that radical change is possible.[29] For Moylan, the critical dystopia thereby resists the ideological closure of capitalism and "reaches toward utopia not by delineation of fully detailed better places, but by dropping in on decidedly worse places" while still allowing for spaces for resistance.[30] It is in this way that the critical dystopia is dialectically revealed to be a utopian figuration in disguise within a world in which radical praxis is seemingly impossible or, as Phillip Wegner argues, has "itself been relocated to the utopian horizon."[31] Building on Moylan's analysis, Jameson proposes that the dystopia therefore can be reclaimed as "a negative cousin of utopia proper," rather than its opposite, by virtue of its "positive conception of human social possibilities" and "politically enabling stance."[32] While this distinction between the older dystopian (or, more precisely, "anti-utopian") form and the critical dystopia might seem largely semantic, it is ideologically important as it reinforces the radical—and, indeed, ultimately utopian—nature of the dystopia. It is in this sense that I argue the radical hopes and possibilities of *Dollhouse* are most active within the dystopian "Epitaph" episodes that conclude each season.

Furthermore, the generic category of dystopia also helps explain why these radical hopes and possibilities are not aimed at any specific reorganization of society or comprehensive "blueprint" for change but simply reflect the belief that some kind of radical change is still possible. As Phillip Wegner argues, "The subject of utopian desire in the dystopia is politics itself—of agency and of kick-starting the engine of history in a moment when it seems to be terminally stalled."[33] As I have argued, the world of biopolitical late capitalism represented in *Dollhouse* by Rossum is precisely the sort of situation in which a praxis powerful enough to radically transform the system appears to be impossible. Yet Caroline and then Echo both reject this kind of defeatism and continue to struggle forward into an unknown future, holding out hope "in the conditions of dimmest radical political possibility . . . that such a politics might emerge."[34] Instead of any specific plan, the focus of *Dollhouse*'s utopian project is simply political praxis itself, or the beginnings of a utopian form of political organization that might be capable of reordering the global field to bring about a more just world. Importantly, this belief in the possibility of radical change is coupled with an uncompromising rejection of nonutopian forms of political organization as both ineffective and ultimately incapable of overturning late capitalist domination.

This unwavering commitment to utopian praxis and concomitant rejection of an insufficiently utopian politics can best be apprehended in the conclusion to "The Hollow Men," in which the destruction of the Rossum mainframe serves as a potent negative diagnostic of a political act that does not enable the struggle toward a utopian future. Immediately after the Rossum mainframe is destroyed, Paul remarks to Echo that "we saved the world," which is a rather interesting repetition of the moment in the last episode of *Buffy the Vampire Slayer*, in which Xander says exactly the same thing. However, in *Buffy*, Willow adds that "we changed the world," implying that a much more radical transformation has taken place.[35] The absence of this second line of dialogue in *Dollhouse* is telling. While Echo, Paul, Adelle, and Topher have seemingly stopped Rossum's plans for the time being, the system in which Rossum is able to operate is itself still intact; there has been no radical break that might usher in hope for a utopian world. Indeed, the mise-en-scène of this ending effectively illustrates the continuation of the status quo: the camera pulls back from Paul and Echo to slowly track across the largely untouched Rossum skyscraper that they have just escaped. Then, when two sun-drenched skyscrapers are in the frame, the camera tilts up so that the tall buildings dwarf our viewing position, reminding us that what Rossum represents—no less than multinational capital itself—remains unchallenged. Following this shot, the episode abruptly cuts to the year 2019, revealing that the postapocalyptic world of "Epitaph One" came about despite the group's efforts. The brightness of the final sequence of "The Hollow Men" is thus immediately plunged back into the darkness of the postapocalyptic world. I argue that the political act that ends "The Hollow Men" is insufficient precisely because it is not truly utopian; it does not open unto the possibility of a better future but produces a rather facile and compensatory fantasy of political action. Significantly, by rejecting this false "ending," *Dollhouse* is also rejecting a simple narrative resolution in favor of a more complex open narrative in which a properly utopian desire is possible.

Indeed, the act of blowing up Rossum's computer mainframe is an anti-utopian narrative resolution because it displaces the contradictions of the entire system of late capitalism onto a single problem (the "mainframe" itself) that can be "solved" easily. Moylan notes that the truly utopian text avoids precisely this kind of narrative closure in which the reader is "reinscribed within the dominant order" through a "satisfying, quieting ending" wherein real political problems are "resolved" through the formal resolution of the narrative or "plot."[36] Likewise, Jameson has long contended that "the narrative construction of imaginary resolutions" is a way in which texts can attempt to repress true social contradictions.[37] In *Dollhouse*, the destruction of the Rossum mainframe, were the series to end there, would create this kind of simplistic ending in which the contradictions of the current system could be narratively resolved through a single destructive act, ostensibly solving the central conflict while leaving the larger oppressive structures of late capitalism intact. Yet, to the show's credit, *Dollhouse* rejects this kind of narrative resolution. The deeper implication of the abrupt cut to the dark postapocalyptic world, then, is that nothing less than a radical restructuring of society is capable of saving us from corporations like Rossum.

Furthermore, the refusal of the false narrative resolution in "The Hollow Men" is also a rejection of what both Phillip Wegner and Kathi Weeks, drawing on the writings of the Marxist philosopher Ernest Bloch, describe as a compensatory "abstract utopia," which they differentiate from the truly radical "concrete utopia."[38] Wegner describes Bloch's "abstract utopia" as one that "operates in any historical conjuncture primarily in a compensatory fashion, projecting a simple fulfillment of those specific needs that are felt to be lacking there"[39] It is not "something radically new" but rather a "repaired present," which "risks defusing the hope for some fully other existence."[40] The destruction of the Rossum mainframe is precisely this kind of compensatory utopian fantasy that eliminates (or "repairs") one specific problem but leaves the overall structure of society intact, thereby disarming any truly utopian praxis aimed at radical transformation. Interestingly, there is another fantastic illustration of this kind of "compensatory" utopia in the *Dollhouse* episode "Needs" (1.8), in which Echo, Sierra, Victor, and November must act out their fantasies so that they can gain "closure." Echo's fantasy, which is the utopic impulse to, as Boyd says, "free us all," is thereby managed by allowing her to do so within a fantasy setting. Indeed, Dr. Saunders suggests that their problems controlling the Actives could be solved by "giving them what they need: closure." She theorizes that they will return to their docile Doll states if they "get some sense of resolve" for their "desires," which is precisely what happens (for the time being at least). This episode effectively demonstrates the way that potentially radical forms of resistance can be temporarily assuaged or controlled through the production of fantasies that allow one to imagine a resolution to a specific social problem or need while, in actuality, leaving the overall system untouched. I argue that the end of "The Hollow Men" functions similarly: Echo is able to finally fulfill Caroline's fantasy of hurting Rossum, which gives her a kind of closure. The problem is that merely destroying a part of a Rossum building—which is essentially just an act of domestic terrorism—does nothing to bring about a radically new future.

Instead of providing a temporary fulfillment of desire, then, the properly utopian impulse must remain radically open-ended, since the specific contours of a "concrete" utopia are always unimaginable within our historical present and can therefore only be glimpsed in brief moments of what Bloch calls "anticipatory illumination."[41] Likewise, Weeks's reminder that "wanting a different future and making it may not hinge on knowing what it might be" once again suggests that the task is not simply to propose solutions to problems that exist in the present but rather to struggle for an unimaginable future.[42] Utopia becomes, in this sense, radically antiteleological. It is in this way that I argue that the power of the two "Epitaph" episodes lies in their refusal of any specific endpoint in favor of an open-ended struggle toward an unimaginable future: *Dollhouse*, like *Buffy*, ends at the beginning of something new. While the dystopian 2019 segments appear to be a depressing vision of Echo's failure to change the system, I argue that we need to read these apocalyptic sequences dialectically as an "end" that is simultaneously pregnant with the utopic possibility for a radical reorganization of the status quo.[43] This dialectical reading of Rossum's apocalypse is unlocked by the show's uncompromising utopian hope that such a reorganization might be possible.

It is, of course, Echo who serves as the primary figure for this radical utopic hope in the two "Epitaph" episodes, that is, in the dystopic space of 2019. Echo's utopian struggle can be apprehended visually in the conclusion to "Epitaph One": in the episode's final moments, Echo (imprinted on Iris) leads Mag and Zone out of the Dollhouse toward "Safe Haven" (which is at this point in the series a more or less empty signifier that suggests some kind of utopia), climbing up past Adelle's old office and out into the sunlight. The prominence of sunlight in the frame prefigures a similar brightness in the final moments of "The Hollow Men" analyzed earlier. Yet now, in the properly dystopic space of 2019, the skyscrapers are all decayed and broken. This visual difference underscores that the utopic desire that ends this episode—literally climbing out into the sunlight—cannot ever be folded back into the status quo: the world has been radically changed, albeit temporarily for the worse, and there is quite obviously no going back. However, this moment is not antiutopian: as Echo, Zone, and Mag climb upward out of the frame, Echo leads them in the struggle for a better world, reinforcing the utopian impulse that has sustained the show all along. Indeed, the visual conceit that closes "Epitaph One" is made literal at the beginning of "Epitaph Two": after

Zone dismisses Echo's claim that they "still have a chance to make a true world," Echo confidently replies, "We're lost. We are not gone." Zone's pessimism here is in opposition with Echo's unwavering hope that a "true world" might still be in the offing, a hope that sustains not only her utopian struggle but the narrative trajectory of the entire show.

Echo's utopianism in "Epitaph Two" can likewise be contrasted with two antiutopian ideological positions that are registered through the binary opposition of technology/antitechnology, a debate prefigured in "Epitaph One" through Zone. Zone's extreme antitechnology stance is signaled early in the episode when he blasts a radio to pieces and, more explicitly, when he reminds the others that they have decided to forgo the use of all technology. Crucially, this antitechnology position is also associated with a radical ahistoricity, or a retreat from history as such: when the others propose that they might be able to find out what happened by using the imprinting chair, Zone dismissively replies that they would find "no answers, only history" ("Epitaph One"). Zone, or the antitechnology position in general, then denies the importance of the diachronic progression of history that Connelly and Rees argue is so central to the show.[44] A rejection of history is, of course, also an implicit rejection of utopianism, which is by its very definition focused on the future. Similarly, when Priya, who is even more opposed to technology than Zone, tells Anthony that "hope seems almost cruel at this point," she indirectly associates her antitechnology position with a kind of antiutopianism, or a refusal of radical hope ("Epitaph One").

Indeed, Safe Haven itself turns out to be a kind of agrarian enclave that represents a retreat not only from technology but also, at least for Priya, from political engagement. After reminding Anthony that she "stopped fighting long ago," Priya resists Echo's plan to return to the Dollhouse and restore everyone's original personas because she does not want to leave Safe Haven, which she calls her "home" ("Epitaph Two"). As Jameson points out in a different context, this sort of "regression to the enclave is thereby distinguished from the transformation of the system itself, its replacement by a wholly different one."[45] Indeed, the agrarian lifestyle within Safe Haven explicitly looks backward into a nostalgic precapitalist past and denies any kind of utopian transformative politics. The most apparent alternative to the antiutopian agrarian enclave, or the other side of the binary opposition, is Anthony's cyberpunk crew of "tech heads," and yet they, too, are associated with a fear of utopia. When it becomes apparent that Topher can radically remake the world by returning everyone's original persona to them, Rome, a member of Anthony's gang, leads a coup, telling them, "We didn't come down here to let you take our world away" ("Epitaph Two"). Both of these positions then—the return to nature espoused by Priya and the cyberpunk technofascism of Rome—are unmasked as antiutopian. Yet, as Jameson proposes in an analysis of what he calls the city-country opposition in science fiction, these two positions are still valuable within a text as "an ideological critique of its opposite number."[46] Thus, the value of these two positions in the dystopic world of "Epitaph Two" is precisely to remind us that neither one is the answer: that the truly utopic position must neither seek to find redemption in the nostalgic past nor hold on to the present.

The alternative to all these antiutopian positions—that is, the authentically utopian impulse represented by Echo—is therefore both historically centered and radically open-ended. When Topher's plan to return everyone's original identity is revealed, Echo immediately reminds them that they cannot all have their original selves returned, because then they would "forget everything" ("Epitaph Two"). Their plan to hide out in the Dollhouse for a year is therefore borne out of a need to remember history, not to retreat into it as Priya wishes to do, but so that it can never repeat itself. Indeed, the pulse is not simply a "reset" button on history; they also have to contend with the concrete reality of the world that currently exists. After all, as Echo asks Paul, "you think after the pulse everything's gonna be hunky-dory?" ("Epitaph Two"). Similarly, Topher reminds Adelle that, while he can return original memories to everyone, it is up to her to "fix what we did to the rest of the world" ("Epitaph Two"). Each of these statements functions as a reminder that history can never regress or return to some previous high point; the struggle for utopia is a struggle to remake the future from the given historical moment. After the blast, in the last moment of the episode before Echo's reunion with Paul, Adelle and Zone help the people who were in the Dollhouse to become acquainted with the new world. The wreckage in this scene—much like the giant crater that ends the final episode of *Buffy*—signals that there is no turning back. The show ends with a radically open historical moment in which anything seems possible. Indeed, it is at this moment that the hope for radical praxis that has sustained Echo becomes actualized.

It is also through this actualization of a revolutionary event capable of radically changing the world that fantasy itself, which has always been one of the Dollhouse's key strategies of power, is rejected in favor of concrete praxis: after Adelle notes the irony that "the last fantasy the Dollhouse would fulfill" is Echo's dream of "leading them into the light for good," Echo simply responds, "I don't have any fantasies" ("Epitaph Two"). If, as I argue, the "fantasy" of liberating the Dollhouse as depicted in "Needs" was nothing more than a compensatory fiction aimed at containment rather than true change, then Echo's rejection of "fantasies" here signals a shift from an ideological narrative resolution to true radical praxis. What Adelle never understands is that people do not need just fantasies; they also can challenge the very structure that produces the need for these fantasies. Echo's actualization of what was previously only a compensatory fiction actually cancels out the need for that fantasy in the first place. Indeed, Phillip Wegner argues that this "canceling" is a feature of the dystopic narrative structure itself: "the critical dystopia becomes a self-consuming text, one that narrates the desired abolition of the very conditions of its own emergence, such that politics once again replaces the desire for politics."[47] *Dollhouse* therefore ends on the cusp of a radically different world, one in which the kinds of fantasies the Dollhouse once provided are no longer necessary.

It is also through this kind of an open ending that utopian texts are, according to Moylan, able to invite the reader "to carry over the tensions of the text to their own struggles . . . in their actual social reality."[48] In our own world, then, *Dollhouse* provides us with a timely reminder that hope itself can be a radical political project, especially given our current historical position in a world that appears incapable of being radically changed. Despite its bleak vision of where humanity might be headed, the show remains stubbornly committed to a utopian impulse bent on nothing less than the systematic reworking of the status quo. Not only does *Dollhouse* reject the hegemonic ideology of "capitalist realism," but it also consistently demonstrates the dangers of both compensatory utopian fantasies and antiutopian retreats from history. The fact that this powerful utopian desire can exist alongside or, indeed, even because of the apparent bleakness of the series also suggests the continued importance of Moylan's generic category of "critical dystopia."[49] Moreover, by situating its utopic impulse against the backdrop of global capitalism and biopolitical power, *Dollhouse* also places this utopic hope within the trajectory of our own historical moment, giving it a profound political relevance. Despite how unlikely the prospects for bringing about a radically different world might appear from our current vantage point, today's global left would perhaps do well to remember Echo's militant optimism: "We're lost. We are not gone."

BIBLIOGRAPHY

Agamben, Giorgio. *Homo Sacer: Sovereign Power and Bare Life*. Translated by Daniel Heller-Roazen. Stanford, CA: Stanford University Press, 1998.

Bennett, Eve. "Deconstructing the Dream Factory: Personal Fantasy and Corporate Manipulation in Joss Whedon's *Dollhouse*." *Slayage: The Journal of the Whedon Studies Association* 9, no. 1 (2011).

Connelly, Tom, and Shelley S. Rees. "Alienation and the Dialectics of History in Joss Whedon's *Dollhouse*." *Slayage: The Journal of the Whedon Studies Association* 8, nos. 2–3 (2010).

Fisher, Mark. *Capitalist Realism: Is There No Alternative?* Washington, DC: Zero Books, 2009.

Foucault, Michel. *Security, Territory, and Population: Lectures at the Collège de France 1977–1978*. Translated by Graham Burchell. New York: Picador, 2009.

Ginn, Sherry. *Power and Control in the Television Worlds of Joss Whedon*. Jefferson, NC: McFarland, 2012.

Hardt, Michael, and Antonio Negri. *Empire*. Cambridge, MA: Harvard University Press, 2000.

Horkheimer, Max, and Theodor W. Adorno. "The Culture Industry: Enlightenment as Mass Deception." In *Dialectic of Enlightenment,* edited by Gunzelin Schmid Noerr, translated by Edmund Jephcott. Stanford, CA: Stanford University Press, 2002.

Jameson, Fredric. *Archaeologies of the Future: The Desire Called Utopia and Other Science Fictions*. New York: Verso, 2005.

———. *The Cultural Turn: Selected Writings on the Postmodern 1983–1998*. New York: Verso, 1998.

———. "The Politics of Utopia." *New Left Review* 25 (2004): 35–54.

———. *Postmodernism; or, The Cultural Logic of Late Capitalism*. Durham, NC: Duke University Press, 1991.

———. "Reification and Utopia in Mass Culture." *Signatures of the Visible*. New York: Routledge, 2007.

———. *Valences of the Dialectic*. New York: Verso, 2010.

Marx, Karl. *Capital*. Vol. 1. Translated by Ben Fowkes. New York: Vintage, 1976.

Moylan, Tom. *Scraps of the Untainted Sky: Science Fiction, Utopia, Dystopia*. Boulder, CO: Westview Press, 2000.

Orwell, George. *1984*. New York: Signet Classics, 1950.

Vinci, Tony M. "Not an Apocalypse, the Apocalypse: Existential Proletarisation and the Possibility of a Soul in Joss Whedon's *Dollhouse*." *Science Fiction Film and Television* 4, no. 2 (2011): 225–48.

Weeks, Kathi. *The Problem with Work: Feminism, Marxism, Antiwork Politics, and Postwork Imaginaries*. Durham, NC: Duke University Press, 2011.

Wegner, Phillip E. "Horizons, Figures, and Machines: The Dialectic of Utopia in the Works of Fredric Jameson." *Utopian Studies* 9, no. 2 (1998): 58–77.

———. *Life between Two Deaths, 1989–2001: US Culture in the Long Nineties*. Durham, NC: Duke University Press, 2009.

Žižek, Slavoj. *Living in the End Times*. New York: Verso, 2010.

FILMOGRAPHY

"Chosen." *Buffy the Vampire Slayer*. Season 7, episode 22. Written and directed by Joss Whedon. First Broadcast May 20, 2003.

NOTES

1. Vinci, "Not an Apocalypse."
2. Bennett, "Deconstructing."
3. The term "late capitalism" is used by Jameson in reference to Ernest Mandel's book of the same name but should be understood as interchangeable with "global," "multinational," or "finance" capitalism for the purposes of my chapter here. As Jameson explains in the introduction to *Postmodernism; or, The Cultural Logic of Late Capitalism*, the point of the phrase is simply to indicate the "sense that something has changed," that "things are different" within this era of capitalism, but that it is still a version of capitalism (xxi). Jameson's framework thus allows a periodizing analysis that can help unlock the specific historical and cultural dynamics of this era of capitalism and how they are reflected within *Dollhouse.*
4. Jameson, *Postmodernism*, xviii–xvx.
5. Vinci, "Not an Apocalypse," 227.
6. Vinci, "Not an Apocalypse," 225–27.
7. Jameson, *Postmodernism*, 36.
8. Bennett, "Deconstructing."
9. Jameson, *Postmodernism*, xx.
10. Jameson, *Postmodernism*, 37–38.
11. Hardt and Negri, *Empire.*

12. Foucault, *Security,* 1.
13. Agamben, *Homo Sacer*, 6.
14. Marx, *Capital* Volume 1.
15. Hardt and Negri, *Empire,* 29–32.
16. Hardt and Negri, *Empire,* 292.
17. Weeks, *Problem with Work*.
18. Weeks, *Problem with Work*, 69–70.
19. Weeks, *Problem with Work*, 72.
20. Fisher, *Capitalist Realism*, 2 (emphasis in original).
21. Jameson, *Cultural Turn*, 90.
22. Jameson, *Cultural Turn*, 91.
23. Connelly and Rees, "Alienation."
24. Ginn, *Power*.
25. Connelly and Rees, "Alienation."
26. Connelly and Rees, "Alienation."
27. Connelly and Rees, "Alienation."
28. Orwell, *1984*; Jameson, *Archaeologies*, 199.
29. Moylan, *Scraps*.
30. Moylan, *Scraps*, 106.
31. Quoted in Moylan, *Scraps*, 106.
32. Jameson, *Archaeologies*, 198.
33. Wegner, *Life*, 119.
34. Wegner, *Life*, 119.
35. *Buffy the Vampire Slayer*, "Chosen." For a detailed reading of Willow and Xander's exchange and the utopian possibilities of these final moments of *Buffy*, see Wegner's *Life between Two Deaths*, 215–17.
36. Moylan, *Scraps*, 55.
37. Jameson, "Reification," 34.
38. Wegner, "Horizons," 61; Weeks, *Problem with Work*, 195–97.
39. Wegner, "Horizons," 61.
40. Wegner, "Horizons," 61.
41. Wegner, "Horizons," 61.
42. Weeks, *Problem with Work*, 197.
43. Toward a similar end, Slavoj Žižek uses Elisabeth Kübler-Ross's famous five stages of grief to describe our collective experience toward the possibility of apocalypse, concluding that "after passing through this zero-point [at the completion of the previous four stages], the subject no longer perceives the situation as a threat, but as the chance of a new beginning" (xii).
44. Connelly and Rees, "Alienation."
45. Jameson, *Valences*, 578.
46. Jameson, "Politics of Utopia," 50.
47. Wegner, *Life*, 119–20.
48. Moylan, *Scraps*, 55.
49. Moylan, *Scraps*.

Chapter Twelve

Welcome to the *Dollhouse*

Reading Its Opening Title Sequences

David Kociemba

> Echo [as Margaret]: Will I see my whole life flash before my eyes?
> DeWitt: Every single moment . . .
>
> —"Haunted," 1.10

> Producers feel [that show openers] are a signature. It's part of the identity of the show.
>
> —Joanne Curley-Kerner, producer of *The Cosby Show* (quoted in Dupree, 34)

Opening title sequences play an important role in understanding television series. They are "a film within a film" whose "very purpose is to serve a whole array of functions: copyright law, economics, certification of employment in the context of careers, movie title, entertainment, commercials, fashion, and art."[1] Opening title sequences are but one example of what Gérard Genette terms *paratexts*, which are secondary texts that influence the reading of the primary text, such as book covers, title pages, author names, interviews, ads, reviews, and public responses.[2] Genette likened paratexts to thresholds or vestibules, in line with Jorge Luis Borges's description of prefaces.[3] In film and television, opening title and closing credit sequences often function as the equivalent to the covers of a book, marking its beginning and end. Yet those metaphors are inexact. Some films eliminate the opening titles entirely.[4] On television, title sequences are increasingly contained within the text after a cold open yet still serve to mark its borders. *Lost* (2004–2010) and *Heroes* (2006–2010) popularized the practice of showing only the series's title, highlighting this function.[5] Yet borders do more than mark territory: they also guide travelers. Like other paratexts, opening title sequences

> guide the reader's attention, influence how a text is read, and communicate such information as to give a text its first contours, its manageable identity so to speak. And paratexts are important not only for the process of textual reception but also for text production; they function as indicators to be aimed for, as structures of literary expectations.[6]

They are a film within a film that is about the narrative and an independent creation. Even *Lost*'s sparse title sequence tells a story about the series's themes of mystery, displacement, trauma, and insight through the rich cult television history of the score's glissando and its tonal destabilization, the title's swirl evoking the series's iconic plane crash as it moves from background to fall off-screen in the foreground, and its letters that come into momentary focus before clarity is once again lost.[7] Opening title sequences are a form of narration, a pleasure in their own right, a genre of short filmmaking, and a mark of authorship,[8] but not that of the auteur—as "it is the work of specialists, title sequence designers, themselves readers foremost" creating with incomplete access to the series's final form and contingent on network and creator approval.[9]

One can learn a great deal about a series through studying the visual, musical, and editing rhetorics of a series's opening title sequences. With the *Dollhouse* montages, the primary story is whether a woman in a noirish world can have agency under predatory corporate surveillance and male objectifying gazes. The wordless voice of the theme song may belong to this woman, Echo, but it may, at times, be read as Caroline's or a little girl's; but all three are the same person. And Echo is one of us, picked from the crowd. Her story is ours, as we are all Dolls. Yet, we are simultaneously implicated in our own oppression by our consumption of the Dream Factory's products.

As a signature of ownership, the credits within the opening title sequence have meaning for artists, executives, and viewers. They establish the identity of the authors by declaring who is responsible for what work. Early silent film producers fought copyright fraud by including in each film a board with the title, company name, and/or trademark. Movie casts started being listed in 1911, with the screenwriter position the year following. In 1913, D. W. Griffith left Biograph and announced his credits as its sole director in a *New York Dramatic Mirror* ad.[10] Director Frank Capra considered the hierarchy implicit in credits so emblematic of his struggles for artistic control for himself and his peers that he named his autobiography *The Name above the Title*.[11] Credits communicate the relative industry power of various parties involved in the creation and distribution of the film or series.

Joss Whedon's signature on *Dollhouse* was especially important to both viewers and network executives. The auteur as a marketing brand of quality television[12] rose with the increased importance of demographics in advertising; with the splintering of the mass audience due to alternatives to broadcast television provided by cable and satellite television, DVDs, and video-streaming services, as well as competition from videogames and other digital entertainments; and as a way to create separation from the legion of reality series. In 2009, *Dollhouse* went up against CBS's army of procedurals (*Numb3rs*, *Medium*, *Ghost Whisperer*), shows on the way out (*Ugly Betty*, *Law & Order*), reality programs across the dial (*Wife Swap*, *Supernanny*, *Shark Tank*, *Dateline NBC*), and reruns. *Dollhouse* stood out as quality television and as science fiction; its only genre competitors were *Smallville* and CBS's contemporary fantasy-procedural hybrids. Fox's strategy was to win Friday-night audience demographics and, more important, minimize the ratings damage from airing a niche auteur's cult show that would weekly change the genre, style, and identity of the protagonist.[13] Airing the first season's episodes of *Dollhouse* with half the usual commercials and promos to retain viewers and present the series as quality television, Fox hoped to leverage Whedon's brand name to win Friday nights with advertisers through demographics rather than mass audiences.[14]

For fans, Whedon's signature in the credits showcased their hope that it was his show, not the network's. Viewers understood that the Friday-night slotting of *Dollhouse* was disastrous for the future of the show and as another betrayal of Joss Whedon by Fox. One fan community started a website to save *Dollhouse* from cancellation before the show even aired. With Mutant Enemy's prior show, *Firefly* (2002), Fox had demanded a new pilot episode, did not air several episodes, and aired every episode out of its intended order. History repeated itself with *Dollhouse*. The network required Joss Whedon to write a new pilot, did not air its thirteenth first-season episode, renewed it very late, and combined episodes to burn more quickly through the soon-canceled second season. Whedon said that "the network pretty much wanted to back away from the concept five minutes after they bought it."[15] He sarcastically summarized the network's instructions to the creative team as follows: "Up the stakes, make the episodes more stand-alone, stop talking about relationships, and cut to the chase. Oh, and add a chase. That you can cut to."[16] Whedon's signature on a Fox show airing Friday night immediately signaled an auteur embattled to fans and in need of active support to survive.

Yet the "Created by Joss Whedon" credit obscures the real identities of the artists behind the series as much as it asserts it against network interference. During *Dollhouse*'s first season on air, Whedon was busy on another project, cowriting *The Cabin in the Woods* (2012) with Drew Goddard. The film's principal photography concluded just two months prior to when filming began on the series's second season. Thus, Whedon shared showrunner duties with Elizabeth Craft and Sarah Fain, who consequently had influence on set and in the writer's room. They were under contract to Fox, not to Mutant Enemy. And Whedon has noted that Fox pushed the series to be more action oriented with clearer plot arcs featuring Rossum and less of his original intent to engage in an ambivalent study of the nature of fantasy and its effects on its producers and consumers.[17] The creative role of the Fox executive in charge of the program, Beth Miyares, is obscured as well by the auteur's brand. The affiliate logo, sometimes flashed when Whedon's credit is broadcast, testifies to the influence of the network. This mingling of signature and logo on the border between show and advertising break illustrates how opening title sequences regulate the "defining characteristic" of television for Raymond Williams, which is the planned flow among programs, ads, and channel identification.[18] The opening credit sequence's imagery, however, focuses on another important creator: Eliza Dushku. Her production company, Diva Productions, is the third involved in making *Dollhouse*, along with Whedon's Mutant Enemy and Twentieth Century Fox Television. *Dollhouse* sprang from her discussions with Whedon about the nature of acting and performance's impact on identity and fantasy.[19] Stanitzek observes,

> Paratexts have the effect of promoting the unity of a text, but they can only accomplish this without hindrance when they are not *read* in the strict sense of the word as such, that is, when no questions are asked about details, when there are no inquiries into how they function, how they make references to circumstances of production or distribution or to other aspects.[20]

However, Genette suggested that paratexts promote "a more pertinent reading . . . in the eyes of the author and his allies."[21] He seems to not have realized how much multiple authors can struggle for control over a text and how different their viewpoints of what is pertinent can be; indeed, he uses the singular term *author* exclusively. While it is indisputable that Whedon is a primary author of the series, the idea that he is the only one is a carefully crafted fantasy created by the opening title sequence and problematized by the closing credits.

The image that Whedon chose to accompany his *Dollhouse* credit is significant. With *Buffy the Vampire Slayer* (1997–2003) and *Angel* (1999–2004), Whedon uses the last shot of the montage to forge a connection between him and the title character; it is his "created by" credit that runs over this shot. He signs the shot of Buffy (the heroine standing alone against the forces of darkness) and Angel (the noir hero stalking down a dark alleyway, trenchcoat billowing) as explicitly his creation and his responsibility.[22] With *Firefly*, however, he signs an image iconic to the world he built: the spaceship Serenity swooping down to drive a herd of cattle. With *Dollhouse*, he signs an overhead shot of the sleeping pods from an alternate setup in "Needs" (1.8), where Echo, Sierra, November, and Victor awaken with some memories of their preimprint selves. Whedon emphasizes place rather than character, genre, or world. Whedon's name fits snugly among the pinwheel of sleeping pods, emphasizing the importance of Stuart Blatt's design of the Los Angeles Dollhouse branch. Physical space grounds the series—a necessary task given the lack of character continuity due to its premise of unmoored psyches. But this place is not as emblematic of the series as Adelle's office, the imprinting chair, or the central hub of the Dollhouse. This shot emphasizes theme more than the Dollhouse itself. The pods emphasize the series's motifs of sleep and fantasy.[23] Here, Echo hides her book and her notes—cunningly scratched into the sliding-glass lid—that help her remember. It is the one place suggestive of solidarity in an opening title sequence unique in Whedon's career for its exclusive focus on a single character, Echo, rather than the ensemble. But it is thwarted in that purpose for all but this episode, as the sleepers are forcibly separated. This shot connects the artist to ideas of forgetting, memory, fantasy, isolation, and connection.

The typography of the credits themselves can have expressive effects. The first opening title sequence to animate type in a subject-appropriate manner was *Gone with the Wind* (1939), whose credits gust on and off the screen, seemingly made italic by the wind.[24] In *Dr. Strangelove; or, How I Learned to Stop Worrying and Love the Bomb* (1964), title designer Pablo Ferro's iconic opening sequence makes a typographic game out of such contractual obligations. While the actors' names were drawn to the size required, unimportant connecting words like "and" were made very tall or wide, sometimes dwarfing the actors' names. Kyle Cooper and Imaginary Forces are often credited with revitalizing the art of opening title sequences and credits in 1990s film.[25] Their best-known work is *Se7en* (1995), whose "degraded typography, jittery jump-cutting and twitching, hand-hewn type ushered in a new era in film titles."[26] Opening title sequences were made newly prominent in the industry just before Whedon began his career as an executive producer with *Buffy the Vampire Slayer* in 1997.[27] *Dollhouse*'s use of all capital letters evokes the underlying conformity imposed by the Dollhouse on its Dolls, as lowercase provides greater variety in letter shape. The "doll" is bolded with light, emphasizing the figure within the environment.

The theme music of opening title sequences can establish a setting, provide exposition, and establish the tone of the series. *The Cosby Show* (1984–1992) reinforces its tone of sweet quirkiness through dance, which features characters grooving to jazz music alone or with Bill Cosby. Scotty Dupree notes the importance of theme music for catching the attention of viewers, citing "*The Brady Bunch*'s popping squares, *Hawaii Five-O*'s raging waves and jolting music, to the snapping fingers of the *Addams Family*."[28] The opening credits of *Twin Peaks* (1990–1991) are unusual in that they neither provide background plot nor introduce characters. Instead, establishing mood and tone is the primary goal of its title montage. The saturated primary colors of the images of the Pacific Northwest, the languorous rhythms of its montage, and Angelo Badalamenti's theme music create a tone teetering on the edge of intense, banal, and fascinating. It evokes the oneiric experience of watching this strange series to prepare the viewer for another episode. Ferro's title sequence for *Dr. Strangelove* sets the film's subversive tone of black satire by showing refueling B-52 airplanes bump and grind to "Try a Little Tenderness." Design writer Ken Coupland suggests that "the first few minutes of a film can be compared to the curious stage of consciousness that marks the transition between wakefulness and sleep."[29] The opening title sequence is when we begin to get some sense of what dreams may come.[30]

Janet K. Halfyard, in her conference paper on Jonatha Brooke's theme song for *Dollhouse*, argues that the theme song foreshadows important developments in the series's central themes of identity and loss.[31] Whedon seems to have played an unusually active role in selecting the theme, as he lobbied for Jonatha Brooke with Fox executives, asked for a glockenspiel sound, and spliced pieces of "What You Don't Know" to make the final piece.[32] At first, viewers make the connection between the voice and Echo, as "the song [is] stripped of its words as Echo herself is stripped of memory," and there is no more prominent character in the opening than her.[33] But the omitted lyrics provide a suggestive exposition of basic Dollhouse procedures, a neat little reward for active viewers who do some research (which *Buffy the Vampire Slayer* fans, for instance, were wont to do with music; cf. Dechert).[34] Halfyard goes further in reading the lyrics as Caroline explaining her surrender to Rossum Corporation to herself, to Ballard, and to us, with the coda sung by Echo.[35] The omitted song lyrics reference the moral question of whether Echo or Caroline has the right to her body, a moral quandary first begun in "Omega" (1.12) and, Halfyard argues, embedded in the opening title sequence's image of a partially nude Echo facing her kimono-clad reflection, which she reads as Caroline.[36] Halfyard writes, "Jonatha Brooke's song itself, absent from the *Dollhouse* text, is in effect Caroline's missing narrative voice, filling us in on what she has done, informing us of her absence."[37]

The theme music (as opposed to the song that the theme music was drawn from) presents another use of foreshadowing. The theme music of Whedon's prior shows prepares viewers to hear *Dollhouse*'s theme music in particular ways. The wordless themes of Whedon's prior signature series clearly inscribe Buffy as a teen outsider hero and Angel as a noir hero wont to brood romantically. *Firefly*'s Whedon-written theme establishes the frontier setting of that 'verse. Past history primes Whedon's fans to hear the wordless song as being about Echo or the Dollhouse setting. The light and breathy timbre and glockenspiel suggest both a child and an innocence. But Echo is not a child. Halfyard reads the voice of the theme music as better heard in season 2 as Caroline's future voice, embedded in the body of a little girl after the "thought-pocalypse."[38] We meet this Caroline in the unaired season 1 finale, "Epitaph One" (1.13), and again (and yet for the first time for broadcast-only viewers) in "Epitaph Two: Return" (2.13), the series finale. I would add that this childlike voice is sung by an adult woman, furthering establishing the posthuman nature of Caroline's identity.[39] The broadly descending pattern of the notes and mournful tone further suggest the early wisdom of a child who survives desperate times, like the girl beneath the Iris and Caroline imprints in the two season finales. The voice of the *Dollhouse* theme music can be productively read as Caroline's, Echo's, and a future little girl's, an identity shifting during the series, adding identities as it goes.

Subservient to the film or series that contains them, the opening title sequences can do work that serves the needs of the narrative. Jonathan Gray observes that

> paratexts tell us what to expect, and in doing so, they shape the reading strategies that we will take with us "into" the text, as they provide the all-important early frames through which we will examine, react to, and evaluate textual consumption. As such, the study of paratexts is the study of how meaning is created, and of how texts begin.[40]

Early functions of these openings were to indicate genre and establish setting and period. American silent films often did this literally. Andrea Codrington writes that their titles "played against the backdrop of billowing sails in swashbuckler adventures, or on brown-edged parchment in historical dramas."[41] Opening title sequences provide concise reminders of plot points essential to understanding the current narrative. The lyrics of the theme song to *Gilligan's Island* (1964–1967) provide the necessary exposition of the storm and shipwreck that ruined the protagonists' three-hour tour. The opening titles can foreshadow important events and themes, as well as emphasizing existing ones. Stephen Frankfurt's opening sequence for *To Kill a Mockingbird* was the first to slowly pan "across details like scissors, crayons, dolls and pens—small, precious objects that grow in significance later in the film."[42] Graphic designer David Peters and design writer Ken Coupland note that Saul Bass's opening to *Vertigo* connects to the film's important themes. Bass's close-up of a woman's face suggests the movie's interest in voyeurism. The spinning spiral in the iris of her eye represents both the title and the vertigo of its central character.[43] Opening title sequences that work in these ways function as synecdoche, providing a "concise story about the story."[44] In addition, opening title sequences function quite differently in serial television than in films. The first alteration is the result of the different uses of the film and television media in the United States. The website of Randy Balsmeyer's Big Film Design observes that "broadcast design is the concentrated version of film design. Television title sequences must work on a smaller screen and in a shorter time frame. The audience is not captive. Television titles have to make people want to watch them, instead of reaching for the remote. Everything needs to happen bigger, faster, clearer."[45]

Creators take advantage of the serial nature of American television fiction programming. Varying the opening encourages active viewing positions. *Ellen* (1994–1998) uses this "variations on a theme" approach to subvert the season-long stasis of television title sequences through featuring a rhythmic gymnast routine given color commentary by John Tesh, clogging, Wolfgang Puck, the Captain & Tennille, and flamenco dancing.[46] A different message is left on Rockford's answering machine in each episode of *The Rockford Files* (1974–1980). *The Simpsons* (1989–present) varies Bart's chalkboard punishment and has the entire family sit on the couch with different results. Sometimes their difference is highlighted as a spectacle independent of its series, as in the case of Banksy's redesign of *The Simpsons*'s sequence. In film, the opening credits are designed to declare ownership, state narrative details, make implicit promises, evoke moods, and suggest what is to come. One act of communication accomplishes many goals. In serial television, it can be much more complex than that one-step process.[47]

Whedon's previous use of opening title sequences on *Buffy the Vampire Slayer* was quite complex, priming *Dollhouse* viewers to interpret its opening in light of the auteur's prior use of them. The wacky title of *Buffy the Vampire Slayer* was intentionally chosen to foreground the series's sense of humor, genre savvy, and feminist themes, just as *Dollhouse*'s title provides a theatrical and feminist frame through its reference to Ibsen. (How would we read *Dollhouse* if it were not called *Dollhouse*?) As with *Dollhouse*'s use of font, the *Buffy* logo rewarded close attention to design as it evolved along with the series. During the first two seasons, the chalk-like font for Buffy's name suggests that her very identity has been written for her by authority figures like teachers, another key theme of *Dollhouse*. During the final five seasons, the logo changes to emphasize a new theme: finding a balance between public and personal identities. (*Dollhouse* would express this theme through the trinity of original consciousness, Doll state, and Active state.) The credits for *Buffy*'s performers express the series's understanding of identity. The words themselves are dynamic and in flux, suggesting that people contain multitudes of identities, as Echo in *Dollhouse* does. At times, the credits for Anthony Stewart Head, Amber Benson, Dushku, and Alyson Hannigan inextricably link these performers to their *Buffy* roles. This technique prepares the audience to understand identity not as a natural essence and thus permanent but rather as a performance. These sequences promise stories that feature action, monsters, and romance to viewers using a media-literate and artistically innovative approach for viewers with an affinity for youth culture and B-movies. (*Dollhouse* similarly establishes its own intended audience and seductive pleasures.)

Each season's opening title sequence functions differently, which encourages *Dollhouse* viewers to compare the opening sequences of its two seasons for suggestive changes. *Buffy*'s montage in the first two seasons helps viewers adjust to the genre of the series by establishing the iconography of fantasy horror, while later seasons cut these shots to devote more time to establishing increasingly complex characters. The first season's opening rewarded attentive viewing through extensive use of foreshadowing, which *Dollhouse* would do as well. *Buffy*'s opening title sequence from the second season emphasizes a shared history to deepen an already existing commitment on the part of the viewer. In the third season's opening, the creators tease out the complexity of that shared history. A number of images lend themselves to multiple readings, depending on the extent of one's knowledge of the narrative; *Dollhouse* uses this technique from the beginning. The rapid pace and layering of the images in both series' openings required great attention—and often a pause button—to unpack their meanings. Finally, "Superstar" (4.17), among other *Buffy* episodes, demonstrates explicitly that the series creators always used the opening title sequences expressively. When Jonathan's magic spell allows him to alter the opening montage in "Superstar," it reveals that the opening title sequences are not exterior to the narrative world. Whedon acknowledges explicitly his implicit presence in the narrative as narrator of the series, just as *Dollhouse* works as an extended meditation on his own culpability as a purveyor of fantasies. The opening title sequences of *Buffy the Vampire Slayer* function as a microcosm of the series itself. They reveal the influence of the creators' perception of their audience and their own work, the medium's narrative and artistic conventions, and the media industry's own practices. They construct the series's past, shape the viewer's present experience of the episode, and prepare the way for future narratives.[48] Whedon's past use of opening title sequences critically informs how viewers read the ones for *Dollhouse*.

The two opening title sequences of *Dollhouse* similarly influence viewers through narrative references, rapid editing rhythms, mise-en-scène, and the evolution of the sequence over time. Their meaning changes as the audience progresses through each season, from establishing theme and the basics of its narrative to foreshadowing, to shared memory. The low ratings of the first season are reflected in how transparent a narrative the second-season sequence provides when compared to the first. Echo sits, looks up, sees herself in the imprinting chair, then we get images of fifteen separate engagements. That is followed by the television screen vortex shot of memory draining away, another flash in the chair, then Echo lies down to sleep in her pod. This is the only opening title sequence of Whedon's career that attempts to tell a linear narrative summarizing the story. The first season's titles assumed a fan base eager to decode the images. The opening title sequence of the last season of *Dollhouse* instead strove toward clarity to gain audience, establish the action and science fiction genres, and provide exposition as the series went more heavily into serial narrative to finish the story before cancellation.

Whereas the opening title sequences of both seasons start with cityscapes, they are rendered very differently. The first season's cityscape presents L.A. at night, from an overpass, with the sickly yellows and browns of night street lighting edging out of the murky blacks. Contrast this shot with the second season's aerial daytime view above the skyscrapers whose steel gray hues are quickly overwhelmed by a flare of white light. The former shot evokes the noirish moral corruption and limited perspective shared by all the characters in the first season. (Even Adelle does not know the extent of Rossum's evil, which she will learn through Sierra's background, Daniel Perrin's rise to power, and the remote wipe technology.) That white light marking the first season's transition to Echo strutting through the Dollhouse flares mostly from the two skyscrapers, linking the corporate world with the harsh glare that rewrites Dolls' minds many times in the series. Light from the horizon overwhelms the LA cityscape of the second-season title sequence, foreshadowing artistically the thought-pocalypse of the two "Epitaph" episodes and Topher's device that resolves it in the finale, exploding from Adelle's top-floor office.

In addition, the first cityscape is one of several expressive uses of tilt-shift photography. Either through the use of a special lens or digital postproduction techniques, tilt-shift photography creates a narrow depth of field that makes figures look like miniatures, with Echo being followed down a country road an excellent example. The first season's title sequence is the only one that uses this technique. This technique shows that we are all Dolls, programmed by society if not imprinted in a Dollhouse. As Topher observes to Boyd in the unaired pilot: "Why do you wear a tie? It's just what grown-up men do in our culture. They put a piece of cloth around their necks so they can assert their status, and so they can recognize each other as non-threatening kindred. . . . You wear a tie because it never occurred to you not to. . . . Everybody's programmed, Boyd." Each tilt-shift shot after the first is paired with a zoom and narrowing targeting brackets that suggest a digital surveillance system. It picks Echo out from doll-sized figures on a city street, a rural road, and from among the other Dolls in the Dollhouse's main traffic area. Each of these shots ends with a small superimposed title in red with small caps lettering that read, in order, "active located," "active secure," "active engaged." Rossum's controlling panoptic gaze hijacks the first opening title sequence, making it as much of a contested arena within the narrative as the "created by" credit is outside it.

Yet these three images foreshadow the possibility of resistance, subvert the rhetorical positioning of the Dolls as innocents, and encode Echo's agency. When the camera digitally zooms in to Echo whirling around listening to a cell phone, marking it "active located," she is being remotely wiped by

Alpha in "Gray Hour" (1.4). This episode presents a key early moment of moral agency by Echo, who, in her Doll state, picks which robber to save. The image of Echo stretching with the red title "active secure" had voice-over dialogue: "Happy? No, this is something quite apart from happiness. Call it a kind of bliss, an unquestioning serenity. True happiness requires some measure of self-awareness" ("True Believer," 1.5). Senator Boxbaum is talking about a religious cult that exploits its believers, but laying the line over Echo's image forges the connection between the cult and the Dollhouse that the senator and Adelle acknowledge later in the scene. With "active engaged," Echo whirls to face her attacker in "The Target" with gun drawn. She saves her severely wounded Handler, Boyd, inverting the chivalrous ritual of the imprinting ceremony, which has the passive Active confirm her or his trust that everything is going to be alright now that the Handler is by her or his side. Here, Boyd is the passive trusting one to be saved by the active female agent.

Rossum's surveillance is not the only predatory gaze to threaten Echo. She is heavily objectified by the camera in both seasons' sequences. Her costumes include thigh-high white lace stockings (twice framed so that Dushku's legs are the dominant focus of the shot), a low-cut tanktop, leather pants, and a dominatrix outfit. Furthermore, Fox's known interference in the series can lead critical viewers to suggest that these shots illustrate the network's vision of the series as Science Fiction Action Barbie, which becomes even more apt when one remembers that Barbie's inspiration was a sex doll. (What is interesting is that while the opening title sequence sells the series through titillation, the network was quite uncomfortable with stories exploring sexuality complexly.)[49] The camera of the opening title sequence is strongly reminiscent of the controlling male gaze, made infamous through the critiques of feminist film studies theorists like Laura Mulvey, which serves to soothe the male creator's and audience's (purported) castration anxiety through voyeurism, sadism, and the creation of fetishes.[50]

Yet these images were authored with Dushku's informed consent and participation, as her producer credit and production company's involvement in the series ensures a *more* equal—yet still hierarchical—power dynamic than the narratives that fueled 1970s' feminist gaze theory.[51] This negotiation between female objectification and female agency might be signified by the next shot, cutting in to the second, a differently costumed version of Echo reflected between her legs to show this other Echo smiling slyly to the camera. (In addition, this outfit is worn in "Echoes" [1.7], which marks the return of Caroline, the campus activist turned terrorist, for the first time since she has been wiped. The episode links Doll and radical, which is anticipated in these shots.) These two moments work simultaneously to objectify Echo and present her subversive masquerade as one source of Echo's power. Traditional feminist gaze theory, as critiqued in Mary Anne Doane's classic essay advancing masquerade as a covert source of power, gave the female spectator two options: the masochism of overidentification with the frequently punished sexually active female figure or the narcissism involved in becoming one's own object of desire (the desiring female gaze directed at the attractive female object of desire).[52]

Masquerade manufactures a critical distance from the image. It makes the image able to be manipulated, producible, and readable by the female spectator. Performers such as Mae West, Marilyn Monroe, and Madonna acknowledge femininity as constructed mask, a decorative layer that conceals agency. Mae West's masquerade flaunts her femininity, while her performance reveals it as such and holds it at a distance. Such masquerade, an excess of femininity, is often aligned with the femme fatale, necessarily regarded as evil, to be destroyed or tamed (via a reintegration into the mainstream, often via marriage). Destabilizing the image confounds the masculine structure of the look and defamiliarizes female iconography, making it threatening to male systems of knowing and pleasure. (Several times a mirror reflects a different image of Echo than the woman standing before it, confusing viewers as to which one is real or, better, if both are, as Echo's image is partially independent of her.) In essence, such female performers might be considered as demonstrating just how difficult a line all women have to walk, or strut, when it comes to the presentation of their sexuality.[53] Coker argues, "The feminist ethos of *Dollhouse* is a thorough explication of what makes society an enemy of women, and how women can fight society and hopefully make it better."[54] That message is embedded in the opening title sequences. Predatory gazes, panoptic and objectifying, seek to suppress the covert agency of masquerade in the opening title sequences, just as the imprinting technology in the Dollhouses is overtly oppressive and controlling in its construction of identity, while latently being a source of revolutionary power through the networked individualism of Echo-Caroline's reconstructed posthuman identities.

The series increasingly codes these contested images as exploitative, spoiling the sexual charge, as the series labels these "romance engagements" as sexual slavery and human trafficking in both seasons. Even the "best" encounters have these erotic fantasies originate from mourning, lonely, or ridiculous clients such as Tempura Joe. The dynamic is hardly the erotics of power and control with which the purveyors of the male gaze flatter their consumers. These images move between being coded as conventionally sexy to corporate sexism to rather pathetic fantasies to those emblematic of global gendered exploitation. Rhonda Wilcox argues the series meditates on the complicity of the protagonist, the male heroes Boyd and Ballard, the Dollhouse clients and employees, and the audience itself.[55] Catherine Coker and Eve Bennett examine how *Dollhouse* implicates its creators in crafting the sexual and heroic fantasies that are the foundation of real-life sexual objectification and sexism that ultimately support and promote prostitution and sexual slavery.[56] As an important producer in the television branch of the Dream Factory, Whedon works as an inspiration for Topher Brink, consciousness writer, but Whedon claims to see himself more in Adelle DeWitt, head of the consciousness production house.[57] The fact that Olivia Williams's credit runs over Echo going to sleep in the pods certainly implies DeWitt's role in the L.A. Dollhouse and further recommends DeWitt as an author figure to the viewer. Whedon is trapped in the Dollhouse, as is the audience.

This reevaluation of images is a foundational function of opening title sequences on serial television. As each episode airs, these images change from being an elliptical form of foreshadowing to a cue to the viewer's memory that binds them in a community with other knowledgeable viewers. The reference to saving Boyd when Echo whirls around with a gun plays very differently once you know of Boyd's betrayal. Her fight with Ballard is referenced in both seasons' sequences. To an uninitiated viewer, the shots of Echo with a knife (in the first title sequence) and her blocked punches (in season 2) merely convey the promise of action, but a fan knows that these shots are early in her history with Ballard, one of her most important and most positive relationships. Opening title sequences forge and reflect fan communities.

The opening title sequences serve as a codex to cue memory and emotion for both Echo and the audience. In the first season's sequence, Echo wipes off a steamy mirror to see her own reflection. This image is found at the end of "Gray Hour," where Echo draws in the steamed surface an image of the Picasso painting that led her to decide that she is not broken. This painting is a crucial early indicator of memory and her progress toward agency. The two images from the beginning of the second season's sequence deal with memory as well. They come from her discussion with Ballard in the second-season pilot, "Vows," the last episode written by Whedon:

> Echo: I remember everything. Sometimes I'm someone else, and then I come back but I still feel them. All of them. I've been many people. I can hear them, sometimes suddenly. I'm all of them, but none of them is me. Do you know who's real?
>
> Paul Ballard: Caroline.
>
> Echo: I want to find her. I want to find all of them. Real them. They can be found. We are lost, but we are not gone. Will you help me?
>
> Paul Ballard: I swear. No matter what, I'm with you. Everything's gonna be alright.

Echo: Now that you're here.

Paul Ballard: Do you trust me?

Echo: With my life.

Echo and Ballard forge their alliance based on her memory. His real chivalry replaces Boyd's tainted patriarchal concern. Their two goals of rescuing the Dolls and finding the "last soldier," Caroline, are decided on. The shot selected from their conversation evokes her ability to remember her former engagements, as her look upward is a typical sign of someone trying to remember something.

Fanvids of *Dollhouse*'s opening title sequences show active audiences creating and promulgating their own preferred readings. Some sequences make minor changes, by emphasizing the ensemble rather than the star. (Indeed, one recast the series by subtracting Eliza Dushku.) Another fanvid emphasized Eliza Dushku by repurposing *Dollhouse* images to make a title sequence homage to *Tru Calling*. One vidder chose Kimberly Cole's "Superstar (Smash It)" from "Stage Fright" (1.3) as the theme song. Other montages emphasize particular themes and content, such as highlighting digital manipulation and surveillance (through painterly editing effects) or the series's violence (by selecting Topher covered in blood or Alpha's bloody knife). Some fan title sequences emphasize the continuity in Whedon's work, linking Echo-Caroline and Buffy Summers (and Adelle DeWitt and Rupert Giles) through switching title songs of each series, changing the font and its animation, and selecting images that echo the other series's sequence. (Indeed, these changes implicitly suggest a desire for a lighter, more hopeful tone to the rather bleak *Dollhouse*.) One fanvid provides an opener for a hypothetical third season of *Dollhouse*, summarizing the second season, then ending on an image of Topher's wipe of the mass imprints.

There have been many arguments about whether preferred readings come from the text, authors, culture, industry, critics, or audiences.[58] The study of paratexts such as opening title sequences shows them to be a site of contestation of all those sources in a complex negotiation of interests, desires, and ideologies. The opening title sequences of *Dollhouse* shape the preferred readings of the series's narrative, show the network and production companies' perceptions of the audience, and express its creators' understanding of the essentials of the text. The opening titles present a series of promises about what narrative forms (such as the science fiction, noir, and action genres), visual pleasures (such as the erotic display of Eliza Dushku's body in a variety of fetishized costumes), formal approaches (such as heavily stylized editing, tilt-shift photography, and a suggestive tonal clash between sound and image), and themes (identity, fantasy, trauma, sexism, corporate power, surveillance, and the possibility of agency) to expect. They cue memory for audiences returning to the series, functioning as an intellectual and emotional codex. The sequence serves as a signature that highlights the presence of some authors (creator Whedon, producer and star Dushku) and hides others (screenwriters, showrunners, network executives, cinematographers, and designers). This signature becomes even more important in an industrial context increasingly reliant on auteur branding, especially in the case of *Dollhouse*, where control over the series is sharply contested between network and production company. The opening title sequence of *Dollhouse* is a many-layered text whose meanings operate differently based on whether one takes the perspective of a network or an auteur, a fan or a viewer, a man or a woman. The whole series flashes before your eyes.

BIBLIOGRAPHY

Balsmeyer, Randy. "Index." Big Film Design.

Barthes, Roland. "The Death of the Author." In *Theories of Authorship*, edited by John Caughie, 208–13. New York: Routledge, 1996.

Bennett, Eve. "Deconstructing the Dream Factory: Personal Fantasy and Corporate Manipulation in Joss Whedon's *Dollhouse*." *Slayage: The Journal of the Whedon Studies Association* 9, no. 1 (2011).

Boxer, Sarah. "Film Titles Offer a Peek at the Future in More Ways Than One." *New York Times*. April 22, 2000: 9.

Capra, Frank. *The Name above the Title: An Autobiography*. New York: Macmillan, 1971.

Codrington, Andrea. *Kyle Cooper (Monographics)*. New Haven, CT: Yale University Press, 2003.

Coker, Catherine. "Exploitation of Minds and Bodies in Season One of *Dollhouse*." In *Sexual Rhetoric in the Works of Joss Whedon*, edited by Erin B. Waggoner, 226–38. Jefferson, NC: McFarland, 2010.

Cook, David A. *A History of Narrative Film*. 4th ed. New York: Norton, 2004.

Counts, James. "Just the Beginning: The Art of Film Titles." http://casaforte.blogspot.com/2007/03/just-beginning-art-of-film-titles.html. December 7, 2006.

Dechert, S. Renee. "'My Boyfriend's in the Band!' *Buffy* and the Rhetoric of Music." In *Fighting the Forces: What's at Stake in "Buffy the Vampire Slayer,"* edited by Rhonda V. Wilcox and David Lavery, 218–26. New York: Rowman & Littlefield, 2002.

Doane, Mary Ann. "Film and the Masquerade: Theorising the Female Spectator." *Screen* 23, nos. 3–4 (1982): 74–88.

Dupree, Scotty. "Stylish Sequences Make a Comeback." *Media Week*. September 23, 1996: 34.

Fish, Stanley. *Is There a Text in This Class? The Authority of Interpretative Communities*. Cambridge, MA: Harvard University Press, 1980.

Genette, Gérard (Marie MacLean, trans.). "Introduction to the Paratext." *New Literary History* 22, no. 2 (1991): 261–72.

Gray, Jonathan. *Show Sold Separately: Promos, Spoilers, and Other Media Paratexts*. New York: New York University Press, 2010.

Halfyard, Janet K. "Music in and out of the *Dollhouse*." Presented at the Slayage Conference on the Whedonverses 5, Vancouver, BC, July 2012.

Hall, Stuart. "Encoding, Decoding." In *Culture, Media, Language: Working Papers in Cultural Studies, 1972–1979*, edited by Stuart Hall et al., 128–38. London: Unwin Hyman, 1980.

Heller, Steven. "Quick Cuts, Coarse Letter, Multiple Screens." https://www.typotheque.com/site/articles.php?&id=48. 1999.

Kociemba, David. "'Actually, It Explains a Lot': Reading the Opening Title Sequences of *BtVS*." *Slayage: The Journal of the Whedon Studies Association* 6, no. 2 (2006).
Lewis, Justin. "The Encoding/Decoding Model: Criticism and Redevelopments for Research on Decoding." *Media, Culture and Society* 5 (1983): 179–97.
McCabe, Janet, and Kim Akass, ed. *Quality TV: Contemporary American Television and Beyond*. London: Tauris, 2007.
Morley, David. *Television, Audiences and Cultural Studies*. New York: Routledge, 1997.
Mulvey, Laura. "Visual Pleasure and Narrative Cinema." *Screen* 16, no. 3 (1975): 6–18.
Ryan, Maureen. "Sex, Secrets and *Dollhouse*: Joss Whedon Talks about the End of His Fox Show." *Chicago Tribune*. December 3, 2009.
Schneider, Michael. "New Fox Dramas to Limit Commercials." May 15, 2008.
St. Louis, Renee, and Miriam Riggs. "'A Painful, Bleeding Sleep': Sleeping Beauty in the *Dollhouse*." *Slayage: The Journal of the Whedon Studies Association* 8, nos. 2–3 (2010).
Stanitzek, Georg (Ellen Klein, trans.). "Texts and Paratexts in Media." *Critical Inquiry* 32, no. 1 (2005): 27–42.
Sullivan, Brad Ford. "Interview: *Dollhouse* Creator Joss Whedon." *The Futon Critic*. January 6, 2009.
Thompson, Robert. *Television's Second Golden Age: From "Hill Street Blues: to "ER."* New York: Syracuse University Press, 1997.
van Elferen, Isabella. "Music of Other Spheres: Diagonal Time and Metaphysics in *Lost*." *Science Fiction Film and Television* 3, no. 2 (2010): 253–70.
"What Happened When the Lights Went Out." Posting to Whedonesque.com. October 26, 2008.
Wilcox, Rhonda. "Echoes of Complicity: Reflexivity and Identity in Joss Whedon's *Dollhouse*." *Slayage: The Journal of the Whedon Studies Association* 8, nos. 2–3 (2010).
Williams, Raymond. *Television: Technology and Cultural Form*. London: Fontana/Collins, 1974.

FILMOGRAPHY

Addams Family. Created by Charles Addams. Developed by David Levy. MGM Television through the Program Exchange, 1964–1966. DVD.
Bourne Ultimatum. Directed by Paul Greengrass. Universal Pictures, 2007. DVD.
The Cabin in the Woods. Directed by Drew Goddard. Lionsgate Films, 2012. DVD.
Citizen Kane. Directed by Orson Welles. RKO Radio Pictures, 1941. DVD.
Dr. Strangelove; or, How I Learned to Stop Worrying and Love the Bomb. Directed by Stanley Kubrick. Columbia Pictures, 1964. DVD.
Gone with the Wind. Directed by Victor Fleming, George Cukor, and Sam Wood. Loew's, 1939. DVD.
"Man on the Street." Commentary. Written by Joss Whedon. Directed by David Straiton. *Dollhouse*. Season 1, episode 6. Twentieth Century Fox Home Entertainment, 2009. DVD.
Se7en. Directed by David Fincher. New Line Cinema, 1995. DVD.
Star Wars Episode IV: A New Hope. Directed by George Lucas. Twentieth Century Fox, 1977. DVD.
"Superstar." *Buffy the Vampire Slayer*. Season 4, episode 17. Written by Jane Espenson. Directed by David Grossman. First broadcast April 4, 2000.
To Kill a Mockingbird. Directed by Robert Mulligan. Universal Pictures, 1962. DVD.
Vertigo. Directed by Alfred Hitchcock. Paramount Pictures, 1958. DVD.
Whedon, Joss. "Making *Dollhouse*." *Joss Whedon's Dollhouse: Season One*. Twentieth Century Fox Home Entertainment, 2009. DVD.
———. "Vows." Commentary. Season 2, episode 1. *Joss Whedon's Dollhouse: The Complete Season Two*. Twentieth Century Fox Home Entertainment, 2009. DVD.

NOTES

1. Stanitzek, "Texts and Paratexts," 45, 46.
2. Ginette, "Introduction to the Paratext."
3. Ginette, "Introduction to the Paratext," 261.
4. For example, *Citizen Kane* (1941), *Star Wars* (1977), and *The Bourne Ultimatum* (2007).
5. Gray, *Show Sold Separately*, 74.
6. Stanitzek, "Texts and Paratexts," 32.
7. van Elferen, "Music of Other Spheres," 254–58.
8. Indeed, it is unknown at this time who designed the opening credit sequence of *Dollhouse,* and it remains uncredited in the trade presses and industry online sites.
9. Stanitzek, "Texts and Paratexts," 48.
10. Cook, *History Narrative Film*, 63.
11. Capra, *Name.*
12. Quality television defies or transforms genre and seeks out taboo narrative subjects for the industry's typical use of the medium; it creates new genres through mixing old ones, such as the dramedy; it is created by an auteur who survives or wins a battle against unappreciative audiences or network interference; it emphasizes the writer through narrative complexity and character depth (story and character arcs, show mythology, large ensemble cast, long-form continuity); it uses self-conscious allusions as a frequent technique; it is watched by upscale, well-educated viewers; and it is television's version of the art film or prides itself on its realism or authenticity. Readers interested in learning more about quality television could start with Robert Thompson's *Television's Second Golden Age* and the anthology edited by Janet McCabe and Kim Akass, *Quality TV: Contemporary American Television and Beyond.* "Cult television" is often the term used for quality television using the science fiction and fantasy genres.
13. Whedon's pitch for the show was "*Alias* meets *Quantum Leap.*" *Quantum Leap* (1989–1993) never aired on the same day in consecutive seasons, which suggests the network's lack of confidence in its mass audience appeal. *Alias* (2001–2006) ranked 60th, 72nd, 78th, 37th, and 90th in the ratings. Guess which season *Alias* aired immediately following *Lost*?
14. Schneider, "New Fox dramas."
15. Ryan, "Sex, Secrets and *Dollhouse.*"
16. "What Happened."
17. Sullivan, "Interview: *Dollhouse* Creator"; Ryan, "Sex, Secrets and *Dollhouse*"; "Man on the Street."
18. Williams, *Television: Technology*, 86.
19. Whedon, "Making *Dollhouse.*"
20. Stanitzek, "Texts and Paratexts," 34.
21. Ginette, "Introduction to the Paratext," 262.
22. Kociemba, "'Actually, It Explains a Lot.'"
23. St. Louis and Riggs, "'Painful, Bleeding Sleep.'"
24. Codrington, *Kyle Cooper*, 8.
25. Counts, "Just the Beginning."
26. Boxer, "Film Titles."
27. Kociemba, "'Actually, It Explains a Lot.'"
28. Dupree, "Stylish Sequences," 34.
29. Boxer, "Film Titles."
30. Kociemba, "'Actually, It Explains a Lot.'"
31. Halfyard, "Music."
32. Halfyard, "Music."
33. Halfyard, "Music."
34. Dechert, "'My Boyfriend.'"
35. Halfyard, "Music."
36. Halfyard, "Music."
37. Halfyard, "Music."

38. Halfyard, "Music."

39. Editors' note: The fact that some women put on a childish demeanor in concert with sexual desire (and that some men like this) adds another possible layer to this reading of the theme song and the schoolgirl outfit featured in the opening title sequence.

40. Gray, *Show Sold Separately*, 26.

41. Codrington, *Kyle Cooper*, 8.

42. Boxer, "Film Titles."

43. Boxer, "Film Titles."

44. Heller, "Quick Cuts," 92.

45. Balsmeyer, Big Film Design.

46. Dupree, "Stylish Sequences," 34.

47. Kociemba, "'Actually, It Explains a Lot.'"

48. Kociemba, "'Actually, It Explains a Lot.'"

49. Bennett, "Deconstructing."

50. Mulvey, "Visual Pleasure."

51. Two of the showrunners and the executive in charge were both women as well, although the presence of powerful media women executives, stars, producers, and writers in creating *Dollhouse* is obviously no guarantee that the series is free from gendered, sexualized exploitation. Indeed, my reading of the series through its opening title sequences depends on the viewer being periodically uneasy at the critical, creative, and profitable use of objectification by the series and even feeling implicated in watching it.

52. Doane, "Film and the Masquerade."

53. As a film studies major in college, it is extremely likely that Whedon encountered the gendered gaze and masquerade theories essential to this reading of the series and its opening title sequences. Editors' note: Actually Whedon addresses this issue in the commentary to "Vows" on the season 2 DVD.

54. Coker, "Exploitation," 237.

55. Wilcox, "Echoes."

56. Coker, "Exploitation"; Bennett, "Deconstructing."

57. Bennett, "Deconstructing"; Whedon, "Vows," commentary.

58. Compare Barthes, "The Death of the Author"; Fish, *Is There a Text*; Hall, "Encoding, Decoding"; Lewis, "The Encoding/Decoding Model"; Morley, *Television, Audiences*.

Chapter Thirteen

Ritual, Rebirth, and the Rising Tide

Water and the Transcendent Self

Ian G. Klein

The first trailer for *Dollhouse* begins not with the gunfights, motorcycle chases, torrid romances, or nightmarish visions of the imprint room that came to visually define the series; instead, it begins with an image of utter stillness and isolation. Show star and producer Eliza Dushku sits at the bottom of a swimming pool in a black one-piece swimsuit as light refracts around her and a single muted note drones in the background. With arms wrapped around her legs, this portrait could hardly be more womblike. The silky voice of the person whom we later learn is Adelle DeWitt—the director of the Dollhouse—pours over this evocative image: "The world is a very simple place at first. Then, as we grow up," she says while views of a city began to interlace with the woman in the pool, "it grows around us, a dense thicket of complication and disappointment," each word clearly enunciated in a posh British accent. Along with flashes of a door, fingerprint, and syringe, the first twenty seconds of the trailer establish a tension between a sound protected interior and a potentially dangerous exterior.[1] With Adelle's soothing voice and the image of the lone woman in the pool, the trailer reflects the concerns of a mother for her child. What challenges will she have to face out in the world? What kind of person will she become? When the woman in the pool opens her eyes, the viewer, who will be witness to her development, is meant to ask, "Who will she be?" The trailer's depiction of this woman—codenamed Echo—in a fetal position is one of potential, both her own and that of the series. It lets prospective viewers know that, when this woman breaks the surface of the water—symbolizing her birth—she can be anything. Over the course of the series, Echo becomes something extraordinary. In *Dollhouse*, the process of becoming for Echo and other characters is, like birth, mediated by water. From the bottom of the Dollhouse pool and the serenity found therein to the pool of blood in the showers during Alpha's rampage to its crystallized form as snow in "The Attic" (2.10), water is that symbolic medium through which characters both navigate the "dense thicket of complication and disappointment" that pervades their world and achieve a transcendent self.

In many ways, water makes life possible. This reality has ensured water's position in the human psyche as a powerful force that both shapes and sustains. It is central to the process of human birth and makes up more of the human body and Earth's surface than any other substance. Nothing else has the capacity to affect people and the landscapes they inhabit quite like water, which is why it possesses countless associations with the metaphysical and has featured at the center of myth, ritual, and philosophy across time and cultures, from Achilles's dip in the River Styx, which brought him great strength, to Sigmund Freud's theories of the unconscious. That many, regardless of background, recognize water as something that soothes, satisfies, and cleanses makes it a fitting element to be emblematic of spiritual birth and rebirth. Its ubiquity suggests that personal transformation is attainable for all, although the results of such metamorphoses are often complicated. Whether the result of a conscious decision or otherwise, interactions with water in *Dollhouse* offer challenges to the new identity.

Apart from the show's trailer, the earliest suggestion of water as an agent of change in *Dollhouse* comes as early as the first episode of the series to air. "You ever try to clean an actual slate?" a soon-to-be-wiped Echo asks Adelle. "You always see what was on it before" ("Ghost," 1.1). Though the line is thematically heavy-handed, it calls into question the ability of water to cleanse absolutely; to act in line with new paradigms, however, does not mean that one must completely forget. The second episode, "The Target" (1.2), suggests a more nuanced reading of the role of water as a trigger for psychological growth and discovery, resulting in the coalescing of a new identity. In these instances and in its various uses and forms, water propels and supports meaningful change in Echo and other characters throughout the series.

Whereas water appears in many scripted instances, there is one significant use of water in the show that came about more or less by accident. In the wake of the 2008 writer's strike, Fox decided that, in lieu of a pilot, it would invest in the construction of a "life-size dollhouse" to serve as the heart of the show in Stage 19 on the Fox Studios lot.[2] The 14,000-square-foot, 35-foot-tall soundstage is identical to adjacent ones on the lot from the outside but possesses what proved to be both a boon to storytelling and a challenge to production: a 28- by 30-foot pit nearly in the middle of the floor.[3] According to a *Los Angeles Times* article on the *Dollhouse* set, this unique feature prompted Whedon and production designer Stuart Blatt to "create a conversation pit with a fountain that would serve as the set's focal point."[4] Two weeks before shooting was to begin on the ultimately unaired pilot, Blatt asked Whedon, "What if I could build you a Japanese reflecting pool with a deck over it that our actors could do yoga and tai chi on?" Blatt says that Whedon "was immediately sold" and, from the beginning, he had "wanted to have a water feature that would be silent and yet always be there to look at. Water is a calming influence. It's sexy and mysterious."[5] Not only does its presence ensure that water appears in every episode of the show, but its centrality makes it a key reference point for water's transformative effects.

On a tour of the dollhouse, Whedon explains, "We thought that water element would help the sort of natural feeling of the dollhouse, which was the most important thing."[6] By bringing the outside in, Whedon and Blatt were able to capture the primordial essence of water, bringing with it timeless principles that thematically support the evolution of those who inhabit the space. Because it is located in the middle of the Dollhouse's main room, Actives would traverse the pool daily. The act of crossing a body of water to signal a change has deep roots in literature, typically marking advancement in characters' personal journeys. As a place of constant crossing, the deck above the pool functions as a symbolic microcosm of persistent change for the Actives. That the design of the Dollhouse is much informed by an Asian aesthetic also invites interpretation of it through the lens of the cultures from which it draws.[7] For the Japanese, water "prefigures the purity and pliant simplicity of life" and, in its still form, beckons contemplation in its surface.[8] Water mediates life on planes both cerebral and physiological. In Chinese philosophy, the softer, passive yin of the yin-yang dichotomy is commonly aligned with water. Peter Merel's interpretation of chapter 43 of the ancient Chinese text the *Tao Te Ching*, by Lao Tse, begins, "Water overcomes the stone."[9] Given enough time, even something so "yielding" as water can change the most stubborn, seemingly fixed aspects of the world.

The other water feature in the Dollhouse is a swimming pool, the one in which a meditative Echo appears at the bottom of in the trailer. In any other "luxurious, spa-like environment" ("Omega," 1.12), the swimming pool would be a site of relaxation and rejuvenation. Its purpose in the Dollhouse is ostensibly similar; however, like everything else in the Dollhouse, it must serve a more critical role in the lives of the Actives. The Actives' Doll state precludes the total embrace of such sensations as "relaxation" and "leisure." Nothing in the Dollhouse is for fun. Every activity therein helps Actives maintain a healthy balance of mind and body. For instance, immediately following one tai chi session, Echo tells the instructor, "I think I'd like to swim now." The tai chi master bows, and in this moment, the audience gets a glimpse of what life for an Active is like—moving from one mode of conditioning to another ("The Target"). Later, in "Needs" (1.8), Mike, an Active who regresses into a Doll state after he and others awake with their original personalities more or less intact, tells his quite conscious peers, "I'm going to swim in the pool. . . . I do thirty laps every day—helps me be my best." Not only is swimming part of the Dolls' regular exercise program, but the Actives also are programmed to associate it with the ideal of themselves. Perhaps daily immersion in water also has the psychological effect of evoking a return to a prenatal state. Just as a baptism within the Christian Church is a sacrament of rebirth—an external testimony to internal change—water therapies may complement the Dollhouse's manipulations of the brain. As if undergoing a daily baptism, the Active is made new, ready for imprint.

Early on in the series, the audience discovers that one particular Active has a harder time with the process of total identity cleansing than others. "The Target" features an adrenaline-fueled plot driven by a shadowy scheme manufactured to challenge Echo's abilities through a man's twisted desire to hunt a fellow human. Written and directed by longtime Whedon collaborator Steven DeKnight, the episode contains many allusions to the short story "The Most Dangerous Game," by Richard Connell, which, in "The Target," is the name of the man who hunts Echo, in her persona as Jenny.[10]

In "The Most Dangerous Game," hunted man Sanger Rainsford achieves his freedom by jumping into the ocean. Water also plays a key role in the salvation of the would-be victim in "The Target," though a much more active one. Here, water directly inspires Echo's actions and serves as a touchstone for her gradual change across the entire series. In her flight from Connell, Echo comes across a park ranger station. She finds a canteen and drinks from it. Whereas river imagery is prominent in this episode, Echo's drink from the canteen in the park ranger station is a more noteworthy narrative use of water in light of Echo's evolution. Within moments, she begins coughing while speaking with Connell on the radio. He observes, "You don't sound too good. Hey, you didn't drink from a canteen in there did you? 'Cause . . . that would be bad." Echo looks over in fear to the canteen lying on its side, large and foreboding in the foreground, visually overtaking the frame, just as the water will overtake Echo.

When Echo raises her voice in defiance, Connell encourages her, "That's it! Shoulder to the wheel. Prove you're not just an echo." Of course, the audience hears the double meaning here. Over the course of the episode, Echo will indeed prove that she is more than Caroline, more than Jenny, and more than the active named Echo—all of this spurred by the effects of the poisoned water. Later, Echo is visibly disoriented as she wanders the woods. The voice of Connell tells her that the drugs will not in fact kill her but, according to the guy that he got them from, will "put a spin on things." Connell provokes her, "So what do you say? You spinning yet?" At this moment, Echo sees another woman with long brown hair. Just as she reaches the figure, the woman turns around. It is the youthful and carefree Caroline that the viewer has seen before in pictures and video. What Echo is experiencing are no ordinary hallucinations; they are much more familiar. Each is tied to a specific persona that together forms the composite identity that the audience will come to know—independent of any natural or imprinted personality—as Echo.

Faced with a perilous situation, Echo has no choice but to confront these fragments of self and reconstruct them into something new to have any chance of survival. Echo's first hallucination catches her off-guard so that she steps backward and trips over a branch, tumbling into a rushing river, itself a familiar motif for constant change. Before she awakens, she experiences a flashback of herself, coughing up water in the Dollhouse shower, surrounded by naked, severely wounded bodies: the result of Alpha's bloody rampage through the Dollhouse. Echo has been spared due to Alpha's particular affinity for her, as is revealed in season 2. That "The Target" largely comprises flashbacks tracing the aftermath of Alpha's "composite event" validates Echo's deeply personal flashback here, which effectively bridges reality and fantasy. Like the episode's structure, which interweaves past and present, so, too, must Echo reconcile her visions with memory. Her Handler, Boyd Langton, is forced to acknowledge the authenticity of her experience when she recounts what she has seen in a subsequent scene:

> There was something in the water. Richard put something in my water. . . . He said it wouldn't kill me, but I've been seeing things. . . . A girl that looks like me but isn't. Dead bodies in a shower. And a man standing over me holding a knife, but I can't see his face; I don't think he has one. Are you here? Is this—is this real?

"I'm here, and, yeah," Langton responds, "this is real" ("The Target"). The tainted water has beckoned reality—and, with it, identity—to the surface, and Echo must not just face the consequences but also embrace them.

This is no easy task, but Echo is unique. Up until this moment, the only thing that Echo has been certain of since drinking the poisoned water is that a man is hunting her. A few visions later, she understandably has doubts as to her ability to defeat him. She panics: "He's going to kill us. He's going to find us and he's . . ." "Us," here, could refer to Langton and her as well as the multiple selves that she holds within. Langton interrupts her, "Hey. Hey! Everything's going to be alright." This elicits a clear assertion from Echo, but it is not the scripted response programmed between Active and Handler to establish complete trust on the part of the imprinted. Instead of following Langton's prompt with the expected, "Now that you're here," Echo declares, "No, it isn't." Langton is stunned. "Everything's not going to be alright," says a newly confident Echo. "You don't get to live just because you deserve to. You have to prove it. You have to put your shoulder to the wheel." In her recollection of Connell's earlier speech, Echo becomes the hunter. Langton pleads with her, unsure of whom she is: "You can't go after this guy. You don't have the right imprint. You don't have the right training." With ferocity in her eyes, Echo responds, "I'm a fast learner." Realizing that he is losing this argument, Langton, with bated breath, tries using the imprint name to reason with her: "Jenny." Echo ignores this final appeal and simply asks, "Do you trust me?" Langton is confused, so Echo repeats the question, and, in a reversal of the Active/Handler script—the "neural lock-and-key" behind their bond—Langton replies, "With my life." At this point, Echo takes over, prioritizing herself over any unique identity, imprinted or otherwise. Jenny is all but forgotten; Caroline is a fleeting vision; and the Active named Echo is unable to be reached via the verbal triggers. They are part of her, but she is greater. This Echo is new and empowered. Echo takes the gun that Langton gives her and begins her pursuit of Connell.

Before she has an opportunity to fire on him, Echo encounters a driven, more severe version of Caroline, who exclaims, "I just wanted to make a difference," linking her to the actions that brought her to the Dollhouse. Echo is distracted long enough for Connell to find her. After a tense standoff, they fire their weapons, but neither deals a fatal blow. Echo charges at Connell, and the two are embroiled in hand-to-hand combat. Just when Echo seems to be weakening under Connell's two-handed grip around her throat, she sees three versions of herself looking down at her, two Carolines from two time frames and Echo in her Doll state. The latter one says, "I try to be my best," and casts her eyes downward. Following her gaze, Echo sees an arrow lying within her reach. She snatches it from the ground and jabs it into Connell's neck. Connell collapses on a log. Blood spilling from his mouth and neck, Connell manages, "He was right about you. You really are special." The episode suggests that "he" is Alpha and that he masterminded this matchup.

Could Alpha have been the one who gave Connell the drugs to put in the water? If, indeed, the goal of the engagement was to encourage Echo's evolution, Alpha's instincts were correct. The drugs disrupt Echo's programming as an imprinted Doll but grant her an altered means of perception that, however fractured, becomes the catalyst for fighting back. By counting on her need of water, Connell thought that he would gain an advantage. To his misfortune, what was in the canteen satisfied not her physical thirst but her thirst for self, thus demonstrating water's capacity to effect great change from the inside out.

"The Target"—unconsciously or otherwise—evokes a pair of ancient Greek cults whose rites parallel Echo's thirst for self as depicted in the episode. The first of these is the cult of Demeter and Persephone. Held annually for some two thousand years, the Eleusinian mysteries—or initiation ceremonies—celebrated the return of Demeter's daughter, Persephone, from Hades's clutches in the underworld. They gain their name from the city Eleusis, the destination for the thousands of men and woman who took part in the festivities that began in Athens.[11] Once the procession reached Eleusis, the initiates fasted for one day. To break their fast, they drank a sacramental concoction, *kykeon*—after the Greek word "to mix" or "stir"—to encourage an elevation of consciousness.[12] As William Savage notes in "Quest of the Soul: The Eleusinian Mysteries," the goal of the mysteries "was to activate the *nous* or higher mind."[13] While there is no consensus among scientists regarding what specific chemical agent Greeks would have used to induce such revelatory states—though ergot fungi are a common hypothesis—chemist and independent psychedelic researcher Peter Webster states, "The *kykeon* surely contained a powerful psychoactive substance similar to one of the major vision-inducing and psychedelic plant alkaloids that Western science has only recently rediscovered."[14]

Like the drugged water that Echo drinks, the "stuff's not gonna kill you . . . just put a spin on things" ("The Target"). In partaking of the *kykeon*, initiates are said to have experienced mystic visions connected to the "mysteries of life and death."[15] The term for one who had seen these was "epoptes," which Savage defines as an "overseer, eyewitness, one who perceives things as they really are."[16] The water from the canteen in the ranger station grants Echo the very same ability. Everything that Echo sees is not just real but also part of her own self rising from the depths of her consciousness. Savage explains, "The mysteries take the view that the soul wants to regain her freedom," which is made possible by rites that include the drinking of the *kykeon*.[17] Both this vital rite of the Eleusinian mysteries and Echo's experience in "The Target" recall the ascent of Persephone from the Underworld. Through her *kykeon* substitute, Echo becomes an *epopte*, a seer of truth. What was intended to harm becomes sacred, the key to the liberation of her soul from that which has been imprinted in her flesh.

The second Greek cult to provide a meaningful touch point for the role of water in *Dollhouse* is that of Orphism. As in the Eleusinian mysteries, the directive to drink within this cult plays a key role in self-revelation. Several gold-leaf tablets found in graves across the Mediterranean and attributed to Orphism are inscribed with instructions for the soul when it reaches the afterlife. One of the Orphic tablets, found tied around the neck of a skeleton in a tomb in Petelia, Southern Italy, dating back to fourth century BCE is translated as such:

> Thou shalt find on the left of the House of Hades a Well-spring,
> And by the side thereof standing a white cypress.
> To this Well-spring approach not near.
> But thou shalt find another by the Lake of Mnemosyne,
> Cold Water flowing forth, and there are guardians before it.
> Say: "I am a child of Earth and of Starry Heaven;
> But my race is of Heaven (alone). This ye know yourselves.
> And lo, I am parched with thirst and I perish. Give me quickly

> The cold water flowing forth from the Lake of Mnemosyne."
> And of themselves they will give thee to drink from the holy Well-spring.
> And thereafter among the other Heroes thou shalt have lordship.[18]

Mnemosyne is the Greek goddess of memory, but as Jules Cashford discusses in her thorough article on this figure, particularly as it relates to Orphic ritual texts, she is much more than a figurehead for memory; she is the embodiment of the memory of the cosmos.[19] In Cashford's translation of an Orphic hymn written to Mnemosyne, the goddess is said to hold "all things together in the same dwelling place, in the mind and soul of mortals, she strengthens the powerful ability of humans to think."[20] The pool of Mnemosyne offers those who drink of it direct access to their origins. Drinking of Mnemosyne's well grants initiates "consciousness of the whole. . . . They remember what they once knew."[21] The myth recalls William Wordsworth's words from *Ode: Intimations of Immortality from Recollections of Early Childhood*, which aim to capture the essence of man's crisis of origin: "Our birth is but a sleep and a forgetting / The Soul that rises with us, our life's Star / Hath had elsewhere its setting."[22] The passage speaks to the longing of the self to reclaim one's eternal soul through remembrance. This, too, is the goal of those who reach the underworld in the Orphic tradition, and, in *Dollhouse*, it is Echo who embarks on a quest for her soul.

The Petelia tablet presumes that the initiates are aware of their deficiency of knowledge, as illustrated by their instructions to approach the well's guardians with a conscious plea to satisfy this void. Unlike the initiates, Echo initially lacks such awareness but, in drinking from the canteen, essentially drinks from the pool of memory, which places her on a trajectory to become whole. Mythopoeic author and Jungian analyst Marion Woodman contends, "One symbol, perhaps above all others, holds a crucial significance for everyone in search of wholeness: the well."[23] Indeed, in Jungian tradition, the pool of water represents the collective unconscious. Jung says that water itself "is the commonest symbol for the unconscious."[24] Thanks to the effects of the psychotropic drugs dissolved in the water, Echo's collective unconscious surfaces in the form of unique personas that inform her sense of self. The visions bestow upon her the beginning of understanding the manner of her creation. Cashford writes, "the myth of Mnemosyne express [*sic*] an archetypal image, that there is a knowing in us—a wiser knowing—which does not come from us, which we receive as a gift of grace—and to this Jung gives the name of the Self."[25] In "The Target," Connell unwittingly gives Echo the gift of self, proving to Alpha and him that she is indeed more than "just an echo." In his notes on *The Mystical Hymns of Orpheus*, translator Thomas Taylor confirms Orpheus's descriptions of the role of the goddess of memory: "to conjoin the soul with intellect."[26] The *Dollhouse* trailer asks viewers to ponder the question, "You can wipe away a memory . . . but can you wipe away a soul?"[27] The implied answer, of course, is a certain "no."

Throughout the remaining twenty-four episodes, Echo's soul—that eternal part of her identity—persists, but memories are not as easily dismissed as the show's premise might have one believe.[28] Even while Echo remembers, however, others try—or are made—to forget.

In Greek mythology, there is a place for forgetting too. A cross-comparison of other gold tablets reveals the "well-spring," marked by a white cypress that initiates are told to avoid upon entry to the underworld, to be Lethe—the pool of forgetfulness—which was named after Lesmosyne, Mnemosyne's sister. The equivalent of visiting this pool in *Dollhouse* is the House's showers, which were only intended to be used in the pilot but kept appearing in scripts for subsequent episodes. "Those wacky writers!" joked Whedon on a press tour of the *Dollhouse* sets. "Sometimes," he acknowledged, the showers were included "for a very good reason, because, as innocence starts to fall away, other things start to become noticeable."[29] Given this understanding, it makes sense, then, that Whedon would choose the showers as the site of a major challenge to Echo's innocence: watching as Alpha struck down her fellow Actives, leaving her sitting in pools of their mingled blood, a trauma that would play a part in shaping her identity. Whedon further said of the showers, "The whole point about this environment is that it's completely innocent; [it is] as somebody said, 'a Garden of Eden.'"[30]

Like the Garden, the Dollhouse showers are meant to be a place largely devoid of self-knowledge. Epitomizing this is Echo's matter-of-fact statement "I'm wet," in the episode "Belle Chose" (2.3), as she stands before Paul Ballard in her Doll state, naked in body and pure of mind, like Eve before the fall. That the Dollhouse is closed off from the rest of the world reinforces it as a seeming proxy for unadulterated paradise, though that purity is an illusion. Alpha, pretending to be Stephen Keplar—the "environmental systems consultant" for the Dollhouse—tells Ballard, "They told me this was going to be the new Eden." It is a thought that would have vastly divergent meanings for Keplar and Alpha. When approached for the job, Keplar might have seen the construction of such a state-of-the-art facility as the career opportunity of a lifetime. Alpha's perception of this Eden, however, is the setting for the rise of the *übermensch*, a new race. When Keplar describes his trade as creating "the self-contained, renewable everything" ("Briar Rose," 1.11), it takes on a sinister connotation through the mouth of Alpha. The Dollhouse's claim to "renewable everything" includes the people in it.

"You're always new!" one of the Dollhouse's stylists tells Echo when she visits post shower ("Belle Chose"). To uphold its business of dealing "in fantasy," the Dollhouse must continue in its efforts to wipe the slate clean ("Man on the Street," 1.6; "Ghost"). Just as their minds are scrubbed, their bodies must be too. For Rossum, an Active's daily shower is akin to a ritual cleansing, a forced return to innocence. The showers also provide the Dollhouse an ideal opportunity to assess the neutrality of its Actives. When

Victor experiences a "man reaction" in the shower in the vicinity of Sierra, the Dollhouse staff takes immediate action to combat this character "growth." Despite the Dollhouse's best efforts to make the Actives forget, their innocence fades away with the increase of not only self-awareness but awareness of the individuality of others.

Because of the showers' psychosomatic capacity to induce the sensation of a prelapsarian state, they are a place where characters come of their own volition, to feel—however briefly—that life can begin anew. The visual of a character taking a shower has often been used as narrative shorthand for cleansing in the wake of trauma, extreme guilt, or overwhelming sadness. In *Dollhouse*, showering as an emotionally driven act can occur only with a non-Active or Active with a burgeoning unique identity. Each comes to the shower to willingly forget. Whereas Orphic texts would speak of forgetfulness as something to be avoided, other poets frame it as a blessing. As Cashford notes, in Orestes's Euripides, the poet calls the pool of Lethe a "wise physician."[31] In this way, the showers are the real "treatment," as opposed to that associated with being placed in the chair, which, rather than "treating" any ailment, destroys identity.

The showers in the Dollhouse never have such immediate effects as the chair, but true healing involves considerable time and self-reflection. In "Belonging" (2.4), Victor goes to the showers to wash away the black paint that Sierra had been using to create foreboding drawings, visual manifestations of her inner turmoil. Victor tells her, "Now you don't have to use this color anymore." They begin to paint each other's faces. Jed Whedon, who cowrote this episode with his wife, Maurissa Tancharoen, says, "When we came up with this scene it felt so right because it was just like play—it was very innocent."[32] When Victor sees the black paint on Sierra's face, he experiences a wartime flashback of Tony's life. If this is Eden, it is Eden after the Fall, and, in it, two of its occupants attempt to recapture some of the innocence they once had. Sierra comforts Victor as he lies on the floor beneath the falling water, the black paint washing away toward the drain, taking their dark pasts along with it.[33] Confronting their trauma together in this way is an early step in a long process of recovery.

Even the one tasked with keeping the Actives oblivious to reality finds her way to this well of forgetfulness. Her authority dissolving in front of her eyes, Adelle visits the showers in a symbolic pronouncement of her morality. In "Stop-Loss" (2.9), Langton accuses her of being "a drunk" and of doing anything that she can "to keep Harding smiling." "We may not have agreed in the past," he says, "but at least we put this house and the Actives first," suggesting that her priorities have shifted from protecting the Actives to protecting herself. "What happened to that woman?" he asks. Adelle insists, "She's standing in front of you." It is clear, though, to the viewer that Langton is right in saying, "No, she's not" ("Stop-Loss"). When the episode returns to Adelle, she showers among the Actives. She, too, must wipe the slate clean, not of the darkness that others have brought upon her, but of her own wrongdoings. Indeed, in Cashford's conceptualization of Lethe, the pool would eradicate the memory of all sins.[34] By showering with those whom she was meant to control, Adelle cleanses herself of selfishness and strengthens her resolve to help overthrow Rossum on their behalf. Though her first act immediately following the shower is sending Echo to the Attic, this decision proves to be the key to salvation for the Actives as well as her own conscience.

Until this point, the Attic had been wrapped in enigma. It is only when the narrative follows Echo's passage into it that the audience gets a thorough sense of its underlying purpose: using adrenaline-fueled human brains as living processors for Rossum's network. The Actives' interaction with the physical framework that allows this to take place exemplifies the role of water as a facilitator of transformation. In the preparation of the bodies, a thick, blue-green liquid is poured around them, likely intended to help prolong the body's value as a serviceable unit, acting as coolant for the human machine. In many ways, the process resembles that of embalming, which, even today, maintains a highly ritualistic quality. The Dollhouse employees who facilitate this procedure are like Charon from Greek mythology, ferrying the souls of the deceased across the River Styx, the division between the world of the living and the dead.

Thanks to Adelle's cunning, Echo, Priya, and Anthony—all sent to the Attic at the end of "Stop-Loss"—return from this underworld with valuable insight into their destinies, not unlike their epic hero counterparts Odysseus, Aeneas, and Heracles. In "The Attic," emergence from the plastic slab is visually poignant. As it turns out, coming out of the Attic is a lot like birth. The rooms housing the bodies are dark, and the basins that hold the bodies simultaneously evoke coffin and womb. The plastic wrap covering the body and fluid is like an amniotic sac—a "bag of waters," as the membrane is called—broken when the Actives are "disengaged" from the network and regain consciousness within the real world. For Echo, Priya, and Anthony, coming into the real world via this dramatic process represents a symbolic rebirth and a physical affirmation of their freedom. In the beginning of the episode, which depicts Echo's repeated fantasy of escaping the Attic, a member of the Attic staff tells Echo, "We don't use Active codes here. We use your real names." Echo retorts, "My real name's Echo" ("The Attic").

In *Dollhouse's* final few episodes, Echo's process of becoming, incited to a great degree by events in "The Target," accelerates. With only a thirteen-episode order for its second season, *Dollhouse* writers had a great deal of ground to cover between the events of "Omega" and the future depicted in "Epitaph One" (1.13). What originally had been written as a stand-alone coda for the series suddenly became a target for the resolution of many of the show's plotlines, resulting in hastened character development. In "Meet Jane Doe" (2.7), Echo has been absent from the Dollhouse for a lengthy but indeterminate amount of time, wandering the country. When Echo steals food from a convenience store for a hungry woman, a policeman corners Echo and pulls a gun. Flashes of previous imprints flood the screen as the words "I'm not broken" repeat in Echo's voice from "A Spy in the House of Love" (1.9). The images coalesce, and Echo knocks the man to the ground. Three months later, Echo recounts the incident to Ballard:

> When I took out that deputy, it wasn't like a glitch, the skills, it was like . . . a wave. Just everything rushing over me and then the wave pulls back and you see what it washed up, picking up the shells and smooth stones from the shore. Collecting myself. ("Meet Jane Doe")

That Echo—the individual, not the Active—is a product of many imprints does not mean that she is fragmented but that she exists across a sea of consciousness, a sea in which great tidal forces are at work. With its every rise and fall, Echo is able to identify each change and addition made to the shore of her psyche. The more of these details she recognizes, the closer she gets to achieving self-transcendence.

Echo's maturation into a true individual is well represented in "The Attic." In between her verbal declaration of her "real name," Echo, and her metaphorical rebirth, Echo manages to break her nightmare cycle and climb out of a hole. She emerges from the trunk of a snow-covered tree standing in the center of the Dollhouse. That it is snowing is significant. What is snow but crystallized water? In the same way, Echo's identity is no longer so fluid. In this environment, Echo stands as an objective witness to where she came from and where she is. The tree is "from one of my childhood homes," she tells Dominic ("The Attic"). The girl who circles the tree is a young Caroline. Living vignettes associated with events and imprints that have made her who she is appear around the room. Here, Echo is able to recognize what is both "other" and part of her. Her history may be a collective one, but Echo is more than the sum of her parts. Echo's gradual transformation has heretofore been mediated by water's liquid phase—ever changing—but here, her identity, like the snow, is crystallized.

"The Attic" is also the first episode to make an explicit connection between the events of the show's present and those associated with the future depicted in "Epitaph One," the postapocalyptic future that Echo and her cohorts are unable to avert. Regarding the torn Los Angeles skyline that Mag, Zone, and the girl imprinted as Caroline—clearly an evolved Echo in the context of the show's narrative—see at the end of "Epitaph One," the girl remarks, "Kids playing with matches . . . and they burned the house down" ("Epitaph One"). This world, to echo Robert Frost's poem "Fire and Ice," ends not in ice—nor water for that matter—but in fire.[35] It is appropriate, then, that the landscapes of Tucson, Arizona, and Southern California depicted in "Epitaph One" and the final episodes of season 2 are of a parched earth—natural, surely, but striking when every episode has had an image of water, several in most. As it is in most postapocalyptic visions, water is a precious commodity, a requirement for humanity to thrive.

Another feature common to these narratives is the presence of lost souls wandering desolate environments. In *Dollhouse*, these lost souls are the empty "dumbshows," violent "butchers," and "actuals," men and women untouched by imprinting technology. Each of them thirsts for self, even the actuals who are compelled to find purpose in the context of this strange new world. In this way, those who roam the desert landscape are like the initiates described in the Orphic tablets. The ritual texts instruct souls to say to the guardians of the well of memory, "And lo, I am parched with thirst and I perish. Give me quickly / The cold water flowing forth from the Lake of Mnemosyne."[36] The arid future that *Dollhouse* paints is a world in need of memory. At the beginning of "Epitaph Two: Return" (2.13), "butchers"—those programmed to kill the unimprinted—chase a man toward a Jeep in which a hooded Zone is sleeping in the bright sun. The environment is clearly dry, the film saturated with yellow. Caroline, in the child's body, and Mag carry buckets of water. There is a close-up of one of the buckets before the women drop them and run. The buckets of water, on one hand, represent their real physical need but, on the other, signify their self-appointed mission to bring knowledge of self back to humanity, one lost soul at a time. It is as if Caroline and Mag have just come from the well of Mnemosyne, bearing the sacred element that would bring memory back to the world.

As Alpha puts together clues for Ballard that will guide him to the Dollhouse in the show's pilot, Caroline speaks on the screen in front of him, "I'd like to take my place in the world, like Mrs. Dundee taught us. . . . The world is in need of some serious saving" ("Ghost"). As part of Echo, she does, in a way, get to save the world. Echo helps facilitate the pulse that restores humanity's original personalities by way of a digital deluge. Like Noah and his ark in the biblical story of the flood, Echo and other Actives shelter themselves in the Dollhouse to protect their acquired memories. The pulse rids the world of Rossum's sins by way of a global baptism, and the would-be-victims of the Dollhouse become victors.

Many of them also have achieved personal victories in the realization of their unique selves, a process marked by water and thereof including metaphorical baptism. Each of these transformations signifies a radical step toward wholeness, and no other transformation is as radical as that undergone by Echo, whose individual evolution extends far beyond human understanding of "self." "I've been many people. . . . I'm all of them, but none of them is me," she tells Ballard at the beginning of season 2 ("Vows," 2.1). Through water, she and others develop the ability to grow, nurture, and shape identity in profound ways. Together, they restore identity to the world.

BIBLIOGRAPHY

Cashford, Jules. "Mnemosyne: Greek Goddess of Memory." Mnemosyne Foundation.

Connell, Richard. *The Most Dangerous Game*. EServer.

Dana, Rebecca. "Post-strike, Networks Revamp Pilot Season." *Wall Street Journal*. March 31, 2008.

"Designing the Perfect Dollhouse." *Dollhouse: Season One*. Created by Joss Whedon and Eliza Dushku. Twentieth Century Fox Home Entertainment, 2009. DVD.

"*Dollhouse* Trailer." Fox Broadcasting Company. YouTube. May 19, 2008.

Fernandez, Maria Elena. "For *Dollhouse* on Fox, the Set Is One of the Stars." *Los Angeles Times*. February 1, 2009.

Frost, Robert. "Fire and Ice." In *Robert Frost: Collected Poems, Prose, and Plays*, edited by Mark Richardson, 204. New York: Library of America, 1995.

Hall, Manly Palmer. *The Secret Teachings of All Ages: An Encyclopedic Outline of Masonic, Hermetic, Qabbalistic and Rosicrucian Symbolical Philosophy*. San Francisco: Crocker, 1928.

Jung., C. G. *The Archetypes and the Collective Unconscious*. Vol. 9, part 1, of *Collected Works of C. G. Jung*. Translated by R. F. C. Hull. Princeton, NJ: Princeton University Press, 1981.

Lo, Melinda. "Touring the Dollhouse of *Dollhouse*." AfterEllen.com. July 29, 2008.

Merel, Peter, trans. *The GNL Tao Te Ching*. By Lao Tse. Brooklyn College.

Orpheus. *Hymns and Initiations: The Mystical Hymns of Orpheus*. Translated by Thomas Taylor. London: Bertram Dobell and Reeves and Turner, 1896.

Savage, William A. "Quest of the Soul: The Eleusinian Mysteries." Theosophical Society.

"Stage 19." Fox Studios Production Services.

Webster, Peter. "Mixing the *Kykeon*: Part 1." *ELEUSIS: Journal of Psychoactive Plants and Compounds* 4 (2000): 1–8.

Whedon, Jed, and Maurissa Tancharoen. "Belonging." Commentary. *Dollhouse: Season Two*. Created by Joss Whedon and Eliza Dushku. Twentieth Century Fox Home Entertainment, 2010. DVD.

Witcombe, Christopher. "Water and the Sacred." Sweet Briar College.

Woodman, Marion. *Sitting by the Well: Bringing the Feminine to Consciousness through Language, Dreams, and Metaphor*. Sounds True, 2007. CD.

Wordsworth, William. "Ode, Intimations on Immortality." In *The Poetical Works of William Wordsworth*, edited by William Knight, 8:190–98. London: Macmillan, 1896.

NOTES

1. "*Dollhouse* Trailer."
2. Dana, "Post-strike."
3. "Stage 19."
4. Fernandez, "For Dollhouse."
5. Fernandez, "For Dollhouse."
6. "Designing."
7. A *Los Angeles Times* article notes, "Before settling on a design, Whedon pored over architecture books, focusing on traditional and modern Asian building styles and elegant spas. Much of his inspiration came from *Thai Style* by Luca Invernizzi Tettoni and *William Warren and Architecture in China* by Philip Jodidio." Blatt described the approach as a "world-class spa" with "a minimalist Japanese feel" (Fernandez).
8. Witcombe, "Water and the Sacred."
9. Merel, *The GNL Tao Te Ching*.

10. The 1924 story "The Most Dangerous Game" is the account of Sanger Rainsford, a big-game hunter who falls from his ship into the Caribbean Sea and swims to a nearby island, where he meets a well-to-do man, General Zaroff, who, bored with hunting animals, has taken up a new hobby: hunting humans. Zaroff gives Rainsford a three-hour head start before taking up pursuit. Rainsford narrowly escapes the general and his pack of dogs by diving off a cliff into the sea. The end of the story implies that, upon Rainsford's return to Zaroff's estate, he kills Zaroff in a fight (Connell).

11. Hall, *Secret Teachings,* 52.

12. According to the Homeric Hymn to Demeter, "kykeon" was a blend of water, barley, and a kind of mint, but all accounts of its dramatic effects suggest that this description is insufficient, possibly intentionally so to keep its contents a secret and thus make its replication impossible (Webster, "Mixing the *Kykeon*," 3).

13. Savage, "Quest of the Soul."

14. Webster, "Mixing the *Kykeon*," 2.

15. Savage, "Quest of the Soul."

16. Savage, "Quest of the Soul."

17. Savage, "Quest of the Soul."

18. Cashford, "Mnemosyne."

19. Cashford, "Mnemosyne."

20. Mnemosyne's offspring, the Nine Muses, have all but eclipsed their mother in popular consciousness for their much-celebrated capacity to inspire (Cashford, "Mnemosyne").

21. Cashford, "Mnemosyne."

22. Wordsworth, *Ode*, 58–60.

23. Woodman, *Sitting by the Well.*

24. Jung, *Archetypes*, 18.

25. Cashford, "Mnemosyne."

26. Orpheus, *Hymns and Initiations*, 146.

27. "*Dollhouse* Trailer"

28. The use of "Echo's soul" here is deliberate in that it is clear that the persona that develops out of this individual's experiences is unique. Caroline is but one persona in a sea of others.

29. Lo, "Touring."

30. Lo, "Touring."

31. Cashford, "Mnemosyne."

32. Whedon and Tancharoen, "Belonging."

33. It is also worth noting Topher's near-futile attempts to wash Nolan's blood off his hands in the sink after he cuts up the body later in this episode ("Belonging"), perhaps a conscious allusion to Lady Macbeth's similar efforts to wash her hands of Duncan's blood in the Shakespearean tragedy *Macbeth.*

34. Cashford, "Mnemosyne."

35. Frost, "Fire and Ice."

36. Cashford, "Mnemosyne."

Dollhouse Episode List

Key: Number / Title / Writer / Director / Air Date

SEASON 1

1.0 / Echo / Joss Whedon / Joss Whedon
1.1 / Ghost / Joss Whedon / Joss Whedon / February 13, 2009
1.2 / The Target / Stephen S. DeKnight / Stephen S. DeKnight / February 20, 2009
1.3 / Stage Fright / Maurissa Tancharoen and Jed Whedon / David Solomon / February 27, 2009
1.4 / Gray Hour / Sarah Fain and Elizabeth Craft / Rod Hardy / March 6, 2009
1.5 / True Believer / Tim Minear / Allan Kroeker / March 13, 2009
1.6 / Man on the Street / Joss Whedon / David Straiton / March 20, 2009
1.7 / Echoes / Elizabeth Craft and Sarah Fain / James A. Contner / March 27, 2009
1.8 / Needs / Tracy Bellomo / Felix E. Alcalá / April 3, 2009
1.9 / A Spy in the House of Love / Andrew Chambliss / David Solomon / April 10, 2009
1.10 / Haunted / Jane Espenson, Maurissa Tancharoen and Jed Whedon / Elodie Keene / April 24, 2009
1.11 / Briar Rose / Jane Espenson / Dwight Little / May 1, 2009
1.12 / Omega / Tim Minear / Tim Minear / May 8, 2009
1.13 / Epitaph One / Maurissa Tancharoen and Jed Whedon / David Solomon[1]

SEASON 2

2.1 / Vows / Joss Whedon / Joss Whedon / September 25, 2009
2.2 / Instinct / Michele Fazekas and Tara Butters / Marita Grabiak / October 2, 2009
2.3 / Belle Chose / Tim Minear / David Solomon / October 9, 2009
2.4 / Belonging / Maurissa Tancharoen and Jed Whedon / Jonathan Frakes / October 23, 2009
2.5 / The Public Eye / Andrew Chambliss / David Solomon / December 4, 2009
2.6 / The Left Hand / Tracy Bellomo / Wendey Stanzler / December 4, 2009
2.7 / Meet Jane Doe / Maurissa Tancharoen, Jed Whedon, and Andrew Chambliss / Dwight Little / December 11, 2009
2.8 / A Love Supreme / Jenny DeArmitt / David Straiton / December 11, 2009
2.9 / Stop-Loss / Andrew Chambliss / Felix E. Alcalá / December 18, 2009
2.10 / The Attic / Maurissa Tancharoen and Jed Whedon / John Cassaday / December 18, 2009
2.11 / Getting Closer / Tim Minear / Tim Minear / January 8, 2010
2.12 / The Hollow Men / Michele Fazekas, Tara Butters, and Tracy Bellomo / Terrence O'Hara / January 15, 2010
2.13 / Epitaph Two: Return / Maurissa Tancharoen, Jed Whedon, and Andrew Chambliss / David Solomon / January 29, 2010

NOTES

The episodes listed herein are the episodes that were filmed and/or broadcast for each season. Dates listed are for the first broadcast of the episode on the Fox Broadcasting Company. We have listed the episodes in the order in which they are indicated on their respective DVDs. All the information in this list was verified by the DVDs for each series and a multitude of Internet sites.

1. The unaired pilot "Echo" and the last episode of season 1, "Epitaph One," were not broadcast in the United States.

Index

About the Contributors

Alyson R. Buckman earned an MA (1991) and PhD (1999) in American studies from Purdue University. She teaches American studies, film, popular culture, and multiculturalism in the humanities and religious studies department at California State University, Sacramento. Outside the Whedonverse, she has published essays on Alice Walker, Octavia Butler, and *The Gilmore Girls*, among others. Her work in the Whedonverse has appeared in the journal *Slayage* as well as the anthologies *Investigating "Firefly" and "Serenity," Sexual Rhetoric in the Works of Joss Whedon*, and *The Joss Whedon Reader*, and she won the Mr. Pointy Award for the best essay published in Whedon Studies in 2010. She founded the Joss Whedon subarea within the Science Fiction and Fantasy Area of the Southwest/Texas Popular Culture/ American Culture conference and was a chair of that area until 2012. She is involved actively with the Whedon Studies Association.

Bronwen Calvert is a senior lecturer in English literature at Sunderland University, working at the North East Centre for Lifelong Learning in Newcastle upon Tyne, and an associate lecturer with the Open University in the North of England. She researches on aspects of embodiment in science fiction and fantasy narratives. She has published work on various television series, including *Buffy the Vampire Slayer, Battlestar Galactica, Supernatural, Heroes*, and, most recently, "William Gibson's Cyberpunk *X-Files*" in *Science Fiction Film and Television* (2013) and "'The Shell I'm In': Illyria and Monstrous Embodiment" (on *Angel*) in *Joss Whedon: The Complete Companion* (2012).

Margo Collins is a visiting assistant professor of English at DeVry University. She holds a PhD in eighteenth-century British literature from the University of North Texas and in recent years has focused on the study of gothic literature and popular culture. She coedits the journal *Supernatural Studies* and is currently editing collections on the television shows *Farscape* and *The Vampire Diaries* for Scarecrow Press. Her recent publications include articles on zombies in popular culture, Lois Lane in *Smallville*, memory and monstrosity in *Buffy the Vampire Slayer*, posthumanism in the *Resident Evil* movies, and the created family in supernatural teen dramas. She is a member of the Whedon Studies Association, the Supernatural Studies Association, and the Southwest Popular/American Culture Association. She is also the author of the paranormal mystery *Waking Up Dead*, due out in 2013, and *Legally Undead*, first in an urban fantasy series, forthcoming in 2014.

Tom Garbett has degrees in law, literature, and linguistics and is currently pursuing a PhD in the design of regulation at the University of British Columbia. He is assistant editor of the law and theater journal *Masks* and is interested in the intersection of law and popular culture in all forms. Current projects include a collection of poetry based on six months living in Stockholm, a debut novel, and the completion of his PhD. He is a practicing lawyer based in the UK.

Sherry Ginn earned both her MA (1984) and PhD (1988) in general-experimental psychology from the University of South Carolina. She has published numerous research articles in the fields of neuroscience and psychology but currently focuses her extracurricular work on the intersection of popular culture with psychology and neuroscience. Her book *Our Space, Our Place: Women in the Worlds of Science Fiction Television* was published in 2005, followed by *Power and Control in the Television Worlds of Joss Whedon* (2012); *The Sex Is out of This World: Essays on the Carnal Side of Science Fiction*, which she coedited with Michael Cornelius (2012); and *The Worlds of "Farscape": Essays on the Groundbreaking Series*, which she edited (2013). Current projects include collections on the award-winning television series *Fringe* with coeditors Tanya R. Cochran and Paul Zinder, time travel in television, the companions of *Doctor Who*, and an examination of neuroscience in the fiction of Anglo-Irish author Joseph Sheridan Le Fanu.

Joel Hawkes completed a master in literature at the University of Glasgow in 2005 and a PhD in literature (with a dose of anthropology) at the University of Bristol in 2011. He has published a number of articles relating to his research areas of literary romanticism and modernism and is near completion of a forthcoming edited collection of articles by the modernist author Mary Butts. He is in the process of rewriting his PhD thesis into a publishable monograph. He is particularly interested in the practices and performances that create physical and literary spaces, and the theory of this research has lent itself to a number of journal articles, including the forthcoming "Rituals of Madness in the Practices of Place," in the *Journal of Medical Humanities*. He also has a great interest in gothic and science fiction literature, television, and film. Joel is currently teaching at Thompson Rivers University in British Columbia.

Lorna Jowett is a reader in television studies at the University of Northampton, where she teaches some of her favorite things, including horror, science fiction, and television, sometimes all at once. She is the coauthor with Stacey Abbott of *TV Horror: Investigating the Dark Side of the Small Screen* and the author of *Sex and the Slayer: A Gender Studies Primer for the "Buffy" Fan*, and she is on the editorial board of *Slayage: The Journal of the Whedon Studies Association* and *Intensities: The Journal of Cult Media*. She has published many articles on television, film, and popular culture and is particularly interested in genre and representation. Currently, she is working on an edited collection of essays about time in television.

Derrick King is currently a PhD student and graduate assistant in the English department at the University of Florida. He holds a BA (2009) and an MA (2012) in English from Missouri State University, where he studied literature and film. His research interests include Marxism, utopian studies, queer theory, and contemporary American literature, film, and culture. He has presented papers at several conferences, including the Pop Culture Association, Slayage, and the Marxist Reading Group.

Ian G. Klein holds an MFA in dramaturgy (2012) from Columbia University in the City of New York. His research interests in narrative development span from posttraumatic identity formation to the subject of his master's thesis: the storytelling potential of created environments. He contributes regularly to popular culture studies as a writer of numerous essays, including a chapter in *Inside Joss' "Dollhouse": From Alpha to Rossum*, edited by Jane Espenson and published in 2010. He also frequently presents at both the biennial Slayage Conference on the Whedonverses and the annual meetings of the Popular Culture Association and American Culture Association. Ian resides in Los Angeles, where he is a producer and freelance dramaturg.

David Kociemba has taught courses in media history, television studies, disability studies, digital media and culture, video art, and fandom studies. He currently serves as the president of the Affiliated Faculty of Emerson College union. His writing focuses on the work of Joss Whedon and Jane Espenson, disability representation in *Glee* and the films of Todd Haynes, and paratexts such as spoilers and opening title sequences. He has written for the journals *In Medias Res*, *Slayage*, and *Transformative Works and Cul-*

tures, along with numerous anthologies. Along with his writing on film and television, Kociemba has appeared in *Playing Columbine*, the 2008 documentary directed, edited, and produced by Danny Ledonne that chronicles the history of the game "Super Columbine Massacre RPG!" Kociemba currently serves as the editor of *Watcher Junior*, a peer-reviewed journal of undergraduate scholarship on Joss Whedon. He recently served as an expert on Internet cat videos for *Animal Planet*.

Ananya Mukherjea is associate professor in sociology at the CUNY College of Staten Island and in public health at the CUNY Graduate Center. In public health, she has published on HIV, avian influenza, and the social politics of epidemics and their prevention, as well as editing the volume *Understanding Emerging Epidemics: Social and Political Approaches* (2010). In popular culture studies, she has published on *Buffy the Vampire Slayer*, the *Twilight* series, and the genre of vampire romance more generally. Ananya has attended six international conferences on the works of Joss Whedon, and her 2008 paper "When You Kiss Me, I Want to Die" (published in *Slayage*) won the Mr. Pointy paper award from the Whedon Studies Association for that year. She has served on the boards of the Prison Moratorium Project, the Center for Lesbian and Gay Studies at CUNY, and the Foucault Society in New York. She is currently a member of the editorial board for the Popular Culture Association in the South's journal *Studies in Popular Culture*.

Samira Nadkarni is currently earning a PhD at the University of Aberdeen by writing a cross-disciplinary thesis on changing theories of an ethics of reading and its applications with regard to digital texts. Her publications trace her interest in British postmodern poetry and performance, Whedon studies, hermeneutics, ethics, postcolonialism and neocolonialism, fan studies, feminism, and digital texts. She served on the poetry selection board of the creative writing publication *New Writing Dundee* (2008–2011) and has had her creative work published in *New Writing Dundee*, *Grund Lit*, and *Causeway Magazine*. She spends her free time fighting with her cat about whether the laptop is for writing or being slept on.

Meg Saint Clair Pearson is a graduate of Southern Illinois University Edwardsville, earning both her undergraduate degree (2008) and master of arts (2013). She is currently working as an adjunct English instructor for St. Louis University and teaching English at Althoff Catholic High School in Belleville, Illinois. Academically, Megan has focused much of her work on television and film studies, including pieces on *Doctor Who*, the 1973 film *The Exorcist*, *Star Wars*, and the rest of the Whedonverse. Most of her free time is divided among grading papers, watching *Mystery Science Theater 3000*, and spoiling her dog, Bowie.

Heather M. Porter received her MS in television from Boston University. She is an accomplished line producer and coordinating producer in reality television. A member of the Producers Guild and the Academy of Television Arts and Sciences, she has worked on five seasons of *Hell's Kitchen* and three seasons of *RuPaul's Drag Race*. A Whedon scholar and charter member of the Whedon Studies Association, Porter has presented at all five Slayage conferences. Her research interests are in quantitative analyses of images of violence, intelligence, and sex in the science fiction and fantasy genres. Her paper "'They Teach You That in Whore Academy?' A Quantitative Examination of Sex and Sex Workers in Joss Whedon's *Firefly* and *Dollhouse*" was recently included in *The Sex Is out of This World: The Carnal Side of Science Fiction*, coedited by Sherry Ginn and Michael G. Cornelius.

Michael Starr lectures in media, English, and cultural studies at the University of Northampton, England. His academic research is concerned with science fiction and philosophy, with particular emphasis on poststructural conceptions of the posthuman. He has co-organized international conferences at his institution concerning vampires on television and Alan Moore, and he recently edited a themed collection dedicated to the work of the latter. Current research projects include spatial conceptions of intertextuality in the Whedonverse, the articulation of horror in splatterpunk literature, and philosophical conceptions of time travel narratives.